Lecture Notes in Mathematics

Volume 2345

This series reports on new developments in all areas of mathematics and their applications - quickly, informally and at a high level. Mathematical texts analysing new developments in modelling and numerical simulation are welcome. The type of material considered for publication includes:

1. Research monographs
2. Lectures on a new field or presentations of a new angle in a classical field
3. Summer schools and intensive courses on topics of current research.

Texts which are out of print but still in demand may also be considered if they fall within these categories. The timeliness of a manuscript is sometimes more important than its form, which may be preliminary or tentative. Please visit the LNM Editorial Policy (https://drive.google.com/file/d/1MOg4TbwOSokRnFJ3ZR3ciEeKs9hOnNX_/view?usp=sharing)

Titles from this series are indexed by Scopus, Web of Science, Mathematical Reviews, and zbMATH.

Ahmed Abbes • Michel Gros

The p-adic Simpson Correspondence and Hodge-Tate Local Systems

Springer

Ahmed Abbes
CNRS & IHES
Institut des Hautes Études Scientifiques
Bures Sur Yvette, France

Michel Gros
CNRS & IRMAR
University of Rennes 1
Rennes Cedex, France

ISSN 0075-8434 ISSN 1617-9692 (electronic)
Lecture Notes in Mathematics
ISBN 978-3-031-55913-6 ISBN 978-3-031-55914-3 (eBook)
https://doi.org/10.1007/978-3-031-55914-3

Mathematics Subject Classification (2020): 14F30, 14F20, 14D07, 14G20, 14G45, 14A21, 14B10

This Springer imprint is published by the registered company Springer Nature Switzerland AG
The registered company address is: Gewerbestrasse 11, 6330 Cham, Switzerland

In memory of Pierre Berthelot

Preface

Continuing a lineage of pioneering works initiated by Weil in 1958 and developed mainly by Narashiman–Seshadri, Donaldson, Hitchin and Corlette, Simpson made a significant breakthrough in the 1980s [50, 51, 52, 53]. He established a correspondence now known as the Simpson correspondence, linking the category of local systems on a smooth complex projective variety to the category of semi-stable *Higgs bundles* with vanishing Chern classes. This correspondence boasts remarkable properties, such as functoriality under proper higher direct images [13, 35], and finds numerous significant applications. Quickly becoming a central topic in (complex) Hodge theory, the Simpson correspondence continues to inspire a wealth of research. Furthermore, its scope has expanded to include the study of singularities, notably through the major contributions of Sabbah and Mochizuki on wild harmonic bundles, thus enhancing its relevance and impact.

In a short and dense article [18], Faltings proposed in 2005 a partial p-adic analogue of Simpson's results. The constructions use several tools that he developed to establish the existence of Hodge–Tate decompositions [16], in particular his theory of almost étale extensions [17]. The *p-adic Simpson correspondence* that he then established for a proper and smooth algebraic variety over a p-adic field applies in fact to objects more general than local systems, which he calls *generalized representations*. Despite the formal analogy, the p-adic and complex correspondences operate in significantly different contexts and employ entirely different techniques. The p-adic correspondence, the subject of this book, can be more appropriately regarded as a generalization of Sen's theory. The construction of Faltings is done in two stages. We first establish a correspondence between generalized representations and Higgs bundles which are both, according to his terminology, *small*, i.e., p-adically close to trivial objects. We then extend the construction to all objects, not necessarily small, by descent. This second stage was only sketched by Faltings for curves.

The p-adic Simpson correspondence has, since its introduction, been the subject of various works, following different approaches, including those that we pursued in [3] where the notion of smallness considered by Faltings is replaced by that, essentially equivalent, of *admissible* objects (in the sense of Fontaine) for certain period rings that we have introduced; these are *Dolbeault representations* and *solvable Higgs bundles*. This volume is a continuation of loc. cit. and is enriched by our investigations on the

Hodge–Tate spectral sequences in [2]. Two major themes are addressed: *Hodge–Tate local systems* and the *functoriality of the correspondence by proper higher direct images*. Along the way, we expand the scope of our previous constructions. However, we do not investigate in this book the question of descent, which will be the subject of future work.

Hodge–Tate local systems were introduced by Hyodo [36] as a p-adic analogue of *variations of Hodge structures*. His construction, based on Faltings' approach to p-adic Hodge theory, came at too early a stage and some of his arguments are valid only in the affine case. In this book, we develop an approach to deal with the general case. For a small affine scheme, Hyodo introduced a ring of periods to define the Hodge–Tate local systems. Our main ring of periods, *the Higgs–Tate algebra*, constructed by a different approach, based on deformation theory, is an integral model of his ring. The integral structure allows the ring to be enlarged by weak completion, which gives rise to a ring having adequate cohomological properties. This weak completion is the ring of periods that we used to define the p-adic Simpson correspondence for admissible (or small) objects. The integral structure also allows us to glue our construction in the Faltings topos, which seems difficult to envisage for Hyodo's construction. Inspired by a classical characterization of complex variations of Hodge structures in terms of Higgs bundles, we propose the p-adic analog as a general definition of Hodge–Tate local systems: these are the Dolbeault (generalized) representations whose associated Higgs bundles are *nilpotent*. We recover in the small affine case the notion introduced by Hyodo.

The second theme addressed in this book is the functoriality of the p-adic Simpson correspondence by higher direct images by a proper and smooth morphism. We show on the one hand that the higher direct images of a Dolbeault generalized representation are essentially Dolbeault, and on the other hand that the associated Higgs bundles by the p-adic Simpson correspondence are obtained by a construction "à la Katz–Oda", analogous to that of the Gauss–Manin connection. These results inspired by our construction of the relative Hodge–Tate spectral sequence for the constant sheaf, give rise to generalizations of this spectral sequence for other local systems, in particular the Hodge–Tate ones.

Ahmed Abbes
CNRS & IHES
Bures-sur-Yvette, France
abbes@ihes.fr

Michel Gros
CNRS & Université de Rennes 1
Rennes, France
michel.gros@univ-rennes1.fr

Contents

Chapter 1
An Overview

1.1 Introduction

1.1.1 Initiated by Faltings [18] and developed following different approaches including those by Tsuji and by the authors [3], the p-adic Simpson correspondence provides an equivalence of categories between certain *p-adic étale local systems* over an algebraic variety defined over a p-adic field and certain *Higgs bundles*. The key idea behind its construction comes from Faltings' approach in p-adic Hodge theory, more precisely from his strategy for the computation of the cohomology of a p-adic étale local system.

1.1.2 Let K be a complete discrete valuation field of characteristic 0, with *algebraically closed* residue field k of characteristic $p > 0$, $\mathscr{O}_K$ the valuation ring of K, $\overline{K}$ an algebraic closure of K, $\mathscr{O}_{\overline{K}}$ the integral closure of $\mathscr{O}_K$ in $\overline{K}$, G_K the Galois group of $\overline{K}$ over K, $\mathscr{O}_C$ the p-adic Hausdorff completion of $\mathscr{O}_{\overline{K}}$, $\mathfrak{m}_C$ its maximal ideal, C its field of fractions. We set $S = \mathrm{Spec}(\mathscr{O}_K)$, $\overline{S} = \mathrm{Spec}(\mathscr{O}_{\overline{K}})$ and we denote by s (resp. η, resp. $\overline{\eta}$) the closed point of S (resp. generic point of S, resp. generic point of $\overline{S}$).

Let X be a proper smooth S-scheme, L a locally constant constructible sheaf of $\mathbb{Z}/p^n\mathbb{Z}$-modules of $X_{\overline{\eta},\text{ét}}$ for an integer $n \geq 0$. To compute the cohomology of L, Faltings introduced a ringed topos $(\widetilde{E}, \overline{\mathscr{B}})$ endowed with two morphisms of topos

$$X_{\overline{\eta},\text{ét}} \xrightarrow{\psi} \widetilde{E} \xrightarrow{\sigma} X_{\text{ét}}\,. \tag{1.1.2.1}$$

For every integer $j \geq 1$, we have $\mathrm{R}^j\psi_*(L) = 0$. In particular, for every $i \geq 0$, we have a canonical isomorphism

$$\mathrm{H}^i(X_{\overline{\eta},\text{ét}}, L) \xrightarrow{\sim} \mathrm{H}^i(\widetilde{E}, \psi_*(L)). \tag{1.1.2.2}$$

Using Artin–Schreier theory, Faltings proved a refinement of this result, namely that the canonical morphism

$$\mathrm{H}^i(X_{\overline{\eta},\text{ét}}, L) \otimes_{\mathbb{Z}_p} \mathscr{O}_C \to \mathrm{H}^i(\widetilde{E}, \psi_*(L) \otimes_{\mathbb{Z}_p} \overline{\mathscr{B}}) \tag{1.1.2.3}$$

A. Abbes, M. Gros, *The p-adic Simpson Correspondence and Hodge-Tate Local Systems*, Lecture Notes in Mathematics 2345, https://doi.org/10.1007/978-3-031-55914-3_1

is an *almost isomorphism*, i.e., its kernel and its cokernel are annihilated by $\mathfrak{m}_C$. Setting $\mathscr{M} = \psi_*(L) \otimes_{\mathbb{Z}_p} \overline{\mathscr{B}}$, we have a canonical isomorphism

$$\mathrm{R}\Gamma(\widetilde{E}, \mathscr{M}) \xrightarrow{\sim} \mathrm{R}\Gamma(X_{\text{ét}}, \mathrm{R}\sigma_*(\mathscr{M})). \tag{1.1.2.4}$$

This computation extends naturally to $\mathbb{Q}_p$-local systems by passing to the inverse limit on n and inverting p. For certain $\mathbb{Q}_p$-local systems L, the corresponding complex $\mathrm{R}\sigma_*(\mathscr{M})$ is the Dolbeault complex of a Higgs bundle canonically associated with L by the *p-adic Simpson correspondence* (see 2.8.1 for the terminology on Higgs modules). The Cartan–Leray spectral sequence for σ (1.1.2.4) and the almost isomorphism (1.1.2.3) lead then to a generalization of the Hodge–Tate spectral sequence [2].

1.1.3 To describe our construction of the p-adic Simpson correspondence, we consider the ring $\breve{\overline{\mathscr{B}}} = (\overline{\mathscr{B}}/p^n\overline{\mathscr{B}})_{n\geq 0}$ of the topos $\widetilde{E}^{\mathbb{N}^\circ}$ of inverse systems of objects of $\widetilde{E}$, indexed by the ordered set $\mathbb{N}$. We work in the *category of $\breve{\overline{\mathscr{B}}}$-modules up to isogeny*, that we call the category of $\breve{\overline{\mathscr{B}}}_{\mathbb{Q}}$-modules. It is a counterpart on the side of the Faltings topos of the category of $\mathbb{Q}_p$-local systems of $X_{\overline{\eta},\text{ét}}$.

1.1.4 Taking up the classical pattern of correspondences constructed by Fontaine, we have established in [3] a correspondence between certain $\breve{\overline{\mathscr{B}}}_{\mathbb{Q}}$-modules and certain Higgs bundles involving a period ring of $\widetilde{E}$ that we call the *Higgs–Tate algebra*. It is an integral model of Hyodo's ring B_{HT}, that we construct using deformation theory, inspired by Faltings' original approach and the work of Ogus–Vologodsky on the Cartier transform in characteristic p [46]. A "weak" p-adic completion of the Higgs–Tate algebra has nice cohomological properties that lead to an equivalence between the categories of admissible objects, namely *Dolbeault $\breve{\overline{\mathscr{B}}}_{\mathbb{Q}}$-modules* and *solvable Higgs bundles*.

1.1.5 We review in this new monograph the construction of this correspondence, enlarging along the way its scope, to prove the following important new features:

(i) We characterize the *Hodge–Tate* $\breve{\overline{\mathscr{B}}}_{\mathbb{Q}}$-modules among the Dolbeault $\breve{\overline{\mathscr{B}}}_{\mathbb{Q}}$-modules, namely those whose associated Higgs bundle is nilpotent (4.11.1).
(ii) We prove the functoriality of the p-adic Simpson correspondence by proper direct image (6.5.24), which leads to a generalization of the relative Hodge–Tate spectral sequence (6.5.28).

Another new feature worth mentioning is the cohomological descent of Dolbeault modules in the small affine case (4.8.16) which leads to a description of the category of Dolbeault modules in terms of their global sections, and allows us to compare the global and the local theories (4.8.31 and 4.8.32).

1.1.6 The p-adic Simpson correspondence requires the existence of a smooth deformation of the scheme $X \otimes_{\mathscr{O}_K} \mathscr{O}_C$ over Fontaine's universal p-adic $\mathrm{W}(k)$-infinitesimal thickening $\mathscr{A}_2(\mathscr{O}_{\overline{K}})$ (2.3.5 and 2.3.8), and it depends on the choice of such a deformation, while the theory of Hodge–Tate $\breve{\overline{\mathscr{B}}}_{\mathbb{Q}}$-modules does not depend on it. This apparent paradox is resolved by introducing a new version of our construction of the p-adic Simpson correspondence which this time depends on the choice of a deformation of the scheme $X \otimes_{\mathscr{O}_K} \mathscr{O}_C$ over a logarithmic version $\mathscr{A}_2^*(\mathscr{O}_{\overline{K}}/\mathscr{O}_K)$ of Fontaine's universal p-adic $\mathscr{O}_K$-infinitesimal thickening $\mathscr{A}_2(\mathscr{O}_{\overline{K}}/\mathscr{O}_K)$ ((2.3.2.4) and (2.4.5.1)). We refer to this context as the *relative case* and to the former context as the *absolute case*. We deal with both cases simultaneously, each having its advantages and disadvantages. Since $\mathscr{A}_2^*(\mathscr{O}_{\overline{K}}/\mathscr{O}_K)$ is naturally an $\mathscr{O}_K$-algebra, $X \otimes_{\mathscr{O}_K} \mathscr{O}_C$ has a canonical smooth $\mathscr{A}_2^*(\mathscr{O}_{\overline{K}}/\mathscr{O}_K)$-deformation, namely the base change of X. However, the theory in the relative case has its limits. Indeed, we know in the affine case that Dolbeault modules are small (3.4.31); roughly speaking, they are "trivial" modulo a prescribed power of p. This power of p is independent of K only in the absolute case. Its dependence on K limits drastically the scope of the theory in the relative case, in particular if we would like to extend the correspondence to all $\breve{\overline{\mathscr{B}}}_{\mathbb{Q}}$-modules by descent. However, Hodge–Tate $\breve{\overline{\mathscr{B}}}_{\mathbb{Q}}$-modules are Dolbeault both in the absolute case and in the relative case, which makes our definition of Hodge–Tate $\breve{\overline{\mathscr{B}}}_{\mathbb{Q}}$-modules independent of any deformation.

1.1.7 Compared to [3], we also extend the theory developed for the category of $\breve{\overline{\mathscr{B}}}_{\mathbb{Q}}$-modules to the larger category of *ind-$\breve{\overline{\mathscr{B}}}$-modules*. The latter admits filtered direct limits and has better properties (2.6, 2.7 and [40]). This enlargement, which seems at first sight technical, turns out to be quite useful and necessary for the study of the functoriality of the p-adic Simpson correspondence by proper direct image.

1.1.8 In this first chapter, we give a detailed overview of the content of this monograph. We treat the case of schemes with toric singularities using logarithmic geometry but, for simplicity, we restrict ourselves in this overview to the smooth case.

1.1.9 The p-adic Simpson correspondence presented here is inspired by both the complex Simpson correspondence and that in characteristic $p > 0$. These fall within substantially different frameworks regarding the nature of the objects considered and the techniques employed. Going beyond the analogies already explored in this work would thus be premature at this stage. In particular, we only mention references from these distant domains that play a direct role in our study or have inspired it.

Acknowledgements We extend our heartfelt thanks to T. Tsuji for the valuable advice he generously provided throughout this work. Our deep appreciation also goes to the referees for their meticulous review and insightful comments concerning potential connections with other works.

1.2 Faltings Topos

1.2.1 Let X be a smooth S-scheme, E the category of morphisms $(V \to U)$ over the canonical morphism $X_{\overline{\eta}} \to X$, i.e., commutative diagrams

$$\begin{array}{ccc} V & \longrightarrow & U \\ \downarrow & & \downarrow \\ X_{\overline{\eta}} & \longrightarrow & X \end{array} \tag{1.2.1.1}$$

such that U is étale over X and the canonical morphism $V \to U_{\overline{\eta}}$ is *finite étale*. It is useful to consider the category E as fibered by the functor

$$\pi \colon E \to \mathbf{\acute{E}t}_{/X}, \quad (V \to U) \mapsto U, \tag{1.2.1.2}$$

over the étale site of X.

The fiber of π over an object U of $\mathbf{\acute{E}t}_{/X}$ is canonically equivalent to the category $\mathbf{\acute{E}t}_{\mathrm{f}/U_{\overline{\eta}}}$ of finite étale morphisms over $U_{\overline{\eta}}$. We endow it with the étale topology and denote by $U_{\overline{\eta},\mathrm{f\acute{e}t}}$ the associated topos. If $U_{\overline{\eta}}$ is connected and if $\overline{y}$ is a geometric point of $U_{\overline{\eta}}$, the topos $U_{\overline{\eta},\mathrm{f\acute{e}t}}$ is then equivalent to the classifying topos of the profinite group $\pi_1(U_{\overline{\eta}}, \overline{y})$, i.e., to the category of discrete sets endowed with a continuous left action of $\pi_1(U_{\overline{\eta}}, \overline{y})$.

1.2.2 We endow E with the *covanishing* topology ([3] VI.10.1), that is to say with the topology generated by the coverings $\{(V_i \to U_i) \to (V \to U)\}_{i \in I}$ of the following two types:

(v) $U_i = U$ for all $i \in I$ and $(V_i \to V)_{i \in I}$ is a covering;
(c) $(U_i \to U)_{i \in I}$ is a covering and $V_i = V \times_U U_i$ for all $i \in I$.

We denote by $\widetilde{E}$ the topos of sheaves of sets on E.

To give a sheaf F on E is equivalent to give:

(i) for any object U of $\mathbf{\acute{E}t}_{/X}$, a sheaf F_U of $U_{\overline{\eta},\mathrm{f\acute{e}t}}$, namely the restriction of F to the fiber of π over U;
(ii) for any morphism $f \colon U' \to U$ of $\mathbf{\acute{E}t}_{/X}$, a morphism $\gamma_f \colon F_U \to f_{\overline{\eta}*}(F_{U'})$.

These data should satisfy a cocycle condition (for the composition of morphisms) and a gluing condition (for the coverings of $\mathbf{\acute{E}t}_{/X}$). We write $F = \{U \mapsto F_U\}$ and think of it as a sheaf on $\mathbf{\acute{E}t}_{/X}$ with values in finite étale topos.

1.2.3 Any specialization map $\overline{y} \rightsquigarrow \overline{x}$ from a geometric point $\overline{y}$ of $X_{\overline{\eta}}$ to a geometric point $\overline{x}$ of X determines a point of $\widetilde{E}$ that we denote by $\rho(\overline{y} \rightsquigarrow \overline{x})$. *The collection of these points of* $\widetilde{E}$ *is conservative* ([3] VI.10.21).

1.2.4 There exist three morphisms of topos ([3] VI.10.6 and VI.10.7)

$$\begin{array}{ccccc} X_{\overline{\eta},\text{ét}} & \xrightarrow{\psi} & \widetilde{E} & \xrightarrow{\sigma} & X_{\text{ét}} \\ & & \downarrow{\scriptstyle\beta} & & \\ & & X_{\overline{\eta},\text{fét}} & & \end{array} \tag{1.2.4.1}$$

defined by

$$(V \mapsto U) \in \text{Ob}(E) \mapsto \psi^*(V \to U) = V, \tag{1.2.4.2}$$

$$U \in \text{Ob}(\mathbf{\acute{E}t}_{/X}) \mapsto \sigma^*(U) = (U_{\overline{\eta}} \to U)^a, \tag{1.2.4.3}$$

$$V \in \text{Ob}(\mathbf{\acute{E}t}_{\text{f}/X_{\overline{\eta}}}) \mapsto \beta^*(V) = (V \to X)^a, \tag{1.2.4.4}$$

where the exponent a denotes the associated sheaf.

1.2.5 The higher direct images of σ sheafify Galois cohomology ([3] VI.10.40): if $F = \{U \mapsto F_U\}$ is an abelian group of $\widetilde{E}$, for each integer $i \geq 0$, $\text{R}^i\sigma_*(F)$ is canonically isomorphic to the sheaf associated with the presheaf

$$U \mapsto \text{H}^i(U_{\overline{\eta},\text{fét}}, F_U). \tag{1.2.5.1}$$

Proposition 1.2.6 ([2] 4.4.2) *For every locally constant constructible torsion abelian sheaf F of $X_{\overline{\eta},\text{ét}}$, we have $\text{R}^i\psi_*(F) = 0$ for all $i \geq 1$.*

This statement is a consequence of the fact that for every geometric point $\overline{x}$ of X over s, denoting by $\underline{X}$ the strict localization of X at $\overline{x}$, $\underline{X}_{\overline{\eta}}$ is a $K(\pi, 1)$ scheme. This property was proved by Faltings ([16] Lemma 2.3 page 281), generalizing results of Artin ([5] XI). It was further generalized by Achinger in the log-smooth case ([4], [2] 4.3.6).

1.2.7 For any object $(V \to U)$ of E, we denote by $\overline{U}^V$ the integral closure of $\overline{U} = U \times_S \overline{S}$ in V and we set

$$\overline{\mathscr{B}}(V \to U) = \Gamma(\overline{U}^V, \mathscr{O}_{\overline{U}^V}). \tag{1.2.7.1}$$

The presheaf on E thus defined is in fact a sheaf. As for any sheaf, we can write $\overline{\mathscr{B}} = \{U \mapsto \overline{\mathscr{B}}_U\}$.

Let $U = \text{Spec}(R)$ be an étale X-scheme, $\overline{y}$ a geometric point of $U_{\overline{\eta}}$. The stalk $\overline{\mathscr{B}}_{U,\overline{y}}$ can be described as follows. We denote by $(V_i)_{i \in I}$ the universal cover of $U_{\overline{\eta}}$ at $\overline{y}$. For each $i \in I$, let $U_i = \text{Spec}(R_i)$ be the normalization of $\overline{U}$ in V_i

$$\begin{array}{ccc} V_i & \longrightarrow & U_i \\ \downarrow & & \downarrow \\ U_{\overline{\eta}} & \longrightarrow & \overline{U}. \end{array} \tag{1.2.7.2}$$

The stalk $\overline{\mathscr{B}}_{U,\overline{y}}$ is then isomorphic to the following $\mathscr{O}_{\overline{K}}$-representation of $\pi_1(U_{\overline{\eta}}, \overline{y})$:

$$\overline{R} = \lim_{\substack{\longrightarrow \\ i \in I}} R_i. \tag{1.2.7.3}$$

Using Artin–Schreier theory, Faltings proved the following refinement of 1.2.6:

Theorem 1.2.8 ([17], [2] 4.8.13) *For every locally constant constructible sheaf of $(\mathbb{Z}/p^n\mathbb{Z})$-modules F of $X_{\overline{\eta},\text{ét}}$, the canonical morphism*

$$\mathrm{H}^i(X_{\overline{\eta},\text{ét}}, F) \otimes_{\mathbb{Z}_p} \mathscr{O}_C \to \mathrm{H}^i(\widetilde{E}, \psi_*(F) \otimes_{\mathbb{Z}_p} \overline{\mathscr{B}}) \tag{1.2.8.1}$$

is an almost isomorphism, *i.e., its kernel and its cokernel are annihilated by* $\mathfrak{m}_C$.

Faltings derived all the comparison theorems between p-adic étale cohomology and the other p-adic cohomologies from this fundamental comparison theorem.

1.3 Local Theory. The Torsor of Deformations

1.3.1 First we review the local variant of the p-adic Simpson correspondence for *small* affine S-schemes. Our approach uses a period ring, the *Higgs–Tate algebra*, that we introduced in [3].

1.3.2 Let us first recall (2.2.3, [20] 1.2.1, [55] 1.1) that Fontaine associated functorially with any $\mathbb{Z}_{(p)}$-algebra A such that $A/pA \neq 0$ the ring

$$A^\flat = \lim_{\substack{\longleftarrow \\ \mathbb{N}}} A/pA, \tag{1.3.2.1}$$

where the transition morphisms are the absolute Frobenius morphisms of A/pA, together with the ring homomorphism

$$\theta \colon \mathrm{W}(A^\flat) \to \widehat{A}, \tag{1.3.2.2}$$

from the ring of Witt vectors of $A^\flat$ to the p-adic Hausdorff completion of A, defined for $x = (x_0, x_1, \dots) \in \mathrm{W}(A^\flat)$, by

$$\theta(x) = \lim_{m \to +\infty} (\widetilde{x}_{0m}^{p^m} + p\widetilde{x}_{1m}^{p^{m-1}} + \dots + p^m \widetilde{x}_{mm}), \tag{1.3.2.3}$$

where, for each $n \geq 0$, we write $x_n = (x_{nm})_{m \geq 0} \in A^\flat$ and where, for $x \in A/pA$, $\widetilde{x}$ denotes a lifting in A.

The ring $A^\flat$ is perfect of characteristic p, and the homomorphism θ is surjective if the absolute Frobenius morphism of A/pA is surjective.

1.3.3 The ring $\mathscr{O}_{\overline{K}^\flat} = (\mathscr{O}_{\overline{K}})^\flat$ is a complete non-discrete valuation ring of height 1. We denote by $\overline{K}^\flat$ its field of fractions. We choose a sequence $(p_n)_{n\geq 0}$ of elements of $\mathscr{O}_{\overline{K}}$ such that $p_0 = p$ and $p_{n+1}^p = p_n$ for all $n \geq 0$. We denote by ϖ the element of $\mathscr{O}_{\overline{K}^\flat}$ defined by (p_n) and we set

$$\xi = [\varpi] - p \in \mathrm{W}(\mathscr{O}_{\overline{K}^\flat}). \tag{1.3.3.1}$$

It is a generator of the kernel of θ (1.3.2.2). We set

$$\mathscr{A}_2(\mathscr{O}_{\overline{K}}) = \mathrm{W}(\mathscr{O}_{\overline{K}^\flat})/\ker(\theta)^2. \tag{1.3.3.2}$$

Then, we have an exact sequence (3.1.4.5)

$$0 \longrightarrow \mathscr{O}_C \xrightarrow{\cdot\xi} \mathscr{A}_2(\mathscr{O}_{\overline{K}}) \xrightarrow{\theta} \mathscr{O}_C \longrightarrow 0. \tag{1.3.3.3}$$

We also have a canonical homomorphism $\mathbb{Z}_p(1) \to \mathscr{O}^\times_{\overline{K}^\flat}$. For every $\zeta \in \mathbb{Z}_p(1)$, we have $\theta([\zeta] - 1) = 0$. We deduce from this a group homomorphism

$$\mathbb{Z}_p(1) \to \mathscr{A}_2(\mathscr{O}_{\overline{K}}), \quad \zeta \mapsto \log([\zeta]) = [\zeta] - 1, \tag{1.3.3.4}$$

whose image generates the ideal $p^{\frac{1}{p-1}}\xi\mathscr{O}_C$ of $\mathscr{A}_2(\mathscr{O}_{\overline{K}})$. It induces an $\mathscr{O}_C$-linear isomorphism (3.1.5.3)

$$\mathscr{O}_C(1) \xrightarrow{\sim} p^{\frac{1}{p-1}}\xi\mathscr{O}_C. \tag{1.3.3.5}$$

1.3.4 We introduce in 3.1.6 a logarithmic relative version of the extension $\mathscr{A}_2(\mathscr{O}_{\overline{K}})$ (1.3.3.3) over $\mathscr{O}_K$. For this we fix a uniformizer π of $\mathscr{O}_K$, a sequence $(\pi_n)_{n\geq 0}$ of elements of $\mathscr{O}_{\overline{K}}$ such that $\pi_0 = \pi$ and $\pi_{n+1}^p = \pi_n$ (for all $n \geq 0$) and we denote by $\underline{\pi}$ the associated element of $\mathscr{O}_{\overline{K}^\flat}$. We set

$$\mathrm{W}_{\mathscr{O}_K}(\mathscr{O}_{\overline{K}^\flat}) = \mathrm{W}(\mathscr{O}_{\overline{K}^\flat}) \otimes_{\mathrm{W}(k)} \mathscr{O}_K. \tag{1.3.4.1}$$

We denote by $\mathrm{W}^*_{\mathscr{O}_K}(\mathscr{O}_{\overline{K}^\flat})$ the sub-$\mathrm{W}_{\mathscr{O}_K}(\mathscr{O}_{\overline{K}^\flat})$-algebra of $\mathrm{W}_K(\mathscr{O}_{\overline{K}^\flat}) = \mathrm{W}(\mathscr{O}_{\overline{K}^\flat})\otimes_{\mathrm{W}(k)} K$ generated by $[\underline{\pi}]/\pi$ and we set

$$\xi^*_\pi = \frac{[\underline{\pi}]}{\pi} - 1 \in \mathrm{W}^*_{\mathscr{O}_K}(\mathscr{O}_{\overline{K}^\flat}). \tag{1.3.4.2}$$

It is a generator of the kernel of the homomorphism $\theta^*_{\mathscr{O}_K} : \mathrm{W}^*_{\mathscr{O}_K}(\mathscr{O}_{\overline{K}^\flat}) \to \mathscr{O}_C$ induced by θ (1.3.2.2). Observe that this algebra depends on $(\pi_n)_{n\geq 0}$ (3.1.6). We set

$$\mathscr{A}^*_2(\mathscr{O}_{\overline{K}}/\mathscr{O}_K) = \mathrm{W}^*_{\mathscr{O}_K}(\mathscr{O}_{\overline{K}^\flat})/(\xi^*_\pi)^2\mathrm{W}^*_{\mathscr{O}_K}(\mathscr{O}_{\overline{K}^\flat}). \tag{1.3.4.3}$$

Then, we have an exact sequence (3.1.6.9)

$$0 \longrightarrow \mathscr{O}_C \xrightarrow{\cdot\xi^*_\pi} \mathscr{A}^*_2(\mathscr{O}_{\overline{K}}/\mathscr{O}_K) \xrightarrow{\theta^*_{\mathscr{O}_K}} \mathscr{O}_C \longrightarrow 0\,. \tag{1.3.4.4}$$

Denoting by K_0 the field of fractions of $\mathrm{W}(k)$ and by $\mathfrak{d}$ the different of K/K_0, the canonical homomorphism $\mathscr{A}_2(\mathscr{O}_{\overline{K}}) \to \mathscr{A}_2^*(\mathscr{O}_{\overline{K}}/\mathscr{O}_K)$ induces an $\mathscr{O}_C$-linear isomorphism (3.1.10)

$$\xi\mathscr{O}_C \xrightarrow{\sim} \pi\mathfrak{d}\xi_\pi^*\mathscr{O}_C. \tag{1.3.4.5}$$

1.3.5 Let $X = \mathrm{Spec}(R)$ be an affine smooth S-scheme which is *small* in the sense of Faltings (i.e., which admits an étale S-morphism $X \to \mathbb{G}_{m,S}^d = \mathrm{Spec}(\mathscr{O}_K[T_1^{\pm 1},\dots,T_d^{\pm 1}])$ for an integer $d \geq 0$) and such that $X_s \neq \emptyset$. We fix a geometric point $\overline{y}$ of $X_{\overline{\eta}}$ and we denote by $\overline{X}^\star$ (resp. $X_{\overline{\eta}}^\star$) the connected component of $\overline{X} = X \times_S \overline{S}$ (resp. $X_{\overline{\eta}}$) containing the image of $\overline{y}$; we have $X_{\overline{\eta}}^\star = \overline{X}^\star \times_{\overline{S}} \overline{\eta}$. We set $\Delta = \pi_1(X_{\overline{\eta}}^\star, \overline{y})$, $\overline{R} = \overline{\mathscr{B}}_{X,\overline{y}}$ (1.2.7.3) and

$$R_1 = \Gamma(\overline{X}^\star, \mathscr{O}_{\overline{X}}). \tag{1.3.5.1}$$

Let $\widehat{R_1}$ (resp. $\widehat{\overline{R}}$) be the p-adic Hausdorff completion of R_1 (resp. $\overline{R}$). Recall that Δ acts naturally on $\overline{R}$ by ring homomorphisms. We denote by $\mathbf{Rep}_{\widehat{\overline{R}}}(\Delta)$ the category of $\widehat{\overline{R}}$-representations of Δ (2.1.2).

Applying the constructions of 1.3.2–1.3.4 to the algebra $\overline{R}$ (3.2.8 and 3.2.10), we set

$$\mathscr{A}_2(\overline{R}) = \mathrm{W}(\overline{R}^\flat)/\xi^2\mathrm{W}(\overline{R}^\flat), \tag{1.3.5.2}$$

$$\mathscr{A}_2^*(\overline{R}/\mathscr{O}_K) = \mathrm{W}_{\mathscr{O}_K}^*(\overline{R}^\flat)/(\xi_\pi^*)^2\mathrm{W}_{\mathscr{O}_K}(\overline{R}^\flat). \tag{1.3.5.3}$$

We have exact sequences

$$0 \longrightarrow \widehat{\overline{R}} \xrightarrow{\cdot\xi} \mathscr{A}_2(\overline{R}) \longrightarrow \widehat{\overline{R}} \longrightarrow 0, \tag{1.3.5.4}$$

$$0 \longrightarrow \widehat{\overline{R}} \xrightarrow{\cdot\xi_\pi^*} \mathscr{A}_2^*(\overline{R}/\mathscr{O}_K) \longrightarrow \widehat{\overline{R}} \longrightarrow 0. \tag{1.3.5.5}$$

1.3.6 Let $\widetilde{S}$ be one of the schemes $\mathrm{Spec}(\mathscr{A}_2(\mathscr{O}_{\overline{K}}))$ or $\mathrm{Spec}(\mathscr{A}_2^*(\mathscr{O}_{\overline{K}}/\mathscr{O}_K))$; the first case will be called *absolute* and the second one *relative*. We denote by

$$i_S \colon \mathrm{Spec}(\mathscr{O}_C) \to \widetilde{S} \tag{1.3.6.1}$$

the closed immersion defined by the ideal of square zero generated by $\widetilde{\xi} = \xi$ in the absolute case and by $\widetilde{\xi} = \xi_\pi^*$ in the relative case.

Note that in the relative case, $\widetilde{S}$ is naturally an S-scheme, since $\mathscr{A}_2^*(\mathscr{O}_{\overline{K}}/\mathscr{O}_K)$ is an $\mathscr{O}_K$-algebra.

1.3.7 Depending on whether we are in the absolute or relative case, we denote by $\widetilde{\mathbb{X}}$ one of the schemes $\mathrm{Spec}(\mathscr{A}_2(\overline{R}))$ or $\mathrm{Spec}(\mathscr{A}_2^*(\overline{R}/\mathscr{O}_K))$. There exists a canonical closed immersion

$$i_X \colon \mathrm{Spec}(\widehat{\overline{R}}) \to \widetilde{\mathbb{X}} \tag{1.3.7.1}$$

over the closed immersion i_S defined by the ideal of $\mathscr{O}_{\widetilde{\mathbb{X}}}$ generated by $\widetilde{\xi}$.

For any $\widehat{R_1}$-algebra A, we consider Higgs A-modules with coefficients in $\widetilde{\xi}^{-1}\Omega^1_{R/\mathscr{O}_K} \otimes_R A$ (2.5.1). We say abusively that they have coefficients in $\widetilde{\xi}^{-1}\Omega^1_{R/\mathscr{O}_K}$. The category of these modules will be denoted by $\mathbf{HM}(A, \widetilde{\xi}^{-1}\Omega^1_{R/\mathscr{O}_K})$.

1.3.8 The p-adic Simpson correspondence depends on the choice of a smooth $\widetilde{S}$-deformation $\widetilde{X}$ of $X \otimes_{\mathscr{O}_K} \mathscr{O}_C$,

$$\begin{array}{ccc} X \otimes_{\mathscr{O}_K} \mathscr{O}_C & \longrightarrow & \widetilde{X} \\ \downarrow & \square & \downarrow \\ \mathrm{Spec}(\mathscr{O}_C) & \longrightarrow & \widetilde{S}. \end{array} \tag{1.3.8.1}$$

Since X is affine, such a deformation always exists and is unique up to a non-unique isomorphism. We fix one in the following.

Let U be an open subscheme of $\mathrm{Spec}(\widehat{\overline{R}})$, $\widetilde{U}$ the open subscheme of $\widetilde{\mathbb{X}}$ defined by U (1.3.7.1). We denote by $\mathscr{L}(U)$ the set of morphisms represented by the dotted arrows completing the diagram

$$\begin{array}{ccc} U & \longrightarrow & \widetilde{U} \\ \downarrow & & \vdots \\ X \otimes_{\mathscr{O}_K} \mathscr{O}_C & \longrightarrow & \widetilde{X} \\ \downarrow & \square & \downarrow \\ \mathrm{Spec}(\mathscr{O}_C) & \longrightarrow & \widetilde{S} \end{array} \tag{1.3.8.2}$$

in such a way that it remains commutative. The functor $U \mapsto \mathscr{L}(U)$ is a torsor for the Zariski topology of $\mathrm{Spec}(\widehat{\overline{R}})$ under the $\widehat{\overline{R}}$-module $\mathrm{Hom}_R(\Omega^1_{R/\mathscr{O}_K}, \xi\widehat{\overline{R}})$ (3.2.14). Such a torsor is very easy to describe. Let $\mathscr{F}$ be the $\widehat{\overline{R}}$-module of affine functions on $\mathscr{L}$ ([3] II.4.9). The latter fits into a canonical exact sequence

$$0 \to \widehat{\overline{R}} \to \mathscr{F} \to \widetilde{\xi}^{-1}\Omega^1_{R/\mathscr{O}_K} \otimes_R \widehat{\overline{R}} \to 0. \tag{1.3.8.3}$$

We consider the $\widehat{\overline{R}}$-algebra

$$\mathscr{C} = \varinjlim_{n \geq 0} \mathrm{Sym}^n_{\widehat{\overline{R}}}(\mathscr{F}), \tag{1.3.8.4}$$

where the transition morphisms are defined by sending $x_1 \otimes \cdots \otimes x_n$ to $1 \otimes x_1 \otimes \cdots \otimes x_n$. Then, the functor $\mathscr{L}$ is represented by $\mathrm{Spec}(\mathscr{C})$ ([3] II.4.10).

The natural action of Δ on $\overline{R}$ induces an action on the scheme $\widetilde{\mathbb{X}}$, and hence an $\widehat{\overline{R}}$-semi-linear action on $\mathscr{F}$, such that the morphisms of (1.3.8.3) are Δ-equivariant (3.2.15). We deduce from this an action of Δ on $\mathscr{C}$ by ring homomorphisms. These actions are continuous for the p-adic topology ([3] II.12.4). We call the $\widehat{\overline{R}}$-algebra $\mathscr{C}$

endowed with this action of Δ the *Higgs–Tate algebra* associated with the deformation $\widetilde{X}$.

1.3.9 The Higgs–Tate algebra $\mathscr{C}$ is an integral model of Hyodo's ring ([3] II.15.6). We introduce a "weak" p-adic completion that serves as a period ring for the p-adic Simpson correspondence. For any rational number $r \geq 0$, we denote by $\mathscr{F}^{(r)}$ the $\widehat{\overline{R}}$-representation of Δ deduced from $\mathscr{F}$ by pullback by the morphism of multiplication by p^r on $\widetilde{\xi}^{-1}\Omega^1_{R/\mathscr{O}_K} \otimes_R \widehat{\overline{R}}$, so that we have an exact sequence

$$0 \to \widehat{\overline{R}} \to \mathscr{F}^{(r)} \to \widetilde{\xi}^{-1}\Omega^1_{R/\mathscr{O}_K} \otimes_R \widehat{\overline{R}} \to 0. \tag{1.3.9.1}$$

We consider the $\widehat{\overline{R}}$-algebra

$$\mathscr{C}^{(r)} = \varinjlim_{n\geq 0} \mathrm{Sym}^n_{\widehat{\overline{R}}}(\mathscr{F}^{(r)}), \tag{1.3.9.2}$$

where the transition morphisms are defined by sending $x_1 \otimes \cdots \otimes x_n$ to $1 \otimes x_1 \otimes \cdots \otimes x_n$. The action of Δ on $\mathscr{F}^{(r)}$ induces an action on $\mathscr{C}^{(r)}$ by ring automorphisms, compatible with its action on $\widehat{\overline{R}}$. We denote by $\widehat{\mathscr{C}}^{(r)}$ the p-adic Hausdorff completion of $\mathscr{C}^{(r)}$.

For all rational numbers $r' \geq r \geq 0$, we have a canonical injective and Δ-equivariant $\widehat{\overline{R}}$-homomorphism $\alpha^{r,r'} : \mathscr{C}^{(r')} \to \mathscr{C}^{(r)}$. One easily verifies that the induced homomorphism $\widehat{\alpha}^{r,r'} : \widehat{\mathscr{C}}^{(r')} \to \widehat{\mathscr{C}}^{(r)}$ is injective. We set

$$\widehat{\mathscr{C}}^{(r+)} = \varinjlim_{t\in\mathbb{Q}_{>r}} \widehat{\mathscr{C}}^{(t)}, \tag{1.3.9.3}$$

which we identify with a sub-$\widehat{\overline{R}}$-algebra of $\widehat{\mathscr{C}} = \widehat{\mathscr{C}}^{(0)}$. The group Δ acts naturally on $\widehat{\mathscr{C}}^{(r+)}$ by ring automorphisms in a manner compatible with its actions on $\widehat{\overline{R}}$ and on $\widehat{\mathscr{C}}$.

We denote by

$$d_{\mathscr{C}^{(r)}} : \mathscr{C}^{(r)} \to \widetilde{\xi}^{-1}\Omega^1_{R/\mathscr{O}_K} \otimes_R \mathscr{C}^{(r)} \tag{1.3.9.4}$$

the $\widehat{\overline{R}}$-universal derivation of $\mathscr{C}^{(r)}$ (3.2.19) and by

$$d_{\widehat{\mathscr{C}}^{(r)}} : \widehat{\mathscr{C}}^{(r)} \to \widetilde{\xi}^{-1}\Omega^1_{R/\mathscr{O}_K} \otimes_R \widehat{\mathscr{C}}^{(r)} \tag{1.3.9.5}$$

its extension to the p-adic completions. These derivations are clearly Δ-equivariant, and are Higgs $\widehat{\overline{R}}$-fields with coefficients in $\widetilde{\xi}^{-1}\Omega^1_{R/\mathscr{O}_K}$ by 2.5.26(i) (see 3.2.19).

For all rational numbers $r' \geq r \geq 0$, we have

$$p^{r'}(\mathrm{id} \times \alpha^{r,r'}) \circ d_{\mathscr{C}^{(r')}} = p^r d_{\mathscr{C}^{(r)}} \circ \alpha^{r,r'}. \tag{1.3.9.6}$$

Therefore, the derivations $(p^t d_{\widehat{\mathscr{C}}^{(t)}})_{t\in\mathbb{Q}_{>r}}$ induce an $\widehat{\overline{R}}$-derivation

$$d^{(r)}_{\widehat{\mathscr{C}}^{(r+)}} : \widehat{\mathscr{C}}^{(r+)} \to \widetilde{\xi}^{-1}\Omega^1_{R/\mathscr{O}_K} \otimes_R \widehat{\mathscr{C}}^{(r+)}, \tag{1.3.9.7}$$

which is also the restriction of $p^r d_{\widehat{\mathscr{C}}^{(r)}}$ to $\widehat{\mathscr{C}}^{(r+)}$.

1.3.10 For any $\widehat{\overline{R}}$-representation M of Δ (2.1.2), we denote by $\mathbb{H}(M)$ the $\widehat{R_1}$-module defined by

$$\mathbb{H}(M) = (M \otimes_{\widehat{\overline{R}}} \widehat{\mathscr{C}}^{(0+)})^{\Delta}. \tag{1.3.10.1}$$

We endow it with the Higgs $\widehat{R_1}$-field with coefficients in $\widetilde{\xi}^{-1}\Omega^1_{R/\mathscr{O}_K}$ induced by $d^{(0)}_{\widehat{\mathscr{C}}^{(0+)}}$ (1.3.9.7). We thus define a functor

$$\mathbb{H}\colon \mathbf{Rep}_{\widehat{\overline{R}}}(\Delta) \to \mathbf{HM}(\widehat{R_1}, \widetilde{\xi}^{-1}\Omega^1_{R/\mathscr{O}_K}). \tag{1.3.10.2}$$

1.3.11 For any Higgs $\widehat{R_1}$–module (N, θ) with coefficients in $\widetilde{\xi}^{-1}\Omega^1_{R/\mathscr{O}_K}$, we denote by $\mathbb{V}(N)$ the $\widehat{\overline{R}}$-module defined by

$$\mathbb{V}(N) = (N \otimes_{\widehat{R_1}} \widehat{\mathscr{C}}^{(0+)})^{\theta_{\rm tot}=0}, \tag{1.3.11.1}$$

where $\theta_{\rm tot} = \theta \otimes {\rm id} + {\rm id} \otimes d^{(0)}_{\widehat{\mathscr{C}}^{(0+)}}$ is the total Higgs $\widehat{R_1}$-field on $N \otimes_{\widehat{R_1}} \widehat{\mathscr{C}}^{(0+)}$. We endow it with the $\widehat{\overline{R}}$-semi-linear action of Δ induced by its natural action on $\widehat{\mathscr{C}}^{(0+)}$. We thus define a functor

$$\mathbb{V}\colon \mathbf{HM}(\widehat{R_1}, \widetilde{\xi}^{-1}\Omega^1_{R/\mathscr{O}_K}) \to \mathbf{Rep}_{\widehat{\overline{R}}}(\Delta). \tag{1.3.11.2}$$

Definition 1.3.12 (cf. 3.3.9) We say that an $\widehat{\overline{R}}[\frac{1}{p}]$-representation M of Δ is *Dolbeault* if the following conditions are satisfied:

(i) $\mathbb{H}(M)$ is a projective $\widehat{R_1}[\frac{1}{p}]$-module of finite type;
(ii) the canonical morphism

$$\mathbb{H}(M) \otimes_{\widehat{R_1}} \widehat{\mathscr{C}}^{(0+)} \to M \otimes_{\widehat{\overline{R}}} \widehat{\mathscr{C}}^{(0+)} \tag{1.3.12.1}$$

is an isomorphism.

We prove that every Dolbeault $\widehat{\overline{R}}[\frac{1}{p}]$-representation of Δ is continuous for the p-adic topology (3.3.12).

Definition 1.3.13 (cf. 3.3.10) We say that a Higgs $\widehat{R_1}[\frac{1}{p}]$-module (N, θ) with coefficients in $\widetilde{\xi}^{-1}\Omega^1_{R/\mathscr{O}_K}$ is *solvable* if the following conditions are satisfied:

(i) N is a projective $\widehat{R_1}[\frac{1}{p}]$-module of finite type;
(ii) the canonical morphism

$$\mathbb{V}(N) \otimes_{\widehat{\overline{R}}} \widehat{\mathscr{C}}^{(0+)} \to N \otimes_{\widehat{R_1}} \widehat{\mathscr{C}}^{(0+)} \tag{1.3.13.1}$$

is an isomorphism.

The notions 1.3.12 and 1.3.13 do not depend on the choice of the $\widetilde{S}$-deformation $\widetilde{X}$ (3.3.8). Nevertheless they depend a priori on the framework, absolute or relative, considered in 1.3.6. They are equivalent to the *smallness* conditions introduced by Faltings [18], i.e., triviality conditions modulo prescribed powers of p (3.4.30–3.4.32).

Proposition 1.3.14 (cf. 3.3.16) *The functors* $\mathbb{H}$ (1.3.10.1) *and* $\mathbb{V}$ (1.3.11.1) *induce equivalences of categories quasi-inverse to each other between the category of Dolbeault* $\widehat{\overline{R}}[\frac{1}{p}]$*-representations of* Δ *and that of solvable Higgs* $\widehat{R_1}[\frac{1}{p}]$*-modules with coefficients in* $\widetilde{\xi}^{-1}\Omega^1_{R/\mathscr{O}_K}$.

Proposition 1.3.15 (cf. 3.3.17) *For every Dolbeault* $\widehat{\overline{R}}[\frac{1}{p}]$*-representation* M *of* Δ*, there exists a canonical functorial isomorphism in* $\mathbf{D}^+(\mathbf{Mod}(\widehat{R_1}[\frac{1}{p}]))$

$$\mathbf{C}^\bullet_{\mathrm{cont}}(\Delta, M) \xrightarrow{\sim} \mathbb{K}^\bullet(\mathbb{H}(M)), \tag{1.3.15.1}$$

where $\mathbf{C}^\bullet_{\mathrm{cont}}(\Delta, M)$ *is the complex of continuous cochains of* Δ *with values in* M *and* $\mathbb{K}^\bullet(\mathbb{H}(M))$ *is the Dolbeault complex of the Higgs* $\widehat{R_1}[\frac{1}{p}]$*-module* $\mathbb{H}(M)$ *associated with* M (1.3.10.2).

Proposition 1.3.16 (cf. 3.5.5) *Let* M *be a projective* $\widehat{\overline{R}}[\frac{1}{p}]$*-module of finite type endowed with an* $\widehat{\overline{R}}[\frac{1}{p}]$*-semi-linear action of* Δ*. Then, the following properties are equivalent:*

(i) *The* $\widehat{\overline{R}}[\frac{1}{p}]$*-representation* M *of* Δ *is Dolbeault and the associated Higgs* $\widehat{R_1}[\frac{1}{p}]$*-module* $(\mathbb{H}(M), \theta)$ (1.3.10.1) *is nilpotent, i.e., there exists a finite decreasing filtration* $(\mathbb{H}_i)_{0 \le i \le n}$ *of* $\mathbb{H}(M)$ *by sub-*$\widehat{R_1}[\frac{1}{p}]$*-modules such that* $\mathbb{H}_0 = \mathbb{H}(M)$, $\mathbb{H}_n = 0$ *and that for every* $0 \le i \le n-1$*, we have*

$$\theta(\mathbb{H}_i) \subset \widetilde{\xi}^{-1}\Omega^1_{R/\mathscr{O}_K} \otimes_R \mathbb{H}_{i+1}. \tag{1.3.16.1}$$

(ii) *There exists a projective* $\widehat{R_1}[\frac{1}{p}]$*-module of finite type* N*, a Higgs* $\widehat{R_1}[\frac{1}{p}]$*-field* θ *on* N *with coefficients in* $\widetilde{\xi}^{-1}\Omega^1_{R/\mathscr{O}_K}$ *and a* $\mathscr{C}$*-linear and* Δ*-equivariant isomorphism of Higgs* $\widehat{\overline{R}}[\frac{1}{p}]$*-modules*

$$N \otimes_{\widehat{R_1}} \mathscr{C} \xrightarrow{\sim} M \otimes_{\widehat{\overline{R}}} \mathscr{C}. \tag{1.3.16.2}$$

Moreover, under these conditions, we have an isomorphism of Higgs $\widehat{R_1}[\frac{1}{p}]$*-modules*

$$\mathbb{H}(M) \xrightarrow{\sim} (N, \theta). \tag{1.3.16.3}$$

Definition 1.3.17 (cf. 3.5.6) We say that an $\widehat{\overline{R}}[\frac{1}{p}]$-representation M of Δ is *Hodge–Tate* if it satisfies the equivalent conditions 1.3.16.

This notion does not depend on the choice of the $\widetilde{S}$-deformation $\widetilde{X}$, not even on the absolute or relative case (1.3.6).

Remark 1.3.18 Tsuji [57] developed an arithmetic version of the local p-adic Simpson correspondence. He associates with a p-adic $\widehat{\overline{R}}[\frac{1}{p}]$-representation of $\Gamma = \pi_1(X^\star_{\overline{\eta}}, \overline{y})$ (1.3.5) a Higgs field and an arithmetic Sen operator satisfying a compatibility relation that forces the Higgs field to be nilpotent (see also [31]). This explains the relation between the work of Liu and Zhu [44] and the point of view developed here. The reader should beware, however, that our notion of Hodge–Tate local systems does not correspond to that of Liu and Zhu since we consider geometric local systems while they consider arithmetic local systems.

1.4 Global Theory. Dolbeault Modules

1.4.1 The local p-adic Simpson correspondence described in § 1.3 cannot be easily glued into a global correspondence for schemes that are not small affine. Instead, we sheafify the Higgs–Tate algebra in Faltings topos and then use it to build a global p-adic Simpson correspondence parallel to the local picture. Then, we prove by cohomological descent that the global correspondence is equivalent to the local one for small affine schemes.

1.4.2 Let X be a smooth S-scheme. With the notation of 1.3.6, we assume that there exists a smooth $\widetilde{S}$-deformation $\widetilde{X}$ of $X \otimes_{\mathscr{O}_K} \mathscr{O}_C$ that we fix in this section:

$$\begin{array}{ccc} X \otimes_{\mathscr{O}_K} \mathscr{O}_C & \longrightarrow & \widetilde{X} \\ \downarrow & \square & \downarrow \\ \mathrm{Spec}(\mathscr{O}_C) & \longrightarrow & \widetilde{S}. \end{array} \tag{1.4.2.1}$$

Note that the condition is superfluous in the *relative case*; we can indeed take $\widetilde{X} = X \times_S \widetilde{S}$.

We take again the notation of Section 1.2. Recall in particular that we have the ring $\overline{\mathscr{B}} = \{U \mapsto \overline{\mathscr{B}}_U\}$ of Faltings topos $\widetilde{E}$ (1.2.7). For every $n \geq 0$, we set $\overline{\mathscr{B}}_n = \overline{\mathscr{B}}/p^n\overline{\mathscr{B}}$ and for every $U \in \mathrm{Ob}(\mathbf{\acute{E}t}_{/X})$, $\overline{\mathscr{B}}_{U,n} = \overline{\mathscr{B}}_U/p^n\overline{\mathscr{B}}_U$, which is a ring of $U_{\overline{\eta},\mathrm{f\acute{e}t}}$.

For every small affine étale X-scheme U (1.3.5), there exists a canonical exact sequence of $\overline{\mathscr{B}}_{U,n}$-modules of $U_{\overline{\eta},\mathrm{f\acute{e}t}}$

$$0 \to \overline{\mathscr{B}}_{U,n} \to \mathscr{F}_{U,n} \to \widetilde{\xi}^{-1}\Omega^1_{X/S}(U) \otimes_{\mathscr{O}_X(U)} \overline{\mathscr{B}}_{U,n} \to 0, \tag{1.4.2.2}$$

such that for every geometric point $\overline{y}$ of $U_{\overline{\eta}}$, we have a canonical isomorphism of $\overline{\mathscr{B}}_{U,\overline{y}}$-representations of $\pi_1(U_{\overline{\eta}}, \overline{y})$

$$(\mathscr{F}_{U,n})_{\overline{y}} \xrightarrow{\sim} \mathscr{F}^{\overline{y}}_U/p^n\mathscr{F}^{\overline{y}}_U, \tag{1.4.2.3}$$

where $\mathscr{F}^{\overline{y}}_U$ is the Higgs–Tate $\overline{\mathscr{B}}_{U,\overline{y}}$-extension (1.3.8.3) relative to the restriction of $\widetilde{X}$ over U (4.4.5). For every rational number $r \geq 0$, let

$$0 \to \overline{\mathscr{B}}_{U,n} \to \mathscr{F}^{(r)}_{U,n} \to \widetilde{\xi}^{-1}\Omega^1_{X/S}(U) \otimes_{\mathscr{O}_X(U)} \overline{\mathscr{B}}_{U,n} \to 0 \tag{1.4.2.4}$$

be the extension of $\overline{\mathscr{B}}_{U,n}$-modules of $U_{\overline{\eta},\text{fét}}$ obtained from $\mathscr{F}_{U,n}$ by pullback by the multiplication by p^r on $\widetilde{\xi}^{-1}\Omega^1_{X/S}(U) \otimes_{\mathscr{O}_X(U)} \overline{\mathscr{B}}_{U,n}$, and let

$$\mathscr{C}^{(r)}_{U,n} = \varinjlim_{m\geq 0} \mathrm{Sym}^m_{\overline{\mathscr{B}}_{U,n}}(\mathscr{F}^{(r)}_{U,n}) \tag{1.4.2.5}$$

be the associated $\overline{\mathscr{B}}_{U,n}$-algebra of $U_{\overline{\eta},\text{fét}}$, where the transition morphisms are locally defined by sending $x_1 \otimes \cdots \otimes x_m$ to $1 \otimes x_1 \otimes \cdots \otimes x_m$.

The formation of $\mathscr{F}^{(r)}_{U,n}$ being functorial in U, the correspondences

$$\{U \mapsto \mathscr{F}^{(r)}_{U,n}\} \quad \text{and} \quad \{U \mapsto \mathscr{C}^{(r)}_{U,n}\} \tag{1.4.2.6}$$

define presheaves on the subcategory E^{sm} of E made up of objects $(V \to U)$ such that U is small affine (1.2.1). The latter is topologically generating of E. Therefore, taking the associated sheaves, we get a $\overline{\mathscr{B}}_n$-module $\mathscr{F}^{(r)}_n$ and a $\overline{\mathscr{B}}_n$-algebra $\mathscr{C}^{(r)}_n$ of $\widetilde{E}$ (4.4.10).

1.4.3 We set $\mathscr{S} = \mathrm{Spf}(\mathscr{O}_C)$ and we denote by $\mathfrak{X}$ the formal scheme p-adic completion of $\overline{X} = X \times_S \overline{S}$. Similarly, to take into account the p-adic topology, we consider the formal p-adic completion of the ringed topos $(\widetilde{E}, \overline{\mathscr{B}})$. First, we define the *special fiber* $\widetilde{E}_s$ of $\widetilde{E}$, a topos that fits into a commutative diagram

$$\begin{array}{ccc} \widetilde{E}_s & \xrightarrow{\sigma_s} & X_{s,\text{ét}} \\ {\scriptstyle\delta}\downarrow & & \downarrow{\scriptstyle\iota} \\ \widetilde{E} & \xrightarrow{\sigma} & X_{\text{ét}} \end{array} \tag{1.4.3.1}$$

where ι denotes the canonical injection (4.3.3). Concretely, $\widetilde{E}_s$ is the full subcategory of $\widetilde{E}$ made up of objects F such that $F|\sigma^*(X_\eta)$ is the final object of $\widetilde{E}_{/\sigma^*(X_\eta)}$, and $\delta_*\colon \widetilde{E}_s \to \widetilde{E}$ is the canonical injection functor.

For every integer $n \geq 0$, $\overline{\mathscr{B}}_n$ is an object of $\widetilde{E}_s$. We denote by $\overline{X}_n$ and $\overline{S}_n$ the reductions of $\overline{X}$ and $\overline{S}$ modulo p^n. Then, the morphism σ_s is underlying a canonical morphism of ringed topos

$$\sigma_n\colon (\widetilde{E}_s, \overline{\mathscr{B}}_n) \to (X_{s,\text{ét}}, \mathscr{O}_{\overline{X}_n}), \tag{1.4.3.2}$$

where we identified the étale topos of X_s and $\overline{X}_n$, since k is algebraically closed.

The formal p-adic completion of $(\widetilde{E}, \overline{\mathscr{B}})$ is the ringed topos $(\widetilde{E}_s^{\mathbb{N}^\circ}, \breve{\overline{\mathscr{B}}})$, where $\widetilde{E}_s^{\mathbb{N}^\circ}$ denotes the topos of inverse systems of $\widetilde{E}_s$ indexed by the ordered set $\mathbb{N}$ ([3] III.7) and $\breve{\overline{\mathscr{B}}} = (\overline{\mathscr{B}}_n)_{n\geq 0}$. The morphisms σ_n induce a morphism of topos

$$\widehat{\sigma}\colon (\widetilde{E}_s^{\mathbb{N}^\circ}, \breve{\overline{\mathscr{B}}}) \to (X_{s,\text{zar}}, \mathscr{O}_{\mathfrak{X}}). \tag{1.4.3.3}$$

We work in the category $\mathbf{Mod}_{\mathbb{Q}}(\breve{\overline{\mathscr{B}}})$ of *$\breve{\overline{\mathscr{B}}}$-modules up to isogeny*, i.e., the category having for objects $\breve{\overline{\mathscr{B}}}$-modules, and for any $\breve{\overline{\mathscr{B}}}$-modules $\mathscr{F}$ and $\mathscr{G}$,

$$\mathrm{Hom}_{\mathbf{Mod}_{\mathbb{Q}}(\breve{\overline{\mathscr{B}}})}(\mathscr{F},\mathscr{G}) = \mathrm{Hom}_{\mathbf{Mod}(\breve{\overline{\mathscr{B}}})}(\mathscr{F},\mathscr{G}) \otimes_{\mathbb{Z}} \mathbb{Q}. \tag{1.4.3.4}$$

We denote the localization functor $\mathbf{Mod}(\breve{\overline{\mathscr{B}}}) \to \mathbf{Mod}_{\mathbb{Q}}(\breve{\overline{\mathscr{B}}})$ by $\mathscr{F} \mapsto \mathscr{F}_{\mathbb{Q}}$. We call *$\breve{\overline{\mathscr{B}}}_{\mathbb{Q}}$-modules* the objects of $\mathbf{Mod}_{\mathbb{Q}}(\breve{\overline{\mathscr{B}}})$.

1.4.4 For every rational number $r \geq 0$ and every integer $n \geq 0$, we have a canonical locally split exact sequence

$$0 \to \overline{\mathscr{B}}_n \to \mathscr{F}_n^{(r)} \to \sigma_n^*(\widetilde{\xi}^{-1}\Omega^1_{\overline{X}_n/\overline{S}_n}) \to 0 \tag{1.4.4.1}$$

and a canonical isomorphism of $\overline{\mathscr{B}}_n$-algebras

$$\mathscr{C}_n^{(r)} \xrightarrow{\sim} \varinjlim_{m\geq 0} \mathrm{Sym}^m_{\overline{\mathscr{B}}_n}(\mathscr{F}_n^{(r)}), \tag{1.4.4.2}$$

where the transition morphisms are locally defined by sending $x_1 \otimes \cdots \otimes x_m$ to $1 \otimes x_1 \otimes \cdots \otimes x_m$ (4.4.11). We set $\breve{\mathscr{F}}^{(r)} = (\mathscr{F}_n^{(r)})_{n\geq 0}$, which is a $\breve{\overline{\mathscr{B}}}$-module, and $\breve{\mathscr{C}}^{(r)} = (\mathscr{C}_n^{(r)})_{n\geq 0}$, which is a $\breve{\overline{\mathscr{B}}}$-algebra. We have a canonical exact sequence of $\breve{\overline{\mathscr{B}}}$-modules

$$0 \to \breve{\overline{\mathscr{B}}} \to \breve{\mathscr{F}}^{(r)} \to \widehat{\sigma}^*(\widetilde{\xi}^{-1}\Omega^1_{\mathfrak{X}/\mathcal{S}}) \to 0. \tag{1.4.4.3}$$

The universal $\breve{\overline{\mathscr{B}}}$-derivation of $\breve{\mathscr{C}}^{(r)}$ can be identified with a derivation

$$d_{\breve{\mathscr{C}}^{(r)}} : \breve{\mathscr{C}}^{(r)} \to \widehat{\sigma}^*(\widetilde{\xi}^{-1}\Omega^1_{\mathfrak{X}/\mathcal{S}}) \otimes_{\breve{\overline{\mathscr{B}}}} \breve{\mathscr{C}}^{(r)}. \tag{1.4.4.4}$$

It is a Higgs $\breve{\overline{\mathscr{B}}}$-field. We denote by $\mathbb{K}^\bullet(\breve{\mathscr{C}}^{(r)})$ the Dolbeault complex of the Higgs $\breve{\overline{\mathscr{B}}}$-module $(\breve{\mathscr{C}}^{(r)}, p^r d_{\breve{\mathscr{C}}^{(r)}})$.

For all rational numbers $r \geq r' \geq 0$, we have a canonical homomorphism of $\breve{\overline{\mathscr{B}}}$-algebras $\breve{\mathscr{C}}^{(r)} \to \breve{\mathscr{C}}^{(r')}$. Observe that the restriction of the derivation $p^{r'} d_{\breve{\mathscr{C}}^{(r')}}$ is $p^r d_{\breve{\mathscr{C}}^{(r)}}$. Hence, we have a morphism of complexes

$$\mathbb{K}^\bullet(\breve{\mathscr{C}}^{(r)}) \to \mathbb{K}^\bullet(\breve{\mathscr{C}}^{(r')}). \tag{1.4.4.5}$$

Proposition 1.4.5 (cf. 4.4.32, [3] III.11.18) *The canonical homomorphism*

$$\mathscr{O}_{\mathfrak{X}}[\frac{1}{p}] \to \varinjlim_{r\in\mathbb{Q}_{>0}} \widehat{\sigma}_*(\breve{\mathscr{C}}^{(r)})[\frac{1}{p}] \tag{1.4.5.1}$$

is an isomorphism and, for every $q \geq 1$,

$$\varinjlim_{r\in\mathbb{Q}_{>0}} \mathrm{R}^q\widehat{\sigma}_*(\breve{\mathscr{C}}^{(r)})[\frac{1}{p}] = 0. \tag{1.4.5.2}$$

This result is a sheafification of the computation of the Galois cohomology of the Higgs–Tate algebra over a small affine scheme ([3] II.12.5). This computation relies on Faltings' almost purity result and the sheafication requires to prove a version modulo p^n, up to some bounded defect (3.2.24, [3] II.12.7).

Proposition 1.4.6 (cf. 4.4.36, [3] III.11.24) *The canonical morphism of* $\mathbf{Mod}_{\mathbb{Q}}(\overline{\breve{\mathscr{B}}})$

$$\overline{\breve{\mathscr{B}}}_{\mathbb{Q}} \to \varinjlim_{r\in\mathbb{Q}_{>0}} \mathrm{H}^0(\mathbb{K}^{\bullet}_{\mathbb{Q}}(\breve{\mathscr{C}}^{(r)})) \tag{1.4.6.1}$$

is an isomorphism, and for every $q \geq 1$,

$$\varinjlim_{r\in\mathbb{Q}_{>0}} \mathrm{H}^q(\mathbb{K}^{\bullet}_{\mathbb{Q}}(\breve{\mathscr{C}}^{(r)})) = 0. \tag{1.4.6.2}$$

This result is a sheafification of the computation of the de Rham cohomology of the Higgs–Tate algebra over a small affine scheme (3.2.22, [3] II.12.3).

1.4.7 Filtered direct limits are not a priori representable in the category $\mathbf{Mod}_{\mathbb{Q}}(\overline{\breve{\mathscr{B}}})$. However, we can naturally embed this category into the abelian category $\mathbf{Ind\text{-}Mod}(\overline{\breve{\mathscr{B}}})$ of ind-$\overline{\breve{\mathscr{B}}}$-modules where filtered direct limits are representable and which has better properties (2.6, 2.7, [40]). In the same way, one can naturally embed the category of coherent $\mathscr{O}_{\mathfrak{X}}[\frac{1}{p}]$-modules into the category $\mathbf{Ind\text{-}Mod}(\mathscr{O}_{\mathfrak{X}})$ of ind-$\mathscr{O}_{\mathfrak{X}}$-modules (4.3.16). The morphism $\widehat{\sigma}$ (1.4.3.3) induces two adjoint functors

$$\mathbf{Ind\text{-}Mod}(\overline{\breve{\mathscr{B}}}) \underset{\mathrm{I}\widehat{\sigma}^*}{\overset{\mathrm{I}\widehat{\sigma}_*}{\rightleftarrows}} \mathbf{Ind\text{-}Mod}(\mathscr{O}_{\mathfrak{X}}) \tag{1.4.7.1}$$

that extend the adjoint functors $\widehat{\sigma}^*$ and $\widehat{\sigma}_*$.

Definition 1.4.8 We call a *Higgs* $\mathscr{O}_{\mathfrak{X}}[\frac{1}{p}]$*-bundle with coefficients in* $\widetilde{\xi}^{-1}\Omega^1_{\mathfrak{X}/\mathscr{S}}$ any locally projective $\mathscr{O}_{\mathfrak{X}}[\frac{1}{p}]$-module of finite type $\mathscr{N}$ (2.1.11) endowed with an $\mathscr{O}_{\mathfrak{X}}$-linear morphism $\theta\colon \mathscr{N} \to \widetilde{\xi}^{-1}\Omega^1_{\mathfrak{X}/\mathscr{S}} \otimes_{\mathscr{O}_{\mathfrak{X}}} \mathscr{N}$ such that $\theta \wedge \theta = 0$.

Observe that since the stalks of the ring $\mathscr{O}_{\mathfrak{X}}[\frac{1}{p}]$ are not necessarily local rings, locally projective $\mathscr{O}_{\mathfrak{X}}[\frac{1}{p}]$-modules of finite type are not necessarily locally free.

Definition 1.4.9 (cf. 4.5.4) Let $\mathscr{M}$ be an ind-$\overline{\breve{\mathscr{B}}}$-module, $\mathscr{N}$ a Higgs $\mathscr{O}_{\mathfrak{X}}[\frac{1}{p}]$-bundle with coefficients in $\widetilde{\xi}^{-1}\Omega^1_{\mathfrak{X}/\mathscr{S}}$.

(i) We say that $\mathscr{M}$ and $\mathscr{N}$ are *r-associated* (for $r \in \mathbb{Q}_{>0}$) if there exists an isomorphism of ind-$\breve{\mathscr{C}}^{(r)}$-modules

$$\mathscr{M} \otimes_{\overline{\breve{\mathscr{B}}}} \breve{\mathscr{C}}^{(r)} \xrightarrow{\sim} \mathrm{I}\widehat{\sigma}^*(\mathscr{N}) \otimes_{\overline{\breve{\mathscr{B}}}} \breve{\mathscr{C}}^{(r)}, \tag{1.4.9.1}$$

compatible with the total Higgs $\overline{\breve{\mathscr{B}}}$-fields with coefficients in $\widehat{\sigma}^*(\widetilde{\xi}^{-1}\Omega^1_{\mathfrak{X}/\mathscr{S}})$, where $\mathscr{M}$ is endowed with the zero Higgs field and $\breve{\mathscr{C}}^{(r)}$ with the Higgs field $p^r d_{\breve{\mathscr{C}}^{(r)}}$.

(ii) We say that $\mathscr{M}$ and $\mathscr{N}$ are *associated* if they are r-associated for a rational number $r > 0$.

In fact, (1.4.9.1) is an isomorphism of *ind-$\breve{\mathscr{C}}^{(r)}$-modules with p^r-connection with respect to the extension* $\breve{\mathscr{C}}^{(r)}/\breve{\overline{\mathscr{B}}}$ (4.5.1). In particular, for all rational numbers $r \geq r' > 0$, if $\mathscr{M}$ and $\mathscr{N}$ are r-associated, they are r'-associated.

Definition 1.4.10 (cf. 4.5.5)

(i) We say that an ind-$\breve{\overline{\mathscr{B}}}$-module is *Dolbeault* if it is associated with a Higgs $\mathscr{O}_{\mathfrak{X}}[\frac{1}{p}]$-bundle with coefficients in $\widetilde{\xi}^{-1}\Omega^1_{\mathfrak{X}/\mathscr{S}}$.

(ii) We say that a Higgs $\mathscr{O}_{\mathfrak{X}}[\frac{1}{p}]$-bundle with coefficients in $\widetilde{\xi}^{-1}\Omega^1_{\mathfrak{X}/\mathscr{S}}$ is *solvable* if it is associated with an ind-$\breve{\overline{\mathscr{B}}}$-module.

The property for an ind-$\breve{\overline{\mathscr{B}}}$-module to be Dolbeault does not depend on the choice of the deformation $\widetilde{X}$ (1.4.2) provided that we stay in one of the settings, absolute or relative (1.3.6) (see 4.10.4). The property for a Higgs $\mathscr{O}_{\mathfrak{X}}[\frac{1}{p}]$-bundle to be solvable depends a priori on the deformation $\widetilde{X}$ (see however 4.10.12).

The notion of being Dolbeault applies to $\breve{\overline{\mathscr{B}}}_{\mathbb{Q}}$-modules (1.4.7). We call the associated Higgs bundles *rationally solvable*.

In ([3] III.12.11), we considered only $\breve{\overline{\mathscr{B}}}_{\mathbb{Q}}$-modules and we requested moreover that they are adic of finite type. We renamed them, in 4.6.6, *strongly Dolbeault* $\breve{\overline{\mathscr{B}}}_{\mathbb{Q}}$-modules and renamed the associated Higgs bundles *strongly solvable*. The finiteness condition is important for matching the global and local theories for small affine schemes (1.4.20).

Theorem 1.4.11 (cf. 4.5.20) *There are explicit equivalences of categories quasi-inverse to each other*

$$\mathbf{Ind\text{-}Mod}^{\mathrm{Dolb}}(\breve{\overline{\mathscr{B}}}) \underset{\mathscr{V}}{\overset{\mathscr{H}}{\rightleftarrows}} \mathbf{HM}^{\mathrm{sol}}(\mathscr{O}_{\mathfrak{X}}[\tfrac{1}{p}], \widetilde{\xi}^{-1}\Omega^1_{\mathfrak{X}/\mathscr{S}}) \qquad (1.4.11.1)$$

between the category of Dolbeault ind-$\breve{\overline{\mathscr{B}}}$-modules and the category of solvable Higgs $\mathscr{O}_{\mathfrak{X}}[\frac{1}{p}]$-bundles with coefficients in $\xi^{-1}\Omega^1_{\mathfrak{X}/\mathscr{S}}$.

These functors are explicitly defined in 4.5.7 as follows. Let $\breve{\sigma}_*$ be the composed functor

$$\mathbf{Ind\text{-}Mod}(\breve{\overline{\mathscr{B}}}) \xrightarrow{\mathrm{I}\breve{\sigma}_*} \mathbf{Ind\text{-}Mod}(\mathscr{O}_{\mathfrak{X}}) \xrightarrow{\kappa_{\mathscr{O}_{\mathfrak{X}}}} \mathbf{Mod}(\mathscr{O}_{\mathfrak{X}}), \qquad (1.4.11.2)$$

where $\mathrm{I}\breve{\sigma}_*$ is defined in (1.4.7.1) and

$$\kappa_{\mathscr{O}_{\mathfrak{X}}}(\text{“}\underset{\longrightarrow}{\lim}\text{”}\alpha) = \underset{\longrightarrow}{\lim}\,\alpha. \qquad (1.4.11.3)$$

Then, the functor $\mathscr{H}$ can in fact be defined for any ind-$\breve{\overline{\mathscr{B}}}$-module $\mathscr{M}$, by

$$\mathscr{H}(\mathscr{M}) = \varinjlim_{r\in\mathbb{Q}_{>0}} \vec{\sigma}_*(\mathscr{M} \otimes_{\breve{\overline{\mathscr{B}}}} \breve{\mathscr{C}}^{(r)}, p^r \mathrm{id} \otimes d_{\breve{\mathscr{C}}^{(r)}}). \qquad (1.4.11.4)$$

We have a similar definition for $\mathscr{V}$. The functors $\mathscr{H}$ and $\mathscr{V}$ depend a priori on the deformation $\widetilde{X}$.

Remark 1.4.12 Faltings [18] outlined a strategy to extend the correspondence (1.4.11.1) for semi-stable proper curves over S that are (strict and) smooth over η, to all $\breve{\overline{\mathscr{B}}}_{\mathbb{Q}}$-modules and all Higgs $\mathscr{O}_{\mathfrak{X}}[\frac{1}{p}]$-bundles. For this purpose, he uses descent: any generalized representation (resp. Higgs bundle) becomes Dolbeault (resp. solvable), or equivalently small, on a finite étale covering of the geometric generic fiber of the curve. The descent of the correspondence (1.4.11.1) defined on a semi-stable integral model of the covering to a correspondence defined on the original curve is not immediate because it depends on the choice of a deformation (1.4.2.1), and the finite étale covering does not necessarily extend to the given deformations of the integral models. Faltings showed that after twisting the inverse image of the Higgs bundle by the obstruction to the existence of such an extension, the correspondence descends properly. This twisted inverse image is closely related to the *Hitchin fibration* [34]. There have been recent interesting developments concerning this aspect of the theory, due independently to Heuer [33] and Xu [58]. We will study the descent in the context of our approach in an upcoming joint work with T. Tsuji.

Theorem 1.4.13 (cf. 4.7.4) *For every Dolbeault ind-$\breve{\overline{\mathscr{B}}}$-module $\mathscr{M}$ and every integer $q \geq 0$, we have a canonical functorial isomorphism of* $\mathbf{D}^+(\mathbf{Mod}(\mathscr{O}_{\mathfrak{X}}))$

$$\mathrm{R}\vec{\sigma}_*(\mathscr{M}) \xrightarrow{\sim} \mathbb{K}^\bullet(\mathscr{H}(\mathscr{M})), \qquad (1.4.13.1)$$

where $\mathbb{K}^\bullet(\mathscr{H}(\mathscr{M}))$ is the Dolbeault complex of $\mathscr{H}(\mathscr{M})$.

This is a global analog of 1.3.15.

1.4.14 The morphism ψ (1.2.4.1) induces a morphism of topos

$$\breve{\psi}\colon X_{\overline{\eta},\text{ét}}^{\mathbb{N}^\circ} \to \widetilde{E}^{\mathbb{N}^\circ}, \qquad (1.4.14.1)$$

where $X_{\overline{\eta},\text{ét}}^{\mathbb{N}^\circ}$ is the topos of the inverse systems of objects of $X_{\overline{\eta},\text{ét}}$, indexed by the ordered set $\mathbb{N}$. We denote by $\breve{\mathbb{Z}}_p$ the ring $(\mathbb{Z}/p^n\mathbb{Z})_{n\geq 0}$ of $X_{\overline{\eta},\text{ét}}^{\mathbb{N}^\circ}$.

We say that a $\breve{\mathbb{Z}}_p$-module $M = (M_n)_{n\in\mathbb{N}}$ of $X_{\overline{\eta},\text{ét}}^{\mathbb{N}^\circ}$ is a *local system* if the following two conditions are satisfied:

(a) M is p-adic, i.e., for all integers $n \geq m \geq 0$, the morphism $M_n/p^m M_n \to M_m$ deduced from the transition morphism $M_n \to M_m$ is an isomorphism;
(b) for all integer $n \geq 0$, the $\mathbb{Z}/p^n\mathbb{Z}$-module M_n of $X_{\overline{\eta},\text{ét}}$ is locally constant constructible.

Corollary 1.4.15 (cf. 4.12.6) *Let $M = (M_n)_{n\geq 0}$ be a $\check{\mathbb{Z}}_p$-local system of $X^{\mathbb{N}^\circ}_{\overline{\eta},\text{ét}}$, $\mathscr{M} = \check{\psi}_*(M) \otimes_{\check{\mathbb{Z}}_p} \overline{\check{\mathscr{B}}}$. Assume that X is proper over S and that the $\overline{\check{\mathscr{B}}}_{\mathbb{Q}}$-module $\mathscr{M}_{\mathbb{Q}}$ is Dolbeault. Then, there exists a canonical spectral sequence*

$$\mathrm{E}_2^{i,j} = \mathrm{H}^i(X_s, \mathrm{H}^j(\mathbb{K}^\bullet)) \Rightarrow \mathrm{H}^{i+j}(X^{\mathbb{N}^\circ}_{\overline{\eta},\text{ét}}, M) \otimes_{\mathbb{Z}_p} C, \tag{1.4.15.1}$$

where $\mathbb{K}^\bullet$ is the Dolbeault complex of $\mathscr{H}(\mathscr{M}_{\mathbb{Q}})$.

This follows from 1.2.8 and 1.4.13.

Remark 1.4.16 In 1.4.15, if we take $M = \check{\mathbb{Z}}_p$, then $\mathscr{M} = \overline{\check{\mathscr{B}}}$, the $\overline{\check{\mathscr{B}}}_{\mathbb{Q}}$-module $\overline{\check{\mathscr{B}}}_{\mathbb{Q}}$ is Dolbeault and $\mathscr{H}(\overline{\check{\mathscr{B}}}_{\mathbb{Q}})$ is equal to $\mathscr{O}_{\mathfrak{X}}[\frac{1}{p}]$ endowed with the zero Higgs field (4.6.10). The spectral sequence (1.4.15.1) is the Hodge–Tate spectral sequence ([2] 6.4.6). Observe that the construction 1.4.15 of this spectral sequence shows directly that it degenerates at E_2 and that the abutment filtration is split without using Tate's theorem on the Galois cohomology of $C(j)$. This construction applies in particular by taking for $\widetilde{X}$ in the relative case (1.3.6) the trivial deformation (1.4.2).

Definition 1.4.17 (cf. 4.11.1) We call a *Hodge–Tate $\overline{\check{\mathscr{B}}}_{\mathbb{Q}}$-module* any Dolbeault $\overline{\check{\mathscr{B}}}_{\mathbb{Q}}$-module $\mathscr{M}$ (1.4.10) whose associated Higgs $\mathscr{O}_{\mathfrak{X}}[\frac{1}{p}]$-bundle $(\mathscr{H}(\mathscr{M}), \theta)$ (1.4.11.1) is nilpotent, i.e., there exists a decreasing finite filtration $(\mathscr{H}_i(\mathscr{M}))_{0\leq i\leq n}$ of $\mathscr{H}(\mathscr{M})$ by coherent sub-$\mathscr{O}_{\mathfrak{X}}[\frac{1}{p}]$-modules such that $\mathscr{H}_0(\mathscr{M}) = \mathscr{H}(\mathscr{M})$, $\mathscr{H}_n(\mathscr{M}) = 0$ and that for every $0 \leq i \leq n-1$, we have

$$\theta(\mathscr{H}_i(\mathscr{M})) \subset \widetilde{\xi}^{-1}\Omega^1_{\mathfrak{X}/\mathscr{S}} \otimes_{\mathscr{O}_{\mathfrak{X}}} \mathscr{H}_{i+1}(\mathscr{M}). \tag{1.4.17.1}$$

This notion does not depend on the choice of the $\widetilde{S}$-deformation $\widetilde{X}$, not even on the absolute or relative case (1.3.6) (see 4.11.9).

Proposition 1.4.18 (cf. 4.11.2) *The functors $\mathscr{H}$ and $\mathscr{V}$ (1.4.11.1) induce equivalences of categories quasi-inverse to each other*

$$\mathbf{Mod}^{\mathrm{HT}}_{\mathbb{Q}}(\overline{\check{\mathscr{B}}}) \underset{\mathscr{V}}{\overset{\mathscr{H}}{\rightleftarrows}} \mathbf{HM}^{\mathrm{qsolnilp}}(\mathscr{O}_{\mathfrak{X}}[\tfrac{1}{p}], \widetilde{\xi}^{-1}\Omega^1_{\mathfrak{X}/\mathscr{S}}) \tag{1.4.18.1}$$

between the category of Hodge–Tate $\overline{\check{\mathscr{B}}}_{\mathbb{Q}}$-modules and the category of rationally solvable and nilpotent Higgs $\mathscr{O}_{\mathfrak{X}}[\frac{1}{p}]$-bundles with coefficients in $\widetilde{\xi}^{-1}\Omega^1_{\mathfrak{X}/\mathscr{S}}$ (1.4.10).

1.4.19 We assume in the remaining part of this section that X is small affine (1.3.5), that $X_s \neq \emptyset$ and, for simplicity, that $X_{\overline{\eta}}$ is connected. We fix a geometric point $\overline{y}$ of $X_{\overline{\eta}}$. We set $\Delta = \pi_1(X_{\overline{\eta}}, \overline{y})$, and we denote by $\mathbf{B}_\Delta$ the classifying topos of Δ and by

$$\nu\colon X_{\overline{\eta},\text{fét}} \xrightarrow{\sim} \mathbf{B}_\Delta \tag{1.4.19.1}$$

the fiber functor of $X_{\overline{\eta},\text{fét}}$ at $\overline{y}$ ([3] (VI.9.8.4)). We denote by β the composed functor

$$\beta\colon \widetilde{E} \to \mathbf{B}_\Delta, \quad F \mapsto \nu \circ (\beta_*(F)), \tag{1.4.19.2}$$

where β is the morphism of topos (1.2.4.4). We thus define a functor from the category of abelian sheaves of $\widetilde{E}$ to the category of $\mathbb{Z}[\Delta]$-modules. The latter being left exact, we denote by $\mathrm{R}^q\upbeta$ $(q \geq 0)$ its right derived functors. For every abelian sheaf F of $\widetilde{E}$ and every integer $q \geq 0$, we have a canonical functorial isomorphism

$$\mathrm{R}^q\upbeta(F) \xrightarrow{\sim} \nu \circ (\mathrm{R}^q\beta_*(F)). \tag{1.4.19.3}$$

For any abelian sheaf $F = (F_n)_{n\geq 0}$ of $\widetilde{E}^{\mathbb{N}^\circ}$, we set

$$\widehat{\upbeta}(F) = \lim_{\substack{\longleftarrow \\ n\geq 0}} \upbeta(F_n). \tag{1.4.19.4}$$

We thus define a functor from the category of abelian sheaves of $\widetilde{E}^{\mathbb{N}^\circ}$ to the category of $\mathbb{Z}[\Delta]$-modules. The latter being left exact, we abusively denote by $\mathrm{R}^q\widehat{\upbeta}(F)$ $(q \geq 0)$ its right derived functors. By ([39] 1.6), we have a canonical exact sequence

$$0 \to \mathrm{R}^1\lim_{\substack{\longleftarrow \\ n\geq 0}} \mathrm{R}^{q-1}\upbeta(F_n) \to \mathrm{R}^q\widehat{\upbeta}(F) \to \lim_{\substack{\longleftarrow \\ n\geq 0}} \mathrm{R}^q\upbeta(F_n) \to 0, \tag{1.4.19.5}$$

where we set $\mathrm{R}^{-1}\upbeta(F_n) = 0$ for any $n \geq 0$.

We set $\overline{R} = \nu(\mathscr{B}_X)$, which is nothing but the algebra defined in (1.2.7.3) endowed with the canonical action of Δ. For every integer $q \geq 0$, $\mathrm{R}^q\widehat{\upbeta}$ induces a functor that we also denote by

$$\mathrm{R}^q\widehat{\upbeta}\colon \mathbf{Mod}(\breve{\overline{\mathscr{B}}}) \to \mathbf{Rep}_{\widehat{\overline{R}}}(\Delta). \tag{1.4.19.6}$$

The latter induces a functor that we also denote by

$$\mathrm{R}^q\widehat{\upbeta}\colon \mathbf{Mod}_{\mathbb{Q}}(\breve{\overline{\mathscr{B}}}) \to \mathbf{Rep}_{\widehat{\overline{R}}[\frac{1}{p}]}(\Delta). \tag{1.4.19.7}$$

Theorem 1.4.20 (cf. 4.8.31) *We keep the assumptions and notation of* 1.4.19. *Then,*

(i) *The functor* $\widehat{\upbeta}$ (1.4.19.7) *induces an equivalence of categories*

$$\mathbf{Mod}^{\mathrm{sDolb}}_{\mathbb{Q}}(\breve{\overline{\mathscr{B}}}) \xrightarrow{\sim} \mathbf{Rep}^{\mathrm{Dolb}}_{\widehat{\overline{R}}[\frac{1}{p}]}(\Delta), \tag{1.4.20.1}$$

between the category of strongly Dolbeault $\breve{\overline{\mathscr{B}}}_{\mathbb{Q}}$*-modules* (1.4.10) *and the category of Dolbeault* $\widehat{\overline{R}}[\frac{1}{p}]$*-representations of* Δ (1.3.12).

(ii) *For every strongly Dolbeault* $\breve{\overline{\mathscr{B}}}_{\mathbb{Q}}$*–module* $\mathscr{M}$ *and every integer* $q \geq 1$, *we have*

$$\mathrm{R}^q\widehat{\upbeta}(\mathscr{M}) = 0. \tag{1.4.20.2}$$

This follows from a cohomological descent result for strongly Dolbeault $\breve{\overline{\mathscr{B}}}_{\mathbb{Q}}$-modules 4.8.16 which, *in fine*, reduces to a cohomological descent result for the ring $\breve{\overline{\mathscr{B}}}_{\mathbb{Q}}$ ([2] 4.6.30).

We show that under the equivalence of categories (1.4.20.1), the functors $\mathscr{H}$ (1.4.11.1) and $\mathbb{H}$ (1.3.10.2) (resp. $\mathscr{V}$ (1.4.11.1) and $\mathbb{V}$ (1.3.11.2)) correspond (4.8.32).

1.5 Functoriality of the p-adic Simpson Correspondence by Proper Direct Image

1.5.1 Let $g\colon X' \to X$ be a smooth morphism of smooth S-schemes. We endow with a prime $'$ the objects associated with X'/S. By functoriality of the Faltings topos, g induces a canonical morphism Θ between the corresponding Faltings topos that fits into a commutative diagram

$$\begin{array}{ccccc} X'_{\overline{\eta},\text{ét}} & \xrightarrow{\psi'} & \widetilde{E}' & \xrightarrow{\sigma'} & X'_{\text{ét}} \\ \downarrow{\scriptstyle g_{\overline{\eta}}} & & \downarrow{\scriptstyle \Theta} & & \downarrow{\scriptstyle g} \\ X_{\overline{\eta},\text{ét}} & \xrightarrow{\psi} & \widetilde{E} & \xrightarrow{\sigma} & X_{\text{ét}}. \end{array} \tag{1.5.1.1}$$

We have also a canonical ring homomorphism

$$\overline{\mathscr{B}} \to \Theta_*(\overline{\mathscr{B}}'). \tag{1.5.1.2}$$

Theorem 1.5.2 ([17], [2] 5.7.4) *Assume that $g\colon X' \to X$ is proper, and let F' be a locally constant constructible sheaf of $(\mathbb{Z}/p^n\mathbb{Z})$-modules of $X'_{\overline{\eta},\text{ét}}$ $(n \geq 1)$. Then, for every integer $i \geq 0$, the canonical morphism*

$$\psi_*(\mathrm{R}^i g_{\overline{\eta}*}(F')) \otimes_{\mathbb{Z}_p} \overline{\mathscr{B}} \to \mathrm{R}^i\Theta_*(\psi'_*(F') \otimes_{\mathbb{Z}_p} \overline{\mathscr{B}}') \tag{1.5.2.1}$$

is an almost isomorphism.

Observe that the sheaves $\mathrm{R}^i g_{\overline{\eta}*}(F)$ are locally constant constructible on $X_{\overline{\eta}}$ by the smooth and the proper base change theorems.

Faltings formulated a *relative version* of his main p-adic comparison theorem in [17] and he very roughly sketched a proof in the appendix. Some arguments have to be modified and our actual proof in [2] requires much more work.

In ([2] 5.7.4), we actually required g to be projective. However, as pointed out in ([2] 5.7.6), the result holds under the more general assumption that g is proper by the same proof, replacing the almost finiteness result ([2] 2.8.18) by the recent generalization of He ([32] 1.5).

1.5.3 With the notation of 1.3.6, we assume that there exists a commutative diagram with Cartesian squares

$$\begin{array}{ccc} X' \otimes_{\mathscr{O}_K} \mathscr{O}_C & \longrightarrow & \widetilde{X}' \\ \downarrow{\scriptstyle g\otimes \mathrm{id}} & \square & \downarrow{\scriptstyle \widetilde{g}} \\ X \otimes_{\mathscr{O}_K} \mathscr{O}_C & \longrightarrow & \widetilde{X} \\ \downarrow & \square & \downarrow \\ \mathrm{Spec}(\mathscr{O}_C) & \longrightarrow & \widetilde{S} \end{array} \tag{1.5.3.1}$$

where $\widetilde{X}$ and $\widetilde{X}'$ are smooth $\widetilde{S}$-schemes, *that we fix in this section*. Note that the condition is superfluous in the *relative case*; we can indeed take $\widetilde{g} = g \times_S \widetilde{S}$. Observe that $\widetilde{g}$ is smooth by ([30] 17.11.1).

Diagram (1.5.1.1) induces a commutative diagram of morphisms of ringed topos

$$\begin{array}{ccc} (\widetilde{E}'^{\mathbb{N}^\circ}_s, \breve{\overline{\mathscr{B}}}') & \xrightarrow{\check{\theta}} & (\widetilde{E}^{\mathbb{N}^\circ}_s, \breve{\overline{\mathscr{B}}}) \\ \Big\downarrow{\scriptstyle \widehat{\sigma}'} & & \Big\downarrow{\scriptstyle \widehat{\sigma}} \\ (X'_{s,\mathrm{zar}}, \mathscr{O}_{\mathfrak{X}'}) & \xrightarrow{\mathfrak{g}} & (X_{s,\mathrm{zar}}, \mathscr{O}_{\mathfrak{X}}) \end{array} \tag{1.5.3.2}$$

in which the horizontal arrows are induced by Θ and g, and the vertical arrows are induced by σ' and σ (1.4.3.3). We prove in 6.3.20 that for every rational number $r \geq 0$, the lifting $\widetilde{g}$ (1.5.3.1) induces a homomorphism of Higgs–Tate algebras

$$\check{\theta}^*(\breve{\mathscr{C}}^{(r)}) \to \breve{\mathscr{C}}'^{(r)}, \tag{1.5.3.3}$$

whose construction is rather subtle.

Theorem 1.5.4 (cf. 6.5.24) *Assume that* $g\colon X' \to X$ *is proper. Let* $\mathscr{M}$ *be a Dolbeault ind-*$\breve{\overline{\mathscr{B}}}'$*-module* (1.4.10),

$$\mathscr{H}'(\mathscr{M}) \to \widetilde{\xi}^{-1}\Omega^1_{\mathfrak{X}'/\mathscr{S}} \otimes_{\mathscr{O}_{\mathfrak{X}'}} \mathscr{H}'(\mathscr{M}) \tag{1.5.4.1}$$

the associated Higgs bundle (1.4.11.1),

$$\underline{\mathscr{H}}'(\mathscr{M}) \to \widetilde{\xi}^{-1}\Omega^1_{\mathfrak{X}'/\mathfrak{X}} \otimes_{\mathscr{O}_{\mathfrak{X}'}} \underline{\mathscr{H}}'(\mathscr{M}) \tag{1.5.4.2}$$

the relative Higgs bundle induced by (1.5.4.1), $\underline{\mathbb{K}}^\bullet$ *the Dolbeault complex of* $\underline{\mathscr{H}}'(\mathscr{M})$. *Then, for every integer* $q \geq 0$, *there exists a rational number* $r > 0$ *and a* $\breve{\mathscr{C}}^{(r)}$*-isomorphism*

$$\mathrm{R}^q\mathrm{I}\check{\theta}_*(\mathscr{M}) \otimes_{\breve{\overline{\mathscr{B}}}} \breve{\mathscr{C}}^{(r)} \xrightarrow{\sim} \mathrm{I}\widehat{\sigma}^*(\mathrm{R}^q\mathfrak{g}_*(\underline{\mathbb{K}}^\bullet)) \otimes_{\breve{\overline{\mathscr{B}}}} \breve{\mathscr{C}}^{(r)}, \tag{1.5.4.3}$$

compatible with the total Higgs fields, where $\breve{\mathscr{C}}^{(r)}$ *is endowed with the Higgs field* $p^r d_{\breve{\mathscr{C}}^{(r)}}$, $\mathrm{R}^q\mathrm{I}\check{\theta}_*(\mathscr{M})$ *with the zero Higgs field and* $\mathrm{R}^q\mathfrak{g}_*(\underline{\mathbb{K}}^\bullet)$ *with the Katz–Oda field* (2.5.16).

Observe that the $\mathscr{O}_{\mathfrak{X}}[\frac{1}{p}]$-module $\mathrm{R}^q\mathfrak{g}_*(\underline{\mathbb{K}}^\bullet)$ is coherent, and that the functor

$$\mathrm{I}\widehat{\sigma}^*\colon \mathbf{Mod}^{\mathrm{coh}}(\mathscr{O}_{\mathfrak{X}}[\frac{1}{p}]) \to \mathbf{Ind\text{-}Mod}(\breve{\overline{\mathscr{B}}}) \tag{1.5.4.4}$$

is exact (6.2.6).

Corollary 1.5.5 (cf. 6.5.25) *Under the assumptions of* 1.5.4, *if the* $\mathscr{O}_{\mathfrak{X}}[\frac{1}{p}]$*-module* $\mathrm{R}^q\mathfrak{g}_*(\underline{\mathbb{K}}^\bullet)$ *is locally projective of finite type, then the ind-*$\breve{\overline{\mathscr{B}}}$*-module* $\mathrm{R}^q\mathrm{I}\check{\theta}_*(\mathscr{M})$ *is Dolbeault, and we have an isomorphism*

$$\mathscr{H}(\mathrm{R}^q\mathrm{I}\breve{\theta}_*(\mathscr{M})) \xrightarrow{\sim} \mathrm{R}^q\mathfrak{g}_*(\underline{\mathbb{K}}^\bullet), \tag{1.5.5.1}$$

where $\mathrm{R}^q\mathfrak{g}_*(\underline{\mathbb{K}}^\bullet)$ is endowed with the Katz–Oda field.

Corollary 1.5.6 (cf. 6.5.34) *Assume $g\colon X' \to X$ proper. Let*

$$\mathscr{M}^n = \breve{\psi}_*(\mathrm{R}^n \breve{g}_{\overline{\eta}*}(\breve{\mathbb{Z}}_p)) \otimes_{\breve{\mathbb{Z}}_p} \breve{\overline{\mathscr{B}}}$$

for an integer $n \geq 0$. Then, the $\breve{\overline{\mathscr{B}}}_{\mathbb{Q}}$-module $\mathscr{M}^n_{\mathbb{Q}}$ is Hodge–Tate (1.4.17), and we have an isomorphism

$$\mathscr{H}(\mathscr{M}^n_{\mathbb{Q}}) \xrightarrow{\sim} \oplus_{0\leq i\leq n}\mathrm{R}^i g_*(\widetilde{\xi}^{i-n}\Omega^{n-i}_{X'/X}) \otimes_{\mathscr{O}_X} \mathscr{O}_{\mathfrak{X}}[\frac{1}{p}], \tag{1.5.6.1}$$

where the Higgs field on the right-hand side is induced by the Kodaira–Spencer maps of g

$$\mathrm{R}g_*(\widetilde{\xi}^{-j}\Omega^j_{X'/X}) \to \widetilde{\xi}^{-1}\Omega^1_{X/S} \otimes_{\mathscr{O}_X} \mathrm{R}g_*(\widetilde{\xi}^{1-j}\Omega^{j-1}_{X'/X})[+1]. \tag{1.5.6.2}$$

This follows from 1.5.2 and 1.5.5. Indeed, $\breve{\psi}'_*(\breve{\mathbb{Z}}_p) = \breve{\mathbb{Z}}_p$, the $\breve{\overline{\mathscr{B}}}'_{\mathbb{Q}}$-module $\breve{\overline{\mathscr{B}}}'_{\mathbb{Q}}$ is Dolbeault and $\underline{\mathscr{H}}'(\breve{\overline{\mathscr{B}}}'_{\mathbb{Q}})$ is identified with $\mathscr{O}_{\mathfrak{X}'}[\frac{1}{p}]$ endowed with the zero Higgs field (4.6.10). Therefore, with the notation of 1.5.4, for every $q \geq 0$, the $\mathscr{O}_{\mathfrak{X}}[\frac{1}{p}]$-module $\mathrm{R}^q\mathfrak{g}_*(\underline{\mathbb{K}}^\bullet)$ is locally free of finite type by ([12] 5.5), which completes the proof of the first statement. The second statement follows easily from the definition of the Katz–Oda field (2.5.17, [42] 1.2).

Corollary 1.5.7 (cf. 6.5.28) *Let $M = (M_n)_{n\geq 0}$ be a $\breve{\mathbb{Z}}_p$-local system of $X'^{\mathbb{N}^\circ}_{\overline{\eta},\text{ét}}$ (1.4.14). We set $\mathscr{M} = \breve{\psi}'_*(M) \otimes_{\breve{\mathbb{Z}}_p} \breve{\overline{\mathscr{B}}}'$, we assume that $g\colon X' \to X$ is proper and that the $\breve{\overline{\mathscr{B}}}'_{\mathbb{Q}}$-module $\mathscr{M}_{\mathbb{Q}}$ is Dolbeault. Then, there exists a rational number $r > 0$ and a spectral sequence*

$$\mathrm{E}_2^{i,j} = \widehat{\sigma}^*_{\mathbb{Q}}(\mathrm{R}^i\mathfrak{g}_*(\mathrm{H}^j(\underline{\mathbb{K}}^\bullet))) \otimes_{\breve{\overline{\mathscr{B}}}_{\mathbb{Q}}} \breve{\mathscr{C}}^{(r)}_{\mathbb{Q}} \Rightarrow \breve{\psi}_*(\mathrm{R}^{i+j}\breve{g}_{\overline{\eta}*}(M)) \otimes_{\breve{\mathbb{Z}}_p} \breve{\mathscr{C}}^{(r)}_{\mathbb{Q}}, \tag{1.5.7.1}$$

where $\underline{\mathbb{K}}^\bullet$ is the Dolbeault complex of the relative Higgs $\mathscr{O}_{\mathfrak{X}'}[\frac{1}{p}]$-bundle $\underline{\mathscr{H}}'(\mathscr{M}_{\mathbb{Q}})$ (1.5.4.2).

This follows from 1.5.2 and 1.5.4.

Corollary 1.5.8 (cf. 6.5.29) *Assume that $g\colon X' \to X$ is proper. Then, there exists a rational number $r > 0$ and for every integer $n \geq 0$; a canonical isomorphism of $\breve{\mathscr{C}}^{(r)}_{\mathbb{Q}}$-modules*

$$\breve{\psi}_*(\mathrm{R}^n\breve{g}_{\overline{\eta}*}(\breve{\mathbb{Z}}_p)) \otimes_{\breve{\mathbb{Z}}_p} \breve{\mathscr{C}}^{(r)}_{\mathbb{Q}} \xrightarrow{\sim} \oplus_{0\leq i\leq n}\sigma^*(\mathrm{R}^i g_*(\Omega^{n-i}_{X'/X})) \otimes_{\sigma^*(\mathscr{O}_X)} \breve{\mathscr{C}}^{(r)}_{\mathbb{Q}}(i-n). \tag{1.5.8.1}$$

This follows from 1.5.4 applied to $\mathscr{M} = \breve{\overline{\mathscr{B}}}'_{\mathbb{Q}}$ since the $\breve{\overline{\mathscr{B}}}'_{\mathbb{Q}}$-module $\breve{\overline{\mathscr{B}}}'_{\mathbb{Q}}$ is Dolbeault and $\underline{\mathscr{H}}'(\breve{\overline{\mathscr{B}}}'_{\mathbb{Q}})$ is the trivial bundle $\mathscr{O}_{\mathfrak{X}'}[\frac{1}{p}]$ endowed with the zero Higgs field.

This corollary applies in particular by taking for $\widetilde{g}$ in the relative case (1.3.6) the trivial deformation (1.5.3).

Theorem 1.5.9 ([2] 6.7.5) *Assume that $g\colon X' \to X$ is proper. Then, we have a canonical spectral sequence of $\overline{\breve{\mathscr{B}}}_{\mathbb{Q}}$-modules,* the relative Hodge–Tate spectral sequence,

$$\mathrm{E}_2^{i,j} = \sigma^*(\mathrm{R}^i g_*(\Omega^j_{X'/X})) \otimes_{\sigma^*(\mathscr{O}_X)} \overline{\breve{\mathscr{B}}}_{\mathbb{Q}}(-j) \Rightarrow \breve{\psi}_*(\mathrm{R}^{i+j} \breve{g}_{\overline{\eta}*}(\breve{\mathbb{Z}}_p)) \otimes_{\breve{\mathbb{Z}}_p} \overline{\breve{\mathscr{B}}}_{\mathbb{Q}}. \quad (1.5.9.1)$$

This spectral sequence does not require any deformation (1.5.3.1). It is G_K-equivariant for the natural G_K-equivariant structures on the various topos and objects involved. Hence it degenerates at E_2 but the abutment filtration does not split in general. However, we can check that it splits after base change from $\overline{\breve{\mathscr{B}}}$ to $\breve{\mathscr{C}}^{(r)}$ for a rational number $r > 0$ and that it corresponds to the decomposition (1.5.8.1).

Remark 1.5.10 Theorem 1.5.4 and its corollary 1.5.5 provide p-adic analogs of results from [7] for the complex Simpson correspondence, but we allow for more general logarithmic schemes. Analogous results for the Simpson correspondence in characteristic p were established in [46] for proper smooth morphisms. In the complex setting, statements dealing with more general singularities than logarithmic singularities can be found in [13]. We hope to establish p-adic analogs generalizing our work.

1.5.11 The proof of 1.5.4 can be divided into three steps. First, we compute the relative Galois and Higgs cohomologies of the Higgs–Tate algebra by adapting Faltings' computation in the absolute case. Second, to sheafify these computations, we consider the fiber product of topos

$$\begin{array}{ccc} \widetilde{E}' & & \\ \downarrow \tau & \searrow^{\sigma'} & \\ \widetilde{E} \times_{X_{\text{ét}}} X'_{\text{ét}} & \longrightarrow & X'_{\text{ét}} \\ \downarrow & \square & \downarrow g \\ \widetilde{E} & \xrightarrow{\sigma} & X_{\text{ét}}. \end{array} \quad (1.5.11.1)$$

The local relative Galois cohomology computation can be globalized into a computation of the sheaves $\mathrm{R}^i\tau_*(\mathscr{C}'^{(r)}_n)$. The last step is a base change theorem with respect to the above Cartesian square.

It turns out that there is a very natural site underlying the topos $\widetilde{E} \times_{X_{\text{ét}}} X'_{\text{ét}}$ which is a relative variant of the Faltings topos. Its definition was inspired by that of the oriented products of topos (beyond the covanishing topos which inspired the definition of the Faltings topos).

1.6 Relative Faltings Topos

1.6.1 Let $g\colon X' \to X$ be a morphism of S-schemes. We denote by G the category of morphisms $(W \to U \leftarrow V)$ over the canonical morphisms $X' \to X \leftarrow X_{\overline{\eta}}$, i.e., commutative diagrams

$$\begin{array}{ccccc} W & \longrightarrow & U & \longleftarrow & V \\ \downarrow & & \downarrow & & \downarrow \\ X' & \longrightarrow & X & \longleftarrow & X_{\overline{\eta}} \end{array} \qquad (1.6.1.1)$$

such that W is étale over X', U is étale over X and the canonical morphism $V \to U_{\overline{\eta}}$ is *finite étale*. We endow it with the topology generated by the coverings

$$\{(W_i \to U_i \leftarrow V_i) \to (W \to U \leftarrow V)\}_{i \in I}$$

of the following three types:

(a) $U_i = U$, $V_i = V$ for all $i \in I$ and $(W_i \to W)_{i \in I}$ is a covering;
(b) $W_i = W$, $U_i = U$ for all $i \in I$ and $(V_i \to V)_{i \in I}$ is a covering;
(c) diagrams

$$\begin{array}{ccccc} W' & \longrightarrow & U' & \longleftarrow & V' \\ \| & & \downarrow & \square & \downarrow \\ W & \longrightarrow & U & \longleftarrow & V \end{array} \qquad (1.6.1.2)$$

where $U' \to U$ is any morphism and where the right square is Cartesian.

We denote by $\widetilde{G}$ the topos of sheaves of sets on G and we call it the *relative Faltings topos* associated with the pair of morphisms $(X_{\overline{\eta}} \to X, X' \to X)$.

We have a canonical morphism of topos

$$\pi\colon \widetilde{G} \to X'_{\text{ét}}, \quad W \in \mathrm{Ob}(\mathbf{Ét}_{/X'}) \mapsto \pi^*(W) = (W \to X \leftarrow X_{\overline{\eta}})^a. \qquad (1.6.1.3)$$

1.6.2 If $X' = X$, the topos $\widetilde{G}$ is canonically equivalent to the Faltings topos $\widetilde{E}$. By functoriality of the relative Faltings topos, we then get a natural factorization of the canonical morphism $\Theta\colon \widetilde{E}' \to \widetilde{E}$ that fits into a commutative diagram

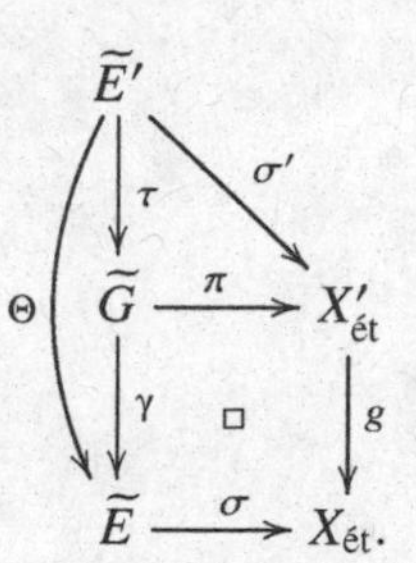

(1.6.2.1)

We prove that the lower square is Cartesian.

We prove first a base change theorem with respect to this square for torsion abelian sheaves of $X'_{\text{ét}}$, inspired by a base change theorem for oriented products due to Gabber ([2] 6.5.5). It reduces to the proper base change theorem for the étale topos. Then, we prove the following result which plays a crucial role in the proofs of both 1.5.4 and 1.5.9:

Theorem 1.6.3 ([2] 6.5.31) *Let $g\colon X' \to X$ be a proper and smooth morphism of smooth S-schemes. Then, there exists an integer $N \geq 0$ such that for all integers $n \geq 1$ and $q \geq 0$, and every quasi-coherent $\mathscr{O}_{X'_n}$-module $\mathscr{F}$, the kernel and the cokernel of the base change morphism*

$$\sigma^*(\mathrm{R}^q g_*(\mathscr{F})) \to \mathrm{R}^q \gamma_*(\pi^*(\mathscr{F})), \tag{1.6.3.1}$$

are annihilated by p^N.

In this statement, π^* and σ^* denote the pullbacks for the morphisms of ringed topos

$$\pi\colon (\widetilde{G}, \tau_*(\overline{\mathscr{B}}')) \to (X'_{\text{ét}}, \mathscr{O}_{X'}), \tag{1.6.3.2}$$

$$\sigma\colon (\widetilde{E}, \overline{\mathscr{B}}) \to (X_{\text{ét}}, \mathscr{O}_X). \tag{1.6.3.3}$$

Chapter 2
Preliminaries

2.1 Notation and Conventions

All rings considered in this book have an identity element; all ring homomorphisms map the identity element to the identity element. We mostly consider commutative rings, and rings are assumed to be commutative unless stated otherwise; in particular, when we take a ringed topos (X, A), the ring A is assumed to be commutative unless stated otherwise.

By monoid *we will mean a commutative monoid with an identity element. Homomorphisms of monoids are assumed to map the identity element to the identity element.*

2.1.1 Let p be a prime number. We endow $\mathbb{Z}_p$ with the p-adic topology and do the same for every adic $\mathbb{Z}_p$-algebra (i.e., every $\mathbb{Z}_p$-algebra that is separated and complete for the p-adic topology). Let A be an adic $\mathbb{Z}_p$-algebra, $i\colon A \to A[\frac{1}{p}]$ the canonical homomorphism. We call the *p-adic topology* on $A[\frac{1}{p}]$ the unique topology compatible with its additive group structure for which the subgroups $i(p^nA)$, for $n \in \mathbb{N}$, form a fundamental system of neighborhoods of 0 ([11] chap. III §1.2, prop. 1). It makes $A[\frac{1}{p}]$ into a topological ring. Let M be an $A[\frac{1}{p}]$-module of finite type, M° a sub-A-module of finite type of M that generates it over $A[\frac{1}{p}]$. We call the *p-adic topology* on M the unique topology compatible with its additive group structure for which the subgroups p^nM°, for $n \in \mathbb{N}$, form a fundamental system of neighborhoods of 0. This topology does not depend on the choice of M°. Indeed, if M' is another sub-A-module of finite type of M which generates it over $A[\frac{1}{p}]$, then there exists an $m \geq 0$ such that $p^mM^\circ \subset M'$ and $p^mM' \subset M^\circ$. It is clear that M is a topological $A[\frac{1}{p}]$-module.

2.1.2 Let G be a profinite group, A a topological ring endowed with a continuous action of G by ring homomorphisms. An *A-representation* of G is an A-module M endowed with a A-semi-linear action of G on M, i.e., such that for all $g \in G$, $a \in A$ and $m \in M$, we have $g(am) = g(a)g(m)$. We say that the A-representation is *continuous* if M is a topological A-module and if the action of G on M is continuous. Let M, N

A. Abbes, M. Gros, *The p-adic Simpson Correspondence and Hodge-Tate Local Systems*, Lecture Notes in Mathematics 2345, https://doi.org/10.1007/978-3-031-55914-3_2

be two A-representations (resp. two continuous A-representations) of G. A morphism from M to N is an A-linear and G-equivariant (resp. an A-linear, continuous and G-equivariant) morphism from M to N. We denote by $\mathbf{Rep}_A(G)$ (resp. $\mathbf{Rep}_A^{\mathrm{cont}}(G)$) the category of A-representations (resp. continuous A-representations) of G. If M and N are two A-representations of G, the A-modules $M \otimes_A N$ and $\mathrm{Hom}_A(M, N)$ are naturally A-representations of G.

2.1.3 Let A be a ring, p a prime number, n an integer ≥ 1. We denote by $\mathrm{W}(A)$ (resp. $\mathrm{W}_n(A)$) the ring of Witt vectors (resp. Witt vectors of length n) with respect to p with coefficients in A. We have a ring homomorphism

$$\begin{aligned} \Phi_n\colon \quad \mathrm{W}_n(A) &\to A, \\ (x_0, \dots, x_{n-1}) &\mapsto x_0^{p^{n-1}} + p x_1^{p^{n-2}} + \cdots + p^{n-1} x_{n-1} \end{aligned} \tag{2.1.3.1}$$

called the nth phantom component. We also have the restriction, the shift and the Frobenius morphisms

$$\mathrm{R}\colon \mathrm{W}_{n+1}(A) \to \mathrm{W}_n(A), \tag{2.1.3.2}$$

$$\mathrm{V}\colon \mathrm{W}_n(A) \to \mathrm{W}_{n+1}(A), \tag{2.1.3.3}$$

$$\mathrm{F}\colon \mathrm{W}_{n+1}(A) \to \mathrm{W}_n(A). \tag{2.1.3.4}$$

When A has characteristic p, F induces an endomorphism of $\mathrm{W}_n(A)$, that we denote also by F.

2.1.4 For any ring R and any monoid M, we denote by $R[M]$ the R-algebra of M and by $e\colon M \to R[M]$ the canonical homomorphism, where $R[M]$ is considered as a multiplicative monoid. For any $x \in M$, we will write e^x instead of $e(x)$.

We denote by $\mathbf{A}_M$ the scheme $\mathrm{Spec}(\mathbb{Z}[M])$ endowed with the logarithmic structure associated with the prelogarithmic structure defined by $e\colon M \to \mathbb{Z}[M]$ ([3] II.5.9). For any homomorphism of monoids $\vartheta\colon M \to N$, we denote by $\mathbf{A}_\vartheta\colon \mathbf{A}_N \to \mathbf{A}_M$ the associated morphism of logarithmic schemes.

2.1.5 For a category $\mathscr{C}$, we denote by $\mathrm{Ob}(\mathscr{C})$ the set of its objects, by $\mathscr{C}^\circ$ the opposite category, and for $X, Y \in \mathrm{Ob}(\mathscr{C})$, by $\mathrm{Hom}_{\mathscr{C}}(X, Y)$ (or $\mathrm{Hom}(X, Y)$ when there is no ambiguity) the set of morphisms from X to Y.

If $\mathscr{C}$ and $\mathscr{C}'$ are two categories, we denote by $\mathrm{Hom}(\mathscr{C}, \mathscr{C}')$ the set of functors from $\mathscr{C}$ to $\mathscr{C}'$, and by $\mathbf{Hom}(\mathscr{C}, \mathscr{C}')$ the category of functors from $\mathscr{C}$ to $\mathscr{C}'$.

Let I be a category, $\mathscr{C}$ and $\mathscr{C}'$ two categories over I ([24] VI 2). We denote by $\mathrm{Hom}_I(\mathscr{C}, \mathscr{C}')$ the set of I-functors from $\mathscr{C}$ to $\mathscr{C}'$ and by $\mathrm{Hom}_{\mathrm{cart}/I}(\mathscr{C}, \mathscr{C}')$ the set of Cartesian functors ([24] VI 5.2). We denote by $\mathbf{Hom}_I(\mathscr{C}, \mathscr{C}')$ the category of I-functors from $\mathscr{C}$ to $\mathscr{C}'$ and by $\mathbf{Hom}_{\mathrm{cart}/I}(\mathscr{C}, \mathscr{C}')$ the full subcategory made up of Cartesian functors.

2.1.6 For any abelian category $\mathscr{C}$, we denote by $\mathbf{D}(\mathscr{C})$ its derived category and by $\mathbf{D}^-(\mathscr{C})$, $\mathbf{D}^+(\mathscr{C})$ and $\mathbf{D}^{\mathrm{b}}(\mathscr{C})$ the full subcategories of $\mathbf{D}(\mathscr{C})$ made up of complexes with cohomology bounded from above, from below and from both sides, respectively

([40] 13.1.2). Unless mentioned otherwise, the complexes of $\mathscr{C}$ have differentials of degree +1, the degree being written as an exponent.

2.1.7 Throughout this monograph, we fix a universe $\mathbb{U}$ with an element of infinite cardinality. A set is said to be $\mathbb{U}$-*small* (or, when no confusion arises, *small*) if it is isomorphic to an element of $\mathbb{U}$. We also use the terminology: *small group*, *small ring*, *small category*... We say that a category $\mathscr{C}$ is a $\mathbb{U}$-*category* if for all objects X, Y of $\mathscr{C}$, the set $\mathrm{Hom}_{\mathscr{C}}(X,Y)$ is $\mathbb{U}$-small ([5] I 1.1). We call the *category of* $\mathbb{U}$-*sets* and we denote by **Sets** the category of sets that are in $\mathbb{U}$. It is a punctual $\mathbb{U}$-topos that we also denote by **Pt** ([5] IV 2.2). We denote by **Sch** the category of schemes that are elements of $\mathbb{U}$. Unless explicitly stated otherwise, it will be understood that the rings and the logarithmic schemes (and in particular the schemes) considered in this monograph are elements of the universe $\mathbb{U}$.

2.1.8 Let $\mathscr{C}$ be a $\mathbb{U}$-category (2.1.7) ([5] I 1.1). We denote by $\widehat{\mathscr{C}}$ the category of presheaves of $\mathbb{U}$-sets on $\mathscr{C}$, that is, the category of contravariant functors on $\mathscr{C}$ with values in **Sets** ([5] I 1.2). If $\mathscr{C}$ is endowed with a topology ([5] II 1.1), we denote by $\widetilde{\mathscr{C}}$ the topos of sheaves of $\mathbb{U}$-sets on $\mathscr{C}$ ([5] II 2.1).

For an object F of $\widehat{\mathscr{C}}$, we denote by $\mathscr{C}_{/F}$ the following category ([5] I 3.4.0). The objects of $\mathscr{C}_{/F}$ are the pairs consisting of an object X of $\mathscr{C}$ and a morphism u from X to F. If (X,u) and (Y,v) are two objects, a morphism from (X,u) to (Y,v) is a morphism $g\colon X \to Y$ such that $u = v \circ g$.

2.1.9 Let X be a $\mathbb{U}$-topos ([5] IV 1.1.2). Inverse systems of objects of X indexed by the ordered set of natural numbers $\mathbb{N}$ form a topos that we denote by $X^{\mathbb{N}^\circ}$. We refer to ([3] III.7) for useful facts on this type of topos. We recall, in particular, that we have a morphism of topos

$$\lambda\colon X^{\mathbb{N}^\circ} \to X, \tag{2.1.9.1}$$

whose pullback functor λ^* associates with any object F of X the constant inverse system of value F, and whose direct image functor λ_* associates with any inverse system its inverse limit ([3] III.7.4).

2.1.10 Let (X,A) be a ringed $\mathbb{U}$-topos ([5] IV 11.1.1). We denote by $\mathbf{Mod}(A)$ or $\mathbf{Mod}(A,X)$ the category of A-modules of X. If M is an A-module, we denote by $\mathrm{S}_A(M)$ (resp. $\wedge_A(M)$, resp. $\Gamma_A(M)$) the symmetric (resp. exterior, resp. divided powers) algebra of M and for any integer $n \geq 0$, by $\mathrm{S}^n_A(M)$ (resp. $\wedge^n_A(M)$, resp. $\Gamma^n_A(M)$) its homogeneous part of degree n. The formations of these algebras commute with localization over an object of X. We will omit the ring A from the notation $\wedge_A(M)$ and $\wedge^n_A(M)$ when there is no risk of ambiguity.

Definition 2.1.11 ([6] I 1.3.1) Let (X, A) be a ringed topos. We say that an A-module M of X is *locally projective of finite type* if the following equivalent conditions are satisfied:

(i) M is of finite type and the functor $\mathscr{H}om_A(M, \cdot)$ is exact;
(ii) M is of finite type and every epimorphism of A-modules $N \to M$ admits locally a section;

(iii) M is locally a direct factor of a free A-module of finite type.

If X has enough points and if for every point x of X, the stalk of A at x is a local ring, then the locally projective A-modules of finite type are the locally free A-modules of finite type ([6] I 2.15.1).

2.1.12 For any scheme X, we denote by $\mathbf{\acute{E}t}_{/X}$ the *étale site* of X, i.e., the full subcategory of $\mathbf{Sch}_{/X}$ (2.1.7) made up of the étale schemes over X, endowed with the étale topology; it is a $\mathbb{U}$-site. We denote by $X_{\text{ét}}$ the *étale topos* of X, that is, the topos of sheaves of $\mathbb{U}$-sets on $\mathbf{\acute{E}t}_{/X}$.

We denote by $\mathbf{\acute{E}t}_{\text{coh}/X}$ the full subcategory of $\mathbf{\acute{E}t}_{/X}$ made up of étale schemes of finite presentation on X, endowed with the topology induced by that of $\mathbf{\acute{E}t}_{/X}$; it is a $\mathbb{U}$-small site. If X is quasi-separated, the restriction functor from $X_{\text{ét}}$ to the topos of sheaves of $\mathbb{U}$-sets on $\mathbf{\acute{E}t}_{\text{coh}/X}$ is an equivalence of categories ([5] VII 3.1 and 3.2).

We denote by $\mathbf{\acute{E}t}_{\text{f}/X}$ the full subcategory of $\mathbf{\acute{E}t}_{/X}$ made up of finite étale schemes on X, endowed with the topology induced by that of $\mathbf{\acute{E}t}_{/X}$; it is a $\mathbb{U}$-small site. We call the *finite étale topos* of X and we denote by $X_{\text{fét}}$ the topos of sheaves of $\mathbb{U}$-sets on $\mathbf{\acute{E}t}_{\text{f}/X}$ (see [3] VI.9.2). The canonical injection $\mathbf{\acute{E}t}_{\text{f}/X} \to \mathbf{\acute{E}t}_{/X}$ induces a morphism of topos

$$\rho_X \colon X_{\text{ét}} \to X_{\text{fét}}. \tag{2.1.12.1}$$

2.1.13 Let X be a scheme. We denote by X_{zar} the Zariski topos of X and by

$$u_X \colon X_{\text{ét}} \to X_{\text{zar}} \tag{2.1.13.1}$$

the canonical morphism ([5] VII 4.2.2). If F is a quasi-coherent $\mathscr{O}_X$-module of X_{zar}, we denote by $\iota(F)$ the sheaf of $X_{\text{ét}}$ defined for any étale X-scheme U by ([5] VII 2 c))

$$\iota(F)(U) = \Gamma(U, F \otimes_{\mathscr{O}_X} \mathscr{O}_U). \tag{2.1.13.2}$$

It is convenient, when there is no risk of confusion, to abusively denote $\iota(F)$ by F. Note that $\iota(\mathscr{O}_X)$ is a ring of $X_{\text{ét}}$ and that $\iota(F)$ is a $\iota(\mathscr{O}_X)$-module ([2] 2.1.18).

Denoting by $\mathbf{Mod}^{\text{qcoh}}(\mathscr{O}_X, X_{\text{zar}})$ the full subcategory of $\mathbf{Mod}(\mathscr{O}_X, X_{\text{zar}})$ made up of quasi-coherent $\mathscr{O}_X$-modules (2.1.10), the correspondence $F \mapsto \iota(F)$ defines a functor

$$\iota \colon \mathbf{Mod}^{\text{qcoh}}(\mathscr{O}_X, X_{\text{zar}}) \to \mathbf{Mod}(\mathscr{O}_X, X_{\text{ét}}). \tag{2.1.13.3}$$

For every quasi-coherent $\mathscr{O}_X$-module F of X_{zar}, we have a canonical isomorphism

$$F \xrightarrow{\sim} u_{X*}(\iota(F)). \tag{2.1.13.4}$$

We therefore consider u_X as a morphism of ringed topos

$$u_X \colon (X_{\text{ét}}, \mathscr{O}_X) \to (X_{\text{zar}}, \mathscr{O}_X). \tag{2.1.13.5}$$

For modules, we use the notation u_X^{-1} to denote the pullback in the sense of abelian sheaves, and we keep the notation u_X^* for the pullback in the sense of modules. By ([2] 2.1.18), the isomorphism (2.1.13.4) induces by adjunction an isomorphism

$$u_X^*(F) \xrightarrow{\sim} \iota(F). \tag{2.1.13.6}$$

2.1.14 Let R be a ring endowed with a non-discrete valuation of height 1, $\mathfrak{m}$ its maximal ideal, $\gamma \in R$, X a $\mathbb{U}$-topos, A an R-algebra of X. We consider the notions of α-algebra (or almost algebra) over R introduced in ([2] 2.6–2.9) (see [2] 2.10.1). We say that a morphism of A-modules is a *γ-isomorphism* (resp. an *α-isomorphism*) if its kernel and cokernel are annihilated by γ (resp. every element of $\mathfrak{m}$) ([2] 2.6.2 and 2.7.2). We say that a complex of A-modules with differential of degree 1, the degree being written as an exponent, is *γ-acyclic* if its cohomology groups are annihilated by γ, and that it is *α-acyclic* if it is δ-acyclic for all $\delta \in \mathfrak{m}$. A short sequence of A-modules is said to be *γ-exact* (resp. *α-exact*) if it is γ-acyclic (resp. α-acyclic) as a complex of A-modules.

We denote by $\mathbf{Mod}(A)$ the category of A-modules of X and by $\mathbf{D}(\mathbf{Mod}(A))$ its derived category. A morphism of complexes of $\mathbf{Mod}(A)$ is said to be a *γ-quasi-isomorphism* (resp. an *α-quasi-isomorphism*) if it induces γ-isomorphisms (resp. α-isomorphisms) on the cohomology modules. Similarly, a morphism of $\mathbf{D}(\mathbf{Mod}(A))$ is said to be a *γ-isomorphism* (resp. an *α-isomorphism*) if it induces γ-isomorphisms (resp. α-isomorphisms) on the cohomology modules. We denote by $\mathscr{N}$ the thick subcategory of $\mathbf{Mod}(A)$ made up of the α-zero A-modules, and by $\mathbf{D}_{\mathscr{N}}(\mathbf{Mod}(A))$ the full subcategory of $\mathbf{D}(\mathbf{Mod}(A))$ made up of the complexes whose cohomology modules are α-zero. The latter is a strictly full and saturated triangulated subcategory of $\mathbf{D}(\mathbf{Mod}(A))$ ([59] 06UQ). The family of α-isomorphisms of $\mathbf{D}(\mathbf{Mod}(A))$ allows a two-sided calculus of fractions ([37] I 1.4.2). The triangulated category obtained by localization of $\mathbf{D}(\mathbf{Mod}(A))$ by the family of α-isomorphisms "represents" the triangulated category quotient of $\mathbf{D}(\mathbf{Mod}(A))$ by $\mathbf{D}_{\mathscr{N}}(\mathbf{Mod}(A))$ ([40] 10.2.3).

2.2 Reminder on a Construction by Fontaine–Grothendieck

2.2.1 The construction recalled in this section was introduced independently by Grothendieck ([26] IV 3.3) and Fontaine ([19] 2.2). We fix a prime number p. All the rings of Witt vectors considered in this monograph are with respect to p (see 2.1.3). Let A be a $\mathbb{Z}_{(p)}$-algebra, n an integer ≥ 1. The ring homomorphism

$$\begin{aligned} \Phi_{n+1}\colon \mathrm{W}_{n+1}(A/p^nA) &\to A/p^nA \\ (x_0,\ldots,x_n) &\mapsto x_0^{p^n} + px_1^{p^{n-1}} + \cdots + p^n x_n \end{aligned} \tag{2.2.1.1}$$

vanishes on $\mathrm{V}^n(A/p^nA)$ and induces therefore, by taking a quotient, a ring homomorphism

$$\begin{aligned} \Phi'_{n+1}\colon \mathrm{W}_n(A/p^nA) &\to A/p^nA \\ (x_0,\ldots,x_{n-1}) &\mapsto x_0^{p^n} + px_1^{p^{n-1}} + \cdots + p^{n-1}x_{n-1}^p. \end{aligned} \tag{2.2.1.2}$$

The latter vanishes on

$$\mathrm{W}_n(pA/p^nA) = \ker(\mathrm{W}_n(A/p^nA) \to \mathrm{W}_n(A/pA)) \tag{2.2.1.3}$$

and in turn factors into a ring homomorphism

$$\theta_n \colon \mathrm{W}_n(A/pA) \to A/p^n A. \tag{2.2.1.4}$$

It immediately follows from the definition that the diagram

$$\begin{array}{ccc} \mathrm{W}_{n+1}(A/pA) & \xrightarrow{\theta_{n+1}} & A/p^{n+1}A \\ \downarrow{\scriptstyle \mathrm{RF}} & & \downarrow \\ \mathrm{W}_n(A/pA) & \xrightarrow{\theta_n} & A/p^n A, \end{array} \tag{2.2.1.5}$$

where R is the restriction morphism (2.1.3.2), F is the Frobenius (2.1.3.4) and the unlabeled arrow is the canonical homomorphism, is commutative.

For every homomorphism of commutative $\mathbb{Z}_{(p)}$-algebras $\varphi \colon A \to B$, the diagram

$$\begin{array}{ccc} \mathrm{W}_n(A/pA) & \longrightarrow & \mathrm{W}_n(B/pB) \\ \downarrow{\scriptstyle \theta_n} & & \downarrow{\scriptstyle \theta_n} \\ A/p^n A & \longrightarrow & B/p^n B \end{array} \tag{2.2.1.6}$$

where the horizontal arrows are the morphisms induced by φ, is commutative.

Proposition 2.2.2 ([3] II.9.2) *Let A be a $\mathbb{Z}_{(p)}$-algebra satisfying the following conditions:*

(i) *A is $\mathbb{Z}_{(p)}$-flat.*
(ii) *A is integrally closed in $A[\frac{1}{p}]$.*
(iii) *The absolute Frobenius of A/pA is surjective.*
(iv) *There exist an integer $N \geq 1$ and a sequence $(p_n)_{0 \leq n \leq N}$ of elements of A such that $p_0 = p$ and $p_{n+1}^p = p_n$ for every $0 \leq n \leq N-1$.*

For any integer $1 \leq n \leq N$, we set

$$\xi_n = [\overline{p}_n] - p \in \mathrm{W}_n(A/pA), \tag{2.2.2.1}$$

where $\overline{p}_n$ is the class of p_n in A/pA and $[\ ]$ denotes the multiplicative representative. Then, for all integers $n \geq 1$ and $i \geq 0$ such that $n+i \leq N$, the sequence

$$\mathrm{W}_n(A/pA) \xrightarrow{\cdot \mathrm{R}^i(\xi_{n+i})} \mathrm{W}_n(A/pA) \xrightarrow{\theta_n \circ \mathrm{F}^i} A/p^n A \longrightarrow 0 \tag{2.2.2.2}$$

is exact.

2.2.3 Let A be a $\mathbb{Z}_{(p)}$-algebra, $\widehat{A}$ the p-adic Hausdorff completion of A. We denote by $A^\flat$ the inverse limit of the inverse system $(A/pA)_{\mathbb{N}}$ whose transition morphisms are the iterates of the Frobenius endomorphism of A/pA.

$$A^\flat = \lim_{\overleftarrow{\mathbb{N}}} A/pA. \tag{2.2.3.1}$$

It is a perfect ring of characteristic p. For every integer $n \geq 1$, the canonical projection $A^\flat \to A/pA$ onto the $(n+1)$th component of the inverse system $(A/pA)_{\mathbb{N}}$ (i.e., the component of index n) induces a homomorphism (2.1.3)

$$\nu_n \colon \mathrm{W}(A^\flat) \to \mathrm{W}_n(A/pA). \qquad (2.2.3.2)$$

As $\nu_n = \mathrm{F} \circ \mathrm{R} \circ \nu_{n+1}$, by taking the inverse limit, we obtain a homomorphism

$$\nu \colon \mathrm{W}(A^\flat) \to \varprojlim_{n\geq 1} \mathrm{W}_n(A/pA), \qquad (2.2.3.3)$$

where the transition morphisms of the inverse limit are the morphisms FR. One immediately verifies that it is bijective. By virtue of (2.2.1.5), by taking the inverse limit, the homomorphisms θ_n induce a homomorphism

$$\theta \colon \mathrm{W}(A^\flat) \to \widehat{A}. \qquad (2.2.3.4)$$

We recover the homomorphism defined by Fontaine ([19] 2.2). For any integer $r \geq 1$, we set

$$\mathscr{A}_r(A) = \mathrm{W}(A^\flat)/\ker(\theta)^r, \qquad (2.2.3.5)$$

and we denote by $\theta_r : \mathscr{A}_r(A) \to \widehat{A}$ the homomorphism induced by θ (see [20] 1.2.2).

For every homomorphism of commutative $\mathbb{Z}_{(p)}$-algebras $\varphi \colon A \to B$, the diagram

$$\begin{array}{ccc} \mathrm{W}(A^\flat) & \longrightarrow & \mathrm{W}(B^\flat) \\ {\scriptstyle\theta}\downarrow & & \downarrow{\scriptstyle\theta} \\ \widehat{A} & \longrightarrow & \widehat{B} \end{array} \qquad (2.2.3.6)$$

where the horizontal arrows are the morphisms induced by φ, is commutative (2.2.1.6). The correspondence $A \mapsto \mathscr{A}_r(A)$ is therefore functorial.

Remark 2.2.4 Let k be a perfect field. The canonical projection $\mathrm{W}(k)^\flat \to k$ onto the first component (i.e., with index 0) is an isomorphism. It therefore induces an isomorphism $\mathrm{W}(\mathrm{W}(k)^\flat) \xrightarrow{\sim} \mathrm{W}(k)$, which we use to identify these two rings. The homomorphism θ then identifies with the identity endomorphism of $\mathrm{W}(k)$.

Lemma 2.2.5 *Let A be a $\mathbb{Z}_{(p)}$-algebra, $\widehat{A}$ the p-adic Hausdorff completion of A, $A^\flat$ the ring defined in* (2.2.3.1), *$\mathbb{A}$ the set of sequences $(x_n)_{\mathbb{N}}$ of $\widehat{A}$ such that $x_{n+1}^p = x_n$ for all $n \geq 0$. Then,*

(i) *The map*

$$\mathbb{A} \to A^\flat, \quad (x_n)_{\mathbb{N}} \mapsto (\overline{x}_n)_{\mathbb{N}}, \qquad (2.2.5.1)$$

where $\overline{x}_n$ is the reduction of x_n modulo p, is an isomorphism of multiplicative monoids.

(ii) *For all $(x_0, x_1, \dots) \in \mathrm{W}(A^\flat)$, we have*

$$\theta(x_0, x_1, \dots) = \sum_{n\geq 0} p^n x_n^{(n)}, \qquad (2.2.5.2)$$

where θ is the homomorphism (2.2.3.4) *and for all* $n \geq 0$, $(x_n^{(m)})_{m\geq 0}$ *is the element of* $\mathbb{A}$ *associated with* $x_n \in A^\flat$ (2.2.5.1).

(i) Denoting by v the valuation of $\mathbb{Z}_{(p)}$ normalized by $v(p) = 1$, for all integers $m, i \geq 1$ such that $i \leq p^m$, we have $v(\binom{p^m}{i}) = m - v(i)$ and therefore $v(\binom{p^m}{i}) + i \geq m$. Let $(x_n)_\mathbb{N}$ and $(y_n)_\mathbb{N}$ be two sequences of elements of $\widehat{A}$ which induce by reduction modulo p the same sequence $(\overline{x}_n)_\mathbb{N} = (\overline{y}_n)_\mathbb{N}$ of A/pA. For all $n, m \geq 0$, we then have

$$x_{n+m}^{p^m} - y_{n+m}^{p^m} \in p^m\widehat{A}. \tag{2.2.5.3}$$

In particular, if $(x_n)_\mathbb{N}$ and $(y_n)_\mathbb{N}$ are elements of $\mathbb{A}$, then $x_n = y_n$ for all $n \geq 0$ since $\widehat{A}$ is separated for the p-adic topology. Consequently, the map (2.2.5.1) is injective. Let us show that it is surjective. Let $(z_n)_\mathbb{N}$ be an element of $A^\flat$, $(y_n)_\mathbb{N}$ a sequence of elements of A which lifts $(z_n)_\mathbb{N}$. Applying (2.2.5.3) to the sequences $(y_n)_{n\geq 0}$ and $(y_{n+1}^p)_{n\geq 0}$, we see that for all $n, m \geq 0$, we have

$$y_{n+m+1}^{p^{m+1}} - y_{n+m}^{p^m} \in p^m\widehat{A}. \tag{2.2.5.4}$$

Hence, for every integer $n \geq 0$, $(y_{n+m}^{p^m})_{m\geq 0}$ converges to an element x_n of $\widehat{A}$. The sequence $(x_n)_\mathbb{N}$ clearly belongs to $\mathbb{A}$ and is sent to $(z_n)_\mathbb{N}$ by (2.2.5.1); the surjectivity follows.

(ii) This follows immediately from the definitions.

2.3 Fontaine Universal p-adic Infinitesimal Thickenings

2.3.1 Let Λ be a ring, A a Λ-algebra, m an integer ≥ 0. A Λ-infinitesimal thickening of order $\leq m$ of A is a pair (D, θ) consisting of a Λ-algebra D and a surjective homomorphism of Λ-algebras $\theta \colon D \to A$ such that $(\ker(\theta))^{m+1} = 0$. The Λ-infinitesimal thickenings of order $\leq m$ of A form a category denoted $\mathbf{E}_m(A/\Lambda)$: if (D_1, θ_1) and (D_2, θ_2) are two Λ-infinitesimal thickenings of order $\leq m$ of A, a morphism from (D_1, θ_1) to (D_2, θ_2) is a homomorphism of Λ-algebras $f \colon D_1 \to D_2$ such that $\theta_1 = \theta_2 \circ f$. Following Fontaine ([20] 1.1.1), if this category admits an initial object, it is called the *universal* Λ*-infinitesimal thickening of order* $\leq m$ *of* A.

Let p be a prime number and suppose, moreover, that A is complete and separated for the p-adic topology. An object (D, θ) of $\mathbf{E}_m(A/\Lambda)$ is said to be *p-adic* if the Λ-algebra D is complete and separated for the p-adic topology. We denote by $\mathbf{E}_m^p(A/\Lambda)$ the full subcategory of $\mathbf{E}_m(A/\Lambda)$ made up of the p-adic Λ-infinitesimal thickenings of order $\leq m$ of A. Following Fontaine ([20] 1.1.3), if the category $\mathbf{E}_m^p(A/\Lambda)$ admits an initial object, it is called the *universal p-adic* Λ*-infinitesimal thickening of order* $\leq m$ *of* A.

2.3.2 Let k be a perfect field of characteristic $p > 0$, W the ring of Witt vectors with coefficients in k, Λ a W-algebra, A a Λ-algebra. All the rings of Witt vectors considered in this monograph are with respect to p (see 2.1.3). We denote by $\widehat{A}$ the p-adic Hausdorff completion of A, $A^\flat$ the ring associated with A by the functor defined in (2.2.3.1) and

$$\theta\colon \mathrm{W}(A^\flat) \to \widehat{A} \tag{2.3.2.1}$$

the canonical homomorphism (2.2.3.4). By 2.2.4, we can canonically endow $A^\flat$ with the structure of a k-algebra, and thus $\mathrm{W}(A^\flat)$ with the structure of a W-algebra. Moreover, θ is a homomorphism of W-algebras. We set

$$\mathrm{W}_\Lambda(A^\flat) = \mathrm{W}(A^\flat) \otimes_W \Lambda. \tag{2.3.2.2}$$

We denote by

$$\theta_\Lambda\colon \mathrm{W}_\Lambda(A^\flat) \to \widehat{A} \tag{2.3.2.3}$$

the homomorphism of Λ-algebras induced by θ, and by J_Λ its kernel. For any integer $r \geq 1$, we set

$$\mathscr{A}_r(A/\Lambda) = \mathrm{W}_\Lambda(A^\flat)/J_\Lambda^r \tag{2.3.2.4}$$

and we denote by $\widehat{\mathscr{A}}_r(A/\Lambda)$ its p-adic Hausdorff completion.

$$\widehat{\mathscr{A}}_r(A/\Lambda) = \varprojlim_{n\geq 0} \frac{\mathrm{W}_\Lambda(A^\flat)}{J_\Lambda^r + p^n\mathrm{W}_\Lambda(A^\flat)}. \tag{2.3.2.5}$$

We denote by $\theta_{\Lambda,r}\colon \mathscr{A}_r(A/\Lambda) \to \widehat{A}$ the homomorphism induced by θ_Λ and by $\widehat{\theta}_{\Lambda,r}\colon \widehat{\mathscr{A}}_r(A/\Lambda) \to \widehat{A}$ its extension to the p-adic completions.

We denote by $\mathrm{W}'_\Lambda(A^\flat)$ the p-adic Hausdorff completion of $\mathrm{W}_\Lambda(A^\flat)$ for the topology defined by the ideal $\theta_\Lambda^{-1}(p\widehat{A})$, by

$$\theta'_\Lambda\colon \mathrm{W}'_\Lambda(A^\flat) \to \widehat{A} \tag{2.3.2.6}$$

the extension of θ_Λ to the completions, by J'_Λ the kernel of θ'_Λ, and for any integer $r \geq 1$, by $\widehat{\mathscr{A}}'_r(A/\Lambda)$ the p-adic Hausdorff completion of $\mathrm{W}'_\Lambda(A^\flat)/J'^r_\Lambda$.

Lemma 2.3.3 *We keep the assumptions and notation of* 2.3.2; *we suppose moreover that the Frobenius endomorphism of A/pA is surjective. Then,*

(i) *The homomorphisms θ and θ_Λ are surjective.*
(ii) *For every $r \geq 1$, we have* $\ker(\widehat{\theta}_{\Lambda,r})^r = 0$.
(iii) *If, moreover, J_Λ is a finitely generated ideal of $\mathrm{W}_\Lambda(A^\flat)$, then, for every integer $r \geq 1$, the canonical homomorphism $\mathrm{W}_\Lambda(A^\flat) \to \mathrm{W}'_\Lambda(A^\flat)$ induces an isomorphism of Λ-algebras*

$$\widehat{\mathscr{A}}_r(A/\Lambda) \to \widehat{\mathscr{A}}'_r(A/\Lambda). \tag{2.3.3.1}$$

(i) The homomorphism θ reduces modulo p to the morphism $A^\flat \to A/pA$ which is the projection onto the second component (2.2.3.1) (i.e., the component of index 1). It is surjective since the Frobenius endomorphism of A/pA is surjective. We deduce from this that θ is surjective because $\mathrm{W}(A^\flat)$ and $\widehat{A}$ are complete and separated for the p-adic topologies ([1] 1.8.5). The same is then true of θ_Λ.

(ii) Indeed, by virtue of (2.3.2.5), the kernel of $\widehat{\theta}_{\Lambda,r}$ identifies with

$$\varprojlim_{n\geq 0} \frac{J_\Lambda + p^n\mathrm{W}_\Lambda(A^\flat)}{J_\Lambda^r + p^n\mathrm{W}_\Lambda(A^\flat)}. \tag{2.3.3.2}$$

We then have $\ker(\widehat{\theta}_{\Lambda,r})^r = 0$.

(iii) Note first that as θ_Λ and θ'_Λ are surjective, for every integer $n \geq 1$, we have

$$\theta_\Lambda^{-1}(p^n\widehat{A}) = J_\Lambda + p^n\mathrm{W}_\Lambda(A^\flat), \tag{2.3.3.3}$$
$$\theta'^{-1}_\Lambda(p^n\widehat{A}) = J'_\Lambda + p^n\mathrm{W}'_\Lambda(A^\flat). \tag{2.3.3.4}$$

Denote by

$$h_n\colon \mathrm{W}'_\Lambda(A^\flat) \to \mathrm{W}_\Lambda(A^\flat)/(J_\Lambda + p\mathrm{W}_\Lambda(A^\flat))^n \tag{2.3.3.5}$$

the canonical homomorphism. The homomorphism θ_Λ induces an injective homomorphism

$$\mathrm{W}_\Lambda(A^\flat)/(J_\Lambda + p\mathrm{W}_\Lambda(A^\flat)) \to A/pA, \tag{2.3.3.6}$$

whose composition with h_1 is induced by θ'_Λ. We deduce from this that $\ker(h_1) = J'_\Lambda + p\mathrm{W}'_\Lambda(A^\flat)$. By ([10] chap. III § 2.11 prop. 14 and cor. 1), as J_Λ is of finite type, for every $n \geq 1$, h_n is surjective and $\ker(h_n) = (J'_\Lambda + p\mathrm{W}'_\Lambda(A^\flat))^n$. Consequently, h_n induces an isomorphism

$$\frac{\mathrm{W}'_\Lambda(A^\flat)}{(J'_\Lambda + p\mathrm{W}'_\Lambda(A^\flat))^n} \xrightarrow{\sim} \frac{\mathrm{W}_\Lambda(A^\flat)}{(J_\Lambda + p\mathrm{W}_\Lambda(A^\flat))^n} \tag{2.3.3.7}$$

that fits into a commutative diagram

$$\begin{array}{ccc} \mathrm{W}'_\Lambda(A^\flat)/(J'_\Lambda + p\mathrm{W}'_\Lambda(A^\flat))^n & \xrightarrow{\sim} & \mathrm{W}_\Lambda(A^\flat)/(J_\Lambda + p\mathrm{W}_\Lambda(A^\flat))^n \\ & \searrow^{\theta'^{(n)}_\Lambda} & \downarrow{\scriptstyle \theta^{(n)}_\Lambda} \\ & & A/p^nA \end{array} \tag{2.3.3.8}$$

where $\theta^{(n)}_\Lambda$ (resp. $\theta'^{(n)}_\Lambda$) is induced by θ_Λ (resp. θ'_Λ). Hence, (2.3.3.7) induces an isomorphism between the kernels of $\theta_{\Lambda,n}$ and $\theta'_{\Lambda,n}$.

$$\frac{J'_\Lambda + p^n\mathrm{W}'_\Lambda(A^\flat)}{(J'_\Lambda + p\mathrm{W}'_\Lambda(A^\flat))^n} \xrightarrow{\sim} \frac{J_\Lambda + p^n\mathrm{W}_\Lambda(A^\flat)}{(J_\Lambda + p\mathrm{W}_\Lambda(A^\flat))^n}. \tag{2.3.3.9}$$

Moreover, for every integer $m \geq 0$, (2.3.3.7) induces an isomorphism

$$\frac{p^m\mathrm{W}'_\Lambda(A^\flat) + (J'_\Lambda + p\mathrm{W}'_\Lambda(A^\flat))^n}{(J'_\Lambda + p\mathrm{W}'_\Lambda(A^\flat))^n} \xrightarrow{\sim} \frac{p^m\mathrm{W}_\Lambda(A^\flat) + (J_\Lambda + p\mathrm{W}_\Lambda(A^\flat))^n}{(J_\Lambda + p\mathrm{W}_\Lambda(A^\flat))^n}. \tag{2.3.3.10}$$

Replace n by $n+r$ and take $m = n$. As we have

$$(J'_\Lambda + p^{n+r}\mathrm{W}'_\Lambda(A^\flat))^r + (J'_\Lambda + p\mathrm{W}'_\Lambda(A^\flat))^{n+r} + p^n\mathrm{W}'_\Lambda(A^\flat) = J'^r_\Lambda + p^n\mathrm{W}'_\Lambda(A^\flat),$$

and similarly for $\mathrm{W}_\Lambda(A^\flat)$ and J_Λ, we deduce from (2.3.3.9) and (2.3.3.10) an isomorphism

$$\frac{J'^r_\Lambda + p^n \mathrm{W}'_\Lambda(A^\flat)}{(J'_\Lambda + p\mathrm{W}'_\Lambda(A^\flat))^{n+r}} \xrightarrow{\sim} \frac{J^r_\Lambda + p^n \mathrm{W}_\Lambda(A^\flat)}{(J_\Lambda + p\mathrm{W}_\Lambda(A^\flat))^{n+r}}. \tag{2.3.3.11}$$

The isomorphism (2.3.3.7) (for $n + r$) induces therefore an isomorphism

$$\frac{\mathrm{W}'_\Lambda(A^\flat)}{J'^r_\Lambda + p^n \mathrm{W}'_\Lambda(A^\flat)} \xrightarrow{\sim} \frac{\mathrm{W}_\Lambda(A^\flat)}{J^r_\Lambda + p^n \mathrm{W}_\Lambda(A^\flat)}. \tag{2.3.3.12}$$

By taking the inverse limit, we obtain an isomorphism

$$\widehat{\mathscr{A}}'_r(A/\Lambda) \xrightarrow{\sim} \widehat{\mathscr{A}}_r(A/\Lambda) \tag{2.3.3.13}$$

whose composition with the canonical morphism (2.3.3.1) is clearly the identity of $\widehat{\mathscr{A}}_r(A/\Lambda)$.

Remark 2.3.4 Under the assumptions of 2.3.2, Fontaine ([20] 1.2.1) asserts that $\mathrm{W}'_\Lambda(A^\flat)$ is complete and separated for the p-adic topology, but his proof has a problem.

Proposition 2.3.5 ([20] 1.2.1) *We keep the assumptions and notation of* 2.3.2; *we suppose moreover that the Frobenius endomorphism of A/pA is surjective. Then, for every integer $r \geq 1$, $(\widehat{\mathscr{A}}_r(A/\Lambda), \widehat{\theta}_{\Lambda,r})$ is a universal p-adic Λ-infinitesimal thickening of order $\leq r-1$ of $\widehat{A}$.*

Note first that $(\widehat{\mathscr{A}}_r(A/\Lambda), \widehat{\theta}_{\Lambda,r})$ is an object of the category $\mathbf{E}^p_{r-1}(\widehat{A}/\Lambda)$ (2.3.1) by virtue of 2.3.3. Let (D, θ_D) be an object of $\mathbf{E}^p_{r-1}(\widehat{A}/\Lambda)$, $I_D = \ker(\theta_D)$. Let $\mathbb{D}$ (resp. $\mathbb{A}$) denote the set of sequences $(x_n)_\mathbb{N}$ of elements of D (resp. $\widehat{A}$) such that $x^p_{n+1} = x_n$. We will show that the map

$$\mathbb{D} \mapsto \mathbb{A}, \quad (x_n)_\mathbb{N} \mapsto (\theta_D(x_n))_\mathbb{N} \tag{2.3.5.1}$$

is bijective. Note first that $I_D + pD$ is an ideal of definition for the p-adic topology of D. Indeed, for every $N \geq r$, we have

$$p^N D \subset (I_D + pD)^N \subset p^{N-r} D. \tag{2.3.5.2}$$

Let $(x_n)_\mathbb{N}$ and $(y_n)_\mathbb{N}$ be two sequences of elements of D such that $\theta_D(x_n) = \theta_D(y_n)$ for all $n \geq 0$. As $x_n \equiv y_n$ modulo I_D and therefore a fortiori modulo $I_D + pD$, proceeding as in the proof of 2.2.5(i), we see that for all $n, m \geq 0$, we have

$$x^{p^m}_{n+m} - y^{p^m}_{n+m} \in (I_D + pD)^m. \tag{2.3.5.3}$$

Consequently, the map (2.3.5.1) is injective. Suppose that $(\theta_D(x_n))_\mathbb{N}$ belongs to $\mathbb{A}$. Applying (2.3.5.3) to the sequences $(x_n)_{n\geq 0}$ and $(x^p_{n+1})_{n\geq 0}$, we deduce that for every $n \geq 0$, the sequence $(x^{p^m}_{n+m})_{m\geq 0}$ converges to an element $\widetilde{x}_n$ of D which depends only on $(\theta_D(x_n))_\mathbb{N}$ but not on $(x_n)_\mathbb{N}$. It is clear that the sequence $(\widetilde{x}_n)_\mathbb{N}$ belongs to $\mathbb{D}$ and that we have $\theta_D(\widetilde{x}_n) = \theta_D(x_n)$ for all $n \geq 0$; hence the surjectivity of (2.3.5.1).

We will then prove that there exists a unique homomorphism

$$\alpha\colon \mathrm{W}(A^\flat) \to D \tag{2.3.5.4}$$

such that $\theta = \theta_D \circ \alpha$. Suppose first given such a morphism and let us show that it is unique. Let $x \in A^\flat$, $(x^{(n)})_{\mathbb{N}}$ be the associated sequence of $\mathbb{A}$ (2.2.5), $(\widetilde{x}^{(n)})_{\mathbb{N}}$ its preimage by the isomorphism (2.3.5.1). For every $n \geq 0$, considering $(x^{(n+m)})_{m\geq 0}$ as an element of $A^\flat$, we have

$$\theta([(x^{(n+m)})_{m\geq 0}]) = x^{(n)}. \tag{2.3.5.5}$$

The sequence $(\alpha([(x^{(n+m)})_{m\geq 0}]))_{n\geq 0}$ being clearly an element of $\mathbb{D}$, we deduce that we have

$$\alpha([(x^{(n+m)})_{m\geq 0}]) = \widetilde{x}^{(n)}. \tag{2.3.5.6}$$

Consequently, we have

$$\alpha(\mathrm{V}^n[x]) = p^n\alpha([(x^{(n+m)})_{m\geq 0}]) = p^n\widetilde{x}^{(n)}; \tag{2.3.5.7}$$

hence the uniqueness of α. To construct α, we proceed as in 2.2.1. For any integer $n \geq 1$, the ring homomorphism (2.1.3.1)

$$\begin{aligned} \Phi_{n+r+1}\colon \mathrm{W}_{n+r+1}(D/p^{n+r}D) &\to D/p^{n+r}D \\ (x_0,\ldots,x_{n+r}) &\mapsto x_0^{p^{n+r}} + px_1^{p^{n+r-1}} + \cdots + p^{n+r}x_{n+r} \end{aligned} \tag{2.3.5.8}$$

induces a ring homomorphism

$$\begin{aligned} \Phi'_{n+r+1}\colon \mathrm{W}_n(D/p^{n+r}D) &\to D/p^nD \\ (x_0,\ldots,x_{n-1}) &\mapsto x_0^{p^{n+r}} + px_1^{p^{n+r-1}} + \cdots + p^{n-1}x_{n-1}^{p^{r+1}}. \end{aligned} \tag{2.3.5.9}$$

For every $0 \leq i \leq n-1$, we have (2.3.5.2)

$$p^i(I_D + pD)^{p^{n+r-i}} \subset p^{p^{n+r-i}+i-r}D \subset p^nD. \tag{2.3.5.10}$$

The homomorphism Φ'_{n+r+1} therefore vanishes on

$$\mathrm{W}_n((I_D + pD)/p^{n+r}D) = \ker(\mathrm{W}_n(D/p^{n+r}D) \to \mathrm{W}_n(A/pA)) \tag{2.3.5.11}$$

and in turn induces a ring homomorphism

$$\alpha_n\colon \mathrm{W}_n(A/pA) \to D/p^nD. \tag{2.3.5.12}$$

It immediately follows from the definition that the diagram

$$\begin{array}{ccc} \mathrm{W}_{n+1}(A/pA) & \xrightarrow{\alpha_{n+1}} & D/p^{n+1}D \\ \Big\downarrow{\scriptstyle \mathrm{RF}} & & \Big\downarrow \\ \mathrm{W}_n(A/pA) & \xrightarrow{\alpha_n} & D/p^nD, \end{array} \tag{2.3.5.13}$$

where R is the restriction morphism (2.1.3.2), F is the Frobenius (2.1.3.4) and the unlabeled arrow is the canonical homomorphism, is commutative. The homomorphisms α_n define, by taking the inverse limit, the desired homomorphism (2.3.5.4).

Let $\alpha_\Lambda : \mathrm{W}(A^\flat) \otimes_W \Lambda \to D$ denote the homomorphism deduced from α. Since D is separated and complete for the p-adic topology, α_Λ induces a homomorphism $\widehat{\mathscr{A}}_r(A/\Lambda) \to D$ satisfying $\widehat{\theta}_{\Lambda,r} = \theta_D \circ \alpha_\Lambda$. The homomorphism $\widehat{\mathscr{A}}_r(A/\Lambda) \to D$ of $\mathbf{E}^p_{r-1}(\widehat{A}/\Lambda)$ thus defined is clearly unique. The proposition follows.

Lemma 2.3.6 *Let A be a ring, $t \in A$, M, M', M'' three complete and separated A-modules for the t-adic topology, $u: M' \to M$, $v: M \to M''$ two A-linear morphisms such that $v \circ u = 0$. We denote by $F \mapsto F_n$ the functor of reduction modulo t^n on the category of A-modules. We assume that t is not a zero divisor in any of the A-modules M, M' and M'' and that the sequence of A_1-modules*

$$0 \longrightarrow M'_1 \xrightarrow{u_1} M_1 \xrightarrow{v_1} M''_1 \longrightarrow 0 \tag{2.3.6.1}$$

is exact. Then, the sequence

$$0 \longrightarrow M' \xrightarrow{u} M \xrightarrow{v} M'' \longrightarrow 0 \tag{2.3.6.2}$$

is exact.

Indeed, the columns of the diagram

$$\begin{array}{ccccccccc} & & 0 & & 0 & & 0 & & \\ & & \downarrow & & \downarrow & & \downarrow & & \\ 0 & \longrightarrow & M'_n & \xrightarrow{u_n} & M_n & \xrightarrow{v_n} & M'_n & \longrightarrow & 0 \\ & & \downarrow{\cdot t} & & \downarrow{\cdot t} & & \downarrow{\cdot t} & & \\ 0 & \longrightarrow & M'_{n+1} & \xrightarrow{u_{n+1}} & M_{n+1} & \xrightarrow{v_{n+1}} & M'_{n+1} & \longrightarrow & 0 \\ & & \downarrow & & \downarrow & & \downarrow & & \\ 0 & \longrightarrow & M'_1 & \xrightarrow{u_1} & M_1 & \xrightarrow{v_1} & M'_1 & \longrightarrow & 0 \\ & & \downarrow & & \downarrow & & \downarrow & & \\ & & 0 & & 0 & & 0 & & \end{array} \tag{2.3.6.3}$$

are exact. We deduce from this by induction that for every $n \geq 1$, the sequence of A_n-modules

$$0 \longrightarrow M'_n \xrightarrow{u_n} M_n \xrightarrow{v_n} M''_n \longrightarrow 0 \tag{2.3.6.4}$$

is exact. The proposition follows by virtue of ([29] 0.13.2.2).

Proposition 2.3.7 ([3] II.9.5; [55] A.1.1 and A.2.2) *Let K be a complete discrete valuation field of characteristic 0, with perfect residue field k of characteristic $p > 0$, $\mathscr{O}_K$ the valuation ring of K, π a uniformizer of $\mathscr{O}_K$, W the ring of Witt vectors with coefficients in k, A an $\mathscr{O}_K$-algebra. Assume that the following conditions hold:*

(i) *A is $\mathscr{O}_K$-flat.*
(ii) *A is integrally closed in $A[\frac{1}{p}]$.*

(iii) *The Frobenius endomorphism of A/pA is surjective.*
(iv) *There exists a sequence $(\pi_n)_{n\geq 0}$ of elements of A such that $\pi_0 = \pi$ and $\pi_{n+1}^p = \pi_n$ for all $n \geq 0$.*

We use the notation of 2.3.2 *with $\Lambda = \mathscr{O}_K$. We denote by $\underline{\pi}$ the element of $A^\flat$ defined by the sequence $(\pi_n)_{n\geq 0}$ and we set*

$$\xi_\pi = [\underline{\pi}] - \pi \in \mathrm{W}_{\mathscr{O}_K}(A^\flat), \tag{2.3.7.1}$$

where $[\ ]$ is the multiplicative representative. Then, the sequence

$$0 \longrightarrow \mathrm{W}_{\mathscr{O}_K}(A^\flat) \xrightarrow{\cdot \xi_\pi} \mathrm{W}_{\mathscr{O}_K}(A^\flat) \xrightarrow{\theta_{\mathscr{O}_K}} \widehat{A} \longrightarrow 0 \tag{2.3.7.2}$$

is exact.

Indeed, we clearly have $\theta_{\mathscr{O}_K}(\xi_\pi) = 0$. The algebra $\mathrm{W}_{\mathscr{O}_K}(A^\flat)$ being free of finite type over $\mathrm{W}(A^\flat)$, it is complete and separated for the p-adic topology or, what amounts to the same, for the π-adic topology. Moreover, $\mathrm{W}_{\mathscr{O}_K}(A^\flat)$ being $\mathscr{O}_K$-flat, π is not a zero divisor in $\mathrm{W}_{\mathscr{O}_K}(A^\flat)$. Similarly, for every $n \geq 1$, the sequence

$$0 \longrightarrow A/\pi^n A \xrightarrow{\cdot \pi} A/\pi^{n+1} A \longrightarrow A/\pi A \rightarrow 0 \tag{2.3.7.3}$$

is exact, and therefore π is not a zero divisor in $\widehat{A}$ ([29] 0.13.2.2). Given 2.3.6, it suffices to show that the sequence

$$0 \longrightarrow A^\flat \xrightarrow{\cdot \underline{\pi}} A^\flat \longrightarrow A/\pi A \longrightarrow 0 \tag{2.3.7.4}$$

where the third arrow is induced by the canonical projection $A^\flat \rightarrow A/pA$ onto the first factor (i.e., of index 0), is exact.

For every $n \geq 0$, there exists a $q_n \in A$ such that $p = \pi_n q_n$. Since π_n is not a zero divisor in A, we have $q_{n+1}^p = p^{p-1} q_n$. Let $y \in A^\flat$ be such that $\underline{\pi} \cdot y = 0$. For any $n \geq 0$, let y_n be the image of y by the canonical projection $A^\flat \rightarrow A/pA$ on the $(n+1)$th factor (i.e., the component of index n), and let $\widetilde{y}_n$ be a lifting of y_n in A. We then have $\pi_n \widetilde{y}_n \in pA$ and, consequently, $\widetilde{y}_n \in q_n A$ and

$$y_n = y_{n+1}^p (= \widetilde{y}_{n+1}^p \mod pA) = 0. \tag{2.3.7.5}$$

We deduce from this that $\underline{\pi}$ is not a zero divisor in $A^\flat$.

Moreover, let $n \geq 0$, $x \in A$ be such that $x^{p^n} \in \pi A$. As $\pi_n^{-1} \in p^{-1} A$, we see from (i) and (ii) that $x \in \pi_n A$. The Frobenius endomorphism φ of A/pA being surjective by (iii), we deduce that the sequence

$$0 \longrightarrow A/q_n A \xrightarrow{\cdot \pi_n} A/pA \xrightarrow{\lambda \circ \varphi^n} A/\pi A \longrightarrow 0\,, \tag{2.3.7.6}$$

where $\lambda\colon A/pA \rightarrow A/\pi A$ is the canonical morphism, is exact. Hence, the sequence (2.3.7.4) is exact by ([29] 0.13.2.2).

Corollary 2.3.8 *Under the assumptions of* 2.3.7, *for every integer* $r \geq 1$, *the rings* $\mathscr{A}_r(A)$ (2.2.3.5) *and* $\mathscr{A}_r(A/\mathscr{O}_K)$ (2.3.2.4) *are complete and separated for the p-adic topologies.*

We first show that $\widehat{A}$ is $\mathscr{O}_K$-flat. By ([10] chap. III §2.11 prop. 14 and cor. 1), for every $n \geq 0$, we have

$$\widehat{A}/p^n\widehat{A} \simeq A/p^nA. \tag{2.3.8.1}$$

Let $x \in \widehat{A}$ be such that $px = 0$, $\overline{x}$ the residue class of x in $\widehat{A}/p^n\widehat{A}$ $(n \geq 1)$. As A is $\mathscr{O}_{\overline{K}}$-flat, it follows from (2.3.8.1) that $\overline{x} \in p^{n-1}\widehat{A}/p^n\widehat{A}$. We deduce from this that $x \in \cap_{n\geq 0} p^n\widehat{A} = \{0\}$. Consequently, p is not a zero divisor in $\widehat{A}$, and therefore $\widehat{A}$ is $\mathscr{O}_K$-flat.

On the other hand, for every integer $r \geq 1$, the sequence (2.3.7.2) induces an exact sequence

$$0 \longrightarrow \mathscr{A}_r(A/\mathscr{O}_K) \xrightarrow{\cdot \xi_\pi} \mathscr{A}_{r+1}(A/\mathscr{O}_K) \xrightarrow{\theta_{\mathscr{O}_K, r+1}} \widehat{A} \longrightarrow 0. \tag{2.3.8.2}$$

We deduce from this by induction on r that $\mathscr{A}_r(A/\mathscr{O}_K)$ is complete and separated for the p-adic topology. Replacing $\mathscr{O}_K$ by W, we obtain that $\mathscr{A}_r(A)$ is complete and separated for the p-adic topology.

2.4 Logarithmic Infinitesimal Thickenings

We refer to ([3] II.5) for a lexicon of logarithmic geometry.

2.4.1 Let $(X, \mathscr{M}_X)$ be a logarithmic scheme, M a monoid, $u \colon M \to \Gamma(X, \mathscr{M}_X)$ a homomorphism, p a prime number. Suppose that the scheme X is affine with ring a $\mathbb{Z}_{(p)}$-algebra A. We denote by $\widehat{A}$ the p-adic Hausdorff completion of A. We will use the notation of 2.2.3 for A. Consider the inverse system of multiplicative monoids $(A)_{n\in\mathbb{N}}$ where the transition morphisms are all equal to the pth power. We denote by Q the fiber product of the diagram of homomorphisms of monoids

$$\begin{array}{ccc} & & M \\ & & \downarrow \\ \varprojlim_{\mathbb{N}} A & \longrightarrow & A \end{array} \tag{2.4.1.1}$$

where the horizontal arrow is the projection onto the first component (i.e., of index 0) and the vertical arrow is induced by u. We denote by τ the composed homomorphism

$$\tau \colon Q \longrightarrow \varprojlim_{x \mapsto x^p} A \longrightarrow A^\flat \xrightarrow{[\;]} \mathrm{W}(A^\flat), \tag{2.4.1.2}$$

where the first and second arrows are the canonical homomorphisms (2.2.3.1) and [ ] is the multiplicative representative. It immediately follows from the definition that the diagram

$$\begin{array}{ccc} Q & \longrightarrow & M \\ {\scriptstyle\tau}\downarrow & & \downarrow \\ \mathrm{W}(A^{\flat}) & \xrightarrow{\theta} & \widehat{A} \end{array} \tag{2.4.1.3}$$

where θ is the canonical homomorphism (2.2.3.4) and the unlabeled arrows are the canonical morphisms, is commutative.

We set $\widehat{X} = \mathrm{Spec}(\widehat{A})$, which we endow with the logarithmic structure $\mathscr{M}_{\widehat{X}}$ pullback of $\mathscr{M}_X$. We endow $\mathrm{Spec}(\mathrm{W}(A^{\flat}))$ with the logarithmic structure $\mathscr{Q}$ associated with the prelogarithmic structure defined by the homomorphism τ (2.4.1.2). By (2.4.1.3), θ induces a morphism of logarithmic schemes

$$(\widehat{X}, \mathscr{M}_{\widehat{X}}) \to (\mathrm{Spec}(\mathrm{W}(A^{\flat})), \mathscr{Q}). \tag{2.4.1.4}$$

Proposition 2.4.2 ([3] II.9.7) *We keep the assumptions of* 2.4.1, *denote by* X° *the maximal open subscheme of* X *where the logarithmic structure* $\mathscr{M}_X$ *is trivial, and suppose that the following conditions are satisfied:*

(a) *A is a normal integral domain.*
(b) *X° is a non-empty simply connected $\mathbb{Q}$-scheme.*
(c) *M is integral and there exist a fine and saturated monoid M' and a homomorphism $\nu\colon M' \to M$ such that the induced homomorphism $M' \to M/M^{\times}$ is an isomorphism.*

Then,

(i) *The monoid Q is integral and the group M'^{gp} is free.*
(ii) *We can complete diagram* (2.4.1.1) *into a commutative diagram*

$$\begin{array}{ccc} M' & \xrightarrow{\nu} & M \\ {\scriptstyle w}\downarrow & & \downarrow \\ \varprojlim\limits_{x\mapsto x^p} A & \longrightarrow & A. \end{array} \tag{2.4.2.1}$$

We denote by $\beta\colon M' \to Q$ the induced homomorphism.

(iii) *The logarithmic structure $\mathscr{Q}$ on* $\mathrm{Spec}(\mathrm{W}(A^{\flat}))$ *is associated with the prelogarithmic structure defined by the composition*

$$M' \xrightarrow{\beta} Q \xrightarrow{\tau} \mathrm{W}(A^{\flat}). \tag{2.4.2.2}$$

In particular, the logarithmic scheme $(\mathrm{Spec}(\mathrm{W}(A^{\flat})), \mathscr{Q})$ *is fine and saturated.*

(iv) *If, moreover, the composed homomorphism $u \circ \nu\colon M' \to \Gamma(X, \mathscr{M}_X)$ is a chart for $(X, \mathscr{M}_X)$, then the morphism* (2.4.1.4) *is strict.*

2.4.3 In the remaining part of this section, K denotes a complete discrete valuation field of characteristic 0, with perfect residue field k of characteristic $p > 0$, $\mathscr{O}_K$ the valuation ring of K, π a uniformizer of $\mathscr{O}_K$, W the ring of Witt vectors (with respect to p) with coefficients in k and A an $\mathscr{O}_K$-algebra satisfying the following conditions:

(L_1) A is $\mathscr{O}_K$-flat.
(L_2) A is integrally closed in $A[\frac{1}{p}]$.
(L_3) The Frobenius endomorphism of A/pA is surjective.
(L_4) There exists a sequence $(\pi_n)_{n\geq 0}$ of elements of A such that $\pi_0 = \pi$ and $\pi_{n+1}^p = \pi_n$ for every $n \geq 0$.

We will use the notation of 2.3.2 with $\Lambda = \mathscr{O}_K$. We denote by $\underline{\pi}$ the element of $A^\flat$ induced by the sequence $(\pi_n)_{n\geq 0}$ and we set

$$\xi_\pi = [\underline{\pi}] - \pi \in \mathrm{W}_{\mathscr{O}_K}(A^\flat), \tag{2.4.3.1}$$

where [ ] is the multiplicative representative. By virtue of 2.3.7, the sequence

$$0 \longrightarrow \mathrm{W}_{\mathscr{O}_K}(A^\flat) \xrightarrow{\cdot\xi_\pi} \mathrm{W}_{\mathscr{O}_K}(A^\flat) \xrightarrow{\theta_{\mathscr{O}_K}} \widehat{A} \longrightarrow 0 \tag{2.4.3.2}$$

is exact. In particular, we have $\ker(\theta) \subset \xi_\pi \mathrm{W}_{\mathscr{O}_K}(A^\flat)$.

It follows from (L_1) that $\widehat{A}$ is $\mathscr{O}_K$-flat (see the proof of 2.3.8). We set

$$\mathrm{W}_K(A^\flat) = \mathrm{W}(A^\flat) \otimes_W K \tag{2.4.3.3}$$

and we denote by $\theta_K \colon \mathrm{W}_K(A^\flat) \to \widehat{A}[\frac{1}{p}]$ the homomorphism induced by θ (2.2.3.4). We denote by $\mathrm{W}^*_{\mathscr{O}_K}(A^\flat)$ the sub-$\mathrm{W}_{\mathscr{O}_K}(A^\flat)$-algebra of $\mathrm{W}_K(A^\flat)$ generated by $[\underline{\pi}]/\pi$ and we set

$$\xi^*_\pi = \frac{\xi_\pi}{\pi} = \frac{[\underline{\pi}]}{\pi} - 1 \in \mathrm{W}^*_{\mathscr{O}_K}(A^\flat). \tag{2.4.3.4}$$

Note that $\mathrm{W}^*_{\mathscr{O}_K}(A^\flat)$ depends on the sequence $(\pi_n)_{n\geq 0}$. The homomorphism θ_K induces a homomorphism

$$\theta^*_{\mathscr{O}_K} \colon \mathrm{W}^*_{\mathscr{O}_K}(A^\flat) \to \widehat{A} \tag{2.4.3.5}$$

such that $\theta^*_{\mathscr{O}_K}(\xi^*_\pi) = 0$.

Lemma 2.4.4 (i) *For every integer $r \geq 0$, we have*

$$(\xi^*_\pi)^r \mathrm{W}^*_{\mathscr{O}_K}(A^\flat) = \mathrm{W}^*_{\mathscr{O}_K}(A^\flat) \cap ((\xi^*_\pi)^r \mathrm{W}_K(A^\flat)). \tag{2.4.4.1}$$

(ii) *The sequence*

$$0 \longrightarrow \mathrm{W}^*_{\mathscr{O}_K}(A^\flat) \xrightarrow{\cdot\xi^*_\pi} \mathrm{W}^*_{\mathscr{O}_K}(A^\flat) \xrightarrow{\theta^*_{\mathscr{O}_K}} \widehat{A} \longrightarrow 0 \tag{2.4.4.2}$$

is exact.

(i) Note first that $\mathrm{W}(A^\flat)$ is W-flat since $A^\flat$ is perfect. Consequently, $\mathrm{W}_{\mathscr{O}_K}(A^\flat)$ is identified with a subalgebra of $\mathrm{W}_K(A^\flat)$. We proceed by induction on r. The assertion is obvious for $r = 0$. Suppose that $r \geq 1$ and the assertion established for $r-1$. Let $x \in \mathrm{W}^*_{\mathcal{O}_K}(A^\flat) \cap ((\xi^*_\pi)^r \mathrm{W}_K(A^\flat))$. By the induction hypothesis, there exists an $a \in \mathrm{W}_{\mathcal{O}_K}(A^\flat)$ such that

$$x - a(\xi^*_\pi)^{r-1} \in (\xi^*_\pi)^r \mathrm{W}^*_{\mathcal{O}_K}(A^\flat). \tag{2.4.4.3}$$

As ξ^*_π is not a zero divisor in $\mathrm{W}_K(A^\flat)$ (2.4.3.2), we deduce that

$$a \in \mathrm{W}_{\mathcal{O}_K}(A^\flat) \cap \xi^*_\pi \mathrm{W}_K(A^\flat). \tag{2.4.4.4}$$

Consequently, $\theta_{\mathcal{O}_K}(a) = 0$ and therefore there exists a $b \in \mathrm{W}_{\mathcal{O}_K}(A^\flat)$ such that $a = \xi_\pi b$ by virtue of (2.4.3.2). It follows that $x \in (\xi^*_\pi)^r \mathrm{W}^*_{\mathcal{O}_K}(A^\flat)$, hence the proposition.

(ii) This follows from (i) and (2.4.3.2).

2.4.5 For any integer $r \geq 1$, we set

$$\mathscr{A}^*_r(A/\mathcal{O}_K) = \mathrm{W}^*_{\mathcal{O}_K}(A^\flat)/(\xi^*_\pi)^r \mathrm{W}^*_{\mathcal{O}_K}(A^\flat) \tag{2.4.5.1}$$

and we denote by $\theta^*_{\mathcal{O}_K,r} : \mathscr{A}^*_r(A/\mathcal{O}_K) \to \widehat{A}$ the homomorphism induced by $\theta^*_{\mathcal{O}_K}$. It follows from 2.4.4(ii) that $\mathscr{A}^*_r(A/\mathcal{O}_K)$ is complete and separated for the p-adic topology. The canonical homomorphism $\mathrm{W}(A^\flat) \to \mathrm{W}^*_{\mathcal{O}_K}(A^\flat)$ induces a homomorphism (2.2.3.5)

$$\mathscr{A}_r(A) \to \mathscr{A}^*_r(A/\mathcal{O}_K). \tag{2.4.5.2}$$

2.4.6 We set $S = \mathrm{Spec}(\mathcal{O}_K)$ and we denote by s (resp. η) the closed point (resp. generic) of S (2.4.3). We endow S with the logarithmic structure $\mathscr{M}_S$ defined by its closed point, that is, $\mathscr{M}_S = u_*(\mathcal{O}^\times_\eta) \cap \mathcal{O}_S$, where $u : \eta \to S$ is the canonical injection. We denote by $\iota : \mathbb{N} \to \Gamma(S, \mathscr{M}_S)$ the homomorphism defined by $\iota(1) = \pi$, which is a chart for $(S, \mathscr{M}_S)$. We set $X = \mathrm{Spec}(A)$ which we suppose endowed with a logarithmic structure $\mathscr{M}_X$. We also fix a morphism $f : (X, \mathscr{M}_X) \to (S, \mathscr{M}_S)$, a monoid M and two homomorphisms $\mathbb{N} \xrightarrow{\gamma} M \xrightarrow{u} \Gamma(X, \mathscr{M}_X)$ that fit into a commutative diagram

$$\begin{array}{ccc} \mathbb{N} & \xrightarrow{\iota} & \Gamma(S, \mathscr{M}_S) \\ {\scriptstyle\gamma}\downarrow & & \downarrow{\scriptstyle f^\flat} \\ M & \xrightarrow{u} & \Gamma(X, \mathscr{M}_X) \end{array} \tag{2.4.6.1}$$

where $f^\flat$ is the homomorphism induced by f. We take again the notation of 2.4.1 for the affine logarithmic scheme $(X, \mathscr{M}_X)$ and the homomorphism u.

For any integer $r \geq 1$, we set (2.4.5.1)

$$\mathscr{A}^*_r(X/S) = \mathrm{Spec}(\mathscr{A}^*_r(A/\mathcal{O}_K)). \tag{2.4.6.2}$$

We endow it with the logarithmic structure $\mathscr{M}_{\mathscr{A}^*_r(X/S)}$ associated with the prelogarithmic structure defined by the homomorphism

$$Q \to \mathscr{A}^*_r(A/\mathcal{O}_K) \tag{2.4.6.3}$$

induced by τ (2.4.1.2). By (2.4.1.3), the homomorphism $\theta^*_{\mathscr{O}_K,r}$ induces a morphism

$$(\widehat{X}, \mathscr{M}_{\widehat{X}}) \to (\mathscr{A}^*_r(X/S), \mathscr{M}_{\mathscr{A}^*_r(X/S)}). \tag{2.4.6.4}$$

With the notation of 2.4.3(L$_4$), we denote by $\widetilde{\pi}$ the element of Q defined by its projections (2.4.1.1)

$$(\pi_n)_{n\geq 0} \in \varprojlim_{\mathbb{N}} A \quad \text{and} \quad \gamma(1) \in M. \tag{2.4.6.5}$$

The element $(\xi^*_\pi + 1)$ being invertible in $\mathscr{A}^*_r(A/\mathscr{O}_K)$, the homomorphism

$$\mathbb{N} \to \Gamma(\mathscr{A}^*_r(X/S), \mathscr{M}_{\mathscr{A}^*_r(X/S)}), \quad 1 \mapsto (\xi^*_\pi + 1)^{-1}\widetilde{\pi} \tag{2.4.6.6}$$

induces a morphism of logarithmic schemes

$$\mathrm{pr}_1 : (\mathscr{A}^*_r(X/S), \mathscr{M}_{\mathscr{A}^*_r(X/S)}) \to (S, \mathscr{M}_S). \tag{2.4.6.7}$$

We set (2.2.3.5)

$$\mathscr{A}_r(X) = \mathrm{Spec}(\mathscr{A}_r(A)), \tag{2.4.6.8}$$

that we endow with the logarithmic structure $\mathscr{M}_{\mathscr{A}_r(X)}$ pullback of the logarithmic structure $\mathscr{Q}$ on $\mathrm{Spec}(\mathrm{W}(A^\flat))$ (2.4.1). We clearly have a morphism of logarithmic schemes

$$\mathrm{pr}_2 : (\mathscr{A}^*_r(X/S), \mathscr{M}_{\mathscr{A}^*_r(X/S)}) \to (\mathscr{A}_r(X), \mathscr{M}_{\mathscr{A}_r(X)}). \tag{2.4.6.9}$$

Proposition 2.4.7 *We keep the assumptions and notation of* 2.4.3 *and* 2.4.6, *denote by X° the maximal open subscheme of X where the logarithmic structure $\mathscr{M}_X$ is trivial and assume that the following conditions are satisfied:*

(L$_5$) *A is a normal integral domain.*

(L$_6$) *X° is a non-empty simply connected η-scheme.*

(L$_7$) *M is integral and there exist a fine and saturated monoid M' and a homomorphism $\nu: M' \to M$ such that the induced homomorphism $M' \to M/M^\times$ is an isomorphism and the composed homomorphism $u \circ \nu: M' \to \Gamma(X, \mathscr{M}_X)$ is a chart for $(X, \mathscr{M}_X)$.*

Then,

(i) *The monoid Q is integral and the group M'^{gp} is free.*

(ii) *We can complete the diagram* (2.4.1.1) *into a commutative diagram*

$$\begin{array}{ccc} M' & \xrightarrow{\nu} & M \\ {\scriptstyle w}\downarrow & & \downarrow \\ \varprojlim_{x\mapsto x^p} A & \longrightarrow & A. \end{array} \tag{2.4.7.1}$$

We denote by $\beta: M' \to Q$ the induced homomorphism.

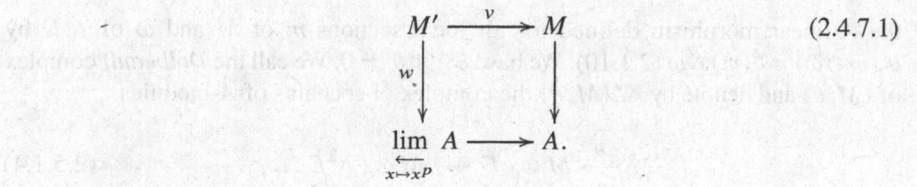

(iii) *For every integer* $r \geq 1$, *the logarithmic structure* $\mathscr{M}_{\mathscr{A}_r^*(X/S)}$ *on* $\mathscr{A}_r^*(X/S)$ *is associated with the prelogarithmic structure defined by the composition*

$$M' \xrightarrow{\beta} Q \longrightarrow \mathscr{A}_r^*(A/\mathscr{O}_K), \tag{2.4.7.2}$$

where the second arrow is the homomorphism (2.4.6.3). *In particular, the logarithmic scheme* $(\mathscr{A}_r^*(X/S), \mathscr{M}_{\mathscr{A}_r^*(X/S)})$ *is fine and saturated and the morphism* (2.4.6.4) *is a strict closed immersion.*

The first two propositions are mentioned as a reminder (2.4.2). Let G be the preimage of $M^\times$ by the canonical homomorphism $Q \to M$. It immediately follows from the definition (2.4.1.1) that G is a subgroup of Q. On the other hand, the composed homomorphism $M' \to Q/G \to M/M^\times$, where the first arrow is deduced from β, is an isomorphism. Hence, $M' \to Q/G$ is an isomorphism. Proposition (iii) then follows from ([55] 1.3.1).

2.5 Higgs Modules and λ-connections

Definition 2.5.1 Let (X, A) be a ringed topos, E an A-module.

(i) We call a *Higgs A-module with coefficients in E* a pair (M, θ) consisting of an A-module M and an A-linear morphism

$$\theta \colon M \to M \otimes_A E \tag{2.5.1.1}$$

such that $\theta \wedge \theta = 0$. We then say that θ is a *Higgs A-field* on M with coefficients in E.

(ii) If (M, θ) and (M', θ') are two Higgs A-modules with coefficients in E, a morphism from (M, θ) to (M', θ') is an A-linear morphism $u \colon M \to M'$ such that $(u \otimes \mathrm{id}_E) \circ \theta = \theta' \circ u$.

The Higgs A-modules with coefficients in E form a category that we denote by $\mathbf{HM}(A, E)$. We can complete the terminology and make the following remarks.

2.5.1.2 Let (M, θ) be a Higgs A-module with coefficients in E. For each $i \geq 1$, we denote by

$$\theta^i \colon M \otimes_A \wedge^i E \to M \otimes_A \wedge^{i+1} E \tag{2.5.1.3}$$

the A-linear morphism defined for all local sections m of M and ω of $\wedge^i E$ by $\theta^i(m \otimes \omega) = \theta(m) \wedge \omega$ (2.1.10). We have $\theta^{i+1} \circ \theta^i = 0$. We call the *Dolbeault* complex of (M, θ) and denote by $\mathbb{K}^\bullet(M, \theta)$ the complex of cochains of A-modules

$$M \xrightarrow{\theta} M \otimes_A E \xrightarrow{\theta^1} M \otimes_A \wedge^2 E \ldots, \tag{2.5.1.4}$$

where M is placed in degree 0 and the differentials are of degree 1.

2.5.1.5 Let (M, θ) be a Higgs A-module with coefficients in E with M a locally free A-module of finite type. Consider, for an integer $i \geq 1$, the composed morphism

$$\wedge^i M \xrightarrow{\wedge^i \theta} \wedge^i (M \otimes_A E) \longrightarrow \wedge^i M \otimes_A \mathrm{S}^i E, \tag{2.5.1.6}$$

where the second arrow is the canonical morphism ([37] V 4.5). We call the *ith characteristic invariant* of θ, and denote by $\lambda_i(\theta)$, the trace of the morphism (2.5.1.6) viewed as a section of $\Gamma(X, \mathrm{S}^i E)$.

2.5.1.7 Let (M, θ), (M', θ') be two Higgs A-modules with coefficients in E. We call the *total* Higgs field on $M \otimes_A M'$ the A-linear morphism

$$\theta_{\mathrm{tot}}\colon M \otimes_A M' \to M \otimes_A M' \otimes_A E \tag{2.5.1.8}$$

defined by

$$\theta_{\mathrm{tot}} = \theta \otimes \mathrm{id}_{M'} + \mathrm{id}_M \otimes \theta'. \tag{2.5.1.9}$$

We say that $(M \otimes_A M', \theta_{\mathrm{tot}})$ is the *tensor product* of (M, θ) and (M', θ').

2.5.1.10 Suppose that E is locally free of finite type over A; let $F = \mathscr{H}om_A(E, A)$ be its dual. For every A-module M, since the canonical morphism

$$\mathscr{E}nd_A(M) \otimes_A E \to \mathscr{H}om_A(M, M \otimes_A E) \tag{2.5.1.11}$$

is an isomorphism, giving a Higgs A-field θ on M is equivalent to giving an $\mathrm{S}_A(F)$-module structure on M that is compatible with its A-module structure (2.1.10). By ([3] II.2.8.10), we then have a canonical isomorphism of complexes of A-modules

$$\mathbb{K}^\bullet(M, \theta) \xrightarrow{\sim} \mathbb{K}^\bullet_{\mathrm{S}(F)}(M), \tag{2.5.1.12}$$

where $\mathbb{K}^\bullet_{\mathrm{S}(F)}(M)$ is the Koszul complex of the $\mathrm{S}(F)$-module M ([3] II.2.7.5).

2.5.2 Let (X, A) be a ringed topos, E a flat A-module. The category $\mathbf{HM}(A, E)$ of Higgs A-modules with coefficients in E is naturally an abelian category. Let (M, θ) be an object of this category. Giving a subobject of (M, θ) in $\mathbf{HM}(A, E)$ is equivalent to giving a sub-A-module N of M such that $\theta(N) \subset N \otimes_A E$. The quotient A-module M/N endowed with the Higgs field induced by θ is a quotient of (M, θ) in $\mathbf{HM}(A, E)$. We will abusively say that N (resp. M/N) is a subobject (resp. quotient) of (M, θ).

Giving a decreasing filtration of (M, θ) in $\mathbf{HM}(A, E)$, indexed by $\mathbb{Z}$, is equivalent to giving a decreasing filtration $(\mathrm{F}^i M)_{i \in \mathbb{Z}}$ of M by sub-A-modules such that $\theta(\mathrm{F}^i M) \subset \mathrm{F}^i M \otimes_A E$ for all $i \in \mathbb{Z}$. We will abusively say that $\mathrm{F}^\bullet M$ is a filtration of (M, θ). For each $i \in \mathbb{Z}$, the Higgs field θ induces on $\mathrm{Gr}^i_{\mathrm{F}} M = \mathrm{F}^i M/\mathrm{F}^{i+1} M$ a Higgs field, and $\mathrm{Gr}^\bullet_{\mathrm{F}} M = \oplus_{i \in \mathbb{Z}} \mathrm{Gr}^i_{\mathrm{F}} M$ is the graded object associated with the filtration $\mathrm{F}^\bullet M$ of (M, θ) in $\mathbf{HM}(A, E)$.

Given a filtration $\mathrm{F}^\bullet M$ of (M, θ) in $\mathbf{HM}(A, E)$ and a subobject N of (M, θ), $(N \cap \mathrm{F}^i M)_{i \in \mathbb{Z}}$ (resp. $((\mathrm{F}^i M + N)/N)_{i \in \mathbb{Z}}$) is a filtration of the subobject N (resp. quotient M/N) of (M, θ), called the *induced* (resp. *quotient*) filtration. Note that for every $i \in \mathbb{Z}$, we have $(N \cap \mathrm{F}^i M) \otimes_A E = (N \otimes_A E) \cap (\mathrm{F}^i M \otimes_A E)$.

Definition 2.5.3 Let (X, A) be a ringed topos, E a flat A-module. We say that a Higgs A-module (M, θ) with coefficients in E is *quasi-nilpotent* if there exist a decreasing filtration $(\mathrm{F}^i M)_{i \in \mathbb{Z}}$ of (M, θ) in $\mathbf{HM}(A, E)$ (2.5.2) and two integers $m \leq n$ such that $\mathrm{F}^n M = 0$, $\mathrm{F}^m M = M$ and that the Higgs field of $\mathrm{Gr}^\bullet_{\mathrm{F}} M$ is zero (or equivalently that we have $\theta(\mathrm{F}^i M) \subset \mathrm{F}^{i+1} M \otimes_A E$ for all $i \in \mathbb{Z}$). We then also say that the Higgs field θ is *quasi-nilpotent*, and that the filtration $\mathrm{F}^\bullet M$ of (M, θ) is *quasi-nilpotent*.

We reserve the terminology *nilpotent* for a slightly stronger condition (2.5.6).

Lemma 2.5.4 *Let (X, A) be a ringed topos, E a flat A-module, (M, θ) a Higgs A-module with coefficients in E, N (resp. P) a subobject (resp. quotient) of (M, θ) in $\mathbf{HM}(A, E)$, $(\mathrm{F}^i M)_{i \in \mathbb{Z}}$ a decreasing filtration of (M, θ) in $\mathbf{HM}(A, E)$ such that $\mathrm{F}^n M = 0$, $\mathrm{F}^m M = M$ for two integers $m \leq n$ (2.5.2).*

(i) *If (M, θ) is quasi-nilpotent, so are N and P. Moreover, if the filtration $\mathrm{F}^\bullet M$ of (M, θ) is quasi-nilpotent, so are the filtrations induced on N and P (2.5.2).*
(ii) *For (M, θ) to be quasi-nilpotent, it is necessary and sufficient that the graded object $\mathrm{Gr}^\bullet_{\mathrm{F}} M$ of the filtration $\mathrm{F}^\bullet M$ is.*

(i) Indeed, denoting by $\mathrm{F}^\bullet N$ (resp. $\mathrm{F}^\bullet P$) the filtration induced by $\mathrm{F}^\bullet M$ on N (resp. P) (2.5.2), $\mathrm{Gr}^\bullet_{\mathrm{F}} N$ (resp. $\mathrm{Gr}^\bullet_{\mathrm{F}} P$) is identified with a subobject (resp. quotient) of $\mathrm{Gr}^\bullet_{\mathrm{F}} M$ in $\mathbf{HM}(A, E)$. More precisely, if $P = M/N$, then we have a canonical exact sequence of $\mathbf{HM}(A, E)$,

$$0 \to \mathrm{Gr}^\bullet_{\mathrm{F}} N \to \mathrm{Gr}^\bullet_{\mathrm{F}} M \to \mathrm{Gr}^\bullet_{\mathrm{F}} P \to 0. \tag{2.5.4.1}$$

(ii) The condition is necessary from (i). Conversely, if $\mathrm{Gr}^\bullet_{\mathrm{F}} M$ is quasi-nilpotent, so is $\mathrm{Gr}^i_{\mathrm{F}} M$ for all $i \in \mathbb{Z}$. There exists hence a filtration $(\mathrm{W}^i M)_{i \in \mathbb{Z}}$ of (M, θ) in $\mathbf{HM}(A, E)$ which refines the filtration $\mathrm{F}^\bullet M$ and such that the Higgs field of the graded object $\mathrm{Gr}^\bullet_{\mathrm{W}} M$ is zero. Hence, (M, θ) is quasi-nilpotent; the condition is therefore sufficient.

Proposition 2.5.5 *Let X be a scheme, $\mathscr{E}$ a locally free $\mathscr{O}_X$-module of finite type, $\mathscr{F} = \mathscr{H}om_{\mathscr{O}_X}(\mathscr{E}, \mathscr{O}_X)$ the dual of $\mathscr{E}$, $\mathscr{M}$ a quasi-coherent $\mathscr{O}_X$-module, $\theta \colon \mathscr{M} \to \mathscr{M} \otimes_{\mathscr{O}_X} \mathscr{E}$ a Higgs $\mathscr{O}_X$-field on $\mathscr{M}$ with coefficients in $\mathscr{E}$. We denote by $T^* = \mathrm{Spec}(\mathrm{S}_{\mathscr{O}_X}(\mathscr{F}))$ the vector bundle on X associated with $\mathscr{F}$ (2.1.10), by $\sigma \colon X \to T^*$ the zero section, by $\mathscr{J}$ the ideal of $\mathscr{O}_{T^*}$ corresponding to the closed immersion σ and by $\mathscr{M}^+$ the $\mathscr{O}_{T^*}$-module associated with $(\mathscr{M}, \theta)$ (2.5.1.10). For any open subset U of X and any section $d \in \Gamma(U, \mathscr{F})$, we denote by θ_d the endomorphism of $\mathscr{M}|U$ deduced from θ and d. Consider the following conditions:*

(i) *The support of $\mathscr{M}^+$ is contained in $\sigma(X)$.*
(ii) *For every quasi-compact open subset U of X and every section $s \in \Gamma(U, \mathscr{M})$, there exists an integer $n \geq 1$ such that for all sections $d_1, \ldots, d_n \in \Gamma(U, \mathscr{F})$, $\theta_{d_n} \circ \cdots \circ \theta_{d_1}(s) = 0$.*
(iii) *There exists a finite decreasing filtration $(\mathscr{M}_i)_{0 \leq i \leq n}$ of $\mathscr{M}$ by quasi-coherent sub-$\mathscr{O}_X$-modules such that $\mathscr{M}_0 = \mathscr{M}$, $\mathscr{M}_n = 0$ and for all $0 \leq i \leq n-1$, we have*

$$\theta(\mathscr{M}_i) \subset \mathscr{E} \otimes_{\mathscr{O}_X} \mathscr{M}_{i+1}. \tag{2.5.5.1}$$

(iv) *There exists an integer $n \geq 1$ such that $\mathscr{J}^n \mathscr{M}^+ = 0$.*

Then we have (iv)$\Rightarrow$(iii)$\Rightarrow$(ii)$\Leftrightarrow$(i). *If, moreover, X is quasi-compact and the $\mathscr{O}_{T^*}$-module $\mathscr{M}^+$ is of finite type, the four conditions are equivalent.*

By "support of $\mathscr{M}^+$" we mean the set (not necessarily closed) of points of T^* where the stalk of $\mathscr{M}^+$ is not zero ([27] 0.3.1.5). It is closed if the $\mathscr{O}_{T^*}$-module $\mathscr{M}^+$ is of finite type ([27] 0.5.2.2); this is the case if the $\mathscr{O}_X$-module $\mathscr{M}$ is of finite type. Note that the $\mathscr{O}_{T^*}$-module $\mathscr{M}^+$ is quasi-coherent.

First assume that condition (iv) is satisfied. For each $0 \le i \le n$, let us denote by $\mathscr{M}_i^+$ the largest sub-$\mathscr{O}_{T^*}$-module of $\mathscr{M}^+$ annihilated by $\mathscr{J}^{n-i}$. We therefore have $\mathscr{M}_0^+ = \mathscr{M}^+$, $\mathscr{M}_n^+ = 0$ and for every $0 \le i \le n-1$, $\mathscr{J}\mathscr{M}_i^+ \subset \mathscr{M}_{i+1}^+$. Denoting by $\pi\colon T^* \to X$ the canonical projection, the filtration of $\mathscr{M}$ defined, for all $0 \le i \le n$, by $\mathscr{M}_i = \pi_*(\mathscr{M}_i^+)$ then verifies condition (iii).

It is clear that we have (iii)$\Rightarrow$(ii).

The implication (ii)$\Rightarrow$(i) follows from the fact that for every open subset U of X, we have $\Gamma(T_U^*, \mathscr{M}^+) = \Gamma(U, \mathscr{M})$, and the action of $\Gamma(T_U^*, \mathscr{O}_{T^*})$ on $\Gamma(T_U^*, \mathscr{M}^+)$ is induced by the map

$$\Gamma(U, \mathscr{F}) \times \Gamma(U, \mathscr{M}) \to \Gamma(U, \mathscr{M}), \quad (d, m) \mapsto \theta_d(m). \tag{2.5.5.2}$$

Let U be a quasi-compact open set of X, $s \in \Gamma(T_U^*, \mathscr{M}^+)$ a section with support contained in $\sigma(X)$. By virtue of ([27] 6.8.4) applied to the submodule $\mathscr{O}_{T_U^*} s$ of $\mathscr{M}^+|T_U^*$, there exists an integer $n \ge 1$ such that $\mathscr{J}^n s = 0$; the implication (i)$\Rightarrow$(ii) follows.

If X is quasi-compact and if the $\mathscr{O}_{T^*}$-module $\mathscr{M}^+$ is of finite type, we have (i)$\Rightarrow$(iv) by virtue of ([27] 6.8.4); the proposition follows.

Definition 2.5.6 Let X be a scheme, $\mathscr{E}$ a locally free $\mathscr{O}_X$-module of finite type. We say that a Higgs $\mathscr{O}_X$-module $(\mathscr{M}, \theta)$ with coefficients in $\mathscr{E}$ is *nilpotent* if the $\mathscr{O}_X$-module $\mathscr{M}$ is quasi-coherent and if the equivalent conditions 2.5.5(i)-(ii) are satisfied. We then also say that the Higgs field θ is *nilpotent*.

For a Higgs $\mathscr{O}_X$-module $(\mathscr{M}, \theta)$ with coefficients in $\mathscr{E}$ to be nilpotent, it is necessary and sufficient that $(\mathscr{M}, \theta)$ is quasi-nilpotent (2.5.3) and admits a quasi-nilpotent filtration consisting of quasi-coherent sub-$\mathscr{O}_X$-modules.

Remark 2.5.7 We take again the assumptions of 2.5.5 and suppose, moreover, that the scheme X is affine of ring R and that the $\mathscr{O}_X$-module $\mathscr{M}$ is of finite type. Let $E = \Gamma(X, \mathscr{E})$, $F = \Gamma(X, \mathscr{F})$ and $M = \Gamma(X, \mathscr{M})$ which are R-modules, and $M^+ = \Gamma(T^*, \mathscr{M}^+)$, which is an $\mathrm{S}_R(F)$-module. Let J be the ideal of the canonical augmentation of $\mathrm{S}_R(F)$. Then, the following properties are equivalent:

(i) The Higgs $\mathscr{O}_X$-module $(\mathscr{M}, \theta)$ is nilpotent.

(ii) For every $s \in M$, there exists an integer $n \ge 1$ such that for all sections $d_1, \ldots, d_n \in F$, $\theta_{d_n} \circ \cdots \circ \theta_{d_1}(s) = 0$.

(iii) There exists a finite decreasing filtration $(M_i)_{0 \le i \le n}$ of M such that $M_0 = M$, $M_n = 0$ and for all $0 \le i \le n-1$, we have

$$\theta(M_i) \subset E \otimes_R M_{i+1}. \tag{2.5.7.1}$$

(iv) There exists an integer $n \ge 1$ such that $J^n M^+ = 0$.

Lemma 2.5.8 *Let X be a scheme, $\mathscr{E}$ a locally free $\mathscr{O}_X$-module of finite type, $\mathscr{M}$ a quasi-coherent $\mathscr{O}_X$-module, θ a Higgs $\mathscr{O}_X$-field on $\mathscr{M}$ with coefficients in $\mathscr{E}$, $\mathscr{N}$ (resp. $\mathscr{P}$) a subobject (resp. quotient) of $(\mathscr{M}, \theta)$ in $\mathbf{HM}(\mathscr{O}_X, \mathscr{E})$ with a quasi-coherent underlying $\mathscr{O}_X$-module, $(\mathrm{F}^i\mathscr{M})_{i \in \mathbb{Z}}$ a decreasing filtration of $(\mathscr{M}, \theta)$ in $\mathbf{HM}(\mathscr{O}_X, \mathscr{E})$* (2.5.2) *such that $\mathrm{F}^n\mathscr{M} = 0$, $\mathrm{F}^m\mathscr{M} = \mathscr{M}$ for two integers $m \leq n$ and for all $i \in \mathbb{Z}$, the $\mathscr{O}_X$-module $\mathscr{M}_i$ is quasi-coherent.*

(i) If $(\mathscr{M}, \theta)$ is nilpotent, so are $\mathscr{N}$ and $\mathscr{P}$.

(ii) For $(\mathscr{M}, \theta)$ to be nilpotent, it is necessary and sufficient that the graded object $\mathrm{Gr}^{\bullet}_{\mathrm{F}}\mathscr{M}$ associated to the filtration $\mathrm{F}^{\bullet}\mathscr{M}$ is nilpotent.

It suffices to copy the proof of 2.5.4.

Lemma 2.5.9 *Let X be a reduced scheme, $\mathscr{E}$, $\mathscr{M}$ two locally free $\mathscr{O}_X$-modules of finite type, θ a nilpotent Higgs $\mathscr{O}_X$-field on $\mathscr{M}$ with coefficients in $\mathscr{E}$. Then, for all $n \geq 1$, the nth characteristic invariant $\lambda_n(\theta)$ of θ vanishes* (2.5.1.5).

Indeed, let $\mathscr{F} = \mathscr{H}om_{\mathscr{O}_X}(\mathscr{E}, \mathscr{O}_X)$ be the dual of $\mathscr{E}$. For any open set U of X and any section $d \in \Gamma(U, \mathscr{F})$, we denote by θ_d the endomorphism of $\mathscr{M}|U$ deduced from θ and d, and by $e_d \colon \mathrm{S}_{\mathscr{O}_X}(\mathscr{E})|U \to \mathscr{O}_X|U$ the homomorphism of $\mathscr{O}_X$-algebras induced by d. For every integer $n \geq 1$, we then have

$$e_d(\lambda_n(\theta)) = \mathrm{Tr}(\wedge^n(\theta_d)). \tag{2.5.9.1}$$

As θ_d is nilpotent, we deduce from this that $e_d(\lambda_n(\theta))$ is nilpotent and therefore it vanishes because X is reduced; the statement follows.

Lemma 2.5.10 *Let X be a scheme, $\mathscr{E}$ a locally free $\mathscr{O}_X$-module of finite type, $\mathscr{M}$ a quasi-coherent $\mathscr{O}_X$-module, $\theta \colon \mathscr{M} \to \mathscr{M} \otimes_{\mathscr{O}_X} \mathscr{E}$ a Higgs $\mathscr{O}_X$-field on $\mathscr{M}$ with coefficients in $\mathscr{E}$. Then, there exists a quasi-coherent sub-$\mathscr{O}_X$-module $\mathscr{N}$ of $\mathscr{M}$ satisfying the following properties:*

(i) $\theta(\mathscr{N}) \subset \mathscr{N} \otimes_{\mathscr{O}_X} \mathscr{E}$ and the Higgs $\mathscr{O}_X$-module $(\mathscr{N}, \theta)$ with coefficients in $\mathscr{E}$ is nilpotent.

(ii) For every morphism $u \colon (\mathscr{M}', \theta') \to (\mathscr{M}, \theta)$ of Higgs $\mathscr{O}_X$-modules with coefficients in $\mathscr{E}$, such that $(\mathscr{M}', \theta')$ is nilpotent, we have $u(\mathscr{M}') \subset \mathscr{N}$.

We say that $(\mathscr{N}, \theta)$ is the maximal nilpotent Higgs sub-$\mathscr{O}_X$-module *of $(\mathscr{M}, \theta)$.*

Indeed, let $\mathscr{F} = \mathscr{H}om_{\mathscr{O}_X}(\mathscr{E}, \mathscr{O}_X)$ be the dual of $\mathscr{E}$, $T^* = \mathrm{Spec}(\mathrm{S}_{\mathscr{O}_X}(\mathscr{F}))$ the vector bundle over X associated with $\mathscr{F}$, $\pi \colon T^* \to X$ the canonical projection, $\sigma \colon X \to T^*$ the zero section, V the open complement of $\sigma(X)$ in T^*, $j \colon V \to T^*$ the canonical injection, $\mathscr{M}^+$ the $\mathscr{O}_{T^*}$-module associated with $(\mathscr{M}, \theta)$ (2.5.1.10). The kernel $\mathscr{N}^+$ of the adjunction morphism $\mathscr{M}^+ \to j_*(j^*\mathscr{M}^+)$ is the maximal sub-$\mathscr{O}_{T^*}$-module of $\mathscr{M}^+$ with support contained in $\sigma(X)$. As the immersion j is coherent (i.e., quasi-compact and quasi-separated), the $\mathscr{O}_{T^*}$-module $\mathscr{N}^+$ is quasi-coherent ([27] 6.7.1). We immediately see that the sub-$\mathscr{O}_X$-module $\mathscr{N} = \pi_*(\mathscr{N}^+)$ of $\mathscr{M}$ answers the question.

2.5.11 Let (X, A) be a ringed topos,

$$0 \to L \to E \to \underline{E} \to 0 \tag{2.5.11.1}$$

an exact sequence of locally free A-modules of finite type. We endow E with the exhaustive decreasing filtration defined by (2.5.11.1), whose associated graded module is $\underline{E} \oplus L$, concentrated in degrees $[0, 1]$. For any integers $i, j \geq 0$, we denote by $\mathrm{W}^i \wedge^j E$ the image of the canonical morphism

$$\wedge^i L \otimes_A \wedge^{j-i} E \to \wedge^j E. \tag{2.5.11.2}$$

We thus define an exhaustive decreasing filtration $(\mathrm{W}^i \wedge^\bullet E)_{i \geq 0}$ of the exterior algebra $\wedge^\bullet E$ of E by ideals, called the *Koszul filtration of* $\wedge^\bullet E$ *associated with the exact sequence* (2.5.11.1). We denote by $\mathrm{Gr}^\bullet_{\mathrm{W}} \wedge^\bullet E$ the associated graded algebra, which is in fact bigraded. The canonical injection $\underline{E} \oplus L \to \mathrm{Gr}^\bullet_{\mathrm{W}} \wedge^\bullet E$ extends canonically into a homomorphism of bigraded algebras

$$\wedge^\bullet(\underline{E} \oplus L) \to \mathrm{Gr}^\bullet_{\mathrm{W}} \wedge^\bullet E. \tag{2.5.11.3}$$

This is an isomorphism by ([37] V 4.1.6). Moreover, by virtue of ([8] III § 7.7 prop. 10), we have a canonical isomorphism of bigraded algebras

$$\wedge^\bullet \underline{E} \,{}^g\!\otimes \wedge^\bullet L \xrightarrow{\sim} \wedge^\bullet(\underline{E} \oplus L), \tag{2.5.11.4}$$

where the symbol ${}^g\!\otimes$ denotes the left tensor product (see [8] III § 4.7 remarks page 49). In particular, for every integer $i \geq 0$, we have a canonical isomorphism

$$\mathrm{Gr}^i_{\mathrm{W}} \wedge^\bullet E \xrightarrow{\sim} (\wedge^i L) \otimes_A (\wedge^{\bullet - i} \underline{E}). \tag{2.5.11.5}$$

Let (M, θ) be a Higgs A-module with coefficients in E. We denote by $\underline{\theta}\colon M \to M \otimes_A \underline{E}$ the Higgs A-field induced by θ, and by $\mathbb{K}^\bullet$ (resp. $\underline{\mathbb{K}}^\bullet$) the Dolbeault complex of (M, θ) (resp. $(M, \underline{\theta})$) (2.5.1.2). We observe that the complex $\mathbb{K}^\bullet$ is a graded (right) module over the graded algebra $\wedge^\bullet E$ and that its differential of degree one is $(\wedge^\bullet E)$-linear. For any integers $i, j \geq 0$, we set

$$\mathrm{W}^i \mathbb{K}^j = M \otimes_A (\mathrm{W}^i \wedge^j E). \tag{2.5.11.6}$$

We thus define an exhaustive decreasing filtration of the $\wedge^\bullet E$-graded module $\mathbb{K}^\bullet$ by graded sub-$\wedge^\bullet E$-modules $(\mathrm{W}^i \mathbb{K}^\bullet)_{i \geq 0}$, stable by the differential of $\mathbb{K}^\bullet$, called the *Koszul filtration of* $\mathbb{K}^\bullet$ *associated with the exact sequence* (2.5.11.1); in particular, it is a filtration of the complex $\mathbb{K}^\bullet$ by subcomplexes. We denote by $\mathrm{Gr}^\bullet_{\mathrm{W}} \mathbb{K}^\bullet$ the associated graded complex.

Taking into account (2.5.11.3) and (2.5.11.4), for every integer $i \geq 0$, we have a canonical isomorphism of complexes

$$\mathrm{Gr}^i_{\mathrm{W}} \mathbb{K}^\bullet \xrightarrow{\sim} \wedge^i L \otimes_A \underline{\mathbb{K}}^\bullet[-i], \tag{2.5.11.7}$$

where the differentials of $\underline{\mathbb{K}}^\bullet[-i]$ are those of $\underline{\mathbb{K}}^\bullet$ multiplied by $(-1)^i$. We therefore have an exact sequence of complexes of A-modules

$$0 \to \wedge^{i+1} L \otimes_A \underline{\mathbb{K}}^\bullet[-i-1] \to \mathrm{W}^i/\mathrm{W}^{i+2}(\mathbb{K}^\bullet) \to \wedge^i L \otimes_A \underline{\mathbb{K}}^\bullet[-i] \to 0. \quad (2.5.11.8)$$

This induces a morphism of the derived category $\mathbf{D}^+(\mathbf{Mod}(A))$,

$$\partial^i\colon \wedge^i L \otimes_A \underline{\mathbb{K}}^\bullet \to \wedge^{i+1} L \otimes_A \underline{\mathbb{K}}^\bullet, \quad (2.5.11.9)$$

which we will call the *boundary map associated with the Koszul filtration of* $\mathbb{K}^\bullet$.

2.5.12 We keep the assumptions and notation of 2.5.11. Let, moreover, (N, κ) be a Higgs A-module with coefficients in L. We denote by κ' the Higgs A-field on N with coefficients in E induced by κ, by $\theta' = \theta \otimes \mathrm{id} + \mathrm{id} \otimes \kappa'$ the total Higgs A-field on $M \otimes_A N$ with coefficients in E, and by $\underline{\theta}'$ the Higgs A-field on $M \otimes_A N$ with coefficients in $\underline{E}$ induced by θ', which is none other than $\underline{\theta} \otimes \mathrm{id}$. We denote by $\mathbb{K}'^\bullet$ (resp. $\underline{\mathbb{K}}'^\bullet$, resp. $\mathscr{K}'^\bullet$) the Dolbeault complex of $(M \otimes_A N, \theta')$ (resp. $(M \otimes_A N, \underline{\theta}')$, resp. (N, κ')) and by $\theta'^\bullet$ (resp. $\underline{\theta}'^\bullet$, resp. $\kappa'^\bullet$) its differentials.

We endow $\mathbb{K}^\bullet$ and $\mathbb{K}'^\bullet$ with the Koszul filtrations associated with the exact sequence (2.5.11.1) and we set $\mathbb{G}^\bullet = \mathrm{W}^0/\mathrm{W}^2(\mathbb{K}^\bullet)$ and $\mathbb{G}'^\bullet = \mathrm{W}^0/\mathrm{W}^2(\mathbb{K}'^\bullet)$ (2.5.11.6). We (abusively) denote the differentials of $\mathbb{G}^\bullet$ (resp. $\mathbb{G}'^\bullet$) also by θ^i (resp. θ'^i), which does not induce any ambiguity. We then have canonical exact sequences of complexes (2.5.11.8)

$$0 \longrightarrow L \otimes_A \underline{\mathbb{K}}^\bullet[-1] \xrightarrow{u^\bullet} \mathbb{G}^\bullet \xrightarrow{v^\bullet} \underline{\mathbb{K}}^\bullet \longrightarrow 0, \quad (2.5.12.1)$$

$$0 \longrightarrow L \otimes_A \underline{\mathbb{K}}'^\bullet[-1] \xrightarrow{u'^\bullet} \mathbb{G}'^\bullet \xrightarrow{v'^\bullet} \underline{\mathbb{K}}'^\bullet \longrightarrow 0, \quad (2.5.12.2)$$

whose associated boundaries in $\mathbf{D}^+(\mathbf{Mod}(A))$ are denoted by

$$\partial\colon \underline{\mathbb{K}}^\bullet \to L \otimes_A \underline{\mathbb{K}}^\bullet, \quad (2.5.12.3)$$

$$\partial'\colon \underline{\mathbb{K}}'^\bullet \to L \otimes_A \underline{\mathbb{K}}'^\bullet. \quad (2.5.12.4)$$

We denote by $\mathrm{C}^\bullet$ (resp. $\mathrm{C}'^\bullet$) the mapping cone of the morphism $u^\bullet$ (resp. $u'^\bullet$) and by $c^\bullet$ (resp. $c'^\bullet$) its differentials. For every integer i, we therefore have $\mathrm{C}^i = (L \otimes_A \underline{\mathbb{K}}^i) \oplus \mathbb{G}^i$ and c^i is defined by the matrix

$$\begin{pmatrix} \mathrm{id} \otimes \underline{\theta}^i & 0 \\ u^{i+1} & \theta^i \end{pmatrix}. \quad (2.5.12.5)$$

We have a similar description of the complex $\mathrm{C}'^\bullet$.

Note that we have $\mathbb{K}'^i = \mathbb{K}^i \otimes_A N$, $\theta'^i = \theta^i \otimes \mathrm{id} + \mathrm{id} \otimes \kappa'^i$, $\underline{\mathbb{K}}'^i = \underline{\mathbb{K}}^i \otimes_A N$, $\underline{\theta}'^i = \underline{\theta}^i \otimes \mathrm{id}$, $\mathbb{G}'^i = \mathbb{G}^i \otimes N$ and $\mathrm{C}'^i = \mathrm{C}^i \otimes N$. We will implicitly use these identifications in the rest of this section.

For all integers $i, j \geq 0$, the morphism $\kappa'^i \otimes \mathrm{id}\colon \mathbb{K}'^i \to \mathbb{K}'^{i+1}$ sends $\mathrm{F}^j\mathbb{K}'^i$ into $\mathrm{F}^{j+1}\mathbb{K}'^{i+1}$. It therefore induces a morphism

$$\delta'^i\colon \mathbb{G}'^i \to L \otimes_A \underline{\mathbb{K}}'^i \quad (2.5.12.6)$$

such that $\delta'^i \circ u'^i = 0$ (2.5.12.2).

Lemma 2.5.13 *We keep the assumptions and notation of 2.5.12. Then,*

(i) *For every integer $i \geq 0$, we have $\delta'^i = (\kappa \otimes \mathrm{id}) \circ v'^i$ (2.5.12.2).*

(ii) *The morphisms $\delta'^\bullet$ define a morphism of complexes $\mathbb{G}'^\bullet \to L \otimes_A \underline{\mathbb{K}}'^\bullet$.*

(iii) *There exists an isomorphism of complexes of A-modules*

$$\omega^\bullet : \mathrm{C}'^\bullet \xrightarrow{\sim} \mathrm{C}^\bullet \otimes_A N \qquad (2.5.13.1)$$

defined in degree i by the automorphism of $\mathrm{C}'^i = (L \otimes_A \underline{\mathbb{K}}'^i) \oplus \mathbb{G}'^i$ which maps (x, y) to $(x + \delta'^i(y), y)$.

Proposition (i) is clear and it immediately implies (ii). To establish (iii), we have to show that for every element (x, y) of $\mathrm{C}'^i = (L \otimes_A \underline{\mathbb{K}}'^i) \oplus \mathbb{G}'^i$, we have

$$(\underline{\theta}^i \otimes \mathrm{id})(x) + \delta'^{i+1}(u'^{i+1}(x) + \theta'^i(y)) = (\underline{\theta}^i \otimes \mathrm{id})(x + \delta'^i(y)), \qquad (2.5.13.2)$$

$$u'^{i+1}(x) + \theta'^i(y) = u'^{i+1}(x + \delta'^i(y)) + (\theta^i \otimes \mathrm{id})(y). \qquad (2.5.13.3)$$

The first equation follows from (ii) and from the fact that $\delta'^{i+1} \circ u'^{i+1} = 0$. The second equation is a consequence of the relations $\theta'^i = \theta^i \otimes \mathrm{id} + \kappa'^i \otimes \mathrm{id}$ and $\kappa'^i \otimes \mathrm{id} = u'^{i+1} \circ \delta'^i$.

Proposition 2.5.14 *Under the assumptions and with the notation of 2.5.12, identifying $\underline{\mathbb{K}}'^\bullet$ with $\underline{\mathbb{K}}^\bullet \otimes_A N$, we have*

$$\partial' = \partial \otimes \mathrm{id} + \mathrm{id} \otimes \kappa. \qquad (2.5.14.1)$$

Indeed, let us denote by $\pi_1'^\bullet$ (resp. $\pi_2'^\bullet$) the projection of $\mathrm{C}'^\bullet = (L \otimes_A \underline{\mathbb{K}}'^\bullet) \oplus \mathbb{G}'^\bullet$ onto $L \otimes_A \underline{\mathbb{K}}'^\bullet$ (resp. $\mathbb{G}'^\bullet$). The composition $v'^\bullet \circ \pi_2'^\bullet : \mathrm{C}'^\bullet \to \underline{\mathbb{K}}'^\bullet$ is then a quasi-isomorphism (2.5.12.2), and $-\partial'$ (2.5.12.4) is the composition in $\mathbf{D}^+(\mathbf{Mod}(A))$ of the inverse of $v'^\bullet \circ \pi_2'^\bullet$ and $\pi_1'^\bullet$ ([59] 09KF). We similarly describe ∂ (2.5.12.3) in terms of the canonical projections $\pi_1^\bullet$ and $\pi_2^\bullet$ of $\mathrm{C}^\bullet$. In view of 2.5.13, we have

$$(\pi_1^\bullet \otimes \mathrm{id}) \circ \omega^\bullet = \pi_1'^\bullet + \delta'^\bullet \circ \pi_2'^\bullet, \qquad (2.5.14.2)$$

$$(\pi_2^\bullet \otimes \mathrm{id}) \circ \omega^\bullet = \pi_2'^\bullet. \qquad (2.5.14.3)$$

We deduce from this that

$$\begin{aligned}
\partial \otimes \mathrm{id} &= \partial' - \delta'^\bullet \circ \pi_2'^\bullet \circ (v'^\bullet \circ \pi_2'^\bullet)^{-1} \\
&= \partial' - (f^*(\kappa) \otimes \mathrm{id}) \circ v'^\bullet \circ \pi_2'^\bullet \circ (v'^\bullet \circ \pi_2'^\bullet)^{-1} \\
&= \partial' - \kappa \otimes \mathrm{id},
\end{aligned}$$

where the second relation follows from 2.5.13(ii); the proposition follows.

2.5.15 Let $f : (X, A) \to (Y, B)$ be a morphism of ringed topos, L a locally free B-module of finite type, E and $\underline{E}$ two locally free A-modules of finite type,

$$0 \longrightarrow f^*(L) \longrightarrow E \xrightarrow{u} \underline{E} \longrightarrow 0 \qquad (2.5.15.1)$$

an exact sequence and (M, θ) a Higgs A-module with coefficients in E. We denote by $\underline{\theta}$ the Higgs A-field on M with coefficients in $\underline{E}$ induced by θ, and by $\mathbb{K}^\bullet$ (resp. $\underline{\mathbb{K}}^\bullet$) the Dolbeault complex of (M, θ) (resp. $(M, \underline{\theta})$). We endow $\mathbb{K}^\bullet$ with the Koszul filtration associated with the exact sequence (2.5.15.1), see (2.5.11.6). We naturally consider

$$\oplus_{i\geq 0}\mathrm{W}^i\mathbb{K}^\bullet \tag{2.5.15.2}$$

as a bigraded (right) module over the bigraded A-algebra $\oplus_{i\geq 0}\mathrm{W}^i \wedge^\bullet E$. Note that its differential of bidegree $(0, 1)$ is $(\oplus_{i\geq 0}\mathrm{W}^i \wedge^\bullet E)$-linear. Moreover, we have a canonical homomorphism of A-graded algebras (graded by i)

$$f^*(\oplus_{i\geq 0} \wedge^i L) \to \oplus_{i\geq 0}\mathrm{W}^i \wedge^\bullet E. \tag{2.5.15.3}$$

For any integer $i \geq 0$, we set

$$\mathrm{E}_0^{i,\bullet} = \mathrm{Gr}^i_{\mathrm{W}}(\mathbb{K}^\bullet). \tag{2.5.15.4}$$

We have an exact sequence of $(\oplus_{i\geq 0}\mathrm{F}^i_{\cdot} \wedge^\bullet E)$-bigraded modules

$$0 \to \oplus_{i\geq 0}\mathrm{E}_0^{i+1,\bullet} \to \oplus_{i\geq 0}\mathrm{W}^i/\mathrm{W}^{i+2}(\mathbb{K}^\bullet) \to \oplus_{i\geq 0}\mathrm{E}_0^{i,\bullet} \to 0. \tag{2.5.15.5}$$

Taking into account (2.5.15.3), we can consider this sequence as an exact sequence of complexes of $f^*(\oplus_{i\geq 0} \wedge^i L)$-graded modules (graded by i).

By virtue of ([59] 015W), there exists a canonical convergent spectral sequence

$$\mathrm{E}_1^{i,j} = \mathrm{R}^{i+j} f_*(\mathrm{E}_0^{i,\bullet}) \Rightarrow \mathrm{R}^{i+j} f_*(\mathbb{K}^\bullet), \tag{2.5.15.6}$$

called the hypercohomology spectral sequence of the filtered complex $\mathbb{K}^\bullet$ ([14] 1.4.5). By (2.5.11.7) and the projection formula ([59] 0B54), we have a canonical isomorphism

$$\mathrm{E}_1^{i,j} \xrightarrow{\sim} \mathrm{R}^j f_*(\underline{\mathbb{K}}^\bullet) \otimes_A \wedge^i L. \tag{2.5.15.7}$$

For every integer $j \geq 0$, we obtain a complex of B-modules

$$\mathrm{R}^j f_*(\underline{\mathbb{K}}^\bullet) \xrightarrow{d_1^{0,j}} \mathrm{R}^j f_*(\underline{\mathbb{K}}^\bullet) \otimes_A L \xrightarrow{d_1^{1,j}} \mathrm{R}^j f_*(\underline{\mathbb{K}}^\bullet) \otimes_A \wedge^2 L \xrightarrow{d_1^{2,j}} \dots \tag{2.5.15.8}$$

whose differentials are none other than the homogeneous components of the boundary maps induced by the exact sequence (2.5.15.5). As the latter is an exact sequence of complexes of graded (right) $f^*(\oplus_{i\geq 0} \wedge^i L)$-modules (graded by i), we obtain the following statement:

Lemma 2.5.16 *Under the assumptions of* 2.5.15 *and with the same notation, for every integer $j \geq 0$, the morphism $d_1^{0,j}$* (2.5.15.8) *is a Higgs B-field on* $\mathrm{R}^j f_*(\underline{\mathbb{K}}^\bullet)$ *with coefficients in L, whose Dolbeault complex is none other than the complex* $(\mathrm{E}_1^{i,j})_{i\geq 0}$.

We call the *Katz–Oda B-field on* $\mathrm{R}^j f_*(\underline{\mathbb{K}}^\bullet)$ *with coefficients in L* the Higgs B-field $d_1^{0,j}$, whose construction recalls that of the Gauss–Manin connection by these authors [43]. It is also called the *Gauss–Manin field* in ([46] § 3.1).

Remark 2.5.17 We keep the assumptions and notation of 2.5.15. We assume further that the Higgs field θ is zero. The Koszul filtration of the exterior A-algebra $\wedge^\bullet E$ associated with the exact sequence (2.5.15.1) induces for every integer $j \geq 0$ an exact sequence (2.5.11.5)

$$0 \to f^*(L) \otimes_{\mathscr{O}_{\mathfrak{X}'}} \wedge^{j-1}\underline{E} \to \mathrm{W}^0/\mathrm{W}^2(\wedge^\bullet E) \to \wedge^j \underline{E} \to 0. \tag{2.5.17.1}$$

Taking into account the projection formula ([59] 0B54), we deduce from this a morphism of $\mathbf{D}^+(\mathbf{Mod}(B))$

$$\mathrm{R}f_*(M \otimes_A \wedge^j \underline{E}) \to L \otimes_B \mathrm{R}f_*(M \otimes_A \wedge^{j-1}\underline{E})[+1]. \tag{2.5.17.2}$$

For every integer $q \geq 0$, we then have a canonical isomorphism

$$\mathrm{R}^q f_*(\underline{\mathbb{K}}^\bullet) \xrightarrow{\sim} \oplus_{0 \leq i \leq q} \mathrm{R}^i f_*(M \otimes_A \wedge^{q-i}\underline{E}), \tag{2.5.17.3}$$

the Katz–Oda field on $\mathrm{R}^q f_*(\underline{\mathbb{K}}^\bullet)$ with coefficients in L (2.5.16) being induced by the morphisms (2.5.17.2). It is therefore quasi-nilpotent (2.5.3).

2.5.18 We keep the assumptions and notation of 2.5.15. Let, moreover, $(\mathrm{F}^i M)_{i \geq 0}$ be a decreasing filtration of (M, θ) in $\mathbf{HM}(A, E)$ (2.5.2) such that $\mathrm{F}^0 M = M$ and $\mathrm{F}^n M = 0$ for an integer $n \geq 0$. This is also a filtration of $(M, \underline{\theta})$ in $\mathbf{HM}(A, \underline{E})$. We denote by $\mathrm{F}^i\mathbb{K}^\bullet = \mathrm{F}^i M \otimes_A \wedge^\bullet E$ and $\mathrm{F}^i\underline{\mathbb{K}}^\bullet = \mathrm{F}^i M \otimes_A \wedge^\bullet \underline{E}$ the Dolbeault complexes of $\mathrm{F}^i M$ considered as an object of $\mathbf{HM}(A, E)$ and $\mathbf{HM}(A, \underline{E})$ respectively. The Higgs fields θ and $\underline{\theta}$ induce Higgs fields on $\mathrm{Gr}^i_{\mathrm{F}} M$. We denote by $\mathrm{Gr}^i_{\mathrm{F}}\mathbb{K}^\bullet = \mathrm{Gr}^i_{\mathrm{F}} M \otimes_A \wedge^\bullet E$ and $\mathrm{Gr}^i_{\mathrm{F}}\underline{\mathbb{K}}^\bullet = \mathrm{Gr}^i_{\mathrm{F}} M \otimes_A \wedge^\bullet \underline{E}$ the associated Dolbeault complexes. We thus obtain a filtration $(\mathrm{F}^i\mathbb{K}^\bullet)_{0 \leq i \leq n}$ (resp. $(\mathrm{F}^i\underline{\mathbb{K}}^\bullet)_{0 \leq i \leq n}$) of the complex $\mathbb{K}^\bullet$ (resp. $\underline{\mathbb{K}}^\bullet$) with associated graded complexes $(\mathrm{Gr}^i_{\mathrm{F}}\mathbb{K}^\bullet)_{0 \leq i \leq n}$ (resp. $(\mathrm{Gr}^i_{\mathrm{F}}\underline{\mathbb{K}}^\bullet)_{0 \leq i \leq n}$).

By ([59] 015W), there exists a canonical convergent spectral sequence

$$\mathrm{E}_1^{i,j} = \mathrm{R}^{i+j} f_*(\mathrm{Gr}^i_{\mathrm{F}}\underline{\mathbb{K}}^\bullet) \Rightarrow \mathrm{R}^{i+j} f_*(\underline{\mathbb{K}}^\bullet). \tag{2.5.18.1}$$

For any integer $q \geq 0$, we denote by

$$\kappa^q \colon \mathrm{R}^q f_*(\underline{\mathbb{K}}^\bullet) \to \mathrm{R}^q f_*(\underline{\mathbb{K}}^\bullet) \otimes_B L \tag{2.5.18.2}$$

the Katz–Oda field associated with the Higgs A-module (M, θ) (2.5.16).

For any integers $i, j \geq 0$, we denote by

$$\kappa_1^{i,j} \colon \mathrm{R}^{i+j} f_*(\mathrm{Gr}^i_{\mathrm{F}}\underline{\mathbb{K}}^\bullet) \to \mathrm{R}^{i+j} f_*(\mathrm{Gr}^i_{\mathrm{F}}\underline{\mathbb{K}}^\bullet) \otimes_B L \tag{2.5.18.3}$$

the Katz–Oda field associated with the Higgs A-module with coefficients in E induced by θ on $\mathrm{Gr}^i_{\mathrm{F}} M$.

Proposition 2.5.19 *Under the assumptions of* 2.5.18, *the morphisms* (2.5.18.2) *and* (2.5.18.3) *define a morphism from the spectral sequence* (2.5.18.1) *to the spectral sequence which is deduced from it by applying the functor* $- \otimes_B L$.

Let j be an integer ≥ 0. Consider the exact sequence

$$0 \longrightarrow \mathrm{Gr}_W^{j+1}(\mathbb{K}^\bullet) \overset{u}{\longrightarrow} W^j/W^{j+2}(\mathbb{K}^\bullet) \overset{v}{\longrightarrow} \mathrm{Gr}_W^j(\mathbb{K}^\bullet) \longrightarrow 0 \qquad (2.5.19.1)$$

induced by the Koszul filtration of $\mathbb{K}^\bullet$ (2.5.11.6), and let $C^\bullet$ denote the mapping cone of the morphism v and

$$\mathrm{Gr}_W^j(\mathbb{K}^\bullet) \overset{\nu}{\longrightarrow} C^\bullet \overset{\mu}{\longleftarrow} \mathrm{Gr}_W^{j+1}(\mathbb{K}^\bullet)[1] \qquad (2.5.19.2)$$

the canonical morphisms. Then, μ is a quasi-isomorphism, and the morphism $\mu^{-1} \circ \nu$ of $\mathbf{D}^+(\mathbf{Mod}(A))$ is identified with the opposite of the boundary map ∂^j associated with the Koszul filtration of $\mathbb{K}^\bullet$ (2.5.11.9).

The filtration F of $\mathbb{K}^\bullet$ (2.5.18) induces a filtration of the complex $W^j\mathbb{K}^\bullet$ ([14] 1.1.8) which in turn induces filtrations of the complexes $\mathrm{Gr}_W^j\mathbb{K}^\bullet$ and $W^j/W^{j+2}(\mathbb{K}^\bullet)$. The morphisms u and v are strict ([14] 1.1.9). In particular, the filtration F of $\mathbb{K}^\bullet$ induces a filtration of $C^\bullet$.

We recall that for any integer $i \geq 0$, $\mathrm{F}^i\mathbb{K}^\bullet$ (resp. $\mathrm{Gr}_F^i\mathbb{K}^\bullet$) is the Dolbeault complex of M_i (resp. Gr_F^iM). By ([10] I §2.6 prop. 7), as for every $\ell \geq 0$, $\wedge^\ell E/W^j \wedge^\ell E$ is A-flat (2.5.11.5), we have a canonical isomorphism

$$\mathrm{F}^i(W^j\mathbb{K}^\bullet) \overset{\sim}{\to} W^j(\mathrm{F}^i\mathbb{K}^\bullet). \qquad (2.5.19.3)$$

This induces an isomorphism

$$\mathrm{Gr}_F^i(W^j\mathbb{K}^\bullet) \overset{\sim}{\to} W^j(\mathrm{Gr}_F^i\mathbb{K}^\bullet). \qquad (2.5.19.4)$$

By ([14] 1.1.11), we deduce from this, for every integer $j' \geq j+1$, a canonical isomorphism

$$\mathrm{Gr}_F^i(W^j/W^{j'}(\mathbb{K}^\bullet)) \overset{\sim}{\to} W^j/W^{j'}(\mathrm{Gr}_F^i\mathbb{K}^\bullet). \qquad (2.5.19.5)$$

The morphism μ (2.5.19.2) is a filtered quasi-isomorphism for the filtrations induced by the filtration F of $\mathbb{K}^\bullet$ ([14] 1.3.6). Indeed, by what precedes and ([14] 1.1.11), the image of the sequence (2.5.19.1) by the functor Gr_F^i is identified with the canonical exact sequence

$$0 \to \mathrm{Gr}_W^{j+1}(\mathrm{Gr}_F^i(\mathbb{K}^\bullet)) \to W^j/W^{j+2}(\mathrm{Gr}_F^i(\mathbb{K}^\bullet)) \to \mathrm{Gr}_W^j(\mathrm{Gr}_F^i(\mathbb{K}^\bullet)) \to 0. \quad (2.5.19.6)$$

The mapping cone of the third arrow is then identified with $\mathrm{Gr}_F^iC^\bullet$, and $\mathrm{Gr}_F^i\mu$ and $\mathrm{Gr}_F^i\nu$ are identified with the morphisms analogous to μ and ν for the sequence (2.5.19.6). Consequently, $\mathrm{Gr}_F^i\mu$ is a quasi-isomorphism; the assertion follows.

Taking $j = 0$, the morphisms μ and ν define morphisms of hypercohomology spectral sequences of filtered complexes (by the filtrations induced by F), the morphism induced by μ being an isomorphism. The proposition follows in view of (2.5.11.7) and the definition of the Katz–Oda field (2.5.16).

Corollary 2.5.20 *Under the assumptions of 2.5.18, the spectral sequence (2.5.18.1) is underlying a spectral sequence in the abelian category* $\mathbf{HM}(B, L)$,

$$\mathrm{E}_1^{i,j} = (\mathrm{R}^{i+j}f_*(\mathrm{Gr}_F^i\underline{\mathbb{K}}^\bullet), \kappa_1^{i,j}) \Rightarrow (\mathrm{R}^{i+j}f_*(\underline{\mathbb{K}}^\bullet), \kappa^{i+j}). \qquad (2.5.20.1)$$

Corollary 2.5.21 *We keep the assumptions of* 2.5.18 *and for any integer $q \geq 0$, we denote by $(\mathrm{N}^i\mathrm{R}^q f_*(\underline{\mathbb{K}}^\bullet))_{0\leq i\leq q}$ the filtration of $\mathrm{R}^q f_*(\underline{\mathbb{K}}^\bullet)$ abutment of the spectral sequence* (2.5.18.1)*. Then, for every $0 \leq i \leq q$, we have*

$$\kappa^q(\mathrm{N}^i\mathrm{R}^q f_*(\underline{\mathbb{K}}^\bullet)) \subset \mathrm{N}^i\mathrm{R}^q f_*(\underline{\mathbb{K}}^\bullet) \otimes_B L. \tag{2.5.21.1}$$

This follows immediately from 2.5.20.

Proposition 2.5.22 *We keep the assumptions of* 2.5.15 *and we also assume that the Higgs A-module (M, θ)* (2.5.3) *is quasi-nilpotent. Then, for every integer $q \geq 0$, the Higgs B-module $(\mathrm{R}^q f_*(\underline{\mathbb{K}}^\bullet), \kappa^q)$, where κ^q is the Katz–Oda field* (2.5.16)*, is quasi-nilpotent.*

Indeed, let $(\mathrm{F}^i M)_{i\geq 0}$ be a quasi-nilpotent decreasing filtration of (M, θ) in $\mathbf{HM}(A, E)$ (2.5.3) such that $\mathrm{F}^0 M = M$ and $\mathrm{F}^n M = 0$ for an integer $n \geq 0$. We take again the notation of 2.5.18. For all integers $i, j \geq 0$, the Higgs field induced by $\underline{\theta}$ on $\mathrm{Gr}^i_{\mathrm{F}} M$ being zero, the Higgs B-module $(\mathrm{R}^{i+j} f_*(\mathrm{Gr}^i_{\mathrm{F}}\underline{\mathbb{K}}^\bullet), \kappa_1^{i,j})$ is quasi-nilpotent by 2.5.17. By virtue of 2.5.4(i), the graded Higgs modules of the filtration of $(\mathrm{R}^q f_*(\underline{\mathbb{K}}^\bullet), \kappa^q)$ in $\mathbf{HM}(B, L)$ abutment of the spectral sequence (2.5.20.1) is therefore quasi-nilpotent. Hence, the Higgs B-module $(\mathrm{R}^q f_*(\underline{\mathbb{K}}^\bullet), \kappa^q)$ is quasi-nilpotent by 2.5.4(ii).

Proposition 2.5.23 *We keep the assumptions of* 2.5.15 *and we suppose moreover that f is a coherent morphism of schemes, that the $\mathscr{O}_X$-modules E, $\underline{E}$ and L are quasi-coherent, and that the Higgs $\mathscr{O}_X$-module (M, θ) is nilpotent* (2.5.6)*. Then, for every integer $q \geq 0$, the Higgs $\mathscr{O}_Y$-module $(\mathrm{R}^q f_*(\underline{\mathbb{K}}^\bullet), \kappa^q)$, where κ^q is the Katz–Oda field* (2.5.16)*, is nilpotent.*

The proof is identical to that of 2.5.22 using 2.5.8 instead of 2.5.4.

2.5.24 Let (X, A) be a ringed topos, B an A-algebra, $\Omega^1_{B/A}$ the B-module of Kähler differentials of B over A ([37] II 1.1.2),

$$\Omega_{B/A} = \oplus_{n\in\mathbb{N}}\Omega^n_{B/A} = \wedge_B(\Omega^1_{B/A}) \tag{2.5.24.1}$$

the exterior algebra of $\Omega^1_{B/A}$. Then, there exists one and only one A-anti-derivation $d\colon \Omega_{B/A} \to \Omega_{B/A}$ of degree 1 and square zero, which extends the universal A-derivation $d\colon B \to \Omega^1_{B/A}$. This follows, for example, from ([9] §2.10 prop. 13) noting that $\Omega^1_{B/A}$ is the sheaf associated with the presheaf $U \mapsto \Omega^1_{B(U)/A(U)}$ $(U \in \mathrm{Ob}(X))$.

Let M be a B-module, $\lambda \in A(X)$. A *λ-connection* on M with respect to the extension B/A is an A-linear morphism

$$\nabla\colon M \to \Omega^1_{B/A} \otimes_B M \tag{2.5.24.2}$$

such that for all local sections x of B and s of M, we have

$$\nabla(xs) = \lambda d(x) \otimes s + x\nabla(s). \tag{2.5.24.3}$$

We also say that (M, ∇) is a B-module with λ-connection with respect to the extension B/A. The extension B/A will be omitted from the terminology when there is no risk of confusion. The morphism ∇ extends into a unique graded A-linear morphism of degree 1 which we denote also by

$$\nabla\colon \Omega_{B/A} \otimes_B M \to \Omega_{B/A} \otimes_B M, \tag{2.5.24.4}$$

such that for all local sections ω of $\Omega^i_{B/A}$ and s of $\Omega^j_{B/A} \otimes_B M$ $(i, j \in \mathbb{N})$, we have

$$\nabla(\omega \wedge s) = \lambda d(\omega) \wedge s + (-1)^i \omega \wedge \nabla(s). \tag{2.5.24.5}$$

Iterating this formula gives

$$\nabla \circ \nabla(\omega \wedge s) = \omega \wedge \nabla \circ \nabla(s). \tag{2.5.24.6}$$

We say that ∇ is *integrable* if $\nabla \circ \nabla = 0$.

Let (M, ∇), (M', ∇') be two B-modules with λ-connection. A morphism from (M, ∇) to (M', ∇') is a B-linear morphism $u\colon M \to M'$ such that $(\mathrm{id} \otimes u) \circ \nabla = \nabla' \circ u$.

Classically, 1-connections are called *connections*. Integrable 0-connections are Higgs B-fields with coefficients in $\Omega^1_{B/A}$ (2.5.1).

2.5.25 Let $f\colon (X', A') \to (X, A)$ be a morphism of ringed topos, B an A-algebra, B' an A'-algebra, $\alpha\colon f^*(B) \to B'$ a homomorphism of A'-algebras, $\lambda \in \Gamma(X, A)$, (M, ∇) a module with λ-connection with respect to the extension B/A (2.5.24). We denote by λ' the canonical image of λ in $\Gamma(X', A')$, by $d'\colon B' \to \Omega^1_{B'/A'}$ the universal A'-derivation of B' and by

$$\gamma\colon f^*(\Omega^1_{B/A}) \to \Omega^1_{B'/A'} \tag{2.5.25.1}$$

the canonical α-linear morphism. We immediately see that $f^*(\nabla)$ is a λ'-connection on $f^*(M)$ with respect to the extension $f^*(B)/A'$. It is integrable if ∇ is. Moreover, there exists a unique A'-linear morphism

$$\nabla'\colon B' \otimes_{f^*(B)} f^*(M) \to \Omega^1_{B'/A'} \otimes_{f^*(B)} f^*(M) \tag{2.5.25.2}$$

such that for all local sections x' of B' and t of $f^*(M)$, we have

$$\nabla'(x' \otimes t) = \lambda' d'(x') \otimes t + x'(\gamma \otimes \mathrm{id})(f^*(\nabla)(t)). \tag{2.5.25.3}$$

This is a λ'-connection on $B' \otimes_{f^*(B)} f^*(M)$ with respect to the extension B'/A'. It is integrable if ∇ is.

2.5.26 Let (X, A) be a ringed topos, B an A-algebra, $\lambda \in \Gamma(X, A)$, (M, ∇) a module with λ-connection with respect to the extension B/A (2.5.24). Suppose that there exists an A-module E and a B-isomorphism $\gamma\colon E \otimes_A B \xrightarrow{\sim} \Omega^1_{B/A}$ such that for every local section ω of E, we have $d(\gamma(\omega \otimes 1)) = 0$. We denote by $\vartheta\colon M \to E \otimes_A M$ the morphism induced by ∇ and γ.

(i) The λ-connection ∇ is integrable if and only if ϑ is a Higgs A-field on M with coefficients in E ([3] II.2.12).

(ii) Let (N, θ) be a Higgs A-module with coefficients in E. The A-linear morphism

$$\nabla' \colon M \otimes_A N \to \Omega^1_{B/A} \otimes_B M \otimes_A N \tag{2.5.26.1}$$

defined by

$$\nabla' = \nabla \otimes_A \mathrm{id}_N + (\gamma \otimes_B \mathrm{id}_{M \otimes_A N})(\mathrm{id}_M \otimes_A \theta) \tag{2.5.26.2}$$

is a λ-connection on $M \otimes_A N$ with respect to the extension B/A. If ∇ is integrable, so is ∇'.

(iii) Let B' be a B-algebra, $\delta' \colon B' \to \Omega^1_{B/A} \otimes_B B'$ an A-derivation such that for every local section b of B, we have $\delta'(b) = \lambda d(b) \otimes 1$. Then, there exists a unique A-linear morphism

$$\nabla' \colon M \otimes_B B' \to \Omega^1_{B/A} \otimes_B M \otimes_B B' \tag{2.5.26.3}$$

such that for all local sections m of M and b' of B', we have

$$\nabla'(m \otimes b') = \nabla(m) \otimes_B b' + m \otimes_B \delta'(b'). \tag{2.5.26.4}$$

We denote by $\vartheta' \colon M \otimes_B B' \to E \otimes_A M \otimes_B B'$ (resp. $\theta' \colon B' \to E \otimes_A B'$) the morphism induced by ∇' (resp. δ') and γ. If ∇ is integrable and if θ' is a Higgs A-field with coefficients in E, then ϑ' is a Higgs A-field with coefficients in E.

2.6 Ind-objects of a Category

2.6.1 Let $\mathscr{C}$ be a $\mathbb{U}$-category ([5] I 1.1) (2.1.7), $\widehat{\mathscr{C}}$ the category of presheaves of $\mathbb{U}$-sets on $\mathscr{C}$ (2.1.8),

$$\mathrm{h}_{\mathscr{C}} \colon \mathscr{C} \to \widehat{\mathscr{C}}, \quad X \mapsto (\mathrm{h}_{\mathscr{C}}(X) \colon Y \mapsto \mathrm{Hom}_{\mathscr{C}}(Y, X)), \tag{2.6.1.1}$$

the canonical functor ([5] I 1.3). The latter is fully faithful ([5] I 1.4). Recall that direct limits in $\widehat{\mathscr{C}}$ are representable and that the evaluation functor over every object of $\mathscr{C}$ commutes with direct limits ([5] I 3.1). Note, however, that even when direct limits in $\mathscr{C}$ are representable, the functor $\mathrm{h}_{\mathscr{C}}$ does not in general commute with direct limits. We will use the notation "$\varinjlim$" to denote direct limits in $\widehat{\mathscr{C}}$ and we will keep the notation $\varinjlim$ to denote direct limits in $\mathscr{C}$.

Let I and J be two small categories, $\alpha \colon I \to \mathscr{C}$ and $\beta \colon J \to \mathscr{C}$ two functors. For every object X of $\mathscr{C}$, we have

$$\mathrm{Hom}_{\widehat{\mathscr{C}}}(X, \text{"}\varinjlim\text{"}\alpha) = \varinjlim_{i \in I} \mathrm{Hom}_{\mathscr{C}}(X, \alpha(i)). \tag{2.6.1.2}$$

We have canonical isomorphisms

$$\mathrm{Hom}_{\widehat{\mathscr{C}}}(\text{“}\varinjlim\text{”}\alpha, \text{“}\varinjlim\text{”}\beta) \xrightarrow{\sim} \varprojlim_{i \in I} \mathrm{Hom}_{\widehat{\mathscr{C}}}(\alpha(i), \text{“}\varinjlim\text{”}\beta) \xrightarrow{\sim} \varprojlim_{i\in I} \varinjlim_{j \in J} \mathrm{Hom}_{\mathscr{C}}(\alpha(i), \beta(j)). \tag{2.6.1.3}$$

Definition 2.6.2 ([5] I 8.1.8) A filtered category I is said to be *essentially small* if I is a $\mathbb{U}$-category and if $\mathrm{Ob}(I)$ admits a cofinal small subset ([5] I 8.1.4).

For a small subset E of $\mathrm{Ob}(I)$ to be cofinal, it is necessary and sufficient that for every object i of I, there exists a morphism $i \to j$ with $j \in E$ ([5] 8.1.3(c)).

Definition 2.6.3 ([40] 6.1.1) Let $\mathscr{C}$ be a $\mathbb{U}$-category.

(i) We call an *ind-object of* $\mathscr{C}$ any object of $\widehat{\mathscr{C}}$ isomorphic to “$\varinjlim$”α for some functor $\alpha: I \to \mathscr{C}$ with I a small filtered category.
(ii) We call the category of *ind-objects of* $\mathscr{C}$ and we denote by $\mathbf{Ind}(\mathscr{C})$ the full subcategory of $\widehat{\mathscr{C}}$ made up of the ind-objects of $\mathscr{C}$. We denote by

$$\iota_{\mathscr{C}}: \mathscr{C} \to \mathbf{Ind}(\mathscr{C}) \tag{2.6.3.1}$$

the functor induced by $h_{\mathscr{C}}$.

We can make the following remarks:

2.6.3.2 For every object F of $\widehat{\mathscr{C}}$, denoting by $\alpha: \mathscr{C}_{/F} \to \mathscr{C}, (X \to F) \mapsto X$, the “source” functor, the canonical morphism

$$\text{“}\varinjlim\text{”}\alpha \to F \tag{2.6.3.3}$$

is an isomorphism ([5] I 3.4)

2.6.3.4 An object F of $\widehat{\mathscr{C}}$ is an ind-object of $\mathscr{C}$ if and only if the category $\mathscr{C}_{/F}$ (2.1.8) is filtered and essentially small ([5] I 8.3.3, [40] 6.1.5).

2.6.3.5 The category $\mathbf{Ind}(\mathscr{C})$ is a $\mathbb{U}$-category ([40] 6.1.2). The canonical functor $\iota_{\mathscr{C}}$ (2.6.3.1) is right exact and right small ([40] 6.1.6) (see [40] 3.3.1 and 3.3.8). In particular, if finite direct limits in $\mathscr{C}$ are representable, $\iota_{\mathscr{C}}$ commutes with these limits ([40] 3.3.2).

2.6.3.6 Small filtered direct limits in $\mathbf{Ind}(\mathscr{C})$ are representable and the canonical injection functor $\mathbf{Ind}(\mathscr{C}) \to \widehat{\mathscr{C}}$ commutes with these limits ([40] 6.1.8).

2.6.3.7 If finite inverse limits (resp. small inverse limits) in $\mathscr{C}$ are representable, then they are also in $\mathbf{Ind}(\mathscr{C})$ and the canonical functors $\iota_{\mathscr{C}}: \mathscr{C} \to \mathbf{Ind}(\mathscr{C})$ and $\mathbf{Ind}(\mathscr{C}) \to \widehat{\mathscr{C}}$ commute with these limits ([40] 6.1.17).

2.6.3.8 If cokernels of double arrows (resp. finite sums, resp. finite direct limits) in $\mathscr{C}$ are representable, then cokernels of the double arrows (resp. small sums, resp. small direct limits) in $\mathbf{Ind}(\mathscr{C})$ are representable ([40] 6.1.18).

2.6.3.9 If finite inverse limits and finite direct limits in $\mathscr{C}$ are representable, then small filtered direct limits in $\mathbf{Ind}(\mathscr{C})$ are exact ([40] 6.1.19).

2.6.4 Let $\phi\colon \mathscr{C} \to \mathscr{C}'$ be a functor between $\mathbb{U}$-categories. By virtue of ([5] I 8.6.3, [40] 6.1.9), there exists essentially a unique functor

$$\mathrm{I}\phi\colon \mathbf{Ind}(\mathscr{C}) \to \mathbf{Ind}(\mathscr{C}') \tag{2.6.4.1}$$

satisfying the following two conditions:

(i) the diagram

$$\begin{array}{ccc} \mathscr{C} & \xrightarrow{\phi} & \mathscr{C}' \\ {\scriptstyle \iota_{\mathscr{C}}}\downarrow & & \downarrow{\scriptstyle \iota_{\mathscr{C}'}} \\ \mathbf{Ind}(\mathscr{C}) & \xrightarrow{\mathrm{I}\phi} & \mathbf{Ind}(\mathscr{C}') \end{array} \tag{2.6.4.2}$$

is commutative up to canonical isomorphism;

(ii) for every small filtered category J and every functor $\alpha\colon J \to \mathbf{Ind}(\mathscr{C})$, the canonical morphism

$$\mathrm{I}\phi(\text{“}\varinjlim\text{”}\alpha) \to \text{“}\varinjlim\text{”}(\phi \circ \alpha) \tag{2.6.4.3}$$

is an isomorphism.

We can make the following remarks:

2.6.4.4 If ϕ is faithful (resp. fully faithful), so is $\mathrm{I}\phi$ ([40] 6.1.10).

2.6.4.5 If $\psi\colon \mathscr{C}' \to \mathscr{C}''$ is a functor with $\mathscr{C}''$ a $\mathbb{U}$-category, then we have a canonical isomorphism between functors from $\mathbf{Ind}(\mathscr{C})$ to $\mathbf{Ind}(\mathscr{C}'')$,

$$\mathrm{I}(\psi \circ \phi) \xrightarrow{\sim} \mathrm{I}\psi \circ \mathrm{I}\phi. \tag{2.6.4.6}$$

2.6.5 Let $\mathscr{C}$ be a $\mathbb{U}$-category in which small filtered direct limits are representable. The canonical functor $\iota_{\mathscr{C}}\colon \mathscr{C} \to \mathbf{Ind}(\mathscr{C})$ admits a left adjoint

$$\kappa_{\mathscr{C}}\colon \mathbf{Ind}(\mathscr{C}) \to \mathscr{C}. \tag{2.6.5.1}$$

For every small filtered category J and every functor $\alpha\colon J \to \mathscr{C}$, we have an isomorphism

$$\kappa_{\mathscr{C}}(\text{“}\varinjlim_{J}\text{”}\alpha) \xrightarrow{\sim} \varinjlim_{J} \alpha. \tag{2.6.5.2}$$

The canonical morphism $\kappa_{\mathscr{C}} \circ \iota_{\mathscr{C}} \to \mathrm{id}_{\mathscr{C}}$ is an isomorphism.

2.6.6 Let $\mathscr{C}$ be a $\mathbb{U}$-category, S a *right multiplicative system of morphisms of* $\mathscr{C}$ ([40] 7.1.5). For any object X of $\mathscr{C}$, we denote by S^X the category defined in the following way. The objects of S^X are the morphisms $s\colon X \to X'$ of $\mathscr{C}$ which belong to S. Let $s\colon X \to X'$, $t\colon X \to X''$ be two objects of S^X. A morphism from $s\colon X \to X'$ to $t\colon X \to X''$ is a morphism $h\colon X' \to X''$ of $\mathscr{C}$ such that $t = h \circ s$. Note that we do not require h to be in S. The category S^X is filtered ([40] 7.1.10). We denote by

$$\alpha^X\colon S^X \to \mathscr{C}, \quad (s\colon X \to X') \mapsto X', \tag{2.6.6.1}$$

the target functor.

For all objects X, Y of $\mathscr{C}$, the canonical map

$$\varprojlim_{(X\to X')\in S^X} \varinjlim_{(Y\to Y')\in S^X} \mathrm{Hom}_{\mathscr{C}}(X', Y') \to \varinjlim_{(Y\to Y')\in S^X} \mathrm{Hom}_{\mathscr{C}}(X, Y') \tag{2.6.6.2}$$

is bijective ([40] 7.1.5).

We denote by $\mathscr{C}_S$ the category having the same objects as $\mathscr{C}$, defined by localization with respect to S ([40] 7.1.11) and by

$$Q\colon \mathscr{C} \to \mathscr{C}_S \tag{2.6.6.3}$$

the canonical functor ([40] 7.1.16).

Suppose that for all $X \in \mathrm{Ob}(\mathscr{C})$, the category S^X is essentially small. Then, $\mathscr{C}_S$ is a $\mathbb{U}$-category ([40] 7.1.14). For all objects X, Y of $\mathscr{C}$, the inverse of the isomorphism (2.6.6.2) induces an isomorphism

$$\mathrm{Hom}_{\mathscr{C}_S}(X, Y) \xrightarrow{\sim} \mathrm{Hom}_{\mathbf{Ind}(\mathscr{C})}(\text{“}\varinjlim\text{”}\alpha^X, \text{“}\varinjlim\text{”}\alpha^Y). \tag{2.6.6.4}$$

The correspondence

$$\alpha_S\colon \mathscr{C}_S \to \mathbf{Ind}(\mathscr{C}), \quad X \mapsto \text{“}\varinjlim\text{”}\alpha^X, \tag{2.6.6.5}$$

therefore defines a fully faithful functor ([40] 7.4.1). Note that the triangle

$$\begin{array}{ccc} \mathscr{C} & \xrightarrow{Q} & \mathscr{C}_S \\ & \searrow^{\iota_{\mathscr{C}}} & \downarrow{\scriptstyle \alpha_S} \\ & & \mathbf{Ind}(\mathscr{C}) \end{array} \tag{2.6.6.6}$$

is not commutative in general. Nevertheless, there exists a canonical morphism

$$\iota_{\mathscr{C}} \to \alpha_S \circ Q, \tag{2.6.6.7}$$

defined for every object X of $\mathscr{C}$ by the canonical morphism

$$\iota_{\mathscr{C}}(X) \to \text{“}\varinjlim\text{”}\alpha^X = (\alpha_S \circ Q)(X). \tag{2.6.6.8}$$

2.6.7 Let $\mathscr{C}$ be an abelian $\mathbb{U}$-category. We denote by $\widehat{\mathscr{C}}^{\mathrm{add}}$ the category of additive functors from $\mathscr{C}^{\circ}$ to $\mathbf{Mod}(\mathbb{Z})$. It is an abelian category and it is a full subcategory of $\widehat{\mathscr{C}}$ by ([40] 8.1.12). The canonical functor

$$\mathscr{C} \to \widehat{\mathscr{C}}^{\mathrm{add}}, \quad X \mapsto \mathrm{Hom}_{\mathscr{C}}(-, X) \tag{2.6.7.1}$$

is fully faithful and left exact, but it is not exact in general.

For any small family of objects $(X_i)_{i\in I}$ of $\widehat{\mathscr{C}}^{\mathrm{add}}$, we denote by "$\bigoplus_{i\in I}$"$X_i$ the object "$\varinjlim$"$(\bigoplus_J X_j)$ of $\widehat{\mathscr{C}}^{\mathrm{add}}$, where the limit is taken over the finite subsets J of I. For every object Y of $\mathscr{C}$, we therefore have a canonical isomorphism

$$\mathrm{Hom}_{\widehat{\mathscr{C}}^{\mathrm{add}}}(Y, \text{“}\bigoplus_{i\in I}\text{”}X_i) \xrightarrow{\sim} \bigoplus_{i\in I} \mathrm{Hom}_{\widehat{\mathscr{C}}^{\mathrm{add}}}(Y, X_i). \tag{2.6.7.2}$$

We can make the following remarks ([40] 8.6.5):

2.6.7.3 The category $\mathbf{Ind}(\mathscr{C})$ is abelian.

2.6.7.4 The canonical functor $\iota_{\mathscr{C}}\colon \mathscr{C} \to \mathbf{Ind}(\mathscr{C})$ is fully faithful and exact, and the canonical functor $\mathbf{Ind}(\mathscr{C}) \to \widehat{\mathscr{C}}^{\mathrm{add}}$ is fully faithful and left exact.

2.6.7.5 The category $\mathbf{Ind}(\mathscr{C})$ admits small direct limits. Moreover, small filtered direct limits are exact.

2.6.7.6 Small sums in $\mathbf{Ind}(\mathscr{C})$ are representable by "$\bigoplus$".

2.6.7.7 If small inverse limits in $\mathscr{C}$ are representable, they are also in $\mathbf{Ind}(\mathscr{C})$.

2.6.7.8 If the category $\mathscr{C}$ is essentially small, $\mathbf{Ind}(\mathscr{C})$ admits a generator and is therefore a Grothendieck category ([40] 8.3.24).

2.6.8 Let $\mathscr{C}$ be an abelian $\mathbb{U}$-category. Note that the category $\mathbf{Ind}(\mathscr{C})$ does not have enough injectives ([40], 15.1.3). Following ([40] 15.2.1), an object F of $\mathbf{Ind}(\mathscr{C})$ is said to be *quasi-injective* if the functor

$$\mathscr{C} \to \mathbf{Mod}(\mathbb{Z}), \quad X \mapsto F(X) = \mathrm{Hom}_{\mathbf{Ind}(\mathscr{C})}(X, F) \tag{2.6.8.1}$$

is exact.

We can make the following remarks:

2.6.8.2 Suppose that $\mathscr{C}$ has enough injectives and let F be an object of $\mathbf{Ind}(\mathscr{C})$. The following properties are then equivalent ([40], 15.2.3):

(i) F is quasi-injective;
(ii) there exists a small filtered category J and a functor $\alpha\colon J \to \mathscr{C}$ such that $F = \text{“}\varinjlim\text{”}\alpha$ and that $\alpha(j)$ is injective for all $j \in \mathrm{Ob}(J)$;

(iii) every morphism $X \to F$, where X is an object of $\mathscr{C}$, factors through an injective object Y of $\mathscr{C}$.

In particular, if $\mathscr{I}$ denotes the full subcategory of injective objects of $\mathscr{C}$, $\mathbf{Ind}(\mathscr{I})$ is canonically identified with the full subcategory of quasi-injective objects of $\mathbf{Ind}(\mathscr{C})$ (2.6.4.4).

2.6.8.3 If $\mathscr{C}$ has enough injectives, $\mathbf{Ind}(\mathscr{C})$ has enough quasi-injectives, i.e., the full subcategory made up of quasi-injective objects is cogenerating in $\mathbf{Ind}(\mathscr{C})$ ([40], 15.2.7).

2.6.9 Let $\phi \colon \mathscr{C} \to \mathscr{C}'$ be a left exact functor between abelian $\mathbb{U}$-categories. The functor $\mathrm{I}\phi \colon \mathbf{Ind}(\mathscr{C}) \to \mathbf{Ind}(\mathscr{C}')$ associated with ϕ is left exact ([40] 8.6.8). Suppose that there exists a ϕ-injective subcategory $\mathscr{J}$ of $\mathscr{C}$ in the sense of ([40] 13.3.4). By virtue of ([40] 13.3.5) and with the notation of 2.1.6, the functor ϕ admits a right derived functor

$$\mathrm{R}\phi \colon \mathbf{D}^+(\mathscr{C}) \to \mathbf{D}^+(\mathscr{C}'). \tag{2.6.9.1}$$

For any integer i, we denote by $\mathrm{R}^i\phi \colon \mathscr{C} \to \mathscr{C}'$ the ith right derived functor of ϕ and by

$$\mathrm{I}(\mathrm{R}^i\phi) \colon \mathbf{Ind}(\mathscr{C}) \to \mathbf{Ind}(\mathscr{C}') \tag{2.6.9.2}$$

the associated functor. By ([40] 15.3.2), the category $\mathbf{Ind}(\mathscr{J})$ is $\mathrm{I}\phi$-injective and the functor $\mathrm{I}\phi$ admits a right derived functor

$$\mathrm{R}(\mathrm{I}\phi) \colon \mathbf{D}^+(\mathbf{Ind}(\mathscr{C})) \to \mathbf{D}^+(\mathbf{Ind}(\mathscr{C}')). \tag{2.6.9.3}$$

Moreover, for every integer i, we have a canonical isomorphism

$$\mathrm{R}^i(\mathrm{I}\phi) \xrightarrow{\sim} \mathrm{I}(\mathrm{R}^i\phi). \tag{2.6.9.4}$$

In particular, $\mathrm{R}^i(\mathrm{I}\phi)$ commutes with small filtered direct limits. Moreover, since the canonical functor $\iota_{\mathscr{C}} \colon \mathscr{C} \to \mathbf{Ind}(\mathscr{C})$ is exact (2.6.3.5 and 2.6.3.7) and $\iota_{\mathscr{C}}(\mathscr{J}) \subset \mathbf{Ind}(\mathscr{J})$, the diagram

$$\begin{array}{ccc} \mathbf{D}^+(\mathscr{C}) & \xrightarrow{\mathrm{R}\phi} & \mathbf{D}^+(\mathscr{C}') \\ \downarrow{\scriptstyle \iota_{\mathscr{C}}} & & \downarrow{\scriptstyle \iota_{\mathscr{C}'}} \\ \mathbf{D}^+(\mathbf{Ind}(\mathscr{C})) & \xrightarrow{\mathrm{R}(\mathrm{I}\phi)} & \mathbf{D}^+(\mathbf{Ind}(\mathscr{C}')), \end{array} \tag{2.6.9.5}$$

is commutative up to canonical isomorphism.

2.6.10 Let $\phi \colon \mathscr{C} \to \mathscr{C}'$ be a left exact functor between abelian $\mathbb{U}$-categories such that $\mathscr{C}$ has enough injectives. We denote by $\mathscr{I}$ the full subcategory of $\mathscr{C}$ made up of the injective objects. By ([40] 13.3.6(iii)), $\mathscr{I}$ is ϕ-injective. By virtue of ([40] 13.3.5), the functor ϕ therefore admits a right derived functor

$$\mathrm{R}\phi \colon \mathbf{D}^+(\mathscr{C}) \to \mathbf{D}^+(\mathscr{C}'). \tag{2.6.10.1}$$

The category $\mathbf{Ind}(\mathscr{J})$ is identified with the full subcategory of $\mathbf{Ind}(\mathscr{C})$ made up of the quasi-injective objects (2.6.8). By virtue of ([40] 15.3.2), $\mathbf{Ind}(\mathscr{J})$ is $\mathrm{I}\phi$-injective and the functor $\mathrm{I}\phi$ admits a right derived functor

$$\mathrm{R}(\mathrm{I}\phi)\colon \mathbf{D}^+(\mathbf{Ind}(\mathscr{C})) \to \mathbf{D}^+(\mathbf{Ind}(\mathscr{C}')). \tag{2.6.10.2}$$

2.6.11 Let I, J, K be three categories, $\varphi\colon I \to K$, $\psi\colon J \to K$ two functors. We denote by L the category of triples (i,j,u) where $i \in \mathrm{Ob}(I)$, $j \in \mathrm{Ob}(J)$ and $u \in \mathrm{Hom}_K(\varphi(i),\psi(j))$, with the obvious morphisms (see [40] 3.4.1). We then have two canonical functors $\mathtt{s}\colon L \to I$ and $\mathtt{t}\colon L \to J$ and a canonical morphism of functors $\varphi \circ \mathtt{s} \to \psi \circ \mathtt{t}$. If the categories I and J are small, so is L. If the categories I and J are filtered and if the functor ψ is cofinal, then the category L is filtered and the functors $\mathtt{s}$ and $\mathtt{t}$ are cofinal by virtue of ([40] 3.4.5).

2.7 Ind-modules

2.7.1 Let (X,A) be a ringed $\mathbb{U}$-topos. We denote by $\mathbf{Mod}(A)$ the category of A-modules of X. It is an abelian $\mathbb{U}$-category ([5] II 4.11). Small inverse (resp. direct) limits are representable in $\mathbf{Mod}(A)$ ([5] II 4.1), and small filtered direct limits are exact ([5] III 4.3(4)). We denote by $\mathbf{Ind}\text{-}\mathbf{Mod}(A)$ the category of *ind-A-modules*, i.e., ind-objects of $\mathbf{Mod}(A)$ (2.6.3). It is an abelian $\mathbb{U}$-category (2.6.7.3), $A(X)$-additive (2.6.1.3). Small direct (resp. inverse) limits are representable in $\mathbf{Ind}\text{-}\mathbf{Mod}(A)$, and small filtered direct limits are exact (2.6.7.5). We have functors

$$\iota_A\colon \mathbf{Mod}(A) \to \mathbf{Ind}\text{-}\mathbf{Mod}(A), \tag{2.7.1.1}$$

$$\kappa_A\colon \mathbf{Ind}\text{-}\mathbf{Mod}(A) \to \mathbf{Mod}(A), \tag{2.7.1.2}$$

defined in (2.6.3.1) and (2.6.5.1), respectively. The functor ι_A is exact and fully faithful and it commutes with small inverse limits (see 2.6.3.5 and 2.6.3.7). The functor κ_A is a left adjoint of ι_A, and the canonical morphism $\kappa_A \circ \iota_A \to \mathrm{id}_{\mathbf{Mod}(A)}$ is an isomorphism (2.6.5).

When there is no risk of ambiguity, we identify $\mathbf{Mod}(A)$ with a full subcategory of $\mathbf{Ind}\text{-}\mathbf{Mod}(A)$ by the functors ι_A that we omit from notation.

Let B, C be two A-algebras. We define the bifunctor

$$\otimes_A\colon \mathbf{Ind}\text{-}\mathbf{Mod}(B) \times \mathbf{Ind}\text{-}\mathbf{Mod}(C) \to \mathbf{Ind}\text{-}\mathbf{Mod}(B \otimes_A C) \tag{2.7.1.3}$$

by setting for any ind-B-module F and any ind-C-module G,

$$F \otimes_A G = \underset{\substack{(M\to F)\in\mathbf{Mod}(B)_{/F} \\ (N\to G)\in\mathbf{Mod}(C)_{/G}}}{\text{“}\varinjlim\text{”}} M \otimes_A N, \tag{2.7.1.4}$$

where $M \otimes_A N$ is the tensor product in $\mathbf{Mod}(B \otimes_A C)$, which is well defined in view of 2.6.3.4. This bifunctor clearly extends the tensor product of modules.

We have a canonical isomorphism of $\mathbf{Mod}(B \otimes_A C)$

$$\kappa_{B\otimes_A C}(F \otimes_A G) \xrightarrow{\sim} \kappa_B(F) \otimes_A \kappa_C(G). \tag{2.7.1.5}$$

The canonical forgetful functor $\mathbf{Mod}(B) \to \mathbf{Mod}(A)$ induces an additive functor

$$\mathbf{Ind}\text{-}\mathbf{Mod}(B) \to \mathbf{Ind}\text{-}\mathbf{Mod}(A). \tag{2.7.1.6}$$

We consider any ind-B-module as an ind-A-module via this functor, which we omit from the notation. For every ind-A-module F and every ind-B-module G, we have a canonical isomorphism

$$\mathrm{Hom}_{\mathbf{Ind}\text{-}\mathbf{Mod}(B)}(B \otimes_A F, G) \xrightarrow{\sim} \mathrm{Hom}_{\mathbf{Ind}\text{-}\mathbf{Mod}(A)}(F, G). \tag{2.7.1.7}$$

By 2.6.4(ii) applied to the functor (2.7.1.6), for every ind-B-modules F, G, we have a canonical morphism of $\mathbf{Ind}\text{-}\mathbf{Mod}(B \otimes_A B)$

$$F \otimes_A G \to F \otimes_B G. \tag{2.7.1.8}$$

Lemma 2.7.2 *Under the assumptions of* 2.7.1, *the functor* κ_A (2.7.1.2) *is exact and it commutes with small direct limits.*

Indeed, κ_A commutes with small direct limits since it admits a right adjoint. Let

$$0 \to F' \to F \to F'' \to 0 \tag{2.7.2.1}$$

be an exact sequence of $\mathbf{Ind}\text{-}\mathbf{Mod}(A)$. By ([40] 8.6.6), there exists a small filtered category J and an exact sequence of functors from J to $\mathbf{Mod}(A)$

$$0 \to \varphi' \to \varphi \to \varphi'' \to 0 \tag{2.7.2.2}$$

which induces the sequence (2.7.2.1) by taking the direct limit in $\mathbf{Ind}\text{-}\mathbf{Mod}(A)$ (2.6.7.5). The following sequence

$$0 \to \varinjlim_J \varphi' \to \varinjlim_J \varphi \to \varinjlim_J \varphi'' \to 0 \tag{2.7.2.3}$$

obtained by taking the direct limit in $\mathbf{Mod}(A)$ is exact ([5] III 4.3(4)). Hence, κ_A is exact.

Lemma 2.7.3 *Under the assumptions of* 2.7.1, *the bifunctor* (2.7.1.3) *commutes with small filtered direct limits in each of its variables. In particular, for all small filtered categories I and J and all functors $\alpha\colon I \to \mathbf{Mod}(B)$ and $\beta\colon J \to \mathbf{Mod}(C)$, we have a canonical isomorphism*

$$\text{“}\varinjlim_{i\in I}\text{”}\,\alpha \otimes_A \text{“}\varinjlim_{j\in J}\text{”}\,\beta \xrightarrow{\sim} \text{“}\varinjlim_{i\in I;\, j\in J}\text{”}\,\alpha(i) \otimes_A \beta(j), \tag{2.7.3.1}$$

where $\alpha(i) \otimes_A \beta(j)$ is the tensor product in $\mathbf{Mod}(B \otimes_A C)$.

Let $\mathbb{V}$ be a universe containing $\mathbb{U}$ and $\mathbf{Mod}(B)$. We denote by $\mathbf{Mod}(B)^{\wedge}_{\mathbb{U}}$ (resp. $\mathbf{Mod}(B)^{\wedge}_{\mathbb{V}}$) the category of presheaves of $\mathbb{U}$-sets (resp. $\mathbb{V}$-sets) on $\mathbf{Mod}(B)$ and by $\mathrm{h}^{\mathbb{V}}_B \colon \mathbf{Mod}(B) \to \mathbf{Mod}(B)^{\wedge}_{\mathbb{V}}$ the canonical functor (2.6.1.1). The canonical injection functor $\mathbf{Ind\text{-}Mod}(B) \to \mathbf{Mod}(B)^{\wedge}_{\mathbb{U}}$ is fully faithful and it commutes with $\mathbb{U}$-small filtered direct limits (2.6.3.6). The canonical injection functor $\mathbf{Mod}(B)^{\wedge}_{\mathbb{U}} \to \mathbf{Mod}(B)^{\wedge}_{\mathbb{V}}$ is fully faithful and it commutes with $\mathbb{V}$-small direct limits ([5] I 3.6).

Let $G \in \mathbf{Ind\text{-}Mod}(C)$. Let us consider the functor

$$\varphi \colon \mathbf{Ind\text{-}Mod}(B) \to \mathbf{Ind\text{-}Mod}(B \otimes_A C), F \mapsto F \otimes_A G, \qquad (2.7.3.2)$$

and denote by ϕ the composed functor

$$\mathbf{Mod}(B) \to \mathbf{Ind\text{-}Mod}(B) \xrightarrow{\varphi} \mathbf{Ind\text{-}Mod}(B \otimes_A C) \to \mathbf{Mod}(B \otimes_A C)^{\wedge}_{\mathbb{V}}, \qquad (2.7.3.3)$$

where the first and last arrows are the canonical injection functors. By ([40] 2.7.1), there exists a functor $\Phi \colon \mathbf{Mod}(B)^{\wedge}_{\mathbb{V}} \to \mathbf{Mod}(B \otimes_A C)^{\wedge}_{\mathbb{V}}$, commuting with $\mathbb{V}$-small direct limits, and an isomorphism $\Phi \circ \mathrm{h}^{\mathbb{V}}_B \xrightarrow{\sim} \phi$. Moreover, it follows from the proof of *loc. cit.* and 2.6.3.4 that the diagram

$$\begin{array}{ccc} \mathbf{Ind\text{-}Mod}(B) & \xrightarrow{\varphi} & \mathbf{Ind\text{-}Mod}(B \otimes_A C) \\ \downarrow & & \downarrow \\ \mathbf{Mod}(B)^{\wedge}_{\mathbb{V}} & \xrightarrow{\Phi} & \mathbf{Mod}(B \otimes_A C)^{\wedge}_{\mathbb{V}}, \end{array} \qquad (2.7.3.4)$$

where the vertical arrows are the canonical injection functors, is commutative up to isomorphism. Consequently, φ commutes with $\mathbb{U}$-small filtered direct limits. The proposition follows since the tensor product (2.7.1.3) is symmetric.

Lemma 2.7.4 *Under the assumptions of* 2.7.1, *the bifunctor* (2.7.1.3) *is right exact.*

This follows from 2.6.7.5, 2.7.3 and ([40] 8.6.6(a)).

2.7.5 We take again the assumptions of 2.7.1. By 2.7.3, for all ind-B-modules F and F' and all ind-C-modules G and G', we have canonical isomorphisms of $\mathbf{Ind\text{-}Mod}(B \otimes_A C)$

$$(F \otimes_A G) \otimes_C G' \xrightarrow{\sim} F \otimes_A (G \otimes_C G'), \qquad (2.7.5.1)$$

$$(F' \otimes_B F) \otimes_A G \xrightarrow{\sim} F' \otimes_B (F \otimes_A G). \qquad (2.7.5.2)$$

By 2.6.4(ii) and 2.7.3, for every ind-B-module F and every ind-C-module G, the image of $F \otimes_A G$ by the canonical functor $\mathbf{Ind\text{-}Mod}(B \otimes_A C) \to \mathbf{Ind\text{-}Mod}(A)$ is nothing but the tensor product of F and G in $\mathbf{Ind\text{-}Mod}(A)$.

When $A = B = C$, the tensor product (2.7.1.3) makes $\mathbf{Ind\text{-}Mod}(A)$ into a symmetric monoidal category, having A as unit object.

Lemma 2.7.6 *Let (X, A) be a ringed $\mathbb{U}$-topos, B an A-algebra, M, N two ind-B-modules, $u\colon M \to N$ a morphism of* $\mathbf{Ind\text{-}Mod}(A)$ (2.7.1.6) *such that the diagram*

$$\begin{array}{ccc} B \otimes_A M & \xrightarrow{\mathrm{id}\otimes u} & B \otimes_A N \\ \downarrow & & \downarrow \\ M & \xrightarrow{u} & N, \end{array} \tag{2.7.6.1}$$

where the vertical arrows are the canonical morphisms (2.7.1.8), *is commutative. Then, there exists a small filtered category J, two functors $\alpha, \beta\colon K \rightrightarrows$* $\mathbf{Mod}(B)$, *a morphism of functors $\sigma\colon \alpha \to \beta$ and two isomorphisms*

$$M \xrightarrow{\sim} \text{“}\varinjlim\text{”}\alpha \quad \text{and} \quad N \xrightarrow{\sim} \text{“}\varinjlim\text{”}\beta, \tag{2.7.6.2}$$

such that the image of the direct limit of σ by the canonical functor $\mathbf{Ind\text{-}Mod}(B) \to \mathbf{Ind\text{-}Mod}(A)$ *is u.*

Write $M = \text{“}\varinjlim\text{”}\varphi$ and $N = \text{“}\varinjlim\text{”}\psi$, where $\varphi\colon I \to \mathbf{Mod}(B)$ and $\psi\colon J \to \mathbf{Mod}(B)$ are two functors such that the categories I and J are small and filtered. We denote by $\widetilde{\varphi}\colon I \to \mathbf{Mod}(A)_{/M}$, $\widetilde{\psi}\colon J \to \mathbf{Mod}(A)_{/N}$ and $\widetilde{u}\colon \mathbf{Mod}(A)_{/M} \to \mathbf{Mod}(A)_{/N}$ the functors induced by φ, ψ and u. We denote by K' the category of triples (i, j, v) where $i \in \mathrm{Ob}(I)$, $j \in \mathrm{Ob}(J)$ and $v \in \mathrm{Hom}_{\mathbf{Mod}(A)_{/N}}(\widetilde{u} \circ \widetilde{\varphi}(i), \widetilde{\psi}(j))$, with the obvious morphisms (2.6.11). Concretely, the objects of K' are the triples (i, j, v) where $i \in \mathrm{Ob}(I)$, $j \in \mathrm{Ob}(J)$ and $v\colon \varphi(i) \to \psi(j)$ is a A-linear morphism such that the diagram of $\mathbf{Ind\text{-}Mod}(A)$

$$\begin{array}{ccc} \varphi(i) & \xrightarrow{v} & \psi(j) \\ \downarrow & & \downarrow \\ M & \xrightarrow{u} & N, \end{array} \tag{2.7.6.3}$$

where the vertical arrows are the canonical morphisms, is commutative.

For every A-module F, we have

$$\mathrm{Hom}_{\mathbf{Ind\text{-}Mod}(A)}(F, N) = \varinjlim_{j \in J} \mathrm{Hom}_{\mathbf{Mod}(A)}(F, \psi(j)). \tag{2.7.6.4}$$

Since the category J is filtered, we deduce from this that the functor $\widetilde{\psi}$ is cofinal by ([5] I 8.1.3(b)). By virtue of ([40] 3.4.5), the category K' is filtered and the canonical projections $\alpha'\colon K' \to I$ and $\beta'\colon K' \to J$ are cofinal. Moreover, K' is clearly small.

We denote by K the full subcategory of K' made up of the triples (i, j, v), where $i \in \mathrm{Ob}(I)$, $j \in \mathrm{Ob}(J)$ and $v\colon \varphi(i) \to \psi(j)$ is a B-linear morphism. It follows from (2.7.6.1), (2.7.6.4) and ([5] I 8.1.3(c)) that the injection functor $\iota\colon K \to K'$ is cofinal and that K is filtered. We take $\alpha = \varphi \circ \alpha' \circ \iota$ and $\beta = \psi \circ \beta' \circ \iota$. We have a canonical morphism $\sigma\colon \beta \to \alpha$ whose direct limit is identified with u by (2.7.6.3); the proposition follows.

Remark 2.7.7 Under the assumptions of 2.7.15, the canonical functor $\mathbf{Ind\text{-}Mod}(B) \to \mathbf{Ind\text{-}Mod}(A)$ being faithful by 2.6.4.4, we can canonically consider u as a morphism of $\mathbf{Ind\text{-}Mod}(B)$.

Lemma 2.7.8 *Let (X, A) be a ringed $\mathbb{U}$-topos, F an ind-A-module. Then, the functor*

$$\mathbf{Mod}(A) \to \mathbf{Ind\text{-}Mod}(A), \quad M \mapsto M \otimes_A F \tag{2.7.8.1}$$

is exact if and only if the functor

$$\mathbf{Ind\text{-}Mod}(A) \to \mathbf{Ind\text{-}Mod}(A), \quad G \mapsto G \otimes_A F \tag{2.7.8.2}$$

is exact.

Indeed, if the functor (2.7.8.2) is exact, so is the functor (2.7.8.1) since the canonical functor $\mathbf{Mod}(A) \to \mathbf{Ind\text{-}Mod}(A)$ is exact (2.6.7.4). The converse implication follows from 2.6.7.5, 2.7.3 and ([40] 8.6.6(1)).

Definition 2.7.9 Let (X, A) be a ringed $\mathbb{U}$-topos. An ind-A-module is said to be *flat* (or *A-flat*) if it satisfies the equivalent conditions of 2.7.8.

This notion extends that of flatness for A-modules ([5] V 1.1). We can make the following remarks:

2.7.9.1 A small filtered direct limit of flat ind-A-modules is a flat ind-A-module by 2.6.7.5 and 2.7.3.

2.7.9.2 Let B be an A-algebra and F a flat ind-A-module. Then the ind-B-module $B \otimes_A F$ is flat. Indeed, the canonical functor $\mathbf{Ind\text{-}Mod}(B) \to \mathbf{Ind\text{-}Mod}(A)$ is exact and faithful by 2.6.4.4 and ([40] 8.6.8), and for every B-module N, we have a canonical isomorphism $N \otimes_B (B \otimes_A F) \xrightarrow{\sim} N \otimes_A F$ of $\mathbf{Ind\text{-}Mod}(B)$ (2.7.5.2).

2.7.9.3 Let B be an A-algebra such that the functor

$$\mathbf{Mod}(A) \to \mathbf{Mod}(B), \ M \mapsto M \otimes_A B, \tag{2.7.9.4}$$

is exact and faithful; in particular B is a flat A-algebra. The functor

$$\mathbf{Ind\text{-}Mod}(A) \to \mathbf{Ind\text{-}Mod}(B), \ F \mapsto F \otimes_A B, \tag{2.7.9.5}$$

is then exact and faithful by 2.6.4.4 and ([40] 8.6.8).

An ind-A-module F is flat if and only if the ind-B-module $B \otimes_A F$ is flat. Indeed, the condition is sufficient by 2.7.9.2 and it is necessary in view of the assumption (2.7.9.5) and the fact that for every A-module M, we have a canonical isomorphism of $\mathbf{Ind\text{-}Mod}(B)$

$$B \otimes_A (M \otimes_A F) \xrightarrow{\sim} (B \otimes_A M) \otimes_B (B \otimes_A F). \tag{2.7.9.6}$$

2.7.10 Let $f\colon (X, A) \to (Y, B)$ be a morphism of ringed $\mathbb{U}$-topos. By virtue of 2.6.4, the adjoint functors f^* and f_* between the categories $\mathbf{Mod}(B)$ and $\mathbf{Mod}(A)$ induce additive functors

$$\mathrm{I}f^*\colon \mathbf{Ind\text{-}Mod}(B) \to \mathbf{Ind\text{-}Mod}(A), \tag{2.7.10.1}$$

$$\mathrm{I}f_*\colon \mathbf{Ind\text{-}Mod}(A) \to \mathbf{Ind\text{-}Mod}(B). \tag{2.7.10.2}$$

By ([40] 8.6.8), the functor $\mathrm{I}f^*$ (resp. $\mathrm{I}f_*$) is right exact (resp. left exact). It immediately follows from (2.6.1.3) that the functor $\mathrm{I}f^*$ is a left adjoint of the functor $\mathrm{I}f_*$.

By 2.6.4(ii) and 2.7.3, for all ind-B-modules F, F', we have a canonical isomorphism

$$\mathrm{I}f^*(F \otimes_B F') \xrightarrow{\sim} \mathrm{I}f^*(F) \otimes_A \mathrm{I}f^*(F'). \tag{2.7.10.3}$$

By 2.6.10, the functor $\mathrm{I}f_*$ admits a right derived functor

$$\mathrm{R}(\mathrm{I}f_*)\colon \mathbf{D}^+(\mathbf{Ind\text{-}Mod}(A)) \to \mathbf{D}^+(\mathbf{Ind\text{-}Mod}(B)). \tag{2.7.10.4}$$

For every integer q, we have a canonical isomorphism (2.6.9.4)

$$\mathrm{R}^q(\mathrm{I}f_*) \xrightarrow{\sim} \mathrm{I}(\mathrm{R}^q f_*). \tag{2.7.10.5}$$

The diagram (2.6.9.5)

$$\begin{array}{ccc} \mathbf{D}^+(\mathbf{Mod}(A)) & \xrightarrow{\mathrm{R}f_*} & \mathbf{D}^+(\mathbf{Mod}(B)) \\ \downarrow{\scriptstyle \iota_A} & & \downarrow{\scriptstyle \iota_B} \\ \mathbf{D}^+(\mathbf{Ind\text{-}Mod}(A)) & \xrightarrow{\mathrm{R}(\mathrm{I}f_*)} & \mathbf{D}^+(\mathbf{Ind\text{-}Mod}(B)), \end{array} \tag{2.7.10.6}$$

is commutative up to canonical isomorphism.

2.7.11 Let $f\colon (X, A) \to (Y, B)$ be a morphism of ringed $\mathbb{U}$-topos, F, G two ind-A-modules, p, q two integers ≥ 0. Then, there exists a canonical cup-product morphism

$$\mathrm{R}^p\mathrm{I}f_*(F) \otimes_B \mathrm{R}^q\mathrm{I}f_*(G) \to \mathrm{R}^{p+q}\mathrm{I}f_*(F \otimes_A G). \tag{2.7.11.1}$$

Indeed, let us write $F = \text{“}\varinjlim\text{”}\alpha$ and $G = \text{“}\varinjlim\text{”}\beta$, where $\alpha\colon I \to \mathbf{Mod}(A)$ and $\beta\colon J \to \mathbf{Mod}(A)$ are two functors such that the categories I and J are small and filtered. In view of 2.7.3, (2.7.10.5) and (2.6.4.3), we take for the morphism (2.7.11.1) the direct limit on α and β of the cup-product morphisms

$$\mathrm{R}^p f_*(\alpha(i)) \otimes_B \mathrm{R}^q f_*(\beta(j)) \to \mathrm{R}^{p+q} f_*(\alpha(i) \otimes_A \beta(j)), \quad (i \in \mathrm{Ob}(I),\ j \in \mathrm{Ob}(J)). \tag{2.7.11.2}$$

The functors $\widetilde{\alpha}\colon I \to \mathbf{Mod}(A)_{/F}$ and $\widetilde{\beta}\colon J \to \mathbf{Mod}(A)_{/G}$ induced by α and β being cofinal by (2.6.1.2) and ([5] I 8.1.3(b)), replacing I (resp. J) by $\mathbf{Mod}(A)_{/F}$ (resp. $\mathbf{Mod}(A)_{/G}$), we deduce from this that (2.7.11.1) does not depend on the choice of α and β.

It immediately follows from the associativity of the cup-product of modules ([59] 0FP4) that for all ind-A-modules F, G, H and all integers $p, q, r \geq 0$, the diagram

$$\begin{array}{ccc} \mathrm{R}^p\mathrm{I}f_*(F) \otimes_B \mathrm{R}^q\mathrm{I}f_*(G) \otimes_B \mathrm{R}^r\mathrm{I}f_*(H) & \longrightarrow & \mathrm{R}^{p+q}\mathrm{I}f_*(F \otimes_A G) \otimes_B \mathrm{R}^r\mathrm{I}f_*(H) \\ \downarrow & & \downarrow \\ \mathrm{R}^p\mathrm{I}f_*(F) \otimes_B \mathrm{R}^{q+r}\mathrm{I}f_*(G \otimes_A H) & \longrightarrow & \mathrm{R}^{p+q+r}\mathrm{I}f_*(F \otimes_A G \otimes_A H) \end{array} \tag{2.7.11.3}$$

defined by (2.7.5.1) and the cup-product morphisms (2.7.11.1) is commutative.

2.7.12 Let $f: (X, A) \to (Y, B)$ be a morphism of ringed $\mathbb{U}$-topos. We denote by $\mathscr{F}_A$ (resp. $\mathscr{F}_B$) the full subcategory of $\mathbf{Mod}(A)$ (resp. $\mathbf{Mod}(B)$) made up of flabby modules ([5] V 4.1). Injective A-modules being flabby, the category $\mathscr{F}_A$ is cogenerating in $\mathbf{Ind}$-$\mathbf{Mod}(A)$. By ([5] V 5.2), the objects of $\mathscr{F}_A$ are acyclic for f_*. Hence, by virtue of ([40] 13.3.8), the category $\mathscr{F}_A$ is f_*-injective in the sense of ([40] 13.3.4). We deduce from this that the category $\mathbf{Ind}(\mathscr{F}_A)$ is $\mathrm{I}f_*$-injective by ([40] 15.3.2). Moreover, since $f_*(\mathscr{F}_A) \subset \mathscr{F}_B$ ([5] V 4.9), we have $\mathrm{I}f_*(\mathbf{Ind}(\mathscr{F}_A)) \subset \mathbf{Ind}(\mathscr{F}_B)$.

Let $g: (Y, B) \to (Z, C)$ be a morphism of ringed $\mathbb{U}$-topos. We set $h = gf: (X, A) \to (Z, C)$. By virtue of the above and of ([40] 13.3.13), we have a canonical isomorphism

$$\mathrm{R}(\mathrm{I}h_*) \xrightarrow{\sim} \mathrm{R}(\mathrm{I}g_*) \circ \mathrm{R}(\mathrm{I}f_*). \tag{2.7.12.1}$$

2.7.13 Let

$$\begin{array}{ccc} (X', A') & \xrightarrow{g'} & (X, A) \\ {\scriptstyle f'}\downarrow & & \downarrow{\scriptstyle f} \\ (Y', B') & \xrightarrow{g} & (Y, B) \end{array} \tag{2.7.13.1}$$

be a diagram of morphisms of ringed topos, commutative up to canonical isomorphism; in other words, we have an isomorphism

$$f_*g'_* \xrightarrow{\sim} g_*f'_* \tag{2.7.13.2}$$

and the diagram

$$\begin{array}{ccccc} f_*(A) & \xleftarrow{f^\#} & B & \xrightarrow{g^\#} & g_*(B') \\ {\scriptstyle f_*(g'^\#)}\downarrow & & & & \downarrow{\scriptstyle g_*(f'^\#)} \\ f_*(g'_*(A')) & & \longrightarrow & & g_*(f'_*(A')), \end{array} \tag{2.7.13.3}$$

where the unlabeled arrow is the isomorphism (2.7.13.2), is commutative. If F is an ind-A-module, we have, for every $q \geq 0$, a canonical morphism of $\mathbf{Ind}$-$\mathbf{Mod}(B')$

$$\mathrm{I}g^*(\mathrm{R}^q\mathrm{I}f_*(F)) \to \mathrm{R}^q\mathrm{I}f'_*(\mathrm{I}g'^*(F)), \tag{2.7.13.4}$$

called *base change morphism*. Indeed, this amounts to giving a morphism

$$\mathrm{R}^q\mathrm{I}f_*(F) \to \mathrm{I}g_*(\mathrm{R}^q\mathrm{I}f'_*(\mathrm{I}g'^*(F))), \tag{2.7.13.5}$$

and we take the composed morphism

$$\begin{aligned} &\mathrm{R}^q\mathrm{I}f_*(F) \to \mathrm{R}^q\mathrm{I}f_*(\mathrm{I}g'_*(\mathrm{I}g'^*(F))) \to \\ &\mathrm{R}^q\mathrm{I}(fg')_*(\mathrm{I}g'^*(F)) \xrightarrow{\sim} \mathrm{R}^q\mathrm{I}(gf')_*(\mathrm{I}g'^*(F)) \to \mathrm{I}g_*(\mathrm{R}^q\mathrm{I}f'_*(g'^*(F))), \end{aligned} \tag{2.7.13.6}$$

where the first arrow comes from the adjunction morphism $\mathrm{id} \to \mathrm{I}g'_*\mathrm{I}g'^*$, the second and last arrows from the Cartan–Leray spectral sequence (2.7.12.1) and the third arrow from the isomorphism (2.7.13.2).

In view of (2.7.10.5), we immediately check that the base change morphism (2.7.13.4) is the direct limit of base change morphisms for modules. Hence, (2.7.13.4) is an isomorphism when the base change morphism for modules is an isomorphism. In particular, (2.7.13.4) is an isomorphism if g is the localization morphism of Y over an object V, $B' = g^{-1}(B)$, g' is the localization morphism of X over $U = f^*(V)$, $A' = g'^{-1}(A)$ and $f' = f_{|V}$ ([5] 5.1(3)).

Lemma 2.7.14 *Let $f\colon (X, A) \to (Y, B)$ be a morphism of ringed $\mathbb{U}$-topos, E a locally free B-module of finite type, $\mathscr{F}$ an object of* $\mathbf{D}^+(\mathbf{Ind\text{-}Mod}(A))$. *Then, there exists a canonical isomorphism of* $\mathbf{D}^+(\mathbf{Ind\text{-}Mod}(B))$, *functorial in $\mathscr{F}$,*

$$E \otimes_A \mathrm{RI}f_*(\mathscr{F}) \to \mathrm{RI}f_*(f^*(E) \otimes_B \mathscr{F}). \tag{2.7.14.1}$$

Note first that the functor $E \otimes_A -$, being exact on the category of ind-A-modules (2.7.8), extends to the category $\mathbf{D}^+(\mathbf{Ind\text{-}Mod}(A))$. We define in the same way the functor $f^*(E) \otimes_B -$ on the category $\mathbf{D}^+(\mathbf{Ind\text{-}Mod}(B))$.

We denote by $\mathscr{J}$ the category of injective A-modules, and by $\mathbf{K}^+(\mathbf{Ind\text{-}Mod}(A))$ (resp. $\mathbf{K}^+(\mathbf{Ind}(\mathscr{J}))$) the category of complexes of ind-A-modules (resp. ind-objects of $\mathscr{J}$) bounded from below up to homotopy ([40] 11.3.7). Note that the category $\mathbf{Ind}(\mathscr{J})$ is identified with the full subcategory of $\mathbf{Ind\text{-}Mod}(A)$ made up of quasi-injective objects (2.6.8). Since $\mathscr{J}$ is f_*-injective by virtue of ([40] 13.3.6(iii)), $\mathbf{Ind}(\mathscr{J})$ is $(\mathrm{I}f_*)$-injective by ([40] 15.3.2), in other words, we have the following properties ([40] 10.3.2 and 13.3.4):

(i) for every complex $\mathscr{F}$ of ind-A-modules bounded from below, there exists an object $\mathscr{G}$ of $\mathbf{K}^+(\mathbf{Ind}(\mathscr{J}))$ and a quasi-isomorphism $\mathscr{F} \to \mathscr{G}$;
(ii) for every exact complex $\mathscr{G}$ of $\mathbf{K}^+(\mathbf{Ind}(\mathscr{J}))$, the complex $\mathrm{I}f_*(\mathscr{G})$ is exact.

Denoting by $\mathbf{N}^+$ the null system of exact complexes of $\mathbf{K}^+(\mathbf{Ind\text{-}Mod}(A))$ (see [40] 13.1.2), the functor $\mathrm{RI}f_*$ is then defined by the following diagram:

$$\begin{array}{ccccc} \mathbf{K}^+(\mathbf{Ind\text{-}Mod}(A)) & \longrightarrow & \mathbf{D}^+(\mathbf{Ind\text{-}Mod}(A)) & \xrightarrow{\mathrm{RI}f_*} & \\ \uparrow & & \uparrow \wr & & \searrow \\ \mathbf{K}^+(\mathbf{Ind}(\mathscr{J})) & \longrightarrow & \mathbf{K}^+(\mathbf{Ind}(\mathscr{J}))/(\mathbf{K}^+(\mathbf{Ind}(\mathscr{J})) \cap \mathbf{N}^+) & \longrightarrow & \mathbf{D}^+(\mathbf{Ind\text{-}Mod}(B)). \end{array}$$
$$\mathbf{K}^+(\mathbf{Ind}(\mathscr{J})) \xrightarrow{\mathrm{I}f_*} \mathbf{D}^+(\mathbf{Ind\text{-}Mod}(B))$$

For every object $\mathscr{G}$ of $\mathbf{K}^+(\mathbf{Ind}(\mathscr{I}))$, $f^*(E) \otimes_A \mathscr{G}$ is an object of $\mathbf{K}^+(\mathbf{Ind}(\mathscr{I}))$ by ([59] 01E7). Moreover, it follows from 2.6.4.3 that we have a canonical isomorphism, functorial in $\mathscr{G}$,

$$E \otimes_A \mathrm{I}f_*(\mathscr{G}) \to \mathrm{I}f_*(f^*(E) \otimes_B \mathscr{G}). \tag{2.7.14.2}$$

The statement follows.

2.7.15 Let (X, A) be a ringed topos, U an object of X. We denote by $j_U : X_{/U} \to X$ the localization morphism of X over U. For any $F \in \mathrm{Ob}(X)$, the sheaf $j_U^*(F)$ will also be denoted by $F|U$. The topos $X_{/U}$ will be ringed by $A|U$. By ([40] 8.6.8), j_U^* induces an exact functor

$$\mathrm{I}j_U^* \colon \mathbf{Ind\text{-}Mod}(A) \to \mathbf{Ind\text{-}Mod}(A|U). \tag{2.7.15.1}$$

For any ind-A-module F, the ind-$(A|U)$-module $\mathrm{I}j_U^*(F)$ will also be denoted $F|U$.

The extension by zero functor $j_{U!} \colon \mathbf{Mod}(A|U) \to \mathbf{Mod}(A)$ is exact and faithful ([5] IV 11.3.1). By 2.6.4.4 and ([40] 8.6.8), it induces an exact and faithful functor

$$\mathrm{I}j_{U!} \colon \mathbf{Ind\text{-}Mod}(A|U) \to \mathbf{Ind\text{-}Mod}(A). \tag{2.7.15.2}$$

This is a left adjoint of the functor $\mathrm{I}j_U^*$.

Lemma 2.7.16 *Let (X, A) be a ringed topos, U an object of X. Then,*

(i) *For every flat ind-$(A|U)$-module P (2.7.9), the ind-A-module $\mathrm{I}j_{U!}(P)$ is flat.*
(ii) *For every flat ind-A-module M, the ind-$(A|U)$-module $\mathrm{I}j_U^*(M)$ is flat.*

Indeed, it immediately follows from ([5] IV 12.11) that for every ind-A-module M and every ind-$(A|U)$-module P, we have a canonical functorial isomorphism

$$\mathrm{I}j_{U!}(P \otimes_{(A|U)} \mathrm{I}j_U^*(M)) \xrightarrow{\sim} \mathrm{I}j_{U!}(P) \otimes_A M. \tag{2.7.16.1}$$

(i) This follows from (2.7.16.1) and the exactness of the functors $\mathrm{I}j_U^*$ and $\mathrm{I}j_{U!}$.
(ii) It follows from (2.7.16.1) and the exactness of the functor $\mathrm{I}j_{U!}$ that the functor

$$\mathbf{Ind\text{-}Mod}(A|U) \to \mathbf{Ind\text{-}Mod}(A), \quad P \mapsto \mathrm{I}j_{U!}(P \otimes_{(A_Q|U)} \mathrm{I}j_U^*(M)) \tag{2.7.16.2}$$

is exact. Since the functor $\mathrm{I}j_{U!}$ is moreover faithful (2.7.15), we deduce from this that the functor $P \mapsto P \otimes_{(A_Q|U)} \mathrm{I}j_U^*(M)$ on the category $\mathbf{Ind\text{-}Mod}(A|U)$ is exact; the statement follows.

Lemma 2.7.17 *Let (X, A) be a ringed topos, $(U_p)_{1 \le p \le n}$ a finite covering of the final object of X, F, G two ind-A-modules. For any $1 \le p, q \le n$, we set $U_{pq} = U_p \times U_q$. We take again the notation of 2.7.15. Then,*

(i) *The following diagram of maps of sets is exact*

$$\begin{aligned}\mathrm{Hom}_{\mathbf{Ind\text{-}Mod}(A)}(F, G) \to &\prod_{1 \le p \le n} \mathrm{Hom}_{\mathbf{Ind\text{-}Mod}(A|U_p)}(F|U_p, G|U_p) \\ \rightrightarrows &\prod_{1 \le p,q \le n} \mathrm{Hom}_{\mathbf{Ind\text{-}Mod}(A|U_{p,q})}(F|U_{pq}, G|U_{pq}).\end{aligned} \tag{2.7.17.1}$$

(ii) *The ind-A-module F vanishes if and only if, for all $1 \leq p \leq n$, the ind-$(A|U_p)$-module $F|U_p$ vanishes.*

(iii) *A morphism $u\colon F \to G$ of* **Ind-Mod**(A) *is an isomorphism if and only if, for all $1 \leq p \leq n$, the morphism $u|U_p$ is an isomorphism.*

(iv) *The ind-A-module F is A-flat (2.7.9) if and only if, for all $1 \leq p \leq n$, the ind-$(A|U_p)$-module $F|U_p$ is $(A|U_p)$-flat.*

(i) Write $F = \text{“}\varinjlim\text{”}\alpha$ and $G = \text{“}\varinjlim\text{”}\beta$ where $\alpha\colon I \to \mathbf{Mod}(A)$ and $\beta\colon J \to \mathbf{Mod}(A)$ are two functors such that the categories I and J are small and filtered. Let $u\colon F \to G$ be a morphism of **Ind-Mod**(A). In view of (2.6.1.3), for every $i \in \mathrm{Ob}(I)$, there exists $j \in \mathrm{Ob}(J)$ and a morphism $u_{ij}\colon \alpha(i) \to \beta(j)$ which represents the image of u in

$$\varinjlim_{j \in J} \mathrm{Hom}_{\mathbf{Mod}(A)}(\alpha(i), \beta(j)). \tag{2.7.17.2}$$

We will say that u_{ij} *represents* u.

Let $u, v\colon F \to G$ be two morphisms of **Ind-Mod**(A) such that for all $1 \leq p \leq n$, we have $u|U_p = v|U_p$. In view of (2.6.1.2), for every $i \in \mathrm{Ob}(I)$, there exists $j \in \mathrm{Ob}(J)$ and representatives $u_{ij}, v_{ij}\colon \alpha(i) \to \beta(j)$ of u and v, respectively, such that $u_{ij}|U_p = v_{ij}|U_p$ for all $1 \leq p \leq n$. We deduce from this that $u_{ij} = v_{ij}$ and consequently that $u = v$ (2.6.1.3). The first map of (2.7.17.1) is therefore injective.

Let, for any $1 \leq p \leq n$, $u_p\colon F|U_p \to G|U_p$ be a morphism of **Ind-Mod**$(A|U_p)$ such that $(u_p)_{1 \leq p \leq n}$ is in the kernel of the double arrow of (2.7.17.1). By (2.6.1.2), for every $i \in \mathrm{Ob}(I)$, there exist $j \in \mathrm{Ob}(J)$ and, for every $1 \leq p \leq n$, a representative $u_{ijp}\colon \alpha(i)|U_p \to \beta(j)|U_p$ of u_p, such that $u_{ijp}|U_{pq} = u_{ijq}|U_{pq}$ for all $1 \leq p, q \leq n$. Then, there exists $u_{ij}\colon \alpha(i) \to \beta(j)$ such that $u_{ij}|U_p = u_{ijp}$ for all $1 \leq p \leq n$. For every morphism $h\colon i' \to i$ of I, the image of u_{ij} by the map

$$\varinjlim_{j \in J} \mathrm{Hom}_{\mathbf{Mod}(A)}(\alpha(i), \beta(j)) \to \varinjlim_{j \in J} \mathrm{Hom}_{\mathbf{Mod}(A)}(\alpha(i'), \beta(j)) \tag{2.7.17.3}$$

induced by h coincides with that of $u_{i'j}$; we check it by restriction to U_p for $1 \leq p \leq n$. The morphisms u_{ij} then define a morphism $u\colon F \to G$ of **Ind-Mod**(A) (2.6.1.2) of image $(u_p)_{1 \leq p \leq n}$; hence the diagram (2.7.17.1) is exact at the center.

(ii) Indeed, F vanishes if and only if $\mathrm{id}_F = 0$. The assertion therefore follows from (i).

(iii) Suppose that for all $1 \leq p \leq n$, $u|U_p$ is an isomorphism and let $v_p\colon G|U_p \to F|U_p$ be its inverse. Since $(v_p)_{1 \leq p \leq n}$ is clearly in the kernel of the double arrow of (2.7.17.1), it is the image of a unique morphism $u\colon G \to F$ of **Ind-Mod**(A). In view of the injectivity of the first arrow of (2.7.17.1), v is the inverse of u.

(iv) Indeed, the condition is necessary by 2.7.16(ii) and it is sufficient in view of (iii) and (2.7.10.3).

2.7.18 Let $f\colon (X, A) \to (Y, B)$ be a morphism of ringed $\mathbb{U}$-topos, F an ind-A-module, G an ind-B-module, q an integer ≥ 0. The adjunction morphism $G \to \mathrm{I}f_*(\mathrm{I}f^*(G))$ and the cup-product (2.7.11.1) induce a bifunctorial morphism

$$G \otimes_A \mathrm{R}^q(\mathrm{I}f_*)(F) \to \mathrm{R}^q(\mathrm{I}f_*)(\mathrm{I}f^*(G) \otimes_A F). \tag{2.7.18.1}$$

We can make the following remarks:

(i) For every ind-B-module G', the composition

$$\begin{array}{ccc} G \otimes_B G' \otimes_B \mathrm{R}^q(\mathrm{I}f_*)(F) & \longrightarrow & G \otimes_B \mathrm{R}^q(\mathrm{I}f_*)(\mathrm{I}f^*(G') \otimes_A F) \\ & \searrow & \downarrow \\ & & \mathrm{R}^q(\mathrm{I}f_*)(\mathrm{I}f^*(G \otimes_B G') \otimes_A F) \end{array} \tag{2.7.18.2}$$

of the morphisms induced by the morphisms (2.7.18.1) relative to G and G', is none other than the morphism (2.7.18.1) relative to $B \otimes_B G'$. This immediately follows from (2.7.11.3).

(ii) When $q = 0$, the morphism (2.7.18.1) is the adjoint of the composed morphism

$$\mathrm{I}f^*(G \otimes_B \mathrm{I}f_*(F)) \xrightarrow{\sim} \mathrm{I}f^*(G) \otimes_A \mathrm{I}f^*(\mathrm{I}f_*(F)) \to \mathrm{I}f^*(G) \otimes_A F, \tag{2.7.18.3}$$

where the first arrow is the isomorphism (2.7.10.3) and the second arrow is induced by the canonical morphism $\mathrm{I}f^*(\mathrm{I}f_*(F)) \to F$. This follows from the analogous statement for modules ([59] 0B68).

Lemma 2.7.19 *Let $f\colon (X, A) \to (Y, B)$ be a morphism of ringed $\mathbb{U}$-topos, N a B-module locally projective of finite type* (2.1.11)*, F an ind-A-module, q an integer ≥ 0. Assume that the final object of Y is quasi-compact. Then, the canonical morphism* (2.7.18.1)

$$N \otimes_B \mathrm{R}^q\mathrm{I}f_*(F) \to \mathrm{R}^q(\mathrm{I}f_*)(\mathrm{I}f^*(N) \otimes_A F) \tag{2.7.19.1}$$

is an isomorphism.

There exists a finite covering $(U_i)_{0 \leq i \leq n}$ of the final object of Y such that for all $0 \leq i \leq n$, the $(B|U_i)$-module $N|U_i$ is a direct factor of a free $(B|U_i)$-module of finite type. In view of 2.7.17(iii) and the fact that $\mathrm{R}^q\mathrm{I}f_*$ commutes with localizations (2.7.13), we can then reduce to the case where N is a direct factor of a free B-module of finite type, and even to the case where N is a free B-module of finite type, in which case the assertion is obvious.

2.8 Higgs Ind-modules and λ-connections

Definition 2.8.1 Let (X, A) be a ringed topos, E an A-module.

(i) We call a *Higgs ind-A-module with coefficients in E* a pair (M, θ) consisting of an ind-A-module M and a morphism of **Ind-Mod**(A)

$$\theta\colon M \to M \otimes_A E \tag{2.8.1.1}$$

such that the composed morphism

$$M \xrightarrow{\theta} M \otimes_A E \xrightarrow{\theta \otimes_A \mathrm{id}_E} M \otimes_A E \otimes_A E \xrightarrow{\mathrm{id}_M \otimes_A w} M \otimes_A \wedge^2 E, \tag{2.8.1.2}$$

where $w\colon E \otimes_A E \to \wedge^2 E$ is the exterior product (2.1.10), vanishes. We then say that θ is a *Higgs A-field* on M with coefficients in E.

(ii) If (M, θ) and (M', θ') are two Higgs ind-A-modules, a morphism from (M, θ) to (M', θ') is a morphism $u\colon M \to M'$ of $\mathbf{Ind\text{-}Mod}(A)$ such that $(u \otimes \mathrm{id}_E) \circ \theta = \theta' \circ u$.

Higgs ind-A-modules with coefficients in E form a category denoted by $\mathbf{Ind\text{-}HM}(A, E)$. Despite the notation, this category is not the category of ind-objects of another category (2.6.3). We can complete the terminology and make the following remarks.

2.8.1.3 Let (M, θ) be a Higgs ind-A-module with coefficients in E. For any $i \geq 1$, we denote by

$$\theta_i\colon M \otimes_A \wedge^i E \to M \otimes_A \wedge^{i+1} E \tag{2.8.1.4}$$

the composed morphism of $\mathbf{Ind\text{-}Mod}(A)$

$$M \otimes_A \wedge^i E \xrightarrow{\theta \otimes_A \mathrm{id}_{\wedge^i E}} M \otimes_A E \otimes_A \wedge^i E \xrightarrow{\mathrm{id}_M \otimes_A w^i} M \otimes_A \wedge^{i+1} E \tag{2.8.1.5}$$

where $w^i\colon E \otimes_A \wedge^i E \to \wedge^{i+1} E$ is the exterior product (2.1.10). We have $\theta_{i+1} \circ \theta_i = 0$. We call the *Dolbeault* complex of (M, θ) and denote by $\mathbb{K}^\bullet(M, \theta)$ the complex of cochains of $\mathbf{Ind\text{-}Mod}(A)$

$$M \xrightarrow{\theta} M \otimes_A E \xrightarrow{\theta_1} M \otimes_A \wedge^2 E \ldots, \tag{2.8.1.6}$$

where M is placed in degree 0 and the differentials are of degree 1.

2.8.1.7 Let (M, θ), (M', θ') be two Higgs ind-A-modules with coefficients in E. We call the *total* Higgs field on $M \otimes_A M'$ the A-linear morphism

$$\theta_{\mathrm{tot}}\colon M \otimes_A M' \to M \otimes_A M' \otimes_A E \tag{2.8.1.8}$$

defined by

$$\theta_{\mathrm{tot}} = \theta \otimes \mathrm{id}_{M'} + \mathrm{id}_M \otimes \theta'. \tag{2.8.1.9}$$

We say that $(M \otimes_A M', \theta_{\mathrm{tot}})$ is the *tensor product* of (M, θ) and (M', θ').

2.8.1.10 For any Higgs ind-A-module M with coefficients in E, $\kappa_A(\theta)$ is identified with a Higgs A-field $\kappa_A(M) \to \kappa_A(M) \otimes_A E$ (2.7.1.5). Denoting by $\mathbb{K}^\bullet(M, \theta)$ the Dolbeault complex of (M, θ), $\kappa_A(\mathbb{K}^\bullet(M, \theta))$ is identified with the Dolbeault complex of $(\kappa_A(M), \kappa_A(\theta))$.

2.8.2 Let (X, A) be a ringed topos,

$$0 \to L \to E \to \underline{E} \to 0 \tag{2.8.2.1}$$

an exact sequence of locally free A-modules of finite type. We endow the exterior algebra $\wedge^\bullet E$ of E with the Koszul filtration $\mathrm{W}^\bullet \wedge^\bullet E$ associated with this sequence (see 2.5.11).

Let (M, θ) be a Higgs ind-A-module with coefficients in E. We denote by $\underline{\theta}\colon M \to M \otimes_A \underline{E}$ the Higgs A-field induced by θ, and by $\mathbb{K}^\bullet$ (resp. $\underline{\mathbb{K}}^\bullet$) the Dolbeault complex of (M, θ) (resp. $(M, \underline{\theta})$) (2.8.1.3). For any integers $i, j \geq 0$, we set

$$\mathrm{F}^i\mathbb{K}^j = M \otimes_A (\mathrm{W}^i \wedge^j E). \tag{2.8.2.2}$$

We thus define an exhaustive decreasing filtration of the $\wedge^\bullet E$-graded module $\mathbb{K}^\bullet$ by graded sub-$\wedge^\bullet E$-modules $(\mathrm{F}^i\mathbb{K}^\bullet)_{i\geq 0}$, stable by the differential of $\mathbb{K}^\bullet$, called the *Koszul filtration of $\mathbb{K}^\bullet$ associated with the exact sequence* (2.8.2.1); in particular, it is a filtration of the complex $\mathbb{K}^\bullet$ by subcomplexes.

In view of (2.5.11.3) and (2.5.11.4), for any integer $i \geq 0$, we have a canonical isomorphism of complexes

$$\mathrm{W}^i\mathbb{K}^\bullet/\mathrm{W}^{i+1}\mathbb{K}^\bullet \xrightarrow{\sim} \wedge^i L \otimes_A \underline{\mathbb{K}}^\bullet[-i], \tag{2.8.2.3}$$

where the differentials of $\underline{\mathbb{K}}^\bullet[-i]$ are those of $\underline{\mathbb{K}}^\bullet$ multiplied by $(-1)^i$. We therefore have a canonical exact sequence of ind-A-modules

$$0 \to \wedge^{i+1} L \otimes_A \underline{\mathbb{K}}^\bullet[-i-1] \to \mathrm{W}^i\mathbb{K}^\bullet/\mathrm{W}^{i+2}\mathbb{K}^\bullet \to \wedge^i L \otimes_A \underline{\mathbb{K}}^\bullet[-i] \to 0. \tag{2.8.2.4}$$

The latter induces a morphism in the derived category $\mathbf{D}^+(\mathbf{Ind\text{-}Mod}(A))$,

$$\partial^i\colon \wedge^i L \otimes_A \underline{\mathbb{K}}^\bullet \to \wedge^{i+1} L \otimes_A \underline{\mathbb{K}}^\bullet, \tag{2.8.2.5}$$

which we will call the *boundary map associated with the Koszul filtration of* $\mathbb{K}^\bullet$.

Proposition 2.8.3 *Let (X, A) be a ringed topos,*

$$0 \to L \to E \to \underline{E} \to 0 \tag{2.8.3.1}$$

an exact sequence of locally free A-modules of finite type, (M, θ) a Higgs ind-A-module with coefficients in E, (N, κ) a Higgs ind-A-module with coefficients in L. We denote by $\underline{\theta}$ the Higgs A-field on M with coefficients in $\underline{E}$ induced by θ, by κ' the Higgs A-field on N with coefficients in E induced by κ, by $\theta' = \theta \otimes \mathrm{id} + \mathrm{id} \otimes \kappa'$ the total Higgs A-field on $M \otimes_A N$ with coefficients in E, and by $\underline{\theta}'$ the Higgs A-field on $M \otimes_A N$ with coefficients in $\underline{E}$ induced by θ', which is none other than $\underline{\theta} \otimes \mathrm{id}$. We denote by $\mathbb{K}^\bullet$ (resp. $\underline{\mathbb{K}}^\bullet$, resp. $\mathbb{K}'^\bullet$, resp. $\underline{\mathbb{K}}'^\bullet$) the Dolbeault complex of (M, θ) (resp. $(M, \underline{\theta})$, resp. $(M \otimes_A N, \theta')$, resp. $(M \otimes_A N, \underline{\theta}')$) (2.8.1.3), and by

$$\partial\colon \underline{\mathbb{K}}^\bullet \to L \otimes_A \underline{\mathbb{K}}^\bullet, \tag{2.8.3.2}$$
$$\partial'\colon \underline{\mathbb{K}}'^\bullet \to L \otimes_A \underline{\mathbb{K}}'^\bullet, \tag{2.8.3.3}$$

the boundaries in $\mathbf{D}^+(\mathbf{Ind\text{-}Mod}(A))$ associated with the Koszul filtrations of $\mathbb{K}^\bullet$ and $\mathbb{K}'^\bullet$ with respect to the exact sequence (2.8.3.1); see (2.8.2.5). Identifying $\underline{\mathbb{K}}'^\bullet$ with $\underline{\mathbb{K}}^\bullet \otimes_A N$, we then have

$$\partial' = \partial \otimes \mathrm{id} - \mathrm{id} \otimes \kappa. \tag{2.8.3.4}$$

It suffices to copy the proof from 2.5.14.

Definition 2.8.4 Let (X, A) be a ringed topos, B an A-algebra, $\lambda \in A(X)$, $\Omega^1_{B/A}$ the B-module of Kähler differentials of B/A, $d\colon B \to \Omega^1_{B/A}$ the universal A-derivation (2.5.24), M an ind-B-module,

$$\nabla\colon M \to \Omega^1_{B/A} \otimes_B M \tag{2.8.4.1}$$

a morphism of **Ind-Mod**(A) (2.7.1.6). We say that ∇ is a *λ-connection on M with respect to the extension B/A* (and that (M, ∇) is an *ind-B-module with λ-connection*) if the composed morphism of **Ind-Mod**(A)

$$B \otimes_A M \longrightarrow M \xrightarrow{\nabla} \Omega^1_{B/A} \otimes_B M\ , \tag{2.8.4.2}$$

where the first arrow is the canonical morphism (2.7.1.8), is the sum of the two composed morphisms of **Ind-Mod**(A)

$$B \otimes_A M \xrightarrow{\mathrm{id}_B \otimes_A \nabla} B \otimes_A \Omega^1_{B/A} \otimes_B M \longrightarrow \Omega^1_{B/A} \otimes_B M\ , \tag{2.8.4.3}$$

$$B \otimes_A M \xrightarrow{\lambda d \otimes_A \mathrm{id}_M} \Omega^1_{B/A} \otimes_A M \longrightarrow \Omega^1_{B/A} \otimes_B M\ , \tag{2.8.4.4}$$

where the second arrows are the canonical morphisms (2.7.5.1).

Let (M, ∇), (M', ∇') be two ind-B-modules with λ-connections. A morphism from (M, ∇) to (M', ∇') is a morphism $u\colon M \to M'$ of **Ind-Mod**(B) such that $(\mathrm{id} \otimes_B u) \circ \nabla = \nabla' \circ u$.

Lemma 2.8.5 *Let (X, A) be a ringed topos, B an A-algebra, $\lambda \in A(X)$, J a small filtered category, $\alpha\colon J \to$ **Mod**(B) a functor, $M =$ "$\varinjlim$"α its direct limit in **Ind-Mod**(B), ∇ a λ-connection on M with respect to the extension B/A. Denote by $\iota\colon$ **Mod**$(B) \to$ **Mod**(A) the forgetful functor, and for any $j \in \mathrm{Ob}(J)$, let $M_j = \alpha(j)$. Then there exist a small filtered category K, two cofinal functors $\beta, \gamma\colon K \rightrightarrows J$ and two morphisms of functors $\sigma\colon \beta \to \gamma$ and $V\colon \iota \circ \alpha \circ \beta \to \iota \circ (\Omega^1_{B/A} \otimes_B \alpha \circ \gamma)$, such that for all $k \in \mathrm{Ob}(K)$, we have a B-linear morphism, functorial in k,*

$$\alpha(\sigma_k)\colon M_{\beta(k)} \to M_{\gamma(k)} \tag{2.8.5.1}$$

and an A-linear morphism, functorial in k,

$$V_k\colon M_{\beta(k)} \to \Omega^1_{B/A} \otimes_B M_{\gamma(k)}, \tag{2.8.5.2}$$

such that the following two properties are satisfied:

(i) *for every $k \in \mathrm{Ob}(K)$ and for every local sections b of B and m of $M_{\beta(k)}$, we have*

$$V_k(bm) = \lambda d(b) \otimes_B \alpha(\sigma_k)(m) + bV_k(m); \tag{2.8.5.3}$$

(ii) *the direct limit of $\alpha(\sigma)$ is the identity of M and the direct limit of V is ∇.*

Let $\mathscr{C}$ denote the filtered category $\mathbf{Mod}(A)_{/\Omega^1_{B/A}\otimes_B M}$ (2.6.3.4) and consider the functors

$$\psi\colon J \to \mathscr{C}, \quad j \mapsto (\Omega^1_{B/A} \otimes_B M_j \to \Omega^1_{B/A} \otimes_B M), \tag{2.8.5.4}$$

$$\varphi\colon J \to \mathscr{C}, \quad i \mapsto (M_i \to M \xrightarrow{\nabla} \Omega^1_{B/A} \otimes_B M), \tag{2.8.5.5}$$

induced by the relation $M = \text{“}\varinjlim\text{”}\alpha$. For every A-module N, we have

$$\mathrm{Hom}_{\mathbf{Ind\text{-}Mod}(A)}(N, \Omega^1_{B/A} \otimes_B M) = \varinjlim_{j\in J} \mathrm{Hom}_{\mathbf{Mod}(A)}(N, \Omega^1_{B/A} \otimes_B M_j). \tag{2.8.5.6}$$

Since J is filtered, we deduce from this that the functor ψ is cofinal by ([5] I 8.1.3(b)).

Consider the functors

$$\psi'\colon J \to J \times \mathscr{C}, \quad j \mapsto (j, \psi(j)), \tag{2.8.5.7}$$

$$\varphi'\colon J \to J \times \mathscr{C}, \quad i \mapsto (i, \varphi(i)). \tag{2.8.5.8}$$

It follows from the above and from ([5] I 8.1.3(b)) that ψ' is cofinal. We denote by K' the category of triples (i, j, u) where $i, j \in \mathrm{Ob}(J)$ and $u \in \mathrm{Hom}_{J\times\mathscr{C}}(\varphi'(i), \psi'(j))$, with obvious morphisms (2.6.11). More explicitly, the objects of K' are the quadruples (i, j, s, v) where $i, j \in \mathrm{Ob}(J)$, $s\colon i \to j$ is a morphism of J and

$$v\colon M_i \to \Omega^1_{B/A} \otimes_A M_j \tag{2.8.5.9}$$

is an A-linear morphism that fits into the commutative diagram

$$\begin{array}{ccc} M_i & \xrightarrow{v} & \Omega^1_{B/A} \otimes_A M_j \\ \downarrow & & \downarrow \\ M & \xrightarrow{\nabla} & \Omega^1_{B/A} \otimes_A M \end{array} \tag{2.8.5.10}$$

where the vertical arrows are induced by the relation $M = \text{“}\varinjlim\text{”}\alpha$. By ([40] 3.4.5), the category K' is filtered and the first and second canonical projections $\beta', \gamma'\colon K' \rightrightarrows J$, respectively, are cofinal. Moreover, K' is clearly small.

Let K be the full subcategory of K' made up of the quadruples (i, j, s, v) such that for all local sections b of B and m of M_i, we have

$$v(bm) = \lambda d(b) \otimes_B \alpha(s)(m) + bv(m). \tag{2.8.5.11}$$

It follows from (2.8.5.6) and ([5] I 8.1.3(c)) that the injection functor $K \to K'$ is cofinal and that K is filtered. We denote by $\beta, \gamma\colon K \rightrightarrows J$ the functors induced by β' and γ'. For every $k \in \mathrm{Ob}(K)$, we write

$$k = (\beta(k), \gamma(k), \sigma_k\colon \beta(k) \to \gamma(k), V_k\colon M_{\beta(k)} \to \Omega^1_{B/A} \otimes_B M_{\gamma(k)}). \tag{2.8.5.12}$$

We therefore have two morphisms of functors $\sigma\colon \beta \to \gamma$ and $V\colon \iota \circ \alpha \circ \beta \to \iota \circ (\Omega^1_{B/A} \otimes_B \alpha \circ \gamma)$. It immediately follows from (2.8.5.10) that the direct limit of V is the morphism ∇. Likewise, for every $k \in \mathrm{Ob}(K)$, we have a commutative diagram

$$\begin{array}{ccc} M_{\beta(k)} & \longrightarrow & M_{\gamma(k)} \\ & \searrow & \downarrow \\ & & M. \end{array} \qquad (2.8.5.13)$$

We deduce from this that the direct limit of $\alpha(\sigma)$ is the identity of M, the proposition follows.

Proposition 2.8.6 *Let (X, A) be a ringed topos, B an A-algebra, $\lambda \in A(X)$, (M, ∇) an ind-B-module with λ-connection with respect to the extension B/A. Then, ∇ extends uniquely into a graded morphism of degree 1 of* **Ind-Mod**(A) *which we also denote by*

$$\nabla\colon \Omega_{B/A} \otimes_B M \to \Omega_{B/A} \otimes_B M, \qquad (2.8.6.1)$$

such that for all integers $p, q \geq 0$, the composed morphism of **Ind-Mod**(A)

$$\Omega^p_{B/A} \otimes_A \Omega^q_{B/A} \otimes_B M \longrightarrow \Omega^{p+q}_{B/A} \otimes_B M \xrightarrow{\nabla} \Omega^{p+q+1}_{B/A} \otimes_B M \qquad (2.8.6.2)$$

is the sum of the two composed morphisms of **Ind-Mod**(A)

$$\Omega^p_{B/A} \otimes_A \Omega^q_{B/A} \otimes_B M \xrightarrow{(-1)^p \mathrm{id} \otimes_A \nabla} \Omega^p_{B/A} \otimes_A \Omega^{q+1}_{B/A} \otimes_B M \longrightarrow \Omega^{p+q+1}_{B/A} \otimes_B M\,, \qquad (2.8.6.3)$$

$$\Omega^p_{B/A} \otimes_A \Omega^q_{B/A} \otimes_B M \xrightarrow{\lambda d \otimes_A \mathrm{id}} \Omega^{p+1}_{B/A} \otimes_A \Omega^q_{B/A} \otimes_B M \longrightarrow \Omega^{p+q+1}_{B/A} \otimes_B M\,, \qquad (2.8.6.4)$$

where the unlabeled arrows are induced by the multiplication of $\Omega_{B/A}$. Moreover, for all integers $p, q \geq 0$, the diagram

$$\begin{array}{ccc} \Omega^p_{B/A} \otimes_A \Omega^q_{B/A} \otimes_B M & \longrightarrow & \Omega^{p+q}_{B/A} \otimes_B M \\ \downarrow{\scriptstyle \mathrm{id} \otimes_A \nabla \circ \nabla} & & \downarrow{\scriptstyle \nabla \circ \nabla} \\ \Omega^p_{B/A} \otimes_A \Omega^{q+2}_{B/A} \otimes_B M & \longrightarrow & \Omega^{p+q+2}_{B/A} \otimes_B M, \end{array} \qquad (2.8.6.5)$$

where the horizontal arrows are induced by the multiplication of $\Omega_{B/A}$, is commutative.

Indeed, the uniqueness of ∇ (2.8.6.1) follows from the surjectivity of the multiplication morphisms $\Omega^p_{B/A} \otimes_A \Omega^q_{B/A} \to \Omega^{p+q}_{B/A}$ $(p, q \geq 0)$. To establish its existence, consider a small filtered category J and a functor $\alpha\colon J \to$ **Mod**(B) such that $M = \text{“}\varinjlim\text{”}\alpha$ in **Ind-Mod**(B). We then use 2.8.5 and keep the same notation. For all $k \in \mathrm{Ob}(K)$, we also denote by

$$\alpha(\sigma_k)\colon \Omega_{B/A} \otimes_B M_{\beta(k)} \to \Omega_{B/A} \otimes_B M_{\gamma(k)} \qquad (2.8.6.6)$$

the B-linear morphism induced by $\alpha(\sigma_k)$ (2.8.5.1). The morphism V_k extends into a unique graded A-linear morphism of degree 1 which we also denote by

$$V_k\colon \Omega_{B/A} \otimes_B M_{\beta(k)} \to \Omega_{B/A} \otimes_B M_{\gamma(k)}, \tag{2.8.6.7}$$

such that for all local sections ω of $\Omega^p_{B/A}$ and m of $\Omega^q_{B/A} \otimes_B M_{\beta(k)}$ $(p, q \in \mathbb{N})$, we have

$$V_k(\omega \wedge m) = \lambda d(\omega) \wedge \alpha(\sigma_k)(m) + (-1)^p \omega \wedge V_k(m). \tag{2.8.6.8}$$

We thus obtain a morphism of functors which we also denote by

$$V\colon \iota \circ (\Omega_{B/A} \otimes_B \alpha \circ \beta) \to \iota \circ (\Omega_{B/A} \otimes_B \alpha \circ \gamma). \tag{2.8.6.9}$$

We take for ∇ (2.8.6.1) the direct limit of V in **Ind-Mod**(A), which satisfies the required property since the direct limit of the morphism of functors $\alpha \circ \sigma$ is the identity of M.

By 2.6.11, there exist a small filtered category L, two cofinal functors $\mathtt{s}, \mathtt{t}\colon L \rightrightarrows K$ and a morphism of functors $\tau\colon \gamma \circ \mathtt{s} \to \beta \circ \mathtt{t}$. For every $\ell \in \mathrm{Ob}(L)$, we therefore have the morphisms of J

$$\beta \circ \mathtt{s}(\ell) \xrightarrow{\sigma_{\mathtt{s}(\ell)}} \gamma \circ \mathtt{s}(\ell) \xrightarrow{\tau_\ell} \beta \circ \mathtt{s}(\ell) \xrightarrow{\sigma_{\mathtt{t}(\ell)}} \gamma \circ \mathtt{t}(\ell)\ . \tag{2.8.6.10}$$

We denote by W_ℓ the A-linear morphism of degree 2, defined by the composed morphism

$$\begin{array}{ccc} \Omega_{B/A} \otimes_B M_{\beta(\mathtt{s}(\ell))} & \xrightarrow{W_\ell} & \Omega_{B/A} \otimes_B M_{\gamma(\mathtt{t}(\ell))} \\ \Big\downarrow V_{\mathtt{s}(\ell)} & & \Big\uparrow V_{\mathtt{t}(\ell)} \\ \Omega_{B/A} \otimes_B M_{\gamma(\mathtt{s}(\ell))} & \xrightarrow{\mathrm{id} \otimes \alpha(\tau_\ell)} & \Omega_{B/A} \otimes_B M_{\beta(\mathtt{t}(\ell))}. \end{array} \tag{2.8.6.11}$$

We thus obtain a morphism of functors that we denote by

$$W\colon \iota \circ (\Omega_{B/A} \otimes_B \alpha \circ \beta \circ \mathtt{s}) \to \iota \circ (\Omega_{B/A} \otimes_B \alpha \circ \gamma \circ \mathtt{t}). \tag{2.8.6.12}$$

Since the direct limit of the morphism of functors $\alpha \circ \tau$ is the identity of M, the direct limit of W is the composed functor $\nabla \circ \nabla$.

For every $\ell \in \mathrm{Ob}(L)$ and every local section ω of $\Omega^p_{B/A}$ and m of $\Omega^q_{B/A} \otimes_B M_{\beta(\mathtt{s}(\ell))}$, we have

$$\begin{aligned} W_\ell(\omega \wedge m) = {} & \omega \wedge W_\ell(m) + (-1)^p \lambda d(\omega) \\ & \wedge [(\alpha(\sigma_{\mathtt{t}(\ell)} \circ \tau_\ell) \otimes \mathrm{id}) \circ V_{\mathtt{s}(\ell)}(m) - V_{\mathtt{t}(\ell)} \circ \alpha(\tau_\ell \circ \sigma_{\mathtt{s}(\ell)})(m)]. \end{aligned}$$

The commutativity of the diagram (2.8.6.5) follows.

Corollary 2.8.7 *Under the assumptions of* 2.8.6, *the morphism* $\nabla \circ \nabla$ (2.8.6.1) *is the image of a graded morphism of degree* 2 *of* **Ind-Mod**(B)

$$\Omega_{B/A} \otimes_B M \to \Omega_{B/A} \otimes_B M. \tag{2.8.7.1}$$

This follows from 2.7.6 and (2.8.6.5).

Definition 2.8.8 Let (X, A) be a ringed topos, B an A-algebra, $\lambda \in A(X)$, (M, ∇) an ind-B-module with λ-connection with respect to the extension B/A. We say that ∇ is *integrable* if the following equivalent conditions are satisfied:

(i) the composed morphism $\nabla \circ \nabla \colon \Omega_{B/A} \otimes_B M \to \Omega_{B/A} \otimes_B M$ vanishes (see 2.8.6);
(ii) the composed morphism

$$M \xrightarrow{\nabla} \Omega^1_{B/A} \otimes_B M \xrightarrow{\nabla} \Omega^2_{B/A} \otimes_B M \tag{2.8.8.1}$$

vanishes.

Lemma 2.8.9 *Let (X, A) be a ringed topos, B an A-algebra, $\lambda \in A(X)$, (M, ∇) an ind-B-module with λ-connection with respect to the extension B/A, B' a B-algebra, $d' \colon B' \to \Omega^1_{B'/A}$ its universal A-derivation. We denote by*

$$u \colon B' \otimes_B \Omega_{B/A} \to \Omega_{B'/A} \tag{2.8.9.1}$$

the canonical morphism of graded B'-algebras (2.5.24.1). Then, there exists a unique morphism of **Ind-Mod**(A)

$$\nabla' \colon B' \otimes_B M \to \Omega^1_{B'/A} \otimes_B M, \tag{2.8.9.2}$$

whose composition with the canonical morphism $B' \otimes_A M \to B' \otimes_B M$ is the sum of the following composed morphisms v_1 and v_2:

$$\begin{array}{ccccc}
M \otimes_A B' & \xrightarrow{\nabla \otimes_A \mathrm{id}_{B'}} & (\Omega^1_{B/A} \otimes_B M) \otimes_A B' & \longrightarrow & \Omega^1_{B/A} \otimes_B M \otimes_B B' \\
 & \searrow^{v_1} & & & \downarrow u^1 \otimes_B \mathrm{id}_M \\
 & & & & \Omega^1_{B'/A} \otimes_B M \\
 & \nearrow_{v_2} & & & \uparrow u^1 \otimes_B \mathrm{id}_M \\
B' \otimes_A M & \xrightarrow[\lambda d' \otimes_A \mathrm{id}_M]{} & (\Omega^1_{B/A} \otimes_B B') \otimes_A M & \longrightarrow & \Omega^1_{B/A} \otimes_B B' \otimes_B M
\end{array} \tag{2.8.9.3}$$

where the unlabeled arrows are the canonical morphisms. It is a λ-connection on $B' \otimes_B M$ with respect to the extension B'/A. If, moreover, the λ-connection ∇ is integrable, so is ∇'.

Indeed, the uniqueness of ∇' (2.8.9.1) follows from the surjectivity of the canonical morphism $B' \otimes_A M \to B' \otimes_B M$. To establish its existence, consider a small filtered category J and a functor $\alpha \colon J \to$ **Mod**(B) such that $M = \text{“}\varinjlim\text{”}\,\alpha$ in **Ind-Mod**(B). We then use 2.8.5 and keep the same notation. For every $k \in \mathrm{Ob}(K)$, there exists a unique A-linear morphism

$$V'_k \colon B' \otimes_B M_{\beta(k)} \to \Omega^1_{B'/A} \otimes_B M_{\gamma(k)} \tag{2.8.9.4}$$

such that for all local sections b' of B' and m of $M_{\beta(k)}$, we have

$$V'_k(b' \otimes m) = \lambda d'(b') \otimes \alpha(\sigma_k)(m) + (u \otimes \mathrm{id}_{M_{\gamma(k)}})(b' \otimes V_k(m)). \qquad (2.8.9.5)$$

We obtain a morphism of functors

$$V' : \iota \circ (B' \otimes_B \alpha \circ \beta) \to \iota \circ (\Omega^1_{B'/A} \otimes_B \alpha \circ \gamma). \qquad (2.8.9.6)$$

Since the direct limit of the morphism of functors $\alpha \circ \sigma$ is the identity of M, by taking the direct limit, we obtain a morphism of **Ind-Mod**(A)

$$\nabla' : B' \otimes_B M \to \Omega^1_{B'/A} \otimes_B M. \qquad (2.8.9.7)$$

It immediately follows from (2.8.9.5) that ∇' is a λ-connection on $B' \otimes_B M$ with respect to the extension B'/A.

For every $k \in \mathrm{Ob}(K)$, V_k extends into a unique graded A-linear morphism of degree 1 which we also denote by

$$V_k : \Omega_{B/A} \otimes_B M_{\beta(k)} \to \Omega_{B/A} \otimes_B M_{\gamma(k)}, \qquad (2.8.9.8)$$

such that for all local sections ω of $\Omega^p_{B/A}$ and m of $\Omega^q_{B/A} \otimes_B M_{\beta(k)}$ $(p, q \in \mathbb{N})$, we have

$$V_k(\omega \wedge m) = \lambda d(\omega) \wedge \alpha(\sigma_k)(m) + (-1)^p \omega \wedge V_k(m). \qquad (2.8.9.9)$$

Likewise, for every $k \in \mathrm{Ob}(K)$, V'_k extends into a unique A-linear morphism of degree 1 which we denote by

$$V'_k : \Omega_{B'/A} \otimes_B M_{\beta(k)} \to \Omega_{B'/A} \otimes_B M_{\gamma(k)}, \qquad (2.8.9.10)$$

such that for all local sections ω' of $\Omega^p_{B'/A}$ and m' of $\Omega^q_{B'/A} \otimes_B M_{\beta(k)}$ $(p, q \in \mathbb{N})$, we have

$$V'_k(\omega' \wedge m') = \lambda d'(\omega') \wedge \alpha(\sigma_k)(m') + (-1)^p \omega' \wedge V'_k(m'). \qquad (2.8.9.11)$$

By 2.6.11, there exist a small filtered category L, two cofinal functors $\mathtt{s}, \mathtt{t} : L \rightrightarrows K$ and a morphism of functors $\tau : \gamma \circ \mathtt{s} \to \beta \circ \mathtt{t}$. For every $\ell \in \mathrm{Ob}(L)$, we therefore have the morphisms of J

$$\beta \circ \mathtt{s}(\ell) \xrightarrow{\sigma_{\mathtt{s}(\ell)}} \gamma \circ \mathtt{s}(\ell) \xrightarrow{\tau_\ell} \beta \circ \mathtt{t}(\ell) \xrightarrow{\sigma_{\mathtt{t}(\ell)}} \gamma \circ \mathtt{t}(\ell). \qquad (2.8.9.12)$$

We denote by W_ℓ and W'_ℓ the A-linear morphisms of degree 2, defined by the composed morphisms

$$\begin{array}{ccc} \Omega_{B/A} \otimes_B M_{\beta(\mathtt{s}(\ell))} & \xrightarrow{W_\ell} & \Omega_{B/A} \otimes_B M_{\gamma(\mathtt{t}(\ell))} \\ \downarrow V_{\mathtt{s}(\ell)} & & \uparrow V_{\mathtt{t}(\ell)} \\ \Omega_{B/A} \otimes_B M_{\gamma(\mathtt{s}(\ell))} & \xrightarrow{\mathrm{id} \otimes \alpha(\tau_\ell)} & \Omega_{B/A} \otimes_B M_{\beta(\mathtt{t}(\ell))} \end{array} \qquad (2.8.9.13)$$

$$\begin{array}{ccc} \Omega_{B'/A} \otimes_B M_{\beta(\mathrm{s}(\ell))} & \xrightarrow{W'_\ell} & \Omega_{B'/A} \otimes_B M_{\gamma(\mathrm{t}(\ell))} \\ \Big\downarrow V'_{\mathrm{s}(\ell)} & & \Big\uparrow V'_{\mathrm{t}(\ell)} \\ \Omega_{B'/A} \otimes_B M_{\gamma(\mathrm{s}(\ell))} & \xrightarrow{\mathrm{id}\otimes\alpha(\tau_\ell)} & \Omega_{B'/A} \otimes_B M_{\beta(\mathrm{t}(\ell))}. \end{array} \tag{2.8.9.14}$$

For every local section m of $M_{\beta(\mathrm{s}(\ell))}$, we have

$$\begin{aligned} W'_\ell(m) &= V'_{\mathrm{t}(\ell)} \circ (\mathrm{id} \otimes \alpha(\tau_\ell)) \circ V'_{\mathrm{s}(\ell)}(m) \\ &= V'_{\mathrm{t}(\ell)} \circ (u \otimes \alpha(\tau_\ell)) \circ V_{\mathrm{s}(\ell)}(m) \\ &= (u \otimes \mathrm{id}_{M_{\gamma(\mathrm{t}(\ell))}}) \circ V_{\mathrm{t}(\ell)} \circ (\mathrm{id} \otimes \alpha(\tau_\ell)) \circ V_{\mathrm{s}(\ell)}(m) \\ &= (u \otimes \mathrm{id}_{M_{\gamma(\mathrm{t}(\ell))}}) \circ W_\ell(m). \end{aligned} \tag{2.8.9.15}$$

We deduce from this that if the λ-connection ∇ is integrable, so is ∇' (see 2.8.7).

2.8.10 Let $f\colon (X', A') \to (X, A)$ be a morphism of ringed topos, B an A-algebra, B' an A'-algebra, $\alpha\colon f^*(B) \to B'$ a homomorphism of A'-algebras, $\lambda \in \Gamma(X, A)$, M an ind-B-module, ∇ a λ-connection on M with respect to the extension B/A (2.8.4). Let λ' be the canonical image of λ in $\Gamma(X', A')$. The functor f^* induces a functor (2.7.10.1)

$$\mathrm{I}f^*\colon \mathbf{Ind\text{-}Mod}(B) \to \mathbf{Ind\text{-}Mod}(f^*(B)), \tag{2.8.10.1}$$

compatible with the functor $\mathrm{I}f^*\colon \mathbf{Ind\text{-}Mod}(A) \to \mathbf{Ind\text{-}Mod}(A')$ via the forgetful functor (2.7.1.6). We immediately check (2.7.10.3) that the morphism of $\mathbf{Ind\text{-}Mod}(A')$

$$\mathrm{I}f^*(\nabla)\colon \mathrm{I}f^*(M) \to \mathrm{I}f^*(M) \otimes_{f^*(B)} \Omega^1_{f^*(B)/A'} \tag{2.8.10.2}$$

is a λ'-connection on the ind-$f^*(B)$-module $\mathrm{I}f^*(M)$ with respect to the extension $f^*(B)/A'$. It follows from 2.8.6 that $(\mathrm{I}f^*(M), \mathrm{I}f^*(\nabla))$ is integrable if (M, ∇) is.

By 2.8.9, $(\mathrm{I}f^*(M), \mathrm{I}f^*(\nabla))$ defines, using α, a λ'-connection on the ind-B'-module $\mathrm{I}f^*(M) \otimes_{f^*(B)} B'$ with respect to the extension B'/A',

$$\nabla'\colon \mathrm{I}f^*(M) \otimes_{f^*(B)} B' \to \mathrm{I}f^*(M) \otimes_{f^*(B)} \Omega^1_{B'/A'}, \tag{2.8.10.3}$$

which is integrable if (M, ∇) is.

2.8.11 Let (X, A) be a ringed topos, B an A-algebra, $\lambda \in \Gamma(X, A)$, (M, ∇) an ind-B-module with λ-connection with respect to the extension B/A (2.8.4). Suppose that there exist an A-module E and a B-isomorphism $\tau\colon E \otimes_A B \xrightarrow{\sim} \Omega^1_{B/A}$ such that for every local section ω of E, we have $d(\tau(\omega \otimes 1)) = 0$. We denote by $\vartheta\colon M \to E \otimes_A M$ the morphism induced by ∇ and τ.

(i) The λ-connection ∇ is integrable (2.8.8) if and only if ϑ is a Higgs A-field on M with coefficients in E (2.8.1). Indeed, the diagram

$$\begin{array}{ccccc} E\otimes_A M & \xrightarrow{-\mathrm{id}\otimes\vartheta} & E\otimes_A E\otimes_A M & \longrightarrow & (\wedge^2 E)\otimes_A M \\ \downarrow & & & & \downarrow \\ \Omega^1_{B/A}\otimes_B M & & \xrightarrow{\quad\nabla\quad} & & \Omega^2_{B/A}\otimes_B M. \end{array} \tag{2.8.11.1}$$

where the vertical arrows are the isomorphisms induced by τ, is clearly commutative.

(ii) Let (N,θ) be a Higgs ind-A-module with coefficients in E. The morphism of **Ind-Mod**(A)

$$\nabla'\colon M\otimes_A N\to\Omega^1_{B/A}\otimes_B M\otimes_A N \tag{2.8.11.2}$$

defined by

$$\nabla'=\nabla\otimes_A\mathrm{id}_N+(\tau\otimes_B\mathrm{id}_{M\otimes_A N})(\mathrm{id}_M\otimes_A\theta), \tag{2.8.11.3}$$

is a λ-connection on $M\otimes_A N$ with respect to the extension B/A. If ∇ is integrable, so is ∇'.

Lemma 2.8.12 *Let (X,A) be a ringed topos, B an A-algebra, $\lambda\in\Gamma(X,A)$, (M,∇) an ind-B-module with λ-connection with respect to the extension B/A, B' a B-algebra, $\delta'\colon B'\to\Omega^1_{B/A}\otimes_B B'$ an A-derivation such that for any local section b of B, we have $\delta'(b)=\lambda d(b)\otimes 1$. Then,*

(i) *There is a unique morphism of* **Ind-Mod**(A)

$$\nabla'\colon B'\otimes_B M\to\Omega^1_{B/A}\otimes_B B'\otimes_B M, \tag{2.8.12.1}$$

such that the composition with the canonical morphism $B'\otimes_A M\to B'\otimes_B M$ is the sum of the morphisms

$$M\otimes_A B'\xrightarrow{\nabla\otimes_A\mathrm{id}_{B'}}(\Omega^1_{B/A}\otimes_B M)\otimes_A B'\longrightarrow\Omega^1_{B/A}\otimes_B M\otimes_B B', \tag{2.8.12.2}$$

$$B'\otimes_A M\xrightarrow{\delta'\otimes_A\mathrm{id}_M}(\Omega^1_{B/A}\otimes_B B')\otimes_A M\longrightarrow\Omega^1_{B/A}\otimes_B B'\otimes_B M, \tag{2.8.12.3}$$

where the second arrows are the canonical morphisms.

(ii) *Suppose, moreover, that there exist an A-module E and a B-isomorphism $\tau\colon E\otimes_A B\xrightarrow{\sim}\Omega^1_{B/A}$ such that for every local section ω of E, we have $d(\tau(\omega\otimes 1))=0$. We denote by $\vartheta'\colon B'\otimes_B M\to E\otimes_A B'\otimes_B M$ (resp. $\theta'\colon B'\to E\otimes_A B'$) the morphism induced by ∇' (resp. δ') and τ. If ∇ is integrable and if θ' is a Higgs A-field with coefficients in E, then ϑ' is a Higgs A-field with coefficients in E.*

(i) Indeed, the uniqueness of ∇' (2.8.12.1) follows from the surjectivity of the canonical morphism $M\otimes_A B'\to M\otimes_B B'$. To establish its existence, consider a small filtered category J and a functor $\alpha\colon J\to$ **Mod**(B) such that $M=\text{“}\varinjlim\text{”}\alpha$ in **Ind-Mod**(B). We then use 2.8.5 and keep the same notation. For every $k\in\mathrm{Ob}(K)$, there exists a unique A-linear morphism

$$V'_k\colon B'\otimes_B M_{\beta(k)}\to\Omega^1_{B/A}\otimes_B B'\otimes_B M_{\gamma(k)} \tag{2.8.12.4}$$

such that for all local sections b' of B' and m of $M_{\beta(k)}$, we have

$$V'_k(b'\otimes_B m)=\delta'(b')\otimes_B \alpha(\sigma_k)(m)+b'\otimes_B V_k(m). \tag{2.8.12.5}$$

We thus define a morphism of functors

$$V'\colon \iota\circ(B'\otimes_B\alpha\circ\beta)\to\iota\circ(\Omega^1_{B/A}\otimes_B B'\otimes_B\alpha\circ\gamma). \tag{2.8.12.6}$$

We take for ∇' (2.8.12.1) the direct limit of V', which clearly satisfies the required property.

(ii) We keep the previous notation. For any $k\in\mathrm{Ob}(K)$, we denote by

$$\vartheta_k\colon M_{\beta(k)}\to M_{\gamma(k)}\otimes_A E, \tag{2.8.12.7}$$

$$\vartheta'_k\colon M_{\beta(k)}\otimes_B B'\to M_{\gamma(k)}\otimes_B B'\otimes_A E, \tag{2.8.12.8}$$

the morphisms induced by V_k, V'_k and τ. The morphism ϑ_k induces a unique A-linear morphism, which we also denote by

$$\vartheta_k\colon M_{\beta(k)}\otimes_A E\to M_{\gamma(k)}\otimes_A\wedge^2 E, \tag{2.8.12.9}$$

such that for all local sections m of $M_{\beta(k)}\otimes_A\wedge^p E$ and ω of $\wedge^q E$, we have $\vartheta_k(m\otimes\omega)=\vartheta_k(m)\wedge\omega$. Likewise, θ' and ϑ'_k canonically induce A-linear morphisms that we also denote by

$$\theta'\colon B'\otimes_A E\to B'\otimes_A\wedge^2 E, \tag{2.8.12.10}$$

$$\vartheta'_k\colon M_{\beta(k)}\otimes_B B'\otimes_A E\to M_{\gamma(k)}\otimes_B B'\otimes_A\wedge^2 E. \tag{2.8.12.11}$$

By 2.6.11, there exist a small filtered category L, two cofinal functors $\mathtt{s},\mathtt{t}\colon L\rightrightarrows K$ and a morphism of functors $\tau\colon\gamma\circ\mathtt{s}\to\beta\circ\mathtt{t}$. For every $\ell\in\mathrm{Ob}(L)$, we therefore have the morphisms of J

$$\beta\circ\mathtt{s}(\ell)\xrightarrow{\sigma_{\mathtt{s}(\ell)}}\gamma\circ\mathtt{s}(\ell)\xrightarrow{\tau_\ell}\beta\circ\mathtt{t}(\ell)\xrightarrow{\sigma_{\mathtt{t}(\ell)}}\gamma\circ\mathtt{t}(\ell)\,. \tag{2.8.12.12}$$

We denote by W_ℓ and W'_ℓ the A-linear morphisms defined by the composed morphisms

$$\begin{array}{ccc} M_{\beta(\mathtt{s}(\ell))} & \xrightarrow{W_\ell} & M_{\gamma(\mathtt{t}(\ell))}\otimes_A\wedge^2 E \\ \Big\downarrow{\scriptstyle\vartheta_{\mathtt{s}(\ell)}} & & \Big\uparrow{\scriptstyle\vartheta_{\mathtt{t}(\ell)}} \\ M_{\gamma(\mathtt{s}(\ell))}\otimes_A E & \xrightarrow{\alpha(\tau_\ell)\otimes\mathrm{id}} & M_{\beta(\mathtt{t}(\ell))}\otimes_A E \end{array} \tag{2.8.12.13}$$

$$\begin{array}{ccc} M_{\beta(\mathtt{s}(\ell))}\otimes_B B' & \xrightarrow{W'_\ell} & M_{\gamma(\mathtt{t}(\ell))}\otimes_B B'\otimes_A\wedge^2 E \\ \Big\downarrow{\scriptstyle\vartheta'_{\mathtt{s}(\ell)}} & & \Big\uparrow{\scriptstyle\vartheta'_{\mathtt{t}(\ell)}} \\ M_{\gamma(\mathtt{s}(\ell))}\otimes_B B'\otimes_A E & \xrightarrow{\alpha(\tau_\ell)\otimes\mathrm{id}} & M_{\beta(\mathtt{t}(\ell))}\otimes_B B'\otimes_A E. \end{array} \tag{2.8.12.14}$$

These are morphisms of functors. Since the direct limit of the morphism of functors $\alpha \circ \tau$ is the identity of M, we see that the direct limit of $\ell \mapsto W_\ell$ vanishes.

Furthermore, for all local sections m of $M_{\beta(\mathsf{s}(\ell))}$ and b' of B', since $\theta'^2(b') = 0$, we have

$$W'_\ell(m \otimes_B b') = W_\ell(m) \otimes b' - (\alpha(\sigma_{\mathsf{t}(\ell)} \circ \tau_\ell) \otimes \mathrm{id}) \circ \vartheta_{\mathsf{s}(\ell)}(m) \wedge \theta'(b') + \vartheta_{\mathsf{t}(\ell)} \circ \alpha(\tau_\ell \circ \sigma_{\mathsf{s}(\ell)})(m) \wedge \theta'(b').$$

Since the direct limit of the morphism of functors $\ell \mapsto \alpha(\tau_\ell \circ \sigma_{\mathsf{s}(\ell)})$ (resp. $\ell \mapsto \alpha(\sigma_{\mathsf{t}(\ell)} \circ \tau_\ell)$) is the identity of M, we deduce from this that the direct limit of the morphism of functors $\ell \mapsto W'_\ell$ vanishes, which proves that ϑ' is a Higgs A-field (2.8.12.14).

2.9 Modules up to Isogeny

Definition 2.9.1 Let $\mathscr{C}$ be an additive category.

(i) A morphism $u\colon M \to N$ of $\mathscr{C}$ is called an *isogeny* if there exist an integer $n \neq 0$ and a morphism $v\colon N \to M$ of $\mathscr{C}$ such that $v \circ u = n \cdot \mathrm{id}_M$ and $u \circ v = n \cdot \mathrm{id}_N$.
(ii) An object M of $\mathscr{C}$ is said to be of *finite exponent* if there exists an integer $n \neq 0$ such that $n \cdot \mathrm{id}_M = 0$.

Let us complete the terminology and make a few remarks:

2.9.1.1 The family of isogenies of $\mathscr{C}$ allows a two-sided calculus of fractions ([37] I 1.4.2). We call the *category of objects of $\mathscr{C}$ up to isogeny*, and denote by $\mathscr{C}_{\mathbb{Q}}$, the localized category of $\mathscr{C}$ with respect to isogenies. We denote by

$$Q\colon \mathscr{C} \to \mathscr{C}_{\mathbb{Q}},\ X \mapsto X_{\mathbb{Q}}, \qquad (2.9.1.2)$$

the localization functor. One easily verifies that for all $X, Y \in \mathrm{Ob}(\mathscr{C})$, we have

$$\mathrm{Hom}_{\mathscr{C}_{\mathbb{Q}}}(X_{\mathbb{Q}}, Y_{\mathbb{Q}}) = \mathrm{Hom}_{\mathscr{C}}(X, Y) \otimes_{\mathbb{Z}} \mathbb{Q}. \qquad (2.9.1.3)$$

In particular, the category $\mathscr{C}_{\mathbb{Q}}$ is additive and the localization functor is additive. An object X of $\mathscr{C}$ is of finite exponent if and only if $X_{\mathbb{Q}}$ is zero. A morphism f of $\mathscr{C}$ is an isogeny if and only if $f_{\mathbb{Q}}$ is an isomorphism of $\mathscr{C}_{\mathbb{Q}}$.

2.9.1.4 If $\mathscr{C}$ is an abelian category, the category $\mathscr{C}_{\mathbb{Q}}$ is abelian and the localization functor $Q\colon \mathscr{C} \to \mathscr{C}_{\mathbb{Q}}$ is exact. In fact, $\mathscr{C}_{\mathbb{Q}}$ identifies canonically with the quotient category of $\mathscr{C}$ by the thick subcategory $\mathscr{E}$ of objects of finite exponent. Indeed, denote by $\mathscr{C}/\mathscr{E}$ the quotient category of $\mathscr{C}$ by $\mathscr{E}$ and by $T\colon \mathscr{C} \to \mathscr{C}/\mathscr{E}$ the canonical functor ([22] III §1). For every $X \in \mathrm{Ob}(\mathscr{E})$, we have $Q(X) = 0$. Consequently, there exists a unique functor $Q'\colon \mathscr{C}/\mathscr{E} \to \mathscr{C}_{\mathbb{Q}}$ such that $Q = Q' \circ T$. On the other hand, for every $X \in \mathrm{Ob}(\mathscr{C})$ and every integer $n \neq 0$, $T(n \cdot \mathrm{id}_X)$ is an isomorphism. Hence there exists a unique functor $T'\colon \mathscr{C}_{\mathbb{Q}} \to \mathscr{C}/\mathscr{E}$ such that $T = T' \circ Q$. We immediately see that T'

and Q' are equivalences of categories that are quasi-inverse to each other. The functor Q, which identifies with the functor T, is therefore exact ([22] III prop. 1).

2.9.1.5 Every additive (resp. exact) functor between additive (resp. abelian) categories $\mathscr{C} \to \mathscr{C}'$ uniquely extends to an additive (resp. exact) functor $\mathscr{C}_{\mathbb{Q}} \to \mathscr{C}'_{\mathbb{Q}}$, compatible with the localization functors.

2.9.2 Let $\mathscr{C}$ be an abelian $\mathbb{U}$-category (2.1.7), I the bilateral multiplicative system of the isogenies of $\mathscr{C}$, $\mathscr{C}_{\mathbb{Q}}$ the category of objects of $\mathscr{C}$ up to isogeny, which is an abelian $\mathbb{U}$-category,

$$Q\colon \mathscr{C} \to \mathscr{C}_{\mathbb{Q}},\ X \mapsto X_{\mathbb{Q}}, \tag{2.9.2.1}$$

the canonical functor. For any object X of $\mathscr{C}$, we denote by I^X the category of isogenies of source X and by

$$\alpha^X\colon I^X \to \mathscr{C}, \quad (X \to X') \mapsto X', \tag{2.9.2.2}$$

the target functor (see 2.6.6). The category I^X is filtered ([40] 7.1.10) and essentially small (2.6.2). Indeed, the endomorphisms of X defined by the multiplication by a nonzero integer form a small cofinal subset of $\mathrm{Ob}(I^X)$. We can therefore consider the fully faithful functor (2.6.6.5)

$$\alpha\colon \mathscr{C}_{\mathbb{Q}} \to \mathbf{Ind}(\mathscr{C}), \quad X \mapsto \text{“}\varinjlim\text{”}\alpha^X. \tag{2.9.2.3}$$

We recall that the canonical morphism (2.6.6.7)

$$\iota_{\mathscr{C}} \to \alpha \circ Q \tag{2.9.2.4}$$

is not an isomorphism in general.

Lemma 2.9.3 *We keep the assumptions and notation of* 2.9.2. *Moreover, let us denote by M the multiplicative monoid $\mathbb{Z} - \{0\}$ and by $\underline{M}$ the filtered category whose objects are the elements of M and the morphisms are determined by the divisibility relation in M. Then,*

(i) *For every object X of $\mathscr{C}$, the functor $\mu^X\colon \underline{M} \to I^X$ which sends a nonzero integer n to the isogeny defined by multiplication by n in X, is cofinal.*
(ii) *The functor α (2.9.2.3) is exact.*
(iii) *For every injective object X of $\mathscr{C}$, $Q(X)$ is injective and $\alpha(Q(X))$ is quasi-injective.*

(i) Indeed, by ([5] I 8.1.3(b)), since the category $\underline{M}$ is filtered, it suffices to show that the functor μ^X satisfies the properties F1) and F2) of ([5] I 8.1.3). Let $u\colon X \to X'$ be an isogeny of $\mathscr{C}$. There exist a morphism $v\colon X' \to X$ of $\mathscr{C}$ and an element n of M such that $v \circ u = n \cdot \mathrm{id}_X$ and $u \circ v = n \cdot \mathrm{id}_{X'}$. Consequently, v defines a morphism from $(u\colon X \to X')$ to $\mu^X([n])$ in I^X; the property F1) follows.

Let $m \in M$ and $v_1, v_2\colon X' \rightrightarrows X$ be two morphisms of $\mathscr{C}$ such that $m \cdot \mathrm{id}_X = v_1 \circ u = v_2 \circ u$. We have $m \cdot v = n \cdot v_1 = n \cdot v_2$ in $\mathscr{C}$. Consequently, we have $\mu^X([m] \to [mn]) \circ v_1 = \mu^X([m] \to [mn]) \circ v_2$ in I^X; the property F2) follows.

(ii) Indeed, it follows from (i) and 2.6.7.5 that the functor $\alpha \circ Q$ is exact. The proposition follows by virtue of ([22] III cor. 1 to prop. 1).

(iii) The first assertion follows from ([22] III cor. 1 to prop. 1) and the second assertion is a consequence of (i) and 2.6.8.2.

2.9.4 Let $\phi\colon \mathscr{C} \to \mathscr{C}'$ be an additive functor between abelian $\mathbb{U}$-categories. We take the notation of 2.9.2 for $\mathscr{C}$ and we consider the analogous notation for $\mathscr{C}'$, which we endow with an exponent $'$. For every object X of $\mathscr{C}$, ϕ induces a functor

$$\phi^X\colon I^X \to I'^{\phi(X)} \tag{2.9.4.1}$$

that fits into a strictly commutative diagram

$$\begin{array}{ccc} I^X & \xrightarrow{\phi^X} & I'^{\phi(X)} \\ {\scriptstyle\alpha^X}\downarrow & & \downarrow{\scriptstyle\alpha'^{\phi(X)}} \\ \mathscr{C} & \xrightarrow{\phi} & \mathscr{C}'. \end{array} \tag{2.9.4.2}$$

Lemma 2.9.5 *We keep the assumptions of* 2.9.4. *Then,*

(i) *For every object X of $\mathscr{C}$, the functor ϕ^X is cofinal.*
(ii) *The diagram*

$$\begin{array}{ccc} \mathscr{C}_{\mathbb{Q}} & \xrightarrow{\phi_{\mathbb{Q}}} & \mathscr{C}'_{\mathbb{Q}} \\ {\scriptstyle\alpha}\downarrow & & \downarrow{\scriptstyle\alpha'} \\ \mathbf{Ind}(\mathscr{C}) & \xrightarrow{\mathrm{I}\phi} & \mathbf{Ind}(\mathscr{C}'), \end{array} \tag{2.9.5.1}$$

where $\phi_{\mathbb{Q}}$ and $\mathrm{I}\phi$ are the additive functors induced by ϕ, and α and α' are the canonical functors (2.9.2.2), *is commutative up to canonical isomorphism.*

(i) Indeed, by ([5] I 8.1.3(b)), since the category I^X is filtered, it suffices to show that the functor ϕ^X satisfies properties F1) and F2) of ([5] I 8.1.3). Let $u\colon \phi(X) \to Y$ be an isogeny of $\mathscr{C}'$. There exist a morphism $v\colon Y \to \phi(X)$ and a nonzero integer n such that $v \circ u = n \cdot \mathrm{id}_{\phi(X)}$ and $u \circ v = n \cdot \mathrm{id}_Y$. Consequently, v defines a morphism from u to $\phi^X(n \cdot \mathrm{id}_X)$ in $I'^{\phi(X)}$; the property F1) follows.

Let $w\colon X \to X'$ be an isogeny of $\mathscr{C}$ and $v_1, v_2\colon Y \rightrightarrows \phi(X')$ two morphisms of $\mathscr{C}'$ such that $\phi(w) = v_1 \circ u = v_2 \circ u$. We have $\phi(w) \circ v = n \cdot v_1 = n \cdot v_2$. Hence, $\phi^X(n \cdot \mathrm{id}_{X'}) \circ v_1 = \phi^X(n \cdot \mathrm{id}_{X'}) \circ v_2$; the property F2) follows.

(ii) Indeed, we have canonical isomorphisms

$$\mathrm{I}\phi(\text{“}\varinjlim\text{”}\alpha^X) \xrightarrow{\sim} \text{“}\varinjlim\text{”}(\phi \circ \alpha^X) \xrightarrow{\sim} \text{“}\varinjlim\text{”}(\alpha'^{\phi(X)} \circ \phi^X) \xrightarrow{\sim} \text{“}\varinjlim\text{”}\alpha'^{\phi(X)}, \tag{2.9.5.2}$$

the first is (2.6.4.3), the second is underlying the commutative diagram (2.9.4.2) and the third is the canonical morphism, which is an isomorphism by (i).

2.9.6 Let $\phi\colon \mathscr{C} \to \mathscr{C}'$ be a left exact functor between abelian $\mathbb{U}$-categories such that $\mathscr{C}$ has enough injectives. By ([22] III cor. 1 to prop. 1), the additive functor $\phi_{\mathrm{Q}}\colon \mathscr{C}_{\mathrm{Q}} \to \mathscr{C}'_{\mathrm{Q}}$, induced by ϕ (2.9.2), is left exact. By 2.9.3(iii), the canonical functor $Q\colon \mathscr{C} \to \mathscr{C}_{\mathrm{Q}}$ (2.9.2.1) transforms injective objects into injective objects. We denote by $\mathscr{I}$ the full subcategory of $\mathscr{C}$ made up of injective objects and by $\mathscr{I}_{\mathrm{Q}}$ the full subcategory of $\mathscr{C}_{\mathrm{Q}}$ made up of the images by Q of the injective objects of $\mathscr{C}$. Then, $\mathscr{I}$ (resp. $\mathscr{I}_{\mathrm{Q}}$) is cogenerating in $\mathscr{C}$ (resp. $\mathscr{C}_{\mathrm{Q}}$), in particular $\mathscr{C}_{\mathrm{Q}}$ has enough injectives. Moreover, $\mathscr{I}$ is ϕ-injective by ([40] 13.3.6(iii)), and $\mathscr{I}_{\mathrm{Q}}$ is ϕ_{Q}-injective by ([40] 13.3.7) and ([22] III cor. 1 to prop. 1). Therefore, by ([40] 13.3.5), the functors ϕ and ϕ_{Q} admit right derived functors

$$\mathrm{R}\phi\colon \mathbf{D}^+(\mathscr{C}) \to \mathbf{D}^+(\mathscr{C}'), \qquad (2.9.6.1)$$

$$\mathrm{R}(\phi_{\mathrm{Q}})\colon \mathbf{D}^+(\mathscr{C}_{\mathrm{Q}}) \to \mathbf{D}^+(\mathscr{C}'_{\mathrm{Q}}). \qquad (2.9.6.2)$$

To justify the existence of $\mathrm{R}(\phi_{\mathrm{Q}})$, we can more simply use the fact that $\mathscr{C}_{\mathrm{Q}}$ has enough injectives. However, the introduction of $\mathscr{I}_{\mathrm{Q}}$ is necessary for (2.9.6.7) below.

Since the functors Q and Q' are exact and transform injective objects into injective objects, we have canonical isomorphisms

$$\mathrm{R}(\phi_{\mathrm{Q}}) \circ Q' \xrightarrow{\sim} \mathrm{R}(\phi_{\mathrm{Q}} \circ Q') \xrightarrow{\sim} \mathrm{R}(Q \circ \phi) \xrightarrow{\sim} Q \circ \mathrm{R}(\phi), \qquad (2.9.6.3)$$

where the first and the last isomorphisms result from ([40] 13.3.13) and the central isomorphism is induced by the definition of ϕ_{Q}. The diagram

$$\begin{array}{ccc} \mathbf{D}^+(\mathscr{C}) & \xrightarrow{\mathrm{R}(\phi)} & \mathbf{D}^+(\mathscr{C}') \\ {\scriptstyle Q}\downarrow & & \downarrow{\scriptstyle Q'} \\ \mathbf{D}^+(\mathscr{C}_{\mathrm{Q}}) & \xrightarrow{\mathrm{R}(\phi_{\mathrm{Q}})} & \mathbf{D}^+(\mathscr{C}'_{\mathrm{Q}}) \end{array} \qquad (2.9.6.4)$$

is therefore commutative up to canonical isomorphism. In particular, for every integer i, the functor $\mathrm{R}^i(\phi_{\mathrm{Q}})$ is obtained by localization from the functor $\mathrm{R}^i\phi$.

The category $\mathbf{Ind}(\mathscr{I})$ is identified with the full subcategory of $\mathbf{Ind}(\mathscr{C})$ made up of the quasi-injective objects (2.6.8). By virtue of ([40] 15.3.2), $\mathbf{Ind}(\mathscr{I})$ is $\mathrm{I}\phi$-injective and the functor $\mathrm{I}\phi$ admits a right derived functor

$$\mathrm{R}(\mathrm{I}\phi)\colon \mathbf{D}^+(\mathbf{Ind}(\mathscr{C})) \to \mathbf{D}^+(\mathbf{Ind}(\mathscr{C}')). \qquad (2.9.6.5)$$

It follows from 2.6.9 that for every integer i, we have a canonical isomorphism $\mathrm{R}^i(\mathrm{I}\phi) \xrightarrow{\sim} \mathrm{I}(\mathrm{R}^i\phi)$.

By 2.9.3, the functor $\alpha\colon \mathscr{C}_{\mathrm{Q}} \to \mathbf{Ind}(\mathscr{C})$ (2.9.2.3) is exact and we have $\alpha(\mathscr{I}_{\mathrm{Q}}) \subset \mathbf{Ind}(\mathscr{I})$. Therefore, we have canonical isomorphisms

$$\mathrm{R}(\mathrm{I}\phi) \circ \alpha \xrightarrow{\sim} \mathrm{R}((\mathrm{I}\phi) \circ \alpha) \xrightarrow{\sim} \mathrm{R}(\alpha' \circ \phi_{\mathrm{Q}}) \xrightarrow{\sim} \alpha' \circ \mathrm{R}(\phi_{\mathrm{Q}}), \qquad (2.9.6.6)$$

where the first and the last isomorphisms follow from ([40] 13.3.13) and the central isomorphism is induced by the isomorphism underlying (2.9.5.1). The diagram

$$\begin{array}{ccc} \mathbf{D}^+(\mathscr{C}_{\mathbb{Q}}) & \xrightarrow{\mathrm{R}(\phi_{\mathbb{Q}})} & \mathbf{D}^+(\mathscr{C}'_{\mathbb{Q}}) \\ \Big\downarrow{\scriptstyle \alpha} & & \Big\downarrow{\scriptstyle \alpha'} \\ \mathbf{D}^+(\mathbf{Ind}(\mathscr{C})) & \xrightarrow{\mathrm{R}(\mathrm{I}\phi)} & \mathbf{D}^+(\mathbf{Ind}(\mathscr{C}')) \end{array} \tag{2.9.6.7}$$

is therefore commutative up to a canonical isomorphism. In particular, for every integer i, the diagram

$$\begin{array}{ccc} \mathscr{C}_{\mathbb{Q}} & \xrightarrow{\mathrm{R}^i(\phi_{\mathbb{Q}})} & \mathscr{C}'_{\mathbb{Q}} \\ \Big\downarrow{\scriptstyle \alpha} & & \Big\downarrow{\scriptstyle \alpha'} \\ \mathbf{Ind}(\mathscr{C}) & \xrightarrow{\mathrm{R}^i(\mathrm{I}\phi)} & \mathbf{Ind}(\mathscr{C}') \end{array} \tag{2.9.6.8}$$

is commutative up to canonical isomorphism.

2.9.7 Let (X, A) be a ringed $\mathbb{U}$-topos. We denote by $\mathbf{Mod}_{\mathbb{Q}}(A)$ the category of A-modules of X up to isogeny (2.9.2) and by

$$Q_A \colon \mathbf{Mod}(A) \to \mathbf{Mod}_{\mathbb{Q}}(A), \quad M \mapsto M_{\mathbb{Q}}, \tag{2.9.7.1}$$

the canonical functor (2.9.2.1). The tensor product of $\mathbf{Mod}(A)$ induces a bifunctor

$$\mathbf{Mod}_{\mathbb{Q}}(A) \times \mathbf{Mod}_{\mathbb{Q}}(A) \to \mathbf{Mod}_{\mathbb{Q}}(A), \quad (M, N) \mapsto M \otimes_{A_{\mathbb{Q}}} N, \tag{2.9.7.2}$$

turning $\mathbf{Mod}_{\mathbb{Q}}(A)$ into a symmetric monoidal category, having $A_{\mathbb{Q}}$ as unit object. Objects in $\mathbf{Mod}_{\mathbb{Q}}(A)$ will also be called $A_{\mathbb{Q}}$-*modules*. This terminology is justified by considering $A_{\mathbb{Q}}$ as a monoid of $\mathbf{Mod}_{\mathbb{Q}}(A)$. The bifunctor (2.9.7.2) is right exact.

We have a canonical functor (2.9.2.3)

$$\alpha_A \colon \mathbf{Mod}_{\mathbb{Q}}(A) \to \mathbf{Ind}\text{-}\mathbf{Mod}(A), \tag{2.9.7.3}$$

which is exact by 2.9.3(ii). In view of 2.9.3(i), for all A-modules M and N, we have canonical isomorphisms

$$\alpha_A(M_{\mathbb{Q}}) \otimes_A N \xrightarrow{\sim} \alpha_A(M_{\mathbb{Q}}) \otimes_A \alpha_A(N_{\mathbb{Q}}) \xrightarrow{\sim} \alpha_A(M_{\mathbb{Q}} \otimes_{A_{\mathbb{Q}}} N_{\mathbb{Q}}). \tag{2.9.7.4}$$

2.9.8 Let $f \colon (Y, B) \to (X, A)$ be a morphism of ringed $\mathbb{U}$-topos. The adjoint functors f^* and f_* between the categories $\mathbf{Mod}(A)$ and $\mathbf{Mod}(B)$ induce adjoint additive functors

$$f^*_{\mathbb{Q}} \colon \mathbf{Mod}_{\mathbb{Q}}(A) \to \mathbf{Mod}_{\mathbb{Q}}(B), \tag{2.9.8.1}$$

$$f_{\mathbb{Q}*} \colon \mathbf{Mod}_{\mathbb{Q}}(B) \to \mathbf{Mod}_{\mathbb{Q}}(A), \tag{2.9.8.2}$$

whose first (resp. second) is right (resp. left) exact. By 2.9.6, the functor $f_{\mathbb{Q}*}$ admits a right derived functor

$$\mathrm{R}f_{\mathbb{Q}*} \colon \mathbf{D}^+(\mathbf{Mod}_{\mathbb{Q}}(B)) \to \mathbf{D}^+(\mathbf{Mod}_{\mathbb{Q}}(A)). \tag{2.9.8.3}$$

The diagram

$$\begin{array}{ccc} \mathbf{D}^+(\mathbf{Mod}(A)) & \xrightarrow{\mathrm{R}f_*} & \mathbf{D}^+(\mathbf{Mod}(B)) \\ \downarrow{\scriptstyle Q_A} & & \downarrow{\scriptstyle Q_B} \\ \mathbf{D}^+(\mathbf{Mod}_{\mathbb{Q}}(A)) & \xrightarrow{\mathrm{R}f_{\mathbb{Q}*}} & \mathbf{D}^+(\mathbf{Mod}_{\mathbb{Q}}(B)) \end{array} \tag{2.9.8.4}$$

where Q_A and Q_B are the canonical functors (2.9.7.1), is commutative up to canonical isomorphism.

By 2.9.5(ii), the diagram

$$\begin{array}{ccc} \mathbf{Mod}_{\mathbb{Q}}(A) & \xrightarrow{f^*_{\mathbb{Q}}} & \mathbf{Mod}_{\mathbb{Q}}(B) \\ \downarrow{\scriptstyle \alpha_A} & & \downarrow{\scriptstyle \alpha_B} \\ \mathbf{Ind\text{-}Mod}(A) & \xrightarrow{\mathrm{I}f^*} & \mathbf{Ind\text{-}Mod}(B) \end{array} \tag{2.9.8.5}$$

where α_A and α_B are the canonical functors (2.9.7.3), is commutative up to canonical isomorphism. By virtue of (2.9.6.7), the diagram

$$\begin{array}{ccc} \mathbf{D}^+(\mathbf{Mod}_{\mathbb{Q}}(B)) & \xrightarrow{\mathrm{R}(f_{\mathbb{Q}*})} & \mathbf{D}^+(\mathbf{Mod}_{\mathbb{Q}}(A)) \\ \downarrow{\scriptstyle \alpha_B} & & \downarrow{\scriptstyle \alpha_A} \\ \mathbf{D}^+(\mathbf{Ind\text{-}Mod}(B)) & \xrightarrow{\mathrm{R}(\mathrm{I}f_*)} & \mathbf{D}^+(\mathbf{Ind\text{-}Mod}(A)) \end{array} \tag{2.9.8.6}$$

is commutative up to canonical isomorphism.

2.9.9 Let (X, A) be a ringed topos, E an A-module. We call a *Higgs A-isogeny with coefficients in E* a quadruple

$$(M, N, u\colon M \to N, \theta\colon M \to N \otimes_A E) \tag{2.9.9.1}$$

consisting of two A-modules M and N and two A-linear morphisms u and θ satisfying the following property: there exist an integer $n \neq 0$ and an A-linear morphism $v\colon N \to M$ such that $v \circ u = n \cdot \mathrm{id}_M$, $u \circ v = n \cdot \mathrm{id}_N$ and such that $(M, (v \otimes \mathrm{id}_E) \circ \theta)$ and $(N, \theta \circ v)$ are Higgs A-modules with coefficients in E. Note that u induces an isogeny of Higgs modules between $(M, (v \otimes \mathrm{id}_E) \circ \theta)$ and $(N, \theta \circ v)$ (2.9.1), which explains the terminology. Let (M, N, u, θ), (M', N', u', θ') be two Higgs A-isogenies with coefficients in E. A morphism from (M, N, u, θ) to (M', N', u', θ') consists of two A-linear morphisms $\alpha\colon M \to M'$ and $\beta\colon N \to N'$ such that $\beta \circ u = u' \circ \alpha$ and $(\beta \otimes \mathrm{id}_E) \circ \theta = \theta' \circ \alpha$. We denote by $\mathbf{HI}(A, E)$ the category of Higgs A-isogenies with coefficients in E. This is an additive category. We denote by $\mathbf{HI}_{\mathbb{Q}}(A, E)$ the category of objects of $\mathbf{HI}(A, E)$ up to isogeny (2.9.1).

Let (M, N, u, θ) be a Higgs A-isogeny with coefficients in E. For any $i \geq 1$, we denote by

$$\theta_i\colon M \otimes_A \wedge^i E \to N \otimes_A \wedge^{i+1} E \tag{2.9.9.2}$$

the A-linear morphism defined for all sections m of M and ω of $\wedge^i E$ by $\theta_i(m \otimes \omega) = \theta(m) \wedge \omega$. We set

$$\overline{\theta}_i = (u_{\mathbb{Q}}^{-1} \otimes \mathrm{id}_{\wedge^{i+1}E}) \circ \theta_i \colon M_{\mathbb{Q}} \otimes_{A_{\mathbb{Q}}} (\wedge^i E)_{\mathbb{Q}} \to M_{\mathbb{Q}} \otimes_{A_{\mathbb{Q}}} (\wedge^{i+1} E)_{\mathbb{Q}}. \tag{2.9.9.3}$$

We immediately show that $\overline{\theta}_{i+1} \circ \overline{\theta}_i = 0$ ([3] III.6.9). We call the *Dolbeault* complex of (M, N, u, θ) and denote by $\mathbb{K}^\bullet(M, N, u, \theta)$ the complex of cochains of $\mathbf{Mod}_{\mathbb{Q}}(A)$

$$M_{\mathbb{Q}} \xrightarrow{\overline{\theta}_0} M_{\mathbb{Q}} \otimes_{A_{\mathbb{Q}}} E_{\mathbb{Q}} \xrightarrow{\overline{\theta}_1} M_{\mathbb{Q}} \otimes_{A_{\mathbb{Q}}} (\wedge^2 E)_{\mathbb{Q}} \to \dots, \tag{2.9.9.4}$$

where $M_{\mathbb{Q}}$ is placed in degree 0 and the differentials are of degree 1. We thus obtain a functor from the category $\mathbf{HI}(A, E)$ to the category of complexes of $\mathbf{Mod}_{\mathbb{Q}}(A)$. Any isogeny of $\mathbf{HI}(A, E)$ induces an isomorphism of the associated Dolbeault complexes. The "Dolbeault complex" functor therefore induces a functor from $\mathbf{HI}_{\mathbb{Q}}(A, E)$ to the category of complexes of $\mathbf{Mod}_{\mathbb{Q}}(A)$.

Let (M, N, u, θ), (M', N', u', θ') be two Higgs A-isogenies with coefficients in E. Setting

$$\theta_{\mathrm{tot}} = \theta \otimes_A u' + u \otimes_A \theta' \colon M \otimes_A M' \to N \otimes_A N' \otimes_A E, \tag{2.9.9.5}$$

we immediately see that $(M \otimes_A M', N \otimes_A N', u \otimes_A u', \theta_{\mathrm{tot}})$ is a Higgs A-isogeny with coefficients in E. We thus define a biadditive functor that we denote by

$$\mathbf{HI}(A, E) \times \mathbf{HI}(A, E) \to \mathbf{HI}(A, E), \quad (I, I') \mapsto I \otimes_A I'. \tag{2.9.9.6}$$

This naturally extends to a biadditive functor that we still denote by

$$\mathbf{HI}_{\mathbb{Q}}(A, E) \times \mathbf{HI}_{\mathbb{Q}}(A, E) \to \mathbf{HI}_{\mathbb{Q}}(A, E), \quad (I, I') \mapsto I \otimes_{A_{\mathbb{Q}}} I'. \tag{2.9.9.7}$$

2.9.10 Let (X, A) be a ringed topos, E an A-module, (M, N, u, θ), (M', N', u', θ') two Higgs A-isogenies with coefficients in E. Since $u_{\mathbb{Q}}$ is an isomorphism of $\mathbf{Mod}_{\mathbb{Q}}(A)$ (2.9.7), we can consider the composed morphism

$$\overline{\theta} = (u_{\mathbb{Q}}^{-1} \otimes \mathrm{id}_E) \circ \theta_{\mathbb{Q}} \colon M_{\mathbb{Q}} \to M_{\mathbb{Q}} \otimes_{A_{\mathbb{Q}}} E_{\mathbb{Q}}. \tag{2.9.10.1}$$

We define in the same way the morphism $\overline{\theta}' \colon M'_{\mathbb{Q}} \to M'_{\mathbb{Q}} \otimes_{A_{\mathbb{Q}}} E_{\mathbb{Q}}$ of $\mathbf{Mod}_{\mathbb{Q}}(A)$. The morphism

$$\begin{array}{ccc} \mathrm{Hom}_{\mathbf{HI}(A,E)}((M, N, u, \theta), (M', N', u', \theta')) & \to & \mathrm{Hom}_A(M, M') \\ (\alpha, \beta) & \mapsto & \alpha \end{array} \tag{2.9.10.2}$$

then induces an isomorphism

$$\begin{array}{l} \mathrm{Hom}_{\mathbf{HI}_{\mathbb{Q}}(A,E)}((M, N, u, \theta), (M', N', u', \theta')) \xrightarrow{\sim} \\ \{\alpha \in \mathrm{Hom}_A(M, M') \otimes_{\mathbb{Z}} \mathbb{Q},\ (\alpha \otimes \mathrm{id}_E) \circ \overline{\theta} = \overline{\theta}' \circ \alpha\}. \end{array} \tag{2.9.10.3}$$

Indeed, since $u_{\mathbb{Q}}$ and $u'_{\mathbb{Q}}$ are isomorphisms of $\mathbf{Mod}_{\mathbb{Q}}(A)$, the morphism (2.9.10.2) induces an injective morphism

$$\mathrm{Hom}_{\mathbf{HI}_{\mathbb{Q}}(A,E)}((M,N,u,\theta),(M',N',u',\theta')) \to \mathrm{Hom}_A(M,M') \otimes_{\mathbb{Z}} \mathbb{Q}. \quad (2.9.10.4)$$

Let $\alpha \in \mathrm{Hom}_A(M,M') \otimes_{\mathbb{Z}} \mathbb{Q}$ such that $(\alpha \otimes \mathrm{id}_E) \circ \overline{\theta} = \overline{\theta}' \circ \alpha$. Let us show that α is in the image of the morphism (2.9.10.3). Since the latter is $\mathbb{Q}$-linear, we may assume that $\alpha \in \mathrm{Hom}_A(M,M')$. There exist two nonzero integers n and n' and two A-linear morphisms $v\colon N \to M$ and $v'\colon N' \to M'$ such that $v \circ u = n \cdot \mathrm{id}_M$, $u \circ v = n \cdot \mathrm{id}_N$, $v' \circ u' = n' \cdot \mathrm{id}_{M'}$, $u' \circ v' = n' \cdot \mathrm{id}_{N'}$, and that $(M, (v \otimes \mathrm{id}_E) \circ \theta)$ and $(M', (v' \otimes id_E) \circ \theta')$ are Higgs A-modules with coefficients in E. After multiplying n and n' if necessary, we may assume that $n = n'$ and that the diagram

$$\begin{array}{ccccc} M & \xrightarrow{\theta} & N \otimes_A E & \xrightarrow{v \otimes \mathrm{id}_E} & M \otimes_A E \\ \downarrow{\scriptstyle \alpha} & & & & \downarrow{\scriptstyle \alpha \otimes \mathrm{id}_E} \\ M' & \xrightarrow{\theta'} & N' \otimes_A E & \xrightarrow{v' \otimes \mathrm{id}_E} & M' \otimes_A E \end{array} \quad (2.9.10.5)$$

is commutative. Hence, (α, α) defines a morphism of $\mathbf{HI}(A,E)$ from $(M, M, \mathrm{id}, (v \otimes \mathrm{id}_E) \circ \theta)$ into $(M', M', \mathrm{id}, (v' \otimes \mathrm{id}_E) \circ \theta')$. Since the morphisms

$$(\mathrm{id}_M, v)\colon (M,N,u,\theta) \to (M,M,\mathrm{id},(v \otimes \mathrm{id}_E) \circ \theta), \quad (2.9.10.6)$$
$$(\mathrm{id}_{M'}, v')\colon (M',N',u',\theta') \to (M',M',\mathrm{id},(v' \otimes \mathrm{id}_E) \circ \theta'), \quad (2.9.10.7)$$

are isogenies of $\mathbf{HI}(A,E)$, we deduce that the morphism (2.9.10.3) is surjective and therefore bijective.

2.9.11 Let (X, A) be a ringed topos, B an A-algebra, $d\colon B \to \Omega^1_{B/A}$ the universal A-derivation, $\lambda \in \Gamma(X,A)$. We call *λ-isoconnection with respect to the extension B/A* (or simply *λ-isoconnection* when there is no risk of confusion) a quadruple

$$(M, N, u\colon M \to N, \nabla\colon M \to \Omega^1_{B/A} \otimes_B N) \quad (2.9.11.1)$$

where M and N are B-modules, u an isogeny of B-modules (2.9.1) and ∇ is an A-linear morphism such that for all local sections x of B and t of M, we have

$$\nabla(xt) = \lambda d(x) \otimes u(t) + x\nabla(t). \quad (2.9.11.2)$$

For every B-linear morphism $v\colon N \to M$ for which there exists an integer n such that $u \circ v = n \cdot \mathrm{id}_N$ and $v \circ u = n \cdot \mathrm{id}_M$, the pairs $(M, (\mathrm{id} \otimes v) \circ \nabla)$ and $(N, \nabla \circ v)$ are modules with $(n\lambda)$-connections (2.5.24), and u is a morphism from $(M, (\mathrm{id} \otimes v) \circ \nabla)$ to $(N, \nabla \circ v)$. We say that the λ-isoconnection (M, N, u, ∇) is *integrable* if there exist a B-linear morphism $v\colon N \to M$ and an integer $n \neq 0$ such that $u \circ v = n \cdot \mathrm{id}_N$, $v \circ u = n \cdot \mathrm{id}_M$ and that the $(n\lambda)$-connections $(\mathrm{id} \otimes v) \circ \nabla$ on M and $\nabla \circ v$ on N are integrable.

Let (M, N, u, ∇), (M', N', u', ∇') be two λ-isoconnections. A morphism from (M, N, u, ∇) to (M', N', u', ∇') consists of two B-linear morphisms $\alpha\colon M \to M'$ and $\beta\colon N \to N'$ such that $\beta \circ u = u' \circ \alpha$ and $(\mathrm{id} \otimes \beta) \circ \nabla = \nabla' \circ \alpha$.

We denote by $\mathbf{IC}^{\lambda}(B/A)$ the category of integrable λ-isoconnections with respect to the extension B/A. It is an additive category. We denote by $\mathbf{IC}^{\lambda}_{\mathbb{Q}}(B/A)$ the category of objects of $\mathbf{IC}^{\lambda}(B/A)$ up to isogeny (2.9.1).

2.9.12 Let (X, A) be a ringed topos, B an A-algebra, $\lambda \in \Gamma(X, A)$ (M, N, u, ∇) and (M', N', u', ∇') two λ-isoconnections with respect to the extension B/A. Since $u_{\mathbb{Q}}$ is an isomorphism of $\mathbf{Mod}_{\mathbb{Q}}(B)$ (2.9.7), we can consider the composed morphism of $\mathbf{Mod}_{\mathbb{Q}}(A)$

$$\overline{\nabla} = (\mathrm{id} \otimes u_{\mathbb{Q}}^{-1}) \circ \nabla\colon M_{\mathbb{Q}} \to (\Omega^1_{B/A})_{\mathbb{Q}} \otimes_{A_{\mathbb{Q}}} M_{\mathbb{Q}}. \tag{2.9.12.1}$$

We define in the same way the morphism $\overline{\nabla}'\colon M'_{\mathbb{Q}} \to (\Omega^1_{B/A})_{\mathbb{Q}} \otimes_{A_{\mathbb{Q}}} M'_{\mathbb{Q}}$ of $\mathbf{Mod}_{\mathbb{Q}}(A)$. Copying the proof of 2.9.10, we show that the morphism

$$\begin{array}{ccc} \mathrm{Hom}_{\mathbf{IC}^{\lambda}(B/A)}((M, N, u, \nabla), (M', N', u', \nabla')) & \to & \mathrm{Hom}_B(M, M') \\ (\alpha, \beta) & \mapsto & \alpha \end{array} \tag{2.9.12.2}$$

induces an isomorphism

$$\begin{array}{l} \mathrm{Hom}_{\mathbf{IC}^{\lambda}_{\mathbb{Q}}(B/A)}((M, N, u, \nabla), (M', N', u', \nabla')) \xrightarrow{\sim} \\ \{\alpha \in \mathrm{Hom}_B(M, M') \otimes_{\mathbb{Z}} \mathbb{Q},\ (\mathrm{id} \otimes \alpha) \circ \overline{\nabla} = \overline{\nabla}' \circ \alpha \text{ in } \mathbf{Mod}_{\mathbb{Q}}(A)\}. \end{array} \tag{2.9.12.3}$$

2.9.13 Let $f\colon (X', A') \to (X, A)$ be a morphism of ringed topos, B an A-algebra, B' an A'-algebra, $\alpha\colon f^*(B) \to B'$ a homomorphism of A'-algebras, $\lambda \in \Gamma(X, A)$, (M, N, u, ∇) a λ-isoconnection with respect to the extension B/A. Denote by λ' be the canonical image of λ in $\Gamma(X', A')$, by $d'\colon B' \to \Omega^1_{B'/A'}$ the universal A'-derivation of B' and by

$$\gamma\colon f^*(\Omega^1_{B/A}) \to \Omega^1_{B'/A'} \tag{2.9.13.1}$$

the canonical α-linear morphism. We see immediately that

$$(f^*(M), f^*(N), f^*(u), f^*(\nabla))$$

is a λ'-isoconnection with respect to the extension $f^*(B)/A'$, which is integrable if (M, N, u, ∇) is.

There exists a unique A'-linear morphism

$$\nabla'\colon B' \otimes_{f^*(B)} f^*(M) \to \Omega^1_{B'/A'} \otimes_{f^*(B)} f^*(N) \tag{2.9.13.2}$$

such that for all local sections x' of B' and t of $f^*(M)$, we have

$$\nabla'(x' \otimes t) = \lambda' d'(x') \otimes f^*(u)(t) + x'(\gamma \otimes \mathrm{id}_{f^*(N)})(f^*(\nabla)(t)). \tag{2.9.13.3}$$

The quadruple $(B' \otimes_{f^*(B)} f^*(M), B' \otimes_{f^*(B)} f^*(N), \mathrm{id}_{B'} \otimes_{f^*(B)} f^*(u), \nabla')$ is a λ'-isoconnection with respect to the extension B'/A', which is integrable if (M, N, u, ∇) is.

2.9.14 Let (X, A) be a ringed topos, B an A-algebra, $\lambda \in \Gamma(X, A)$, (M, N, u, ∇) a λ-integrable isoconnection with respect to the extension B/A. Suppose there exist an A-module E and a B-isomorphism $\gamma\colon E \otimes_A B \xrightarrow{\sim} \Omega^1_{B/A}$ such that for every local section ω of E, we have $d(\gamma(\omega \otimes 1)) = 0$. Denote by $\vartheta\colon M \to E \otimes_A N$ the morphism induced by ∇ and γ. Then,

(i) The quadruple (M, N, u, ϑ) is a Higgs A-isogeny with coefficients in E.
(ii) For every Higgs A-isogeny (M', N', u', θ') with coefficients in E, there exists a unique A-linear morphism

$$\nabla'\colon M \otimes_A M' \to \Omega^1_{B/A} \otimes_B N \otimes_A N' \tag{2.9.14.1}$$

such that for all local sections t of M and t' of M', we have

$$\nabla'(t \otimes t') = \nabla(t) \otimes_A u'(t') + (\gamma \otimes_B \mathrm{id}_{N \otimes_A N'})(u(t) \otimes_A \theta'(t')). \tag{2.9.14.2}$$

The quadruple $(M \otimes_A M', N \otimes_A N', u \otimes u', \nabla')$ is an integrable λ-isoconnection.

2.10 Complement on the Functoriality of Generalized Covanishing Topos

2.10.1 We first recall and clarify the terminology introduced in ([3] § VI.5) on generalized covanishing topos. We call a *covanishing fibered site* a $\mathbb{U}$-site I and a cleaved and normalized fibered category over the category underlying I ([24] VI 7.1)

$$\pi\colon E \to I, \tag{2.10.1.1}$$

satisfying the following conditions:

(i) fiber products are representable in I;
(ii) for every $i \in \mathrm{Ob}(I)$, the fiber category E_i of E over i is endowed with a topology making it into a $\mathbb{U}$-site, and finite inverse limits are representable in E_i. We denote by $\widetilde{E}_i$ the topos of sheaves of $\mathbb{U}$-sets on E_i;
(iii) for every morphism $f\colon i \to j$ of I, the inverse image functor $f^+\colon E_j \to E_i$ is continuous and left exact. It therefore defines a morphism of topos that we (abusively) denote also by $f\colon \widetilde{E}_i \to \widetilde{E}_j$ ([5] IV 4.9.2).

For any $i \in \mathrm{Ob}(I)$, we denote by

$$\alpha_i\colon E_i \to E \tag{2.10.1.2}$$

the canonical inclusion functor. Note that this functor was denoted by $\alpha_{i!}$ in ([3] (VI.5.1.2)).

The functor π is in fact a fibered $\mathbb{U}$-site ([5] VI 7.2.1 and 7.2.4). We denote by

$$\mathscr{F} \to I \tag{2.10.1.3}$$

the fibered $\mathbb{U}$-topos associated with π ([5] VI 7.2.6). The fiber category of $\mathscr{F}$ over any $i \in \mathrm{Ob}(I)$ is canonically equivalent to the topos $\widetilde{E}_i$, and the inverse image functor by any morphism $f: i \to j$ of I is identified with the pullback functor $f^*: \widetilde{E}_j \to \widetilde{E}_i$ of the morphism of topos $f: \widetilde{E}_i \to \widetilde{E}_j$. We denote by

$$\mathscr{F}^\vee \to I^\circ \tag{2.10.1.4}$$

the fibered category obtained by associating with any $i \in \mathrm{Ob}(I)$ the category $\widetilde{E}_i$, and with any morphism $f: i \to j$ of I the direct image functor $f_*: \widetilde{E}_i \to \widetilde{E}_j$ of the morphism of topos $f: \widetilde{E}_i \to \widetilde{E}_j$. We denote by

$$\mathscr{P}^\vee \to I^\circ \tag{2.10.1.5}$$

the fibered category obtained by associating with any $i \in \mathrm{Ob}(I)$ the category $\widehat{E}_i$ of presheaves of $\mathbb{U}$-sets on E_i, and with any morphism $f: i \to j$ of I the functor $\widehat{f}^*: \widehat{E}_i \to \widehat{E}_j$ obtained by composing with the inverse image functor $f^+: E_j \to E_i$. Note that the canonical I°-functor $\mathscr{F}^\vee \to \mathscr{P}^\vee$ is compatible with inverse image functors.

In the rest of this section, we fix the covanishing fibered site $\pi: E \to I$ and the associated fibered categories $\mathscr{F}/I$, $\mathscr{F}^\vee/I^\circ$ and $\mathscr{P}^\vee/I^\circ$.

2.10.2 Note that E is a $\mathbb{U}$-category. We denote by $\widehat{E}$ the category of presheaves of $\mathbb{U}$-sets on E. By ([3] VI.5.2) and with the notation of 2.1.5, we have an equivalence of categories

$$\begin{aligned} \widehat{E} &\xrightarrow{\sim} \mathbf{Hom}_{I^\circ}(I^\circ, \mathscr{P}^\vee) \\ F &\mapsto \{i \mapsto F \circ \alpha_i\}, \end{aligned} \tag{2.10.2.1}$$

where α_i is the functor (2.10.1.2). From now on, we will identify F with the section $\{i \mapsto F \circ \alpha_i\}$ that is associated with it by this equivalence.

Definition 2.10.3 We call a *v-presheaf* of $\mathbb{U}$-sets on E any presheaf F of $\mathbb{U}$-sets on E such that for every $i \in \mathrm{Ob}(I)$, $F \circ \alpha_i$ is a sheaf on E_i (2.10.1.2).

We denote by $\widehat{E}_{\mathrm{v}}$ the category of v-presheaves, i.e., the full subcategory of $\widehat{E}$ made up of the v-presheaves. We then have an equivalence of categories

$$\begin{aligned} \widehat{E}_{\mathrm{v}} &\xrightarrow{\sim} \mathbf{Hom}_{I^\circ}(I^\circ, \mathscr{F}^\vee) \\ F &\mapsto \{i \mapsto F \circ \alpha_i\}. \end{aligned} \tag{2.10.3.1}$$

Note that this notion does not use the condition 2.10.1(i).

2.10.4 Following ([3] VI.5.3), we call the *covanishing* topology of E the topology generated by the families of coverings $(V_n \to V)_{n\in\Sigma}$ of the following two types:

(v) There exists an $i \in \mathrm{Ob}(I)$ such that $(V_n \to V)_{n\in\Sigma}$ is a covering family of E_i.
(c) There exists a covering family of morphisms $(f_n: i_n \to i)_{n\in\Sigma}$ of I such that $\pi(V) = i$ and for every $n \in \Sigma$, V_n is isomorphic to $f_n^+(V)$.

The coverings of type (v) are called *vertical*, and those of type (c) are called *Cartesian*. The resulting site is called the *covanishing site* associated with the covanishing fibered site π (2.10.1.1); it is a $\mathbb{U}$-site. We call the *covanishing topos* associated with the covanishing fibered site π, and denote by $\widetilde{E}$, the topos of sheaves of $\mathbb{U}$-sets on E. For any presheaf F of $\mathbb{U}$-sets on E, we denote by F^a the associated sheaf.

2.10.5 Let J be a topologically generating full subcategory of I, $\psi: J \to I$ the canonical functor, which will be omitted from notation when there is no risk of confusion. We denote by

$$\pi_J: E_J \to J \tag{2.10.5.1}$$

the fibered category deduced from π by base change by ψ, and by

$$\Psi: E_J \to E \tag{2.10.5.2}$$

the canonical projection ([24] VI § 3). For every $j \in \mathrm{Ob}(J)$, the fiber category $(E_J)_j$ is canonically equivalent to the fiber category E_j; they will be identified in the following. The fibered topos $\mathscr{F}_J \to J$ associated with π_J is canonically J-equivalent to the fibered topos deduced from $\mathscr{F}/I$ (2.10.1.3) by base change by ψ. We denote by

$$\mathscr{F}_J^\vee \to J^\circ, \tag{2.10.5.3}$$

$$\mathscr{P}_J^\vee \to J^\circ, \tag{2.10.5.4}$$

the fibered categories deduced from $\mathscr{F}^\vee/I^\circ$ (2.10.1.4) and $\mathscr{P}^\vee/I^\circ$ (2.10.1.5) by base change by ψ°. We denote by $\widehat{E}_J$ the category of presheaves of $\mathbb{U}$-sets on E_J. We immediately check that the diagram of functors

$$\begin{array}{ccc} \widehat{E} & \xrightarrow{\sim} & \mathbf{Hom}_{I^\circ}(I^\circ, \mathscr{P}^\vee) \\ {\scriptstyle\widehat{\Psi}^*}\downarrow & & \downarrow \\ \widehat{E}_J & \xrightarrow{\sim} & \mathbf{Hom}_{J^\circ}(J^\circ, \mathscr{P}_J^\vee), \end{array} \tag{2.10.5.5}$$

where the horizontal arrows are the equivalences of categories (2.10.2.1), $\widehat{\Psi}^*$ is the functor defined by composition with Ψ and the right vertical arrow is the canonical functor ([24] VI § 3), is commutative up to canonical isomorphism. Consequently, for every presheaf $F = \{i \mapsto F_i\}$ on E, we have

$$\widehat{\Psi}^*(F) = \{j \in J^\circ \mapsto F_j\}. \tag{2.10.5.6}$$

We endow E_J with the topology induced by the covanishing topology on E by means of the functor Ψ ([5] III §3), and denote by $\widetilde{E}_J$ the topos of sheaves of $\mathbb{U}$-sets on E_J. Observe that the functor Ψ is fully faithful and that the family E_J is topologically generating of E. Suppose that the categories J and E_J are $\mathbb{U}$-small. By ([5] III 4.1) and its proof, the functor Ψ is continuous and cocontinuous, and the functor $\widehat{\Psi}^*$ induces an equivalence of categories

$$\Psi_s : \widetilde{E} \xrightarrow{\sim} \widetilde{E}_J. \tag{2.10.5.7}$$

Lemma 2.10.6 *We keep the assumptions and notation of* 2.10.5. *Then,*

(i) *For every presheaf G on E, we have a canonical isomorphism*

$$\Psi_s(G^a) \xrightarrow{\sim} \widehat{\Psi}^*(G)^a, \tag{2.10.6.1}$$

where the exponent ^{a} *denotes the associated sheaves.*

(ii) *For every presheaf* $F = \{j \in J^\circ \mapsto F_j\}$ *on* E_J*, there exist a canonical presheaf G on E and an isomorphism* $F \xrightarrow{\sim} \widehat{\Psi}^*(G)$*. The latter induces an isomorphism*

$$\Psi_s(G^a) \xrightarrow{\sim} F^a, \tag{2.10.6.2}$$

where the exponent ^{a} *denotes the associated sheaves.*

(i) This follows from ([5] III 2.3) since the functor $\widehat{\Psi}^*$ is continuous and cocontinuous.

(ii) For any object i of I, we define the category $J_{/i}$ as follows. The objects of $J_{/i}$ are the pairs (j, f) made up of an object j of J and a morphism $f : j \to i$ of I. Let (j, f) and (j', f') be two objects of $J_{/i}$. A morphism from (j, f) to (j', f') is a morphism $g : j \to j'$ of J such that $f = f' \circ g$ in I. For every morphism $h : i \to i'$ of I, we have the functor

$$J_{/h} : J_{/i} \to J_{/i'}, \quad (j, f) \mapsto (j, h \circ f). \tag{2.10.6.3}$$

For every $i \in \mathrm{Ob}(I)$, the presheaves $\widehat{f}^*(F_j)$, for $(j, f) \in \mathrm{Ob}(J^\circ_{/i})$, naturally form an inverse system. We set

$$G_i = \lim_{\substack{\longleftarrow \\ (j,f) \in J^\circ_{/i}}} \widehat{f}^*(F_j). \tag{2.10.6.4}$$

For every morphism $h : i \to i'$ of I, since the functor $\widehat{h}^* : \widehat{E}_i \to \widehat{E}_{i'}$ admits a left adjoint, it commutes with inverse limits. Taking into account (2.10.6.3), we deduce a morphism $G_{i'} \to \widehat{h}^*(G_i)$. We immediately check that the collection $G = \{i \in I^\circ \mapsto G_i\}$ forms a presheaf on E.

For every $j \in \mathrm{Ob}(J)$, (j, id_j) is the final object of $J_{/j}$, which implies that $\widehat{\Psi}^*(G) = F$.

2.10.7 Let $\pi' : E' \to I'$ be a covanishing fibered site,

$$\begin{array}{ccc} E & \xrightarrow{\pi} & I \\ {\scriptstyle \Phi}\downarrow & & \downarrow{\scriptstyle \varphi} \\ E' & \xrightarrow{\pi'} & I' \end{array} \tag{2.10.7.1}$$

a strictly commutative diagram of functors, i.e., such that $\varphi \circ \pi = \pi' \circ \Phi$. We denote by

$$\pi'_I : E'_I \to I \tag{2.10.7.2}$$

the fibered category deduced from π' by base change by φ ([24] VI § 3), and by

$$\phi : E \to E'_I \tag{2.10.7.3}$$

the I-functor induced by Φ (2.10.7.1). We suppose that the functor ϕ is *Cartesian*.

We denote by $\mathscr{F}' \to I'$ the fibered topos associated with the fibered site π' and by

$$\mathscr{F}'^{\vee} \to I'^{\circ}, \tag{2.10.7.4}$$

$$\mathscr{P}'^{\vee} \to I'^{\circ}, \tag{2.10.7.5}$$

the fibered categories associated with the fibered site π', defined in (2.10.1.4) and (2.10.1.5), respectively.

For every $i \in \mathrm{Ob}(I)$, the fiber category $(E'_I)_i$ is canonically equivalent to the fiber category $E'_{\varphi(i)}$; they will be identified in the following. Therefore, π'_I is a covanishing fibered site. The fibered topos $\mathscr{F}'_I \to I$ associated with π'_I is canonically I-equivalent to the fibered topos deduced from $\mathscr{F}'/I'$ by base change by φ. We denote by

$$\mathscr{F}'^{\vee}_I \to I^{\circ}, \tag{2.10.7.6}$$

$$\mathscr{P}'^{\vee}_I \to I^{\circ}, \tag{2.10.7.7}$$

the fibered categories associated with π'_I, defined in (2.10.1.4) and (2.10.1.5), respectively. They are canonically I°-equivalent to the fibered categories deduced from $\mathscr{F}'^{\vee}/I'^{\circ}$ and $\mathscr{P}'^{\vee}/I'^{\circ}$ by base change by φ°.

For any $i \in \mathrm{Ob}(I)$, we denote by $\Phi_i : E_i \to E'_{\varphi(i)}$ the functor induced by Φ, by $\widehat{E}_i$ the category of presheaves of $\mathbb{U}$-sets on E_i, and by $\widehat{\Phi}^*_i : \widehat{E}'_{\varphi(i)} \to \widehat{E}_i$ the functor defined by composition with Φ_i. Let $f : j \to i$ be a morphism of I, $f' = \varphi(f) : \varphi(j) \to \varphi(i)$. We have an isomorphism of functors

$$\Phi_j \circ f^+ \xrightarrow{\sim} f'^+ \circ \Phi_i, \tag{2.10.7.8}$$

where $f^+ : E_i \to E_j$ (resp. $f'^+ : E'_{\varphi(i)} \to E'_{\varphi(j)}$) is the inverse image functor by f of E (resp. f' of E'). It induces an isomorphism of functors

$$\widehat{\Phi}^*_i \circ \widehat{f}'^* \xrightarrow{\sim} \widehat{f}^* \circ \widehat{\Phi}^*_j, \tag{2.10.7.9}$$

where $\widehat{f}^* : \widehat{E}_j \to \widehat{E}_i$ (resp. $\widehat{f}'^* : \widehat{E}'_{\varphi(j)} \to \widehat{E}'_{\varphi(i)}$) is the inverse image functor by f° of $\mathscr{P}^{\vee}$ (resp. f'° of $\mathscr{P}'^{\vee}$) (2.10.1.5). The isomorphisms (2.10.7.8) satisfy a cocycle relation of the type ([1] (1.1.2.2)), which induces an analogous relation for the isomorphisms (2.10.7.9). By ([24] VI 12; see also [1] 1.1.2), the functors $\widehat{\Phi}^*_i$ thus define an I°-Cartesian functor

$$\mathscr{P}'^{\vee}_I \to \mathscr{P}^{\vee}. \tag{2.10.7.10}$$

We denote by

$$\widehat{\Phi}^* : \widehat{E}' \to \widehat{E} \tag{2.10.7.11}$$

the functor defined by composition with Φ. We immediately check that the diagram of functors

$$\begin{array}{ccc} \widehat{E}' & \xrightarrow{\sim} & \mathbf{Hom}_{I'^\circ}(I'^\circ, \mathscr{P}'^\vee) \\ & & \downarrow u \\ \widehat{\Phi}^* \downarrow & & \mathbf{Hom}_{I^\circ}(I^\circ, \mathscr{P}'^\vee_I) \\ & & \downarrow v \\ \widehat{E} & \xrightarrow{\sim} & \mathbf{Hom}_{I^\circ}(I^\circ, \mathscr{P}^\vee), \end{array} \tag{2.10.7.12}$$

where u is the canonical functor ([24] VI § 3), v is the functor defined by composition with the functor (2.10.7.10), and the horizontal arrows are the equivalences of categories (2.10.2.1), is commutative up to canonical isomorphism. Hence, for every presheaf $F' = \{i' \mapsto F'_{i'}\}$ on E', we have

$$\widehat{\Phi}^*(F') = \{i \mapsto \widehat{\Phi}_i^*(F'_{\varphi(i)})\}. \tag{2.10.7.13}$$

2.10.8 We keep the assumptions and notation of 2.10.7; suppose, moreover, that the categories E, I and $(E'_{i'})_{i' \in \mathrm{Ob}(I')}$ are $\mathbb{U}$-small. By ([5] I 5.1), the functor $\widehat{\Phi}^*$ (2.10.7.11) admits a left adjoint

$$\widehat{\Phi}_! \colon \widehat{E} \to \widehat{E}'. \tag{2.10.8.1}$$

Likewise, for all $i \in \mathrm{Ob}(I)$, the category E_i being $\mathbb{U}$-small, the functor $\widehat{\Phi}_i^* \colon \widehat{E}'_{\varphi(i)} \to \widehat{E}_i$ admits a left adjoint

$$\widehat{\Phi}_{i!} \colon \widehat{E}_i \to \widehat{E}'_{\varphi(i)}. \tag{2.10.8.2}$$

For every morphism $f \colon j' \to i'$ of I', the functor $\widehat{f}^* \colon \widehat{E}'_{j'} \to \widehat{E}'_{i'}$ inverse image by f° of $\mathscr{P}'^\vee$ admits a left adjoint

$$\widehat{f}_! \colon \widehat{E}'_{i'} \to \widehat{E}'_{j'}. \tag{2.10.8.3}$$

For all composable morphisms $h \colon k' \to j'$ and $f \colon j' \to i'$ of I', setting $g = f \circ h \colon k' \to i'$, we have a canonical isomorphism $\widehat{g}^* \xrightarrow{\sim} \widehat{f}^* \circ \widehat{h}^*$ which induces by adjunction an isomorphism

$$\widehat{g}_! \xrightarrow{\sim} \widehat{h}_! \circ \widehat{f}_!. \tag{2.10.8.4}$$

For any object i' of I', we define the category $I_\varphi^{i'}$ as follows. The objects of $I_\varphi^{i'}$ are the pairs (i, f) consisting of an object i of I and a morphism $f \colon i' \to \varphi(i)$ of I'. Let (i_1, f_1) and (i_2, f_2) be two objects of $I_\varphi^{i'}$. A morphism from (i_1, f_1) to (i_2, f_2) is a morphism $g \colon i_1 \to i_2$ of I such that $f_2 = \varphi(g) \circ f_1$.

For every morphism $h \colon i' \to j'$ of I', we have the functor

$$I_\varphi^h \colon I_\varphi^{j'} \to I_\varphi^{i'}, \quad (j, g) \mapsto (j, g \circ h). \tag{2.10.8.5}$$

Lemma 2.10.9 *We keep the assumptions and notation of* 2.10.7 *and* 2.10.8. *Let, moreover, $F = \{i \mapsto F_i\}$ be a presheaf on E. Then, for every $i' \in \mathrm{Ob}(I')$, we have an isomorphism*

$$\widehat{\Phi}_!(F) \circ \alpha'_{i'} \xrightarrow{\sim} \varinjlim_{(i,f)\in (I^{i'}_{\varphi})^{\circ}} \widehat{f}_!(\widehat{\Phi}_{i!}(F_i)), \tag{2.10.9.1}$$

where $\alpha'_{i'} : E'_{i'} \to E'$ *is the canonical injection. Moreover, for every morphism* $h : i' \to j'$ *of* I' *and every object* (j, g) *of* $I^{j'}_{\varphi}$*, setting* $f = g \circ h : i' \to \varphi(j)$ *so that* $(j, f) = I^h_{\varphi}(j, g)$ (2.10.8.5), *the diagram*

$$\begin{array}{ccc} \widehat{h}_!(\widehat{g}_!(\widehat{\Phi}_{j!}(F_j))) & \longrightarrow & \widehat{f}_!(\widehat{\Phi}_{j!}(F_j)) \\ \downarrow & & \downarrow \\ \widehat{h}_!(\widehat{\Phi}_!(F) \circ \alpha'_{j'}) & \longrightarrow & \widehat{\Phi}_!(F) \circ \alpha'_{i'}, \end{array} \tag{2.10.9.2}$$

where the vertical arrows are induced by the morphisms (2.10.9.1), *the upper horizontal arrow is the canonical isomorphism* (2.10.8.4) *and the lower horizontal arrow is the adjoint morphism of the canonical morphism*

$$\widehat{\Phi}_!(F) \circ \alpha'_{j'} \to \widehat{h}^*(\widehat{\Phi}_!(F) \circ \alpha'_{i'}), \tag{2.10.9.3}$$

is commutative.

Let $U' \in \mathrm{Ob}(E')$, $i' = \pi'(U')$. We define the category $I^{U'}_{\Phi}$ as follows. The objects of $I^{U'}_{\Phi}$ are the pairs (U, m) consisting of an object U of E and a morphism $m : U' \to \Phi(U)$ of E'. Let (U_1, m_1) and (U_2, m_2) be two objects of $I^{U'}_{\Phi}$. A morphism from (U_1, m_1) to (U_2, m_2) is a morphism $n : U_1 \to U_2$ of E such that $m_2 = \Phi(n) \circ m_1$. By the proof of ([5] I 5.1), we have an isomorphism

$$\widehat{\Phi}_!(F)(U') \xrightarrow{\sim} \varinjlim_{(U,m)\in (I^{U'}_{\Phi})^{\circ}} F(U). \tag{2.10.9.4}$$

For all $(i, f) \in \mathrm{Ob}(I^{i'}_{\varphi})$, we have a canonical isomorphism $(f^+ \circ \Phi_i)^* \xrightarrow{\sim} \widehat{\Phi}^*_i \circ \widehat{f}^*$ which induces by adjunction an isomorphism

$$(f^+ \circ \Phi_i)_! \xrightarrow{\sim} \widehat{f}_! \circ \widehat{\Phi}_{i!}. \tag{2.10.9.5}$$

We define the category $I^{U'}_{\Phi_i}$ as follows. The objects of $I^{U'}_{\Phi_i}$ are the pairs (U, m) consisting of an object U of E_i and a morphism $m : U' \to f^+(\Phi_i(U))$ of $E'_{i'}$. Let (U_1, m_1) and (U_2, m_2) be two objects of $I^{U'}_{\Phi_i}$. A morphism from (U_1, m_1) to (U_2, m_2) is a morphism $n : U_1 \to U_2$ of E_i such that $m_2 = f^+(\Phi_i(n)) \circ m_1$. By (2.10.9.5), we have an isomorphism

$$\widehat{f}_!(\widehat{\Phi}_{i!}(F_i))(U') \xrightarrow{\sim} \varinjlim_{(U,m)\in (I^{U'}_{\Phi_i})^{\circ}} F(U). \tag{2.10.9.6}$$

Moreover, we have the functor

$$\varpi : I^{U'}_{\Phi} \to I^{i'}_{\varphi}, \quad (U, m) \mapsto (\pi(U), \pi'(m)). \tag{2.10.9.7}$$

Indeed, we have $\pi'(m)\colon i' \to \pi'(\Phi(U)) = \varphi(\pi(U))$. Since the functor ϕ (2.10.7.3) is Cartesian, for every $(i, f) \in \mathrm{Ob}(I^{i'}_{\varphi})$, the fiber category of ϖ over (i, f) is canonically identified with the category $I^{U'}_{\Phi_i}$. The proposition follows in view of (2.10.9.5) and (2.10.9.6).

Remark 2.10.10 We keep the assumptions and notation of 2.10.7 and 2.10.8.

(i) For all $i \in \mathrm{Ob}(I)$, the diagram

$$\begin{array}{ccc} E_i & \xrightarrow{\alpha_i} & E \\ {\scriptstyle \Phi_i}\downarrow & & \downarrow{\scriptstyle \Phi} \\ E'_{\varphi(i)} & \xrightarrow{\alpha'_{\varphi(i)}} & E' \end{array} \tag{2.10.10.1}$$

is strictly commutative. We deduce from this that for every presheaf F' on E', we have a canonical isomorphism

$$\widehat{\Phi}^*_i(F' \circ \alpha'_{\varphi(i)}) \xrightarrow{\sim} \alpha_i \circ \widehat{\Phi}^*(F'). \tag{2.10.10.2}$$

(ii) With the assumptions and notation of 2.10.9, the diagram

$$\begin{array}{ccccc} F_i & & \xrightarrow{\quad a \quad} & & F \circ \alpha_i \\ \downarrow & & & & \downarrow \\ \widehat{\Phi}^*_i(\widehat{\Phi}_{i!}(F_i)) & \xrightarrow{\widehat{\Phi}^*_i(b)} & \widehat{\Phi}^*_i(\widehat{\Phi}_!(F) \circ \alpha'_{\varphi(i)}) & \xrightarrow{c} & \widehat{\Phi}^*(\widehat{\Phi}_!(F)) \circ \alpha_i, \end{array} \tag{2.10.10.3}$$

where the vertical arrows are induced by the adjunction morphisms, a is the canonical isomorphism, $b\colon \widehat{\Phi}_{i!}(F_i) \to \widehat{\Phi}_!(F) \circ \alpha'_{\varphi(i)}$ is the morphism induced by the isomorphism (2.10.9.1) and the object $(\varphi(i), \mathrm{id})$ of $I^{\varphi(i)}_{\varphi}$ and c is the isomorphism (2.10.10.2), is commutative. This immediately follows from the proof of 2.10.9, in particular the description of the fiber categories of ϖ (2.10.9.7).

2.10.11 We keep the assumptions and notation of 2.10.7 and 2.10.8. Suppose, moreover, that the following assumptions are satisfied:

(i) the functor ϕ (2.10.7.3) is Cartesian;
(ii) the categories E, I and $(E'_{i'})_{i' \in \mathrm{Ob}(I')}$ are $\mathbb{U}$-small;
(iii) the functor φ is continuous;
(iv) for all $i \in \mathrm{Ob}(I)$, the functor $\Phi_i\colon E_i \to E'_{\varphi(i)}$ is continuous.

The assumptions (i)–(ii) are mentioned as a reminder.

The functor Φ (2.10.7.1) is continuous for the covanishing topologies on E and E', by virtue of (2.10.7.12) and ([3] VI 5.10). Denoting by $\widetilde{E}'$ the topos of sheaves of $\mathbb{U}$-sets on E', the functor $\widehat{\Phi}^*$ induces a functor

$$\Phi_s\colon \widetilde{E}' \to \widetilde{E}. \tag{2.10.11.1}$$

By ([5] III 1.3), the functor Φ_s admits a left adjoint

$$\Phi^s : \widetilde{E} \to \widetilde{E}' \tag{2.10.11.2}$$

such that for every presheaf F on E, we have a canonical functorial isomorphism

$$\Phi^s(F^a) = (\widehat{\Phi}_!(F))^a, \tag{2.10.11.3}$$

where the exponent ^{a} denotes the associated sheaves.

For any $i' \in \mathrm{Ob}(I')$, let $\widetilde{E}'_{i'}$ be the topos of sheaves of $\mathbb{U}$-sets on $E'_{i'}$. The functor $\widehat{\Phi}_i^*$ induces a functor

$$\Phi_{i,s} : \widetilde{E}'_{\varphi(i)} \to \widetilde{E}_i. \tag{2.10.11.4}$$

This functor admits a left adjoint

$$\Phi_i^s : \widetilde{E}_i \to \widetilde{E}'_{\varphi(i)} \tag{2.10.11.5}$$

such that for every presheaf G on E_i, we have a canonical functorial isomorphism

$$\Phi_i^s(G^a) = (\widehat{\Phi}_{i!}(G))^a, \tag{2.10.11.6}$$

where the exponent ^{a} denotes the associated sheaves.

Lemma 2.10.12 *We keep the assumptions and notation of* 2.10.11. *Let, moreover,* $F = \{i \mapsto F_i\}$ *be a v-presheaf on* E (2.10.3). *Then,*

(i) *For all* $i' \in \mathrm{Ob}(I')$, *the sheaves* $f^*(\Phi_i^s(F_i))$, *for* $(i, f) \in \mathrm{Ob}((I_\varphi^{i'})^\circ)$, *naturally form a direct system of* $\widetilde{E}'_{i'}$. *We set*

$$F'_{i'} = \varinjlim_{(i,f)\in(I_\varphi^{i'})^\circ} f^*(\Phi_i^s(F_i)). \tag{2.10.12.1}$$

(ii) *The collection* $F' = \{i' \in I'^\circ \mapsto F'_{i'}\}$ *naturally forms a v-presheaf on* E'. *For any morphism* $h : i' \to j'$ *of* I' *and any object* (j, g) *of* $I_\varphi^{j'}$, *setting* $f = g \circ h : i' \to \varphi(j)$ *so that* $(j, f) = I_\varphi^h(j, g)$ (2.10.8.5), *the diagram*

$$\begin{array}{ccc} h^*(g^*(\Phi_j^s(F_i))) & \longrightarrow & f^*(\Phi_j^s(F_i)) \\ \downarrow & & \downarrow \\ h^*(F'_{j'}) & \longrightarrow & F'_{i'}, \end{array} \tag{2.10.12.2}$$

where the vertical arrows are the canonical morphisms (2.10.12.1), *the upper horizontal arrow is the canonical isomorphism and the lower horizontal arrow is the adjoint morphism of the morphism* $F'_{j'} \to h_*(F'_{i'})$ *defining the presheaf structure on* $\{i' \mapsto F'_{i'}\}$, *is commutative.*

(iii) *We have a canonical functorial isomorphism*

$$\Phi^s(F^a) \xrightarrow{\sim} F'^a, \tag{2.10.12.3}$$

where the exponent ^{a} *denotes the associated sheaves.*

Indeed, propositions (i)–(ii) are immediate and proposition (iii) follows from 2.10.9, (2.10.11.3), (2.10.11.6), ([3] VI 5.17) and the fact that the "associated sheaf" functor commutes with direct limits ([5] II 4.1).

2.10.13 We take again the assumptions and notation of 2.10.7. We denote by $\mathbb{I}$ the following category. The objects of $\mathbb{I}$ are the triples (i, i', f) consisting of an object i of I, an object i' of I' and a morphism $f : i' \to \varphi(i)$ of I'. Let (i_1, i'_1, f_1) and (i_2, i'_2, f_2) be two objects of $\mathbb{I}$. A morphism from (i_1, i'_1, f_1) to (i_2, i'_2, f_2) is a morphism $m: i_1 \to i_2$ of I and a morphism $m' : i'_1 \to i'_2$ of I' such that $\varphi(m) \circ f_1 = f_2 \circ m'$. Consider the functors

$$\mathrm{b}: \mathbb{I} \to I, \quad (i, i', f) \mapsto i, \tag{2.10.13.1}$$

$$\mathrm{s}: \mathbb{I} \to I', \quad (i, i', f) \mapsto i'. \tag{2.10.13.2}$$

Note that for every $i' \in \mathrm{Ob}(I')$, the category $I_\varphi^{i'}$ (2.10.8) is identified with the fiber category of the functor s over i'.

Let $\mathbb{J}$ be a full subcategory of $\mathbb{I}$, J and J' its essential images by the functors b and s respectively.

For any $i \in \mathrm{Ob}(I)$, we denote by $J_{/i}$ the category of pairs (j, m) made up of an object j of J and a morphism $m: j \to i$ of I, the morphisms of $J_{/i}$ being naturally induced by those of J. For any $i' \in \mathrm{Ob}(I')$, we define the category $J'_{/i'}$ in the same way. For any $(i, i', f) \in \mathrm{Ob}(\mathbb{I})$, we denote by $\mathbb{J}_{/(i,i',f)}$ the category of quintuples (j, j', g, m, m') consisting of an object (j, j', g) of $\mathbb{J}$ and a morphism $(m, m'): (j, j', g) \to (i, i', f)$ of $\mathbb{I}$, i.e., a commutative diagram

$$\begin{array}{ccc} j' & \xrightarrow{g} & \varphi(j) \\ {\scriptstyle m'}\downarrow & & \downarrow{\scriptstyle \varphi(m)} \\ i' & \xrightarrow{f} & \varphi(i) \end{array} \tag{2.10.13.3}$$

the morphisms of $\mathbb{J}_{/(i,i',f)}$ being naturally induced by those of $\mathbb{J}$.

We denote by $\pi_J: E_J \to J$ the fibered category deduced from π by base change by the canonical functor $J \to I$, by $\Psi: E_J \to E$ the canonical projection ([24] VI § 3) and by $\widehat{\Psi}^*: \widehat{E}_J \to \widehat{E}$ the functor defined by composition with Ψ. Likewise, we denote by $\pi'_{J'}: E'_{J'} \to J'$ the fibered category deduced from π' by base change by the canonical functor $J' \to I'$, by $\Psi': E'_{J'} \to E'$ the canonical projection and by $\widehat{\Psi}'^*: \widehat{E}'_{J'} \to \widehat{E}'$ the functor defined by composition with Ψ'.

Definition 2.10.14 We keep the assumptions and notation of 2.10.13. Let $F = \{j \in J^\circ \mapsto F_j\}$ be a presheaf on E_J, $F' = \{j' \in J'^\circ \mapsto F'_{j'}\}$ a presheaf on $E'_{J'}$. We call the *$\mathbb{J}$-system of Φ-compatible morphisms from F to F'* the data for any object (j, j', g) of $\mathbb{J}$, of a morphism

$$u_{(j,j',g)}: F_j \to \widehat{\Phi}_j^*(\widehat{g}^*(F'_{j'})) \tag{2.10.14.1}$$

of $\widehat{E}_j$, such that for every morphism $(m, m'): (j_1, j'_1, g_1) \to (j_2, j'_2, g_2)$ of $\mathbb{J}$,

$$\begin{array}{ccc} j'_1 & \xrightarrow{g_1} & \varphi(j_1) \\ {\scriptstyle m'}\downarrow & & \downarrow{\scriptstyle \varphi(m)} \\ j'_2 & \xrightarrow[g_2]{} & \varphi(j_2) \end{array} \tag{2.10.14.2}$$

setting $n = \varphi(m)$, the diagram

$$\begin{array}{ccccc} F_{j_2} & \xrightarrow{u_{(j_2,j'_2,g_2)}} & \widehat{\Phi}^*_{j_2}(hg_2^*(F'_{j'_2})) & \xrightarrow{\widehat{\Phi}^*_{j_2}(\widehat{g}_2^*(a'))} & \widehat{\Phi}^*_{j_2}(\widehat{g}_2^*(\widehat{m}'^*(F'_{j'_1}))) \\ {\scriptstyle a}\downarrow & & & & \downarrow{\scriptstyle v_2} \\ \widehat{m}^*(F_{j_1}) & \xrightarrow{\widehat{m}^*(u_{(j_1,j'_1,g_1)})} & \widehat{m}^*(\widehat{\Phi}^*_{j_1}(\widehat{g}_1^*(F'_{j'_1}))) & \xrightarrow{v_1} & \widehat{\Phi}^*_{j_2}(\widehat{n}^*(\widehat{g}_1^*(F'_{j'_1}))), \end{array} \tag{2.10.14.3}$$

where a (resp. $a' : F'_{j'_2} \to \widehat{m}'^*(F'_{j'_1})$) is the transition morphism of the sheaf F (resp. F'), v_1 is the isomorphism (2.10.7.9) and v_2 is the isomorphism induced by the commutative diagram (2.10.14.1), is commutative.

2.10.15 We keep the assumptions and notation of 2.10.13 and take, moreover, those of 2.10.11. Let $F = \{j \in J^\circ \mapsto F_j\}$ be a v-presheaf on E_J (2.10.3), $F' = \{j' \in J'^\circ \mapsto F'_{j'}\}$ a v-presheaf on $E'_{J'}$. Giving a $\mathbb{J}$-system of Φ-compatible morphisms from F to F' (2.10.14) is equivalent by adjunction to giving, for any object (j, j', g) of $\mathbb{J}$, a morphism of $\widetilde{E}'_{j'}$,

$$u'_{(j,j',g)} : g^*(\Phi^s_j(F_j)) \to F'_{j'}, \tag{2.10.15.1}$$

such that for every morphism $(m, m') : (j_1, j'_1, g_1) \to (j_2, j'_2, g_2)$ of $\mathbb{J}$,

$$\begin{array}{ccc} j'_1 & \xrightarrow{g_1} & \varphi(j_1) \\ {\scriptstyle m'}\downarrow & & \downarrow{\scriptstyle \varphi(m)} \\ j'_2 & \xrightarrow[g_2]{} & \varphi(j_2) \end{array} \tag{2.10.15.2}$$

setting $n = \varphi(m)$, the diagram

$$\begin{array}{ccccc} g_1^*(n^*(\Phi^s_{j_2}(F_{j_2}))) & \xrightarrow{v_2} & m'^*(g_2^*(\Phi^s_{j_2}(F_{j_2}))) & \xrightarrow{m'^*(u'_{(j_2,j'_2,f_2)})} & m'^*(F'_{j'_2}) \\ {\scriptstyle v_1}\downarrow & & & & \downarrow{\scriptstyle a'} \\ g_1^*(\Phi^s_{j_1}(m^*(F_{j_2}))) & \xrightarrow{g_1^*(\Phi^*_{j_1}(a))} & g_1^*(\Phi^s_{j_1}(F_{j_1})) & \xrightarrow{u'_{(j_1,j'_1,f_1)}} & F'_{j'_1}, \end{array} \tag{2.10.15.3}$$

where $a : m^*(F_{j_2}) \to F_{j_1}$ (resp. a') is the transition morphism of the v-presheaf F (resp. F'), v_1 is the adjoint isomorphism of (2.10.7.9) and v_2 is the isomorphism induced by the commutative diagram (2.10.15.2), is commutative.

Lemma 2.10.16 *We keep the assumptions and notation of* 2.10.13; *suppose, furthermore, that the following conditions are satisfied:*

(i) *Fiber products are representable in I and I' and φ commutes with these products;*
(ii) *For every object (i, i', f) of $\mathbb{I}$, every object ℓ' of J' and every morphism $u\colon \ell' \to i'$ of I', there exist an object (j, j', g) of $\mathbb{J}$, a morphism $v\colon \ell' \to j'$ of I' and a morphism $(m, m')\colon (j, j', g) \to (i, i', f)$ of $\mathbb{I}$ such that $u = m' \circ v$;*

$$\begin{array}{ccccc} & & \overset{u}{\frown} & & \\ \ell' & \xrightarrow{v} & j' & \xrightarrow{m'} & i' \\ & & \downarrow{\scriptstyle g} & & \downarrow{\scriptstyle f} \\ & & \varphi(j) & \xrightarrow{\varphi(m)} & \varphi(i). \end{array} \tag{2.10.16.1}$$

Let C be a category where inverse limits are representable, (i, i', f) an object of $\mathbb{I}$, $F\colon J'^{\circ}_{/i'} \to C$ a functor. Then, the canonical morphism

$$\varprojlim_{J'^{\circ}_{/i'}} F \to \varprojlim_{\mathbb{J}^{\circ}_{/(i,i',f)}} F \circ \mathrm{s}^{\circ}, \tag{2.10.16.2}$$

where we again denoted by $\mathrm{s}\colon \mathbb{J}_{/(i,i',f)} \to J'_{/i'}$ the functor induced by s (2.10.13.2), *is an isomorphism.*

Note first that it follows from (i) that the inverse limit of a diagram

$$\begin{array}{ccc} & & (i_1, i'_1, f_1) \\ & & \downarrow \\ (i_2, i'_2, f_2) & \longrightarrow & (i, i', f) \end{array} \tag{2.10.16.3}$$

of morphisms of $\mathbb{I}$ is representable by the object $(i_1 \times_i i_2, i'_1 \times_{i'} i'_2, f_1 \times f_2)$ of $\mathbb{I}$.

By (ii), for every object ℓ' of $J'_{/i'}$, there exist an object (j, j', g) of $\mathbb{J}_{/(i,i',f)}$ and an i'-morphism $v\colon \ell' \to i'$. The composed morphism

$$\varprojlim_{\mathbb{J}^{\circ}_{/(i,i',f)}} F \circ \mathrm{s}^{\circ} \to F(j') \to F(\ell'), \tag{2.10.16.4}$$

where the first arrow is the canonical morphism and the second arrow is induced by v, is independent of the choices of (j, j', g) and v. Indeed, if (j_1, j'_1, g_1) and (j_2, j'_2, g_2) are two objects of $\mathbb{J}_{/(i,i',f)}$ and if $v_1\colon \ell' \to j'_1$ and $v_2\colon \ell' \to j'_2$ are two i'-morphisms, there exists, by (ii), an object (j_3, j'_3, g_3) of $\mathbb{J}$, a morphism $v_3\colon \ell' \to i_3$ of I' and a morphism $(m, m')\colon (j_3, j'_3, g_3) \to (j_1 \times_i j_2, j'_1 \times_{i'} j'_2, g_1 \times g_2)$ of $\mathbb{I}$ such that $v_1 \times v_2 = m' \circ v_3$;

$$\begin{array}{ccc} \ell' \xrightarrow{v_3} j'_3 & \xrightarrow{m'} & j'_1 \times_{i'} j'_2 \\ \downarrow g_3 & & \downarrow g_1\times g_2 \\ \varphi(j_3) & \xrightarrow{\varphi(m)} & \varphi(j_1 \times_i j_2). \end{array} \quad (v_1\times v_2 : \ell' \to j'_1 \times_{i'} j'_2) \tag{2.10.16.5}$$

It immediately follows that the morphism (2.10.16.2) is an isomorphism.

Lemma 2.10.17 *We keep the assumptions and notation of* 2.10.13; *suppose, furthermore, that one of the following two conditions is satisfied:*

(a) *For all* $(i, i', f) \in \mathrm{Ob}(\mathbb{I})$, *the functor* $\mathbb{J}_{/(i,i',f)} \to J'_{/i'}$ *induced by the functor* s (2.10.13.2) *is cofinal* ([5] I 8.1.1)*;*
(b) *The conditions* 2.10.16(i)-(ii) *are satisfied.*

Let $F = \{j \in J^\circ \mapsto F_j\}$ *be a presheaf on* E_J, $F' = \{j' \in J'^\circ \mapsto F'_{j'}\}$ *a presheaf on* $E'_{J'}$,

$$u_{(j,j',g)} \colon F_j \to \widehat{\Phi}^*_j(\widehat{g}^*(F'_{j'})), \quad (j, j', g) \in \mathrm{Ob}(\mathbb{J}), \tag{2.10.17.1}$$

a $\mathbb{J}$*-system of compatible* Φ*-morphisms from* F *to* F'. *Then, there exist a canonical presheaf* $G = \{i \in I^\circ \mapsto G_i\}$ *on* E, *a canonical presheaf* $G' = \{i' \in I'^\circ \mapsto G'_{i'}\}$ *on* E', *a canonical* $\mathbb{I}$*-system of compatible* Φ*-morphisms from* G *to* G',

$$v_{(i,i',f)} \colon G_i \to \widehat{\Phi}^*_i(\widehat{f}^*(G'_{i'})), \quad (i, i', f) \in \mathrm{Ob}(\mathbb{I}), \tag{2.10.17.2}$$

and isomorphisms $h \colon \widehat{\Psi}^*(G) \xrightarrow{\sim} F$ *and* $h' \colon \widehat{\Psi}'^*(G') \xrightarrow{\sim} F'$ *such that for all* $(j, j', g) \in \mathrm{Ob}(\mathbb{J})$, *we have*

$$u_{(j,j',f)} \circ h_j = \widehat{\Phi}^*_i(\widehat{g}^*(h'_{j'})) \circ v_{(j,j',g)}. \tag{2.10.17.3}$$

For any object i of I, let

$$G_i = \lim_{\substack{\longleftarrow \\ (j,m)\in J^\circ_{/i}}} \widehat{m}^*(F_j). \tag{2.10.17.4}$$

The collection $G = \{i \in I^\circ \mapsto G_i\}_{i\in I^\circ}$ then forms a presheaf on E and we have a canonical isomorphism $h \colon \widehat{\Psi}^*(G) \xrightarrow{\sim} F$. For any object i' of I', let

$$G'_{i'} = \lim_{\substack{\longleftarrow \\ (j',m')\in J'^\circ_{/i'}}} \widehat{m}'^*(F'_{j'}). \tag{2.10.17.5}$$

The collection $G' = \{i' \in I'^\circ \mapsto G'_{i'}\}_{i'\in I'^\circ}$ then forms a presheaf on E' and we have a canonical isomorphism $h' \colon \widehat{\Psi}'^*(G') \xrightarrow{\sim} F'$.

Let (i, i', f) be an object of $\mathbb{I}$, (j', j', g, m, m') be an object of $\mathbb{J}_{/(i,i',f)}$;

$$\begin{array}{ccc} j' & \xrightarrow{g} & \varphi(j) \\ \downarrow m' & & \downarrow \varphi(m) \\ i' & \xrightarrow{f} & \varphi(i). \end{array} \tag{2.10.17.6}$$

Setting $n = \varphi(m)$, we define the morphism $v_{(j,j',g,m,m')}$ by the commutative diagram

$$\begin{array}{ccccc} G_i & \xrightarrow{\quad v_{(j,j',g,m,m')} \quad} & & & \widehat{\Phi}_i^*(\widehat{f}^*(\widehat{m}'^*(F'_{j'}))) \\ \downarrow a & & & & \uparrow w_2 \\ \widehat{m}^*(F_j) & \xrightarrow{\widehat{m}^*(u_{(j,j',g)})} & \widehat{m}^*(\widehat{\Phi}_j^*(\widehat{g}^*(F'_{j'}))) & \xrightarrow{w_1} & \widehat{\Phi}_i^*(\widehat{n}^*(\widehat{g}^*(F'_{j'}))) \end{array} \tag{2.10.17.7}$$

where a is the canonical morphism (2.10.17.4), w_1 the isomorphism (2.10.7.9) and w_2 the isomorphism induced by the commutative diagram (2.10.17.6). By the relations (2.10.14.3), the morphisms $v_{(j,j',g,m,m')}$ are compatible for $(j, j', g, m, m') \in \mathrm{Ob}(\mathbb{J}_{/(i,i',f)})$. In view of the assumptions and 2.10.16, taking the inverse limit on $\mathbb{J}^\circ_{/(i,i',f)}$, we obtain a morphism

$$v_{(i,i',f)} : G_i \to \widehat{\Phi}_i^*(\widehat{f}^*(G'_{i'})). \tag{2.10.17.8}$$

Indeed, the functors $\widehat{\Phi}_i^*$ and $\widehat{f}^*$ commute with inverse limits since they admit left adjoints. We then painstakingly verify that the morphisms $v_{(i,i',f)}$ for $(i, i', f) \in \mathrm{Ob}(\mathbb{I})$ form an $\mathbb{I}$-system of Φ-compatible morphisms from G to G' which satisfies the required conditions.

2.10.18 We keep the assumptions and notation of 2.10.17 and take moreover those of 2.10.8. For any object (i, i', f) of $\mathbb{I}$, we denote by

$$v'_{(i,i',f)} : \widehat{f}_!(\Phi_{i!}(G_i)) \to G'_{i'} \tag{2.10.18.1}$$

the adjoint morphism of $v_{(i,i',f)}$ (2.10.17.2). By (2.10.14.3), for every $i' \in \mathrm{Ob}(I')$, the morphisms $v'_{(i,i',f)}$, for $(i, f) \in I^{i'}_\varphi$, are compatible. They therefore induce a morphism

$$v'_{i'} : \varinjlim_{(i,f)\in(I^{i'}_\varphi)^\circ} \widehat{f}_!(\Phi_{i!}(G_i)) \to G'_{i'}. \tag{2.10.18.2}$$

In view of 2.10.9 and (2.10.14.3), the morphisms $v'_{i'}$, for $i' \in \mathrm{Ob}(I')$, define a morphism of $\widehat{E}'$,

$$v' : \Phi_!(G) \to G'. \tag{2.10.18.3}$$

2.10.19 We keep the assumptions and notation of 2.10.17 and take moreover those of 2.10.11. Suppose also that the subcategory J (resp. J') of I (resp. I') is topologically generating. By 2.10.5, the functor $\Psi : E_J \to E$ is continuous and cocontinuous and the functor $\widehat{\Psi}^*$ induces an equivalence of categories

$$\Psi_s : \widetilde{E} \xrightarrow{\sim} \widetilde{E}_J. \tag{2.10.19.1}$$

Likewise, the functor $\Psi' : E'_{J'} \to E'$ is continuous and cocontinuous and the functor $\widehat{\Psi}'^*$ induces an equivalence of categories

$$\Psi'_s : \widetilde{E}' \xrightarrow{\sim} \widetilde{E}'_{J'}. \tag{2.10.19.2}$$

Let $F = \{j \in J^\circ \mapsto F_j\}$ be a v-presheaf on E_J (2.10.3), $F' = \{j' \in J'^\circ \mapsto F'_{j'}\}$ a v-presheaf on $E'_{J'}$,

$$u'_{(j,j',g)} : g^*(\Phi^s_j(F_j)) \to F'_{j'}, \quad (j, j', g) \in \mathrm{Ob}(\mathbb{J}), \tag{2.10.19.3}$$

a $\mathbb{J}$-system of compatible Φ-morphisms from F to F' (2.10.15). By 2.10.17 and its proof, there exist a canonical v-presheaf $G = \{i \in I^\circ \mapsto G_i\}$ on E, a canonical v-presheaf $G' = \{i' \in I'^\circ \mapsto G'_{i'}\}$ on E', a canonical $\mathbb{I}$-system of compatible Φ-morphisms from G to G',

$$v'_{(i,i',f)} : f^*(\Phi^s_i(G_i)) \to G'_{i'}, \quad (i, i', f) \in \mathrm{Ob}(\mathbb{I}), \tag{2.10.19.4}$$

and isomorphisms $h: \widehat{\Psi}^*(G) \xrightarrow{\sim} F$ and $h' : \widehat{\Psi}'^*(G') \xrightarrow{\sim} F'$ such that for every $(j, j', g) \in \mathrm{Ob}(\mathbb{J})$, we have

$$u'_{(j,j',f)} \circ g^*(\Phi^s_j(h_j)) = h'_{j'} \circ v'_{(i,i',f)}. \tag{2.10.19.5}$$

By 2.10.6, we have canonical isomorphisms $\Psi_s(G^a) \xrightarrow{\sim} F^a$ and $\Psi'_s(G'^a) \xrightarrow{\sim} F'^a$, where the exponent ^{a} denotes the associated sheaves. For all $i' \in \mathrm{Ob}(I')$, the morphisms $v'_{(i,i',f)}$, for $(i, f) \in I^{i'}_\varphi$, are compatible (2.10.15.3). They therefore induce a morphism

$$v'_{i'} : \varinjlim_{(i,f)\in(I^{i'}_\varphi)^\circ} f^*(\Phi^s_i(G_i)) \to G'_{i'}. \tag{2.10.19.6}$$

In view of 2.10.12 and (2.10.15.3), the morphisms $v'_{i'}$, for $i' \in \mathrm{Ob}(I')$, induce a morphism of $\widetilde{E}'$,

$$v' : \Phi^s(G^a) \to G'^a. \tag{2.10.19.7}$$

Chapter 3
The p-adic Simpson Correspondence and Hodge–Tate Modules. Local Study

3.1 Assumptions and Notation. p-adic Infinitesimal Deformations

3.1.1 In this chapter, K denotes a complete discrete valuation field of characteristic 0, with perfect residue field k of characteristic $p > 0$, $\mathscr{O}_K$ the valuation ring of K, W the ring of Witt vectors with respect to p and with coefficients in k, $\overline{K}$ an algebraic closure of K, $\mathscr{O}_{\overline{K}}$ the integral closure of $\mathscr{O}_K$ in $\overline{K}$, $\mathfrak{m}_{\overline{K}}$ the maximal ideal of $\mathscr{O}_{\overline{K}}$ and G_K the Galois group of $\overline{K}$ over K. We denote by $\mathscr{O}_C$ the p-adic Hausdorff completion of $\mathscr{O}_{\overline{K}}$, by $\mathfrak{m}_C$ its maximal ideal, by C its field of fractions and by v its valuation, normalized by $v(p) = 1$. We denote by $\widehat{\mathbb{Z}}(1)$ and $\mathbb{Z}_p(1)$ the $\mathbb{Z}[G_K]$-modules

$$\widehat{\mathbb{Z}}(1) = \lim_{\substack{\longleftarrow \\ n\geq 1}} \mu_n(\mathscr{O}_{\overline{K}}), \tag{3.1.1.1}$$

$$\mathbb{Z}_p(1) = \lim_{\substack{\longleftarrow \\ n\geq 0}} \mu_{p^n}(\mathscr{O}_{\overline{K}}), \tag{3.1.1.2}$$

where $\mu_n(\mathscr{O}_{\overline{K}})$ denotes the subgroup of nth roots of unity in $\mathscr{O}_{\overline{K}}$. For any $\mathbb{Z}_p[G_K]$-module M and any integer n, we set $M(n) = M \otimes_{\mathbb{Z}_p} \mathbb{Z}_p(1)^{\otimes n}$.

For any $\mathbb{Z}_p$-module A, we denote by $\widehat{A}$ its p-adic Hausdorff completion.

3.1.2 We set $S = \operatorname{Spec}(\mathscr{O}_K)$, $\overline{S} = \operatorname{Spec}(\mathscr{O}_{\overline{K}})$ and $\check{\overline{S}} = \operatorname{Spec}(\mathscr{O}_C)$. We denote by s (resp. η, resp. $\overline{\eta}$) the closed point of S (resp. generic point of S, resp. generic point of $\overline{S}$). For any integer $n \geq 1$, we set $S_n = \operatorname{Spec}(\mathscr{O}_K/p^n\mathscr{O}_K)$. For any S-scheme X, we set

$$\overline{X} = X \times_S \overline{S}, \quad \check{\overline{X}} = X \times_S \check{\overline{S}} \quad \text{and} \quad X_n = X \times_S S_n. \tag{3.1.2.1}$$

We endow S with the logarithmic structure $\mathscr{M}_S$ defined by its closed point, and $\overline{S}$ and $\check{\overline{S}}$ with the logarithmic structures $\mathscr{M}_{\overline{S}}$ and $\mathscr{M}_{\check{\overline{S}}}$ pullbacks of $\mathscr{M}_S$.

A. Abbes, M. Gros, *The p-adic Simpson Correspondence and Hodge-Tate Local Systems*, Lecture Notes in Mathematics 2345, https://doi.org/10.1007/978-3-031-55914-3_3

3.1.3 Since $\mathscr{O}_{\overline{K}}$ is a non-discrete valuation ring of height 1, it is possible to develop the α-algebra (or almost algebra) over this ring ([2] 2.10.1) (see [2] 2.6-2.10). We choose a compatible system $(\beta_n)_{n>0}$ of nth roots of p in $\mathscr{O}_{\overline{K}}$. For any rational number $\varepsilon > 0$, we set $p^{\varepsilon} = (\beta_n)^{\varepsilon n}$, where n is an integer > 0 such that εn is an integer.

3.1.4 All rings of Witt vectors considered in this chapter are with respect to p (2.1.3). We denote by $\mathscr{O}_{\overline{K}^\flat}$ the inverse limit of the inverse system $(\mathscr{O}_{\overline{K}}/p\mathscr{O}_{\overline{K}})_{\mathbb{N}}$ whose transition morphisms are the iterates of the absolute Frobenius endomorphism of $\mathscr{O}_{\overline{K}}/p\mathscr{O}_{\overline{K}}$;

$$\mathscr{O}_{\overline{K}^\flat} = \varprojlim_{\mathbb{N}} \mathscr{O}_{\overline{K}}/p\mathscr{O}_{\overline{K}}. \tag{3.1.4.1}$$

It is a non-discrete valuation ring of height 1, complete and perfect of characteristic p ([2] 4.8.1 and 4.8.2). We denote by $\overline{K}^\flat$ its field of fractions and by $\mathfrak{m}_{\overline{K}^\flat}$ its maximal ideal.

We fix a sequence $(p_n)_{n\geq 0}$ of elements of $\mathscr{O}_{\overline{K}}$ such that $p_0 = p$ and $p^p_{n+1} = p_n$ (for every $n \geq 0$) and denote by ϖ the associated element of $\mathscr{O}_{\overline{K}^\flat}$. We set

$$\xi = [\varpi] - p \in \mathrm{W}(\mathscr{O}_{\overline{K}^\flat}), \tag{3.1.4.2}$$

where [ ] is the multiplicative representative. We take again the notation of 2.2.3 for $A = \mathscr{O}_{\overline{K}}$, in particular, let

$$\mathscr{A}_2(\mathscr{O}_{\overline{K}}) = \mathrm{W}(\mathscr{O}_{\overline{K}^\flat})/\ker(\theta)^2, \tag{3.1.4.3}$$

and denote by $\theta_2 \colon \mathscr{A}_2(\mathscr{O}_{\overline{K}}) \to \mathscr{O}_C$ the homomorphism induced by θ (2.2.3.4). By 2.3.7, the sequence

$$0 \longrightarrow \mathrm{W}(\mathscr{O}_{\overline{K}^\flat}) \xrightarrow{\cdot\xi} \mathrm{W}(\mathscr{O}_{\overline{K}^\flat}) \xrightarrow{\theta} \mathscr{O}_C \longrightarrow 0 \tag{3.1.4.4}$$

is exact. It therefore induces an exact sequence

$$0 \longrightarrow \mathscr{O}_C \xrightarrow{\cdot\xi} \mathscr{A}_2(\mathscr{O}_{\overline{K}}) \xrightarrow{\theta_2} \mathscr{O}_C \longrightarrow 0, \tag{3.1.4.5}$$

where we have denoted also by $\cdot\xi$ the morphism induced by the multiplication by ξ in $\mathscr{A}_2(\mathscr{O}_{\overline{K}})$. The square zero ideal $\ker(\theta_2)$ of $\mathscr{A}_2(\mathscr{O}_{\overline{K}})$ is a free $\mathscr{O}_C$-module with basis ξ. It will be denoted by $\xi\mathscr{O}_C$. Unlike ξ, it does not depend on the choice of the sequence $(p_n)_{n\geq 0}$. We denote by $\xi^{-1}\mathscr{O}_C$ the dual $\mathscr{O}_C$-module of $\xi\mathscr{O}_C$. For any $\mathscr{O}_C$-module M, we denote the $\mathscr{O}_C$-modules $M \otimes_{\mathscr{O}_C} (\xi\mathscr{O}_C)$ and $M \otimes_{\mathscr{O}_C} (\xi^{-1}\mathscr{O}_C)$ simply by ξM and by $\xi^{-1}M$, respectively.

The Galois group G_K acts naturally on $\mathrm{W}(\mathscr{O}_{\overline{K}^\flat})$ by ring automorphisms, and the homomorphism θ is G_K-equivariant. We deduce an action of G_K on $\mathscr{A}_2(\mathscr{O}_{\overline{K}})$ by ring automorphisms such that the homomorphism θ_2 is G_K-equivariant.

3.1.5 We have a canonical homomorphism

$$\mathbb{Z}_p(1) \to \mathscr{O}^{\times}_{\overline{K}^{\flat}}. \tag{3.1.5.1}$$

For any $\zeta \in \mathbb{Z}_p(1)$, we denote also by ζ its image in $\mathscr{O}^{\times}_{\overline{K}^{\flat}}$. Since $\theta([\zeta] - 1) = 0$, we obtain a homomorphism of groups

$$\mathbb{Z}_p(1) \to \mathscr{A}_2(\mathscr{O}_{\overline{K}}), \quad \zeta \mapsto \log([\zeta]) = [\zeta] - 1, \tag{3.1.5.2}$$

whose image is contained in $\ker(\theta_2) = \xi\mathscr{O}_C$. It is clearly $\mathbb{Z}_p$-linear. By ([3] II.9.18), its image generates the ideal $p^{\frac{1}{p-1}}\xi\mathscr{O}_C$ of $\mathscr{A}_2(\mathscr{O}_{\overline{K}})$, and the induced $\mathscr{O}_C$-linear morphism

$$\mathscr{O}_C(1) \to p^{\frac{1}{p-1}}\xi\mathscr{O}_C \tag{3.1.5.3}$$

is an isomorphism.

3.1.6 The ring $\mathrm{W}(\mathscr{O}_{\overline{K}^{\flat}})$ being naturally a W-algebra (2.2.4), we set

$$\mathrm{W}_{\mathscr{O}_K}(\mathscr{O}_{\overline{K}^{\flat}}) = \mathrm{W}(\mathscr{O}_{\overline{K}^{\flat}}) \otimes_W \mathscr{O}_K \tag{3.1.6.1}$$

and denote by $\theta_{\mathscr{O}_K} \colon \mathrm{W}_{\mathscr{O}_K}(\mathscr{O}_{\overline{K}^{\flat}}) \to \mathscr{O}_C$ the homomorphism induced by θ (3.1.4.4). We fix a uniformizer π of $\mathscr{O}_K$ and a sequence $(\pi_n)_{n\geq 0}$ of elements of $\mathscr{O}_{\overline{K}}$ such that $\pi_0 = \pi$ and $\pi_{n+1}^p = \pi_n$ (for every $n \geq 0$) and denote by $\underline{\pi}$ the associated element of $\mathscr{O}_{\overline{K}^{\flat}}$. We set

$$\xi_\pi = [\underline{\pi}] - \pi \in \mathrm{W}_{\mathscr{O}_K}(\mathscr{O}_{\overline{K}^{\flat}}). \tag{3.1.6.2}$$

By 2.3.7, the sequence

$$0 \longrightarrow \mathrm{W}_{\mathscr{O}_K}(\mathscr{O}_{\overline{K}^{\flat}}) \xrightarrow{\cdot \xi_\pi} \mathrm{W}_{\mathscr{O}_K}(\mathscr{O}_{\overline{K}^{\flat}}) \xrightarrow{\theta_{\mathscr{O}_K}} \mathscr{O}_C \longrightarrow 0 \tag{3.1.6.3}$$

is exact. In particular, we have $\xi \in \xi_\pi \mathrm{W}_{\mathscr{O}_K}(\mathscr{O}_{\overline{K}^{\flat}})$ (3.1.4.2). We set

$$\mathrm{W}_K(\mathscr{O}_{\overline{K}^{\flat}}) = \mathrm{W}(\mathscr{O}_{\overline{K}^{\flat}}) \otimes_W K \tag{3.1.6.4}$$

and denote by $\theta_K \colon \mathrm{W}_K(\mathscr{O}_{\overline{K}^{\flat}}) \to C$ the homomorphism induced by θ (3.1.4.4). We denote by $\mathrm{W}^*_{\mathscr{O}_K}(\mathscr{O}_{\overline{K}^{\flat}})$ the sub-$\mathrm{W}_{\mathscr{O}_K}(\mathscr{O}_{\overline{K}^{\flat}})$-algebra of $\mathrm{W}_K(\mathscr{O}_{\overline{K}^{\flat}})$ generated by $[\underline{\pi}]/\pi$ and we set

$$\xi^*_\pi = \frac{\xi_\pi}{\pi} = \frac{[\underline{\pi}]}{\pi} - 1 \in \mathrm{W}^*_{\mathscr{O}_K}(\mathscr{O}_{\overline{K}^{\flat}}). \tag{3.1.6.5}$$

Note that $\mathrm{W}^*_{\mathscr{O}_K}(\mathscr{O}_{\overline{K}^{\flat}})$ depends on the sequence $(\pi_n)_{n\geq 0}$. The homomorphism θ_K induces a homomorphism

$$\theta^*_{\mathscr{O}_K} \colon \mathrm{W}^*_{\mathscr{O}_K}(\mathscr{O}_{\overline{K}^{\flat}}) \to \mathscr{O}_C \tag{3.1.6.6}$$

such that $\theta^*_{\mathscr{O}_K}(\xi^*_\pi) = 0$. By 2.4.4(ii), the sequence

$$0 \longrightarrow \mathrm{W}^*_{\mathscr{O}_K}(\mathscr{O}_{\overline{K}^\flat}) \xrightarrow{\cdot \xi^*_\pi} \mathrm{W}^*_{\mathscr{O}_K}(\mathscr{O}_{\overline{K}^\flat}) \xrightarrow{\theta^*_{\mathscr{O}_K}} \mathscr{O}_C \longrightarrow 0 \tag{3.1.6.7}$$

is exact. We set

$$\mathscr{A}^*_2(\mathscr{O}_{\overline{K}}/\mathscr{O}_K) = \mathrm{W}^*_{\mathscr{O}_K}(\mathscr{O}_{\overline{K}^\flat})/(\xi^*_\pi)^2\mathrm{W}^*_{\mathscr{O}_K}(\mathscr{O}_{\overline{K}^\flat}) \tag{3.1.6.8}$$

and denote by $\theta^*_{\mathscr{O}_K,2} \colon \mathscr{A}^*_2(\mathscr{O}_{\overline{K}}/\mathscr{O}_K) \to \mathscr{O}_C$ the homomorphism induced by $\theta^*_{\mathscr{O}_K}$. The sequence (3.1.6.7) induces an exact sequence

$$0 \longrightarrow \mathscr{O}_C \xrightarrow{\cdot \xi^*_\pi} \mathscr{A}^*_2(\mathscr{O}_{\overline{K}}/\mathscr{O}_K) \xrightarrow{\theta^*_{\mathscr{O}_K,2}} \mathscr{O}_C \longrightarrow 0\,, \tag{3.1.6.9}$$

where we have denoted also by $\cdot\xi^*_\pi$ the morphism induced by the multiplication by ξ^*_π in $\mathscr{A}^*_2(\mathscr{O}_{\overline{K}}/\mathscr{O}_K)$. The square zero ideal $\ker(\theta^*_{\mathscr{O}_K,2})$ of $\mathscr{A}^*_2(\mathscr{O}_{\overline{K}}/\mathscr{O}_K)$ is a free $\mathscr{O}_C$-module with basis ξ^*_π. We denote it by $\xi^*_\pi\mathscr{O}_C$ and by $(\xi^*_\pi)^{-1}\mathscr{O}_C$ its dual $\mathscr{O}_C$-module. For any $\mathscr{O}_C$-module M, we denote the $\mathscr{O}_C$-modules $M \otimes_{\mathscr{O}_C} (\xi^*_\pi\mathscr{O}_C)$ and $M \otimes_{\mathscr{O}_C} ((\xi^*_\pi)^{-1}\mathscr{O}_C)$ simply by $\xi^*_\pi M$ and $(\xi^*_\pi)^{-1}M$, respectively.

3.1.7 For any integer $n \geq 1$, we denote by $\chi_n \colon G_K \to \mu_{p^n}(\mathscr{O}_{\overline{K}})$ the homomorphism defined for any $g \in G_K$ by

$$g(\pi_n) = \chi_n(g)\pi_n, \tag{3.1.7.1}$$

and by

$$\chi \colon G_K \to \mathbb{Z}_p(1) \tag{3.1.7.2}$$

the inverse limit of the χ_n's (3.1.1.2). We denote again by $\chi \colon G_K \to \mathscr{O}^\times_{\overline{K}^\flat}$ the composition of (3.1.7.2) and the canonical homomorphism $\mathbb{Z}_p(1) \to \mathscr{O}^\times_{\overline{K}^\flat}$ (3.1.5.1). For every $g \in G_K$, we then have $g([\underline{\pi}]) = [\chi(g)].[\underline{\pi}]$. Consequently, the natural action of G_K on $\mathrm{W}_K(\mathscr{O}_{\overline{K}^\flat})$ preserves $\mathrm{W}^*_{\mathscr{O}_K}(\mathscr{O}_{\overline{K}^\flat})$, and the homomorphism $\theta^*_{\mathscr{O}_K}$ is G_K-equivariant. We deduce an action of G_K on $\mathscr{A}^*_2(\mathscr{O}_{\overline{K}}/\mathscr{O}_K)$ by ring automorphisms such that the homomorphism $\theta^*_{\mathscr{O}_K,2}$ is G_K-equivariant.

3.1.8 Let K' be a finite extension of K contained in $\overline{K}$, $\mathscr{O}_{K'}$ the integral closure of $\mathscr{O}_K$ in K', K'_0 the maximal unramified extension of K contained in K', $\mathscr{O}_{K'_0}$ the integral closure of $\mathscr{O}_K$ in K'_0, e the degree of the extension K'/K'_0. We fix a uniformizer π' of $\mathscr{O}_{K'}$ and a sequence $(\pi'_n)_{n\geq 0}$ of elements of $\mathscr{O}_{\overline{K}}$ such that $\pi'_0 = \pi'$ and $\pi'^p_{n+1} = \pi'_n$ (for every $n \geq 0$), and denote by $\underline{\pi}'$ the associated element of $\mathscr{O}_{\overline{K}^\flat}$. We set

$$\xi_{\pi'} = [\underline{\pi}'] - \pi' \in \mathrm{W}_{\mathscr{O}_{K'}}(\mathscr{O}_{\overline{K}^\flat}). \tag{3.1.8.1}$$

Let $f \in \mathscr{O}_{K'_0}[T]$ be an Eisenstein polynomial of degree e such that $f(\pi') = 0$, $f' \in \mathscr{O}_{K'_0}[T]$ its derivative, $g \in \mathscr{O}_{K'}[T]$ the polynomial defined by the relation $f(T) = (T - \pi')g(T)$. We then have

$$\theta_{\mathscr{O}_{K'}}(g([\underline{\pi}'])) = f'(\pi') \in \mathscr{O}_C. \tag{3.1.8.2}$$

Proposition 3.1.9 *Under the assumptions of* 3.1.8, *there exists a unit u of* $\mathrm{W}_{\mathscr{O}_{K'}}(\mathscr{O}_{\overline{K}^\flat})$ *such that*

$$\xi_\pi = u \cdot g([\underline{\pi}']) \cdot \xi_{\pi'} \in \mathrm{W}_{\mathscr{O}_{K'}}(\mathscr{O}_{\overline{K}^\flat}). \tag{3.1.9.1}$$

Indeed, if $K' = K_0'$, we have $g = 1$ and the proposition follows immediately since ξ_π and $\xi_{\pi'}$ are two generators of $\ker(\theta_{\mathscr{O}_{K'}})$ by (3.1.6.3). We may therefore restrict to the case where $K = K_0'$. We have $\theta_{\mathscr{O}_K}(f([\underline{\pi}'])) = 0$ and $f([\underline{\pi}']) \bmod \pi = \underline{\pi}'^e$. So there exists a unit v of $\mathscr{O}_{\overline{K}^\flat}$ such that $f([\underline{\pi}']) \bmod \pi = v\underline{\pi}$. Moreover, by (2.3.7.4), the sequence

$$0 \longrightarrow \mathscr{O}_{\overline{K}^\flat} \xrightarrow{\cdot\underline{\pi}} \mathscr{O}_{\overline{K}^\flat} \longrightarrow \mathscr{O}_{\overline{K}}/\pi\mathscr{O}_{\overline{K}} \longrightarrow 0 \tag{3.1.9.2}$$

where the third arrow is induced by the canonical projection $\mathscr{O}_{\overline{K}^\flat} \to \mathscr{O}_{\overline{K}}/p\mathscr{O}_{\overline{K}}$ onto the first factor of the inverse system (3.1.4.1), is exact. It then follows from 2.3.6 that $f([\underline{\pi}'])$ generates $\ker(\theta_{\mathscr{O}_K})$. The proposition follows in view of (3.1.6.3).

Corollary 3.1.10 *Denoting by K_0 the field of fractions of* W (3.1.1) *and by $\mathfrak{d}$ the different of the extension K/K_0, the canonical homomorphism $\mathscr{A}_2(\mathscr{O}_{\overline{K}}) \to \mathscr{A}_2^*(\mathscr{O}_{\overline{K}}/\mathscr{O}_K)$ induces an $\mathscr{O}_C$-linear isomorphism*

$$\xi\mathscr{O}_C \xrightarrow{\sim} \pi\mathfrak{d}\xi_\pi^*\mathscr{O}_C. \tag{3.1.10.1}$$

This follows from 3.1.9 and (3.1.8.2).

3.1.11 Consider the inverse system of multiplicative monoids $(\mathscr{O}_{\overline{K}})_{n\in\mathbb{N}}$, where the transition morphisms are all equal to the pth power. We denote by Q_S the fiber product of the diagram of homomorphisms of monoids

$$\begin{array}{ccc} & & \mathscr{O}_K - \{0\} \\ & & \downarrow \\ \varprojlim\limits_{x\mapsto x^p} \mathscr{O}_{\overline{K}} & \longrightarrow & \mathscr{O}_{\overline{K}} \end{array} \tag{3.1.11.1}$$

where the vertical arrow is the canonical homomorphism and the horizontal arrow is the projection onto the first component (i.e., of index 0). We denote by τ_S the composed homomorphism

$$\tau_S : Q_S \longrightarrow \varprojlim_{x\mapsto x^p} \mathscr{O}_{\overline{K}} \longrightarrow \mathscr{O}_{\overline{K}^\flat} \xrightarrow{[\,]} \mathrm{W}(\mathscr{O}_{\overline{K}^\flat}), \tag{3.1.11.2}$$

where [ ] is the multiplicative representative and the other arrows are the canonical morphisms. It immediately follows from the definitions that the diagram

$$\begin{array}{ccc} Q_S & \longrightarrow & \mathscr{O}_K - \{0\} \\ {\scriptstyle \tau_S}\downarrow & & \downarrow \\ \mathrm{W}(\mathscr{O}_{\overline{K}^\flat}) & \xrightarrow{\theta} & \mathscr{O}_C, \end{array} \tag{3.1.11.3}$$

where the unlabeled arrows are the canonical morphisms, is commutative. Moreover, the Galois group G_K acts naturally on the monoid Q_S, and the homomorphism τ_S is G_K-equivariant.

3.1.12 We set

$$\mathscr{A}_2(\overline{S}) = \mathrm{Spec}(\mathscr{A}_2(\mathscr{O}_{\overline{K}})) \tag{3.1.12.1}$$

which we endow with the logarithmic structure $\mathscr{M}_{\mathscr{A}_2(\overline{S})}$ associated with the prelogarithmic structure defined by the homomorphism $Q_S \to \mathscr{A}_2(\mathscr{O}_{\overline{K}})$ induced by τ_S (3.1.11.2). By virtue of 2.4.2, $\mathscr{M}_{\mathscr{A}_2(\overline{S})}$ is the logarithmic structure on $\mathscr{A}_2(\overline{S})$ associated with the prelogarithmic structure defined by the homomorphism

$$\mathbb{N} \to \mathscr{A}_2(\mathscr{O}_{\overline{K}}), \quad 1 \mapsto [\underline{\pi}]. \tag{3.1.12.2}$$

Indeed, denoting by $\widetilde{\pi}$ the element of Q_S defined by its projections (3.1.6)

$$(\pi_n)_{n\in\mathbb{N}} \in \varprojlim_{\mathbb{N}} \mathscr{O}_{\overline{K}} \quad \text{and} \quad \pi \in \mathscr{O}_K - \{0\}, \tag{3.1.12.3}$$

we have $\tau_S(\widetilde{\pi}) = [\underline{\pi}] \in \mathrm{W}(\mathscr{O}_{\overline{K}^\flat})$. The logarithmic scheme $(\mathscr{A}_2(\overline{S}), \mathscr{M}_{\mathscr{A}_2(\overline{S})})$ is therefore fine and saturated, and θ_2 (3.1.4.3) induces an exact closed immersion (3.1.1)

$$(\check{\overline{S}}, \mathscr{M}_{\check{\overline{S}}}) \to (\mathscr{A}_2(\overline{S}), \mathscr{M}_{\mathscr{A}_2(\overline{S})}). \tag{3.1.12.4}$$

3.1.13 We set

$$\mathscr{A}_2^*(\overline{S}/S) = \mathrm{Spec}(\mathscr{A}_2^*(\mathscr{O}_{\overline{K}}/\mathscr{O}_K)), \tag{3.1.13.1}$$

which we endow with the logarithmic structure $\mathscr{M}_{\mathscr{A}_2^*(\overline{S}/S)}$ associated with the prelogarithmic structure defined by the homomorphism $Q_S \to \mathscr{A}_2^*(\mathscr{O}_{\overline{K}}/\mathscr{O}_K)$ induced by τ_S (3.1.11.2). By virtue of 2.4.7, $\mathscr{M}_{\mathscr{A}_2^*(\overline{S}/S)}$ is the logarithmic structure on $\mathscr{A}_2^*(\overline{S}/S)$ associated with the prelogarithmic structure defined by the homomorphism

$$\mathbb{N} \to \mathscr{A}_2^*(\mathscr{O}_{\overline{K}}/\mathscr{O}_K), \quad 1 \mapsto [\underline{\pi}]. \tag{3.1.13.2}$$

The logarithmic scheme $(\mathscr{A}_2^*(\overline{S}/S), \mathscr{M}_{\mathscr{A}_2^*(\overline{S}/S)})$ is therefore fine and saturated, and $\theta^*_{\mathscr{O}_K,2}$ (3.1.6.8) induces an exact closed immersion (3.1.1)

$$(\check{\overline{S}}, \mathscr{M}_{\check{\overline{S}}}) \to (\mathscr{A}_2^*(\overline{S}/S), \mathscr{M}_{\mathscr{A}_2^*(\overline{S}/S)}). \tag{3.1.13.3}$$

The element $(\xi_\pi^* + 1)$ being invertible in $\mathscr{A}_2^*(\mathscr{O}_{\overline{K}}/\mathscr{O}_K)$, the homomorphism

$$\mathbb{N} \to \Gamma(\mathscr{A}_2^*(\overline{S}/S), \mathscr{M}_{\mathscr{A}_2^*(\overline{S}/S)}), \quad 1 \mapsto (\xi_\pi^* + 1)^{-1}\widetilde{\pi} \tag{3.1.13.4}$$

induces a strict morphism of logarithmic schemes

$$\mathrm{pr}_1 \colon (\mathscr{A}_2^*(\overline{S}/S), \mathscr{M}_{\mathscr{A}_2^*(\overline{S}/S)}) \to (S, \mathscr{M}_S). \tag{3.1.13.5}$$

Moreover, we clearly have a strict morphism of logarithmic schemes

$$\mathrm{pr}_2 \colon (\mathscr{A}_2^*(\overline{S}/S), \mathscr{M}_{\mathscr{A}_2^*(\overline{S}/S)}) \to (\mathscr{A}_2(\overline{S}), \mathscr{M}_{\mathscr{A}_2(\overline{S})}). \tag{3.1.13.6}$$

3.1.14 In the rest of this chapter, $(\widetilde{S}, \mathscr{M}_{\widetilde{S}})$ denotes one of the two logarithmic schemes

$$(\mathscr{A}_2(\overline{S}), \mathscr{M}_{\mathscr{A}_2(\overline{S})}) \quad \text{or} \quad (\mathscr{A}_2^*(\overline{S}/S), \mathscr{M}_{\mathscr{A}_2^*(\overline{S}/S)}); \tag{3.1.14.1}$$

the first case will be called *absolute* and the second will be called *relative*. We denote by

$$i_S \colon (\check{\overline{S}}, \mathscr{M}_{\check{\overline{S}}}) \to (\widetilde{S}, \mathscr{M}_{\widetilde{S}}) \tag{3.1.14.2}$$

the canonical exact closed immersion (3.1.12.4) or (3.1.13.3). We set

$$\widetilde{\xi} = \xi \quad \text{or} \quad \widetilde{\xi} = \xi_\pi^* \tag{3.1.14.3}$$

depending on whether we are in the absolute or relative case. We deal with both cases simultaneously, each having its advantages and disadvantages. Note that in the relative case, $\widetilde{S}$ is naturally an S-scheme (3.1.13.5).

The closed immersion $\check{\overline{S}} \to \widetilde{S}$ is defined by the square zero ideal $\widetilde{\xi}\mathscr{O}_{\widetilde{S}}$ of $\mathscr{O}_{\widetilde{S}}$, associated with the $\mathscr{O}_C$-module $\widetilde{\xi}\mathscr{O}_C$ (see 3.1.4 and 3.1.6). We denote by $\widetilde{\xi}^{-1}\mathscr{O}_C$ the dual $\mathscr{O}_C$-module of $\widetilde{\xi}\mathscr{O}_C$. For any $\mathscr{O}_C$-module M and any integer $i \geq 1$, we denote the $\mathscr{O}_C$-modules $M \otimes_{\mathscr{O}_C} (\widetilde{\xi}\mathscr{O}_C)^{\otimes i}$ and $M \otimes_{\mathscr{O}_C} (\widetilde{\xi}^{-1}\mathscr{O}_C)^{\otimes i}$ simply by $\widetilde{\xi}^i M$ and $\widetilde{\xi}^{-i} M$, respectively.

The Galois group G_K acts naturally on the left on the logarithmic scheme $(\widetilde{S}, \mathscr{M}_{\widetilde{S}})$ and the closed immersion (3.1.14.2) is G_K-equivariant.

3.1.15 Let $f \colon (X, \mathscr{M}_X) \to (S, \mathscr{M}_S)$ be an *adequate* morphism of logarithmic schemes ([3] III.4.7), having an adequate chart ([3] III.4.4), such that $X = \mathrm{Spec}(R)$ is affine and X_s is non-empty. We denote by X° the maximal open subscheme of X where the logarithmic structure $\mathscr{M}_X$ is trivial; it is an open subscheme of X_η. We denote by $j \colon X^\circ \to X$ the canonical injection. For any X-scheme U, we set

$$U^\circ = U \times_X X^\circ. \tag{3.1.15.1}$$

To lighten the notation, we set

$$\widetilde{\Omega}^1_{X/S} = \Omega^1_{(X,\mathscr{M}_X)/(S,\mathscr{M}_S)}, \tag{3.1.15.2}$$

which is considered as a sheaf of $X_{\rm zar}$ or $X_{\rm \acute{e}t}$, depending on the context (2.1.13), and

$$\widetilde{\Omega}^1_{R/\mathscr{O}_K} = \widetilde{\Omega}^1_{X/S}(X). \tag{3.1.15.3}$$

Moreover, we fix an adequate chart $((P,\gamma),(\mathbb{N},\iota),\vartheta)$ for f, i.e., a chart (P,γ) for $(X,\mathscr{M}_X)$ ([3] II.5.13), a chart $(\mathbb{N},\iota)$ for $(S,\mathscr{M}_S)$ and a homomorphism of monoids $\vartheta\colon \mathbb{N}\to P$ such that the following conditions are satisfied:

(i) The diagram of homomorphisms of monoids

$$\begin{array}{ccc} P & \xrightarrow{\gamma} & \Gamma(X,\mathscr{M}_X) \\ \uparrow{\scriptstyle\vartheta} & & \uparrow{\scriptstyle f^\flat} \\ \mathbb{N} & \xrightarrow{\iota} & \Gamma(S,\mathscr{M}_S) \end{array} \tag{3.1.15.4}$$

is commutative, or equivalently (with the notation of 2.1.4), the associated diagram of morphisms of logarithmic schemes

$$\begin{array}{ccc} (X,\mathscr{M}_X) & \xrightarrow{\gamma^a} & \mathbf{A}_P \\ {\scriptstyle f}\downarrow & & \downarrow{\scriptstyle \mathbf{A}_\vartheta} \\ (S,\mathscr{M}_S) & \xrightarrow{\iota^a} & \mathbf{A}_{\mathbb{N}} \end{array} \tag{3.1.15.5}$$

is commutative.

(ii) The monoid P is toric, i.e., P is fine and saturated and $P^{\rm gp}$ is a free $\mathbb{Z}$-module ([3] II.5.1).

(iii) The homomorphism ϑ is saturated ([3] II.5.2).

(iv) The homomorphism $\vartheta^{\rm gp}\colon \mathbb{Z}\to P^{\rm gp}$ is injective, the order of the torsion subgroup of $\mathrm{coker}(\vartheta^{\rm gp})$ is prime to p and the morphism of usual schemes

$$X \to S\times_{\mathbf{A}_{\mathbb{N}}} \mathbf{A}_P \tag{3.1.15.6}$$

deduced from (3.1.15.5) is étale.

(v) We set $\lambda=\vartheta(1)\in P$,

$$L = \mathrm{Hom}_{\mathbb{Z}}(P^{\rm gp},\mathbb{Z}), \tag{3.1.15.7}$$
$$\mathrm{H}(P) = \mathrm{Hom}(P,\mathbb{N}). \tag{3.1.15.8}$$

Note that $\mathrm{H}(P)$ is a fine, saturated and sharp monoid and that the canonical homomorphism $\mathrm{H}(P)^{\rm gp}\to \mathrm{Hom}((P^\sharp)^{\rm gp},\mathbb{Z})$ is an isomorphism ([45] I 2.2.3). Suppose that there exist $h_1,\dots,h_r\in \mathrm{H}(P)$, which are $\mathbb{Z}$-linearly independent in L, such that

$$\ker(\lambda)\cap \mathrm{H}(P) = \{\sum_{i=1}^r a_ih_i \mid (a_1,\dots,a_r)\in\mathbb{N}^r\}, \tag{3.1.15.9}$$

where we consider λ as a homomorphism $L\to\mathbb{Z}$.

We denote by $\alpha\colon P \to R$ the homomorphism induced by the chart (P, γ). We set $\pi = \iota(1)$ which is a uniformizer of $\mathscr{O}_K$. We then have (2.1.4)

$$S \times_{\mathbf{A}_{\mathbb{N}}} \mathbf{A}_P = \mathrm{Spec}(\mathscr{O}_K[P]/(\pi - e^{\lambda})). \tag{3.1.15.10}$$

We denote by P^{gp} the group associated with P. By virtue of ([2] 4.2.2), the $\mathbb{Z}$-module $P^{\mathrm{gp}}/\mathbb{Z}\lambda$ is free of finite type. By 3.1.15(iv) and ([41] 1.8) or ([45] IV 1.1.4), we have a canonical R-linear isomorphism

$$(P^{\mathrm{gp}}/\mathbb{Z}\lambda) \otimes_{\mathbb{Z}} R \xrightarrow{\sim} \widetilde{\Omega}^1_{R/\mathscr{O}_K}. \tag{3.1.15.11}$$

This R-module is therefore free of rank $d = \dim(X/S)$.

Remark 3.1.16 The assumptions of 3.1.15 correspond to the conditions fixed in ([3] II.6.2), with the exception of the connectedness of X. We reduce to the case where this last condition is satisfied by replacing X by its connected components.

3.1.17 We endow $\check{\overline{X}} = X \times_S \check{\overline{S}}$ (3.1.2.1) with the logarithmic structure $\mathscr{M}_{\check{\overline{X}}}$ pullback of $\mathscr{M}_X$. We then have a canonical isomorphism

$$(\check{\overline{X}}, \mathscr{M}_{\check{\overline{X}}}) \xrightarrow{\sim} (X, \mathscr{M}_X) \times_{(S, \mathscr{M}_S)} (\check{\overline{S}}, \mathscr{M}_{\check{\overline{S}}}), \tag{3.1.17.1}$$

where the fiber product is taken indifferently in the category of logarithmic schemes or in that of fine logarithmic schemes.

In the remainder of this chapter, we fix an $(\widetilde{S}, \mathscr{M}_{\widetilde{S}})$-smooth deformation $(\widetilde{X}, \mathscr{M}_{\widetilde{X}})$ of $(\check{\overline{X}}, \mathscr{M}_{\check{\overline{X}}})$, i.e., a smooth morphism of fine logarithmic schemes $(\widetilde{X}, \mathscr{M}_{\widetilde{X}}) \to (\widetilde{S}, \mathscr{M}_{\widetilde{S}})$ and an $(\check{\overline{S}}, \mathscr{M}_{\check{\overline{S}}})$-isomorphism

$$(\check{\overline{X}}, \mathscr{M}_{\check{\overline{X}}}) \xrightarrow{\sim} (\widetilde{X}, \mathscr{M}_{\widetilde{X}}) \times_{(\widetilde{S}, \mathscr{M}_{\widetilde{S}})} (\check{\overline{S}}, \mathscr{M}_{\check{\overline{S}}}). \tag{3.1.17.2}$$

Such a deformation exists and is unique up to isomorphism by virtue of ([41], 3.14).

3.2 Torsors and Higgs–Tate Algebras

3.2.1 For any integer $n \geq 1$, we set

$$\mathscr{O}_{K_n} = \mathscr{O}_K[\zeta]/(\zeta^n - \pi), \tag{3.2.1.1}$$

which is a discrete valuation ring. We denote by K_n the field of fractions of $\mathscr{O}_{K_n}$ and by π_n the class of ζ in $\mathscr{O}_{K_n}$, which is a uniformizer of $\mathscr{O}_{K_n}$. We set $S^{(n)} = \mathrm{Spec}(\mathscr{O}_{K_n})$, which we endow with the logarithmic structure $\mathscr{M}_{S^{(n)}}$ defined by its closed point. We denote by $\iota_n\colon \mathbb{N} \to \Gamma(S^{(n)}, \mathscr{M}_{S^{(n)}})$ the homomorphism defined by $\iota_n(1) = \pi_n$; it is a chart for $(S^{(n)}, \mathscr{M}_{S^{(n)}})$.

Consider the direct system of monoids $(\mathbb{N}^{(n)})_{n\geq 1}$, indexed by the set $\mathbb{Z}_{\geq 1}$ ordered by the divisibility relation, defined by $\mathbb{N}^{(n)} = \mathbb{N}$ for any $n \geq 1$ and whose transition homomorphism $\mathbb{N}^{(n)} \to \mathbb{N}^{(mn)}$ (for $m, n \geq 1$) is the Frobenius endomorphism of order m of $\mathbb{N}$ (i.e., the morphism defined by taking the mth power). We denote $\mathbb{N}^{(1)}$ simply by $\mathbb{N}$. The logarithmic schemes $(S^{(n)}, \mathscr{M}_{S^{(n)}})_{n\geq 1}$ naturally form an inverse system. For all integers $m, n \geq 1$, with the notation of 2.1.4, we have a Cartesian diagram of morphisms of logarithmic schemes

$$\begin{array}{ccc} (S^{(mn)}, \mathscr{M}_{S^{(mn)}}) & \xrightarrow{\iota^a_{mn}} & \mathbf{A}_{\mathbb{N}^{(mn)}} \\ \downarrow & & \downarrow \\ (S^{(n)}, \mathscr{M}_{S^{(n)}}) & \xrightarrow{\iota^a_{n}} & \mathbf{A}_{\mathbb{N}^{(n)}} \end{array} \tag{3.2.1.2}$$

where ι^a_n (resp. ι^a_{mn}) is the morphism associated with ι_n (resp. ι_{mn}) ([3] II.5.13) .

3.2.2 Consider the direct system of monoids $(P^{(n)})_{n\geq 1}$, indexed by the set $\mathbb{Z}_{\geq 1}$ ordered by the divisibility relation, defined by $P^{(n)} = P$ for any $n \geq 1$ and whose transition homomorphism $i_{n,mn} \colon P^{(n)} \to P^{(mn)}$ (for $m, n \geq 1$) is the Frobenius endomorphism of order m of P (i.e., the morphism defined by taking the mth power) (see 3.1.15). For any $n \geq 1$, we denote by

$$P \xrightarrow{\sim} P^{(n)}, \quad t \mapsto t^{(n)}, \tag{3.2.2.1}$$

the canonical isomorphism. For all $t \in P$ and all $m, n \geq 1$, we therefore have

$$i_{n,mn}(t^{(n)}) = (t^{(mn)})^m. \tag{3.2.2.2}$$

We denote $P^{(1)}$ simply by P.

For any integer $n \geq 1$, we set (with the notation of 2.1.4)

$$(X^{(n)}, \mathscr{M}_{X^{(n)}}) = (X, \mathscr{M}_X) \times_{\mathbf{A}_P} \mathbf{A}_{P^{(n)}}. \tag{3.2.2.3}$$

Note that the canonical projection $(X^{(n)}, \mathscr{M}_{X^{(n)}}) \to \mathbf{A}_{P^{(n)}}$ is strict. Since the diagram (3.2.1.2) is Cartesian, there exists a unique morphism

$$f^{(n)} \colon (X^{(n)}, \mathscr{M}_{X^{(n)}}) \to (S^{(n)}, \mathscr{M}_{S^{(n)}}), \tag{3.2.2.4}$$

that fits into the commutative diagram

$$\begin{array}{ccccccc}
(X^{(n)}, \mathscr{M}_{X^{(n)}}) & & & \longrightarrow & & & \mathbf{A}_{P^{(n)}} \\
\downarrow & \searrow^{f^{(n)}} & & \textcircled{1} & & \swarrow^{\mathbf{A}_\vartheta} & \downarrow \\
\downarrow & & (S^{(n)}, \mathscr{M}_{S^{(n)}}) & \xrightarrow{\iota_n^a} & \mathbf{A}_{\mathbb{N}^{(n)}} & & \downarrow \\
\downarrow & & \downarrow & & \downarrow & & \downarrow \\
\downarrow & & (S, \mathscr{M}_S) & \xrightarrow{\iota^a} & \mathbf{A}_{\mathbb{N}} & & \downarrow \\
\downarrow & \nearrow^{f} & & & & \nwarrow^{\mathbf{A}_\vartheta} & \downarrow \\
(X, \mathscr{M}_X) & & & \longrightarrow & & & \mathbf{A}_P.
\end{array} \qquad (3.2.2.5)$$

In the diagram (3.2.2.5), the two horizontal arrows $(X^{(n)}, \mathscr{M}_{X^{(n)}}) \to \mathbf{A}_{P^{(n)}}$ and $(X, \mathscr{M}_X) \to \mathbf{A}_P$ run along the top and bottom, and the two vertical arrows at the left and right edges point down. $f^{(n)}$ goes from $(X^{(n)}, \mathscr{M}_{X^{(n)}})$ to $(S^{(n)}, \mathscr{M}_{S^{(n)}})$ and f goes from $(X, \mathscr{M}_X)$ to $(S, \mathscr{M}_S)$. Each $\mathbf{A}_\vartheta$ points from the right-hand corner ($\mathbf{A}_{P^{(n)}}$, $\mathbf{A}_P$) in to $\mathbf{A}_{\mathbb{N}^{(n)}}$ or $\mathbf{A}_{\mathbb{N}}$. The face marked (1) is the one bounded by $(X^{(n)}, \mathscr{M}_{X^{(n)}})$, $\mathbf{A}_{P^{(n)}}$, $\mathbf{A}_{\mathbb{N}^{(n)}}$ and $(S^{(n)}, \mathscr{M}_{S^{(n)}})$.

By ([2] 4.2.7(i)), the face (1) of the diagram (3.2.2.5) is an adequate chart for $f^{(n)}$; in particular, $f^{(n)}$ is smooth and saturated.

3.2.3 Let $\overline{y}$ be a geometric point of $\overline{X}^{\circ}$ (see (3.1.2.1) and (3.1.15.1)). The scheme $\overline{X}$ being locally irreducible by ([2] 4.2.7(iii)), it is the sum of the schemes induced on its irreducible components. We denote by $\overline{X}^{\star}$ the irreducible component of $\overline{X}$ containing $\overline{y}$. Likewise, $\overline{X}^{\circ}$ is the sum of the schemes induced on its irreducible components and $\overline{X}^{\star\circ} = \overline{X}^{\star} \times_X X^{\circ}$ is the irreducible component of $\overline{X}^{\circ}$ containing $\overline{y}$. We denote by Δ the profinite group $\pi_1(\overline{X}^{\star\circ}, \overline{y})$ and by $(V_i)_{i\in I}$ the normalized universal cover of $\overline{X}^{\star\circ}$ at $\overline{y}$ ([2] 2.1.20). For any $i \in I$, we denote by $\overline{X}_i$ the integral closure of $\overline{X}$ in V_i. The schemes $(\overline{X}_i)_{i\in I}$ then form a filtered inverse system. We set

$$\overline{R} = \varinjlim_{i\in I} \Gamma(\overline{X}_i, \mathscr{O}_{\overline{X}_i}). \qquad (3.2.3.1)$$

This is a normal integral domain ([2] 4.1.10), on which Δ acts naturally by ring homomorphisms and the action is discrete.

We set

$$\mathbb{X} = \mathrm{Spec}(\overline{R}) \quad \text{and} \quad \widehat{\mathbb{X}} = \mathrm{Spec}(\widehat{\overline{R}}), \qquad (3.2.3.2)$$

which we endow with the logarithmic structures pullbacks of $\mathscr{M}_X$, denoted respectively by $\mathscr{M}_{\mathbb{X}}$ and $\mathscr{M}_{\widehat{\mathbb{X}}}$. The actions of Δ on $\overline{R}$ and $\widehat{\overline{R}}$ induce left actions on the logarithmic schemes $(\mathbb{X}, \mathscr{M}_{\mathbb{X}})$ and $(\widehat{\mathbb{X}}, \mathscr{M}_{\widehat{\mathbb{X}}})$. Endowing $(\check{\overline{X}}, \mathscr{M}_{\check{\overline{X}}})$ with the trivial action of Δ (3.1.17), we have a canonical Δ-equivariant morphism

$$(\widehat{\mathbb{X}}, \mathscr{M}_{\widehat{\mathbb{X}}}) \to (\check{\overline{X}}, \mathscr{M}_{\check{\overline{X}}}). \qquad (3.2.3.3)$$

3.2.4 For all integers $m, n \geq 1$, the canonical morphism $X^{(mn)} \to X^{(n)}$ is finite and surjective. By ([30] 8.3.8(i)), then there exists an X-morphism

$$\overline{y} \to \varprojlim_{n\geq 1} X^{(n)}, \qquad (3.2.4.1)$$

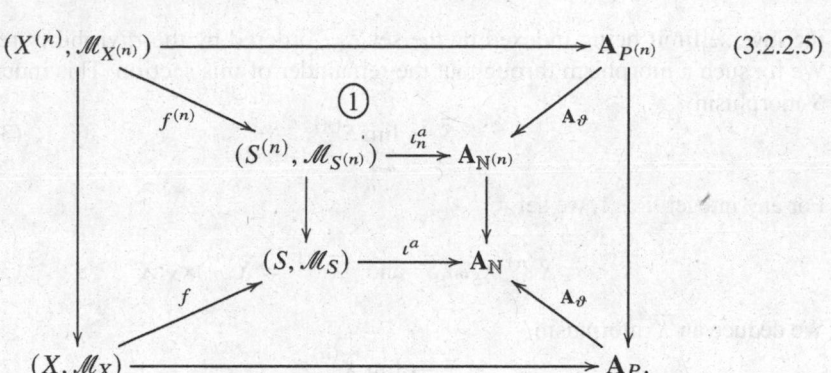

the inverse limit being indexed by the set $\mathbb{Z}_{\geq 1}$ ordered by the divisibility relation. We fix such a morphism throughout the remainder of this section. This induces an S-morphism

$$\overline{S} \to \varprojlim_{n\geq 1} S^{(n)}. \tag{3.2.4.2}$$

For any integer $n \geq 1$, we set

$$\overline{X}^{(n)} = X^{(n)} \times_{S^{(n)}} \overline{S} \quad\text{and}\quad \overline{X}^{(n)\circ} = \overline{X}^{(n)} \times_X X^{\circ}. \tag{3.2.4.3}$$

We deduce an $\overline{X}$-morphism

$$\overline{y} \to \varprojlim_{n\geq 1} \overline{X}^{(n)}. \tag{3.2.4.4}$$

For any integer $n \geq 1$, the scheme $\overline{X}^{(n)}$ being normal and locally irreducible by ([2] 4.2.7(iii)), it is the sum of the schemes induced on its irreducible components. We denote by $\overline{X}^{(n)\star}$ the irreducible component of $\overline{X}^{(n)}$ containing the image of $\overline{y}$ (3.2.4.4). Similarly, $\overline{X}^{(n)\circ}$ is the sum of the schemes induced on its irreducible components and $\overline{X}^{(n)\star\circ} = \overline{X}^{(n)\star} \times_X X^{\circ}$ is the irreducible component of $\overline{X}^{(n)\circ}$ containing the image of $\overline{y}$. Note that $\overline{X}^{(n)}$ being finite over $\overline{X}$ (3.2.2.5), $\overline{X}^{(n)\star}$ is the integral closure of $\overline{X}^{\star}$ in $\overline{X}^{(n)\star\circ}$. We set

$$R_n = \Gamma(\overline{X}^{(n)\star}, \mathscr{O}_{\overline{X}^{(n)}}). \tag{3.2.4.5}$$

By ([2] 4.2.7(v)), the morphism $\overline{X}^{(n)\star\circ} \to \overline{X}^{\star\circ}$ is finite. It follows from the proof of ([3] II.6.8(iv)) that $\overline{X}^{(n)\star\circ}$ is in fact a Galois finite étale cover of $\overline{X}^{\star\circ}$ with group Δ_n canonically isomorphic to a subgroup of $\mathrm{Hom}_{\mathbb{Z}}(P^{\mathrm{gp}}/\mathbb{Z}\lambda, \mu_n(\overline{K}))$. The group Δ_n acts naturally on R_n.

If n is a power of p, the canonical morphism

$$\overline{X}^{(n)\star} \to \overline{X}^{(n)} \times_{\overline{X}} \overline{X}^{\star} \tag{3.2.4.6}$$

is an isomorphism by virtue of ([3] II.6.6(v)), and we therefore have $\Delta_n \simeq \mathrm{Hom}_{\mathbb{Z}}(P^{\mathrm{gp}}/\mathbb{Z}\lambda, \mu_n(\overline{K}))$.

The rings $(R_n)_{n\geq 1}$ naturally form a direct system. We set

$$R_{\infty} = \varinjlim_{n\geq 1} R_n, \tag{3.2.4.7}$$

$$R_{p^{\infty}} = \varinjlim_{n\geq 0} R_{p^n}. \tag{3.2.4.8}$$

They are normal integral domains by ([27] 0.6.1.6(i) and 0.6.5.12(ii)).

The morphism (3.2.4.4) induces an $\overline{X}^{\star}$-morphism

$$\mathrm{Spec}(\overline{R}) \to \varprojlim_{n\geq 1} \overline{X}^{(n)\star}, \tag{3.2.4.9}$$

and therefore injective homomorphisms

$$R_{p^\infty} \to R_\infty \to \overline{R}. \tag{3.2.4.10}$$

Note that the $\mathscr{O}_C$-modules $\widehat{R_1}$ and $\widehat{\overline{R}}$ are flat and that the canonical homomorphism $\widehat{R_1} \to \widehat{\overline{R}}$ is injective ([3] II.6.14).

The groups $(\Delta_n)_{n\geq 1}$ naturally form an inverse system. We set

$$\Delta_\infty = \lim_{\substack{\longleftarrow \\ n\geq 1}} \Delta_n, \tag{3.2.4.11}$$

$$\Delta_{p^\infty} = \lim_{\substack{\longleftarrow \\ n\geq 0}} \Delta_{p^n}. \tag{3.2.4.12}$$

We have canonical homomorphisms

$$\begin{array}{ccc} \Delta_\infty & \hookrightarrow & \mathrm{Hom}_{\mathbb{Z}}(P^{\mathrm{gp}}/\mathbb{Z}\lambda, \widehat{\mathbb{Z}}(1)) \\ \downarrow & & \downarrow \\ \Delta_{p^\infty} & \xrightarrow{\sim} & \mathrm{Hom}_{\mathbb{Z}}(P^{\mathrm{gp}}/\mathbb{Z}\lambda, \mathbb{Z}_p(1)). \end{array} \tag{3.2.4.13}$$

The kernel Σ_0 of the canonical homomorphism $\Delta_\infty \to \Delta_{p^\infty}$ is a profinite group of order prime to p. Moreover, the morphism (3.2.4.4) determines a surjective homomorphism $\Delta \to \Delta_\infty$. We denote by Σ its kernel. The homomorphisms (3.2.4.10) are then Δ-equivariant. The constructions introduced in this subsection are summarized in the following diagram:

$$\xymatrix{ R_1 \ar[r]^{\Delta_{p^\infty}} \ar@/^2pc/[rr]^{\Delta_\infty} \ar@/_2pc/[rrr]_{\Delta} & R_{p^\infty} \ar[r]^{\Sigma_0} & R_\infty \ar[r]^{\Sigma} & \overline{R}. } \tag{3.2.4.14}$$

3.2.5 Consider the inverse system of multiplicative monoids $(\overline{R})_{n\in\mathbb{N}}$, where the transition morphisms are the iterates of the pth power map of $\overline{R}$. We denote by Q_X the fiber product of the diagram of homomorphisms of monoids

$$\begin{array}{ccc} & & \Gamma(X, \mathscr{M}_X) \\ & & \downarrow \\ \lim\limits_{\substack{\longleftarrow \\ \mathbb{N}}} \overline{R} & \longrightarrow & \overline{R} \end{array} \tag{3.2.5.1}$$

where the vertical arrow is the canonical homomorphism (which factorizes through R) and the horizontal arrow is the projection onto the first component (i.e., of index 0). We denote by τ_X the composed homomorphism

$$\tau_X \colon Q_X \longrightarrow \lim_{\substack{\longleftarrow \\ \mathbb{N}}} \overline{R} \longrightarrow \overline{R}^\flat \xrightarrow{[\,]} \mathrm{W}(\overline{R}^\flat), \tag{3.2.5.2}$$

where $\overline{R}^{\flat}$ is the ring defined in (2.2.3.1), [ ] is the multiplicative representative and the other arrows are the canonical morphisms. It immediately follows from the definitions that the diagram

$$\begin{array}{ccc} Q_X & \longrightarrow & \Gamma(X, \mathscr{M}_X) \\ {\scriptstyle \tau_X}\downarrow & & \downarrow \\ \mathrm{W}(\overline{R}^{\flat}) & \xrightarrow{\theta} & \widehat{\overline{R}}, \end{array} \tag{3.2.5.3}$$

where the unlabeled arrows are the canonical morphisms, is commutative. The group Δ acts naturally on the monoid Q_X, and the homomorphism τ_X is Δ-equivariant.

We have a canonical homomorphism $Q_S \to Q_X$ (3.1.11) that fits into a commutative diagram

$$\begin{array}{ccc} Q_S & \longrightarrow & Q_X \\ {\scriptstyle \tau_S}\downarrow & & \downarrow{\scriptstyle \tau_X} \\ \mathrm{W}(\mathscr{O}_{\overline{K}^{\flat}}) & \longrightarrow & \mathrm{W}(\overline{R}^{\flat}). \end{array} \tag{3.2.5.4}$$

3.2.6 Let (3.2.2)

$$P_\infty = \lim_{\substack{\longrightarrow \\ n \geq 1}} P^{(n)}. \tag{3.2.6.1}$$

For any $n \geq 1$, we denote by $\alpha_n \colon P^{(n)} \to R_n$ the homomorphism induced by the canonical strict morphism $(X_n, \mathscr{M}_{X_n}) \to \mathbf{A}_{P^{(n)}}$ (3.2.2.3). For all integers $m, n \geq 1$, the diagram

$$\begin{array}{ccc} P^{(n)} & \xrightarrow{\alpha_n} & R_n \\ {\scriptstyle i_{n,mn}}\downarrow & & \downarrow \\ P^{(mn)} & \xrightarrow{\alpha_{mn}} & R_{mn} \end{array} \tag{3.2.6.2}$$

is commutative. By taking the direct limit, the α_n's therefore define a homomorphism

$$\alpha_\infty \colon P_\infty \to R_\infty. \tag{3.2.6.3}$$

We also denote by $\alpha_\infty \colon P_\infty \to \overline{R}$ the composition of α_∞ with the canonical injection $R_\infty \to \overline{R}$.

For any $t \in P$, we denote by $\widetilde{t}$ the element of Q_X (3.2.5.1) defined by its projections

$$(\alpha_\infty(t^{(p^n)}))_{n \in \mathbb{N}} \in \lim_{\substack{\longleftarrow \\ \mathbb{N}}} \overline{R} \quad \text{and} \quad \gamma(t) \in \Gamma(X, \mathscr{M}_X), \tag{3.2.6.4}$$

where $t^{(p^n)}$ is the image of t in $P^{(p^n)}$ by the isomorphism (3.2.2.1). The map

$$\nu \colon P \to Q_X, \quad t \mapsto \widetilde{t} \tag{3.2.6.5}$$

thus defined is a homomorphism of monoids.

3.2.7 There exists a unique map

$$\langle\ ,\ \rangle \colon \Delta_\infty \times P_\infty \to \mu_\infty(\mathscr{O}_{\overline{K}}) = \varinjlim_{n\geq 1} \mu_n(\mathscr{O}_{\overline{K}}), \tag{3.2.7.1}$$

where the direct limit is indexed by the set $\mathbb{Z}_{\geq 1}$ ordered by the divisibility relation, such that for every $g \in \Delta_\infty$ and every $x \in P^{(n)}$ $(n \geq 1)$, we have $\langle g, x\rangle \in \mu_n(\mathscr{O}_{\overline{K}})$ and

$$g(\alpha_\infty(x)) = \langle g, x\rangle \cdot \alpha_\infty(x), \tag{3.2.7.2}$$

where α_∞ is the homomorphism (3.2.6.3). Indeed, R_∞ is an integral domain, and we have $\alpha_\infty(x)^n \in \alpha(P) \subset R$, which is invertible on X° (3.2.6); hence $\alpha_\infty(x) \neq 0$ and $\alpha_\infty(x)^n$ is invariant by Δ_∞.

The pairing (3.2.7.1) is multiplicative in each of its factors.

Let n be an integer ≥ 1. Recall that we have a canonical isomorphism $P^{(n)} \xrightarrow{\sim} P$ (3.2.2.1) and that Δ_∞ is canonically isomorphic to a subgroup of $\mathrm{Hom}_{\mathbb{Z}}(P^{\mathrm{gp}}/\mathbb{Z}\lambda, \widehat{\mathbb{Z}}(1))$ (3.2.4.13). We therefore have a canonical homomorphism $\Delta_\infty \to \mathrm{Hom}_{\mathbb{Z}}(P^{\mathrm{gp}}/\mathbb{Z}\lambda, \mu_n(\mathscr{O}_{\overline{K}}))$. By ([3] II.6.6(vi)), the diagram

$$\begin{array}{ccc} \Delta_\infty \times P^{(n)} & \xrightarrow{\langle\ ,\ \rangle} & \mu_n(\mathscr{O}_{\overline{K}}) \\ \downarrow & \nearrow & \\ \mathrm{Hom}_{\mathbb{Z}}(P^{\mathrm{gp}}/\mathbb{Z}\lambda, \mu_n(\mathscr{O}_{\overline{K}})) \times P, & & \end{array} \tag{3.2.7.3}$$

where the oblique arrow is the canonical pairing, is commutative.

Let $t \in P$. We denote by

$$\chi_t \colon \Delta_\infty \to \mathbb{Z}_p(1) \tag{3.2.7.4}$$

the map which with any $g \in \Delta_\infty$ associates the element

$$\chi_t(g) = \varprojlim_{n\geq 0} \langle g, t^{(p^n)}\rangle, \tag{3.2.7.5}$$

where $t^{(p^n)}$ is the image of t in $P^{(p^n)}$ by the isomorphism (3.2.2.1) and $\langle g, t^{(p^n)}\rangle \in \mu_{p^n}(\mathscr{O}_{\overline{K}})$ is defined in (3.2.7.1). It is clear that χ_t is a group homomorphism.

We clearly have $\chi_0 = 1$, and for all $t, t' \in P$,

$$\chi_{tt'} = \chi_t \cdot \chi_{t'}. \tag{3.2.7.6}$$

Consequently, the map $P \to \mathrm{Hom}(\Delta_\infty, \mathbb{Z}_p(1))$ defined by $t \mapsto \chi_t$ is a homomorphism of monoids. It therefore induces a homomorphism that we denote again by

$$P^{\mathrm{gp}} \to \mathrm{Hom}(\Delta_\infty, \mathbb{Z}_p(1)), \quad t \mapsto \chi_t. \tag{3.2.7.7}$$

Since $\chi_\lambda = 1$, we deduce from this a homomorphism

$$P^{\mathrm{gp}}/\mathbb{Z}\lambda \to \mathrm{Hom}(\Delta_\infty, \mathbb{Z}_p(1)). \tag{3.2.7.8}$$

By virtue of ([3] (II.7.19.9)), the latter induces an isomorphism

$$(P^{\mathrm{gp}}/\mathbb{Z}\lambda)\otimes_{\mathbb{Z}}\mathbb{Z}_p \xrightarrow{\sim} \mathrm{Hom}(\Delta_\infty,\mathbb{Z}_p(1)). \tag{3.2.7.9}$$

Taking into account (3.1.15.11) and ([3] (II.6.12.2)), we deduce from this an $\widehat{R_1}$-linear isomorphism

$$\widetilde{\Omega}^1_{R/\mathscr{O}_K}\otimes_R \widehat{R_1} \xrightarrow{\sim} \mathrm{Hom}(\Delta_\infty,\widehat{R_1}(1)). \tag{3.2.7.10}$$

3.2.8 We take again the notation of 2.2.3 for $A=\overline{R}$: let

$$\mathscr{A}_2(\overline{R}) = \mathrm{W}(\overline{R}^\flat)/\ker(\theta)^2, \tag{3.2.8.1}$$

and denote by $\theta_2\colon \mathscr{A}_2(\overline{R})\to\widehat{\overline{R}}$ the homomorphism induced by θ (2.2.3.4). By 2.3.7 and ([3] II.9.10), the sequence

$$0\longrightarrow \mathrm{W}(\overline{R}^\flat)\xrightarrow{\cdot\xi}\mathrm{W}(\overline{R}^\flat)\xrightarrow{\theta}\widehat{\overline{R}}\longrightarrow 0, \tag{3.2.8.2}$$

where $\xi\in\mathrm{W}(\mathscr{O}_{\overline{K}^\flat})$ is the element defined in (3.1.4.2), is exact. It therefore induces an exact sequence

$$0\longrightarrow \widehat{\overline{R}}\xrightarrow{\cdot\xi}\mathscr{A}_2(\overline{R})\xrightarrow{\theta_2}\widehat{\overline{R}}\longrightarrow 0, \tag{3.2.8.3}$$

where we denoted also by $\cdot\xi$ the morphism induced by the multiplication by ξ in $\mathscr{A}_2(\overline{R})$.

The group Δ acts naturally on $\mathrm{W}(\overline{R}^\flat)$ by ring automorphisms, and the homomorphism θ is Δ-equivariant. We deduce from this an action of Δ on $\mathscr{A}_2(\overline{R})$ by ring automorphisms such that the homomorphism θ_2 is Δ-equivariant.

We set

$$\mathscr{A}_2(\mathbb{X}) = \mathrm{Spec}(\mathscr{A}_2(\overline{R})) \tag{3.2.8.4}$$

which is endowed with the logarithmic structure $\mathscr{M}'_{\mathscr{A}_2(\mathbb{X})}$ associated with the prelogarithmic structure defined by the homomorphism $Q_X\to\mathscr{A}_2(\overline{R})$ induced by τ_X (3.2.5.2). The homomorphism θ_2 then induces a morphism

$$(\widehat{\mathbb{X}},\mathscr{M}_{\widehat{\mathbb{X}}})\to(\mathscr{A}_2(\mathbb{X}),\mathscr{M}'_{\mathscr{A}_2(\mathbb{X})}). \tag{3.2.8.5}$$

The actions of Δ on the ring $\mathscr{A}_2(\overline{R})$ and the monoid Q_X induce a left action on the logarithmic scheme $(\mathscr{A}_2(\mathbb{X}),\mathscr{M}'_{\mathscr{A}_2(\mathbb{X})})$. The morphism (3.2.8.5) is Δ-equivariant.

Without the conditions of 2.4.2, we do not know if the logarithmic scheme $(\mathscr{A}_2(\mathbb{X}),\mathscr{M}'_{\mathscr{A}_2(\mathbb{X})})$ is fine and saturated and if (3.2.8.5) is an exact closed immersion. This is why we endow $\mathscr{A}_2(\mathbb{X})$ with another logarithmic structure, namely the logarithmic structure $\mathscr{M}_{\mathscr{A}_2(\mathbb{X})}$ associated with the prelogarithmic structure defined by the composed homomorphism

$$\widetilde{\tau}_X\colon P\xrightarrow{\nu}Q_X\xrightarrow{\tau_X}\mathrm{W}(\overline{R}^\flat)\longrightarrow\mathscr{A}_2(\overline{R}), \tag{3.2.8.6}$$

where the first arrow is the homomorphism (3.2.6.5) and the second arrow is the homomorphism (3.2.5.2).

The homomorphism $\nu\colon P \to Q_X$ (3.2.6.5) induces a morphism of logarithmic structures on $\mathscr{A}_2(\mathbb{X})$

$$\mathscr{M}_{\mathscr{A}_2(\mathbb{X})} \longrightarrow \mathscr{M}'_{\mathscr{A}_2(\mathbb{X})}. \tag{3.2.8.7}$$

It is clear that the composed homomorphism $\theta_2 \circ \widetilde{\tau}_X\colon P \to \widehat{\overline{R}}$ is induced by α (see 3.1.15). Hence, θ_2 induces an exact closed immersion

$$(\widehat{\mathbb{X}}, \mathscr{M}_{\widehat{\mathbb{X}}}) \to (\mathscr{A}_2(\mathbb{X}), \mathscr{M}_{\mathscr{A}_2(\mathbb{X})}), \tag{3.2.8.8}$$

which factors through the morphism (3.2.8.5).

3.2.9 The canonical homomorphism $\mathbb{Z}_p(1) \to (\overline{R}^{\flat}, \times)$ and the trivial homomorphism $\mathbb{Z}_p(1) \to \Gamma(X, \mathscr{M}_X)$ (with value 1) induce a homomorphism

$$\mathbb{Z}_p(1) \to Q_X. \tag{3.2.9.1}$$

For every $g \in \Delta$ and every $t \in P$, we have in Q_X

$$g(\nu(t)) = \chi_t(g) \cdot \nu(t), \tag{3.2.9.2}$$

where ν is the map (3.2.6.5) and where we have (abusively) denoted by $\chi_t\colon \Delta \to \mathbb{Z}_p(1)$ the map deduced from (3.2.7.4). We deduce from this the following relation in $\mathscr{A}_2(\overline{R})$

$$g(\widetilde{\tau}_X(t)) = [\chi_t(g)] \cdot \widetilde{\tau}_X(t), \tag{3.2.9.3}$$

where $[\chi_t(g)]$ denotes the image of $\chi_t(g)$ by the composed map

$$\mathbb{Z}_p(1) \longrightarrow \overline{R}^{\flat} \xrightarrow{[\,]} \mathrm{W}(\overline{R}^{\flat}) \longrightarrow \mathscr{A}_2(\overline{R}).$$

For any $g \in \Delta$, we denote by γ_g the automorphism of $\mathscr{A}_2(\mathbb{X})$ induced by the action of g on $\mathscr{A}_2(\overline{R})$. The logarithmic structure $\gamma_g^*(\mathscr{M}_{\mathscr{A}_2(\mathbb{X})})$ on $\mathscr{A}_2(\mathbb{X})$ is associated with the prelogarithmic structure defined by the composed homomorphism $g \circ \widetilde{\tau}_X\colon P \to \mathscr{A}_2(\overline{R})$ (3.2.8.6). The map

$$P \to \Gamma(\mathscr{A}_2(\mathbb{X}), \mathscr{M}_{\mathscr{A}_2(\mathbb{X})}), \quad t \mapsto [\chi_t(g)] \cdot t, \tag{3.2.9.4}$$

is a homomorphism of monoids (3.2.7.6). It therefore induces a morphism of logarithmic structures on $\mathscr{A}_2(\mathbb{X})$

$$a_g\colon \gamma_g^*(\mathscr{M}_{\mathscr{A}_2(\mathbb{X})}) \to \mathscr{M}_{\mathscr{A}_2(\mathbb{X})}. \tag{3.2.9.5}$$

Similarly, the homomorphism of monoids

$$P \to \Gamma(\mathscr{A}_2(\mathbb{X}), \gamma_g^*(\mathscr{M}_{\mathscr{A}_2(\mathbb{X})})), \quad t \mapsto [\chi_t(g^{-1})] \cdot t, \tag{3.2.9.6}$$

induces a morphism of logarithmic structures on $\mathscr{A}_2(\mathbb{X})$

$$b_g : \mathscr{M}_{\mathscr{A}_2(\mathbb{X})} \to \gamma_g^*(\mathscr{M}_{\mathscr{A}_2(\mathbb{X})}). \tag{3.2.9.7}$$

We immediately see that a_g and b_g are isomorphisms inverse to each other, and that the map $g \mapsto (\gamma_{g^{-1}}, a_{g^{-1}})$ is a left action of Δ on the logarithmic scheme $(\mathscr{A}_2(\mathbb{X}), \mathscr{M}_{\mathscr{A}_2(\mathbb{X})})$.

We check immediately that the immersion (3.2.8.8) and the canonical morphism

$$(\mathscr{A}_2(\mathbb{X}), \mathscr{M}_{\mathscr{A}_2(\mathbb{X})}) \to (\mathscr{A}_2(\overline{S}), \mathscr{M}_{\mathscr{A}_2(\overline{S})}) \tag{3.2.9.8}$$

are Δ-equivariant. Moreover, for every $g \in \Delta$, the diagram

$$\begin{array}{ccc} \gamma_g^*(\mathscr{M}_{\mathscr{A}_2(\mathbb{X})}) & \longrightarrow & \gamma_g^*(\mathscr{M}'_{\mathscr{A}_2(\mathbb{X})}) \\ \downarrow{\scriptstyle a_g} & & \downarrow{\scriptstyle a'_g} \\ \mathscr{M}_{\mathscr{A}_2(\mathbb{X})} & \longrightarrow & \mathscr{M}'_{\mathscr{A}_2(\mathbb{X})}, \end{array} \tag{3.2.9.9}$$

where the horizontal arrows are induced by the homomorphism (3.2.8.7) and a'_g is the automorphism of logarithmic structures on $\mathscr{A}_2(\mathbb{X})$ induced by the action of g on Q_X, is commutative.

3.2.10 The ring $\mathrm{W}(\overline{R}^\flat)$ being naturally endowed with a structure of W-algebra (2.2.4), we set

$$\mathrm{W}_{\mathscr{O}_K}(\overline{R}^\flat) = \mathrm{W}(\overline{R}^\flat) \otimes_W \mathscr{O}_K \tag{3.2.10.1}$$

and denote by $\theta_{\mathscr{O}_K} : \mathrm{W}_{\mathscr{O}_K}(\overline{R}^\flat) \to \widehat{\overline{R}}$ the homomorphism induced by θ (3.2.8.2). By 2.3.7 and ([3] II.9.10), the sequence

$$0 \longrightarrow \mathrm{W}_{\mathscr{O}_K}(\overline{R}^\flat) \xrightarrow{\cdot \xi_\pi} \mathrm{W}_{\mathscr{O}_K}(\overline{R}^\flat) \xrightarrow{\theta_{\mathscr{O}_K}} \widehat{\overline{R}} \longrightarrow 0, \tag{3.2.10.2}$$

where $\xi_\pi \in \mathrm{W}_{\mathscr{O}_K}(\mathscr{O}_{\overline{K}^\flat})$ is the element defined in (3.1.6.2), is exact. We set

$$\mathrm{W}_K(\overline{R}^\flat) = \mathrm{W}(\overline{R}^\flat) \otimes_W K \tag{3.2.10.3}$$

and denote by $\theta_K : \mathrm{W}_K(\overline{R}^\flat) \to \widehat{\overline{R}}[\frac{1}{p}]$ the homomorphism induced by θ (3.2.8.2). We denote by $\mathrm{W}^*_{\mathscr{O}_K}(\overline{R}^\flat)$ the sub-$\mathrm{W}_{\mathscr{O}_K}(\overline{R}^\flat)$-algebra of $\mathrm{W}_K(\overline{R}^\flat)$ generated by $\xi^*_\pi = \xi_\pi/\pi$ (3.1.6.5). The homomorphism θ_K induces a homomorphism

$$\theta^*_{\mathscr{O}_K} : \mathrm{W}^*_{\mathscr{O}_K}(\overline{R}^\flat) \to \widehat{\overline{R}} \tag{3.2.10.4}$$

such that $\theta^*_{\mathscr{O}_K}(\xi^*_\pi) = 0$. By 2.4.4(ii), the sequence

$$0 \longrightarrow \mathrm{W}^*_{\mathscr{O}_K}(\overline{R}^\flat) \xrightarrow{\cdot \xi^*_\pi} \mathrm{W}^*_{\mathscr{O}_K}(\overline{R}^\flat) \xrightarrow{\theta^*_{\mathscr{O}_K}} \widehat{\overline{R}} \longrightarrow 0 \tag{3.2.10.5}$$

is exact. We set

$$\mathscr{A}_2^*(\overline{R}/\mathscr{O}_K) = \mathrm{W}^*_{\mathscr{O}_K}(\overrightarrow{R}^{\flat})/(\xi_\pi^*)^2\mathrm{W}^*_{\mathscr{O}_K}(\overrightarrow{R}^{\flat}) \tag{3.2.10.6}$$

and denote by $\theta^*_{\mathscr{O}_K,2}\colon \mathscr{A}_2^*(\overline{R}/\mathscr{O}_K) \to \widehat{\overline{R}}$ the homomorphism induced by $\theta^*_{\mathscr{O}_K}$. The sequence (3.2.10.5) induces an exact sequence

$$0 \longrightarrow \widehat{\overline{R}} \xrightarrow{\cdot\xi_\pi^*} \mathscr{A}_2^*(\overline{R}/\mathscr{O}_K) \xrightarrow{\theta^*_{\mathscr{O}_K,2}} \widehat{\overline{R}} \longrightarrow 0\,, \tag{3.2.10.7}$$

where we denoted also by $\cdot\xi_\pi^*$ the morphism induced by the multiplication by ξ_π^* in $\mathscr{A}_2^*(\overline{R}/\mathscr{O}_K)$.

The group Δ acts naturally on $\mathrm{W}^*_{\mathscr{O}_K}(\overrightarrow{R}^{\flat})$ by ring automorphisms (3.1.7), and the homomorphism $\theta^*_{\mathscr{O}_K}$ is Δ-equivariant. We deduce from this an action of Δ on $\mathscr{A}_2^*(\overline{R}/\mathscr{O}_K)$ by ring automorphisms such that the homomorphism $\theta^*_{\mathscr{O}_K,2}$ is Δ-equivariant.

We set

$$\mathscr{A}_2^*(\mathbb{X}/S) = \mathrm{Spec}(\mathscr{A}_2^*(\overline{R}/\mathscr{O}_K)), \tag{3.2.10.8}$$

which we endow with the logarithmic structure $\mathscr{M}'_{\mathscr{A}_2^*(\mathbb{X}/S)}$ associated with the prelogarithmic structure defined by the homomorphism $Q_X \to \mathscr{A}_2^*(\overline{R}/\mathscr{O}_K)$ induced by τ_X (3.2.5.2). The homomorphism $\theta^*_{\mathscr{O}_K,2}$ induces a morphism

$$(\widehat{\mathbb{X}}, \mathscr{M}_{\widehat{\mathbb{X}}}) \to (\mathscr{A}_2^*(\mathbb{X}/S), \mathscr{M}'_{\mathscr{A}_2^*(\mathbb{X}/S)}). \tag{3.2.10.9}$$

The actions of Δ on the ring $\mathscr{A}_2^*(\overline{R}/\mathscr{O}_K)$ and the monoid Q_X induce a left action on the logarithmic scheme $(\mathscr{A}_2^*(\mathbb{X}/S), \mathscr{M}'_{\mathscr{A}_2^*(\mathbb{X}/S)})$. The morphism (3.2.10.9) is Δ-equivariant.

Without the conditions of 2.4.7, we do not know if the logarithmic scheme $(\mathscr{A}_2^*(\mathbb{X}/S), \mathscr{M}'_{\mathscr{A}_2^*(\mathbb{X}/S)})$ is fine and saturated or if (3.2.10.9) is an exact closed immersion. This is why we endow $\mathscr{A}_2^*(\mathbb{X}/S)$ with another logarithmic structure, namely the logarithmic structure $\mathscr{M}_{\mathscr{A}_2^*(\mathbb{X}/S)}$ associated with the prelogarithmic structure defined by the composed homomorphism

$$\widetilde{\tau}_X^*\colon P \xrightarrow{\ \nu\ } Q_X \xrightarrow{\ \tau_X\ } \mathrm{W}(\overrightarrow{R}^{\flat}) \longrightarrow \mathscr{A}_2^*(\overline{R}/\mathscr{O}_K), \tag{3.2.10.10}$$

where the first arrow is the homomorphism (3.2.6.5) and the second arrow is the homomorphism (3.2.5.2).

The homomorphism $\nu\colon P \to Q_X$ (3.2.6.5) induces a morphism of logarithmic structures on $\mathscr{A}_2^*(\mathbb{X}/S)$

$$\mathscr{M}_{\mathscr{A}_2^*(\mathbb{X}/S)} \to \mathscr{M}'_{\mathscr{A}_2^*(\mathbb{X}/S)}. \tag{3.2.10.11}$$

It is clear that the composed homomorphism $\theta^*_{\mathscr{O}_K,2} \circ \widetilde{\tau}_X\colon P \to \widehat{\overline{R}}$ is induced by α (see 3.1.15). Hence, $\theta^*_{\mathscr{O}_K,2}$ induces an exact closed immersion

$$(\widehat{\mathbb{X}}, \mathscr{M}_{\widehat{\mathbb{X}}}) \to (\mathscr{A}_2^*(\mathbb{X}/S), \mathscr{M}_{\mathscr{A}_2^*(\mathbb{X}/S)}), \tag{3.2.10.12}$$

which factors through the morphism (3.2.10.9).

We immediately see that the morphisms (3.2.9.5) induce a left action of Δ on the logarithmic scheme $(\mathscr{A}_2^*(\mathbb{X}/S), \mathscr{M}_{\mathscr{A}_2^*(\mathbb{X}/S)})$, compatible with the morphism of logarithmic structures (3.2.10.11). The closed immersion (3.2.10.12) is Δ-equivariant.

3.2.11 In the remainder of this section, depending on whether we are in the absolute or relative case (3.1.14), we denote by $\widetilde{\mathbb{X}}$ one of the two schemes

$$\mathscr{A}_2(\mathbb{X}) \quad \text{or} \quad \mathscr{A}_2^*(\mathbb{X}/S), \tag{3.2.11.1}$$

and by $\mathscr{M}_{\widetilde{\mathbb{X}}}$ (resp. $\mathscr{M}'_{\widetilde{\mathbb{X}}}$) one of the two logarithmic structures on $\widetilde{\mathbb{X}}$,

$$\mathscr{M}_{\mathscr{A}_2(\mathbb{X})} \quad \text{or} \quad \mathscr{M}_{\mathscr{A}_2^*(\mathbb{X}/S)} \quad (\text{resp.} \quad \mathscr{M}'_{\mathscr{A}_2(\mathbb{X})} \quad \text{or} \quad \mathscr{M}'_{\mathscr{A}_2^*(\mathbb{X}/S)}). \tag{3.2.11.2}$$

The logarithmic scheme $(\widetilde{\mathbb{X}}, \mathscr{M}_{\widetilde{\mathbb{X}}})$ thus defined is fine and saturated. The homomorphism $\nu\colon P \to Q_X$ (3.2.6.5) induces a morphism of logarithmic structures on $\widetilde{\mathbb{X}}$

$$\mathscr{M}_{\widetilde{\mathbb{X}}} \to \mathscr{M}'_{\widetilde{\mathbb{X}}}. \tag{3.2.11.3}$$

We have a canonical morphism (3.2.8.5) or (3.2.10.9)

$$i'_X\colon (\widehat{\mathbb{X}}, \mathscr{M}_{\widehat{\mathbb{X}}}) \to (\widetilde{\mathbb{X}}, \mathscr{M}'_{\widetilde{\mathbb{X}}}), \tag{3.2.11.4}$$

and a canonical exact closed immersion (3.2.8.8) or (3.2.10.12)

$$i_X\colon (\widehat{\mathbb{X}}, \mathscr{M}_{\widehat{\mathbb{X}}}) \to (\widetilde{\mathbb{X}}, \mathscr{M}_{\widetilde{\mathbb{X}}}), \tag{3.2.11.5}$$

which factors through i'_X. The closed immersion $\widehat{\mathbb{X}} \to \widetilde{\mathbb{X}}$ is defined by the square zero ideal $\widetilde{\xi}\mathscr{O}_{\widetilde{\mathbb{X}}}$ of $\mathscr{O}_{\widetilde{\mathbb{X}}}$, associated with the $\widehat{\overline{R}}$-module $\widetilde{\xi}\widehat{\overline{R}}$ (see 3.1.14 for the notation).

By (3.2.5.4), we have a canonical morphism

$$(\widetilde{\mathbb{X}}, \mathscr{M}_{\widetilde{\mathbb{X}}}) \to (\widetilde{S}, \mathscr{M}_{\widetilde{S}}) \tag{3.2.11.6}$$

that fits into a commutative diagram (3.1.14.2)

$$\begin{array}{ccc} (\widehat{\mathbb{X}}, \mathscr{M}_{\widehat{\mathbb{X}}}) & \xrightarrow{i_X} & (\widetilde{\mathbb{X}}, \mathscr{M}_{\widetilde{\mathbb{X}}}) \\ \downarrow & & \downarrow \\ (\check{\overline{S}}, \mathscr{M}_{\check{\overline{S}}}) & \xrightarrow{i_S} & (\widetilde{S}, \mathscr{M}_{\widetilde{S}}). \end{array} \tag{3.2.11.7}$$

The group Δ acts on the left on the logarithmic schemes $(\widetilde{\mathbb{X}}, \mathscr{M}_{\widetilde{\mathbb{X}}})$ and $(\widetilde{\mathbb{X}}, \mathscr{M}'_{\widetilde{\mathbb{X}}})$, and the morphisms i_X, i'_X and (3.2.11.6) are Δ-equivariant.

Proposition 3.2.12 ([3] II.9.13) *Suppose that there exists a fine and saturated chart $M \to \Gamma(X, \mathscr{M}_X)$ for $(X, \mathscr{M}_X)$ inducing an isomorphism*

$$M \xrightarrow{\sim} \Gamma(X, \mathscr{M}_X)/\Gamma(X, \mathscr{O}_X^\times). \tag{3.2.12.1}$$

Then, the morphism $\mathscr{M}_{\widetilde{\mathbb{X}}} \to \mathscr{M}'_{\widetilde{\mathbb{X}}}$ (3.2.11.3) is an isomorphism. In particular, the homomorphism $P \to \Gamma(\widetilde{\mathbb{X}}, \mathscr{M}'_{\widetilde{\mathbb{X}}})$ induced by (3.2.6.5) is a chart for $(\widetilde{\mathbb{X}}, \mathscr{M}'_{\widetilde{\mathbb{X}}})$, and the morphism i'_X (3.2.11.4) is an exact closed immersion.

By virtue of 2.4.2 and 2.4.7, the logarithmic scheme $(\widetilde{\mathbb{X}}, \mathscr{M}'_{\widetilde{\mathbb{X}}})$ is fine and saturated and the morphism i'_X (3.2.11.4) is an exact closed immersion. For any geometric point $\overline{z}$ of $\widehat{\mathbb{X}}$, denoting the geometric point $i'_X(\overline{z}) = i_X(\overline{z})$ of $\widetilde{\mathbb{X}}$ also by $\overline{z}$, the homomorphism

$$\mathscr{M}_{\widetilde{\mathbb{X}},\overline{z}}/\mathscr{O}^\times_{\widetilde{\mathbb{X}},\overline{z}} \to \mathscr{M}'_{\widetilde{\mathbb{X}},\overline{x}}/\mathscr{O}^\times_{\widetilde{\mathbb{X}},\overline{z}} \tag{3.2.12.2}$$

is an isomorphism. Since $\mathscr{M}'_{\widetilde{\mathbb{X}},\overline{z}}$ is integral, we deduce from this that the morphism $\mathscr{M}_{\widetilde{\mathbb{X}},\overline{z}} \to \mathscr{M}'_{\widetilde{\mathbb{X}},\overline{z}}$ is an isomorphism; the statement follows.

Remark 3.2.13 Unlike $\mathscr{M}_{\widetilde{\mathbb{X}}}$, the logarithmic structure $\mathscr{M}'_{\widetilde{\mathbb{X}}}$ on $\widetilde{\mathbb{X}}$ does not depend on the adequate chart chosen on $(X, \mathscr{M}_X)$. But without the condition of 3.2.12, we do not know if the logarithmic scheme $(\widetilde{\mathbb{X}}, \mathscr{M}'_{\widetilde{\mathbb{X}}})$ is fine and saturated or if the morphism i'_X (3.2.11.4) is an exact closed immersion. Note however that the condition of 3.2.12 is satisfied on an open affine covering of X by ([3] II.5.17).

3.2.14 We set (3.1.15.3)

$$\mathrm{T} = \mathrm{Hom}_{\widehat{\overline{R}}}(\widetilde{\Omega}^1_{R/\mathscr{O}_K} \otimes_R \widehat{\overline{R}}, \widehat{\xi\overline{R}}). \tag{3.2.14.1}$$

We identify the dual $\widehat{\overline{R}}$-module with $\xi^{-1}\widetilde{\Omega}^1_{R/\mathscr{O}_K} \otimes_R \widehat{\overline{R}}$ (see 3.1.14) and denote by $\mathscr{G}$ the associated symmetric $\widehat{\overline{R}}$-algebra (2.1.10)

$$\mathscr{G} = \mathrm{S}_{\widehat{\overline{R}}}(\xi^{-1}\widetilde{\Omega}^1_{R/\mathscr{O}_K} \otimes_R \widehat{\overline{R}}). \tag{3.2.14.2}$$

We denote by $\widehat{\mathbb{X}}_{\mathrm{zar}}$ the Zariski topos of $\widehat{\mathbb{X}} = \mathrm{Spec}(\widehat{\overline{R}})$ (3.2.3.2), by $\widetilde{\mathrm{T}}$ the $\mathscr{O}_{\widehat{\mathbb{X}}}$-module associated with T and by $\mathbf{T}$ the $\widehat{\mathbb{X}}$-vector bundle associated with its dual, in other words,

$$\mathbf{T} = \mathrm{Spec}(\mathscr{G}). \tag{3.2.14.3}$$

Let U be a Zariski open subscheme of $\widehat{\mathbb{X}}$, $\widetilde{U}$ the corresponding open subscheme of $\widetilde{\mathbb{X}}$ (see 3.2.11). We denote by $\mathscr{L}(U)$ the set of morphisms represented by the dotted arrows which complete the canonical diagram

$$\begin{array}{ccc}
(U, \mathscr{M}_{\widetilde{\mathbb{X}}}|U) & \xrightarrow{i_X \times_{\widetilde{\mathbb{X}}} \widetilde{U}} & (\widetilde{U}, \mathscr{M}_{\widetilde{\mathbb{X}}}|\widetilde{U}) \\
\downarrow & & \vdots \\
(\check{\overline{X}}, \mathscr{M}_{\check{\overline{X}}}) & \longrightarrow & (\widetilde{X}, \mathscr{M}_{\widetilde{X}}) \\
\downarrow & & \downarrow \\
(\check{\overline{S}}, \mathscr{M}_{\check{\overline{S}}}) & \xrightarrow{i_S} & (\widetilde{S}, \mathscr{M}_{\widetilde{S}})
\end{array} \tag{3.2.14.4}$$

in such a way that it remains commutative (3.1.17.2). By ([3] II.5.23), the functor $U \mapsto \mathscr{L}(U)$ is a $\widetilde{\mathrm{T}}$-torsor of $\widehat{\mathbb{X}}_{\mathrm{zar}}$. We call it the *Higgs–Tate torsor* associated with $(\widetilde{X}, \mathscr{M}_{\widetilde{X}})$. We denote by $\mathscr{F}$ the $\widehat{\overline{R}}$-module of affine functions on $\mathscr{L}$ (see [3] II.4.9). It fits into a canonical exact sequence

$$0 \to \widehat{\overline{R}} \to \mathscr{F} \to \widetilde{\xi}^{-1}\widetilde{\Omega}^1_{R/\mathscr{O}_K} \otimes_R \widehat{\overline{R}} \to 0. \tag{3.2.14.5}$$

By ([37] I 4.3.1.7), this sequence induces for every integer $n \geq 1$ an exact sequence (2.1.10)

$$0 \to \mathrm{S}^{n-1}_{\widehat{\overline{R}}}(\mathscr{F}) \to \mathrm{S}^{n}_{\widehat{\overline{R}}}(\mathscr{F}) \to \mathrm{S}^{n}_{\widehat{\overline{R}}}(\widetilde{\xi}^{-1}\widetilde{\Omega}^1_{R/\mathscr{O}_K} \otimes_R \widehat{\overline{R}}) \to 0. \tag{3.2.14.6}$$

The $\widehat{\overline{R}}$-modules $(\mathrm{S}^n_{\widehat{\overline{R}}}(\mathscr{F}))_{n \in \mathbb{N}}$ therefore form a filtered direct system, whose direct limit

$$\mathscr{C} = \varinjlim_{n \geq 0} \mathrm{S}^n_{\widehat{\overline{R}}}(\mathscr{F}) \tag{3.2.14.7}$$

is naturally endowed with an $\widehat{\overline{R}}$-algebra structure. By ([3] II.4.10), the $\widehat{\mathbb{X}}$-scheme

$$\mathbf{L} = \mathrm{Spec}(\mathscr{C}) \tag{3.2.14.8}$$

is naturally a homogeneous principal $\mathbf{T}$-bundle over $\widehat{\mathbb{X}}$ which canonically represents $\mathscr{L}$. Note that $\mathscr{L}$, $\mathscr{F}$, $\mathscr{C}$ and $\mathbf{L}$ depend on $(\widetilde{X}, \mathscr{M}_{\widetilde{X}})$.

3.2.15 We endow $\widehat{\mathbb{X}}$ with the natural left action of Δ; for any $g \in \Delta$, the automorphism of $\widehat{\mathbb{X}}$ defined by g, also denoted by g, is induced by the automorphism g^{-1} of $\widehat{\overline{R}}$. We consider $\widetilde{\mathrm{T}}$ as a Δ-equivariant $\mathscr{O}_{\widehat{\mathbb{X}}}$-module using the descent datum corresponding to the $\widehat{R_1}$-module $\mathrm{Hom}_{\widehat{R_1}}(\widetilde{\Omega}^1_{R/\mathscr{O}_K} \otimes_R \widehat{R_1}, \widehat{\xi R_1})$ (see [3] II.4.18). For every $g \in \Delta$, we therefore have a canonical isomorphism of $\mathscr{O}_{\widehat{\mathbb{X}}}$-modules

$$\tau_g^{\widetilde{\mathrm{T}}} \colon \widetilde{\mathrm{T}} \xrightarrow{\sim} g^*(\widetilde{\mathrm{T}}). \tag{3.2.15.1}$$

This induces an isomorphism of $\widehat{\mathbb{X}}$-group schemes

$$\tau_g^{\mathbf{T}} \colon \mathbf{T} \xrightarrow{\sim} g^{\bullet}(\mathbf{T}), \tag{3.2.15.2}$$

where $g^\bullet$ denotes the base change functor by the automorphism g of $\widehat{\mathbb{X}}$. We thus obtain a Δ-equivariant structure on the $\widehat{\mathbb{X}}$-group scheme $\mathbf{T}$ (see [3] II.4.17) and consequently a left action of Δ on $\mathbf{T}$ compatible with its action on $\widehat{\mathbb{X}}$; the automorphism of $\mathbf{T}$ defined by an element g of Δ is the composition of $\tau_g^{\mathbf{T}}$ with the canonical projection $g^\bullet(\mathbf{T}) \to \mathbf{T}$. We deduce from this an action of Δ on $\mathscr{G}$ (3.2.14.2) by ring automorphisms, compatible with its action on $\widehat{\overline{R}}$, that we call the *canonical action*. The latter is concretely induced by the trivial action on $\mathrm{S}_{\widehat{R_1}}(\widetilde{\xi}^{-1}\widetilde{\Omega}^1_{R/\mathscr{O}_K} \otimes_R \widehat{R_1})$.

The natural left action of Δ on the logarithmic scheme $(\widetilde{\mathbb{X}}, \mathscr{M}_{\widetilde{\mathbb{X}}})$ (3.2.8 or 3.2.10) induces on the $\widetilde{\mathrm{T}}$-torsor $\mathscr{L}$ a Δ-equivariant structure (see [3] II.4.18), in other words, it induces for every $g \in \Delta$, a $\tau_g^{\widetilde{\mathrm{T}}}$-equivariant isomorphism

$$\tau_g^{\mathscr{L}} : \mathscr{L} \xrightarrow{\sim} g^*(\mathscr{L}); \tag{3.2.15.3}$$

these isomorphisms being subject to compatibility relations (see [3] II.4.16). Indeed, for any Zariski open subscheme U of $\widehat{\mathbb{X}}$, we take for

$$\tau_g^{\mathscr{L}}(U) : \mathscr{L}(U) \xrightarrow{\sim} \mathscr{L}(g(U)) \tag{3.2.15.4}$$

the isomorphism defined as follows. Let $\widetilde{U}$ be the open subscheme of $\widetilde{\mathbb{X}}$ corresponding to U, $\mu \in \mathscr{L}(U)$ which we consider as a morphism

$$\mu : (\widetilde{U}, \mathscr{M}_{\widetilde{\mathbb{X}}}|\widetilde{U}) \to (\widetilde{X}, \mathscr{M}_{\widetilde{X}}). \tag{3.2.15.5}$$

Since i_X (3.2.11.5) and the morphism (3.2.3.3) are Δ-equivariant, the composed morphism

$$(g(\widetilde{U}), \mathscr{M}_{\widetilde{\mathbb{X}}}|g(\widetilde{U})) \xrightarrow{g^{-1}} (\widetilde{U}, \mathscr{M}_{\widetilde{\mathbb{X}}}|\widetilde{U}) \xrightarrow{\mu} (\widetilde{X}, \mathscr{M}_{\widetilde{X}}) \tag{3.2.15.6}$$

extends the canonical morphism $(g(U), \mathscr{M}_{\widetilde{\mathbb{X}}}|g(U)) \to (\widetilde{X}, \mathscr{M}_{\widetilde{X}})$. It corresponds to the image of μ by $\tau_g^{\mathscr{L}}(U)$. We immediately check that the morphism $\tau_g^{\mathscr{L}}$ thus defined is a $\tau_g^{\widetilde{\mathrm{T}}}$-equivariant isomorphism and that these isomorphisms verify the compatibility relations required in ([3] II.4.16).

By ([3] II.4.21), the Δ-equivariant structures on $\widetilde{\mathrm{T}}$ and $\mathscr{L}$ induce a Δ-equivariant structure on the $\mathscr{O}_{\widetilde{\mathbb{X}}}$-module associated with $\mathscr{F}$, or, what amounts to the same, an $\widehat{\overline{R}}$-semi-linear action of Δ on $\mathscr{F}$, such that the morphisms of the sequence (3.2.14.5) are Δ-equivariant. We deduce on $\mathbf{L}$ (3.2.14.8) the structure of a Δ-equivariant homogeneous principal $\mathbf{T}$-bundle on $\widehat{\mathbb{X}}$ (see [3] II.4.20). For every $g \in \Delta$, we therefore have a $\tau_g^{\mathbf{T}}$-equivariant isomorphism

$$\tau_g^{\mathbf{L}} : \mathbf{L} \xrightarrow{\sim} g^\bullet(\mathbf{L}). \tag{3.2.15.7}$$

This structure determines a left action of Δ on $\mathbf{L}$ compatible with its action on $\widehat{\mathbb{X}}$; the automorphism of $\mathbf{L}$ defined by an element g of Δ is the composition of $\tau_g^{\mathbf{L}}$ and the canonical projection $g^\bullet(\mathbf{L}) \to \mathbf{L}$. We thus obtain an action of Δ on $\mathscr{C}$ (3.2.14.7) by ring automorphisms, compatible with its action on $\widehat{\overline{R}}$, that we call the *canonical action*. The latter is concretely induced by the action of Δ on $\mathscr{F}$.

For any $g \in \Delta$, we denote by

$$\mathbf{L}(\widehat{\mathbb{X}}) \xrightarrow{\sim} \mathbf{L}(\widehat{\mathbb{X}}), \quad \mu \mapsto {}^g\mu \tag{3.2.15.8}$$

the composition of the isomorphisms

$$\tau_g^{\mathbf{L}} \colon \mathbf{L}(\widehat{\mathbb{X}}) \xrightarrow{\sim} g^{\bullet}(\mathbf{L})(\widehat{\mathbb{X}}), \tag{3.2.15.9}$$

$$g^{\bullet}(\mathbf{L})(\widehat{\mathbb{X}}) \xrightarrow{\sim} \mathbf{L}(\widehat{\mathbb{X}}), \quad \mu \mapsto \mathrm{pr} \circ \mu \circ g^{-1}, \tag{3.2.15.10}$$

where g^{-1} acts on $\widehat{\mathbb{X}}$ and $\mathrm{pr} \colon g^{\bullet}(\mathbf{L}) \to \mathbf{L}$ is the canonical projection, so that the diagram

$$\begin{array}{ccc} \mathbf{L} & \xrightarrow{g} & \mathbf{L} \\ \uparrow{\scriptstyle \mu} & & \uparrow{\scriptstyle {}^g\mu} \\ \widehat{\mathbb{X}} & \xrightarrow{g} & \widehat{\mathbb{X}} \end{array} \tag{3.2.15.11}$$

is commutative. In particular, for every $\mu \in \mathscr{L}(\widehat{\mathbb{X}})$ and every $\beta \in \mathscr{F}$, we have

$$(g^{-1}(\beta))(\mu) = g^{-1}(\beta({}^g\mu)). \tag{3.2.15.12}$$

Moreover, ${}^g\mu$ is defined by the composed morphism

$$(\widetilde{\mathbb{X}}, \mathscr{M}_{\widetilde{\mathbb{X}}}) \xrightarrow{g^{-1}} (\widetilde{\mathbb{X}}, \mathscr{M}_{\widetilde{\mathbb{X}}}) \xrightarrow{\mu} (\widetilde{X}, \mathscr{M}_{\widetilde{X}}). \tag{3.2.15.13}$$

Definition 3.2.16 The $\widehat{\overline{R}}$-algebra $\mathscr{C}$ (3.2.14.7), endowed with the canonical action of Δ (3.2.15), is called the *Higgs–Tate algebra* associated with $(\widetilde{X}, \mathscr{M}_{\widetilde{X}})$. The $\widehat{\overline{R}}$-representation $\mathscr{F}$ (3.2.14.5) of Δ is called the *Higgs–Tate extension* associated with $(\widetilde{X}, \mathscr{M}_{\widetilde{X}})$.

3.2.17 For any rational number $r \geq 0$, we denote by $\mathscr{F}^{(r)}$ the $\widehat{\overline{R}}$-representation of Δ deduced from $\mathscr{F}$ (3.2.14.5) by pullback by the multiplication by p^r on $\widetilde{\xi}^{-1}\widetilde{\Omega}^1_{R/\mathscr{O}_K} \otimes_R \widehat{\overline{R}}$, so that we have a split exact sequence of $\widehat{\overline{R}}$-modules

$$0 \to \widehat{\overline{R}} \longrightarrow \mathscr{F}^{(r)} \to \widetilde{\xi}^{-1}\widetilde{\Omega}^1_{R/\mathscr{O}_K} \otimes_R \widehat{\overline{R}} \to 0. \tag{3.2.17.1}$$

By ([37] I 4.3.1.7), this sequence induces for any integer $n \geq 1$, an exact sequence

$$0 \to \mathrm{S}^{n-1}_{\widehat{\overline{R}}}(\mathscr{F}^{(r)}) \to \mathrm{S}^{n}_{\widehat{\overline{R}}}(\mathscr{F}^{(r)}) \to \mathrm{S}^{n}_{\widehat{\overline{R}}}(\widetilde{\xi}^{-1}\widetilde{\Omega}^1_{R/\mathscr{O}_K} \otimes_R \widehat{\overline{R}}) \to 0. \tag{3.2.17.2}$$

The $\widehat{\overline{R}}$-modules $(\mathrm{S}^{n}_{\widehat{\overline{R}}}(\mathscr{F}^{(r)}))_{n \in \mathbb{N}}$ thus form a filtered direct system, whose direct limit

$$\mathscr{C}^{(r)} = \varinjlim_{n \geq 0} \mathrm{S}^{n}_{\widehat{\overline{R}}}(\mathscr{F}^{(r)}) \tag{3.2.17.3}$$

is naturally endowed with an $\widehat{\overline{R}}$-algebra structure. The action of Δ on $\mathscr{F}^{(r)}$ induces an action on $\mathscr{C}^{(r)}$ by ring automorphisms, compatible with its action on $\widehat{\overline{R}}$, that we call the *canonical action*. The $\widehat{\overline{R}}$-algebra $\mathscr{C}^{(r)}$ endowed with this action is called the *Higgs–Tate algebra of thickness r associated with* $(\widetilde{X}, \mathscr{M}_{\widetilde{X}})$.

We denote by $\widehat{\mathscr{C}}^{(r)}$ (resp. $\widehat{\mathscr{C}}$) the p-adic Hausdorff completion of $\mathscr{C}^{(r)}$ (resp. $\mathscr{C}$) that we always assume endowed with the p-adic topology. We endow $\widehat{\mathscr{C}}^{(r)} \otimes_{\mathbb{Z}_p} \mathbb{Q}_p$ with the p-adic topology (2.1.1). We check immediately that $\mathscr{C}^{(r)}$ and $\widehat{\mathscr{C}}^{(r)}$ are $\mathscr{O}_C$-flat. For all rational numbers $r' \geq r \geq 0$, we have an injective and Δ-equivariant canonical $\widehat{\overline{R}}$-homomorphism $\alpha^{r,r'} : \mathscr{C}^{(r')} \to \mathscr{C}^{(r)}$. The induced homomorphism $\widehat{\alpha}^{r,r'} : \widehat{\mathscr{C}}^{(r')} \to \widehat{\mathscr{C}}^{(r)}$ is also injective. We set

$$\widehat{\mathscr{C}}^{(r+)} = \varinjlim_{t \in \mathbb{Q}_{>r}} \widehat{\mathscr{C}}^{(t)}, \tag{3.2.17.4}$$

which is identified with a sub-$\widehat{\overline{R}}$-algebra of $\widehat{\mathscr{C}}^{(r)}$ by the direct limit of the homomorphisms $(\widehat{\alpha}^{r,t})_{t \in \mathbb{Q}_{>r}}$. The actions of Δ on the rings $(\widehat{\mathscr{C}}^{(t)})_{t \in \mathbb{Q}_{>r}}$ induce an action on $\widehat{\mathscr{C}}^{(r+)}$ by ring automorphisms, compatible with its actions on $\widehat{\overline{R}}$ and on $\widehat{\mathscr{C}}$. The algebra $\widehat{\mathscr{C}}^{(0+)}$ was denoted $\mathscr{C}^{\dagger}$ in ([3] (II.12.1.6)).

Remark 3.2.18 For an $\widehat{R_1}$-algebra A, we consider in the remainder of this chapter Higgs A-modules with coefficients in $\widetilde{\xi}^{-1}\widetilde{\Omega}^1_{R/\mathscr{O}_K} \otimes_R A$ (see 2.5.1 and 3.1.14). We will (abusively) say that they have coefficients in $\widetilde{\xi}^{-1}\widetilde{\Omega}^1_{R/\mathscr{O}_K}$. The category of these modules will be denoted by $\mathbf{HM}(A, \widetilde{\xi}^{-1}\widetilde{\Omega}^1_{R/\mathscr{O}_K})$.

3.2.19 We have a canonical $\mathscr{G}$-linear isomorphism (3.2.14.2)

$$\Omega^1_{\mathscr{G}/\widehat{\overline{R}}} \xrightarrow{\sim} \widetilde{\xi}^{-1}\widetilde{\Omega}^1_{R/\mathscr{O}_K} \otimes_R \mathscr{G}. \tag{3.2.19.1}$$

This induces an isomorphism

$$\Omega^1_{\mathscr{C}/\widehat{\overline{R}}} \xrightarrow{\sim} \widetilde{\xi}^{-1}\widetilde{\Omega}^1_{R/\mathscr{O}_K} \otimes_R \mathscr{C}. \tag{3.2.19.2}$$

We denote by

$$d_{\mathscr{C}} : \mathscr{C} \to \widetilde{\xi}^{-1}\widetilde{\Omega}^1_{R/\mathscr{O}_K} \otimes_R \mathscr{C} \tag{3.2.19.3}$$

the $\widehat{\overline{R}}$-universal derivation of $\mathscr{C}$ and by

$$d_{\widehat{\mathscr{C}}} : \widehat{\mathscr{C}} \to \widetilde{\xi}^{-1}\widetilde{\Omega}^1_{R/\mathscr{O}_K} \otimes_R \widehat{\mathscr{C}} \tag{3.2.19.4}$$

its extension to the p-adic completions (note that the R-module $\widetilde{\Omega}^1_{R/\mathscr{O}_K}$ is free of finite type). For any $x \in \mathscr{F}$, $d_{\mathscr{C}}(x)$ is the canonical image of x in $\widetilde{\xi}^{-1}\widetilde{\Omega}^1_{R/\mathscr{O}_K} \otimes_R \widehat{\overline{R}}$ (3.2.14.5).

Similarly, for any rational number $r \geq 0$, we denote by

$$d_{\mathscr{C}^{(r)}} : \mathscr{C}^{(r)} \to \widetilde{\xi}^{-1}\widetilde{\Omega}^1_{R/\mathscr{O}_K} \otimes_R \mathscr{C}^{(r)} \tag{3.2.19.5}$$

the $\overline{R}$-universal derivation of $\mathscr{C}^{(r)}$ and by

$$d_{\widehat{\mathscr{C}}^{(r)}} : \widehat{\mathscr{C}}^{(r)} \to \widetilde{\xi}^{-1}\widetilde{\Omega}^1_{R/\mathscr{O}_K} \otimes_R \widehat{\mathscr{C}}^{(r)} \tag{3.2.19.6}$$

its extension to the p-adic completions. We immediately see that the derivations $d_{\mathscr{C}^{(r)}}$ and $d_{\widehat{\mathscr{C}}^{(r)}}$ are Δ-equivariant. Since $\widetilde{\xi}^{-1}\widetilde{\Omega}^1_{R/\mathscr{O}_K} \otimes_R \overline{R} = d_{\mathscr{C}^{(r)}}(\mathscr{F}^{(r)}) \subset d_{\mathscr{C}^{(r)}}(\mathscr{C}^{(r)})$, $d_{\mathscr{C}^{(r)}}$ and $d_{\widehat{\mathscr{C}}^{(r)}}$ are also Higgs $\widehat{\overline{R}}$-fields with coefficients in $\widetilde{\xi}^{-1}\widetilde{\Omega}^1_{R/\mathscr{O}_K}$ by 2.5.26(i). We denote by $\mathbb{K}^\bullet(\widehat{\mathscr{C}}^{(r)})$ the Dolbeault complex of $(\widehat{\mathscr{C}}^{(r)}, p^r d_{\widehat{\mathscr{C}}^{(r)}})$ (2.5.1.2) and by $\widetilde{\mathbb{K}}^\bullet(\widehat{\mathscr{C}}^{(r)})$ the augmented Dolbeault complex

$$\widehat{\overline{R}} \to \mathbb{K}^0(\widehat{\mathscr{C}}^{(r)}) \to \mathbb{K}^1(\widehat{\mathscr{C}}^{(r)}) \to \cdots \to \mathbb{K}^n(\widehat{\mathscr{C}}^{(r)}) \to \dots, \tag{3.2.19.7}$$

where $\widehat{\overline{R}}$ is placed in degree -1 and the differential $\widehat{\overline{R}} \to \widehat{\mathscr{C}}^{(r)}$ is the canonical homomorphism.

For all rational numbers $r' \geq r \geq 0$, we have

$$p^{r'}(\mathrm{id} \times \alpha^{r,r'}) \circ d_{\mathscr{C}^{(r')}} = p^r d_{\mathscr{C}^{(r)}} \circ \alpha^{r,r'}. \tag{3.2.19.8}$$

Consequently, $\widehat{\alpha}^{r,r'}$ induces a morphism of complexes

$$\iota^{r,r'} : \widetilde{\mathbb{K}}^\bullet(\widehat{\mathscr{C}}^{(r')}) \to \widetilde{\mathbb{K}}^\bullet(\widehat{\mathscr{C}}^{(r)}). \tag{3.2.19.9}$$

By (3.2.19.8), the derivations $(p^t d_{\widehat{\mathscr{C}}^{(t)}})_{t \in \mathbb{Q}_{>r}}$ induce an $\widehat{\overline{R}}$-derivation

$$d^{(r)}_{\widehat{\mathscr{C}}^{(r+)}} : \widehat{\mathscr{C}}^{(r+)} \to \widetilde{\xi}^{-1}\widetilde{\Omega}^1_{R/\mathscr{O}_K} \otimes_R \widehat{\mathscr{C}}^{(r+)}, \tag{3.2.19.10}$$

which is nothing but the restriction of $p^r d_{\widehat{\mathscr{C}}^{(r)}}$ to $\widehat{\mathscr{C}}^{(r+)}$. It is also a Higgs $\widehat{\overline{R}}$-field with coefficients in $\widetilde{\xi}^{-1}\widetilde{\Omega}^1_{R/\mathscr{O}_K}$. We denote by $\mathbb{K}^\bullet(\widehat{\mathscr{C}}^{(r+)})$ the Dolbeault complex of $(\widehat{\mathscr{C}}^{(r+)}, d^{(r)}_{\widehat{\mathscr{C}}^{(r+)}})$. Since $\widehat{\overline{R}}$ is $\mathscr{O}_C$-flat, for every rational number $r \geq 0$, we have

$$\ker(d^{(r)}_{\widehat{\mathscr{C}}^{(r+)}}) = \ker(d_{\widehat{\mathscr{C}}^{(r)}}) = \widehat{\overline{R}}. \tag{3.2.19.11}$$

3.2.20 Consider a second $(\widetilde{S}, \mathscr{M}_{\widetilde{S}})$-smooth deformation $(\widetilde{X}', \mathscr{M}_{\widetilde{X}'})$ of $(\check{\overline{X}}, \mathscr{M}_{\check{\overline{X}}})$ and equip with an exponent $'$ the associated objects (3.2.14). By ([41] 3.14), there exists an isomorphism of $(\widetilde{S}, \mathscr{M}_{\widetilde{S}})$-deformations

$$h : (\widetilde{X}, \mathscr{M}_{\widetilde{X}}) \xrightarrow{\sim} (\widetilde{X}', \mathscr{M}_{\widetilde{X}'}). \tag{3.2.20.1}$$

The isomorphism of $\widetilde{\mathrm{T}}$-torsors $\mathscr{L} \xrightarrow{\sim} \mathscr{L}'$, $\psi \mapsto h \circ \psi$ (3.2.14.4) induces an $\widehat{\overline{R}}$-linear and Δ-equivariant isomorphism

$$\mathscr{F}' \xrightarrow{\sim} \mathscr{F}, \tag{3.2.20.2}$$

that fits into a commutative diagram (3.2.14.5)

$$\begin{array}{ccccccccc} 0 & \longrightarrow & \widehat{\overline{R}} & \longrightarrow & \mathscr{F}' & \longrightarrow & \widetilde{\xi}^{-1}\widetilde{\Omega}^1_{R/\mathscr{O}_K} \otimes_R \widehat{\overline{R}} & \longrightarrow & 0 \\ & & \| & & \downarrow & & \| & & \\ 0 & \longrightarrow & \widehat{\overline{R}} & \longrightarrow & \mathscr{F} & \longrightarrow & \widetilde{\xi}^{-1}\widetilde{\Omega}^1_{R/\mathscr{O}_K} \otimes_R \widehat{\overline{R}} & \longrightarrow & 0. \end{array} \tag{3.2.20.3}$$

We deduce from this a Δ-equivariant $\widehat{\overline{R}}$-isomorphism

$$\mathscr{C}' \xrightarrow{\sim} \mathscr{C}. \tag{3.2.20.4}$$

3.2.21 Let us consider the absolute case, i.e., $(\widetilde{S}, \mathscr{M}_{\widetilde{S}}) = (\mathscr{A}_2(\overline{S}), \mathscr{M}_{\mathscr{A}_2(\overline{S})})$ (3.1.14) and set (3.1.17)

$$(\widetilde{X}', \mathscr{M}_{\widetilde{X}'}) = (\widetilde{X}, \mathscr{M}_{\widetilde{X}}) \times_{(\mathscr{A}_2(\overline{S}), \mathscr{M}_{\mathscr{A}_2(\overline{S})})} (\mathscr{A}_2^*(\overline{S}/S), \mathscr{M}_{\mathscr{A}_2^*(\overline{S}/S)}) \tag{3.2.21.1}$$

where the base change is defined by the morphism pr_2 (3.1.13.6), the product being indifferently taken in the category of logarithmic schemes or in that of fine logarithmic schemes. We equip with an exponent $'$ the objects associated with the $(\mathscr{A}_2^*(\overline{S}/S), \mathscr{M}_{\mathscr{A}_2^*(\overline{S}/S)})$-deformation $(\widetilde{X}', \mathscr{M}_{\widetilde{X}'})$ (3.2.14). We denote by K_0 the field of the fractions of W (3.1.1), by $\mathfrak{d}$ the different of the extension K/K_0 and we set $\rho = v(\pi\mathfrak{d})$. We denote by

$$\iota \colon \xi\mathscr{O}_C \xrightarrow{\sim} \xi_\pi^*\mathscr{O}_C \tag{3.2.21.2}$$

the $\mathscr{O}_C$-linear isomorphism such that the composed morphism

$$\xi\mathscr{O}_C \xrightarrow{\iota} \xi_\pi^*\mathscr{O}_C \xrightarrow{\cdot p^\rho} p^\rho\xi_\pi^*\mathscr{O}_C \tag{3.2.21.3}$$

coincides with the isomorphism (3.1.10.1). The isomorphism $\iota \otimes_{\mathscr{O}_C} \widehat{\overline{R}}$ induces an $\widehat{\overline{R}}$-linear isomorphism $u \colon \mathrm{T} \xrightarrow{\sim} \mathrm{T}'$ (3.2.14.1).

A chase into the commutative diagram

$$\begin{array}{ccccc}
(\widehat{\mathbb{X}}, \mathscr{M}_{\widehat{\mathbb{X}}}) & \longrightarrow & (\mathscr{A}_2^*(\mathbb{X}/S), \mathscr{M}_{\mathscr{A}_2^*(\mathbb{X}/S)}) & \longrightarrow & (\mathscr{A}_2(\mathbb{X}), \mathscr{M}_{\mathscr{A}_2(\mathbb{X})}) \\
\downarrow & & & & \\
(\check{\overline{X}}, \mathscr{M}_{\check{\overline{X}}}) & \longrightarrow & (\widetilde{X}', \mathscr{M}_{\widetilde{X}'}) & \longrightarrow & (\widetilde{X}, \mathscr{M}_{\widetilde{X}}) \\
\downarrow & & \downarrow & \square & \downarrow \\
(\check{\overline{S}}, \mathscr{M}_{\check{\overline{S}}}) & \longrightarrow & (\mathscr{A}_2^*(\overline{S}/S), \mathscr{M}_{\mathscr{A}_2^*(\overline{S}/S)}) & \longrightarrow & (\mathscr{A}_2(\overline{S}), \mathscr{M}_{\mathscr{A}_2(\overline{S})})
\end{array} \qquad (3.2.21.4)$$

induces a $(p^\rho u)$-equivariant and Δ-equivariant morphism $\mathscr{L} \to \mathscr{L}'$ of $\widehat{\mathbb{X}}_{\text{zar}}$. By ([3] II.4.12), we deduce from this an $\widehat{\overline{R}}$-linear and Δ-equivariant morphism

$$v\colon \mathscr{F}' \to \mathscr{F} \qquad (3.2.21.5)$$

that fits into a commutative diagram (3.2.14.5)

$$\begin{array}{ccccccccc}
0 & \longrightarrow & \widehat{\overline{R}} & \longrightarrow & \mathscr{F}' & \longrightarrow & (\xi_\pi^*)^{-1}\widetilde{\Omega}^1_{R/\mathscr{O}_K} \otimes_R \widehat{\overline{R}} & \longrightarrow & 0 \\
 & & \| & & \downarrow v & & \downarrow p^\rho u^\vee & & \\
0 & \longrightarrow & \widehat{\overline{R}} & \longrightarrow & \mathscr{F} & \longrightarrow & \xi^{-1}\widetilde{\Omega}^1_{R/\mathscr{O}_K} \otimes_R \widehat{\overline{R}} & \longrightarrow & 0.
\end{array} \qquad (3.2.21.6)$$

where $u^\vee$ is the dual morphism of u. We deduce from this a Δ-equivariant $\widehat{\overline{R}}$-morphism

$$w\colon \mathscr{C}' \to \mathscr{C}. \qquad (3.2.21.7)$$

It is clear that $v \otimes_{\mathbb{Z}_p} \mathbb{Q}_p$ and $w \otimes_{\mathbb{Z}_p} \mathbb{Q}_p$ are isomorphisms (3.2.21.2) .

By (3.2.21.6), for every rational number $r \geq 0$, the morphism v (3.2.21.5) induces an $\widehat{\overline{R}}$-linear and Δ-equivariant morphism

$$v^{(r)}\colon \mathscr{F}'^{(r)} \to \mathscr{F}^{(r)} \qquad (3.2.21.8)$$

that fits into a commutative diagram (3.2.17.1)

$$\begin{array}{ccccccccc}
0 & \longrightarrow & \widehat{\overline{R}} & \longrightarrow & \mathscr{F}'^{(r)} & \longrightarrow & (\xi_\pi^*)^{-1}\widetilde{\Omega}^1_{R/\mathscr{O}_K} \otimes_R \widehat{\overline{R}} & \longrightarrow & 0 \\
 & & \| & & \downarrow v^{(r)} & & \downarrow p^\rho u^\vee & & \\
0 & \longrightarrow & \widehat{\overline{R}} & \longrightarrow & \mathscr{F}^{(r)} & \longrightarrow & \xi^{-1}\widetilde{\Omega}^1_{R/\mathscr{O}_K} \otimes_R \widehat{\overline{R}} & \longrightarrow & 0.
\end{array} \qquad (3.2.21.9)$$

Since $u^\vee$ is an isomorphism, we deduce from this a canonical $\widehat{\overline{R}}$-linear and Δ-equivariant isomorphism

$$\mathscr{F}'^{(r)} \xrightarrow{\sim} \mathscr{F}^{(r+\rho)} \qquad (3.2.21.10)$$

that fits into a commutative diagram

$$\begin{array}{ccccccccc} 0 & \longrightarrow & \widehat{\overline{R}} & \longrightarrow & \mathscr{F}'^{(r)} & \longrightarrow & (\xi_\pi^*)^{-1}\widetilde{\Omega}^1_{R/\mathscr{O}_K} \otimes_R \widehat{\overline{R}} & \longrightarrow & 0 \\ & & \| & & \downarrow & & \downarrow{\scriptstyle u^\vee} & & \\ 0 & \longrightarrow & \widehat{\overline{R}} & \longrightarrow & \mathscr{F}^{(r+\rho)} & \longrightarrow & \xi^{-1}\widetilde{\Omega}^1_{R/\mathscr{O}_K} \otimes_R \widehat{\overline{R}} & \longrightarrow & 0 \end{array} \qquad (3.2.21.11)$$

We deduce from this a Δ-equivariant isomorphism of $\widehat{\overline{R}}$-algebras

$$\mathscr{C}'^{(r)} \xrightarrow{\sim} \mathscr{C}^{(r+\rho)}. \qquad (3.2.21.12)$$

We therefore have a canonical Δ-equivariant isomorphism of $\widehat{\overline{R}}$-algebras

$$\widehat{\mathscr{C}}'^{(r+)} \xrightarrow{\sim} \widehat{\mathscr{C}}^{(r+\rho+)}. \qquad (3.2.21.13)$$

This identifies $\widehat{\mathscr{C}}'^{(r+)}$ with a sub-$\widehat{\overline{R}}$-algebra of $\widehat{\mathscr{C}}^{(r+)}$.

For every rational number $r \geq 0$, the diagram

$$\begin{array}{ccc} \mathscr{C}'^{(r)} & \xrightarrow{d_{\mathscr{C}'^{(r)}}} & (\xi_\pi^*)^{-1}\widetilde{\Omega}^1_{R/\mathscr{O}_K} \otimes_R \mathscr{C}'^{(r)} \\ \downarrow & & \downarrow{\scriptstyle u^\vee} \\ \mathscr{C}^{(r+\rho)} & \xrightarrow{d_{\mathscr{C}^{(r+\rho)}}} & \xi^{-1}\widetilde{\Omega}^1_{R/\mathscr{O}_K} \otimes_R \mathscr{C}^{(r+\rho)} \end{array} \qquad (3.2.21.14)$$

is commutative (3.2.19.5). We deduce from this that the diagram

$$\begin{array}{ccc} \widehat{\mathscr{C}}'^{(r+)} & \xrightarrow{d^{(r)}_{\widehat{\mathscr{C}}'^{(r+)}}} & (\xi_\pi^*)^{-1}\widetilde{\Omega}^1_{R/\mathscr{O}_K} \otimes_R \widehat{\mathscr{C}}'^{(r+)} \\ \downarrow & & \downarrow{\scriptstyle p^\rho u^\vee} \\ \widehat{\mathscr{C}}^{(r+)} & \xrightarrow{d^{(r)}_{\widehat{\mathscr{C}}^{(r+)}}} & \xi^{-1}\widetilde{\Omega}^1_{R/\mathscr{O}_K} \otimes_R \widehat{\mathscr{C}}^{(r+)} \end{array} \qquad (3.2.21.15)$$

is commutative (3.2.19.10).

Proposition 3.2.22 *Let r, r' be two rational numbers such that $r' > r > 0$. Then,*

(i) *There exists a rational number $\alpha \geq 0$ depending on r and r' but not on the morphism f satisfying the conditions of* 3.1.15 *nor on the adequate chart, such that*

$$p^\alpha \iota^{r,r'} : \widetilde{\mathbb{K}}^\bullet(\widehat{\mathscr{C}}^{(r')}) \to \widetilde{\mathbb{K}}^\bullet(\widehat{\mathscr{C}}^{(r)}), \qquad (3.2.22.1)$$

where $\iota^{r,r'}$ is the morphism (3.2.19.9), is homotopic to 0 by an $\widehat{\overline{R}}$-linear homotopy.

(ii) *The canonical morphism*

$$\iota^{r,r'} \otimes_{\mathbb{Z}_p} \mathbb{Q}_p \colon \widetilde{\mathbb{K}}^\bullet(\widehat{\mathscr{C}}^{(r')}) \otimes_{\mathbb{Z}_p} \mathbb{Q}_p \to \widetilde{\mathbb{K}}^\bullet(\widehat{\mathscr{C}}^{(r)}) \otimes_{\mathbb{Z}_p} \mathbb{Q}_p \tag{3.2.22.2}$$

is homotopic to 0 *by a continuous homotopy.*

(iii) *The complex* $\mathbb{K}^\bullet(\widehat{\mathscr{C}}^{(0+)}) \otimes_{\mathbb{Z}_p} \mathbb{Q}_p$ *is a resolution of* $\widehat{\overline{R}}[\frac{1}{p}]$ (3.2.19.10).

Indeed, a section $\psi \in \mathscr{L}(\widehat{\mathbb{X}})$ induces an isomorphism of $\widehat{\overline{R}}$-algebras $\mathscr{G} \xrightarrow{\sim} \mathscr{C}$ (3.2.14.2). It then suffices to adapt the proofs of ([3] II.11.2, II.11.3 and II.11.4), which correspond to the absolute case (3.1.14) (see [3] II.12.3).

Proposition 3.2.23 *For every rational number* $r \geq 0$*, the actions of* Δ *on* $\mathscr{F}^{(r)}$, $\mathscr{C}^{(r)}$ *and* $\widehat{\mathscr{C}}^{(r)}$ *are continuous for the p-adic topologies.*

The absolute case (3.2.11) was proved in ([3] II.12.4) and the relative case follows from the absolute case, in view of 3.2.20 and 3.2.21, especially (3.2.21.12).

Proposition 3.2.24 *Let* r, r' *be two rational numbers such that* $r' > r > 0$*. Then,*

(i) *For every integer* $n \geq 1$*, the canonical homomorphism*

$$R_1/p^n R_1 \to (\mathscr{C}^{(r)}/p^n\mathscr{C}^{(r)})^\Delta \tag{3.2.24.1}$$

is α*-injective* (3.1.3)*. Let* $\mathscr{H}_n^{(r)}$ *be its cokernel.*

(ii) *There exists an integer* $\alpha \geq 0$*, depending on* r*,* r' *and* $d = \dim(X/S)$*, but not on the morphism* f *satisfying the conditions of* 3.1.15 *nor on the adequate chart, such that for every integer* $n \geq 1$*, the canonical morphism* $\mathscr{H}_n^{(r')} \to \mathscr{H}_n^{(r)}$ *is annihilated by* p^α*.*

(iii) *There exists an integer* $\gamma \geq 0$*, depending on* r*,* r' *and* d*, but not on the morphism* f *satisfying the conditions of* 3.1.15 *nor on the adequate chart, such that for all integers* $n, q \geq 1$*, the canonical morphism*

$$\mathrm{H}^q(\Delta, \mathscr{C}^{(r')}/p^n\mathscr{C}^{(r')}) \to \mathrm{H}^q(\Delta, \mathscr{C}^{(r)}/p^n\mathscr{C}^{(r)}) \tag{3.2.24.2}$$

is annihilated by p^γ*.*

The absolute case (3.2.11) was proved in ([3] II.12.7) and the relative case follows from the absolute case, in view of 3.2.20 and 3.2.21, especially (3.2.21.12).

Proposition 3.2.25 *Let* r *be a rational number* > 0*. Then:*

(i) *The canonical morphism*

$$\widehat{R_1} \otimes_{\mathbb{Z}_p} \mathbb{Q}_p \to (\widehat{\mathscr{C}}^{(r)} \otimes_{\mathbb{Z}_p} \mathbb{Q}_p)^\Delta \tag{3.2.25.1}$$

is an isomorphism.

(ii) *For every integer* $i \geq 1$*, we have*

$$\varinjlim_{r \in \mathbb{Q}_{>0}} \mathrm{H}^i_{\mathrm{cont}}(\Delta, \widehat{\mathscr{C}}^{(r)} \otimes_{\mathbb{Z}_p} \mathbb{Q}_p) = 0. \tag{3.2.25.2}$$

(i) The absolute case (3.2.11) has been proved in ([3] II.12.5(i)) and the relative case follows from the absolute case, in view of 3.2.20 and 3.2.21, especially (3.2.21.12).

(ii) For any topological $\mathscr{O}_C$-Δ-module M, let $\mathrm{C}^\bullet_{\mathrm{cont}}(\Delta, M)$ be the complex of non-homogeneous continuous cochains of Δ with values in M ([3] II.3.8). Since Δ is compact, for every rational number $r > 0$, the canonical morphism

$$\mathrm{C}^\bullet_{\mathrm{cont}}(\Delta, \widehat{\mathscr{C}}^{(r)}) \otimes_{\mathbb{Z}_p} \mathbb{Q}_p \to \mathrm{C}^\bullet_{\mathrm{cont}}(\Delta, \widehat{\mathscr{C}}^{(r)} \otimes_{\mathbb{Z}_p} \mathbb{Q}_p) \tag{3.2.25.3}$$

is an isomorphism. Moreover, by virtue of ([3] (II.3.10.4) and (II.3.10.5)), we have an exact sequence

$$0 \to \mathrm{R}^1\varprojlim_n \mathrm{H}^{i-1}(\Delta, \mathscr{C}^{(r)}/p^n\mathscr{C}^{(r)}) \to \mathrm{H}^i_{\mathrm{cont}}(\Delta, \widehat{\mathscr{C}}^{(r)}) \tag{3.2.25.4}$$
$$\to \varprojlim_n \mathrm{H}^i(\Delta, \mathscr{C}^{(r)}/p^n\mathscr{C}^{(r)}) \to 0.$$

It follows from 3.2.24(iii) that

$$\varinjlim_{r\in\mathbb{Q}_{>0}} (\varprojlim_{n\in\mathbb{N}} \mathrm{H}^i(\Delta, \mathscr{C}^{(r)}/p^n\mathscr{C}^{(r)})) \otimes_{\mathbb{Z}_p} \mathbb{Q}_p = 0. \tag{3.2.25.5}$$

If $i \geq 2$, we have, similarly,

$$\varinjlim_{r\in\mathbb{Q}_{>0}} (\mathrm{R}^1\varprojlim_{n\in\mathbb{N}} \mathrm{H}^{i-1}(\Delta, \mathscr{C}^{(r)}/p^n\mathscr{C}^{(r)})) \otimes_{\mathbb{Z}_p} \mathbb{Q}_p = 0. \tag{3.2.25.6}$$

For any rational number $r > 0$, let $\mathscr{H}_n^{(r)}$ be the cokernel of the canonical morphism $R_1/p^nR_1 \to (\mathscr{C}^{(r)}/p^n\mathscr{C}^{(r)})^\Delta$. By virtue of ([39] 1.15) and the fact that $\mathrm{R}^2\varprojlim_{n\in\mathbb{N}} = 0$, the morphism

$$\mathrm{R}^1\varprojlim_{n\in\mathbb{N}} (\mathscr{C}^{(r)}/p^n\mathscr{C}^{(r)})^\Delta \to \mathrm{R}^1\varprojlim_{n\in\mathbb{N}} \mathscr{H}_n^{(r)} \tag{3.2.25.7}$$

is an isomorphism. It thus follows from 3.2.24(ii) that

$$\varinjlim_{r\in\mathbb{Q}_{>0}} (\mathrm{R}^1\varprojlim_{n\in\mathbb{N}} (\mathscr{C}^{(r)}/p^n\mathscr{C}^{(r)})^\Delta) \otimes_{\mathbb{Z}_p} \mathbb{Q}_p = 0. \tag{3.2.25.8}$$

The proposition then follows from (3.2.25.4).

Corollary 3.2.26 *For every rational number $r > 0$, we have $(\widehat{\mathscr{C}}^{(0+)})^\Delta = (\widehat{\mathscr{C}}^{(r)})^\Delta = \widehat{R_1}$.*

This follows from 3.2.25 by copying the proof of ([3] II.11.8).

3.3 Dolbeault Representations

3.3.1 For any rational number $r \geq 0$, we denote by $\mathbf{MC}^r$ the category of $\mathscr{C}^{(r)}$-modules with integrable p^r-connection with respect to the extension $\mathscr{C}^{(r)}/\widehat{\overline{R}}$ (see 2.5.24). Since $\widetilde{\xi}^{-1}\widetilde{\Omega}^1_{R/\mathscr{O}_K} \otimes_R \widehat{\overline{R}} = d_{\mathscr{C}^{(r)}}(\mathscr{F}^{(r)}) \subset d_{\mathscr{C}^{(r)}}(\mathscr{C}^{(r)})$ (3.2.19), every object of $\mathbf{MC}^r$ defines a Higgs $\widehat{\overline{R}}$–module with coefficients in $\widetilde{\xi}^{-1}\widetilde{\Omega}^1_{R/\mathscr{O}_K}$ by virtue of 2.5.26(i). We have the functor

$$\mathbf{Mod}(\widehat{\overline{R}}) \to \mathbf{MC}^r, \quad M \mapsto (\mathscr{C}^{(r)} \otimes_{\widehat{\overline{R}}} M, p^r d_{\mathscr{C}^{(r)}} \otimes \mathrm{id}_M). \tag{3.3.1.1}$$

In view of 2.5.26(ii) and with the notation of 3.2.18, we have the functor

$$\mathbf{HM}(\widehat{R_1}, \widetilde{\xi}^{-1}\widetilde{\Omega}^1_{R/\mathscr{O}_K}) \to \mathbf{MC}^r, \quad (N, \theta) \mapsto (\mathscr{C}^{(r)} \otimes_{\widehat{R_1}} N, p^r d_{\mathscr{C}^{(r)}} \otimes \mathrm{id}_N + \mathrm{id}_{\mathscr{C}^{(r)}} \otimes \theta). \tag{3.3.1.2}$$

Let r' be a rational number such that $r \geq r' \geq 0$, (M, ∇) an object of $\mathbf{MC}^r$. In view of (3.2.19.8), there exists one and only one $\widehat{\overline{R}}$-linear morphism

$$\nabla' \colon \mathscr{C}^{(r')} \otimes_{\mathscr{C}^{(r)}} M \to \widetilde{\xi}^{-1}\widetilde{\Omega}^1_{R/\mathscr{O}_K} \otimes_R \mathscr{C}^{(r')} \otimes_{\mathscr{C}^{(r)}} M \tag{3.3.1.3}$$

such that for all $t \in \mathscr{C}^{(r')}$ and $x \in M$, we have

$$\nabla'(t \otimes_{\mathscr{C}^{(r)}} x) = p^{r'} d_{\mathscr{C}^{(r')}}(t) \otimes_{\mathscr{C}^{(r)}} x + t \otimes_{\mathscr{C}^{(r)}} \nabla(x). \tag{3.3.1.4}$$

Hence, ∇' is an integrable $p^{r'}$-connection on $\mathscr{C}^{(r')} \otimes_{\mathscr{C}^{(r)}} M$ relative to the extension $\mathscr{C}^{(r')}/\widehat{\overline{R}}$. Note that the canonical morphism $M \to \mathscr{C}^{(r')} \otimes_{\mathscr{C}^{(r)}} M$ is a morphism of Higgs $\widehat{\overline{R}}$-modules with coefficients in $\widetilde{\xi}^{-1}\widetilde{\Omega}^1_{R/\mathscr{O}_K}$. We thus obtain a functor

$$\mathbf{MC}^r \to \mathbf{MC}^{r'}, \quad (M, \nabla) \mapsto (\mathscr{C}^{(r')} \otimes_{\mathscr{C}^{(r)}} M, \nabla'). \tag{3.3.1.5}$$

Remark 3.3.2 The objects of $\mathbf{MC}^0$ are the $\mathscr{C}$-modules with integrable connection with respect to the extension $\mathscr{C}/\widehat{\overline{R}}$. We define in the same way the category of $\mathscr{G}$-modules with integrable connection relative to the extension $\mathscr{G}/\widehat{\overline{R}}$ (3.2.14.2) and the analogous functors to (3.3.1.1) and (3.3.1.2). Every object of this category defines a Higgs $\widehat{\overline{R}}$-module with coefficients in $\widetilde{\xi}^{-1}\widetilde{\Omega}^1_{R/\mathscr{O}_K}$ by (3.2.19.1) and 2.5.26(i).

3.3.3 Let r be a rational number ≥ 0, $\lambda \in \widehat{\overline{R}}$, M a $\widehat{\mathscr{C}}^{(r)}$-module. A *$p$-adic λ-connection on M relative to the extension* $\widehat{\mathscr{C}}^{(r)}/\widehat{\overline{R}}$ is an $\widehat{\overline{R}}$-linear morphism

$$\nabla \colon M \to \widetilde{\xi}^{-1}\widetilde{\Omega}^1_{R/\mathscr{O}_K} \otimes_R M \tag{3.3.3.1}$$

such that for all $t \in \widehat{\mathscr{C}}^{(r)}$ and $x \in M$, we have

$$\nabla(tx) = \lambda d_{\widehat{\mathscr{C}}^{(r)}}(t) \otimes x + t\nabla(x). \tag{3.3.3.2}$$

We say that ∇ is *integrable* if it is a Higgs $\widehat{\overline{R}}$-field with coefficients in $\widetilde{\xi}^{-1}\widetilde{\Omega}^1_{R/\mathscr{O}_K}$ (2.5.1). If M is complete and separated for the p-adic topology, we recover the notion introduced in ([3] II.2.14) (see [3] II.2.16).

3.3.4 Let r be a rational number ≥ 0. We denote by $\mathbf{MC}^r_p$ the category of $\widehat{\mathscr{C}}^{(r)}$-modules with integrable p-adic p^r-connection with respect to the extension $\widehat{\mathscr{C}}^{(r)}/\widehat{\overline{R}}$ (3.3.3). We have the functor

$$\mathbf{Mod}(\widehat{\overline{R}}) \to \mathbf{MC}^r_p, \quad M \mapsto (\widehat{\mathscr{C}}^{(r)} \otimes_{\widehat{\overline{R}}} M, p^r d_{\widehat{\mathscr{C}}^{(r)}} \otimes \mathrm{id}_M). \tag{3.3.4.1}$$

Let (M, ∇) be an object of $\mathbf{MC}^r_p$, (N, θ) a Higgs $\widehat{R_1}$-module with coefficients in $\widetilde{\xi}^{-1}\widetilde{\Omega}^1_{R/\mathscr{O}_K}$ (see 3.2.18). There is one and only one $\widehat{\overline{R}}$-linear morphism

$$\nabla' : M \otimes_{\widehat{R_1}} N \to \widetilde{\xi}^{-1}\widetilde{\Omega}^1_{R/\mathscr{O}_K} \otimes_R M \otimes_{\widehat{R_1}} N \tag{3.3.4.2}$$

such that for all $x \in M$ and $y \in N$, we have

$$\nabla'(x \otimes_{\widehat{R_1}} y) = \nabla(x) \otimes_{\widehat{R_1}} y + x \otimes_{\widehat{R_1}} \theta(y). \tag{3.3.4.3}$$

It is an integrable p-adic p^r-connection on $M \otimes_{\widehat{R_1}} N$ with respect to the extension $\widehat{\mathscr{C}}^{(r)}/\widehat{\overline{R}}$ (2.5.1.7). In particular, we have a functor

$$\mathbf{HM}(\widehat{R_1}, \widetilde{\xi}^{-1}\widetilde{\Omega}^1_{R/\mathscr{O}_K}) \to \mathbf{MC}^r_p, \quad (N, \theta) \mapsto (\widehat{\mathscr{C}}^{(r)} \otimes_{\widehat{R_1}} N, p^r d_{\widehat{\mathscr{C}}^{(r)}} \otimes \mathrm{id}_N + \mathrm{id}_{\widehat{\mathscr{C}}^{(r)}} \otimes \theta). \tag{3.3.4.4}$$

For every object (M, ∇) of $\mathbf{MC}^r$ (3.3.1), there exists one and only one $\widehat{\overline{R}}$-linear morphism

$$\widehat{\nabla} : \widehat{\mathscr{C}}^{(r)} \otimes_{\mathscr{C}^{(r)}} M \to \widetilde{\xi}^{-1}\widetilde{\Omega}^1_{R/\mathscr{O}_K} \otimes_R \widehat{\mathscr{C}}^{(r)} \otimes_{\mathscr{C}^{(r)}} M \tag{3.3.4.5}$$

such that for all $t \in \widehat{\mathscr{C}}^{(r)}$ and $x \in M$, we have

$$\widehat{\nabla}(t \otimes_{\mathscr{C}^{(r)}} x) = p^r d_{\widehat{\mathscr{C}}^{(r)}}(t) \otimes_{\mathscr{C}^{(r)}} x + t \otimes_{\mathscr{C}^{(r)}} \nabla(x). \tag{3.3.4.6}$$

It is an integrable p-adic p^r-connection on $\widehat{\mathscr{C}}^{(r)} \otimes_{\mathscr{C}^{(r)}} M$ with respect to the extension $\widehat{\mathscr{C}}^{(r)}/\widehat{\overline{R}}$. We thus obtain a functor

$$\mathbf{MC}^r \to \mathbf{MC}^r_p, \quad (M, \nabla) \mapsto (\widehat{\mathscr{C}}^{(r)} \otimes_{\mathscr{C}^{(r)}} M, \widehat{\nabla}). \tag{3.3.4.7}$$

Let r' be a rational number such that $r \geq r' \geq 0$, (M, ∇) an object of $\mathbf{MC}^r_p$. In view of (3.2.19.8), there exists one and only one $\widehat{\overline{R}}$-linear morphism

$$\nabla' : \widehat{\mathscr{C}}^{(r')} \otimes_{\widehat{\mathscr{C}}^{(r)}} M \to \widetilde{\xi}^{-1}\widetilde{\Omega}^1_{R/\mathscr{O}_K} \otimes_R \widehat{\mathscr{C}}^{(r')} \otimes_{\widehat{\mathscr{C}}^{(r)}} M \tag{3.3.4.8}$$

such that for all $t \in \widehat{\mathscr{C}}^{(r')}$ and $x \in M$, we have

$$\nabla'(t \otimes_{\widehat{\mathscr{C}}^{(r)}} x) = p^{r'} d_{\widehat{\mathscr{C}}^{(r')}}(t) \otimes_{\widehat{\mathscr{C}}^{(r)}} x + t \otimes_{\widehat{\mathscr{C}}^{(r)}} \nabla(x). \tag{3.3.4.9}$$

It is an integrable p-adic $p^{r'}$-connection on $\widehat{\mathscr{C}}^{(r')} \otimes_{\widehat{\mathscr{C}}^{(r)}} M$ with respect to the extension $\widehat{\mathscr{C}}^{(r')}/\widehat{\overline{R}}$. Note that the canonical morphism $M \to \widehat{\mathscr{C}}^{(r')} \otimes_{\widehat{\mathscr{C}}^{(r)}} M$ is a morphism of Higgs $\widehat{\overline{R}}$-modules with coefficients in $\widetilde{\xi}^{-1}\widetilde{\Omega}^1_{R/\mathscr{O}_K}$. We thus obtain a functor

$$\mathbf{MC}^r_p \to \mathbf{MC}^{r'}_p, \quad (M, \nabla) \mapsto (\widehat{\mathscr{C}}^{(r')} \otimes_{\widehat{\mathscr{C}}^{(r)}} M, \nabla'). \tag{3.3.4.10}$$

Suppose $r > 0$ and let (M, ∇) be an object of $\mathbf{MC}^r_p$. We obtain, by taking the direct limit of the morphisms (3.3.4.8) over the rational numbers $0 < r' \leq r$ tending to 0, a Higgs $\widehat{\overline{R}}$-field

$$\nabla^{(0+)} : \widehat{\mathscr{C}}^{(0+)} \otimes_{\widehat{\mathscr{C}}^{(r)}} M \to \widetilde{\xi}^{-1}\widetilde{\Omega}^1_{R/\mathscr{O}_K} \otimes_R \widehat{\mathscr{C}}^{(0+)} \otimes_{\widehat{\mathscr{C}}^{(r)}} M. \tag{3.3.4.11}$$

Lemma 3.3.5 *Let M be a projective $\widehat{\overline{R}}[\frac{1}{p}]$-module of finite type, r a rational number ≥ 0. We denote by $\mathscr{C}^{(r)} \otimes_{\widehat{\overline{R}}} M$ the $\mathscr{C}^{(r)}$-module with integrable p^r-connection with respect to the extension $\mathscr{C}^{(r)}/\widehat{\overline{R}}$ associated with M (3.3.1) and by $\widehat{\mathscr{C}}^{(r)} \otimes_{\widehat{\overline{R}}} M$ the $\widehat{\mathscr{C}}^{(r)}$-module with integrable p-adic p^r-connection with respect to the extension $\widehat{\mathscr{C}}^{(r)}/\widehat{\overline{R}}$ associated with M (3.3.4). Then, $\mathscr{C}^{(r)} \otimes_{\widehat{\overline{R}}} M$ is the maximal nilpotent Higgs sub-$\widehat{\overline{R}}[\frac{1}{p}]$-module of $\widehat{\mathscr{C}}^{(r)} \otimes_{\widehat{\overline{R}}} M$ (2.5.10).*

Indeed, $\mathscr{C}^{(r)} \otimes_{\widehat{\overline{R}}} M$ is nilpotent, as the restriction of $d_{\mathscr{C}^{(r)}}$ (3.2.19.5) to $\mathscr{F}^{(r)}$ is the canonical morphism $\mathscr{F}^{(r)} \to \widetilde{\xi}^{-1}\widetilde{\Omega}^1_{R/\mathscr{O}_K} \otimes_R \widehat{\overline{R}}$ (3.2.17.1). Since M is a direct factor of a free $\widehat{\overline{R}}[\frac{1}{p}]$-module of finite type, we may restrict to the case where $M = \widehat{\overline{R}}[\frac{1}{p}]$ (see the proof of 2.5.10), in which case the assertion is immediate.

3.3.6 For any $\widehat{\overline{R}}$-representation M of Δ (2.1.2), we denote by $\mathbb{H}(M)$ the $\widehat{R_1}$-module defined by

$$\mathbb{H}(M) = (M \otimes_{\widehat{\overline{R}}} \widehat{\mathscr{C}}^{(0+)})^{\Delta}. \tag{3.3.6.1}$$

We endow it with the Higgs $\widehat{R_1}$-field with coefficients in $\widetilde{\xi}^{-1}\widetilde{\Omega}^1_{R/\mathscr{O}_K}$ induced by $d^{(0)}_{\widehat{\mathscr{C}}^{(0+)}}$ (3.2.19.10) (see 3.2.18). We thus define a functor

$$\mathbb{H} \colon \mathbf{Rep}_{\widehat{\overline{R}}}(\Delta) \to \mathbf{HM}(\widehat{R_1}, \widetilde{\xi}^{-1}\widetilde{\Omega}^1_{R/\mathscr{O}_K}). \tag{3.3.6.2}$$

3.3.7 For any Higgs $\widehat{R_1}$-module (N, θ) with coefficients in $\widetilde{\xi}^{-1}\widetilde{\Omega}^1_{R/\mathscr{O}_K}$ (3.2.18), we denote by $\mathbb{V}(N)$ the $\widehat{\overline{R}}$-module defined by

$$\mathbb{V}(N) = (N \otimes_{\widehat{R_1}} \widehat{\mathscr{C}}^{(0+)})^{\theta_{\rm tot}=0}, \tag{3.3.7.1}$$

where $\theta_{\rm tot} = \theta \otimes {\rm id} + {\rm id} \otimes d^{(0)}_{\widehat{\mathscr{C}}^{(0+)}}$ is the total Higgs $\widehat{R_1}$-field on $N \otimes_{\widehat{R_1}} \widehat{\mathscr{C}}^{(0+)}$ (2.5.1.7). We endow it with the $\widehat{\overline{R}}$-semi-linear action of Δ induced by its natural action on $\widehat{\mathscr{C}}^{(0+)}$.

We thus define a functor

$$\mathbb{V}\colon \mathbf{HM}(\widehat{R_1}, \widetilde{\xi}^{-1}\widetilde{\Omega}^1_{R/\mathscr{O}_K}) \to \mathbf{Rep}_{\widehat{\overline{R}}}(\Delta). \tag{3.3.7.2}$$

Remarks 3.3.8

(i) It follows from 3.2.20 that the functors $\mathbb{H}$ and $\mathbb{V}$ do not depend on the choice of the $(\widetilde{S}, \mathscr{M}_{\widetilde{S}})$-deformation $(\widetilde{X}, \mathscr{M}_{\widetilde{X}})$ (3.1.17), up to *non-canonical* isomorphism.

(ii) For every $\widehat{\overline{R}}$-representation M of Δ, the canonical morphism

$$\mathbb{H}(M) \otimes_{\widehat{R_1}} \widehat{R_1}[\frac{1}{p}] \to \mathbb{H}(M \otimes_{\widehat{\overline{R}}} \widehat{\overline{R}}[\frac{1}{p}]) \tag{3.3.8.1}$$

is an isomorphism.

(iii) For every Higgs $\widehat{R_1}$-module (N, θ) with coefficients in $\widetilde{\xi}^{-1}\widetilde{\Omega}^1_{R/\mathscr{O}_K}$, the canonical morphism

$$\mathbb{V}(N) \otimes_{\widehat{\overline{R}}} \widehat{\overline{R}}[\frac{1}{p}] \to \mathbb{V}(N \otimes_{\widehat{R_1}} \widehat{R_1}[\frac{1}{p}]) \tag{3.3.8.2}$$

is an isomorphism.

Definition 3.3.9 ([3] II.12.11) We say that an $\widehat{\overline{R}}[\frac{1}{p}]$-representation M of Δ is *Dolbeault* if the following conditions are satisfied:

(i) $\mathbb{H}(M)$ is a projective $\widehat{R_1}[\frac{1}{p}]$-module of finite type;

(ii) the canonical morphism

$$\mathbb{H}(M) \otimes_{\widehat{R_1}} \widehat{\mathscr{C}}^{(0+)} \to M \otimes_{\widehat{\overline{R}}} \widehat{\mathscr{C}}^{(0+)} \tag{3.3.9.1}$$

is an isomorphism.

This notion does not depend on the choice of the $(\widetilde{S}, \mathscr{M}_{\widetilde{S}})$-deformation $(\widetilde{X}, \mathscr{M}_{\widetilde{X}})$ by 3.3.8(i). Nevertheless, it depends a priori on the relative or absolute case considered in 3.1.14. When it is necessary to specify, we will say that the $\widehat{\overline{R}}[\frac{1}{p}]$-representation M is *absolute Dolbeault* or *Dolbeault relative to* $\mathscr{O}_K$.

This definition is in fact equivalent to ([3] II.12.11), even if it seems more general, see 3.3.12.

Definition 3.3.10 ([3] II.12.12) We say that a Higgs $\widehat{R_1}[\frac{1}{p}]$-module (N, θ) with coefficients in $\widetilde{\xi}^{-1}\widetilde{\Omega}^1_{R/\mathscr{O}_K}$ is *solvable* if the following conditions are satisfied:

(i) N is a projective $\widehat{R_1}[\frac{1}{p}]$-module of finite type;

(ii) the canonical morphism

$$\mathbb{V}(N) \otimes_{\widehat{\overline{R}}} \widehat{\mathscr{C}}^{(0+)} \to N \otimes_{\widehat{R_1}} \widehat{\mathscr{C}}^{(0+)} \tag{3.3.10.1}$$

is an isomorphism.

This notion does not depend on the choice of the $(\widetilde{S}, \mathscr{M}_{\widetilde{S}})$-deformation $(\widetilde{X}, \mathscr{M}_{\widetilde{X}})$ by 3.3.8(i). Nevertheless it depends a priori on the relative or absolute case considered in 3.1.14. The coefficients of the Higgs module specify this. There is therefore no need to specify it in the terminology.

This definition is in fact equivalent to ([3] II.12.12), even if it seems more general, see 3.3.13.

Proposition 3.3.11 *Let M be an $\widehat{\overline{R}}[\frac{1}{p}]$-representation of Δ, N a $\widehat{R_1}[\frac{1}{p}]$-projective module of finite type, endowed with a Higgs $\widehat{R_1}$-field θ with coefficients in $\widetilde{\xi}^{-1}\widetilde{\Omega}^1_{R/\mathscr{O}_K}$, r a rational number > 0,*

$$N \otimes_{\widehat{R_1}} \widehat{\mathscr{C}}^{(r)} \xrightarrow{\sim} M \otimes_{\widehat{\overline{R}}} \widehat{\mathscr{C}}^{(r)} \tag{3.3.11.1}$$

a $\widehat{\mathscr{C}}^{(r)}$-linear and Δ-equivariant isomorphism of Higgs $\widehat{\overline{R}}[\frac{1}{p}]$-modules with coefficients in $\widetilde{\xi}^{-1}\widetilde{\Omega}^1_{R/\mathscr{O}_K}$, where N is endowed with the trivial action of Δ, M is endowed with the zero Higgs field and $\widehat{\mathscr{C}}^{(r)}$ is endowed with the canonical action of Δ and with the Higgs field $p^r d_{\widehat{\mathscr{C}}^{(r)}}$. Then,

(i) *The $\widehat{\overline{R}}[\frac{1}{p}]$-module M is projective of finite type, and the action of Δ on M is continuous for the p-adic topology* (2.1.1).

(ii) *The $\widehat{\overline{R}}[\frac{1}{p}]$-representation M of Δ is Dolbeault, and we have an isomorphism of Higgs $\widehat{R_1}[\frac{1}{p}]$-modules with coefficients in $\widetilde{\xi}^{-1}\widetilde{\Omega}^1_{R/\mathscr{O}_K}$*

$$\mathbb{H}(M) \xrightarrow{\sim} (N, \theta). \tag{3.3.11.2}$$

(iii) *The Higgs $\widehat{R_1}[\frac{1}{p}]$-module (N, θ) is solvable, and we have an isomorphism of $\widehat{\overline{R}}[\frac{1}{p}]$-representations of Δ*

$$\mathbb{V}(N, \theta) \xrightarrow{\sim} M. \tag{3.3.11.3}$$

Any splitting $\mathscr{F}^{(r)} \to \widehat{\overline{R}}$ of the extension (3.2.17.1), considered as an exact sequence of $\widehat{\overline{R}}$-modules without actions of Δ, defines a retraction of the $\widehat{\overline{R}}$-algebra $\widehat{\mathscr{C}}^{(r)}$. The isomorphism (3.3.11.1) then implies that the $\widehat{\overline{R}}[\frac{1}{p}]$-module M is projective of finite type. The isomorphism (3.3.11.1) is in fact an isomorphism of $\widehat{\mathscr{C}}^{(r)}$-modules with integrable p-adic p^r-connection with respect to the extension $\widehat{\mathscr{C}}^{(r)}/\widehat{\overline{R}}$ (3.3.4). We deduce from it a $\widehat{\mathscr{C}}^{(0+)}$-linear and Δ-equivariant isomorphism of Higgs $\widehat{\overline{R}}[\frac{1}{p}]$-modules with coefficients in $\widetilde{\xi}^{-1}\widetilde{\Omega}^1_{R/\mathscr{O}_K}$ (3.3.4.11)

$$N \otimes_{\widehat{R_1}} \widehat{\mathscr{C}}^{(0+)} \xrightarrow{\sim} M \otimes_{\widehat{\overline{R}}} \widehat{\mathscr{C}}^{(0+)}. \tag{3.3.11.4}$$

Since N is a direct factor of a free $\widehat{R_1}[\frac{1}{p}]$-module of finite type, we have $(N \otimes_{\widehat{R_1}} \widehat{\mathscr{C}}^{(0+)})^{\Delta} = N$ by virtue of 3.2.26. We then obtain from (3.3.11.4) an isomorphism of Higgs $\widehat{R_1}[\frac{1}{p}]$-modules $\mathbb{H}(M) \xrightarrow{\sim} (N, \theta)$. Similarly, in view of (3.2.19.11), we obtain from (3.3.11.4) an isomorphism of $\overline{\widehat{R}}[\frac{1}{p}]$-representations of Δ, $\mathbb{V}(N,\theta) \xrightarrow{\sim} M$. We deduce from this that the Higgs $\widehat{R_1}[\frac{1}{p}]$-module (N,θ) is solvable and that the $\overline{\widehat{R}}[\frac{1}{p}]$-representation M of Δ is Dolbeault. Furthermore, the proof of ([3] II.12.22) shows that the $\overline{\widehat{R}}[\frac{1}{p}]$-representation M of Δ is continuous, proving the proposition. The careful reader will note that ([3] II.12.22) concerns the absolute case (3.2.11), but the same proof holds for the relative case.

Corollary 3.3.12 *An $\overline{\widehat{R}}[\frac{1}{p}]$-representation M of Δ is Dolbeault if and only if there exists a projective $\widehat{R_1}[\frac{1}{p}]$-module of finite type N, a Higgs $\widehat{R_1}$-field θ on N with coefficients in $\widetilde{\xi}^{-1}\widetilde{\Omega}^1_{R/\mathscr{O}_K}$, a rational number $r > 0$ and a $\widehat{\mathscr{C}}^{(r)}$-linear and Δ-equivariant isomorphism of Higgs $\overline{\widehat{R}}[\frac{1}{p}]$-modules with coefficients in $\widetilde{\xi}^{-1}\widetilde{\Omega}^1_{R/\mathscr{O}_K}$*

$$N \otimes_{\widehat{R_1}} \widehat{\mathscr{C}}^{(r)} \xrightarrow{\sim} M \otimes_{\overline{\widehat{R}}} \widehat{\mathscr{C}}^{(r)}, \tag{3.3.12.1}$$

where N is endowed with the trivial action of Δ, M is endowed with the zero Higgs field and $\widehat{\mathscr{C}}^{(r)}$ is endowed with the canonical action of Δ and with the Higgs field $p^r d_{\widehat{\mathscr{C}}^{(r)}}$. Moreover, in this case, we have the following properties:

(i) *The $\overline{\widehat{R}}[\frac{1}{p}]$-module M is projective of finite type, and the action of Δ on M is continuous for the p-adic topology (2.1.1).*
(ii) *We have an isomorphism of Higgs $\widehat{R_1}[\frac{1}{p}]$-modules with coefficients in $\widetilde{\xi}^{-1}\widetilde{\Omega}^1_{R/\mathscr{O}_K}$*

$$\mathbb{H}(M) \xrightarrow{\sim} (N,\theta). \tag{3.3.12.2}$$

Indeed, the condition is sufficient by virtue of 3.3.11. Let us show that it is necessary. Suppose that M is Dolbeault. Any splitting $\mathscr{F} \to \overline{\widehat{R}}$ of the extension (3.2.14.5), considered as an exact sequence of $\overline{\widehat{R}}$-modules without actions of Δ, defines a retraction of the $\overline{\widehat{R}}$-algebra $\widehat{\mathscr{C}} = \widehat{\mathscr{C}}^{(0)}$, and consequently a retraction of the sub-$\overline{\widehat{R}}$-algebra $\widehat{\mathscr{C}}^{(0+)}$ (3.2.17.4). The isomorphism (3.3.9.1) then implies that the $\overline{\widehat{R}}[\frac{1}{p}]$-module M is projective of finite type. Hence, for every rational number $r > 0$, the canonical morphism $M \otimes_{\overline{\widehat{R}}} \widehat{\mathscr{C}}^{(r)} \to M \otimes_{\overline{\widehat{R}}} \widehat{\mathscr{C}}^{(0+)}$ is injective. The $\widehat{R_1}[\frac{1}{p}]$-module $\mathbb{H}(M)$ being of finite type, there exists a rational number $r > 0$ such that $\mathbb{H}(M)$ is contained in $M \otimes_{\overline{\widehat{R}}} \widehat{\mathscr{C}}^{(r)}$. Since M is of finite type over $\overline{\widehat{R}}[\frac{1}{p}]$, up to decreasing r, the canonical morphism

$$\mathbb{H}(M) \otimes_{\widehat{R_1}} \widehat{\mathscr{C}}^{(r)} \to M \otimes_{\overline{\widehat{R}}} \widehat{\mathscr{C}}^{(r)} \tag{3.3.12.3}$$

is surjective. Moreover, $\mathbb{H}(M)$ being $\widehat{R_1}$-flat, the canonical morphism

$$\mathbb{H}(M) \otimes_{\widehat{R_1}} \widehat{\mathscr{C}}^{(r)} \to \mathbb{H}(M) \otimes_{\widehat{R_1}} \widehat{\mathscr{C}}^{(0+)} \tag{3.3.12.4}$$

is injective. We deduce from this that (3.3.12.3) is an isomorphism; hence the required condition. Properties (i) and (ii) result from 3.3.11.

Corollary 3.3.13 *A Higgs $\widehat{R_1}[\frac{1}{p}]$-module (N, θ) with coefficients in $\widetilde{\xi}^{-1}\widetilde{\Omega}^1_{R/\mathscr{O}_K}$ is solvable if and only it satisfies the following conditions:*

(i) *the $\widehat{R_1}[\frac{1}{p}]$-module N is projective of finite type;*

(ii) *there exists an $\widehat{\overline{R}}[\frac{1}{p}]$-representation M of Δ, a rational number $r > 0$ and a $\widehat{\mathscr{C}}^{(r)}$-linear and Δ-equivariant isomorphism of Higgs $\widehat{\overline{R}}[\frac{1}{p}]$-modules with coefficients in $\widetilde{\xi}^{-1}\widetilde{\Omega}^1_{R/\mathscr{O}_K}$*

$$M \otimes_{\widehat{\overline{R}}} \widehat{\mathscr{C}}^{(r)} \xrightarrow{\sim} N \otimes_{\widehat{R_1}} \widehat{\mathscr{C}}^{(r)}, \tag{3.3.13.1}$$

where N is endowed with the trivial action of Δ, M is endowed with the zero Higgs field and $\widehat{\mathscr{C}}^{(r)}$ is endowed with the canonical action of Δ and with the Higgs field $p^r d_{\widehat{\mathscr{C}}^{(r)}}$.

Moreover, in this case, the $\widehat{\overline{R}}[\frac{1}{p}]$-module M is projective of finite type, and we have an isomorphism of $\widehat{\overline{R}}[\frac{1}{p}]$-representations of Δ

$$\mathbb{V}(N, \theta) \xrightarrow{\sim} M. \tag{3.3.13.2}$$

The proof is similar to that of 3.3.12 and is left to the reader.

3.3.14 We take again the assumptions and notation of 3.2.21, and we further denote by

$$\nu\colon (\xi_\pi^*)^{-1}\widehat{R_1} \otimes_R \widetilde{\Omega}^1_{R/\mathscr{O}_K} \xrightarrow{\sim} \xi^{-1}\widehat{R_1} \otimes_R \widetilde{\Omega}^1_{R/\mathscr{O}_K} \tag{3.3.14.1}$$

the isomorphism induced by (3.2.21.2). Note that $\nu \otimes_{\widehat{R_1}} \widehat{\overline{R}} = u^\vee$ is the dual morphism of u (3.2.21). We denote by

$$\varepsilon\colon \mathbf{HM}(\widehat{R_1}, (\xi_\pi^*)^{-1}\widetilde{\Omega}^1_{R/\mathscr{O}_K}) \to \mathbf{HM}(\widehat{R_1}, \xi^{-1}\widetilde{\Omega}^1_{R/\mathscr{O}_K}) \tag{3.3.14.2}$$

the functor induced by $p^\rho \nu$. We denote by

$$\mathbf{Rep}_{\widehat{\overline{R}}}(\Delta) \underset{\mathbb{V}}{\overset{\mathbb{H}}{\rightleftarrows}} \mathbf{HM}(\widehat{R_1}, \xi^{-1}\widetilde{\Omega}^1_{R/\mathscr{O}_K}) \tag{3.3.14.3}$$

$$\mathbf{Rep}_{\widehat{\overline{R}}}(\Delta) \underset{\mathbb{V}'}{\overset{\mathbb{H}'}{\rightleftarrows}} \mathbf{HM}(\widehat{R_1}, (\xi_\pi^*)^{-1}\widetilde{\Omega}^1_{R/\mathscr{O}_K}) \tag{3.3.14.4}$$

the functors (3.3.6.2) and (3.3.7.2) associated with the deformations $(\widetilde{X}, \mathscr{M}_{\widetilde{X}})$ and $(\widetilde{X}', \mathscr{M}_{\widetilde{X}'})$, respectively.

For every $\widehat{\overline{R}}$-representation M of Δ, the canonical homomorphism $\widehat{\mathscr{C}}'^{(0+)} \to \widehat{\mathscr{C}}^{(0+)}$ (3.2.21.13) induces an $\widehat{R_1}$-linear morphism

$$a_M\colon \mathbb{H}'(M) \to \mathbb{H}(M). \tag{3.3.14.5}$$

In view of (3.2.21.15), this fits into a commutative diagram

$$\begin{array}{ccc} \mathbb{H}'(M) & \xrightarrow{\theta'} & (\xi_\pi^*)^{-1}\widetilde{\Omega}^1_{R/\mathscr{O}_K} \otimes_R \mathbb{H}'(M) \\ \Big\downarrow{\scriptstyle a_M} & & \Big\downarrow{\scriptstyle p^\rho \nu \otimes_{\widehat{R_1}} a_M} \\ \mathbb{H}(M) & \xrightarrow{\theta} & \xi^{-1}\widetilde{\Omega}^1_{R/\mathscr{O}_K} \otimes_R \mathbb{H}(M) \end{array} \tag{3.3.14.6}$$

where θ and θ' are the canonical Higgs fields. Consequently, a_M defines a morphism of functors

$$\varepsilon \circ \mathbb{H}' \to \mathbb{H}. \tag{3.3.14.7}$$

Similarly, the diagram (3.2.21.15) induces a canonical morphism of functors

$$\mathbb{V}' \to \mathbb{V} \circ \varepsilon. \tag{3.3.14.8}$$

Proposition 3.3.15 *The assumptions being those of* 3.3.14, *let M be an $\widehat{\overline{R}}[\frac{1}{p}]$-representation of Δ which is Dolbeault relative to $\mathscr{O}_K$* (3.3.9)*, N a solvable Higgs $\widehat{R_1}[\frac{1}{p}]$-module with coefficients in $(\xi_\pi^*)^{-1}\widetilde{\Omega}^1_{R/\mathscr{O}_K}$* (3.3.10)*. Then,*

(i) *The $\widehat{\overline{R}}[\frac{1}{p}]$-representation M of Δ is absolute Dolbeault and the canonical morphism*

$$\varepsilon(\mathbb{H}'(M)) \to \mathbb{H}(M) \tag{3.3.15.1}$$

is an isomorphism.

(ii) *The Higgs $\widehat{R_1}[\frac{1}{p}]$-module $\varepsilon(N)$ with coefficients in $\xi^{-1}\widetilde{\Omega}^1_{R/\mathscr{O}_K}$ is solvable and the canonical morphism*

$$\mathbb{V}'(N) \to \mathbb{V}(\varepsilon(N)) \tag{3.3.15.2}$$

is an isomorphism.

Suppose there exists a rational number $r > 0$ and a $\widehat{\mathscr{C}}'^{(r)}$-linear and Δ-equivariant isomorphism of Higgs $\widehat{\overline{R}}[\frac{1}{p}]$-modules with coefficients in $(\xi_\pi^*)^{-1}\widetilde{\Omega}^1_{R/\mathscr{O}_K}$

$$N \otimes_{\widehat{R_1}} \widehat{\mathscr{C}}'^{(r)} \xrightarrow{\sim} M \otimes_{\widehat{\overline{R}}} \widehat{\mathscr{C}}'^{(r)}, \tag{3.3.15.3}$$

where N is endowed with the trivial action of Δ, M is endowed with the zero Higgs field and $\widehat{\mathscr{C}}'^{(r)}$ is endowed with the Higgs field $p^r d_{\widehat{\mathscr{C}}'^{(r)}}$ and with the canonical action of Δ. This induces an isomorphism of Higgs $\widehat{R_1}$-modules with coefficients in $(\xi_\pi^*)^{-1}\widetilde{\Omega}^1_{R/\mathscr{O}_K}$ (3.3.12.2)

$$N \xrightarrow{\sim} \mathbb{H}'(M), \tag{3.3.15.4}$$

and an $\widehat{\overline{R}}$-linear and Δ-equivariant isomorphism (3.3.13.2)

$$\mathbb{V}'(N) \xrightarrow{\sim} M. \tag{3.3.15.5}$$

In view of (3.2.21.12) and (3.2.21.14), we deduce from (3.3.15.3) a $\widehat{\mathscr{C}}^{(r+\rho)}$-linear and Δ-equivariant isomorphism of Higgs $\widehat{\overline{R}}[\frac{1}{p}]$-modules with coefficients in $\xi^{-1}\widetilde{\Omega}^1_{R/\mathscr{O}_K}$

$$\varepsilon(N) \otimes_{\widehat{R_1}} \widehat{\mathscr{C}}^{(r+\rho)} \xrightarrow{\sim} M \otimes_{\widehat{\overline{R}}} \widehat{\mathscr{C}}^{(r+\rho)}, \tag{3.3.15.6}$$

where $\varepsilon(N)$ is endowed with the trivial action of Δ, M is endowed with the zero Higgs field and $\widehat{\mathscr{C}}^{(r+\rho)}$ is endowed with the Higgs field $p^{r+\rho} d_{\widehat{\mathscr{C}}^{(r+\rho)}}$ and with the canonical action of Δ. By virtue of 3.3.11, M is absolute Dolbeault, $\varepsilon(N)$ is solvable, and we have an isomorphism of Higgs $\widehat{R_1}$-modules with coefficients in $\xi^{-1}\widetilde{\Omega}^1_{R/\mathscr{O}_K}$

$$\varepsilon(N) \xrightarrow{\sim} \mathbb{H}(M), \tag{3.3.15.7}$$

and an $\widehat{\overline{R}}$-linear and Δ-equivariant isomorphism

$$\mathbb{V}(\varepsilon(N)) \xrightarrow{\sim} M. \tag{3.3.15.8}$$

We check immediately that the isomorphisms (3.3.15.4) and (3.3.15.7) are compatible via the morphism (3.3.14.7), and the isomorphisms (3.3.15.5) and (3.3.15.8) are compatible via the morphism (3.3.14.8). The proposition follows.

Proposition 3.3.16 *The functors* $\mathbb{H}$ (3.3.6.2) *and* $\mathbb{V}$ (3.3.7.2) *induce equivalences of categories quasi-inverse to each other between the category of Dolbeault* $\widehat{\overline{R}}[\frac{1}{p}]$*-representations of* Δ *and that of solvable Higgs* $\widehat{R_1}[\frac{1}{p}]$*-modules with coefficients in* $\widetilde{\xi}^{-1}\widetilde{\Omega}^1_{R/\mathscr{O}_K}$.

The absolute case (3.1.14) was proved in ([3] II.12.24). The relative case follows in view 3.2.20 and 3.3.15.

Proposition 3.3.17 *Let* M *be a Dolbeault* $\widehat{\overline{R}}[\frac{1}{p}]$*-representation of* Δ, $(\mathbb{H}(M), \theta)$ *the associated Higgs* $\widehat{R_1}[\frac{1}{p}]$*-module with coefficients in* $\widetilde{\xi}^{-1}\widetilde{\Omega}^1_{R/\mathscr{O}_K}$*. Then, we have a canonical functorial isomorphism in* $\mathbf{D}^+(\mathbf{Mod}(\widehat{R_1}[\frac{1}{p}]))$

$$\mathrm{C}^\bullet_{\mathrm{cont}}(\Delta, M) \xrightarrow{\sim} \mathbb{K}^\bullet(\mathbb{H}(M), \theta), \tag{3.3.17.1}$$

where $\mathrm{C}^\bullet_{\mathrm{cont}}(\Delta, M)$ *is the complex of continuous cochains of* Δ *with values in* M *and* $\mathbb{K}^\bullet(\mathbb{H}(M), \theta)$ *is the Dolbeault complex of* $(\mathbb{H}(M), \theta)$ (2.5.1.2).

The absolute case (3.1.14) was proved in ([3] II.12.26) and the relative case follows in view of 3.2.20 and 3.3.15.

3.4 Small Higgs Modules

3.4.1 We denote by K_0 the field of the fractions of W (3.1.1) and by $\mathfrak{d}$ the different of the extension K/K_0. We set $\rho = 0$ in the absolute case (3.1.14) and $\rho = v(\pi\mathfrak{d})$ in the relative case. By (3.1.10.1) and (3.1.5.3), we have a canonical $\mathscr{O}_C$-linear isomorphism

$$\mathscr{O}_C(1) \xrightarrow{\sim} p^{\rho + \frac{1}{p-1}} \widetilde{\xi} \mathscr{O}_C. \tag{3.4.1.1}$$

Definition 3.4.2 ([3] II.13.1) Let G be a topological group, A an $\mathscr{O}_C$-algebra that is complete and separated for the p-adic topology, endowed with a continuous action of G (by homomorphisms of $\mathscr{O}_C$-algebras), α a rational number > 0, M a continuous A-representation of G (2.1.2), endowed with the p-adic topology.

(i) We say that M is *α-quasi-small* if the A-module M is complete and separated for the p-adic topology, and is generated by a finite number of elements that are G-invariant modulo $p^\alpha M$.
(ii) We say that M is *quasi-small* if it is α'-quasi-small for a rational number $\alpha' > \rho + \frac{2}{p-1}$ (3.4.1).

We denote by $\mathbf{Rep}_A^{\alpha\text{-qsf}}(G)$ (resp. $\mathbf{Rep}_A^{\mathrm{qsf}}(G)$) the full subcategory of $\mathbf{Rep}_A^{\mathrm{cont}}(G)$ made up of the α-quasi-small (resp. quasi-small) A-representations of G whose underlying A-module is $\mathscr{O}_C$-flat.

Definition 3.4.3 ([3] II.13.4) Let ε be a rational number > 0, (N, θ) a Higgs $\widehat{R_1}$-module with coefficients in $\widetilde{\xi}^{-1}\widetilde{\Omega}^1_{R/\mathscr{O}_K}$ (3.2.18).

(i) We say that (N, θ) is *ε-quasi-small* if N is of finite type over $\widehat{R_1}$ and if θ is a multiple of p^ε in $\widetilde{\xi}^{-1}\mathrm{End}_{\widehat{R_1}}(N) \otimes_R \widetilde{\Omega}^1_{R/\mathscr{O}_K}$ (2.5.1.10). We then also say that the Higgs $\widehat{R_1}$-field θ is *ε-quasi-small*.
(ii) We say that (N, θ) is *quasi-small* if it is ε'-quasi-small for a rational number $\varepsilon' > \frac{1}{p-1}$. We then also say that the Higgs $\widehat{R_1}$-field θ is *quasi-small*.

We denote by $\mathbf{HM}^{\varepsilon\text{-qsf}}(\widehat{R_1}, \widetilde{\xi}^{-1}\widetilde{\Omega}^1_{R/\mathscr{O}_K})$ (resp. $\mathbf{HM}^{\mathrm{qsf}}(\widehat{R_1}, \widetilde{\xi}^{-1}\widetilde{\Omega}^1_{R/\mathscr{O}_K})$) the full subcategory of $\mathbf{HM}(\widehat{R_1}, \widetilde{\xi}^{-1}\widetilde{\Omega}^1_{R/\mathscr{O}_K})$ made up of the ε-quasi-small (resp. quasi-small) Higgs $\widehat{R_1}$-modules whose underlying $\widehat{R_1}$-module is $\mathscr{O}_C$-flat.

3.4.4 We take again the notation of 3.1.15. We recall that the $\mathbb{Z}$-module $P^{\mathrm{gp}}/\mathbb{Z}\lambda$ is free of finite type by virtue of ([2] 4.2.2). Let $t_1, \dots, t_d$ be elements of P^{gp} such that their images in $P^{\mathrm{gp}}/\mathbb{Z}\lambda$ form a $\mathbb{Z}$-basis. For all $1 \leq i \leq d$, let $d\log(t_i)$ be the image of t_i by the morphism (3.1.15.11), and $y_i = \widetilde{\xi}^{-1} d\log(t_i) \in \widetilde{\xi}^{-1}\widetilde{\Omega}^1_{R/\mathscr{O}_K} \otimes_R \widehat{R_1}$, so that $(y_i)_{1 \leq i \leq d}$ is a basis of $\widetilde{\xi}^{-1}\widetilde{\Omega}^1_{R/\mathscr{O}_K} \otimes_R \widehat{R_1}$ over $\widehat{R_1}$. Let χ_{t_i} be the image of t_i by the homomorphism (3.2.7.7), and χ_i the composed homomorphism

$$\Delta_\infty \xrightarrow{\chi_{t_i}} \mathbb{Z}_p(1) \xrightarrow{\log([\,])} p^{\rho + \frac{1}{p-1}} \widetilde{\xi} \mathscr{O}_C, \tag{3.4.4.1}$$

where the second arrow is induced by the isomorphism (3.4.1.1). We also denote by $\chi_i \colon \Delta \to p^{\rho+\frac{1}{p-1}}\widetilde{\xi}\mathscr{O}_C$ the induced homomorphism.

3.4.5 Let M be an $\widehat{R_1}$-module of finite type that is $\mathscr{O}_C$-flat, ε a rational number > 0, $\alpha = \varepsilon + \rho + \frac{1}{p-1}$ (3.4.1). We denote by Ψ_M the composed isomorphism

$$\begin{array}{ccc} \mathrm{Hom}_{\mathbb{Z}}(\Delta_\infty, p^\alpha \mathrm{End}_{\widehat{R_1}}(M)) & \xrightarrow{\sim} & p^\varepsilon \widetilde{\xi}^{-1}\mathrm{End}_{\widehat{R_1}}(M) \otimes_{\widehat{R_1}} \mathrm{Hom}_{\mathbb{Z}}(\Delta_\infty, \widehat{R_1}(1)) \\ & \searrow^{\Psi_M} & \downarrow \\ & & p^\varepsilon \widetilde{\xi}^{-1}\mathrm{End}_{\widehat{R_1}}(M) \otimes_R \widetilde{\Omega}^1_{R/\mathscr{O}_K} \end{array} \tag{3.4.5.1}$$

where the vertical isomorphism is induced by the isomorphism (3.2.7.10) and the horizontal isomorphism comes from the isomorphism (3.4.1.1) and from ([3] (II.6.12.2)). Note that the $\widehat{R_1}$-module $\mathrm{End}_{\widehat{R_1}}(M)$ is complete and separated for the p-adic topology and $\mathscr{O}_C$-flat ([3] II.13.9).

With the notation of 3.4.4, for every homomorphism $\phi\colon \Delta_\infty \to p^\alpha \mathrm{End}_{\widehat{R_1}}(M)$, there exists a $\phi_i \in p^\varepsilon \mathrm{End}_{\widehat{R_1}}(M)$ $(1 \leq i \leq d)$ such that

$$\phi = \sum_{i=1}^{d} \widetilde{\xi}^{-1}\phi_i \otimes \chi_i. \tag{3.4.5.2}$$

We then have ([3] II.13.10)

$$\Psi_M(\phi) = \sum_{i=1}^{d} \widetilde{\xi}^{-1}\phi_i \otimes d\log(t_i). \tag{3.4.5.3}$$

Let φ be an α-quasi-small $\widehat{R_1}$-representation of Δ_∞ over M (3.4.2). Since Δ_∞ acts trivially on $\widehat{R_1}$, φ is a homomorphism

$$\varphi\colon \Delta_\infty \to \mathrm{Aut}_{\widehat{R_1}}(M) \tag{3.4.5.4}$$

whose image is contained in the subgroup $\mathrm{id} + p^\alpha \mathrm{End}_{\widehat{R_1}}(M)$ of $\mathrm{Aut}_{\widehat{R_1}}(M)$. Since Δ_∞ is abelian, we can define the homomorphism ([3] II.13.9)

$$\log(\varphi)\colon \Delta_\infty \to p^\alpha \mathrm{End}_{\widehat{R_1}}(M). \tag{3.4.5.5}$$

We immediately see that $\Psi_M(\log(\varphi)) \wedge \Psi_M(\log(\varphi)) = 0$ ([3] II.13.10), in other words, $\Psi_M(\log(\varphi))$ is an ε-quasi-small Higgs $\widehat{R_1}$-field on M with coefficients in $\widetilde{\xi}^{-1}\widetilde{\Omega}^1_{R/\mathscr{O}_K}$ (3.4.3). We thus obtain a functor

$$\begin{array}{ccc} \mathbf{Rep}^{\alpha\text{-qsf}}_{\widehat{R_1}}(\Delta_\infty) & \to & \mathbf{HM}^{\varepsilon\text{-qsf}}(\widehat{R_1}, \widetilde{\xi}^{-1}\widetilde{\Omega}^1_{R/\mathscr{O}_K}) \\ (M, \varphi) & \mapsto & (M, \Psi_M(\log(\varphi))). \end{array} \tag{3.4.5.6}$$

Let θ be a ε-quasi-small Higgs $\widehat{R_1}$-field on M with coefficients in $\widetilde{\xi}^{-1}\widetilde{\Omega}^1_{R/\mathscr{O}_K}$. Since $\theta \wedge \theta = 0$, the image of the homomorphism $\Psi_M^{-1}(\theta)\colon \Delta_\infty \to p^\alpha \mathrm{End}_{\widehat{R_1}}(M)$ consists of endomorphisms of M that commute pairwise (see [3] 13.10). We can therefore define the homomorphism ([3] II.13.9)

$$\exp(\Psi_M^{-1}(\theta))\colon \Delta_\infty \to \mathrm{Aut}_{\widehat{R_1}}(M), \tag{3.4.5.7}$$

which is clearly an α-quasi-small $\widehat{R_1}$-representation of Δ_∞ over M. We thus define a functor

$$\begin{array}{rcl} \mathbf{HM}^{\varepsilon\text{-qsf}}(\widehat{R_1}, \xi^{-1}\widetilde{\Omega}^1_{R/\mathscr{O}_K}) & \to & \mathbf{Rep}^{\alpha\text{-qsf}}_{\widehat{R_1}}(\Delta_\infty) \\ (M,\theta) & \mapsto & (M, \exp(\Psi_M^{-1}(\theta))). \end{array} \tag{3.4.5.8}$$

3.4.6 By the condition 3.1.15(iv), there exists essentially a unique étale morphism

$$(\widetilde{X}_0, \mathscr{M}_{\widetilde{X}_0}) \to (\widetilde{S}, \mathscr{M}_{\widetilde{S}}) \times_{\mathbf{A}_{\mathbb{N}}} \mathbf{A}_P \tag{3.4.6.1}$$

that fits into a commutative diagram with Cartesian squares

$$\begin{array}{ccccc} (\check{\overline{X}}, \mathscr{M}_{\check{\overline{X}}}) & \longrightarrow & (\check{\overline{S}}, \mathscr{M}_{\check{\overline{S}}}) \times_{\mathbf{A}_{\mathbb{N}}} \mathbf{A}_P & \longrightarrow & (\check{\overline{S}}, \mathscr{M}_{\check{\overline{S}}}) \\ \downarrow & & \downarrow & & \downarrow i_S \\ (\widetilde{X}_0, \mathscr{M}_{\widetilde{X}_0}) & \longrightarrow & (\widetilde{S}, \mathscr{M}_{\widetilde{S}}) \times_{\mathbf{A}_{\mathbb{N}}} \mathbf{A}_P & \longrightarrow & (\widetilde{S}, \mathscr{M}_{\widetilde{S}}) \\ & & \downarrow & & \downarrow a \\ & & \mathbf{A}_P & \longrightarrow & \mathbf{A}_{\mathbb{N}} \end{array} \tag{3.4.6.2}$$

where the morphism a is defined by the chart $\mathbb{N} \to \Gamma(\widetilde{S}, \mathscr{M}_{\widetilde{S}})$, $1 \mapsto [\underline{\pi}]$ ((3.1.12.2) or (3.1.13.2)). We say that $(\widetilde{X}_0, \mathscr{M}_{\widetilde{X}_0})$ *is the* $(\widetilde{S}, \mathscr{M}_{\widetilde{S}})$*-smooth deformation of* $(\check{\overline{X}}, \mathscr{M}_{\check{\overline{X}}})$ *defined by the adequate chart* $((P,\gamma), (\mathbb{N}, \iota), \vartheta)$. We denote by $\mathscr{L}_0$ the Higgs–Tate torsor associated with $(\widetilde{X}_0, \mathscr{M}_{\widetilde{X}_0})$ (3.2.14), by $\mathscr{C}_0$ the Higgs–Tate $\widehat{\overline{R}}$-algebra and by $\mathscr{F}_0$ the Higgs–Tate $\widehat{\overline{R}}$-extension associated with $(\widetilde{X}_0, \mathscr{M}_{\widetilde{X}_0})$ (3.2.16).

We check immediately that the diagram

$$\begin{array}{ccccc} (\widehat{\mathbb{X}}, \mathscr{M}_{\widehat{\mathbb{X}}}) & \xrightarrow{i_X} & (\widetilde{\mathbb{X}}, \mathscr{M}_{\widetilde{\mathbb{X}}}) & \xrightarrow{b} & \mathbf{A}_P \\ \downarrow & & \downarrow & & \downarrow \\ (\check{\overline{S}}, \mathscr{M}_{\check{\overline{S}}}) & \xrightarrow{i_S} & (\widetilde{S}, \mathscr{M}_{\widetilde{S}}) & \xrightarrow{a} & \mathbf{A}_{\mathbb{N}}, \end{array} \tag{3.4.6.3}$$

where the morphism b is the canonical homomorphism (3.2.11), is commutative. We deduce from this a morphism ϕ_0 that fits into the commutative diagram (without the dotted arrow)

$$\begin{array}{ccccc} (\widehat{\mathbb{X}}, \mathscr{M}_{\widehat{\mathbb{X}}}) & \longrightarrow & (\check{\overline{X}}, \mathscr{M}_{\check{\overline{X}}}) & \longrightarrow & (\widetilde{X}_0, \mathscr{M}_{\widetilde{X}_0}) \\ \downarrow i_X & & & \overset{\psi_0}{\nearrow} & \downarrow \\ (\widetilde{\mathbb{X}}, \mathscr{M}_{\widetilde{\mathbb{X}}}) & & \xrightarrow{\phi_0} & & (\widetilde{S}, \mathscr{M}_{\widetilde{S}}) \times_{\mathbf{A}_{\mathbb{N}}} \mathbf{A}_P. \end{array} \tag{3.4.6.4}$$

The latter can be completed with a unique dotted arrow $\psi_0 \in \mathscr{L}_0(\widehat{\mathbb{X}})$ in such a way that it remains commutative. We say that ψ_0 *is the section of* $\mathscr{L}_0(\widehat{\mathbb{X}})$ *defined by the chart* $((P,\gamma),(\mathbb{N},\iota),\vartheta)$.

3.4.7 Let us temporarily distinguish in this subsection the absolute case from the relative case (3.1.14): we denote by $(\widetilde{X}_0, \mathscr{M}_{\widetilde{X}_0})$ (resp. $(\widetilde{X}'_0, \mathscr{M}_{\widetilde{X}'_0})$) the smooth deformation of $(\check{\overline{X}}, \mathscr{M}_{\check{\overline{X}}})$ above

$$(\mathscr{A}_2(\overline{S}), \mathscr{M}_{\mathscr{A}_2(\overline{S})}) \quad (\text{resp. } (\mathscr{A}_2^*(\overline{S}/S), \mathscr{M}_{\mathscr{A}_2^*(\overline{S}/S)}))$$

defined by the adequate chart $((P,\gamma),(\mathbb{N},\iota),\vartheta)$ (3.4.6), by $\mathscr{L}_0$ (resp. $\mathscr{L}'_0$) the associated Higgs–Tate torsor (3.2.14) and by $\psi_0 \in \mathscr{L}_0(\widehat{\mathbb{X}})$ (resp. $\psi'_0 \in \mathscr{L}'_0(\widehat{\mathbb{X}})$) the section defined by the same adequate chart. We check immediately that the diagram (3.2.11)

$$\begin{array}{ccc} (\mathscr{A}_2^*(\mathbb{X}/S), \mathscr{M}_{\mathscr{A}_2^*(\mathbb{X}/S)}) & \longrightarrow & (\mathscr{A}_2(\mathbb{X}), \mathscr{M}_{\mathscr{A}_2(\mathbb{X})}) \\ \Big\downarrow{\scriptstyle \psi'_0} & & \Big\downarrow{\scriptstyle \psi_0} \\ (\widetilde{X}'_0, \mathscr{M}_{\widetilde{X}'_0}) & \longrightarrow & (\widetilde{X}_0, \mathscr{M}_{\widetilde{X}_0}) \\ \Big\downarrow & \Box & \Big\downarrow \\ (\mathscr{A}_2^*(\overline{S}/S), \mathscr{M}_{\mathscr{A}_2^*(\overline{S}/S)}) & \longrightarrow & (\mathscr{A}_2(\overline{S}), \mathscr{M}_{\mathscr{A}_2(\overline{S})}) \end{array} \tag{3.4.7.1}$$

is commutative and that the lower square is Cartesian. In particular, ψ'_0 is induced by ψ_0.

Proposition 3.4.8 *We keep the notation of* 3.4.6. *For all* $t \in P^{\mathrm{gp}}$ *and* $g \in \Delta$, *we have*

$$(\psi_0 - {}^g\psi_0)(d\log(t)) = -\log([\chi_t(g)]), \tag{3.4.8.1}$$

where ${}^g\psi_0$ *is the image of* ψ_0 *under the isomorphism* (3.2.15.8), $\psi_0 - {}^g\psi_0$ *is viewed as an element of* $\mathbf{T}(\widehat{\mathbb{X}}) = \mathrm{T}$ (3.2.14.1), $d\log(t)$ *is the canonical image of* t *in* $\widetilde{\Omega}^1_{R/\mathscr{O}_K}$ *and* $\log([\chi_t])$ *denotes the composed homomorphism*

$$\Delta \longrightarrow \Delta_\infty \xrightarrow{\chi_t} \mathbb{Z}_p(1) \xrightarrow{\log([\ ])} p^{\rho+\frac{1}{p-1}}\xi\widehat{\overline{R}} \longrightarrow \xi\widehat{\overline{R}}, \tag{3.4.8.2}$$

where the first and the last arrows are the canonical morphisms, χ_t *is the image of* t *by the homomorphism* (3.2.7.7) *and the third arrow is induced by the isomorphism* (3.4.1.1).

Since the two sides of the equation (3.4.8.1) are homomorphisms from P^{gp} to $\xi\widehat{\overline{R}}$, we may restrict to the case where $t \in P$. The morphisms ϕ_0 and $\phi_0 \circ g^{-1}$, where ϕ_0 is the morphism defined in (3.4.6.4) and g^{-1} acts on $(\widetilde{\mathbb{X}}, \mathscr{M}_{\widetilde{\mathbb{X}}})$, extend the same morphism

$$(\widehat{\mathbb{X}}, \mathscr{M}_{\widehat{\mathbb{X}}}) \to (\widetilde{S}, \mathscr{M}_{\widetilde{S}}) \times_{\mathbf{A}_{\mathbb{N}}} \mathbf{A}_P.$$

By the definition and condition 3.1.15(iv), the difference $\phi_0 - \phi_0 \circ g^{-1}$ corresponds to the morphism $\psi_0 - {}^g\psi_0 \in \mathbf{T}$. On the other hand, we have $g(\nu(t)) = [\chi_t(g)] \cdot \nu(t)$ in $\Gamma(\widetilde{\mathbb{X}}, \mathscr{M}_{\widetilde{\mathbb{X}}})$ (3.2.9.2). The proposition follows from this in view of (3.1.5.2) and ([3] II.5.23).

3.4.9 We set $\mathbf{L}_0 = \mathrm{Spec}(\mathscr{C}_0)$ (3.2.14.8), which is naturally a principal homogeneous $\mathbf{T}$-bundle on $\widehat{\mathbb{X}}$ that represents $\mathscr{L}_0$. Consider the isomorphism of principal homogeneous $\mathbf{T}$-bundles on $\widehat{\mathbb{X}}$

$$\mathtt{t}_0 \colon \mathbf{T} \xrightarrow{\sim} \mathbf{L}_0, \nu \mapsto \nu + \psi_0. \tag{3.4.9.1}$$

The structure of a Δ-equivariant principal homogeneous $\mathbf{T}$-bundle over $\mathbf{L}_0$ (3.2.15.7) transfers through $\mathtt{t}_0$ to the structure of a Δ-equivariant principal homogeneous $\mathbf{T}$-bundle over $\mathbf{T}$ (see [3] II.4.20). For every $g \in \Delta$, we therefore have an equivariant $\tau_g^{\mathbf{T}}$-isomorphism

$$\tau_{g,\psi_0}^{\mathbf{T}} : \mathbf{T} \xrightarrow{\sim} g^{\bullet}(\mathbf{T}). \tag{3.4.9.2}$$

This structure determines a left action of Δ on $\mathbf{T}$ compatible with its action on $\widehat{\mathbb{X}}$. We deduce from this an action

$$\varphi_0 \colon \Delta \to \mathrm{Aut}_{\widehat{R_1}}(\mathscr{G}) \tag{3.4.9.3}$$

of Δ on $\mathscr{G}$ (3.2.14.2) by ring automorphisms, compatible with its action on $\widehat{\overline{R}}$; for every $g \in \Delta$, $\varphi_0(g)$ is induced by the automorphism of $\mathbf{T}$ defined by g^{-1}.

Proposition 3.4.10 *With the notation of* 3.4.4, *for every* $g \in \Delta$, *we have* (3.4.9.3)

$$\varphi_0(g) = \exp(-\sum_{i=1}^{d} \widetilde{\xi}^{-1} \frac{\partial}{\partial y_i} \otimes \chi_i(g)) \circ g. \tag{3.4.10.1}$$

This follows from 3.4.8 and ([3] (II.10.11.6)), see ([3] II.10.17) where the absolute case (3.1.14) has been proved. We can also deduce the relative case from the absolute case as follows. We take again the notation of 3.4.7 and set

$$\mathscr{G} = \mathrm{S}_{\widehat{R}}(\xi^{-1}\widetilde{\Omega}^1_{R/\mathscr{O}_K} \otimes_R \widehat{\overline{R}}), \tag{3.4.10.2}$$

$$\mathscr{G}' = \mathrm{S}_{\widehat{R}}((\xi^*_\pi)^{-1}\widetilde{\Omega}^1_{R/\mathscr{O}_K} \otimes_R \widehat{\overline{R}}). \tag{3.4.10.3}$$

We denote by φ_0 (resp. φ_0') the action of Δ on $\mathscr{G}$ (resp. $\mathscr{G}'$) induced by the section ψ_0 (resp. ψ_0'). By 3.1.10, we have a canonical isomorphism

$$(\xi^*_\pi)^{-1}\mathscr{O}_C \xrightarrow{\sim} p^{\rho}\xi^{-1}\mathscr{O}_C. \tag{3.4.10.4}$$

We deduce from this an injective homomorphism of $\widehat{\overline{R}}$-algebras

$$\mathscr{G}' \to \mathscr{G}. \tag{3.4.10.5}$$

Let $\mathbf{T} = \mathrm{Spec}(\mathscr{G})$ and $\mathbf{T}' = \mathrm{Spec}(\mathscr{G}')$ and denote by $h \colon \mathbf{T} \to \mathbf{T}'$ the morphism of $\widehat{\mathbb{X}}$-vector bundles deduced from (3.4.10.5). By (3.2.21.4), we have a canonical h-equivariant and Δ-equivariant morphism $\mathscr{L}_0 \to \mathscr{L}_0'$. It transforms ψ_0 into ψ_0' in view of (3.4.7.1). It follows that the homomorphism (3.4.10.5) is Δ-equivariant when we endow $\mathscr{G}$ and $\mathscr{G}'$ with the actions φ_0 and φ'_0 respectively. The proposition in the relative case therefore follows from the absolute case.

3.4.11 For any rational number $r \geq 0$, we denote by $\mathscr{G}^{(r)}$ the sub-$\widehat{\overline{R}}$-algebra of $\mathscr{G}$ (3.2.14.2) defined by (2.1.10)

$$\mathscr{G}^{(r)} = \mathrm{S}_{\widehat{\overline{R}}}(p^r \widetilde{\xi}^{-1} \widetilde{\Omega}^1_{R/\mathscr{O}_K} \otimes_R \widehat{\overline{R}}) \tag{3.4.11.1}$$

and by $\widehat{\mathscr{G}}^{(r)}$ its p-adic Hausdorff completion. In view of ([3] II.6.14) and its proof, $\mathscr{G}^{(r)}$ and $\widehat{\mathscr{G}}^{(r)}$ are $\mathscr{O}_C$-flat. For all rational numbers $r' \geq r \geq 0$, we have a canonical injective homomorphism $a^{r,r'} : \mathscr{G}^{(r')} \to \mathscr{G}^{(r)}$. We immediately check that the induced homomorphism $\widehat{a}^{r,r'} : \widehat{\mathscr{G}}^{(r')} \to \widehat{\mathscr{G}}^{(r)}$ is injective. We have a canonical $\mathscr{G}^{(r)}$-isomorphism

$$\Omega^1_{\mathscr{G}^{(r)}/\widehat{\overline{R}}} \xrightarrow{\sim} \widetilde{\xi}^{-1} \widetilde{\Omega}^1_{R/\mathscr{O}_K} \otimes_R \mathscr{G}^{(r)}. \tag{3.4.11.2}$$

We denote by

$$d_{\mathscr{G}^{(r)}} : \mathscr{G}^{(r)} \to \widetilde{\xi}^{-1} \widetilde{\Omega}^1_{R/\mathscr{O}_K} \otimes_R \mathscr{G}^{(r)} \tag{3.4.11.3}$$

the universal $\widehat{\overline{R}}$-derivation of $\mathscr{G}^{(r)}$ and by

$$d_{\widehat{\mathscr{G}}^{(r)}} : \widehat{\mathscr{G}}^{(r)} \to \widetilde{\xi}^{-1} \widetilde{\Omega}^1_{R/\mathscr{O}_K} \otimes_R \widehat{\mathscr{G}}^{(r)} \tag{3.4.11.4}$$

its extension to the completions (note that the R-module $\widetilde{\Omega}^1_{R/\mathscr{O}_K}$ is free of finite type). Since $\widetilde{\xi}^{-1} \widetilde{\Omega}^1_{R/\mathscr{O}_K} \otimes_R \widehat{\overline{R}} \subset d_{\mathscr{G}^{(r)}}(\mathscr{G}^{(r)})$, $d_{\mathscr{G}^{(r)}}$ and $d_{\widehat{\mathscr{G}}^{(r)}}$ are also Higgs $\widehat{\overline{R}}$-fields with coefficients in $\widetilde{\xi}^{-1} \widetilde{\Omega}^1_{R/\mathscr{O}_K}$ by 2.5.26(i). For all rational numbers $r' \geq r \geq 0$, we have

$$p^{r'-r}(\mathrm{id} \times a^{r,r'}) \circ d_{\mathscr{G}^{(r')}} = d_{\mathscr{G}^{(r)}} \circ a^{r,r'}. \tag{3.4.11.5}$$

The section $\psi_0 \in \mathscr{L}_0(\widehat{\mathbb{X}})$ defined by the adequate chart $((P, \gamma), (\mathbb{N}, \iota), \vartheta)$ (3.4.6) induces an isomorphism of $\widehat{\overline{R}}$-algebras

$$\mathscr{G} \xrightarrow{\sim} \mathscr{C}_0 \tag{3.4.11.6}$$

that is Δ-equivariant when we endow $\mathscr{G}$ with the action φ_0 (3.4.9.3) and $\mathscr{C}_0$ with the canonical action.

By 3.4.12 below, the sub-$\widehat{\overline{R}}$-algebra $\mathscr{G}^{(r)}$ of $\mathscr{G}$ is stable under the action φ_0 of Δ on $\mathscr{G}$. Denoting by $\mathscr{C}_0^{(r)}$ the Higgs–Tate $\widehat{\overline{R}}$-algebra of thickness r associated with $(\widetilde{X}_0, \mathscr{M}_{\widetilde{X}_0})$ (3.2.17.3), we prove that $\mathrm{Spec}(\mathscr{C}_0^{(r)})$ is naturally a principal homogeneous $\mathrm{Spec}(\mathscr{G}^{(r)})$-bundle over $\widehat{\mathbb{X}}$ (see [3] II.12.1). The section ψ_0 induces an $\widehat{\overline{R}}$-homomorphism $\mathscr{C}_0^{(r)} \to \widehat{\overline{R}}$ and consequently an isomorphism of $\widehat{\overline{R}}$-algebras

$$\mathscr{G}^{(r)} \xrightarrow{\sim} \mathscr{C}_0^{(r)}. \tag{3.4.11.7}$$

It is Δ-equivariant when we endow $\mathscr{G}^{(r)}$ with the action induced by φ_0 and $\mathscr{C}_0^{(r)}$ with the canonical action, and it is compatible with the derivations $d_{\mathscr{G}^{(r)}}$ and $d_{\mathscr{C}_0^{(r)}}$.

Proposition 3.4.12 ([3] II.11.6) *For every rational number $r \geq 0$, the sub-$\widehat{\overline{R}}$-algebra $\mathscr{G}^{(r)}$ (3.4.11.1) of $\mathscr{G}$ is stable under the action φ_0 of Δ on $\mathscr{G}$ (3.4.9.3), and the induced actions of Δ on $\mathscr{G}^{(r)}$ and $\widehat{\mathscr{G}}^{(r)}$ are continuous for the p-adic topologies.*

We take again the assumptions and notation of 3.4.4. By 3.4.10, for every $g \in \Delta$ and every $1 \leq i \leq d$, we have

$$\varphi_0(g)(y_i) = y_i - \widetilde{\xi}^{-1}\chi_i(g). \tag{3.4.12.1}$$

Since $\widetilde{\xi}^{-1}\chi_i(g) \in p^{\rho+\frac{1}{p-1}}\mathscr{O}_C$ (3.4.4.1), $\mathscr{G}^{(r)}$ is stable under $\varphi_0(g)$. Let ζ be a generator of $\mathbb{Z}_p(1)$. There exists an $a_g \in \mathbb{Z}_p$ such that $\chi_i(g) = [\zeta^{a_g}] - 1 \in \mathscr{A}_2(\mathscr{O}_{\overline{K}})$. By linearity, we have $\log([\zeta^{a_g}]) \in p^{v_p(a_g)}\xi\mathscr{O}_C$, and consequently $\varphi_0(g)(y_i) - y_i \in p^{\rho+v_p(a_g)}\mathscr{G}$ (3.1.10.1). For every integer $n \geq 0$, the set of $g \in \Delta$ such that $\rho + v_p(a_g) \geq n$ being an open subgroup of Δ, we deduce from this that the stabilizer of the class of $p^r y_i$ in $\mathscr{G}^{(r)}/p^n\mathscr{G}^{(r)}$ is open in Δ. The second assertion follows because the action of Δ on $\overline{R}/p^n\overline{R}$ is continuous for the discrete topology.

3.4.13 We take again the notation of 3.4.7 and for any rational number $r \geq 0$, we set

$$\mathscr{G}^{(r)} = \mathrm{S}_{\widehat{\overline{R}}}(p^r\xi^{-1}\widetilde{\Omega}^1_{R/\mathscr{O}_K} \otimes_R \widehat{\overline{R}}), \tag{3.4.13.1}$$

$$\mathscr{G}'^{(r)} = \mathrm{S}_{\widehat{\overline{R}}}(p^r(\xi^*_\pi)^{-1}\widetilde{\Omega}^1_{R/\mathscr{O}_K} \otimes_R \widehat{\overline{R}}). \tag{3.4.13.2}$$

We then have a canonical isomorphism

$$\mathscr{G}'^{(r)} \xrightarrow{\sim} \mathscr{G}^{(r+\rho)}. \tag{3.4.13.3}$$

It is compatible with the actions of Δ defined by the sections ψ_0 and ψ'_0 by 3.4.10. Moreover, the diagram

$$\begin{array}{ccc} \mathscr{G}'^{(r)} & \xrightarrow{d_{\mathscr{G}'^{(r)}}} & (\xi^*_\pi)^{-1}\widetilde{\Omega}^1_{R/\mathscr{O}_K} \otimes_R \mathscr{G}'^{(r)} \\ \downarrow & & \downarrow{\scriptstyle \nu} \\ \mathscr{G}^{(r+\rho)} & \xrightarrow{d_{\mathscr{G}^{(r+\rho)}}} & \xi^{-1}\widetilde{\Omega}^1_{R/\mathscr{O}_K} \otimes_R \mathscr{G}^{(r+\rho)}, \end{array} \tag{3.4.13.4}$$

where $d_{\mathscr{G}'^{(r)}}$ and $d_{\mathscr{G}^{(r+\rho)}}$ are the universal $\widehat{R_1}$-derivations (3.4.11.3) and ν is the isomorphism induced by (3.2.21.2), is commutative

3.4.14 For any rational number $r \geq 0$, we denote by $\mathfrak{S}^{(r)}$ the sub-$\widehat{R_1}$-algebra of $\mathscr{G}^{(r)}$ (3.4.11.1) defined by (2.1.10)

$$\mathfrak{S}^{(r)} = \mathrm{S}_{\widehat{R_1}}(p^r\widetilde{\xi}^{-1}\widetilde{\Omega}^1_{R/\mathscr{O}_K} \otimes_R \widehat{R_1}), \tag{3.4.14.1}$$

and by $\widehat{\mathfrak{S}}^{(r)}$ its p-adic Hausdorff completion. We set $\mathfrak{S} = \mathfrak{S}^{(0)}$ and $\widehat{\mathfrak{S}} = \widehat{\mathfrak{S}}^{(0)}$. Note that $\widehat{\mathfrak{S}}^{(r)}$ is $\widehat{R_1}$-flat by virtue of ([1] 1.12.4) and is therefore $\mathscr{O}_C$-flat ([3] II. 6.14). For all rational numbers $r' \geq r \geq 0$, we have a canonical injective homomorphism $\mathrm{a}^{r,r'}\colon \mathfrak{S}^{(r')} \to \mathfrak{S}^{(r)}$. We immediately check that the induced homomorphism $\widehat{\mathrm{a}}^{r,r'}\colon \widehat{\mathfrak{S}}^{(r')} \to \widehat{\mathfrak{S}}^{(r)}$ is injective. We have a canonical $\mathfrak{S}^{(r)}$-isomorphism

$$\Omega^1_{\mathfrak{S}^{(r)}/\widehat{R_1}} \xrightarrow{\sim} \xi^{-1}\widetilde{\Omega}^1_{R/\mathscr{O}_K} \otimes_R \mathfrak{S}^{(r)}. \tag{3.4.14.2}$$

We denote by

$$d_{\mathfrak{S}^{(r)}}\colon \mathfrak{S}^{(r)} \to \xi^{-1}\widetilde{\Omega}^1_{R/\mathscr{O}_K} \otimes_R \mathfrak{S}^{(r)} \tag{3.4.14.3}$$

the universal $\widehat{R_1}$-derivation of $\mathfrak{S}^{(r)}$ and by

$$d_{\widehat{\mathfrak{S}}^{(r)}}\colon \widehat{\mathfrak{S}}^{(r)} \to \xi^{-1}\widetilde{\Omega}^1_{R/\mathscr{O}_K} \otimes_R \widehat{\mathfrak{S}}^{(r)} \tag{3.4.14.4}$$

its extension to the completions. Since $\xi^{-1}\widetilde{\Omega}^1_{R/\mathscr{O}_K} \otimes_R \widehat{R_1} \subset d_{\mathfrak{S}^{(r)}}(\mathfrak{S}^{(r)})$, $d_{\mathfrak{S}^{(r)}}$ and $d_{\widehat{\mathfrak{S}}^{(r)}}$ are also Higgs $\widehat{R_1}$-fields with coefficients in $\xi^{-1}\widetilde{\Omega}^1_{R/\mathscr{O}_K}$ by 2.5.26(i). For all rational numbers $r' \geq r \geq 0$, we have

$$p^{r'-r}(\mathrm{id} \times \mathrm{a}^{r,r'}) \circ d_{\mathfrak{S}^{(r')}} = d_{\mathfrak{S}^{(r)}} \circ \mathrm{a}^{r,r'}. \tag{3.4.14.5}$$

Copying the proof of 3.4.12, we show that for every rational number $r \geq 0$, the action φ_0 of Δ on $\mathscr{G}$ (3.4.9.3) preserves the sub-$\widehat{R_1}$-algebra $\mathfrak{S}^{(r)}$, that the induced action of Δ on $\mathfrak{S}^{(r)}$ factors through Δ_∞ and that the action of Δ_∞ on $\mathfrak{S}^{(r)}$ thus defined is continuous for the p-adic topology.

3.4.15 Let r be a rational number ≥ 0, $\lambda \in \widehat{\mathfrak{S}}^{(r)}$, M a $\widehat{\mathfrak{S}}^{(r)}$-module. We call a *$p$-adic λ-connection on M with respect to the extension* $\widehat{\mathfrak{S}}^{(r)}/\widehat{R_1}$ an $\widehat{R_1}$-linear morphism

$$\nabla\colon M \to \xi^{-1}\widetilde{\Omega}^1_{R/\mathscr{O}_K} \otimes_R M \tag{3.4.15.1}$$

such that for all $t \in \widehat{\mathfrak{S}}^{(r)}$ and $x \in M$, we have

$$\nabla(tx) = \lambda d_{\widehat{\mathfrak{S}}^{(r)}}(t) \otimes x + t\nabla(x). \tag{3.4.15.2}$$

We say that ∇ is *integrable* if it is a Higgs $\widehat{R_1}$-field with coefficients in $\xi^{-1}\widetilde{\Omega}^1_{R/\mathscr{O}_K}$. If M is complete and separated for the p-adic topology, we recover the notion introduced in ([3] II.2.14) (see [3] II.2.16).

Let ∇ be an integrable p-adic λ-connection on M with respect to the extension $\widehat{\mathfrak{S}}^{(r)}/\widehat{R_1}$, (N,θ) a Higgs $\widehat{R_1}$-module with coefficients in $\xi^{-1}\widetilde{\Omega}^1_{R/\mathscr{O}_K}$. There is one and only one $\widehat{R_1}$-linear morphism

$$\nabla'\colon M \otimes_{\widehat{R_1}} N \to \xi^{-1}\widetilde{\Omega}^1_{R/\mathscr{O}_K} \otimes_R M \otimes_{\widehat{R_1}} N \tag{3.4.15.3}$$

such that for all $x \in M$ and $y \in N$, we have

$$\nabla'(x \otimes_{\widehat{R_1}} y) = \nabla(x) \otimes_{\widehat{R_1}} y + x \otimes_{\widehat{R_1}} \theta(y). \tag{3.4.15.4}$$

It is an integrable p-adic λ-connection on $M \otimes_{\widehat{R_1}} N$ with respect to the extension $\widehat{\mathfrak{S}}^{(r)}/\widehat{R_1}$.

We define in the same way the notion of an integrable p-adic λ-connection on a $\widehat{\mathscr{G}}^{(r)}$-module relative to the extension $\widehat{\mathscr{G}}^{(r)}/\widehat{\overline{R}}$. If ∇ is an integrable p-adic λ-connection on M with respect to the extension $\widehat{\mathfrak{S}}^{(r)}/\widehat{R_1}$, there exists one and only one $\widehat{\overline{R}}$-linear morphism

$$\nabla' \colon \widehat{\mathscr{G}}^{(r)} \otimes_{\widehat{\mathfrak{S}}^{(r)}} M \to \widetilde{\xi}^{-1}\widetilde{\Omega}^1_{R/\mathscr{O}_K} \otimes_R \widehat{\mathscr{G}}^{(r)} \otimes_{\widehat{\mathfrak{S}}^{(r)}} M \qquad (3.4.15.5)$$

such that for all $t \in \widehat{\mathscr{G}}^{(r)}$ and $x \in M$, we have

$$\nabla'(t \otimes_{\widehat{\mathfrak{S}}^{(r)}} x) = \lambda d_{\widehat{\mathscr{G}}^{(r)}}(t) \otimes_{\widehat{\mathfrak{S}}^{(r)}} x + t \otimes_{\widehat{\mathfrak{S}}^{(r)}} \nabla(x). \qquad (3.4.15.6)$$

It is an integrable p-adic λ-connection on $\widehat{\mathscr{G}}^{(r)} \otimes_{\widehat{\mathfrak{S}}^{(r)}} M$ with respect to the extension $\widehat{\mathscr{G}}^{(r)}/\widehat{\overline{R}}$.

3.4.16 Let r, ε be two rational numbers such that $r \geq 0$ and $\varepsilon > r + \frac{1}{p-1}$, (N, θ) an ε-quasi-small Higgs $\widehat{R_1}$-module with coefficients in $\widetilde{\xi}^{-1}\widetilde{\Omega}^1_{R/\mathscr{O}_K}$ (3.4.3) such that N is $\mathscr{O}_C$-flat. We can write uniquely (3.4.4)

$$\theta = \sum_{i=1}^{d} \theta_i \otimes y_i, \qquad (3.4.16.1)$$

where the θ_i are endomorphisms of N belonging to $p^{\varepsilon}\mathrm{End}_{\widehat{R_1}}(N)$ and pairwise commuting. For any $\underline{n} = (n_1, \ldots, n_d) \in \mathbb{N}^d$, we set $|\underline{n}| = \sum_{i=1}^d n_i$, $\underline{n}! = \prod_{i=1}^d n_i!$, $\underline{\theta}^{\underline{n}} = \prod_{i=1}^d \theta_i^{n_i} \in \mathrm{End}_{\widehat{R_1}}(N)$ and $\underline{y}^{\underline{n}} = \prod_{i=1}^d y_i^{n_i} \in \mathfrak{S}$. Note that $N \otimes_{\widehat{R_1}} \widehat{\mathfrak{S}}^{(r)}$ is complete and separated for the p-adic topology ([1] 1.10.2), and that it is $\mathscr{O}_C$-flat since $\widehat{\mathfrak{S}}^{(r)}$ is $\widehat{R_1}$-flat . Hence, for every $z \in N \otimes_{\widehat{R_1}} \widehat{\mathfrak{S}}^{(r)}$, the series

$$\sum_{\underline{n} \in \mathbb{N}^d} \frac{1}{\underline{n}!}(\underline{\theta}^{\underline{n}} \otimes \underline{y}^{\underline{n}})(z) \qquad (3.4.16.2)$$

converges in $N \otimes_{\widehat{R_1}} \widehat{\mathfrak{S}}^{(r)}$, and defines an $\widehat{\mathfrak{S}}^{(r)}$-linear endomorphism of $N \otimes_{\widehat{R_1}} \widehat{\mathfrak{S}}^{(r)}$, which we denote by

$$\exp_r(\theta) \colon N \otimes_{\widehat{R_1}} \widehat{\mathfrak{S}}^{(r)} \to N \otimes_{\widehat{R_1}} \widehat{\mathfrak{S}}^{(r)}. \qquad (3.4.16.3)$$

For every rational number r' such that $0 \leq r' \leq r$, the diagram

$$\begin{array}{ccc} N \otimes_{\widehat{R_1}} \widehat{\mathfrak{S}}^{(r)} & \xrightarrow{\exp_r(\theta)} & N \otimes_{\widehat{R_1}} \widehat{\mathfrak{S}}^{(r)} \\ \downarrow{\scriptstyle \mathrm{id} \otimes \widehat{\mathrm{a}}^{r',r}} & & \downarrow{\scriptstyle \mathrm{id} \otimes \widehat{\mathrm{a}}^{r',r}} \\ N \otimes_{\widehat{R_1}} \widehat{\mathfrak{S}}^{(r')} & \xrightarrow{\exp_{r'}(\theta)} & N \otimes_{\widehat{R_1}} \widehat{\mathfrak{S}}^{(r')} \end{array} \qquad (3.4.16.4)$$

is commutative. We can therefore omit the index r from the notation $\exp_r(\theta)$ without causing any ambiguity.

Proposition 3.4.17 *Let r, ε be two rational numbers such that $r \geq 0$ and $\varepsilon > r + \frac{1}{p-1}$, N an $\widehat{R_1}$-module of finite type that is $\mathscr{O}_C$-flat, θ an ε-quasi-small Higgs $\widehat{R_1}$-field over N with coefficients in $\widetilde{\xi}^{-1}\widetilde{\Omega}^1_{R/\mathscr{O}_K}$, φ the quasi-small $\widehat{R_1}$-representation of Δ_∞ on N associated with θ by the functor* (3.4.5.8). *Then, the endomorphism* (3.4.16.3)

$$\exp_r(\theta)\colon N \otimes_{\widehat{R_1}} \widehat{\mathfrak{S}}^{(r)} \to N \otimes_{\widehat{R_1}} \widehat{\mathfrak{S}}^{(r)} \tag{3.4.17.1}$$

is a Δ_∞-equivariant isomorphism of $\widehat{\mathfrak{S}}^{(r)}$-modules with integrable p-adic p^r-connection with respect to the extension $\widehat{\mathfrak{S}}^{(r)}/\widehat{R_1}$ (3.4.15), *where $\widehat{\mathfrak{S}}^{(r)}$ is endowed with the action of Δ_∞ induced by φ_0, the module N on the left-hand side is endowed with the trivial action of Δ_∞ and the Higgs $\widehat{R_1}$-field θ, and the module N on the right-hand side is endowed with the action φ of Δ_∞ and the zero Higgs $\widehat{R_1}$-field. In particular, $\exp_r(\theta)$ is an isomorphism of Higgs $\widehat{R_1}$-modules with coefficients in $\widetilde{\xi}^{-1}\widetilde{\Omega}^1_{R/\mathscr{O}_K}$.*

The absolute case (3.1.14) has been proved in ([3] II.13.15) and the relative case follows in view of 3.4.13 and the fact that $N \otimes_{\widehat{R_1}} \widehat{\mathfrak{S}}^{(r)}$ is $\mathscr{O}_C$-flat.

Corollary 3.4.18 *Under the assumptions of* 3.4.17, *we have a functorial Δ-equivariant isomorphism of* $\mathbf{MC}^r_p$ (3.3.4)

$$N \otimes_{\widehat{R_1}} \widehat{\mathscr{C}}^{(r)} \xrightarrow{\sim} N \otimes_{\widehat{R_1}} \widehat{\mathscr{C}}^{(r)}, \tag{3.4.18.1}$$

where $\widehat{\mathscr{C}}^{(r)}$ is endowed with the canonical action of Δ, the module N on the left-hand side is endowed with the trivial action of Δ_∞ and the Higgs $\widehat{R_1}$-field θ, and the module N on the right-hand side is endowed with the action φ of Δ_∞ and the zero Higgs $\widehat{R_1}$-field (3.3.4.4). *If, moreover, the deformation $(\widetilde{X}, \mathscr{M}_{\widetilde{X}})$ is defined by the adequate chart $((P, \gamma), (\mathbb{N}, \iota), \vartheta)$* (3.4.6), *the isomorphism is canonical.*

Indeed, by 3.4.17 and in view of 3.4.15, $\exp_r(\theta)$ induces a functorial and Δ-equivariant isomorphism of $\widehat{\mathscr{G}}^{(r)}$-modules with integrable p-adic p^r-connection with respect to the extension $\widehat{\mathscr{G}}^{(r)}/\widehat{\overline{R}}$

$$N \otimes_{\widehat{R_1}} \widehat{\mathscr{G}}^{(r)} \xrightarrow{\sim} N \otimes_{\widehat{R_1}} \widehat{\mathscr{G}}^{(r)}, \tag{3.4.18.2}$$

where $\widehat{\mathscr{G}}^{(r)}$ is endowed with the action of Δ induced by φ_0 (3.4.11), the module N on the left-hand side is endowed with the trivial action of Δ_∞ and the Higgs $\widehat{R_1}$-field θ, and the module N on the right-hand side is endowed with the action φ of Δ_∞ and the zero Higgs $\widehat{R_1}$-field. The proposition follows from 3.2.20 and (3.4.11.7).

Definition 3.4.19 Let G be a topological group, A an $\mathscr{O}_C$-algebra that is complete and separated for the p-adic topology, endowed with a continuous action of G (by homomorphisms of $\mathscr{O}_C$-algebras). We endow $A[\frac{1}{p}]$ with the p-adic topology (2.1.1). We say that a continuous $A[\frac{1}{p}]$-representation M of G is *small* if the following conditions are satisfied:

(i) M is a projective $A[\frac{1}{p}]$-module of finite type, endowed with the p-adic topology (2.1.1);
(ii) There exists a rational number $\alpha > \rho + \frac{2}{p-1}$ (3.4.1) and a sub-A-module of finite type M° of M that is stable under G, generated by a finite number of elements that are G-invariant modulo $p^\alpha M^\circ$, and that generates M over $A[\frac{1}{p}]$.

We denote by $\mathbf{Rep}^{\mathrm{s}}_{A[\frac{1}{p}]}(G)$ the full subcategory of $\mathbf{Rep}^{\mathrm{cont}}_{A[\frac{1}{p}]}(G)$ made up of the small A-representations of G (2.1.2). Note that this definition corresponds to the one given in ([3] II.13.2) in the absolute case (3.1.14).

Remarks 3.4.20 Let G be a topological group and A an $\mathscr{O}_C$-algebra that is complete and separated for the p-adic topology, endowed with a continuous action of G (by homomorphisms of $\mathscr{O}_C$-algebras).

(i) Let M be a projective $A[\frac{1}{p}]$-module of finite type and M° a sub-A-module of finite type of M. Then M° is complete and separated for the p-adic topology. Indeed, M° is complete by virtue of ([10] chap. III §2.12 cor. 1 of prop. 16). On the other hand, after adding a direct summand to M, if necessary, we may assume that it is free of finite type over $A[\frac{1}{p}]$. Consequently, there exists an integer $m \geq 0$ such that $p^m M^\circ$ is contained in a free A-module of finite type N. Hence $\cap_{n\geq 0} p^n M^\circ \subset \cap_{n \geq 0} p^n N = 0$.
(ii) Let M be a small $A[\frac{1}{p}]$-representation of G, M° a sub-A-module of finite type of M satisfying condition 3.4.19(ii). It then follows from (i) that M° is a quasi-small A-representation of G (3.4.2).

3.4.21 Let G be a topological group, A an $\mathscr{O}_C$-algebra that is complete and separated for the p-adic topology, endowed with a continuous action of G (by homomorphisms of $\mathscr{O}_C$-algebras). We denote by $\mathbf{Rep}'^{\mathrm{qsf}}_A(G)$ the full subcategory of the category $\mathbf{Rep}^{\mathrm{qsf}}_A(G)$ (3.4.2) made up of the A-representations M of G such that the $A[\frac{1}{p}]$-module underlying $M[\frac{1}{p}]$ is projective of finite type. It is an additive category. We denote by $\mathbf{Rep}'^{\mathrm{qsf}}_{A,\mathbb{Q}}(G)$ the category of objects of $\mathbf{Rep}'^{\mathrm{qsf}}_A(G)$ up to isogeny. The functor

$$\mathbf{Rep}'^{\mathrm{qsf}}_A(G) \to \mathbf{Rep}^{\mathrm{s}}_{A[\frac{1}{p}]}(G), \quad M \mapsto M[\frac{1}{p}] \tag{3.4.21.1}$$

then induces a functor

$$\mathbf{Rep}'^{\mathrm{qsf}}_{A,\mathbb{Q}}(G) \to \mathbf{Rep}^{\mathrm{s}}_{A[\frac{1}{p}]}(G). \tag{3.4.21.2}$$

Lemma 3.4.22 *The functor* (3.4.21.2) *is an equivalence of categories.*

Indeed, this functor is essentially surjective by 3.4.20. Let M, N be two A-modules of finite type that are $\mathscr{O}_C$-flat. Then, the canonical morphism

$$\mathrm{Hom}_A(M, N) \otimes_{\mathbb{Z}_p} \mathbb{Q}_p \to \mathrm{Hom}_{A[\frac{1}{p}]}(M[\frac{1}{p}], N[\frac{1}{p}]) \tag{3.4.22.1}$$

is an isomorphism. Suppose that M and N are endowed with A-semi-linear actions of G. The canonical morphism

$$\mathrm{Hom}_{A\langle G\rangle}(M,N)\otimes_{\mathbb{Z}_p}\mathbb{Q}_p \to \mathrm{Hom}_{A[\frac{1}{p}]\langle G\rangle}(M[\frac{1}{p}],N[\frac{1}{p}]) \tag{3.4.22.2}$$

where the source (resp. the target) denotes the set of morphisms of A-representations (resp. $A[\frac{1}{p}]$-representations) of G is then an isomorphism: the injectivity immediately follows from that of (3.4.22.1) and the surjectivity from that of (3.4.22.1) and from the fact that M and N are $\mathscr{O}_C$-flat. The proposition follows.

Definition 3.4.23 We say that a Higgs $\widehat{R_1}[\frac{1}{p}]$-module (N,θ) with coefficients in $\widetilde{\xi}^{-1}\widetilde{\Omega}^1_{R/\mathscr{O}_K}$ is *small* if the following conditions are satisfied:

(i) N is a projective $\widehat{R_1}[\frac{1}{p}]$-module of finite type;
(ii) There exist a rational number $\varepsilon > \frac{1}{p-1}$ and a sub-$\widehat{R_1}$-module of finite type N° of N, that generates it over $\widehat{R_1}[\frac{1}{p}]$, such that we have

$$\theta(N^\circ)\subset p^{\varepsilon}\widetilde{\xi}^{-1}N^\circ\otimes_R\widetilde{\Omega}^1_{R/\mathscr{O}_K}. \tag{3.4.23.1}$$

We denote by $\mathbf{HM}^{\mathrm{s}}(\widehat{R_1}[\frac{1}{p}],\widetilde{\xi}^{-1}\widetilde{\Omega}^1_{R/\mathscr{O}_K})$ the full subcategory of the category $\mathbf{HM}(\widehat{R_1}[\frac{1}{p}],\widetilde{\xi}^{-1}\widetilde{\Omega}^1_{R/\mathscr{O}_K})$ made up of the small Higgs $\widehat{R_1}[\frac{1}{p}]$-modules. Note that this definition corresponds to the one given in ([3] II.13.5) in the absolute case (3.1.14).

Lemma 3.4.24 ([3] II.13.7) *Let (N,θ) be a Higgs $\widehat{R_1}[\frac{1}{p}]$-module with coefficients in $\widetilde{\xi}^{-1}\widetilde{\Omega}^1_{R/\mathscr{O}_K}$ such that the following conditions are satisfied:*

(i) N is a projective $\widehat{R_1}[\frac{1}{p}]$-module of finite type;
(ii) There exists a rational number $\varepsilon > \frac{1}{p-1}$ such that for every $i\geq 1$, the ith characteristic invariant of θ belongs to $p^{i\varepsilon}\widetilde{\xi}^{-i}\mathrm{S}^i_R(\widetilde{\Omega}^1_{R/\mathscr{O}_K})\otimes_R\widehat{R_1}$ (2.5.1.5).

Then (N,θ) is small.

3.4.25 We denote by $\mathbf{HM}'^{\mathrm{qsf}}(\widehat{R_1},\widetilde{\xi}^{-1}\widetilde{\Omega}^1_{R/\mathscr{O}_K})$ the full subcategory of the category $\mathbf{HM}^{\mathrm{qsf}}(\widehat{R_1},\widetilde{\xi}^{-1}\widetilde{\Omega}^1_{R/\mathscr{O}_K})$ (3.4.3) made up of the Higgs $\widehat{R_1}$-modules (N,θ) such that the $\widehat{R_1}[\frac{1}{p}]$-module $N[\frac{1}{p}]$ is projective of finite type. It is an additive category. We denote by $\mathbf{HM}'^{\mathrm{qsf}}_{\mathbb{Q}}(\widehat{R_1},\widetilde{\xi}^{-1}\widetilde{\Omega}^1_{R/\mathscr{O}_K})$ the category of objects of $\mathbf{HM}'^{\mathrm{qsf}}(\widehat{R_1},\widetilde{\xi}^{-1}\widetilde{\Omega}^1_{R/\mathscr{O}_K})$ up to isogeny. The functor

$$\begin{array}{ccc}\mathbf{HM}'^{\mathrm{qsf}}(\widehat{R_1},\widetilde{\xi}^{-1}\widetilde{\Omega}^1_{R/\mathscr{O}_K}) & \to & \mathbf{HM}^{\mathrm{s}}(\widehat{R_1}[\frac{1}{p}],\widetilde{\xi}^{-1}\widetilde{\Omega}^1_{R/\mathscr{O}_K}),\\ (N,\theta) & \mapsto & (N\otimes_{\mathbb{Z}_p}\mathbb{Q}_p,\theta\otimes_{\mathbb{Z}_p}\mathbb{Q}_p)\end{array} \tag{3.4.25.1}$$

then induces a functor

$$\mathbf{HM}'^{\mathrm{qsf}}_{\mathbb{Q}}(\widehat{R_1},\widetilde{\xi}^{-1}\widetilde{\Omega}^1_{R/\mathscr{O}_K})\to\mathbf{HM}^{\mathrm{s}}(\widehat{R_1}[\frac{1}{p}],\widetilde{\xi}^{-1}\widetilde{\Omega}^1_{R/\mathscr{O}_K}). \tag{3.4.25.2}$$

Lemma 3.4.26 *The functor* (3.4.25.2) *is an equivalence of categories.*

Indeed, it immediately follows from the definitions that this functor is essentially surjective. Let M, N be two $\widehat{R_1}$-modules of finite type that are $\mathscr{O}_C$-flat. Then, the canonical morphism

$$\mathrm{Hom}_{\widehat{R_1}}(M,N)\otimes_{\mathbb{Z}_p}\mathbb{Q}_p \to \mathrm{Hom}_{\widehat{R_1}[\frac{1}{p}]}(M\otimes_{\mathbb{Z}_p}\mathbb{Q}_p, N\otimes_{\mathbb{Z}_p}\mathbb{Q}_p) \tag{3.4.26.1}$$

is an isomorphism. Suppose that M and N are endowed with Higgs $\widehat{R_1}$-fields with coefficients in $\widetilde{\xi}^{-1}\widetilde{\Omega}^1_{R/\mathscr{O}_K}$. The canonical morphism

$$\mathrm{Hom}_{\mathbf{HM}}(M,N)\otimes_{\mathbb{Z}_p}\mathbb{Q}_p \to \mathrm{Hom}_{\mathbf{HM}}(M\otimes_{\mathbb{Z}_p}\mathbb{Q}_p, N\otimes_{\mathbb{Z}_p}\mathbb{Q}_p), \tag{3.4.26.2}$$

where the source (resp. target) denotes the set of morphisms of Higgs $\widehat{R_1}$-modules (resp. $\widehat{R_1}[\frac{1}{p}]$-modules), is then an isomorphism: the injectivity immediately follows from that of (3.4.26.1) and the surjectivity from that of (3.4.26.1) and from the fact that M and N are $\mathscr{O}_C$-flat and that $\widetilde{\Omega}^1_{R/\mathscr{O}_K}$ is R-flat. The proposition follows.

3.4.27 In view of 3.4.22 and 3.4.26, the functors (3.4.5.6) induce a functor

$$\mathbf{Rep}^{\mathrm{s}}_{\widehat{R_1}[\frac{1}{p}]}(\Delta_\infty) \to \mathbf{HM}^{\mathrm{s}}(\widehat{R_1}[\frac{1}{p}], \widetilde{\xi}^{-1}\widetilde{\Omega}^1_{R/\mathscr{O}_K}), \tag{3.4.27.1}$$

and the functors (3.4.5.8) induce a functor

$$\mathbf{HM}^{\mathrm{s}}(\widehat{R_1}[\frac{1}{p}], \widetilde{\xi}^{-1}\widetilde{\Omega}^1_{R/\mathscr{O}_K}) \to \mathbf{Rep}^{\mathrm{s}}_{\widehat{R_1}[\frac{1}{p}]}(\Delta_\infty). \tag{3.4.27.2}$$

Proposition 3.4.28 *Let (N,θ) be a small Higgs $\widehat{R_1}[\frac{1}{p}]$-module with coefficients in $\widetilde{\xi}^{-1}\widetilde{\Omega}^1_{R/\mathscr{O}_K}$* (3.4.23) *and φ the small $\widehat{R_1}[\frac{1}{p}]$-representation of Δ_∞ on N associated with θ by the functor* (3.4.27.2). *Then,*

(i) *We have a functorial Δ-equivariant $\widehat{\mathscr{C}}^{(0+)}$-isomorphism of Higgs $\widehat{\overline{R}}$-modules with coefficients in $\widetilde{\xi}^{-1}\widetilde{\Omega}^1_{R/\mathscr{O}_K}$*

$$N\otimes_{\widehat{R_1}}\widehat{\mathscr{C}}^{(0+)} \xrightarrow{\sim} N\otimes_{\widehat{R_1}}\widehat{\mathscr{C}}^{(0+)}, \tag{3.4.28.1}$$

where $\widehat{\mathscr{C}}^{(0+)}$ is endowed with the canonical action of Δ and the Higgs $\widehat{\overline{R}}$-field $d^{(0)}_{\widehat{\mathscr{C}}^{(0+)}}$ (3.2.19.10), *the module N on the left-hand side is endowed with the trivial action of Δ_∞ and the Higgs $\widehat{R_1}$-field θ, and the module N on the right-hand side is endowed with the action φ of Δ_∞ and the zero Higgs $\widehat{R_1}$-field. If, moreover, the deformation $(\widetilde{X}, \mathscr{M}_{\widetilde{X}})$ is defined by the adequate chart $((P,\gamma),(\mathbb{N},\iota),\vartheta)$* (3.4.6), *then the isomorphism is canonical.*

(ii) *The Higgs $\widehat{R_1}[\frac{1}{p}]$-module (N,θ) is solvable* (3.3.10), *and we have a functorial Δ-equivariant $\widehat{\overline{R}}[\frac{1}{p}]$-isomorphism*

$$\mathbb{V}(N) \xrightarrow{\sim} (N,\varphi)\otimes_{\widehat{R_1}}\widehat{\overline{R}}, \tag{3.4.28.2}$$

where $\mathbb{V}$ *is the functor* (3.3.7.2). *If, moreover, the deformation* $(\widetilde{X}, \mathscr{M}_{\widetilde{X}})$ *is defined by the adequate chart* $((P, \gamma), (\mathbb{N}, \iota), \vartheta)$ (3.4.6), *then the isomorphism is canonical.*

(iii) *The* $\widehat{\overline{R}}[\frac{1}{p}]$*-representation* $\mathbb{V}(N)$ *of* Δ *is small and Dolbeault* (3.3.9), *and we have a functorial isomorphism of Higgs* $\widehat{R_1}[\frac{1}{p}]$*-modules*

$$\mathbb{H}(\mathbb{V}(N)) \xrightarrow{\sim} (N, \theta), \tag{3.4.28.3}$$

where $\mathbb{H}$ *is the functor* (3.3.6.2). *If, moreover, the deformation* $(\widetilde{X}, \mathscr{M}_{\widetilde{X}})$ *is defined by the adequate chart* $((P, \gamma), (\mathbb{N}, \iota), \vartheta)$ (3.4.6), *then the isomorphism is canonical.*

The isomorphism (3.4.28.1) follows from 3.4.18. The other assertions follow from the first one since $\ker(d^{(0)}_{\widehat{\mathscr{C}}^{(0+)}}) = \widehat{\overline{R}}$ and $(\widehat{\mathscr{C}}^{(0+)})^{\Delta} = \widehat{R_1}$ (3.2.26).

Proposition 3.4.29 ([3] II.13.24, IV.5.3.10) *Let* N *be a projective* $\widehat{R_1}[\frac{1}{p}]$*-module of finite type,* θ *a Higgs* $\widehat{R_1}[\frac{1}{p}]$*-field on* N *with coefficients in* $\widetilde{\xi}^{-1}\widetilde{\Omega}^1_{R/\mathscr{O}_K}$, M *an* $\widehat{\overline{R}}[\frac{1}{p}]$*-module endowed with the zero Higgs field,* r *a rational number* > 0,

$$N \otimes_{\widehat{R_1}} \widehat{\mathscr{C}}^{(r)} \xrightarrow{\sim} M \otimes_{\widehat{\overline{R}}} \widehat{\mathscr{C}}^{(r)} \tag{3.4.29.1}$$

an isomorphism of $\mathbf{MC}^r_p$ (3.3.4). *Then,* (N, θ) *is a small Higgs* $\widehat{R_1}[\frac{1}{p}]$*-module with coefficients in* $\widetilde{\xi}^{-1}\widetilde{\Omega}^1_{R/\mathscr{O}_K}$ (3.4.23).

In fact, the propositions ([3] II.13.24 and IV.5.3.10) are formulated in the absolute case (3.1.14), but the proof applies *mutatis mutandis* to the relative case.

Corollary 3.4.30 *A Higgs* $\widehat{R_1}[\frac{1}{p}]$*-module with coefficients in* $\widetilde{\xi}^{-1}\widetilde{\Omega}^1_{R/\mathscr{O}_K}$ *is solvable* (3.3.10) *if and only if it is small* (3.4.23).

This follows from 3.4.28(ii) and 3.4.29.

Corollary 3.4.31 *Every Dolbeault* $\widehat{\overline{R}}[\frac{1}{p}]$*-representation of* Δ (3.3.9) *is small* (3.4.19).

This follows from 3.3.16, 3.4.30 and 3.4.28(iii).

Proposition 3.4.32 *In the absolute case* (3.1.14), *an* $\widehat{\overline{R}}[\frac{1}{p}]$*-representation of* Δ *is Dolbeault* (3.3.9) *if and only if it is small* (3.4.19).

This follows from ([3] II.14.8) and ([57] 13.7). Note that the assumptions of ([57] § 2 and § 13) are satisfied by 3.1.15, ([2] 4.2.2(ii)) and ([56] I.5.1).

3.5 Hodge–Tate Representations

We keep the assumptions and notation of 3.2, 3.3 and 3.4 in this section.

3.5.1 We take again the notation of 3.4.4. Let N be a projective $\widehat{R_1}[\frac{1}{p}]$-module of finite type, θ a *nilpotent* Higgs $\widehat{R_1}[\frac{1}{p}]$-field with coefficients in $\widetilde{\xi}^{-1}\widetilde{\Omega}^1_{R/\mathscr{O}_K}$ (2.5.6). There exist $\widehat{R_1}[\frac{1}{p}]$-linear endomorphisms $\theta_1, \ldots, \theta_d$ of N that pairwise commute, such that

$$\theta = \sum_{i=1}^{d} \widetilde{\xi}^{-1}\theta_i \otimes d\log(t_i). \tag{3.5.1.1}$$

The θ_i's being nilpotent, we can define a linear $\widehat{R_1}[\frac{1}{p}]$-representation of Δ_∞ over N by the formula

$$\varphi = \exp\left(\sum_{i=1}^{d} \widetilde{\xi}^{-1}\theta_i \otimes \chi_i\right), \tag{3.5.1.2}$$

where χ_i is defined in (3.4.4.1). Moreover, the Higgs field θ induces a $\mathscr{G}[\frac{1}{p}]$-linear endomorphism of $N \otimes_{\widehat{R_1}} \mathscr{G}$ (3.2.14.2) that we also denote by θ, defined for all $h \in \mathscr{G}$ and $x \in N$ by

$$\theta(x \otimes h) = \sum_{1 \le i \le d} \theta_i(x) \otimes y_i h. \tag{3.5.1.3}$$

This endomorphism being clearly nilpotent, we can define its exponential

$$\exp(\theta)\colon N \otimes_{\widehat{R_1}} \mathscr{G} \to N \otimes_{\widehat{R_1}} \mathscr{G}. \tag{3.5.1.4}$$

Proposition 3.5.2 *We keep the assumptions of* 3.5.1.

(i) *For every rational number $\varepsilon > \frac{1}{p-1}$, there exists a sub-$\widehat{R_1}$-module of finite type N° of N, that generates it over $\widehat{R_1}[\frac{1}{p}]$, such that we have*

$$\theta(N^\circ) \subset p^\varepsilon \widetilde{\xi}^{-1} N^\circ \otimes_R \widetilde{\Omega}^1_{R/\mathscr{O}_K}. \tag{3.5.2.1}$$

In particular, the Higgs $\widehat{R_1}[\frac{1}{p}]$-module (N, θ) is small (3.4.23).

(ii) *The $\widehat{R_1}[\frac{1}{p}]$-representation (N, φ) of Δ_∞* (3.5.1.2) *is the image of the Higgs $\widehat{R_1}[\frac{1}{p}]$-module (N, θ) by the functor* (3.4.27.2).

(iii) *Let r be a rational number ≥ 0, ε a rational number $> r + \frac{1}{p-1}$, N° a sub-$\widehat{R_1}$-module of finite type of N satisfying the conclusions of* (i), *$\widehat{\mathfrak{S}}^{(r)}$ the ring defined in* 3.4.14, $\exp_r(\theta)$ *the $\widehat{\mathfrak{S}}^{(r)}$-linear endomorphism of $N^\circ \otimes_{\widehat{R_1}} \widehat{\mathfrak{S}}^{(r)}$ defined in* (3.4.16.3). *Then, the $\widehat{\mathscr{G}}$-linear endomorphisms of $N \otimes_{\widehat{R_1}} \widehat{\mathscr{G}}$ induced by* $\exp_r(\theta)$ *and* $\exp(\theta)$ (3.5.1.4) *coincide.*

(iv) *The endomorphism* $\exp(\theta)$ (3.5.1.4) *is a Δ-equivariant isomorphism of $\mathscr{G}$-modules with integrable connection with respect to the extension $\mathscr{G}/\widehat{\overline{R}}$* (3.3.2), *where the module N on the left-hand side is endowed with the trivial action of Δ and the Higgs $\widehat{R_1}$-field θ, the module N on the right-hand side is endowed with the action*

φ of Δ and the zero Higgs $\widehat{R_1}$-field and $\mathscr{G}$ is endowed with the action φ_0 of Δ (3.4.9.3).

(i) Since the ring $\widehat{R_1}$ is normal by ([3] II.6.15) and therefore reduced, for every $n \geq 1$, the nth characteristic invariant $\lambda_n(\theta)$ of θ vanishes by virtue of 2.5.9. The proposition then follows from ([3] II.13.7).

(ii) This immediately follows from (i) and 3.4.5.

(iii) This immediately follows from the definitions (cf. 3.4.16).

(iv) This follows from (ii), (iii) and 3.4.17 since the canonical morphism $N \otimes_{\widehat{R_1}} \mathscr{G} \to N \otimes_{\widehat{R_1}} \widehat{\mathscr{G}}$ is injective.

Corollary 3.5.3 *Under the assumptions of* 3.5.1, *there exists a Δ-equivariant isomorphism of $\mathscr{C}$-modules with integrable connection with respect to the extension $\mathscr{C}/\widehat{\overline{R}}$* (3.3.2)

$$N \otimes_{\widehat{R_1}} \mathscr{C} \to N \otimes_{\widehat{R_1}} \mathscr{C}, \tag{3.5.3.1}$$

where $\mathscr{C}$ is endowed with the canonical action of Δ (3.2.16), *the module N on the left-hand side is endowed with the trivial action of Δ and the Higgs $\widehat{R_1}$-field θ and the module N on the right-hand side is endowed with the action φ of Δ* (3.5.1.2) *and the zero Higgs $\widehat{R_1}$-field.*

We may assume that the deformation $(\widetilde{X}, \mathscr{M}_{\widetilde{X}})$ is defined by the adequate chart $((P,\gamma),(\mathbb{N},\iota),\vartheta)$ (3.4.6), in which case the proposition follows from 3.5.2(iv) and (3.4.11.6).

Corollary 3.5.4 *Let N be a projective $\widehat{R_1}[\frac{1}{p}]$-module of finite type, θ a Higgs $\widehat{R_1}[\frac{1}{p}]$-field with coefficients in $\widetilde{\xi}^{-1}\widetilde{\Omega}^1_{R/\mathscr{O}_K}$. Then, the following conditions are equivalent:*

(i) *The Higgs field θ is nilpotent* (2.5.6).

(ii) *There exist an $\widehat{\overline{R}}[\frac{1}{p}]$-representation M of Δ and a $\mathscr{C}$-linear and Δ-equivariant isomorphism of Higgs $\widehat{\overline{R}}[\frac{1}{p}]$-modules with coefficients in $\widetilde{\xi}^{-1}\widetilde{\Omega}^1_{R/\mathscr{O}_K}$*

$$N \otimes_{\widehat{R_1}} \mathscr{C} \xrightarrow{\sim} M \otimes_{\widehat{\overline{R}}} \mathscr{C}, \tag{3.5.4.1}$$

where $\mathscr{C}$ is endowed with the canonical action of Δ and the Higgs field $d_{\mathscr{C}}$, N is endowed with the trivial action of Δ and M is endowed with the zero Higgs field.

Moreover, in this case, the $\widehat{\overline{R}}[\frac{1}{p}]$-module M is projective of finite type, the Higgs $\widehat{R_1}[\frac{1}{p}]$-module (N,θ) is solvable and we have an isomorphism of $\widehat{\overline{R}}[\frac{1}{p}]$-representations of Δ,

$$\mathbb{V}(N,\theta) \xrightarrow{\sim} M. \tag{3.5.4.2}$$

The implication (i)⇒(ii) follows from 3.5.3. Let us show the converse implication. Assume that condition (ii) is satisfied. Any splitting $\mathscr{F} \to \widehat{\overline{R}}$ of the extension (3.2.14.5), seen as an exact sequence of $\widehat{\overline{R}}$-modules without actions of Δ, defines a retraction of the $\widehat{\overline{R}}$-algebra $\mathscr{C}$. The isomorphism (3.5.4.1) then implies that the $\widehat{\overline{R}}[\frac{1}{p}]$-module M is projective of finite type. The $\widehat{R_1}[\frac{1}{p}]$-module N being projective of finite type, there exists an integer $n \geq 1$ such that the isomorphism (3.5.4.1) sends N to $M \otimes_{\widehat{\overline{R}}} \mathrm{S}^n_{\widehat{\overline{R}}}(\mathscr{F})$. We immediately deduce from this that θ is nilpotent. The isomorphism (3.5.4.1) is in fact an isomorphism of $\mathscr{C}$-modules with integrable connection with respect to the extension $\mathscr{C}/\widehat{\overline{R}}$ (3.3.2). For every rational number $r \geq 0$, the canonical morphism $\alpha^{0,r} : \mathscr{C}^{(r)} \to \mathscr{C}$ (3.2.17) induces an isomorphism of $\widehat{\overline{R}}$-algebras

$$\mathscr{C}^{(r)}[\frac{1}{p}] \xrightarrow{\sim} \mathscr{C}[\frac{1}{p}]. \tag{3.5.4.3}$$

In view of (3.2.19.8) and (3.3.4.7), the isomorphism (3.5.4.1) then induces a $\widehat{\mathscr{C}}^{(r)}$-linear and Δ-equivariant isomorphism of Higgs $\widehat{\overline{R}}[\frac{1}{p}]$-modules with coefficients in $\widetilde{\xi}^{-1}\widetilde{\Omega}^1_{R/\mathscr{O}_K}$

$$N \otimes_{\widehat{R_1}} \widehat{\mathscr{C}}^{(r)} \xrightarrow{\sim} M \otimes_{\widehat{\overline{R}}} \widehat{\mathscr{C}}^{(r)}, \tag{3.5.4.4}$$

where $\widehat{\mathscr{C}}^{(r)}$ is endowed with the canonical action of Δ and the Higgs field $p^r d_{\widehat{\mathscr{C}}^{(r)}}$, N is endowed with the trivial action of Δ and M is endowed with the zero Higgs field. It then follows from 3.3.11 that the Higgs $\widehat{R_1}[\frac{1}{p}]$-module (N, θ) is solvable and that we have an isomorphism of $\widehat{\overline{R}}[\frac{1}{p}]$-representations of Δ

$$\mathbb{V}(N, \theta) \xrightarrow{\sim} M. \tag{3.5.4.5}$$

Corollary 3.5.5 *For any $\widehat{\overline{R}}[\frac{1}{p}]$-representation M of Δ, the following conditions are equivalent:*

(i) *The $\widehat{\overline{R}}[\frac{1}{p}]$-representation M of Δ is Dolbeault and the associated Higgs $\widehat{R_1}[\frac{1}{p}]$-module $\mathbb{H}(M)$ with coefficients in $\widetilde{\xi}^{-1}\widetilde{\Omega}^1_{R/\mathscr{O}_K}$ is nilpotent (2.5.6).*

(ii) *There exist a projective $\widehat{R_1}[\frac{1}{p}]$-module of finite type N, a Higgs $\widehat{R_1}[\frac{1}{p}]$-field θ with coefficients in $\widetilde{\xi}^{-1}\widetilde{\Omega}^1_{R/\mathscr{O}_K}$ over N and a $\mathscr{C}$-linear and Δ-equivariant isomorphism of Higgs $\widehat{\overline{R}}[\frac{1}{p}]$-modules with coefficients in $\widetilde{\xi}^{-1}\widetilde{\Omega}^1_{R/\mathscr{O}_K}$*

$$N \otimes_{\widehat{R_1}} \mathscr{C} \xrightarrow{\sim} M \otimes_{\widehat{\overline{R}}} \mathscr{C}, \tag{3.5.5.1}$$

where $\mathscr{C}$ is endowed with the canonical action of Δ and the Higgs field $d_{\mathscr{C}}$, N is endowed with the trivial action of Δ and M is endowed with the zero Higgs field.

Moreover, in this case, we have an isomorphism of Higgs $\widehat{R_1}[\frac{1}{p}]$-modules with coefficients in $\widetilde{\xi}^{-1}\widetilde{\Omega}^1_{R/\mathscr{O}_K}$

$$\mathbb{H}(M) \xrightarrow{\sim} (N, \theta). \tag{3.5.5.2}$$

The implication (i)⇒(ii) follows from 3.3.16 and 3.5.4 applied to $\mathbb{H}(M)$. Let us show the converse implication. Assume that condition (ii) is satisfied. For every rational number $r \geq 0$, the isomorphism (3.5.5.1) induces a $\widehat{\mathscr{C}}^{(r)}$-linear and Δ-equivariant isomorphism of Higgs $\widehat{\overline{R}}[\frac{1}{p}]$-modules with coefficients in $\widetilde{\xi}^{-1}\widetilde{\Omega}^1_{R/\mathscr{O}_K}$

$$N \otimes_{\widehat{R_1}} \widehat{\mathscr{C}}^{(r)} \xrightarrow{\sim} M \otimes_{\widehat{\overline{R}}} \widehat{\mathscr{C}}^{(r)}, \tag{3.5.5.3}$$

where $\widehat{\mathscr{C}}^{(r)}$ is endowed with the canonical action of Δ and the Higgs field $p^r d_{\widehat{\mathscr{C}}^{(r)}}$, N is endowed with the trivial action of Δ and M is endowed with the zero Higgs field (cf. the proof of 3.5.4). It then follows from 3.3.12 that M is Dolbeault and that we have an isomorphism of Higgs $\widehat{R_1}[\frac{1}{p}]$-modules with coefficients in $\widetilde{\xi}^{-1}\widetilde{\Omega}^1_{R/\mathscr{O}_K}$

$$\mathbb{H}(M) \xrightarrow{\sim} (N, \theta). \tag{3.5.5.4}$$

The $\widehat{R_1}[\frac{1}{p}]$-module N (resp. $\widehat{\overline{R}}[\frac{1}{p}]$-module M) being projective of finite type, there exists an integer $n \geq 1$ such that the isomorphism (3.5.5.1) sends N to $M \otimes_{\widehat{\overline{R}}} \mathrm{S}^n_{\widehat{\overline{R}}}(\mathscr{F})$. We immediately deduce from this that θ is nilpotent; condition (i) follows.

Definition 3.5.6 We say that an $\widehat{\overline{R}}[\frac{1}{p}]$-representation M of Δ is *Hodge–Tate* if it satisfies the equivalent conditions of 3.5.5.

This notion does not depend on the choice of the $(\widetilde{S}, \mathscr{M}_{\widetilde{S}})$-deformation $(\widetilde{X}, \mathscr{M}_{\widetilde{X}})$ by 3.3.8(i), nor on the absolute or relative case (3.1.14) by virtue of 3.2.21. When it is necessary to specify, we will say that the $\widehat{\overline{R}}[\frac{1}{p}]$-representation M of Δ is *geometrically Hodge–Tate* to avoid any confusion with the notion of Hodge–Tate representation considered in ([36] 2.1) (cf. [3] II.15).

Proposition 3.5.7 *The functors $\mathbb{H}$ (3.3.6.2) and $\mathbb{V}$ (3.3.7.2) induce equivalences of categories that are quasi-inverse to each other, between the category of Hodge–Tate $\widehat{\overline{R}}[\frac{1}{p}]$-representations of Δ and that of nilpotent Higgs $\widehat{R_1}[\frac{1}{p}]$-modules with coefficients in $\widetilde{\xi}^{-1}\widetilde{\Omega}^1_{R/\mathscr{O}_K}$ whose underlying $\widehat{R_1}[\frac{1}{p}]$-module is projective of finite type.*

This follows from 3.3.16, 3.5.4 and 3.5.5.

Chapter 4
The p-adic Simpson Correspondence and Hodge–Tate Modules. Global Study

4.1 Assumptions and Notation

4.1.1 In this chapter, K denotes a complete discrete valuation field of characteristic 0, with *algebraically closed* residue field k of characteristic $p > 0$, $\mathscr{O}_K$ the valuation ring of K, W the ring of Witt vectors with respect to p with coefficients in k, $\overline{K}$ an algebraic closure of K, $\mathscr{O}_{\overline{K}}$ the integral closure of $\mathscr{O}_K$ in $\overline{K}$, $\mathfrak{m}_{\overline{K}}$ the maximal ideal of $\mathscr{O}_{\overline{K}}$ and G_K the Galois group of $\overline{K}$ over K. We denote by $\mathscr{O}_C$ the p-adic Hausdorff completion of $\mathscr{O}_{\overline{K}}$, $\mathfrak{m}_C$ its maximal ideal, C its field of fractions and v its valuation, normalized by $v(p) = 1$. We denote by $\mathbb{Z}_p(1)$ the $\mathbb{Z}[G_K]$-module

$$\mathbb{Z}_p(1) = \varprojlim_{n\geq 1} \mu_{p^n}(\mathscr{O}_{\overline{K}}), \tag{4.1.1.1}$$

where $\mu_{p^n}(\mathscr{O}_{\overline{K}})$ denotes the subgroup of p^nth roots of unity in $\mathscr{O}_{\overline{K}}$. For any $\mathbb{Z}_p[G_K]$-module M and any integer n, we set $M(n) = M \otimes_{\mathbb{Z}_p} \mathbb{Z}_p(1)^{\otimes n}$.

We set $S = \operatorname{Spec}(\mathscr{O}_K)$, $\overline{S} = \operatorname{Spec}(\mathscr{O}_{\overline{K}})$ and $\check{\overline{S}} = \operatorname{Spec}(\mathscr{O}_C)$. We denote by s (resp. η, resp. $\overline{\eta}$) the closed point of S (resp. generic point of S, resp. generic point of $\overline{S}$). For any integer $n \geq 1$, we set $S_n = \operatorname{Spec}(\mathscr{O}_K/p^n\mathscr{O}_K)$. For any S-scheme X, we set

$$\overline{X} = X \times_S \overline{S}, \quad \check{\overline{X}} = X \times_S \check{\overline{S}} \quad \text{and} \quad X_n = X \times_S S_n. \tag{4.1.1.2}$$

We endow S with the logarithmic structure $\mathscr{M}_S$ defined by its closed point, and $\overline{S}$ and $\check{\overline{S}}$ with the logarithmic structures $\mathscr{M}_{\overline{S}}$ and $\mathscr{M}_{\check{\overline{S}}}$ pullbacks of $\mathscr{M}_S$.

4.1.2 Since $\mathscr{O}_{\overline{K}}$ is a non-discrete valuation ring of height 1, we are allowed to consider an α-algebra (or almost algebra) over this ring ([2] 2.10.1) (see [2] 2.6-2.10). We choose a compatible system $(\beta_n)_{n>0}$ of nth roots of p in $\mathscr{O}_{\overline{K}}$. For any rational number $\varepsilon > 0$, we set $p^\varepsilon = (\beta_n)^{\varepsilon n}$, where n is a positive integer such that εn is an integer.

A. Abbes, M. Gros, *The p-adic Simpson Correspondence and Hodge-Tate Local Systems*, Lecture Notes in Mathematics 2345, https://doi.org/10.1007/978-3-031-55914-3_4

4.1.3   Let $f\colon (X, \mathscr{M}_X) \to (S, \mathscr{M}_S)$ be an *adequate* morphism of logarithmic schemes ([3] III.4.7). We denote by X° the maximal open subscheme of X where the logarithmic structure $\mathscr{M}_X$ is trivial; it is an open subscheme of X_η. We denote by $j\colon X^\circ \to X$ the canonical injection. For any X-scheme U, we set

$$U^\circ = U \times_X X^\circ. \tag{4.1.3.1}$$

We denote by $\hbar\colon \overline{X} \to X$ and $h\colon \overline{X}^\circ \to X$ the canonical morphisms (4.1.1.2), so that we have $h = \hbar \circ j_{\overline{X}}$. To lighten the notation, we set

$$\widetilde{\Omega}^1_{X/S} = \Omega^1_{(X, \mathscr{M}_X)/(S, \mathscr{M}_S)}, \tag{4.1.3.2}$$

that we consider as a sheaf of X_{zar} or $X_{\mathrm{\acute{e}t}}$, depending on the context (see 2.1.13).

4.1.4   For any integer $n \geq 1$, we denote by $a\colon X_s \to X$, $a_n\colon X_s \to X_n$, $\iota_n\colon X_n \to X$ and $\bar\iota_n\colon \overline{X}_n \to \overline{X}$ the canonical injections (4.1.1.2). Since k is algebraically closed, there exists a unique S-morphism $s \to \overline{S}$. It induces closed immersions $\bar{a}\colon X_s \to \overline{X}$ and $\bar{a}_n\colon X_s \to \overline{X}_n$ lifting a and a_n, respectively.

$$
\begin{array}{ccccc}
& & \overset{\bar{a}}{\overbrace{}} & & \\
X_s & \xrightarrow{\ \bar{a}_n\ } & \overline{X}_n & \xrightarrow{\ \bar\iota_n\ } & \overline{X} \\
\big\| & & \downarrow{\hbar_n} & & \downarrow{\hbar} \\
X_s & \xrightarrow{\ a_n\ } & X_n & \xrightarrow{\ \iota_n\ } & X. \\
& & \underset{a}{\underbrace{}} & &
\end{array}
\tag{4.1.4.1}
$$

Since $\hbar$ is integral and $\hbar_n$ is a universal homeomorphism, for every sheaf $\mathscr{F}$ of $\overline{X}_{\mathrm{\acute{e}t}}$, the base change morphism

$$a^*(\hbar_*(\mathscr{F})) \to \bar{a}^*(\mathscr{F}) \tag{4.1.4.2}$$

is an isomorphism ([5] VIII 5.6). Moreover, $\bar{a}_n$ being a universal homeomorphism, we can consider $\mathscr{O}_{\overline{X}_n}$ as a sheaf of $X_{s,\mathrm{zar}}$ or $X_{s,\mathrm{\acute{e}t}}$, depending on the context (2.1.13). We set (4.1.3.2)

$$\widetilde{\Omega}^1_{\overline{X}_n/\overline{S}_n} = \widetilde{\Omega}^1_{X/S} \otimes_{\mathscr{O}_X} \mathscr{O}_{\overline{X}_n}, \tag{4.1.4.3}$$

that we also consider as a sheaf of $X_{s,\mathrm{zar}}$ or $X_{s,\mathrm{\acute{e}t}}$, depending on the context.

4.1.5   We take again the notation of 3.1.14. We set

$$\breve{\xi}^{-1}\widetilde{\Omega}^1_{\overline{X}_n/\overline{S}_n} = \breve{\xi}^{-1}\mathscr{O}_C \otimes_{\mathscr{O}_C} \widetilde{\Omega}^1_{\overline{X}_n/\overline{S}_n}. \tag{4.1.5.1}$$

We endow $\breve{\overline{X}} = X \times_S \breve{\overline{S}}$ (4.1.1.2) with the logarithmic structure $\mathscr{M}_{\breve{\overline{X}}}$ pullback of $\mathscr{M}_X$. We then have a canonical isomorphism

$$(\breve{\overline{X}}, \mathscr{M}_{\breve{\overline{X}}}) \xrightarrow{\sim} (X, \mathscr{M}_X) \times_{(S, \mathscr{M}_S)} (\breve{\overline{S}}, \mathscr{M}_{\breve{\overline{S}}}), \tag{4.1.5.2}$$

the fiber product being taken indifferently in the category of logarithmic schemes or in that of fine logarithmic schemes.

We assume in this chapter that there exists an $(\widetilde{S}, \mathscr{M}_{\widetilde{S}})$*-smooth deformation* $(\widetilde{X}, \mathscr{M}_{\widetilde{X}})$ *of* $(\check{\overline{X}}, \mathscr{M}_{\check{\overline{X}}})$ *that we fix*; i.e., a smooth morphism of fine logarithmic schemes $(\widetilde{X}, \mathscr{M}_{\widetilde{X}}) \to (\widetilde{S}, \mathscr{M}_{\widetilde{S}})$ and an $(\check{\overline{S}}, \mathscr{M}_{\check{\overline{S}}})$-isomorphism

$$(\check{\overline{X}}, \mathscr{M}_{\check{\overline{X}}}) \xrightarrow{\sim} (\widetilde{X}, \mathscr{M}_{\widetilde{X}}) \times_{(\widetilde{S}, \mathscr{M}_{\widetilde{S}})} (\check{\overline{S}}, \mathscr{M}_{\check{\overline{S}}}). \tag{4.1.5.3}$$

Remark 4.1.6 In the relative case (3.1.14), there exists a canonical smooth $(\widetilde{S}, \mathscr{M}_{\widetilde{S}})$-deformation of $(\check{\overline{X}}, \mathscr{M}_{\check{\overline{X}}})$, namely

$$(X, \mathscr{M}_X) \times_{(S, \mathscr{M}_S)} (\widetilde{S}, \mathscr{M}_{\widetilde{S}}) \tag{4.1.6.1}$$

where we consider $(\widetilde{S}, \mathscr{M}_{\widetilde{S}})$ as a logarithmic scheme over $(S, \mathscr{M}_S)$ via pr_1 (3.1.13.5), the fiber product being indifferently taken in the category of logarithmic schemes or in that of fine logarithmic schemes.

4.2 Higgs Modules

4.2.1 We set $\mathscr{S} = \mathrm{Spf}(\mathscr{O}_C)$ and we denote by $\mathfrak{X}$ the formal scheme p-adic completion of $\overline{X}$ (4.1.3). This is a formal $\mathscr{S}$-scheme of finite presentation ([1] 2.3.15). It is therefore idyllic ([1] 2.6.13). We denote by $\mathbf{Mod}(\mathscr{O}_{\mathfrak{X}})$ (resp. $\mathbf{Mod}(\mathscr{O}_{\mathfrak{X}}[\frac{1}{p}])$) the category of $\mathscr{O}_{\mathfrak{X}}$-modules (resp. $\mathscr{O}_{\mathfrak{X}}[\frac{1}{p}]$-modules) of $X_{s,\mathrm{zar}}$, by $\mathbf{Mod}^{\mathrm{coh}}(\mathscr{O}_{\mathfrak{X}})$ (resp. $\mathbf{Mod}^{\mathrm{coh}}(\mathscr{O}_{\mathfrak{X}}[\frac{1}{p}])$) the full subcategory made up of coherent $\mathscr{O}_{\mathfrak{X}}$-modules (resp. $\mathscr{O}_{\mathfrak{X}}[\frac{1}{p}]$-modules) and by $\mathbf{Mod}^{\mathrm{coh}}_{\mathbb{Q}}(\mathscr{O}_{\mathfrak{X}})$ the category of coherent $\mathscr{O}_{\mathfrak{X}}$-modules up to isogeny ([3] III.6.1.1). By ([3] III.6.16), the canonical functor

$$\mathbf{Mod}^{\mathrm{coh}}(\mathscr{O}_{\mathfrak{X}}) \to \mathbf{Mod}^{\mathrm{coh}}(\mathscr{O}_{\mathfrak{X}}[\frac{1}{p}]), \quad \mathscr{N} \mapsto \mathscr{N}_{\mathbb{Q}_p} \tag{4.2.1.1}$$

induces an equivalence of abelian categories

$$\mathbf{Mod}^{\mathrm{coh}}_{\mathbb{Q}}(\mathscr{O}_{\mathfrak{X}}) \xrightarrow{\sim} \mathbf{Mod}^{\mathrm{coh}}(\mathscr{O}_{\mathfrak{X}}[\frac{1}{p}]). \tag{4.2.1.2}$$

4.2.2 With the conventions and notation of 3.1.14 and (4.1.3.2), we denote by $\widetilde{\xi}^{-1}\widetilde{\Omega}^1_{\mathfrak{X}/\mathscr{S}}$ the p-adic completion of the $\mathscr{O}_{\overline{X}}$-module ([1] 2.5.1)

$$\widetilde{\xi}^{-1}\widetilde{\Omega}^1_{\check{\overline{X}}/\check{\overline{S}}} = \widetilde{\xi}^{-1}\widetilde{\Omega}^1_{X/S} \otimes_{\mathscr{O}_X} \mathscr{O}_{\check{\overline{X}}}. \tag{4.2.2.1}$$

We denote by $\mathbf{HM}(\mathscr{O}_{\mathfrak{X}}, \widetilde{\xi}^{-1}\widetilde{\Omega}^1_{\mathfrak{X}/\mathscr{S}})$ the category of Higgs $\mathscr{O}_{\mathfrak{X}}$-modules with coefficients in $\widetilde{\xi}^{-1}\widetilde{\Omega}^1_{\mathfrak{X}/\mathscr{S}}$ (2.5.1). Such a Higgs module is said to be *coherent* if the underlying

$\mathscr{O}_{\mathfrak{X}}$-module is coherent. We denote by $\mathbf{HM}^{\mathrm{coh}}(\mathscr{O}_{\mathfrak{X}}, \widetilde{\xi}^{-1}\widetilde{\Omega}^1_{\mathfrak{X}/\mathscr{S}})$ the full subcategory of $\mathbf{HM}(\mathscr{O}_{\mathfrak{X}}, \widetilde{\xi}^{-1}\widetilde{\Omega}^1_{\mathfrak{X}/\mathscr{S}})$ made up of coherent Higgs modules.

By a *Higgs* $\mathscr{O}_{\mathfrak{X}}[\frac{1}{p}]$*-module with coefficients in* $\widetilde{\xi}^{-1}\widetilde{\Omega}^1_{\mathfrak{X}/\mathscr{S}}$, we mean a Higgs $\mathscr{O}_{\mathfrak{X}}[\frac{1}{p}]$-module with coefficients in $\widetilde{\xi}^{-1}\widetilde{\Omega}^1_{\mathfrak{X}/\mathscr{S}}[\frac{1}{p}]$. Such a Higgs module is said to be *coherent* if the underlying $\mathscr{O}_{\mathfrak{X}}[\frac{1}{p}]$-module is coherent. We denote by $\mathbf{HM}(\mathscr{O}_{\mathfrak{X}}[\frac{1}{p}], \widetilde{\xi}^{-1}\widetilde{\Omega}^1_{\mathfrak{X}/\mathscr{S}})$ the abelian category of Higgs $\mathscr{O}_{\mathfrak{X}}[\frac{1}{p}]$-modules with coefficients in $\widetilde{\xi}^{-1}\widetilde{\Omega}^1_{\mathfrak{X}/\mathscr{S}}$ and by $\mathbf{HM}^{\mathrm{coh}}(\mathscr{O}_{\mathfrak{X}}[\frac{1}{p}], \widetilde{\xi}^{-1}\widetilde{\Omega}^1_{\mathfrak{X}/\mathscr{S}})$ the full subcategory made up of coherent Higgs modules; it is an abelian subcategory.

In the remainder of this chapter, we will omit the Higgs field from the notation of a Higgs module when we do not explicitly need it.

We denote by $\mathbf{HI}(\mathscr{O}_{\mathfrak{X}}, \widetilde{\xi}^{-1}\widetilde{\Omega}^1_{\mathfrak{X}/\mathscr{S}})$ the category of Higgs $\mathscr{O}_{\mathfrak{X}}$-isogenies with coefficients in $\widetilde{\xi}^{-1}\widetilde{\Omega}^1_{\mathfrak{X}/\mathscr{S}}$ (2.9.9) and by $\mathbf{HI}^{\mathrm{coh}}(\mathscr{O}_{\mathfrak{X}}, \widetilde{\xi}^{-1}\widetilde{\Omega}^1_{\mathfrak{X}/\mathscr{S}})$ the full subcategory made up of the quadruples $(\mathscr{M}, \mathscr{N}, u, \theta)$ such that $\mathscr{M}$ and $\mathscr{N}$ are coherent $\mathscr{O}_{\mathfrak{X}}$-modules. These are additive categories. We denote by $\mathbf{HI}_{\mathbb{Q}}(\mathscr{O}_{\mathfrak{X}}, \widetilde{\xi}^{-1}\widetilde{\Omega}^1_{\mathfrak{X}/\mathscr{S}})$ (resp. $\mathbf{HI}^{\mathrm{coh}}_{\mathbb{Q}}(\mathscr{O}_{\mathfrak{X}}, \widetilde{\xi}^{-1}\widetilde{\Omega}^1_{\mathfrak{X}/\mathscr{S}})$) the category of objects of $\mathbf{HI}(\mathscr{O}_{\mathfrak{X}}, \widetilde{\xi}^{-1}\widetilde{\Omega}^1_{\mathfrak{X}/\mathscr{S}})$ (resp. $\mathbf{HI}^{\mathrm{coh}}(\mathscr{O}_{\mathfrak{X}}, \widetilde{\xi}^{-1}\widetilde{\Omega}^1_{\mathfrak{X}/\mathscr{S}})$) up to isogeny ([3] III.6.1.1). The functor

$$\begin{aligned} \mathbf{HI}(\mathscr{O}_{\mathfrak{X}}, \widetilde{\xi}^{-1}\widetilde{\Omega}^1_{\mathfrak{X}/\mathscr{S}}) &\to \mathbf{HM}(\mathscr{O}_{\mathfrak{X}}[\tfrac{1}{p}], \widetilde{\xi}^{-1}\widetilde{\Omega}^1_{\mathfrak{X}/\mathscr{S}}) \\ (\mathscr{M}, \mathscr{N}, u, \theta) &\mapsto (\mathscr{M}_{\mathbb{Q}_p}, (\mathrm{id} \otimes u^{-1}_{\mathbb{Q}_p}) \circ \theta_{\mathbb{Q}_p}) \end{aligned} \tag{4.2.2.2}$$

induces a functor

$$\mathbf{HI}_{\mathbb{Q}}(\mathscr{O}_{\mathfrak{X}}, \widetilde{\xi}^{-1}\widetilde{\Omega}^1_{\mathfrak{X}/\mathscr{S}}) \to \mathbf{HM}(\mathscr{O}_{\mathfrak{X}}[\frac{1}{p}], \widetilde{\xi}^{-1}\widetilde{\Omega}^1_{\mathfrak{X}/\mathscr{S}}). \tag{4.2.2.3}$$

By ([3] III.6.21), the latter induces an equivalence of categories

$$\mathbf{HI}^{\mathrm{coh}}_{\mathbb{Q}}(\mathscr{O}_{\mathfrak{X}}, \widetilde{\xi}^{-1}\widetilde{\Omega}^1_{\mathfrak{X}/\mathscr{S}}) \xrightarrow{\sim} \mathbf{HM}^{\mathrm{coh}}(\mathscr{O}_{\mathfrak{X}}[\frac{1}{p}], \widetilde{\xi}^{-1}\widetilde{\Omega}^1_{\mathfrak{X}/\mathscr{S}}). \tag{4.2.2.4}$$

Definition 4.2.3 We call a *Higgs* $\mathscr{O}_{\mathfrak{X}}[\frac{1}{p}]$*-bundle with coefficients in* $\widetilde{\xi}^{-1}\widetilde{\Omega}^1_{\mathfrak{X}/\mathscr{S}}$ any Higgs $\mathscr{O}_{\mathfrak{X}}[\frac{1}{p}]$- module with coefficients in $\widetilde{\xi}^{-1}\widetilde{\Omega}^1_{\mathfrak{X}/\mathscr{S}}$ whose underlying $\mathscr{O}_{\mathfrak{X}}[\frac{1}{p}]$-module is locally projective of finite type (2.1.11).

Lemma 4.2.4 *Let $(\mathscr{M}, \theta)$ be a coherent Higgs $\mathscr{O}_{\mathfrak{X}}[\frac{1}{p}]$-module with coefficients in $\widetilde{\xi}^{-1}\widetilde{\Omega}^1_{\mathfrak{X}/\mathscr{S}}$. We set $\mathscr{T} = \mathscr{H}om_{\mathscr{O}_{\mathfrak{X}}}(\widetilde{\Omega}^1_{\mathfrak{X}/\mathscr{S}}, \widetilde{\xi}\mathscr{O}_{\mathfrak{X}})$ and we endow $\mathscr{M}$ with the $\mathrm{S}_{\mathscr{O}_{\mathfrak{X}}}(\mathscr{T})$-module structure induced by θ (2.5.1.10). We denote by $\mathscr{J}$ the kernel of the canonical augmentation $\mathrm{S}_{\mathscr{O}_{\mathfrak{X}}}(\mathscr{T}) \to \mathscr{O}_{\mathfrak{X}}$. For any open set U of $\mathfrak{X}$ and any section $d \in \Gamma(U, \mathscr{T})$, we denote by θ_d the endomorphism of $\mathscr{M}|U$ deduced from θ and d. Then, the following conditions are equivalent:*

(i) *There exists an integer $n \geq 1$ such that $\mathscr{J}^n\mathscr{M} = 0$.*
(ii) *There exists an integer $n \geq 1$ such that for all open sets U of $\mathfrak{X}$ and for all sections $s \in \Gamma(U, \mathscr{M})$ and $d_1, \dots, d_n \in \Gamma(U, \mathscr{T})$, we have $\theta_{d_n} \circ \cdots \circ \theta_{d_1}(s) = 0$.*

(iii) *There exists a finite decreasing filtration* $(\mathscr{M}_i)_{0\le i\le n}$ *of* $\mathscr{M}$ *by coherent sub-*$\mathscr{O}_{\mathfrak{X}}[\frac{1}{p}]$*-modules such that* $\mathscr{M}_0 = \mathscr{M}$, $\mathscr{M}_n = 0$ *and that for all* $0 \le i \le n-1$, *we have*

$$\theta(\mathscr{M}_i) \subset \widetilde{\xi}^{-1}\widetilde{\Omega}^1_{\mathfrak{X}/\mathscr{S}} \otimes_{\mathscr{O}_{\mathfrak{X}}} \mathscr{M}_{i+1}. \qquad (4.2.4.1)$$

Indeed, we clearly have (iii)⇒(ii)⇔(i). Assume that condition (i) is satisfied. For any $0 \le i \le n$, we denote by $\mathscr{M}_i$ the largest sub-$\mathrm{S}_{\mathscr{O}_{\mathfrak{X}}}(\mathscr{T})$-module of $\mathscr{M}$ annihilated by $\mathscr{J}^{n-i}$. We therefore have $\mathscr{M}_0 = \mathscr{M}$, $\mathscr{M}_n = 0$ and for every $0 \le i \le n-1$, $\mathscr{J}\mathscr{M}_i \subset \mathscr{M}_{i+1}$. Since the $\mathscr{O}_{\mathfrak{X}}$-module $\mathscr{T}$ is of finite type, the $\mathscr{O}_{\mathfrak{X}}[\frac{1}{p}]$-modules $\mathscr{M}_i$ are coherent ([27] 0.5.3.4). Therefore, the filtration of $\mathscr{M}$ thus defined satisfies condition (iii). The proposition follows.

Definition 4.2.5 We say that a coherent Higgs $\mathscr{O}_{\mathfrak{X}}[\frac{1}{p}]$-module $(\mathscr{M}, \theta)$ with coefficients in $\widetilde{\xi}^{-1}\widetilde{\Omega}^1_{\mathfrak{X}/\mathscr{S}}$ is *nilpotent* if it satisfies the equivalent conditions of 4.2.4. We then also say that the Higgs field θ is *nilpotent*.

A Higgs $\mathscr{O}_{\mathfrak{X}}[\frac{1}{p}]$-module $(\mathscr{M}, \theta)$ with coefficients in $\widetilde{\xi}^{-1}\widetilde{\Omega}^1_{\mathfrak{X}/\mathscr{S}}$ is nilpotent if and only if it is quasi-nilpotent and $(\mathscr{M}, \theta)$ admits a quasi-nilpotent filtration consisting of coherent sub-$\mathscr{O}_{\mathfrak{X}}[\frac{1}{p}]$-modules (2.5.3).

Remark 4.2.6 We keep the assumptions of 4.2.4 and suppose, moreover, $\mathfrak{X}$ is affine of ring A. Let $T = \mathscr{T}(\mathfrak{X})$ and $M = \mathscr{M}(\mathfrak{X})$, which is an $A[\frac{1}{p}]$-module of finite type. We denote by θ the Higgs $A[\frac{1}{p}]$-field on M with coefficients in $\widetilde{\xi}^{-1}\widetilde{\Omega}^1_{\mathfrak{X}/\mathscr{S}}(\mathfrak{X})$ induced by θ ([1] 2.7.2.3 and 2.10.24). Then, the following conditions are equivalent:

(a) The Higgs $\mathscr{O}_{\mathfrak{X}}[\frac{1}{p}]$-module $(\mathscr{M}, \theta)$ is nilpotent in the sense of 4.2.5.

(b) There exists a finite decreasing filtration $(M_i)_{0\le i\le n}$ of M by sub-$A[\frac{1}{p}]$-modules such that $M_0 = M$, $M_n = 0$ and for every $0 \le i \le n-1$, we have

$$\theta(M_i) \subset \widetilde{\xi}^{-1}\widetilde{\Omega}^1_{\mathfrak{X}/\mathscr{S}}(\mathfrak{X}) \otimes_A M_{i+1}. \qquad (4.2.6.1)$$

(c) For every $s \in M$, there exists an integer $n \ge 1$ such that for all sections $d_1, \ldots, d_n \in T$, $\theta_{d_n} \circ \cdots \circ \theta_{d_1}(s) = 0$.

(d) The Higgs $A[\frac{1}{p}]$-module (M, θ) is nilpotent in the sense of 2.5.6.

Indeed, the ring $A[\frac{1}{p}]$ being Noetherian ([1] 1.10.12(i)), conditions (b) and 4.2.4(iii) are equivalent by virtue of ([1] 2.7.2.3 and 2.10.24). On the other hand, conditions (b), (c) and (d) are equivalent by 2.5.7.

Remark 4.2.7 Under the assumptions of 4.2.6, the Higgs $\mathscr{O}_{\mathfrak{X}}[\frac{1}{p}]$-module $(\mathscr{M}, \theta)$ is nilpotent if and only if the following condition is satisfied:

(ii') For any open set U of $\mathfrak{X}$, there exists an integer $n \ge 1$ such that for all sections $s \in \Gamma(U, \mathscr{M})$ and $d_1, \ldots, d_n \in \Gamma(U, \mathscr{T})$, we have $\theta_{d_n} \circ \cdots \circ \theta_{d_1}(s) = 0$.

Moreover, it suffices that it holds for open affine subschemes U covering $\mathfrak{X}$.

Lemma 4.2.8 *Let* $(\mathscr{M}, \theta)$ *be a Higgs* $\mathscr{O}_{\mathfrak{X}}[\frac{1}{p}]$*-module with coefficients in* $\widetilde{\xi}^{-1}\widetilde{\Omega}^1_{\mathfrak{X}/\mathscr{S}}$, $\mathscr{N}$ *(resp.* $\mathscr{P}$*) a subobject (resp. quotient) of* $(\mathscr{M}, \theta)$ *in* $\mathbf{HM}^{\mathrm{coh}}(\mathscr{O}_{\mathfrak{X}}[\frac{1}{p}], \widetilde{\xi}^{-1}\widetilde{\Omega}^1_{\mathfrak{X}/\mathscr{S}})$, $(\mathrm{F}^i\mathscr{M})_{i\in\mathbb{Z}}$ *a decreasing filtration of* $(\mathscr{M}, \theta)$ *in* $\mathbf{HM}^{\mathrm{coh}}(\mathscr{O}_{\mathfrak{X}}[\frac{1}{p}], \widetilde{\xi}^{-1}\widetilde{\Omega}^1_{\mathfrak{X}/\mathscr{S}})$ (2.5.2) *such that* $\mathrm{F}^n\mathscr{M} = 0$, $\mathrm{F}^m\mathscr{M} = \mathscr{M}$ *for two integers* $m \le n$. *Then,*

(i) *If* $(\mathscr{M},\theta)$ *is nilpotent, so are* $\mathscr{N}$ *and* $\mathscr{P}$.
(ii) *The Higgs module* $(\mathscr{M},\theta)$ *is nilpotent if and only if the graded Higgs module* $\mathrm{Gr}^{\bullet}_{\mathrm{F}}\mathscr{M}$ *of the filtration* $\mathrm{F}^{\bullet}\mathscr{M}$ *is nilpotent.*

It suffices to copy the proof of 2.5.4.

Definition 4.2.9 Let ε be a rational number > 0, $(\mathscr{N},\theta)$ a Higgs $\mathscr{O}_{\mathfrak{X}}$-module with coefficients in $\widetilde{\xi}^{-1}\widetilde{\Omega}^1_{\mathfrak{X}/\mathscr{S}}$.

(i) We say that $(\mathscr{N},\theta)$ is *ε-quasi-small* if the $\mathscr{O}_{\mathfrak{X}}$-module $\mathscr{N}$ is coherent and if θ is a multiple of p^{ε} as a section of $\widetilde{\xi}^{-1}\mathscr{E}nd_{\mathscr{O}_{\mathfrak{X}}}(\mathscr{N})\otimes_R\widetilde{\Omega}^1_{\mathfrak{X}/\mathscr{S}}$ (2.5.1.10). We then also say that the Higgs $\mathscr{O}_{\mathfrak{X}}$-field θ is *ε-quasi-small.*
(ii) We say that $(\mathscr{N},\theta)$ is *quasi-small* if it is ε'-quasi-small for a rational number $\varepsilon' > \frac{1}{p-1}$. We then also say that the Higgs $\mathscr{O}_{\mathfrak{X}}$-field θ is *quasi-small.*

We denote by $\mathbf{HM}^{\varepsilon\text{-qsf}}(\mathscr{O}_{\mathfrak{X}},\widetilde{\xi}^{-1}\widetilde{\Omega}^1_{\mathfrak{X}/\mathscr{S}})$ (resp. $\mathbf{HM}^{\mathrm{qsf}}(\mathscr{O}_{\mathfrak{X}},\widetilde{\xi}^{-1}\widetilde{\Omega}^1_{\mathfrak{X}/\mathscr{S}})$) the full subcategory of $\mathbf{HM}(\mathscr{O}_{\mathfrak{X}},\widetilde{\xi}^{-1}\widetilde{\Omega}^1_{\mathfrak{X}/\mathscr{S}})$ made up of ε-quasi-small (resp. quasi-small) Higgs $\mathscr{O}_{\mathfrak{X}}$-modules whose underlying $\mathscr{O}_{\mathfrak{X}}$-module is $\mathscr{S}$-flat.

Remark 4.2.10 Let ε be a rational number > 0, $\mathscr{N}$ a coherent $\mathscr{O}_{\mathfrak{X}}$-module that is $\mathscr{S}$-flat, θ a Higgs $\mathscr{O}_{\mathfrak{X}}$-field with coefficients in $\widetilde{\xi}^{-1}\widetilde{\Omega}^1_{\mathfrak{X}/\mathscr{S}}$. The Higgs module $(\mathscr{N},\theta)$ is ε-quasi-small if and only if

$$\theta(\mathscr{N})\subset p^{\varepsilon}\widetilde{\xi}^{-1}\widetilde{\Omega}^1_{\mathfrak{X}/\mathscr{S}}\otimes_{\mathscr{O}_{\mathfrak{X}}}\mathscr{N}. \tag{4.2.10.1}$$

Definition 4.2.11 Let $(\mathscr{N},\theta)$ be a Higgs $\mathscr{O}_{\mathfrak{X}}[\frac{1}{p}]$-bundle with coefficients in $\widetilde{\xi}^{-1}\widetilde{\Omega}^1_{\mathfrak{X}/\mathscr{S}}$ (4.2.3).

(i) We say that $(\mathscr{N},\theta)$ is *small* if there exist a coherent sub-$\mathscr{O}_{\mathfrak{X}}$-module $\mathfrak{N}$ of $\mathscr{N}$ that generates it over $\mathscr{O}_{\mathfrak{X}}[\frac{1}{p}]$ and a rational number $\varepsilon > \frac{1}{p-1}$ such that

$$\theta(\mathfrak{N})\subset p^{\varepsilon}\widetilde{\xi}^{-1}\widetilde{\Omega}^1_{\mathfrak{X}/\mathscr{S}}\otimes_{\mathscr{O}_{\mathfrak{X}}}\mathfrak{N}. \tag{4.2.11.1}$$

(ii) We say that $(\mathscr{N},\theta)$ is *locally small* if there exists an open covering $(U_i)_{i\in I}$ of X_s such that for every $i\in I$, $(\mathscr{N}|U_i,\theta|U_i)$ is small.

We denote by $\mathbf{HM}^{\mathrm{s}}(\mathscr{O}_{\mathfrak{X}}[\frac{1}{p}],\widetilde{\xi}^{-1}\widetilde{\Omega}^1_{\mathfrak{X}/\mathscr{S}})$ (resp. $\mathbf{HM}^{\mathrm{locs}}(\mathscr{O}_{\mathfrak{X}}[\frac{1}{p}],\widetilde{\xi}^{-1}\widetilde{\Omega}^1_{\mathfrak{X}/\mathscr{S}})$) the full subcategory of $\mathbf{HM}(\mathscr{O}_{\mathfrak{X}}[\frac{1}{p}],\widetilde{\xi}^{-1}\widetilde{\Omega}^1_{\mathfrak{X}/\mathscr{S}})$ made up of small (resp. locally small) Higgs $\mathscr{O}_{\mathfrak{X}}[\frac{1}{p}]$-bundles.

Remark 4.2.12 Assume that X is affine and the $\mathscr{O}_X$-module $\widetilde{\Omega}^1_{X/S}$ is free of finite type. We set $R=\Gamma(X,\mathscr{O}_X)$ and $R_1=R\otimes_{\mathscr{O}_K}\mathscr{O}_{\overline{K}}$ and denote by $\widehat{R}$ and $\widehat{R_1}$ their p-adic Hausdorff completions. Let $(\mathscr{N},\theta)$ be a Higgs $\mathscr{O}_{\mathfrak{X}}[\frac{1}{p}]$-bundle with coefficients in $\widetilde{\xi}^{-1}\widetilde{\Omega}^1_{\mathfrak{X}/\mathscr{S}}$. We set $N=\Gamma(\mathfrak{X},\mathscr{N})$, which is a projective $\widehat{R_1}[\frac{1}{p}]$-module of finite type by ([3] III.6.17), and we denote by θ the Higgs $\widehat{R_1}[\frac{1}{p}]$-field over N with coefficients in $\widetilde{\xi}^{-1}\widetilde{\Omega}^1_{X/S}(X)$ induced by θ ([1] 2.7.2.3 and 2.10.24). The Higgs $\mathscr{O}_{\mathfrak{X}}[\frac{1}{p}]$-bundle

$(\mathscr{N}, \theta)$ is small in the sense of 4.2.11 if and only if the Higgs $\widehat{R_1}[\frac{1}{p}]$-module (N, θ) is small in the sense of 3.4.23. Indeed, the condition is necessary by virtue of ([1] (2.10.5.1)) and it is sufficient in view of ([1] 1.10.2).

Proposition 4.2.13 *Suppose that X is affine, that X_s is non-empty and that f* (4.1.3) *admits an adequate chart* (3.1.15). *Then, every nilpotent Higgs $\mathscr{O}_{\mathfrak{X}}[\frac{1}{p}]$-bundle* (4.2.5) *with coefficients in $\widetilde{\xi}^{-1}\widetilde{\Omega}^1_{\mathfrak{X}/\mathscr{S}}$ is small* (4.2.11).

We set $R = \Gamma(X, \mathscr{O}_X)$ and $R_1 = R \otimes_{\mathscr{O}_K} \mathscr{O}_{\overline{K}}$ and we denote by $\widehat{R_1}$ the p-adic Hausdorff completion of R_1. Let $(\mathscr{N}, \theta)$ be a nilpotent Higgs $\mathscr{O}_{\mathfrak{X}}[\frac{1}{p}]$-bundle with coefficients in $\widetilde{\xi}^{-1}\widetilde{\Omega}^1_{\mathfrak{X}/\mathscr{S}}$. We set $N = \Gamma(\mathfrak{X}, \mathscr{N})$, which is a projective $\widehat{R_1}[\frac{1}{p}]$-module of finite type by ([3] III.6.17), and we denote by θ the Higgs $\widehat{R_1}[\frac{1}{p}]$-field over N with coefficients in $\widetilde{\xi}^{-1}\widetilde{\Omega}^1_{X/S}(X)$ induced by θ. By 4.2.6, the Higgs $\widehat{R_1}[\frac{1}{p}]$-module (N, θ) is nilpotent in the sense of 2.5.6. It is therefore small in the sense of 3.4.23, by virtue of 3.5.2(i). Consequently, the Higgs $\mathscr{O}_{\mathfrak{X}}[\frac{1}{p}]$-bundle $(\mathscr{N}, \theta)$ is small by 4.2.12.

Corollary 4.2.14 *Every nilpotent Higgs $\mathscr{O}_{\mathfrak{X}}[\frac{1}{p}]$-bundle* (4.2.5) *with coefficients in $\widetilde{\xi}^{-1}\widetilde{\Omega}^1_{\mathfrak{X}/\mathscr{S}}$ is locally small* (4.2.11).

4.2.15 We denote by $\mathbf{HM}'^{\mathrm{qsf}}(\mathscr{O}_{\mathfrak{X}}, \widetilde{\xi}^{-1}\widetilde{\Omega}^1_{\mathfrak{X}/\mathscr{S}})$ the full subcategory of the category $\mathbf{HM}^{\mathrm{qsf}}(\mathscr{O}_{\mathfrak{X}}, \widetilde{\xi}^{-1}\widetilde{\Omega}^1_{\mathfrak{X}/\mathscr{S}})$ made up of Higgs $\mathscr{O}_{\mathfrak{X}}$-modules $(\mathscr{N}, \theta)$ such that the $\mathscr{O}_{\mathfrak{X}}[\frac{1}{p}]$-module $\mathscr{N}_{\mathbb{Q}_p}$ is locally projective of finite type. It is an additive category. We denote by $\mathbf{HM}'^{\mathrm{qsf}}_{\mathbb{Q}}(\mathscr{O}_{\mathfrak{X}}, \widetilde{\xi}^{-1}\widetilde{\Omega}^1_{\mathfrak{X}/\mathscr{S}})$ the category of objects of $\mathbf{HM}'^{\mathrm{qsf}}(\mathscr{O}_{\mathfrak{X}}, \widetilde{\xi}^{-1}\widetilde{\Omega}^1_{\mathfrak{X}/\mathscr{S}})$ up to isogeny ([3] III.6.1.1). The functor

$$\begin{array}{ccc} \mathbf{HM}'^{\mathrm{qsf}}(\mathscr{O}_{\mathfrak{X}}, \widetilde{\xi}^{-1}\widetilde{\Omega}^1_{\mathfrak{X}/\mathscr{S}}) & \rightarrow & \mathbf{HM}^{\mathrm{s}}(\mathscr{O}_{\mathfrak{X}}[\frac{1}{p}], \widetilde{\xi}^{-1}\widetilde{\Omega}^1_{\mathfrak{X}/\mathscr{S}}) \\ (\mathscr{N}, \theta) & \mapsto & (\mathscr{N}_{\mathbb{Q}_p}, \theta) \end{array} \tag{4.2.15.1}$$

induces a functor

$$\mathbf{HM}'^{\mathrm{qsf}}_{\mathbb{Q}}(\mathscr{O}_{\mathfrak{X}}, \widetilde{\xi}^{-1}\widetilde{\Omega}^1_{\mathfrak{X}/\mathscr{S}}) \rightarrow \mathbf{HM}^{\mathrm{s}}(\mathscr{O}_{\mathfrak{X}}[\frac{1}{p}], \widetilde{\xi}^{-1}\widetilde{\Omega}^1_{\mathfrak{X}/\mathscr{S}}). \tag{4.2.15.2}$$

Lemma 4.2.16 *The functor* (4.2.15.2) *is an equivalence of categories.*

Indeed, it immediately follows from the definitions that this functor is essentially surjective. Let $\mathscr{M}, \mathscr{N}$ be two coherent $\mathscr{O}_{\mathfrak{X}}$-modules that are $\mathscr{S}$-flat. By (4.2.1.2), the canonical morphism

$$\mathrm{Hom}_{\mathscr{O}_{\mathfrak{X}}}(\mathscr{M}, \mathscr{N}) \otimes_{\mathbb{Z}_p} \mathbb{Q}_p \rightarrow \mathrm{Hom}_{\mathscr{O}_{\mathfrak{X}}[\frac{1}{p}]}(\mathscr{M} \otimes_{\mathbb{Z}_p} \mathbb{Q}_p, \mathscr{N} \otimes_{\mathbb{Z}_p} \mathbb{Q}_p) \tag{4.2.16.1}$$

is an isomorphism. Suppose that $\mathscr{M}$ and $\mathscr{N}$ are endowed with Higgs $\mathscr{O}_{\mathfrak{X}}$-fields with coefficients in $\widetilde{\xi}^{-1}\widetilde{\Omega}^1_{\mathfrak{X}/\mathscr{S}}$. The canonical morphism

$$\mathrm{Hom}_{\mathbf{HM}}(\mathscr{M}, \mathscr{N}) \otimes_{\mathbb{Z}_p} \mathbb{Q}_p \rightarrow \mathrm{Hom}_{\mathbf{HM}}(\mathscr{M} \otimes_{\mathbb{Z}_p} \mathbb{Q}_p, \mathscr{N} \otimes_{\mathbb{Z}_p} \mathbb{Q}_p), \tag{4.2.16.2}$$

where the source (resp. the target) denotes the module of morphisms of Higgs $\mathscr{O}_{\mathfrak{X}}$-modules (resp. $\mathscr{O}_{\mathfrak{X}}[\frac{1}{p}]$-modules), is then an isomorphism: the injectivity immediately follows from that of (4.2.16.1) and the surjectivity from that of (4.2.16.1) and from the facts that $\mathscr{M}$ and $\mathscr{N}$ are $\mathscr{S}$-flat and that $\widetilde{\Omega}^1_{\mathfrak{X}/\mathscr{S}}$ is $\mathfrak{X}$-flat. The proposition follows.

4.3 Faltings Topos

4.3.1 We denote by $\mathscr{R}$ the category of finite étale morphisms (i.e., the full subcategory of the category of morphisms of schemes, made up of finite étale morphisms) and by

$$\mathscr{R} \to \mathbf{Sch} \tag{4.3.1.1}$$

the "target functor", which makes $\mathscr{R}$ into a cleaved and normalized fibered category over **Sch** (2.1.7) ([24] VI): the fiber category over a scheme X is canonically equivalent to the category $\mathbf{\acute{E}t}_{\mathrm{f}/X}$ of finite étale schemes over X, and for any morphism of schemes $f: Y \to X$, the inverse image functor $f^+: \mathbf{\acute{E}t}_{\mathrm{f}/X} \to \mathbf{\acute{E}t}_{\mathrm{f}/Y}$ is none other than the base change functor by f (2.1.12). We consider $\mathscr{R}/\mathbf{Sch}$ as a fibered $\mathbb{U}$-site by endowing each fiber with the étale topology ([5] VI 7.2.1).

4.3.2 With the notation of 2.1.12, we denote by

$$\pi: E \to \mathbf{\acute{E}t}_{/X} \tag{4.3.2.1}$$

the Faltings fibered $\mathbb{U}$-site associated with the morphism $h: \overline{X}^{\circ} \to X$ (4.1.3.1) ([3] VI.10.1), i.e., the fibered $\mathbb{U}$-site deduced from the fibered site of finite étale morphisms $\mathscr{R}/\mathbf{Sch}$ (4.3.1.1) by base change by the functor

$$\mathbf{\acute{E}t}_{/X} \to \mathbf{Sch}, \quad U \mapsto \overline{U}^{\circ} = U \times_X \overline{X}^{\circ}. \tag{4.3.2.2}$$

For any $U \in \mathrm{Ob}(\mathbf{\acute{E}t}_{/X})$, we denote by

$$\iota_U: \mathbf{\acute{E}t}_{\mathrm{f}/\overline{U}^{\circ}} \to E \tag{4.3.2.3}$$

the canonical functor. This functor was denoted by $\alpha_{U!}$ in ([3] (VI.5.1.2)).

We endow E with the covanishing topology associated with π ([3] VI.5.3), that is, with the topology generated by the coverings $\{(V_i \to U_i) \to (V \to U)\}_{i \in I}$ of the following two types:

(v) $U_i = U$ for every $i \in I$, and $(V_i \to V)_{i \in I}$ is an étale covering.
(c) $(U_i \to U)_{i \in I}$ is an étale covering and $V_i = U_i \times_U V$ for every $i \in I$.

The resulting covanishing site E is also called the *Faltings site* associated with h; it is a $\mathbb{U}$-site. We denote by $\widehat{E}$ (resp. $\widetilde{E}$) the category of presheaves (resp. the topos of sheaves) of $\mathbb{U}$-sets on E. We call $\widetilde{E}$ the *Faltings topos* associated with h ([3] VI.10.1).

We denote by

$$\mathfrak{F} \to \mathbf{\acute{E}t}_{/X} \tag{4.3.2.4}$$

the fibered $\mathbb{U}$-topos associated with π. The fiber category of $\mathfrak{F}$ over any $U \in \mathrm{Ob}(\mathbf{\acute{E}t}_{/X})$ is canonically equivalent to the finite étale topos $\overline{U}^{\circ}_{\mathrm{fét}}$ of $\overline{U}^{\circ}$ and the pullback functor for any morphism $\mu\colon U' \to U$ of $\mathbf{\acute{E}t}_{/X}$ is identified with the functor $\overline{\mu}^{\circ *}_{\mathrm{fét}}\colon \overline{U}^{\circ}_{\mathrm{fét}} \to \overline{U}'^{\circ}_{\mathrm{fét}}$ pullback by the morphism of topos $\overline{\mu}^{\circ}_{\mathrm{fét}}\colon \overline{U}'^{\circ}_{\mathrm{fét}} \to \overline{U}^{\circ}_{\mathrm{fét}}$ ([3] VI.9.3). We denote by

$$\mathfrak{F}^{\vee} \to (\mathbf{\acute{E}t}_{/X})^{\circ} \tag{4.3.2.5}$$

the fibered category obtained by associating with any $U \in \mathrm{Ob}(\mathbf{\acute{E}t}_{/X})$ the category $\overline{U}^{\circ}_{\mathrm{fét}}$, and with any morphism $\mu\colon U' \to U$ of $\mathbf{\acute{E}t}_{/X}$ the functor $\overline{\mu}^{\circ}_{\mathrm{fét}*}\colon \overline{U}'^{\circ}_{\mathrm{fét}} \to \overline{U}^{\circ}_{\mathrm{fét}}$ direct image by the morphism of topos $\overline{\mu}^{\circ}_{\mathrm{fét}}$. We denote by

$$\mathscr{P}^{\vee} \to (\mathbf{\acute{E}t}_{/X})^{\circ} \tag{4.3.2.6}$$

the fibered category obtained by associating with any $U \in \mathrm{Ob}(\mathbf{\acute{E}t}_{/X})$ the category $(\mathbf{\acute{E}t}_{\mathrm{f}/\overline{U}^{\circ}})^{\wedge}$ of presheaves of $\mathbb{U}$-sets on $\mathbf{\acute{E}t}_{\mathrm{f}/\overline{U}^{\circ}}$, and with any morphism $\mu\colon U' \to U$ of $\mathbf{\acute{E}t}_{/X}$ the functor

$$\overline{\mu}^{\circ}_{\mathrm{fét}*}\colon (\mathbf{\acute{E}t}_{\mathrm{f}/\overline{U}'^{\circ}})^{\wedge} \to (\mathbf{\acute{E}t}_{\mathrm{f}/\overline{U}^{\circ}})^{\wedge} \tag{4.3.2.7}$$

obtained by composing with the inverse image functor $\overline{\mu}^{\circ +}\colon \mathbf{\acute{E}t}_{\mathrm{f}/\overline{U}^{\circ}} \to \mathbf{\acute{E}t}_{\mathrm{f}/\overline{U}'^{\circ}}$.

We have an equivalence of categories ([3] VI.5.2)

$$\begin{aligned} \widehat{E} &\to \mathbf{Hom}_{(\mathbf{\acute{E}t}_{/X})^{\circ}}((\mathbf{\acute{E}t}_{/X})^{\circ}, \mathscr{P}^{\vee}) \\ F &\mapsto \{U \mapsto F \circ \iota_U\}. \end{aligned} \tag{4.3.2.8}$$

From now on, we will identify F with the section $\{U \mapsto F \circ \iota_U\}$ that is associated with it by this equivalence.

By ([3] VI.5.11), the functor (4.3.2.8) induces a fully faithful functor

$$\widetilde{E} \to \mathbf{Hom}_{(\mathbf{\acute{E}t}_{/X})^{\circ}}((\mathbf{\acute{E}t}_{/X})^{\circ}, \mathfrak{F}^{\vee}), \tag{4.3.2.9}$$

whose essential image is made up of the sections $\{U \mapsto F_U\}$ satisfying a gluing condition.

The functors

$$\sigma^{+}\colon \mathbf{\acute{E}t}_{/X} \to E,\ U \mapsto (\overline{U}^{\circ} \to U), \tag{4.3.2.10}$$

$$\iota_X\colon \mathbf{\acute{E}t}_{\mathrm{f}/\overline{X}^{\circ}} \to E,\ V \mapsto (V \to X), \tag{4.3.2.11}$$

are continuous and left exact ([3] VI.10.6). Hence they define two morphisms of topos

$$\sigma\colon \widetilde{E} \to X_{\mathrm{\acute{e}t}}, \tag{4.3.2.12}$$

$$\beta\colon \widetilde{E} \to \overline{X}^{\circ}_{\mathrm{fét}}. \tag{4.3.2.13}$$

For every sheaf $F = \{U \mapsto F_U\}$ on E, we have $\beta_*(F) = F_X$.

The functor

$$\psi^+ \colon E \to \mathbf{\acute{E}t}_{/\overline{X}^\circ}, \quad (V \to U) \mapsto V \tag{4.3.2.14}$$

is continuous and left exact ([3] VI.10.7). It therefore defines a morphism of topos

$$\psi \colon \overline{X}^\circ_{\text{ét}} \to \widetilde{E}. \tag{4.3.2.15}$$

We change here the notation compared to *loc. cit.*

4.3.3 Since X_η is an open object of $X_{\text{ét}}$, i.e., a subobject of the final object X ([5] IV 8.3), $\sigma^*(X_\eta)$ is an open object of $\widetilde{E}$. We denote by

$$\gamma \colon \widetilde{E}_{/\sigma^*(X_\eta)} \to \widetilde{E} \tag{4.3.3.1}$$

the localization morphism of $\widetilde{E}$ at $\sigma^*(X_\eta)$ ([5] IV 5.2). We denote by $\widetilde{E}_s$ the closed subtopos of $\widetilde{E}$ complement of the open object $\sigma^*(X_\eta)$, that is, the full subcategory of $\widetilde{E}$ made up of the sheaves F such that $\gamma^*(F)$ is a final object of $\widetilde{E}_{/\sigma^*(X_\eta)}$ ([5] IV 9.3.5), by

$$\delta \colon \widetilde{E}_s \to \widetilde{E} \tag{4.3.3.2}$$

the canonical embedding and by

$$\sigma_s \colon \widetilde{E}_s \to X_{s,\text{ét}} \tag{4.3.3.3}$$

the morphism of topos induced by σ ([3] (III.9.8.3)). The diagram of morphisms of topos

$$\begin{array}{ccc} \widetilde{E}_s & \xrightarrow{\sigma_s} & X_{s,\text{ét}} \\ {\scriptstyle\delta}\downarrow & & \downarrow{\scriptstyle a} \\ \widetilde{E} & \xrightarrow{\sigma} & X_{\text{ét}} \end{array} \tag{4.3.3.4}$$

is commutative up to isomorphism.

4.3.4 We denote by $X_{\text{ét}} \overleftarrow{\times}_{X_{\text{ét}}} \overline{X}^\circ_{\text{ét}}$ the covanishing topos of the morphism $f_{\text{ét}} \colon \overline{X}^\circ_{\text{ét}} \to X_{\text{ét}}$ ([3] VI.3.12) and by

$$\rho \colon X_{\text{ét}} \overleftarrow{\times}_{X_{\text{ét}}} \overline{X}^\circ_{\text{ét}} \to \widetilde{E} \tag{4.3.4.1}$$

the canonical morphism ([3] VI.10.15). By ([3] VI.4.20) and ([5] VIII 7.9), giving a point of $X_{\text{ét}} \overleftarrow{\times}_{X_{\text{ét}}} \overline{X}^\circ_{\text{ét}}$ is equivalent to giving a pair of geometric points $\overline{x}$ of X and $\overline{y}$ of $\overline{X}^\circ$ and a specialization morphism from $\hbar(\overline{y})$ to $\overline{x}$, that is, an X-morphism $\overline{y} \to X_{(\overline{x})}$, where $X_{(\overline{x})}$ denotes the strict localization of X in $\overline{x}$. Such a point will be denoted by $(\overline{y} \rightsquigarrow \overline{x})$, and its image by ρ will be denoted by $\rho(\overline{y} \rightsquigarrow \overline{x})$, which is therefore a point of $\widetilde{E}$. The family of fiber functors of $\widetilde{E}$ associated with these points is conservative by ([3] VI.10.21).

4.3.5 Let $\overline{x}$ be a geometric point of X, $\underline{X}$ the strict localization of X in $\overline{x}$. We set $\underline{\overline{X}} = \underline{X} \times_S \overline{S}$ (4.1.1.2) and $\underline{\overline{X}}^{\circ} = \underline{\overline{X}} \times_X X^{\circ}$ (4.1.3.1). We denote by $\underline{\widetilde{E}}$ the Faltings topos associated with the canonical morphism $\underline{\overline{X}}^{\circ} \to \underline{X}$, by

$$\Phi \colon \underline{\widetilde{E}} \to \widetilde{E} \tag{4.3.5.1}$$

the functoriality morphism induced by the canonical morphism $\underline{X} \to X$, by

$$\underline{\beta} \colon \underline{\widetilde{E}} \to \underline{\overline{X}}^{\circ}_{\text{fét}} \tag{4.3.5.2}$$

the canonical morphism (4.3.2.13) and by

$$\theta \colon \underline{\overline{X}}^{\circ}_{\text{fét}} \to \underline{\widetilde{E}} \tag{4.3.5.3}$$

the canonical section of $\underline{\beta}$ defined in ([3] VI.10.23). We denote by

$$\varphi_{\overline{x}} \colon \widetilde{E} \to \underline{\overline{X}}^{\circ}_{\text{fét}} \tag{4.3.5.4}$$

the composed functor $\theta^* \circ \Phi^*$ ([3] VI.10.29).

The composed functor $\varphi_{\overline{x}} \circ \beta^*$ is canonically isomorphic to the pullback functor by the canonical morphism $\underline{\overline{X}}^{\circ}_{\text{fét}} \to \overline{X}^{\circ}_{\text{fét}}$, by ([3] (VI.10.24.3) and (VI.10.12.6)). For every object F of $X_{\text{ét}}$, we have a canonical and functorial isomorphism

$$\varphi_{\overline{x}}(\sigma^*(F)) \xrightarrow{\sim} F_{\overline{x}}, \tag{4.3.5.5}$$

whose target is the constant sheaf of $\underline{\overline{X}}^{\circ}_{\text{fét}}$ with value $F_{\overline{x}}$, by ([3] VI.10.24 and (VI.10.12.6)).

We denote by $\mathfrak{V}_{\overline{x}}$ the category of $\overline{x}$-pointed étale X-schemes, or equivalently, the category of neighborhoods of the point of $X_{\text{ét}}$ associated with $\overline{x}$ in the site $\mathbf{Ét}_{/X}$ ([5] IV 6.8.2 and VIII 3.9). For any object $(U, \mathfrak{p} \colon \overline{x} \to U)$ of $\mathfrak{V}_{\overline{x}}$, we also denote by $\mathfrak{p} \colon \underline{X} \to U$ the morphism deduced from $\mathfrak{p}$ ([5] VIII 7.3) and we set

$$\overline{\mathfrak{p}}^{\circ} = \mathfrak{p} \times_X \overline{X}^{\circ} \colon \underline{\overline{X}}^{\circ} \to \overline{U}^{\circ}. \tag{4.3.5.6}$$

For any object $F = \{U \mapsto F_U\}$ of $\widetilde{E}$ (4.3.2.8), we denote by F^{a} the sheaf of $\widetilde{E}$ associated with F, and for any $U \in \text{Ob}(\mathbf{Ét}_{/X})$, by F^{a}_U the sheaf of $\overline{U}^{\circ}_{\text{fét}}$ associated with F_U. By ([3] VI.10.37), we have a canonical and functorial isomorphism

$$\varphi_{\overline{x}}(F^{\text{a}}) \xrightarrow{\sim} \lim_{\substack{\longrightarrow \\ (U,\mathfrak{p}) \in \mathfrak{V}^{\circ}_{\overline{x}}}} (\overline{\mathfrak{p}}^{\circ})^*_{\text{fét}}(F^{\text{a}}_U). \tag{4.3.5.7}$$

By virtue of ([3] VI.10.30), for every abelian group F of $\widetilde{E}$ and every $q \geq 0$, we have a canonical and functorial isomorphism

$$\mathrm{R}^q \sigma_*(F)_{\overline{x}} \xrightarrow{\sim} \mathrm{H}^q(\underline{\overline{X}}^{\circ}_{\text{fét}}, \varphi_{\overline{x}}(F)). \tag{4.3.5.8}$$

4.3.6 The scheme $\overline{X}$ is normal and locally irreducible by ([3] III.4.2(iii)). Moreover, the immersion $j\colon X^\circ \to X$ is quasi-compact since X is Noetherian. For any object $(V \to U)$ of E, we denote by $\overline{U}^V$ the integral closure of $\overline{U}$ in V. For every morphism $(V' \to U') \to (V \to U)$ of E, we have a canonical morphism $\overline{U}'^{V'} \to \overline{U}^V$ which fits into a commutative diagram

$$\begin{array}{ccccccc} V' & \longrightarrow & \overline{U}'^{V'} & \longrightarrow & \overline{U}' & \longrightarrow & U' \\ \downarrow & & \downarrow & & \downarrow & & \downarrow \\ V & \longrightarrow & \overline{U}^V & \longrightarrow & \overline{U} & \longrightarrow & U. \end{array} \tag{4.3.6.1}$$

We denote by $\overline{\mathscr{B}}$ the presheaf on E defined for any $(V \to U) \in \mathrm{Ob}(E)$, by

$$\overline{\mathscr{B}}((V \to U)) = \Gamma(\overline{U}^V, \mathscr{O}_{\overline{U}^V}). \tag{4.3.6.2}$$

It is a sheaf for the covanishing topology on E by virtue of ([3] III.8.16). For any $U \in \mathrm{Ob}(\mathbf{\acute{E}t}_{/X})$, we set

$$\overline{\mathscr{B}}_U = \overline{\mathscr{B}} \circ \iota_U. \tag{4.3.6.3}$$

By ([3] III.8.17), we have a canonical homomorphism

$$\sigma^*(\hbar_*(\mathscr{O}_{\overline{X}})) \to \overline{\mathscr{B}}. \tag{4.3.6.4}$$

Unless explicitly stated otherwise, we consider σ (4.3.2.12) as a morphism of ringed topos

$$\sigma\colon (\widetilde{E}, \overline{\mathscr{B}}) \to (X_{\text{ét}}, \hbar_*(\mathscr{O}_{\overline{X}})). \tag{4.3.6.5}$$

We denote again by $\mathscr{O}_{\overline{K}}$ the constant sheaf of $\overline{X}^\circ_{\text{fét}}$ with value $\mathscr{O}_{\overline{K}}$. We will also consider β (4.3.2.13) as a morphism of ringed topos

$$\beta\colon (\widetilde{E}, \overline{\mathscr{B}}) \to (\overline{X}^\circ_{\text{fét}}, \mathscr{O}_{\overline{K}}). \tag{4.3.6.6}$$

4.3.7 Let $U \in \mathrm{Ob}(\mathbf{\acute{E}t}_{/X})$, $\overline{y}$ be a geometric point of $\overline{U}^\circ$ (4.1.3.1). The scheme $\overline{U}$ being locally irreducible by ([3] III.3.3 and III.4.2(iii)), it is the sum of the schemes induced on its irreducible components. We denote by $\overline{U}^\star$ the irreducible component of $\overline{U}$ containing $\overline{y}$. Similarly, $\overline{U}^\circ$ is the sum of the schemes induced on its irreducible components and $\overline{U}^{\star\circ} = \overline{U}^\star \times_X X^\circ$ is the irreducible component of $\overline{U}^\circ$ containing $\overline{y}$. We denote by $\mathbf{B}_{\pi_1(\overline{U}^{\star\circ}, \overline{y})}$ the classifying topos of the profinite group $\pi_1(\overline{U}^{\star\circ}, \overline{y})$ and by

$$\nu_{\overline{y}}\colon \overline{U}^{\star\circ}_{\text{fét}} \xrightarrow{\sim} \mathbf{B}_{\pi_1(\overline{U}^{\star\circ}, \overline{y})} \tag{4.3.7.1}$$

the fiber functor of $\overline{U}^{\star\circ}_{\text{fét}}$ at $\overline{y}$ ([3] VI.9.8). We set

$$\overline{R}^{\overline{y}}_U = \nu_{\overline{y}}(\overline{\mathscr{B}}_U | \overline{U}^{\star\circ}). \tag{4.3.7.2}$$

4.3.8 Let $(\overline{y} \rightsquigarrow \overline{x})$ be a point of $X_{\text{ét}} \overleftarrow{\times}_{X_{\text{ét}}} \overline{X}^{\circ}_{\text{ét}}$ (4.3.4), $\underline{X}$ the strict localization of X at $\overline{x}$, $\mathfrak{V}_{\overline{x}}$ the category of $\overline{x}$-pointed étale X-schemes (see 4.3.5). We set $\underline{\overline{X}} = \underline{X} \times_S \overline{S}$ (4.1.1.2) and $\underline{\overline{X}}^{\circ} = \underline{\overline{X}} \times_X X^{\circ}$ (4.1.3.1). The X-morphism $u \colon \overline{y} \to \underline{X}$ defining $(\overline{y} \rightsquigarrow \overline{x})$ lifts into an $\overline{X}^{\circ}$-morphism $v \colon \overline{y} \to \underline{\overline{X}}^{\circ}$ and therefore induces a geometric point of $\underline{\overline{X}}^{\circ}$ which we (abusively) denote also by $\overline{y}$. For any object $(U, \mathfrak{p} \colon \overline{x} \to U)$ of $\mathfrak{V}_{\overline{x}}$, we also denote by $\mathfrak{p} \colon \underline{X} \to U$ the morphism deduced from $\mathfrak{p}$. We (abusively) denote also by $\overline{y}$ the geometric point $\overline{\mathfrak{p}}^{\circ}(v(\overline{y}))$ of $\overline{U}^{\circ}$ (4.3.5.6).

For any object $F = \{U \mapsto F_U\}$ of $\widehat{E}$ (4.3.2.8), we denote by F^{a} the sheaf of $\widetilde{E}$ associated with F, and for any $U \in \mathrm{Ob}(\mathbf{\acute{E}t}_{/X})$, by F^{a}_U the sheaf of $\overline{U}^{\circ}_{\text{fét}}$ associated with F_U. By ([3] VI.10.36 and (VI.9.3.4)), we have a canonical and functorial isomorphism

$$(F^{\mathrm{a}})_{\rho(\overline{y} \rightsquigarrow \overline{x})} \xrightarrow{\sim} \varinjlim_{(U,\mathfrak{p}) \in \mathfrak{V}^{\circ}_{\overline{x}}} (F^{\mathrm{a}}_U)_{\rho_{\overline{U}^{\circ}}(\overline{y})}, \tag{4.3.8.1}$$

where ρ is the morphism (4.3.4.1) and $\rho_{\overline{U}^{\circ}} \colon \overline{U}^{\circ}_{\text{ét}} \to \overline{U}^{\circ}_{\text{fét}}$ is the canonical morphism (2.1.12.1). In view of (4.3.7.2) and ([3] VI.9.9), we deduce from this a canonical isomorphism of $\Gamma(\underline{\overline{X}}, \mathscr{O}_{\underline{\overline{X}}})$-algebras

$$\overline{\mathscr{B}}_{\rho(\overline{y} \rightsquigarrow \overline{x})} \xrightarrow{\sim} \varinjlim_{(U,\mathfrak{p}) \in \mathfrak{V}^{\circ}_{\overline{x}}} \overline{R}^{\overline{y}}_U. \tag{4.3.8.2}$$

4.3.9 We keep the assumptions and notation of 4.3.8; we suppose furthermore that $\overline{x}$ *is above* s. By ([3] III.3.7), $\underline{\overline{X}}$ is normal and strictly local (and in particular is integral). For any object $(U, \mathfrak{p} \colon \overline{x} \to U)$ of $\mathfrak{V}_{\overline{x}}$, we denote by $\overline{U}^{\star}$ the irreducible component of $\overline{U}$ containing $\overline{y}$ and we sèt $\overline{U}^{\star\circ} = \overline{U}^{\star} \times_X X^{\circ}$, which is the irreducible component of $\overline{U}^{\circ}$ containing $\overline{y}$ (see 4.3.7). The morphism $\overline{\mathfrak{p}}^{\circ} \colon \underline{\overline{X}}^{\circ} \to \overline{U}^{\circ}$ (4.3.5.6) factors therefore through $\overline{U}^{\star\circ}$. We denote by $\mathbf{B}_{\pi_1(\underline{\overline{X}}^{\circ}, \overline{y})}$ the classifying topos of the profinite group $\pi_1(\underline{\overline{X}}^{\circ}, \overline{y})$ and by

$$\nu_{\overline{y}} \colon \underline{\overline{X}}^{\circ}_{\text{fét}} \xrightarrow{\sim} \mathbf{B}_{\pi_1(\underline{\overline{X}}^{\circ}, \overline{y})} \tag{4.3.9.1}$$

the fiber functor of $\underline{\overline{X}}^{\circ}_{\text{fét}}$ at $\overline{y}$ ([3] (VI.9.8.4)). By ([3] VI.10.31 and VI.9.9), the composed functor

$$\widetilde{E} \xrightarrow{\varphi_{\overline{x}}} \underline{\overline{X}}^{\circ}_{\text{fét}} \xrightarrow{\nu_{\overline{y}}} \mathbf{B}_{\pi_1(\underline{\overline{X}}^{\circ}, \overline{y})} \longrightarrow \mathbf{Sets}, \tag{4.3.9.2}$$

where $\varphi_{\overline{x}}$ is the canonical functor (4.3.5.4) and thè last arrow is the functor "forgetting the action of $\pi_1(\underline{\overline{X}}^{\circ}, \overline{y})$", is canonically isomorphic to the fiber functor associated with the point $\rho(\overline{y} \rightsquigarrow \overline{x})$ of $\widetilde{E}$ (4.3.4.1).

We define the $\Gamma(\underline{\overline{X}}, \mathscr{O}_{\underline{\overline{X}}})$-algebra $\overline{R}^{\overline{y}}_{\underline{X}}$ of $\mathbf{B}_{\pi_1(\underline{\overline{X}}^{\circ}, \overline{y})}$ by the formula

$$\overline{R}^{\overline{y}}_{\underline{X}} = \varinjlim_{(U,\mathfrak{p}) \in \mathfrak{V}^{\circ}_{\overline{x}}} \overline{R}^{\overline{y}}_U, \tag{4.3.9.3}$$

where we consider $\overline{R}_U^{\overline{y}}$ as a $\Gamma(\overline{U}, \mathscr{O}_{\overline{U}})$-algebra of $\mathbf{B}_{\pi_1(\overline{U}^{\star\circ}, \overline{y})}$ (4.3.7.2). The isomorphism (4.3.5.7) induces an isomorphism of $\Gamma(\underline{\overline{X}}, \mathscr{O}_{\underline{\overline{X}}})$-algebras of $\mathbf{B}_{\pi_1(\underline{\overline{X}}^\circ, \overline{y})}$

$$\nu_{\overline{y}}(\varphi_{\overline{x}}(\overline{\mathscr{B}})) \xrightarrow{\sim} \overline{R}_{\underline{X}}^{\overline{y}}, \tag{4.3.9.4}$$

whose underlying isomorphism of $\Gamma(\underline{\overline{X}}, \mathscr{O}_{\underline{\overline{X}}})$-algebras is (4.3.8.2).

4.3.10 For any integer $n \geq 0$, we set

$$\overline{\mathscr{B}}_n = \overline{\mathscr{B}}/p^n\overline{\mathscr{B}}. \tag{4.3.10.1}$$

By ([3] III.9.7), $\overline{\mathscr{B}}_n$ is a ring of $\widetilde{E}_s$. For any $U \in \mathrm{Ob}(\mathbf{\acute{E}t}_{/X})$, we set (4.3.6.3)

$$\overline{\mathscr{B}}_{U,n} = \overline{\mathscr{B}}_U/p^n\overline{\mathscr{B}}_U. \tag{4.3.10.2}$$

We have a canonical homomorphism $\sigma_s^*(\mathscr{O}_{\overline{X}_n}) \to \overline{\mathscr{B}}_n$ ([3] (III.9.9.3)). The morphism σ_s (4.3.3.3) is therefore underlying a morphism of ringed topos that we denote by

$$\sigma_n \colon (\widetilde{E}_s, \overline{\mathscr{B}}_n) \to (X_{s,\text{ét}}, \mathscr{O}_{\overline{X}_n}). \tag{4.3.10.3}$$

For modules, we use the notation σ_n^{-1} (or σ_s^*) to denote the pullback in the sense of abelian sheaves, and we keep the notation σ_n^* for the pullback in the sense of modules.

We have a canonical homomorphism $\overline{\mathscr{B}}_{X,n} \to \beta_*(\overline{\mathscr{B}}_n)$, which is not an isomorphism in general ([2] 4.1.8). The composed morphism $\beta \circ \delta$ is therefore underlying a morphism of ringed topos that we denote by

$$\beta_n \colon (\widetilde{E}_s, \overline{\mathscr{B}}_n) \to (\overline{X}^\circ_{\text{fét}}, \overline{\mathscr{B}}_{X,n}). \tag{4.3.10.4}$$

We denote by

$$\Sigma_n \colon (\widetilde{E}_s, \overline{\mathscr{B}}_n) \to (X_{s,\text{zar}}, \mathscr{O}_{\overline{X}_n}) \tag{4.3.10.5}$$

the composition of the morphism σ_n and the canonical morphism (2.1.13.5)

$$u_n \colon (X_{s,\text{ét}}, \mathscr{O}_{\overline{X}_n}) \to (X_{s,\text{zar}}, \mathscr{O}_{\overline{X}_n}). \tag{4.3.10.6}$$

For modules, we use the notation Σ_n^{-1} to denote the pullback in the sense of abelian sheaves, and we keep the notation Σ_n^* for the pullback in the sense of modules.

4.3.11 For any $\mathbb{U}$-topos T, we denote by $T^{\mathbb{N}^\circ}$ the topos of inverse systems of T, indexed by the ordered set $\mathbb{N}$ of natural numbers (2.1.9). We denote by $\breve{\overline{\mathscr{B}}}$ the ring $(\overline{\mathscr{B}}_{n+1})_{n\in\mathbb{N}}$ of $\widetilde{E}_s^{\mathbb{N}^\circ}$ (4.3.10.1), by $\mathscr{O}_{\breve{\overline{X}}}$ the ring $(\mathscr{O}_{\overline{X}_{n+1}})_{n\in\mathbb{N}}$ of $X_{s,\text{ét}}^{\mathbb{N}^\circ}$ or $X_{s,\text{zar}}^{\mathbb{N}^\circ}$, depending on the context, and by $\widetilde{\xi}^{-1}\widetilde{\Omega}^1_{\breve{\overline{X}}/\breve{\overline{S}}}$ the $\mathscr{O}_{\breve{\overline{X}}}$-module $(\widetilde{\xi}^{-1}\widetilde{\Omega}^1_{\overline{X}_{n+1}/\overline{S}_{n+1}})_{n\in\mathbb{N}}$ (4.1.5.1). We will take care not to confuse $\mathscr{O}_{\breve{\overline{X}}}$ and $\mathscr{O}_{\breve{X}}$ (4.1.1.2).

We denote by

$$\breve{\sigma}\colon (\widetilde{E}_s^{\mathbb{N}^\circ}, \breve{\overline{\mathscr{B}}}) \to (X_{s,\text{ét}}^{\mathbb{N}^\circ}, \mathscr{O}_{\breve{\overline{X}}}) \quad (4.3.11.1)$$

the morphism of ringed topos induced by $(\sigma_{n+1})_{n\in\mathbb{N}}$ (4.3.10.3). For modules, we use the notation $\breve{\sigma}^{-1}$ to denote the pullback in the sense of abelian sheaves, and we keep the notation $\breve{\sigma}^*$ for the pullback in the sense of modules.

We denote by

$$\breve{\Sigma}\colon (\widetilde{E}_s^{\mathbb{N}^\circ}, \breve{\overline{\mathscr{B}}}) \to (X_{s,\text{zar}}^{\mathbb{N}^\circ}, \mathscr{O}_{\breve{\overline{X}}}) \quad (4.3.11.2)$$

the morphism of ringed topos defined by the $(\Sigma_{n+1})_{n\in\mathbb{N}}$ (4.3.10.5). For modules, we use the notation $\breve{\Sigma}^{-1}$ to denote the pullback in the sense of abelian sheaves, and we keep the notation $\breve{\Sigma}^*$ for the pullback in the sense of modules.

Recall that $\mathfrak{X}$ denotes the formal scheme p-adic completion of $\overline{X}$ (4.2.1). We denote by

$$\lambda\colon (X_{s,\text{zar}}^{\mathbb{N}^\circ}, \mathscr{O}_{\breve{\overline{X}}}) \to (X_{s,\text{zar}}, \mathscr{O}_{\mathfrak{X}}) \quad (4.3.11.3)$$

the morphism of ringed topos for which the functor λ_* is the inverse limit functor (2.1.9.1). We denote by

$$\widehat{\sigma}\colon (\widetilde{E}_s^{\mathbb{N}^\circ}, \breve{\overline{\mathscr{B}}}) \to (X_{s,\text{zar}}, \mathscr{O}_{\mathfrak{X}}) \quad (4.3.11.4)$$

the composed morphism $\lambda \circ \breve{\Sigma}$. For modules, we use the notation $\widehat{\sigma}^{-1}$ to denote the pullback in the sense of abelian sheaves, and we keep the notation $\widehat{\sigma}^*$ for the pullback in the sense of modules. Note that the morphism $\widehat{\sigma}$ has been denoted by $\top$ in ([3] (III.11.1.11)).

For every $\mathscr{O}_{\mathfrak{X}}$-module $\mathscr{F}$ of $X_{s,\text{zar}}$, we have a canonical isomorphism

$$\widehat{\sigma}^*(\mathscr{F}) \xrightarrow{\sim} \breve{\Sigma}^*((\mathscr{F}/p^{n+1}\mathscr{F})_{n\in\mathbb{N}}). \quad (4.3.11.5)$$

In particular, $\widehat{\sigma}^*(\mathscr{F})$ is adic ([3] III.7.18).

4.3.12 We denote by $\mathbf{Mod}(\breve{\overline{\mathscr{B}}})$ the category of $\breve{\overline{\mathscr{B}}}$-modules of $\widetilde{E}_s^{\mathbb{N}^\circ}$, by $\mathbf{Ind\text{-}Mod}(\breve{\overline{\mathscr{B}}})$ the category of *ind-$\breve{\overline{\mathscr{B}}}$-modules* (2.7.1) and by

$$\iota_{\breve{\overline{\mathscr{B}}}}\colon \mathbf{Mod}(\breve{\overline{\mathscr{B}}}) \to \mathbf{Ind\text{-}Mod}(\breve{\overline{\mathscr{B}}}) \quad (4.3.12.1)$$

the canonical functor, which is exact and fully faithful (2.7.1). We will identify $\mathbf{Mod}(\breve{\overline{\mathscr{B}}})$ with a full subcategory of $\mathbf{Ind\text{-}Mod}(\breve{\overline{\mathscr{B}}})$ by the functor $\iota_{\breve{\overline{\mathscr{B}}}}$, which we will omit from the notation.

We denote by $\mathbf{Mod}(\mathscr{O}_{\mathfrak{X}})$ the category of $\mathscr{O}_{\mathfrak{X}}$-modules of $X_{s,\text{zar}}$ (4.2.1), by $\mathbf{Ind\text{-}Mod}(\mathscr{O}_{\mathfrak{X}})$ the category of ind-$\mathscr{O}_{\mathfrak{X}}$-modules and by

$$\iota_{\mathscr{O}_{\mathfrak{X}}}\colon \mathbf{Mod}(\mathscr{O}_{\mathfrak{X}}) \to \mathbf{Ind\text{-}Mod}(\mathscr{O}_{\mathfrak{X}}) \quad (4.3.12.2)$$

the canonical functor, which is exact and fully faithful. We will identify $\mathbf{Mod}(\mathscr{O}_{\mathfrak{X}})$ with a full subcategory of $\mathbf{Ind\text{-}Mod}(\mathscr{O}_{\mathfrak{X}})$ by the functor $\iota_{\mathscr{O}_{\mathfrak{X}}}$, which we will omit from the notation.

By 2.7.10, the morphism of ringed topos $\widehat{\sigma}$ (4.3.11.4) induces two adjoint additive functors

$$\mathrm{I}\widehat{\sigma}^*\colon \mathbf{Ind\text{-}Mod}(\mathscr{O}_{\mathfrak{X}}) \to \mathbf{Ind\text{-}Mod}(\breve{\overline{\mathscr{B}}}), \tag{4.3.12.3}$$

$$\mathrm{I}\widehat{\sigma}_*\colon \mathbf{Ind\text{-}Mod}(\breve{\overline{\mathscr{B}}}) \to \mathbf{Ind\text{-}Mod}(\mathscr{O}_{\mathfrak{X}}). \tag{4.3.12.4}$$

The functor $\mathrm{I}\widehat{\sigma}^*$ (resp. $\mathrm{I}\widehat{\sigma}_*$) is right (resp. left) exact. The functor $\mathrm{I}\widehat{\sigma}_*$ admits a right derived functor

$$\mathrm{RI}\widehat{\sigma}_*\colon \mathbf{D}^+(\mathbf{Ind\text{-}Mod}(\breve{\overline{\mathscr{B}}})) \to \mathbf{D}^+(\mathbf{Ind\text{-}Mod}(\mathscr{O}_{\mathfrak{X}})). \tag{4.3.12.5}$$

The diagram

$$\begin{array}{ccc} \mathbf{D}^+(\mathbf{Mod}(\breve{\overline{\mathscr{B}}})) & \xrightarrow{\mathrm{R}\widehat{\sigma}_*} & \mathbf{D}^+(\mathbf{Mod}(\mathscr{O}_{\mathfrak{X}})) \\ \Big\downarrow{\scriptstyle \iota_{\breve{\overline{\mathscr{B}}}}} & & \Big\downarrow{\scriptstyle \iota_{\mathscr{O}_{\mathfrak{X}}}} \\ \mathbf{D}^+(\mathbf{Ind\text{-}Mod}(\breve{\overline{\mathscr{B}}})) & \xrightarrow{\mathrm{RI}\widehat{\sigma}_*} & \mathbf{D}^+(\mathbf{Ind\text{-}Mod}(\mathscr{O}_{\mathfrak{X}})) \end{array} \tag{4.3.12.6}$$

is commutative up to canonical isomorphism (2.7.10.6).

By 2.6.5, the canonical functor $\iota_{\mathscr{O}_{\mathfrak{X}}}$ admits a left adjoint

$$\kappa_{\mathscr{O}_{\mathfrak{X}}}\colon \mathbf{Ind\text{-}Mod}(\mathscr{O}_{\mathfrak{X}}) \to \mathbf{Mod}(\mathscr{O}_{\mathfrak{X}}), \tag{4.3.12.7}$$

such that for every small filtered category J and every functor $\alpha\colon J \to \mathbf{Mod}(\mathscr{O}_{\mathfrak{X}})$, we have an isomorphism

$$\kappa_{\mathscr{O}_{\mathfrak{X}}}(\text{“}\varinjlim_{J}\text{”}\alpha) \xrightarrow{\sim} \varinjlim_{J} \alpha. \tag{4.3.12.8}$$

The canonical morphism $\kappa_{\mathscr{O}_{\mathfrak{X}}} \circ \iota_{\mathscr{O}_{\mathfrak{X}}} \to \mathrm{id}_{\mathbf{Mod}(\mathscr{O}_{\mathfrak{X}})}$ is an isomorphism. The functor $\kappa_{\mathscr{O}_{\mathfrak{X}}}$ is exact by 2.7.2.

We denote by $\vec{\sigma}_*$ the composed functor

$$\vec{\sigma}_* = \kappa_{\mathscr{O}_{\mathfrak{X}}} \circ \mathrm{I}\widehat{\sigma}_*\colon \mathbf{Ind\text{-}Mod}(\breve{\overline{\mathscr{B}}}) \to \mathbf{Mod}(\mathscr{O}_{\mathfrak{X}}). \tag{4.3.12.9}$$

It is left exact. It admits a right derived functor

$$\mathrm{R}\vec{\sigma}_*\colon \mathbf{D}^+(\mathbf{Ind\text{-}Mod}(\breve{\overline{\mathscr{B}}})) \to \mathbf{D}^+(\mathbf{Mod}(\mathscr{O}_{\mathfrak{X}})), \tag{4.3.12.10}$$

canonically isomorphic to $\kappa_{\mathscr{O}_{\mathfrak{X}}} \circ \mathrm{RI}\widehat{\sigma}_*$ ([40] 13.3.13).

4.3.13 We denote by $\mathbf{Mod}_{\mathbb{Q}}(\mathscr{O}_{\mathfrak{X}})$ (resp. $\mathbf{Mod}_{\mathbb{Q}}(\breve{\overline{\mathscr{B}}})$) the category of $\mathscr{O}_{\mathfrak{X}}$-modules (resp. $\breve{\overline{\mathscr{B}}}$-modules) up to isogeny (2.9.7) and by

$$\mathbf{Mod}(\mathscr{O}_{\mathfrak{X}}) \to \mathbf{Mod}_{\mathbb{Q}}(\mathscr{O}_{\mathfrak{X}}), \quad \mathscr{F} \mapsto \mathscr{F}_{\mathbb{Q}}, \tag{4.3.13.1}$$

$$\mathbf{Mod}(\breve{\overline{\mathscr{B}}}) \to \mathbf{Mod}_{\mathbb{Q}}(\breve{\overline{\mathscr{B}}}), \quad \mathscr{M} \mapsto \mathscr{M}_{\mathbb{Q}}, \tag{4.3.13.2}$$

the canonical functors. These are two abelian symmetric monoidal categories, having $\mathscr{O}_{\mathfrak{X},\mathbb{Q}}$ (resp. $\breve{\overline{\mathscr{B}}}_{\mathbb{Q}}$) as unit object. The objects of $\mathbf{Mod}_{\mathbb{Q}}(\breve{\overline{\mathscr{B}}})$ will also be called $\breve{\overline{\mathscr{B}}}_{\mathbb{Q}}$-*modules*.

By 2.9.2, we have canonical functors (2.9.2.3)

$$\alpha_{\mathscr{O}_{\mathfrak{X}}} : \mathbf{Mod}_{\mathbb{Q}}(\mathscr{O}_{\mathfrak{X}}) \to \mathbf{Ind\text{-}Mod}(\mathscr{O}_{\mathfrak{X}}), \tag{4.3.13.3}$$

$$\alpha_{\breve{\overline{\mathscr{B}}}} : \mathbf{Mod}_{\mathbb{Q}}(\breve{\overline{\mathscr{B}}}) \to \mathbf{Ind\text{-}Mod}(\breve{\overline{\mathscr{B}}}). \tag{4.3.13.4}$$

These functors are fully faithful (2.6.6.5) and exact (2.9.3). We will identify $\mathbf{Mod}_{\mathbb{Q}}(\mathscr{O}_{\mathfrak{X}})$ (resp. $\mathbf{Mod}_{\mathbb{Q}}(\breve{\overline{\mathscr{B}}})$) with a full subcategory of $\mathbf{Ind\text{-}Mod}(\mathscr{O}_{\mathfrak{X}})$ (resp. $\mathbf{Ind\text{-}Mod}(\breve{\overline{\mathscr{B}}})$) by the functor $\alpha_{\mathscr{O}_{\mathfrak{X}}}$ (resp. $\alpha_{\breve{\overline{\mathscr{B}}}}$), which we will omit from the notation. We will therefore consider any $\breve{\overline{\mathscr{B}}}_{\mathbb{Q}}$-module as an ind-$\breve{\overline{\mathscr{B}}}$-module.

By 2.9.7, the morphism of ringed topos $\widehat{\sigma}$ (4.3.11.4) induces two adjoint additive functors

$$\widehat{\sigma}^*_{\mathbb{Q}} : \mathbf{Mod}_{\mathbb{Q}}(\mathscr{O}_{\mathfrak{X}}) \to \mathbf{Mod}_{\mathbb{Q}}(\breve{\overline{\mathscr{B}}}), \tag{4.3.13.5}$$

$$\widehat{\sigma}_{\mathbb{Q}*} : \mathbf{Mod}_{\mathbb{Q}}(\breve{\overline{\mathscr{B}}}) \to \mathbf{Mod}_{\mathbb{Q}}(\mathscr{O}_{\mathfrak{X}}). \tag{4.3.13.6}$$

The functor $\widehat{\sigma}^*_{\mathbb{Q}}$ (resp. $\widehat{\sigma}_{\mathbb{Q}*}$) is right (resp. left) exact. By 2.9.5(ii), the diagrams

$$\begin{array}{ccc} \mathbf{Mod}_{\mathbb{Q}}(\breve{\overline{\mathscr{B}}}) & \xrightarrow{\widehat{\sigma}_{\mathbb{Q}*}} & \mathbf{Mod}_{\mathbb{Q}}(\mathscr{O}_{\mathfrak{X}}) \\ {\scriptstyle \alpha_{\breve{\overline{\mathscr{B}}}}}\downarrow & & \downarrow{\scriptstyle \alpha_{\mathscr{O}_{\mathfrak{X}}}} \\ \mathbf{Ind\text{-}Mod}(\breve{\overline{\mathscr{B}}}) & \xrightarrow{I\widehat{\sigma}_*} & \mathbf{Ind\text{-}Mod}(\mathscr{O}_{\mathfrak{X}}) \end{array} \tag{4.3.13.7}$$

$$\begin{array}{ccc} \mathbf{Mod}_{\mathbb{Q}}(\mathscr{O}_{\mathfrak{X}}) & \xrightarrow{\widehat{\sigma}^*_{\mathbb{Q}}} & \mathbf{Mod}_{\mathbb{Q}}(\breve{\overline{\mathscr{B}}}) \\ {\scriptstyle \alpha_{\mathscr{O}_{\mathfrak{X}}}}\downarrow & & \downarrow{\scriptstyle \alpha_{\breve{\overline{\mathscr{B}}}}} \\ \mathbf{Ind\text{-}Mod}(\mathscr{O}_{\mathfrak{X}}) & \xrightarrow{I\widehat{\sigma}^*} & \mathbf{Ind\text{-}Mod}(\breve{\overline{\mathscr{B}}}) \end{array} \tag{4.3.13.8}$$

are commutative up to canonical isomorphisms. ∎

The canonical functor

$$\mathbf{Mod}(\mathscr{O}_{\mathfrak{X}}) \to \mathbf{Mod}(\mathscr{O}_{\mathfrak{X}}[\frac{1}{p}]), \quad \mathscr{F} \mapsto \mathscr{F}[\frac{1}{p}] \tag{4.3.13.9}$$

induces an exact functor

$$\mathbf{Mod}_{\mathbb{Q}}(\mathscr{O}_{\mathfrak{X}}) \to \mathbf{Mod}(\mathscr{O}_{\mathfrak{X}}[\frac{1}{p}]). \tag{4.3.13.10}$$

We denote again by

$$\widehat{\sigma}_{\mathbb{Q}*} : \mathbf{Mod}_{\mathbb{Q}}(\breve{\overline{\mathscr{B}}}) \to \mathbf{Mod}(\mathscr{O}_{\mathfrak{X}}[\frac{1}{p}]) \tag{4.3.13.11}$$

the composition of the functor $\widehat{\sigma}_{\mathbb{Q}*}$ (4.3.13.6) and the functor (4.3.13.10).

By 2.9.6, the functor $\widehat{\sigma}_{\mathbb{Q}*}$ (4.3.13.6) admits a right derived functor

$$\mathrm{R}\widehat{\sigma}_{\mathbb{Q}*}\colon \mathbf{D}^+(\mathbf{Mod}_{\mathbb{Q}}(\breve{\overline{\mathscr{B}}})) \to \mathbf{D}^+(\mathbf{Mod}_{\mathbb{Q}}(\mathscr{O}_{\mathfrak{X}})). \tag{4.3.13.12}$$

By virtue of (2.9.6.7), the diagram

$$\begin{array}{ccc} \mathbf{D}^+(\mathbf{Mod}_{\mathbb{Q}}(\breve{\overline{\mathscr{B}}})) & \xrightarrow{\mathrm{R}\widehat{\sigma}_{\mathbb{Q}*}} & \mathbf{D}^+(\mathbf{Mod}_{\mathbb{Q}}(\mathscr{O}_{\mathfrak{X}})) \\ \downarrow{\scriptstyle \alpha_{\breve{\overline{\mathscr{B}}}}} & & \downarrow{\scriptstyle \alpha_{\mathscr{O}_{\mathfrak{X}}}} \\ \mathbf{D}^+(\mathbf{Ind}\text{-}\mathbf{Mod}(\breve{\overline{\mathscr{B}}})) & \xrightarrow{\mathrm{RI}\widehat{\sigma}_*} & \mathbf{D}^+(\mathbf{Ind}\text{-}\mathbf{Mod}(\mathscr{O}_{\mathfrak{X}})) \end{array} \tag{4.3.13.13}$$

is commutative up to a canonical isomorphism.

Similarly, the functor $\widehat{\sigma}_{\mathbb{Q}*}$ (4.3.13.11) admits a right derived functor

$$\mathrm{R}\widehat{\sigma}_{\mathbb{Q}*}\colon \mathbf{D}^+(\mathbf{Mod}_{\mathbb{Q}}(\breve{\overline{\mathscr{B}}})) \to \mathbf{D}^+(\mathbf{Mod}(\mathscr{O}_{\mathfrak{X}}[\frac{1}{p}])) \tag{4.3.13.14}$$

that is none other than the composition of the functor $\mathrm{R}\widehat{\sigma}_{\mathbb{Q}*}$ (4.3.13.12) and the exact functor (4.3.13.10). This abuse of notation does not lead to any confusion. It follows from (4.3.13.13) that the diagram

$$\begin{array}{ccc} \mathbf{D}^+(\mathbf{Mod}_{\mathbb{Q}}(\breve{\overline{\mathscr{B}}})) & \xrightarrow{\mathrm{R}\widehat{\sigma}_{\mathbb{Q}*}} & \mathbf{D}^+(\mathbf{Mod}(\mathscr{O}_{\mathfrak{X}}[\frac{1}{p}])) \\ \downarrow{\scriptstyle \alpha_{\breve{\overline{\mathscr{B}}}}} & & \downarrow \\ \mathbf{D}^+(\mathbf{Ind}\text{-}\mathbf{Mod}(\breve{\overline{\mathscr{B}}})) & \xrightarrow{\mathrm{R}\vec{\sigma}_*} & \mathbf{D}^+(\mathbf{Mod}(\mathscr{O}_{\mathfrak{X}})), \end{array} \tag{4.3.13.15}$$

where $\mathrm{R}\vec{\sigma}_*$ is the functor (4.3.12.10) and the unlabeled arrow is the canonical functor, is commutative up to canonical isomorphism.

Definition 4.3.14 We say that a $\breve{\overline{\mathscr{B}}}_{\mathbb{Q}}$-module is *adic of finite type* if it is isomorphic to $\mathscr{M}_{\mathbb{Q}}$ (4.3.13.2), where $\mathscr{M}$ is an adic $\breve{\overline{\mathscr{B}}}$-module of finite type ([3] III.7.16).

We denote by $\mathbf{Mod}^{\mathrm{aft}}(\breve{\overline{\mathscr{B}}})$ the full subcategory of $\mathbf{Mod}(\breve{\overline{\mathscr{B}}})$ made up of the adic $\breve{\overline{\mathscr{B}}}$-modules of finite type and by $\mathbf{Mod}^{\mathrm{aft}}_{\mathbb{Q}}(\breve{\overline{\mathscr{B}}})$ the category of objects of $\mathbf{Mod}^{\mathrm{aft}}(\breve{\overline{\mathscr{B}}})$ up to isogeny (2.9.1). The canonical functor

$$\mathbf{Mod}^{\mathrm{aft}}_{\mathbb{Q}}(\breve{\overline{\mathscr{B}}}) \to \mathbf{Mod}_{\mathbb{Q}}(\breve{\overline{\mathscr{B}}}) \tag{4.3.14.1}$$

being fully faithful, a $\breve{\overline{\mathscr{B}}}_{\mathbb{Q}}$-module is adic of finite type if and only if it is in the essential image of this functor.

Definition 4.3.15 We say that an ind-$\breve{\overline{\mathscr{B}}}$-module $\mathscr{F}$ is *rational* if the multiplication by p on $\mathscr{F}$ is an isomorphism.

Observe that the essential image of $\alpha_{\breve{\overline{\mathscr{B}}}}$ is made up of rational ind-$\breve{\overline{\mathscr{B}}}$-modules.

4.3.16 We denote by $\mathbf{Mod}^{\mathrm{coh}}(\mathscr{O}_{\mathfrak{X}})$ (resp. $\mathbf{Mod}^{\mathrm{coh}}(\mathscr{O}_{\mathfrak{X}}[\frac{1}{p}])$) the category of coherent $\mathscr{O}_{\mathfrak{X}}$-modules (resp. $\mathscr{O}_{\mathfrak{X}}[\frac{1}{p}]$-modules) of $X_{s,\mathrm{zar}}$ and by $\mathbf{Mod}^{\mathrm{coh}}_{\mathbb{Q}}(\mathscr{O}_{\mathfrak{X}})$ the category of coherent $\mathscr{O}_{\mathfrak{X}}$-modules up to isogeny ([3] III.6.1.1). By ([3] III.6.16), the canonical functor

$$\mathbf{Mod}^{\mathrm{coh}}(\mathscr{O}_{\mathfrak{X}}) \to \mathbf{Mod}^{\mathrm{coh}}(\mathscr{O}_{\mathfrak{X}}[\frac{1}{p}]), \quad \mathscr{F} \mapsto \mathscr{F}[\frac{1}{p}] \tag{4.3.16.1}$$

induces an equivalence of abelian categories

$$\mathbf{Mod}^{\mathrm{coh}}_{\mathbb{Q}}(\mathscr{O}_{\mathfrak{X}}) \xrightarrow{\sim} \mathbf{Mod}^{\mathrm{coh}}(\mathscr{O}_{\mathfrak{X}}[\frac{1}{p}]). \tag{4.3.16.2}$$

Consider the composed functor

$$\mathbf{Mod}^{\mathrm{coh}}(\mathscr{O}_{\mathfrak{X}}[\frac{1}{p}]) \xrightarrow{\sim} \mathbf{Mod}^{\mathrm{coh}}_{\mathbb{Q}}(\mathscr{O}_{\mathfrak{X}}) \longrightarrow \mathbf{Mod}_{\mathbb{Q}}(\mathscr{O}_{\mathfrak{X}}) \xrightarrow{\alpha_{\mathscr{O}_{\mathfrak{X}}}} \mathbf{Ind\text{-}Mod}(\mathscr{O}_{\mathfrak{X}}), \tag{4.3.16.3}$$

where the first arrow is a quasi-inverse of the equivalence of categories (4.3.16.2) and the second arrow is the canonical fully faithful functor. We will identify $\mathbf{Mod}^{\mathrm{coh}}(\mathscr{O}_{\mathfrak{X}}[\frac{1}{p}])$ with a full subcategory of $\mathbf{Ind\text{-}Mod}(\mathscr{O}_{\mathfrak{X}})$ by this composed functor. We will therefore consider any coherent $\mathscr{O}_{\mathfrak{X}}[\frac{1}{p}]$-module also as an ind-$\mathscr{O}_{\mathfrak{X}}$-module.

For every coherent $\mathscr{O}_{\mathfrak{X}}[\frac{1}{p}]$-module $\mathscr{N}$, considered as an ind-$\mathscr{O}_{\mathfrak{X}}$-module, we have a canonical functorial isomorphism

$$\kappa_{\mathscr{O}_{\mathfrak{X}}}(\mathscr{N}) \xrightarrow{\sim} \mathscr{N}. \tag{4.3.16.4}$$

This follows immediately from the definitions and from 2.9.3(i).

Let $\mathscr{N}$ be a coherent $\mathscr{O}_{\mathfrak{X}}[\frac{1}{p}]$-module, $\mathscr{M}$ an ind-$\breve{\overline{\mathscr{B}}}$-module. The adjoint functors $\mathrm{I}\widehat{\sigma}^*$ and $\mathrm{I}\widehat{\sigma}_*$ and the isomorphism (4.3.16.4) induce a map

$$\mathrm{Hom}_{\mathbf{Ind\text{-}Mod}(\breve{\overline{\mathscr{B}}})}(\mathrm{I}\widehat{\sigma}^*(\mathscr{N}), \mathscr{M}) \to \mathrm{Hom}_{\mathbf{Mod}(\mathscr{O}_{\mathfrak{X}})}(\mathscr{N}, \vec{\sigma}_*(\mathscr{M})). \tag{4.3.16.5}$$

It is injective by 4.3.17 below. In view of (2.7.1.5) and (4.3.16.4), for every integer $q \geq 0$, the canonical morphism (2.7.18.1)

$$\mathscr{N} \otimes_{\mathscr{O}_{\mathfrak{X}}} \mathrm{R}^q\mathrm{I}\widehat{\sigma}_*(\mathscr{M}) \to \mathrm{R}^q\mathrm{I}\widehat{\sigma}_*(\mathrm{I}\widehat{\sigma}^*(\mathscr{N}) \otimes_{\breve{\overline{\mathscr{B}}}} \mathscr{M}) \tag{4.3.16.6}$$

induces by applying the functor $\kappa_{\mathscr{O}_{\mathfrak{X}}}$ a morphism

$$\mathscr{N} \otimes_{\mathscr{O}_{\mathfrak{X}}} \mathrm{R}^q\vec{\sigma}_*(\mathscr{M}) \to \mathrm{R}^q\vec{\sigma}_*(\mathrm{I}\widehat{\sigma}^*(\mathscr{N}) \otimes_{\breve{\overline{\mathscr{B}}}} \mathscr{M}). \tag{4.3.16.7}$$

Lemma 4.3.17 *Let $\mathscr{N}$ be a coherent $\mathscr{O}_{\mathfrak{X}}[\frac{1}{p}]$-module, $\mathscr{F}$ an ind-$\mathscr{O}_{\mathfrak{X}}$-module. Then, the functor $\kappa_{\mathscr{O}_{\mathfrak{X}}}$ induces an injective map*

$$\mathrm{Hom}_{\mathbf{Ind\text{-}Mod}(\mathscr{O}_{\mathfrak{X}})}(\mathscr{N}, \mathscr{F}) \to \mathrm{Hom}_{\mathbf{Mod}(\mathscr{O}_{\mathfrak{X}})}(\mathscr{N}, \kappa_{\mathscr{O}_{\mathfrak{X}}}(\mathscr{F})). \tag{4.3.17.1}$$

Let $\mathscr{N}^\circ$ be a coherent $\mathscr{O}_{\mathfrak{X}}$-module such that $\mathscr{N}^\circ[\frac{1}{p}] = \mathscr{N}$ (4.3.16.1), $\alpha\colon J \to$ **Mod**$(\mathscr{O}_{\mathfrak{X}})$ a functor with J a small filtered category such that $\mathscr{F} = \text{“}\varinjlim\text{”}\alpha$. In view of (2.6.1.3), we have to show that the canonical map

$$\varprojlim_{\mathbb{N}} \varinjlim_{j\in J} \mathrm{Hom}_{\mathscr{O}_{\mathfrak{X}}}(\mathscr{N}^\circ, \alpha(j)) \to \mathrm{Hom}_{\mathscr{O}_{\mathfrak{X}}}(\mathscr{N}, \varinjlim_{j\in J} \alpha(j)), \tag{4.3.17.2}$$

where the transition morphisms of the inverse limit are induced by the multiplication by the powers of p, is injective. The question being local, we may suppose $\mathscr{N}^\circ$ is generated by its global sections, and consequently limit ourselves to the case where $\mathscr{N}^\circ$ is free of finite type and even to the case where $\mathscr{N}^\circ = \mathscr{O}_{\mathfrak{X}}$. By ([5] VI 5.3), the canonical morphism

$$\varinjlim_{j\in J} \mathrm{Hom}_{\mathscr{O}_{\mathfrak{X}}}(\mathscr{O}_{\mathfrak{X}}, \alpha(j)) \to \mathrm{Hom}_{\mathscr{O}_{\mathfrak{X}}}(\mathscr{O}_{\mathfrak{X}}, \varinjlim_{j\in J} \alpha(j)) \tag{4.3.17.3}$$

is an isomorphism. Moreover, the canonical morphism

$$\varprojlim_{\mathbb{N}} \mathrm{Hom}_{\mathscr{O}_{\mathfrak{X}}}(\mathscr{O}_{\mathfrak{X}}, \varinjlim_{j\in J} \alpha(j)) \to \mathrm{Hom}_{\mathscr{O}_{\mathfrak{X}}}(\mathscr{O}_{\mathfrak{X}}[\frac{1}{p}], \varinjlim_{j\in J} \alpha(j)), \tag{4.3.17.4}$$

where the transition morphisms of the inverse limit are induced by the multiplication by the powers of p, is an isomorphism, proving the desired statement.

Lemma 4.3.18 *Let $\mathscr{N}$ be a locally projective $\mathscr{O}_{\mathfrak{X}}[\frac{1}{p}]$-module of finite type* (2.1.11), *$\mathscr{M}$ an ind-$\breve{\overline{\mathscr{B}}}$-module, q an integer ≥ 0. Then, the canonical morphism* (4.3.16.6)

$$\mathscr{N} \otimes_{\mathscr{O}_{\mathfrak{X}}} \mathrm{R}^q\mathrm{I}\widehat{\sigma}_*(\mathscr{M}) \to \mathrm{R}^q\mathrm{I}\widehat{\sigma}_*(\mathrm{I}\widehat{\sigma}^*(\mathscr{N}) \otimes_{\breve{\overline{\mathscr{B}}}} \mathscr{M}) \tag{4.3.18.1}$$

is an isomorphism. The induced morphism (4.3.16.7)

$$\mathscr{N} \otimes_{\mathscr{O}_{\mathfrak{X}}} \mathrm{R}^q\vec{\sigma}_*(\mathscr{M}) \to \mathrm{R}^q\vec{\sigma}_*(\mathrm{I}\widehat{\sigma}^*(\mathscr{N}) \otimes_{\breve{\overline{\mathscr{B}}}} \mathscr{M}) \tag{4.3.18.2}$$

is also an isomorphism.

There exists a Zariski open covering $(U_i)_{0\leq i\leq n}$ of X such that for all $0 \leq i \leq n$, the restriction of $\mathscr{N}$ to $(U_i)_s$ is a direct factor of a free $(\mathscr{O}_{\mathfrak{X}}|U_i)[\frac{1}{p}]$-module. By 2.9.5(ii), 2.7.17(iii) and the fact that $\mathrm{R}^q\mathrm{I}\widehat{\sigma}_*$ commutes with localizations (2.7.13), we may then restrict to the case where $\mathscr{N}$ is a direct factor of a free $\mathscr{O}_{\mathfrak{X}}[\frac{1}{p}]$-module of finite type, and even to the case where $\mathscr{N}$ is a free $\mathscr{O}_{\mathfrak{X}}[\frac{1}{p}]$-module of finite type. Since $\mathrm{R}^q\mathrm{I}\widehat{\sigma}_*$ commutes with direct limits (2.7.10.5), the morphism (4.3.18.1) is an isomorphism. It immediately follows that the morphism (4.3.18.2) is an isomorphism.

4.4 Higgs–Tate Algebras in Faltings Topos

4.4.1 We denote by $\mathbf{P}$ the full subcategory of $\mathbf{\acute{E}t}_{/X}$ made up of the affine schemes U such that one of the following two conditions is satisfied:

(i) the scheme U_s is empty; or
(ii) the morphism $(U, \mathscr{M}_X|U) \to (S, \mathscr{M}_S)$ induced by f (4.1.3) admits an adequate chart (3.1.15).

We endow $\mathbf{P}$ with the topology induced by that of $\mathbf{\acute{E}t}_{/X}$. Since X is Noetherian and therefore quasi-separated, any object of $\mathbf{P}$ is coherent over X. Hence, $\mathbf{P}$ is a $\mathbb{U}$-small family topologically generating the site $\mathbf{\acute{E}t}_{/X}$ and is stable by fiber products.

We denote by

$$\pi_{\mathbf{P}} \colon E_{\mathbf{P}} \to \mathbf{P} \tag{4.4.1.1}$$

the fibered site deduced from π (4.3.2.1) by base change by the canonical injection functor $\mathbf{P} \to \mathbf{\acute{E}t}_{/X}$, by

$$\mathscr{P}^{\vee}_{\mathbf{P}} \to \mathbf{P}^{\circ} \tag{4.4.1.2}$$

the fibered category over $\mathbf{P}^{\circ}$ deduced from the fibered category $\mathscr{P}^{\vee}$ (4.3.2.6) by base change by the canonical injection functor $\mathbf{P} \to \mathbf{\acute{E}t}_{/X}$, and by $\widehat{E}_{\mathbf{P}}$ the category of presheaves of $\mathbb{U}$-sets on $E_{\mathbf{P}}$. For any $U \in \mathrm{Ob}(\mathbf{P})$, we denote by $\iota_U \colon \mathbf{\acute{E}t}_{\mathrm{f}/\overline{U}^{\circ}} \to E_{\mathbf{P}}$ the canonical functor (4.3.2.3). We have an equivalence of categories

$$\begin{aligned} \widehat{E}_{\mathbf{P}} &\stackrel{\sim}{\to} \mathbf{Hom}_{\mathbf{P}^{\circ}}(\mathbf{P}^{\circ}, \mathscr{P}^{\vee}_{\mathbf{P}}) \\ F &\mapsto \{U \mapsto F \circ \iota_U\}. \end{aligned} \tag{4.4.1.3}$$

From now on, we will identify F with the section $\{U \mapsto F \circ \iota_U\}$ that is associated with it by this equivalence.

We endow $E_{\mathbf{P}}$ with the covanishing topology defined by $\pi_{\mathbf{P}}$ ([3] VI.5.3) and we denote by $\widetilde{E}_{\mathbf{P}}$ the topos of sheaves of $\mathbb{U}$-sets on $E_{\mathbf{P}}$. By ([3] VI.5.21 and VI.5.22), the topology of $E_{\mathbf{P}}$ is induced by that of E by the canonical projection functor $E_{\mathbf{P}} \to E$, and the latter induces by restriction an equivalence of categories

$$\widetilde{E} \stackrel{\sim}{\to} \widetilde{E}_{\mathbf{P}}. \tag{4.4.1.4}$$

4.4.2 We denote by $\mathbf{Q}$ the full subcategory of $\mathbf{P}$ made up of the affine schemes U such that one of the following conditions is satisfied:

(i) the scheme U_s is empty; or
(ii) there exists a fine and saturated chart $M \to \Gamma(U, \mathscr{M}_X)$ for $(U, \mathscr{M}_X|U)$ inducing an isomorphism

$$M \stackrel{\sim}{\to} \Gamma(U, \mathscr{M}_X)/\Gamma(U, \mathscr{O}_X^{\times}). \tag{4.4.2.1}$$

This chart is a priori independent of the adequate chart required in 4.4.1(ii).

We endow $\mathbf{Q}$ with the topology induced by that of $\mathbf{\acute{E}t}_{/X}$. It follows from ([3] II.5.17) that $\mathbf{Q}$ is a topologically generating subcategory of $\mathbf{\acute{E}t}_{/X}$.

We denote by

$$\pi_{\mathbf{Q}} \colon E_{\mathbf{Q}} \to \mathbf{Q} \tag{4.4.2.2}$$

the fibered site deduced from π (4.3.2.1) by base change by the canonical injection functor $\mathbf{Q} \to \mathbf{\acute{E}t}_{/X}$, by

$$\mathscr{P}^{\vee}_{\mathbf{Q}} \to \mathbf{Q}^{\circ} \tag{4.4.2.3}$$

the fibered category over $\mathbf{Q}^{\circ}$ deduced from the fibered category $\mathscr{P}^{\vee}$ (4.3.2.6) by base change by the canonical injection functor $\mathbf{Q} \to \mathbf{\acute{E}t}_{/X}$, and by $\widehat{E}_{\mathbf{Q}}$ the category of presheaves of $\mathbb{U}$-sets on $E_{\mathbf{Q}}$. For any $U \in \mathrm{Ob}(\mathbf{Q})$, we denote by $\iota_U \colon \mathbf{\acute{E}t}_{\mathrm{f}/\overline{U}^{\circ}} \to E_{\mathbf{Q}}$ the canonical functor (4.3.2.3). We have an equivalence of categories

$$\begin{aligned} \widehat{E}_{\mathbf{Q}} &\xrightarrow{\sim} \mathbf{Hom}_{\mathbf{Q}^{\circ}}(\mathbf{Q}^{\circ}, \mathscr{P}^{\vee}_{\mathbf{Q}}) \\ F &\mapsto \{U \mapsto F \circ \iota_U\}. \end{aligned} \tag{4.4.2.4}$$

From now on, we will identify F with the section $\{U \mapsto F \circ \iota_U\}$ that is associated with it by this equivalence.

The canonical projection functor $E_{\mathbf{Q}} \to E$ is fully faithful and the category $E_{\mathbf{Q}}$ is $\mathbb{U}$-small and topologically generating of the site E. We endow $E_{\mathbf{Q}}$ with the topology induced by that of E. By restriction, the topos $\widetilde{E}$ is then equivalent to the category of sheaves of $\mathbb{U}$-sets on $E_{\mathbf{Q}}$ ([5] III 4.1). Note that in general, $\mathbf{Q}$ not being stable under fiber products, we cannot speak of the covanishing topology on $E_{\mathbf{Q}}$ associated with $\pi_{\mathbf{Q}}$, and even less apply ([3] VI.5.21 and VI.5.22).

Remark 4.4.3

(i) The subcategories $\mathbf{P}$ and $\mathbf{Q}$ of $\mathbf{\acute{E}t}_{/X}$ were introduced in ([3] III.10.3 and III.10.5), but we omitted to include the objects U such that U_s is empty, which is however necessary for these subcategories to be topologically generating. We fix this omission in 4.4.1 and 4.4.2. This erratum has no consequences for the remainder of ([3] III).

(ii) Unlike ([3] III.10.5), we do not assume the schemes in $\mathbf{Q}$ to be connected.

4.4.4 Let Y be an object of $\mathbf{P}$ such that Y_s is nonempty, $((P, \gamma), (\mathbb{N}, \iota), \vartheta)$ an adequate chart for the morphism $f|Y \colon (Y, \mathscr{M}_X|Y) \to (S, \mathscr{M}_S)$ induced by f, $\overline{y}$ a geometric point of $\overline{Y}^{\circ}$. Note that the assumptions of 3.1.15 are satisfied. The scheme $\overline{Y}$ being locally irreducible by ([2] 4.2.7(iii) and [3] III.3.3), it is the sum of the schemes induced on its irreducible components. We denote by $\overline{Y}^{\star}$ the irreducible component of $\overline{Y}$ containing $\overline{y}$. Similarly, $\overline{Y}^{\circ}$ is the sum of the schemes induced on its irreducible components and $\overline{Y}^{\star\circ} = \overline{Y}^{\star} \times_X X^{\circ}$ is the irreducible component of $\overline{Y}^{\circ}$ containing $\overline{y}$. We denote by $\overline{R}^{\overline{y}}_Y$ the discrete representation of $\pi_1(\overline{Y}^{\star\circ}, \overline{y})$ defined in (4.3.7.2) and by $\widehat{\overline{R}}^{\overline{y}}_Y$ its p-adic Hausdorff completion. We set

$$\mathbb{Y}^{\overline{y}} = \mathrm{Spec}(\overline{R}^{\overline{y}}_Y) \quad \text{and} \quad \widehat{\mathbb{Y}}^{\overline{y}} = \mathrm{Spec}(\widehat{\overline{R}}^{\overline{y}}_Y), \tag{4.4.4.1}$$

which we endow with the logarithmic structures pullbacks of $\mathscr{M}_X$, denoted respectively by $\mathscr{M}_{\mathbb{Y}^{\overline{y}}}$ and $\mathscr{M}_{\widehat{\mathbb{Y}}^{\overline{y}}}$. Depending on the case considered in 3.1.14, we denote by

$(\widetilde{\mathbb{Y}}^{\overline{y}}, \mathscr{M}_{\widetilde{\mathbb{Y}}^{\overline{y}}})$ one or the other of the logarithmic schemes

$$(\mathscr{A}_2(\mathbb{Y}^{\overline{y}}), \mathscr{M}_{\mathscr{A}_2(\mathbb{Y}^{\overline{y}})}) \quad \text{or} \quad (\mathscr{A}_2^*(\mathbb{Y}^{\overline{y}}/S), \mathscr{M}_{\mathscr{A}_2^*(\mathbb{Y}^{\overline{y}}/S)}) \tag{4.4.4.2}$$

associated with the morphism $f|Y$ and the adequate chart $((P,\gamma),(\mathbb{N},\iota),\vartheta)$ in 3.2.8 and 3.2.10. We have a canonical exact closed immersion (3.2.11.5)

$$i_Y^{\overline{y}}\colon (\widehat{\mathbb{Y}}^{\overline{y}}, \mathscr{M}_{\widehat{\mathbb{Y}}^{\overline{y}}}) \to (\widetilde{\mathbb{Y}}^{\overline{y}}, \mathscr{M}_{\widetilde{\mathbb{Y}}^{\overline{y}}}). \tag{4.4.4.3}$$

We set (4.1.3.2)

$$\mathrm{T}_Y^{\overline{y}} = \mathrm{Hom}_{\widehat{\overline{R}}_Y^{\overline{y}}}(\widetilde{\Omega}^1_{X/S}(Y) \otimes_{\mathscr{O}_X(Y)} \widehat{\overline{R}}_Y^{\overline{y}}, \widehat{\xi\overline{R}}_Y^{\overline{y}}). \tag{4.4.4.4}$$

We identify the dual $\widehat{\overline{R}}_Y^{\overline{y}}$-module with $\xi^{-1}\widetilde{\Omega}^1_{X/S}(Y) \otimes_{\mathscr{O}_X(Y)} \widehat{\overline{R}}_Y^{\overline{y}}$ (see 3.1.14). We denote by $\widehat{\mathbb{Y}}^{\overline{y}}_{\mathrm{zar}}$ the Zariski topos of $\widehat{\mathbb{Y}}^{\overline{y}}$ and by $\widetilde{\mathrm{T}}_Y^{\overline{y}}$ the $\mathscr{O}_{\widehat{\mathbb{Y}}^{\overline{y}}}$-module associated with $\mathrm{T}_Y^{\overline{y}}$. Let U be a Zariski open subscheme of $\widehat{\mathbb{Y}}^{\overline{y}}$, $\widetilde{U}$ the corresponding open subscheme of $\widetilde{\mathbb{Y}}^{\overline{y}}$. We denote by $\mathscr{L}_Y^{\overline{y}}(U)$ the set of morphisms represented by the dotted arrows that complete the canonical diagram (4.1.5.3)

$$\begin{array}{ccc}
(U, \mathscr{M}_{\widehat{\mathbb{Y}}^{\overline{y}}}|U) & \xrightarrow{i_Y^{\overline{y}} \times_{\widetilde{\mathbb{Y}}^{\overline{y}}} \widetilde{U}} & (\widetilde{U}, \mathscr{M}_{\widetilde{\mathbb{Y}}^{\overline{y}}}|\widetilde{U}) \\
\downarrow & & \vdots \\
(\check{\overline{X}}, \mathscr{M}_{\check{\overline{X}}}) & \longrightarrow & (\widetilde{X}, \mathscr{M}_{\widetilde{X}}) \\
\downarrow & & \downarrow \\
(\check{\overline{S}}, \mathscr{M}_{\check{\overline{S}}}) & \xrightarrow{i_S} & (\widetilde{S}, \mathscr{M}_{\widetilde{S}})
\end{array} \tag{4.4.4.5}$$

in such a way that it remains commutative. By ([3] II.5.23), the functor $U \mapsto \mathscr{L}_Y^{\overline{y}}(U)$ is a $\widetilde{\mathrm{T}}_Y^{\overline{y}}$-torsor of $\widehat{\mathbb{Y}}_{\mathrm{zar}}$. Denoting by $\widetilde{Y} \to \widetilde{X}$ the unique étale morphism which lifts $\check{\overline{Y}} \to \check{\overline{X}}$ ([30] 18.1.2) (see 3.1.17), $\mathscr{L}_Y^{\overline{y}}$ identifies with the Higgs–Tate torsor associated with $(\widetilde{Y}, \mathscr{M}_{\widetilde{X}}|\widetilde{Y})$ (3.2.14). We denote by $\mathscr{F}_Y^{\overline{y}}$ the $\widehat{\overline{R}}_Y^{\overline{y}}$-module of affine functions on $\mathscr{L}_Y^{\overline{y}}$ (see [3] II.4.9). This $\widehat{\overline{R}}_Y^{\overline{y}}$-module fits into a canonical exact sequence (3.2.14.5)

$$0 \to \widehat{\overline{R}}_Y^{\overline{y}} \to \mathscr{F}_Y^{\overline{y}} \to \xi^{-1}\widetilde{\Omega}^1_{X/S}(Y) \otimes_{\mathscr{O}_X(Y)} \widehat{\overline{R}}_Y^{\overline{y}} \to 0. \tag{4.4.4.6}$$

We denote by $\mathscr{C}_Y^{\overline{y}}$ the $\widehat{\overline{R}}_Y^{\overline{y}}$-algebra (3.2.14.7)

$$\mathscr{C}_Y^{\overline{y}} = \lim_{\substack{\longrightarrow \\ n \geq 0}} \mathrm{S}^n_{\widehat{\overline{R}}_Y^{\overline{y}}}(\mathscr{F}_Y^{\overline{y}}). \tag{4.4.4.7}$$

By 3.2.15, the $\widehat{\overline{R}}{}^{\overline{y}}_Y$-module $\mathscr{F}_Y^{\overline{y}}$ is canonically endowed with an $\widehat{\overline{R}}{}^{\overline{y}}_Y$-semi-linear action of $\pi_1(\overline{Y}^{\star\circ}, \overline{y})$ that is continuous for the p-adic topology. The morphisms of the sequence (4.4.4.6) are $\pi_1(\overline{Y}^{\star\circ}, \overline{y})$-equivariant; we thus obtain the Higgs–Tate extension associated with $(\widetilde{Y}, \mathscr{M}_{\widetilde{X}}|\widetilde{Y})$ (3.2.16). We deduce from this an action of $\pi_1(\overline{Y}^{\star\circ}, \overline{y})$ on $\mathscr{C}_Y^{\overline{y}}$ by ring homomorphisms that is continuous for the p-adic topology and which extends the canonical action on $\widehat{\overline{R}}{}^{\overline{y}}_Y$; we thus obtain the Higgs–Tate algebra associated with $(\widetilde{Y}, \mathscr{M}_{\widetilde{X}}|\widetilde{Y})$ (3.2.16). Note that these representations depend on the adequate chart $((P, \gamma), (\mathbb{N}, \iota), \vartheta)$. However, they do not depend on it if Y is an object of $\mathbf{Q}$ by 3.2.12.

4.4.5 Let Y be an object of $\mathbf{P}$ such that Y_s is nonempty, $((P, \gamma), (\mathbb{N}, \iota), \vartheta)$ an adequate chart for the morphism $f|Y \colon (Y, \mathscr{M}_X|Y) \to (S, \mathscr{M}_S)$ induced by f, n an integer ≥ 0. If A is a ring and M an A-module, we denote again by A (resp. M) the constant sheaf with value A (resp. M) of $\overline{Y}^{\circ}_{\text{fét}}$. We recall that $\overline{Y}^{\circ}$ is the sum of the schemes induced on its irreducible components (4.4.4). Let W be an irreducible component of $\overline{Y}^{\circ}$, $\Pi(W)$ its fundamental groupoid ([3] VI.9.10). In view of (4.3.7.2) and ([3] VI.9.11), the sheaf $\overline{\mathscr{B}}_Y|W$ (4.3.6.3) of $W_{\text{fét}}$ defines a functor

$$\Pi(W) \to \mathbf{Sets}, \quad \overline{y} \mapsto \overline{R}{}^{\overline{y}}_Y. \tag{4.4.5.1}$$

We deduce from this a functor

$$\Pi(W) \to \mathbf{Sets}, \quad \overline{y} \mapsto \mathscr{F}_Y^{\overline{y}}/p^n\mathscr{F}_Y^{\overline{y}}. \tag{4.4.5.2}$$

For every geometric point $\overline{y}$ of W, $\mathscr{F}_Y^{\overline{y}}/p^n\mathscr{F}_Y^{\overline{y}}$ is a continuous and discrete representation of $\pi_1(W, \overline{y})$. Consequently, by virtue of ([3] VI.9.11), the functor (4.4.5.2) defines a $(\overline{\mathscr{B}}_{Y,n}|W)$-module $\mathscr{F}_{W,n}$ of $W_{\text{fét}}$, unique up to canonical isomorphism, where $\overline{\mathscr{B}}_{Y,n} = \overline{\mathscr{B}}_Y/p^n\overline{\mathscr{B}}_Y$ (4.3.10.2). By descent ([23] II 3.4.4), there exists a $\overline{\mathscr{B}}_{Y,n}$-module $\mathscr{F}_{Y,n}$ of $\overline{Y}^{\circ}_{\text{fét}}$, unique up to canonical isomorphism, such that for every irreducible component W of $\overline{Y}^{\circ}$, we have $\mathscr{F}_{Y,n}|W = \mathscr{F}_{W,n}$.

The exact sequence (4.4.4.6) induces an exact sequence of $\overline{\mathscr{B}}_{Y,n}$-modules

$$0 \to \overline{\mathscr{B}}_{Y,n} \to \mathscr{F}_{Y,n} \to \widetilde{\xi}^{-1}\widetilde{\Omega}^1_{X/S}(Y) \otimes_{\mathscr{O}_X(Y)} \overline{\mathscr{B}}_{Y,n} \to 0. \tag{4.4.5.3}$$

By ([37] I 4.3.1.7), this induces for every integer $m \geq 1$ an exact sequence

$$0 \to \mathrm{S}^{m-1}_{\overline{\mathscr{B}}_{Y,n}}(\mathscr{F}_{Y,n}) \to \mathrm{S}^{m}_{\overline{\mathscr{B}}_{Y,n}}(\mathscr{F}_{Y,n}) \to \mathrm{S}^{m}_{\overline{\mathscr{B}}_{Y,n}}(\widetilde{\xi}^{-1}\widetilde{\Omega}^1_{X/S}(Y) \otimes_{\mathscr{O}_X(Y)} \overline{\mathscr{B}}_{Y,n}) \to 0.$$

The $\overline{\mathscr{B}}_{Y,n}$-modules $(\mathrm{S}^{m}_{\overline{\mathscr{B}}_{Y,n}}(\mathscr{F}_{Y,n}))_{m\in\mathbb{N}}$ thus form a direct system whose direct limit

$$\mathscr{C}_{Y,n} = \lim_{\substack{\longrightarrow \\ m\geq 0}} \mathrm{S}^{m}_{\overline{\mathscr{B}}_{Y,n}}(\mathscr{F}_{Y,n}) \tag{4.4.5.4}$$

is naturally endowed with the structure of a $\overline{\mathscr{B}}_{Y,n}$-algebra of $\overline{Y}^{\circ}_{\text{fét}}$.

Note that $\mathscr{F}_{Y,n}$ and $\mathscr{C}_{Y,n}$ depend on the choice of the adequate chart $((P,\gamma),(\mathbb{N},\iota),\vartheta)$. However, they do not depend on it if Y is an object of $\mathbf{Q}$ by 3.2.12. Moreover, $\mathscr{F}_{Y,n}$ and $\mathscr{C}_{Y,n}$ depend on the choice of the deformation $(\widetilde{X},\mathscr{M}_{\widetilde{X}})$ fixed in 4.1.5.

4.4.6 Let $g\colon Y\to Z$ be a morphism of $\mathbf{P}$ such that Y_s is non-empty, $((P,\gamma),(\mathbb{N},\iota),\vartheta)$ an adequate chart for the morphism $f|Z\colon (Z,\mathscr{M}_X|Z)\to(S,\mathscr{M}_S)$ induced by f, $\overline{y}$ a geometric point of $\overline{Y}^{\circ}$, $\overline{z}=\overline{g}(\overline{y})$. We endow the morphism $f|Y\colon (Y,\mathscr{M}_X|Y)\to(S,\mathscr{M}_S)$ induced by f with the same adequate chart $((P,\gamma),(\mathbb{N},\iota),\vartheta)$. Recall that $\overline{Y}$ and $\overline{Z}$ are sums of the schemes induced over their irreducible components. We denote by $\overline{Y}^{\star}$ the irreducible component of $\overline{Y}$ containing $\overline{y}$ and by $\overline{Z}^{\star}$ the irreducible component of $\overline{Z}$ containing $\overline{z}$, so that $\overline{g}(\overline{Y}^{\star})\subset\overline{Z}^{\star}$. The morphism $\overline{g}^{\circ}\colon\overline{Y}^{\circ}\to\overline{Z}^{\circ}$ induces a group homomorphism $\pi_1(\overline{Y}^{\star\circ},\overline{y})\to\pi_1(\overline{Z}^{\star\circ},\overline{z})$. The canonical morphism $(\overline{g}^{\circ})^*_{\text{fét}}(\overline{\mathscr{B}}_Z)\to\overline{\mathscr{B}}_Y$ induces a $\pi_1(\overline{Y}^{\star\circ},\overline{y})$-equivariant ring homomorphism

$$\overline{R}^{\overline{z}}_Z\to\overline{R}^{\overline{y}}_Y, \tag{4.4.6.1}$$

and hence a $\pi_1(\overline{Y}^{\star\circ},\overline{y})$-equivariant morphism of schemes $h\colon\widehat{\mathbb{Y}}^{\overline{y}}\to\widehat{\mathbb{Z}}^{\overline{z}}$. Since g is étale, we have a canonical $\mathscr{O}_{\widehat{\mathbb{Z}}^{\overline{z}}}$-linear and $\pi_1(\overline{Y}^{\star\circ},\overline{y})$-equivariant morphism $u\colon\widetilde{\mathrm{T}}^{\overline{z}}_Z\to h_*(\widetilde{\mathrm{T}}^{\overline{y}}_Y)$ such that the adjoint morphism $u^{\sharp}\colon h^*(\widetilde{\mathrm{T}}^{\overline{z}}_Z)\to\widetilde{\mathrm{T}}^{\overline{y}}_Y$ is an isomorphism. It immediately follows from the definitions (4.4.4.5) that we have a canonical u-equivariant and $\pi_1(\overline{Y}^{\star\circ},\overline{y})$-equivariant morphism

$$v\colon\mathscr{L}^{\overline{z}}_Z\to h_*(\mathscr{L}^{\overline{y}}_Y). \tag{4.4.6.2}$$

By ([3] II.4.22), the pair (u,v) induces an $\widehat{\overline{R}}^{\overline{z}}_Y$-linear and $\pi_1(\overline{Y}^{\star\circ},\overline{y})$-equivariant isomorphism

$$\mathscr{F}^{\overline{y}}_Y\xrightarrow{\sim}\mathscr{F}^{\overline{z}}_Z\otimes_{\widehat{\overline{R}}^{\overline{z}}_Z}\widehat{\overline{R}}^{\overline{y}}_Y, \tag{4.4.6.3}$$

and hence an $\widehat{\overline{R}}^{\overline{z}}_Z$-linear and $\pi_1(\overline{Y}^{\star\circ},\overline{y})$-equivariant morphism

$$\mathscr{F}^{\overline{z}}_Z\to\mathscr{F}^{\overline{y}}_Y \tag{4.4.6.4}$$

that fits into a commutative diagram

$$\begin{array}{ccccccccc}
0 & \longrightarrow & \widehat{\overline{R}}^{\overline{z}}_Z & \longrightarrow & \mathscr{F}^{\overline{z}}_Z & \longrightarrow & \widetilde{\xi}^{-1}\widetilde{\Omega}^1_{X/S}(Z)\otimes_{\mathscr{O}_X(Z)}\widehat{\overline{R}}^{\overline{z}}_Z & \longrightarrow & 0 \\
 & & \downarrow & & \downarrow & & \downarrow & & \\
0 & \longrightarrow & \widehat{\overline{R}}^{\overline{y}}_Y & \longrightarrow & \mathscr{F}^{\overline{y}}_Y & \longrightarrow & \widetilde{\xi}^{-1}\widetilde{\Omega}^1_{X/S}(Y)\otimes_{\mathscr{O}_X(Y)}\widehat{\overline{R}}^{\overline{y}}_Y & \longrightarrow & 0.
\end{array} \tag{4.4.6.5}$$

We deduce from this a $\pi_1(\overline{Y}^{\star\circ}, \overline{y})$-equivariant homomorphism of $\widehat{\overline{R}}{}^{\overline{z}}_Z$-algebras

$$\mathscr{C}^{\overline{z}}_Z \to \mathscr{C}^{\overline{y}}_Y. \tag{4.4.6.6}$$

We denote by $\Pi(\overline{Y}^{\star\circ})$ and $\Pi(\overline{Z}^{\star\circ})$ the fundamental groupoids of $\overline{Y}^{\star\circ}$ and $\overline{Z}^{\star\circ}$ and by

$$\gamma\colon \Pi(\overline{Y}^{\star\circ}) \to \Pi(\overline{Z}^{\star\circ}) \tag{4.4.6.7}$$

the functor induced by the pullback functor $\mathbf{\acute{E}t}_{\mathrm{f}/\overline{Z}^{\star\circ}} \to \mathbf{\acute{E}t}_{\mathrm{f}/\overline{Y}^{\star\circ}}$. For any integer $n \geq 0$, we denote by $F_{Y,n}\colon \Pi(\overline{Y}^{\star\circ}) \to \mathbf{Sets}$ and $F_{Z,n}\colon \Pi(\overline{Z}^{\star\circ}) \to \mathbf{Sets}$ the functors associated by ([3] VI.9.11) with the objects $\mathscr{F}_{Y,n}|\overline{Y}^{\star\circ}$ of $\overline{Y}^{\star\circ}_{\mathrm{f\acute{e}t}}$ and $\mathscr{F}_{Z,n}|\overline{Z}^{\star\circ}$ of $\overline{Z}^{\star\circ}_{\mathrm{f\acute{e}t}}$, respectively. The morphism (4.4.6.4) clearly induces a morphism of functors

$$F_{Z,n} \circ \gamma \to F_{Y,n}. \tag{4.4.6.8}$$

We deduce by ([3] VI.9.11) a $(\overline{g}^{\circ})^*_{\mathrm{f\acute{e}t}}(\overline{\mathscr{B}}_{Z,n})$-linear morphism

$$(\overline{g}^{\circ})^*_{\mathrm{f\acute{e}t}}(\mathscr{F}_{Z,n}) \to \mathscr{F}_{Y,n}, \tag{4.4.6.9}$$

and therefore by adjunction, a $\overline{\mathscr{B}}_{Z,n}$-linear morphism

$$\mathscr{F}_{Z,n} \to \overline{g}^{\circ}_{\mathrm{f\acute{e}t}*}(\mathscr{F}_{Y,n}). \tag{4.4.6.10}$$

It follows from (4.4.6.5) that the diagram

$$\begin{array}{ccccccccc} 0 & \longrightarrow & (\overline{g}^{\circ})^*_{\mathrm{f\acute{e}t}}(\overline{\mathscr{B}}_{Z,n}) & \longrightarrow & (\overline{g}^{\circ})^*_{\mathrm{f\acute{e}t}}(\mathscr{F}_{Z,n}) & \longrightarrow & \widetilde{\xi}^{-1}\widetilde{\Omega}^1_{X/S}(Z) \otimes_{\mathscr{O}_X(Z)} (\overline{g}^{\circ})^*_{\mathrm{f\acute{e}t}}(\overline{\mathscr{B}}_{Z,n}) & \longrightarrow & 0 \\ & & \downarrow & & \downarrow & & \downarrow & & \\ 0 & \longrightarrow & \overline{\mathscr{B}}_{Y,n} & \longrightarrow & \mathscr{F}_{Y,n} & \longrightarrow & \widetilde{\xi}^{-1}\widetilde{\Omega}^1_{X/S}(Y) \otimes_{\mathscr{O}_X(Y)} \overline{\mathscr{B}}_{Y,n} & \longrightarrow & 0 \end{array} \tag{4.4.6.11}$$

is commutative. We deduce from this a homomorphism of $(\overline{g}^{\circ})^*_{\mathrm{f\acute{e}t}}(\overline{\mathscr{B}}_{Z,n})$-algebras

$$(\overline{g}^{\circ})^*_{\mathrm{f\acute{e}t}}(\mathscr{C}_{Z,n}) \to \mathscr{C}_{Y,n}, \tag{4.4.6.12}$$

and therefore by adjunction a homomorphism of $\overline{\mathscr{B}}_{Z,n}$-algebras

$$\mathscr{C}_{Z,n} \to \overline{g}^{\circ}_{\mathrm{f\acute{e}t}*}(\mathscr{C}_{Y,n}). \tag{4.4.6.13}$$

Note that the morphisms (4.4.6.10) and (4.4.6.13) depend on the choice of the adequate chart $((P,\gamma), (\mathbb{N}, \iota), \vartheta)$. However, they do not depend on it if Y and Z are objects of $\mathbf{Q}$ by 3.2.12.

4.4.7 We take again the assumptions and notation of 4.4.5. Let r be a rational number ≥ 0, n an integer ≥ 0. We denote by $\mathscr{F}^{(r)}_{Y,n}$ the extension of $\overline{\mathscr{B}}_{Y,n}$-modules of $\overline{Y}^{\circ}_{\mathrm{f\acute{e}t}}$ deduced from $\mathscr{F}_{Y,n}$ (4.4.5.3) by pullback by the morphism of multiplication by p^r on $\widetilde{\xi}^{-1}\widetilde{\Omega}^1_{X/S}(Y) \otimes_{\mathscr{O}_X(Y)} \overline{\mathscr{B}}_{Y,n}$, so that we have a canonical exact sequence of

$\overline{\mathscr{B}}_{Y,n}$-modules

$$0 \to \overline{\mathscr{B}}_{Y,n} \to \mathscr{F}^{(r)}_{Y,n} \to \widetilde{\xi}^{-1}\widetilde{\Omega}^1_{X/S}(Y) \otimes_{\mathscr{O}_X(Y)} \overline{\mathscr{B}}_{Y,n} \to 0. \tag{4.4.7.1}$$

This induces for every integer $m \geq 1$ an exact sequence of $\overline{\mathscr{B}}_{Y,n}$-modules

$$0 \to \mathrm{S}^{m-1}_{\overline{\mathscr{B}}_{Y,n}}(\mathscr{F}^{(r)}_{Y,n}) \to \mathrm{S}^{m}_{\overline{\mathscr{B}}_{Y,n}}(\mathscr{F}^{(r)}_{Y,n}) \to \mathrm{S}^{m}_{\overline{\mathscr{B}}_{Y,n}}(\widetilde{\xi}^{-1}\widetilde{\Omega}^1_{X/S}(Y) \otimes_{\mathscr{O}_X(Y)} \overline{\mathscr{B}}_{Y,n}) \to 0.$$

The $\overline{\mathscr{B}}_{Y,n}$-modules $(\mathrm{S}^{m}_{\overline{\mathscr{B}}_{Y,n}}(\mathscr{F}^{(r)}_{Y,n}))_{m\in\mathbb{N}}$ thus form a direct system whose direct limit

$$\mathscr{C}^{(r)}_{Y,n} = \varinjlim_{m\geq 0} \mathrm{S}^{m}_{\overline{\mathscr{B}}_{Y,n}}(\mathscr{F}^{(r)}_{Y,n}) \tag{4.4.7.2}$$

is naturally endowed with a structure of $\overline{\mathscr{B}}_{Y,n}$-algebra of $\overline{Y}^{\circ}_{\text{fét}}$.

For all rational numbers $r \geq r' \geq 0$, we have a canonical $\overline{\mathscr{B}}_{Y,n}$-linear morphism

$$\mathrm{a}^{r,r'}_{Y,n} \colon \mathscr{F}^{(r)}_{Y,n} \to \mathscr{F}^{(r')}_{Y,n} \tag{4.4.7.3}$$

which lifts the multiplication by $p^{r-r'}$ over $\widetilde{\xi}^{-1}\widetilde{\Omega}^1_{X/S}(Y) \otimes_{\mathscr{O}_X(Y)} \overline{\mathscr{B}}_{Y,n}$ and which extends the identity over $\overline{\mathscr{B}}_{Y,n}$ (4.4.7.1). It induces a homomorphism of $\overline{\mathscr{B}}_{Y,n}$-algebras

$$\alpha^{r,r'}_{Y,n} \colon \mathscr{C}^{(r)}_{Y,n} \to \mathscr{C}^{(r')}_{Y,n}. \tag{4.4.7.4}$$

4.4.8 We take again the assumptions and notation of 4.4.6. Let r be a rational number ≥ 0, n an integer ≥ 0. The diagram (4.4.6.11) induces a $(\overline{g}^{\circ})^*_{\text{fét}}(\overline{\mathscr{B}}_{Z,n})$-linear morphism

$$(\overline{g}^{\circ})^*_{\text{fét}}(\mathscr{F}^{(r)}_{Z,n}) \to \mathscr{F}^{(r)}_{Y,n} \tag{4.4.8.1}$$

which fits into a commutative diagram

$$\begin{array}{ccccccccc} 0 & \to & (\overline{g}^{\circ})^*_{\text{fét}}(\overline{\mathscr{B}}_{Z,n}) & \to & (\overline{g}^{\circ})^*_{\text{fét}}(\mathscr{F}^{(r)}_{Z,n}) & \to & \widetilde{\xi}^{-1}\widetilde{\Omega}^1_{X/S}(Z) \otimes_{\mathscr{O}_X(Z)} (\overline{g}^{\circ})^*_{\text{fét}}(\overline{\mathscr{B}}_{Z,n}) & \to & 0 \\ & & \downarrow & & \downarrow & & \downarrow & & \\ 0 & \to & \overline{\mathscr{B}}_{Y,n} & \to & \mathscr{F}^{(r)}_{Y,n} & \to & \widetilde{\xi}^{-1}\widetilde{\Omega}^1_{X/S}(Y) \otimes_{\mathscr{O}_X(Y)} \overline{\mathscr{B}}_{Y,n} & \to & 0. \end{array} \tag{4.4.8.2}$$

We deduce by adjunction a $\overline{\mathscr{B}}_{Z,n}$-linear morphism

$$\mathscr{F}^{(r)}_{Z,n} \to (\overline{g}^{\circ})_{\text{fét}*}(\mathscr{F}^{(r)}_{Y,n}). \tag{4.4.8.3}$$

We also deduce a morphism of $(\overline{g}^{\circ})^*_{\text{fét}}(\overline{\mathscr{B}}_{Z,n})$-algebras

$$(\overline{g}^{\circ})^*_{\text{fét}}(\mathscr{C}^{(r)}_{Z,n}) \to \mathscr{C}^{(r)}_{Y,n}, \tag{4.4.8.4}$$

and therefore by adjunction a morphism of $\overline{\mathscr{B}}_{Z,n}$-algebras

$$\mathscr{C}^{(r)}_{Z,n} \to (\overline{g}^{\circ})_{\text{fét}*}(\mathscr{C}^{(r)}_{Y,n}). \tag{4.4.8.5}$$

The morphisms (4.4.8.3) and (4.4.8.5) satisfy cocycle relations of type ([1] (1.1.2.2)).

4.4.9 For any rational number $r \geq 0$, any integer $n \geq 0$ and any object Y of $\mathbf{P}$ such that Y_s is empty, we set $\mathscr{C}^{(r)}_{Y,n} = \mathscr{F}^{(r)}_{Y,n} = 0$. The exact sequence (4.4.7.1) still holds in this case, since $\overline{\mathscr{B}}_Y$ is a $\overline{K}$-algebra. The morphisms (4.4.8.3) and (4.4.8.5) are then defined for any morphism of $\mathbf{P}$, and they satisfy cocycle relations of the type ([1] (1.1.2.2)).

4.4.10 Let r be a rational number ≥ 0, n an integer ≥ 0. The correspondences $\{U \mapsto p^n\overline{\mathscr{B}}_U\}$ and $\{U \mapsto \overline{\mathscr{B}}_{U,n}\}$ naturally form presheaves on E (4.3.2.8), and the canonical morphisms

$$\{U \mapsto p^n\overline{\mathscr{B}}_U\}^{\mathrm{a}} \to p^n\overline{\mathscr{B}}, \tag{4.4.10.1}$$
$$\{U \mapsto \overline{\mathscr{B}}_{U,n}\}^{\mathrm{a}} \to \overline{\mathscr{B}}_n, \tag{4.4.10.2}$$

where the terms on the left denote the associated sheaves in $\widetilde{E}$, are isomorphisms by virtue of ([3] VI.8.2 and VI.8.9). By 4.4.8, the correspondences

$$\{Y \in \mathbf{Q}^{\circ} \mapsto \mathscr{F}^{(r)}_{Y,n}\} \quad \text{and} \quad \{Y \in \mathbf{Q}^{\circ} \mapsto \mathscr{C}^{(r)}_{Y,n}\} \tag{4.4.10.3}$$

define presheaves on $E_{\mathbf{Q}}$ (4.4.2.4) of modules and algebras, respectively, relative to the ring $\{Y \in \mathbf{Q}^{\circ} \mapsto \overline{\mathscr{B}}_{Y,n}\}$. We denote by

$$\mathscr{F}^{(r)}_n = \{Y \in \mathbf{Q}^{\circ} \mapsto \mathscr{F}^{(r)}_{Y,n}\}^a, \tag{4.4.10.4}$$
$$\mathscr{C}^{(r)}_n = \{Y \in \mathbf{Q}^{\circ} \mapsto \mathscr{C}^{(r)}_{Y,n}\}^a, \tag{4.4.10.5}$$

the associated sheaves in $\widetilde{E}$ (see 2.10.5 and 4.4.2). By (4.4.10.2) and ([3] (III.10.6.5)), $\mathscr{F}^{(r)}_n$ is a $\overline{\mathscr{B}}_n$-module; it is called the Higgs–Tate $\overline{\mathscr{B}}_n$*-extension of thickness* r associated with $(f, \widetilde{X}, \mathscr{M}_{\widetilde{X}})$. Similarly, $\mathscr{C}^{(r)}_n$ is a $\overline{\mathscr{B}}_n$-algebra; it is called the Higgs–Tate $\overline{\mathscr{B}}_n$*-algebra of thickness* r associated with $(f, \widetilde{X}, \mathscr{M}_{\widetilde{X}})$. We set $\mathscr{F}_n = \mathscr{F}^{(0)}_n$ and $\mathscr{C}_n = \mathscr{C}^{(0)}_n$, that we call the *Higgs–Tate* $\overline{\mathscr{B}}_n$*-extension* and the *Higgs–Tate* $\overline{\mathscr{B}}_n$*-algebra*, respectively, associated with $(f, \widetilde{X}, \mathscr{M}_{\widetilde{X}})$.

For all rational numbers $r \geq r' \geq 0$, the morphisms (4.4.7.3) induce a $\overline{\mathscr{B}}_n$-linear morphism

$$\mathrm{a}^{r,r'}_n : \mathscr{F}^{(r)}_n \to \mathscr{F}^{(r')}_n. \tag{4.4.10.6}$$

The homomorphisms (4.4.7.4) induce a homomorphism of $\overline{\mathscr{B}}_n$-algebras

$$\alpha^{r,r'}_n : \mathscr{C}^{(r)}_n \to \mathscr{C}^{(r')}_n. \tag{4.4.10.7}$$

For all rational numbers $r \geq r' \geq r'' \geq 0$, we have

$$\mathrm{a}^{r,r''}_n = \mathrm{a}^{r',r''}_n \circ \mathrm{a}^{r,r'}_n \quad \text{and} \quad \alpha^{r,r''}_n = \alpha^{r',r''}_n \circ \alpha^{r,r'}_n. \tag{4.4.10.8}$$

Proposition 4.4.11 ([3] III.10.22) *Let r be a rational number ≥ 0, n an integer ≥ 1. Then,*

(i) *The sheaves $\mathscr{F}_n^{(r)}$ and $\mathscr{C}_n^{(r)}$ are objects of $\widetilde{E}_s$.*

(ii) *With the notation of (4.1.5.1) and (4.3.10.3), we have a canonical locally split exact sequence of $\overline{\mathscr{B}}_n$-modules*

$$0 \to \overline{\mathscr{B}}_n \to \mathscr{F}_n^{(r)} \to \sigma_n^*(\widetilde{\xi}^{-1}\widetilde{\Omega}^1_{\overline{X}_n/\overline{S}_n}) \to 0. \tag{4.4.11.1}$$

It induces for any integer $m \geq 1$ an exact sequence of $\overline{\mathscr{B}}_n$-modules

$$0 \to \mathrm{S}^{m-1}_{\overline{\mathscr{B}}_n}(\mathscr{F}_n^{(r)}) \to \mathrm{S}^m_{\overline{\mathscr{B}}_n}(\mathscr{F}_n^{(r)}) \to \sigma_n^*(\mathrm{S}^m_{\mathscr{O}_{\overline{X}_n}}(\widetilde{\xi}^{-1}\widetilde{\Omega}^1_{\overline{X}_n/\overline{S}_n})) \to 0. \tag{4.4.11.2}$$

In particular, the $\overline{\mathscr{B}}_n$-modules $(\mathrm{S}^m_{\overline{\mathscr{B}}_n}(\mathscr{F}_n^{(r)}))_{m\in\mathbb{N}}$ form a filtered direct system.

(iii) *We have a canonical isomorphism of $\overline{\mathscr{B}}_n$-algebras*

$$\mathscr{C}_n^{(r)} \xrightarrow{\sim} \varinjlim_{m\geq 0} \mathrm{S}^m_{\overline{\mathscr{B}}_n}(\mathscr{F}_n^{(r)}). \tag{4.4.11.3}$$

(iv) *For all rational numbers $r \geq r' \geq 0$, the diagram*

$$\begin{array}{ccccccccc} 0 & \longrightarrow & \overline{\mathscr{B}}_n & \longrightarrow & \mathscr{F}_n^{(r)} & \longrightarrow & \sigma_n^*(\widetilde{\xi}^{-1}\widetilde{\Omega}^1_{\overline{X}_n/\overline{S}_n}) & \longrightarrow & 0 \\ & & \| & & \downarrow{\scriptstyle \mathrm{a}_n^{r,r'}} & & \downarrow{\scriptstyle \cdot p^{r-r'}} & & \\ 0 & \longrightarrow & \overline{\mathscr{B}}_n & \longrightarrow & \mathscr{F}_n^{(r')} & \longrightarrow & \sigma_n^*(\widetilde{\xi}^{-1}\widetilde{\Omega}^1_{\overline{X}_n/\overline{S}_n}) & \longrightarrow & 0, \end{array} \tag{4.4.11.4}$$

where the horizontal lines are the exact sequences (4.4.11.1) and the right vertical arrow denotes multiplication by $p^{r-r'}$, is commutative. Moreover, the morphisms $\mathrm{a}_n^{r,r'}$ and $\alpha_n^{r,r'}$ are compatible with the isomorphisms (4.4.11.3) for r and r'.

4.4.12 Suppose that the morphism $f: (X, \mathscr{M}_X) \to (S, \mathscr{M}_S)$ admits an adequate chart $((P, \gamma), (\mathbb{N}, \iota), \vartheta)$ that we fix. Let r be a rational number ≥ 0, n an integer ≥ 0. By 4.4.8, the correspondences

$$\{Y \in \mathbf{P}^{\circ} \mapsto \mathscr{F}_{Y,n}^{(r)}\} \quad \text{and} \quad \{Y \in \mathbf{P}^{\circ} \mapsto \mathscr{C}_{Y,n}^{(r)}\} \tag{4.4.12.1}$$

then define presheaves on $\underline{E}_{\mathbf{P}}$ (4.4.1.3) of modules and algebras, respectively, relative to the ring $\{Y \mapsto \overline{\mathscr{B}}_{Y,n}\}$. These presheaves depend on the adequate chart $((P, \gamma), (\mathbb{N}, \iota), \vartheta)$, but the associated sheaves do not depend on it. Indeed, by virtue of 2.10.6(i), we have canonical isomorphisms

$$\mathscr{F}_n^{(r)} \xrightarrow{\sim} \{Y \in \mathbf{P}^{\circ} \mapsto \mathscr{F}_{Y,n}^{(r)}\}^a, \tag{4.4.12.2}$$

$$\mathscr{C}_n^{(r)} \xrightarrow{\sim} \{Y \in \mathbf{P}^{\circ} \mapsto \mathscr{C}_{Y,n}^{(r)}\}^a, \tag{4.4.12.3}$$

where the $\overline{\mathscr{B}}_n$-module $\mathscr{F}_n^{(r)}$ and the $\overline{\mathscr{B}}_n$-algebra $\mathscr{C}_n^{(r)}$ are defined in 4.4.10.

4.4.13 Let Y be an object of $\mathbf{P}$ such that Y_s is nonempty, $((P,\gamma),(\mathbb{N},\iota),\vartheta)$ an adequate chart for the morphism $f|Y\colon (Y,\mathscr{M}_X|Y)\to(S,\mathscr{M}_S)$ induced by f, $\overline{y}$ a geometric point of $\overline{Y}^{\circ}$. We take again the notation of 4.4.4. For any rational number $r\geq 0$, we denote by $\mathscr{F}_Y^{\overline{y},(r)}$ the extension of $\widehat{\overline{R}}{}^{\overline{y}}_Y$-modules deduced from $\mathscr{F}_Y^{\overline{y}}$ (4.4.4.6) by pullback by the morphism of multiplication by p^r on $\widetilde{\xi}^{-1}\widetilde{\Omega}^1_{X/S}(Y)\otimes_{\mathscr{O}_X(Y)}\widehat{\overline{R}}{}^{\overline{y}}_Y$, so that we have an exact sequence of $\widehat{\overline{R}}{}^{\overline{y}}_Y$-modules

$$0\to\widehat{\overline{R}}{}^{\overline{y}}_Y\to\mathscr{F}_Y^{\overline{y},(r)}\to\widetilde{\xi}^{-1}\widetilde{\Omega}^1_{X/S}(Y)\otimes_{\mathscr{O}_X(Y)}\widehat{\overline{R}}{}^{\overline{y}}_Y\to 0. \tag{4.4.13.1}$$

We denote by $\mathscr{C}_Y^{\overline{y},(r)}$ the $\widehat{\overline{R}}{}^{\overline{y}}_Y$-algebra (3.2.17.3)

$$\mathscr{C}_Y^{\overline{y},(r)}=\varinjlim_{m\geq 0}\mathrm{S}^m_{\widehat{\overline{R}}{}^{\overline{y}}_Y}(\mathscr{F}_Y^{\overline{y},(r)}). \tag{4.4.13.2}$$

It follows from 4.4.6 that the formations of $\mathscr{F}_Y^{\overline{y},(r)}$ and $\mathscr{C}_Y^{\overline{y},(r)}$ are functorial in the pair $(Y,\overline{y})$. More precisely, let $g\colon Z\to Y$ be a morphism of $\mathbf{P}$, $\overline{z}$ a geometric point of $\overline{Z}^{\circ}$ with image $\overline{y}$ by the morphism $\overline{g}^{\circ}\colon\overline{Z}^{\circ}\to\overline{Y}^{\circ}$. We endow the morphism $f|Z\colon(Z,\mathscr{M}_X|Z)\to(S,\mathscr{M}_S)$ induced by f with the same adequate chart $((P,\gamma),(\mathbb{N},\iota),\vartheta)$. It immediately follows from 4.4.6 that the canonical diagram

$$\begin{array}{ccccccccc}
0 & \longrightarrow & \widehat{\overline{R}}{}^{\overline{y}}_Y & \longrightarrow & \mathscr{F}_Y^{\overline{y},(r)} & \longrightarrow & \widetilde{\xi}^{-1}\widetilde{\Omega}^1_{X/S}(Y)\otimes_{\mathscr{O}_X(Y)}\widehat{\overline{R}}{}^{\overline{y}}_Y & \longrightarrow & 0\\
 & & \downarrow & & \downarrow & & \downarrow & & \\
0 & \longrightarrow & \widehat{\overline{R}}{}^{\overline{z}}_Z & \longrightarrow & \mathscr{F}_Z^{\overline{z},(r)} & \longrightarrow & \widetilde{\xi}^{-1}\widetilde{\Omega}^1_{X/S}(Z)\otimes_{\mathscr{O}_X(Z)}\widehat{\overline{R}}{}^{\overline{z}}_Z & \longrightarrow & 0
\end{array} \tag{4.4.13.3}$$

is commutative. The canonical morphism

$$\widetilde{\Omega}^1_{X/S}(Y)\otimes_{\mathscr{O}_X(Y)}\mathscr{O}_X(Z)\to\widetilde{\Omega}^1_{X/S}(Z) \tag{4.4.13.4}$$

being an isomorphism, we deduce that the canonical morphisms

$$\mathscr{F}_Y^{\overline{y},(r)}\otimes_{\widehat{\overline{R}}{}^{\overline{y}}_Y}\widehat{\overline{R}}{}^{\overline{z}}_Z\to\mathscr{F}_Z^{\overline{z},(r)}, \tag{4.4.13.5}$$

$$\mathscr{C}_Y^{\overline{y},(r)}\otimes_{\widehat{\overline{R}}{}^{\overline{y}}_Y}\widehat{\overline{R}}{}^{\overline{z}}_Z\to\mathscr{C}_Z^{\overline{z},(r)}, \tag{4.4.13.6}$$

are isomorphisms.

Since $\overline{Y}$ is locally irreducible (4.4.4), it is the sum of the schemes induced on its irreducible components. We denote by $\overline{Y}^{\star}$ the irreducible component of $\overline{Y}$ containing $\overline{y}$. Similarly, $\overline{Y}^{\circ}$ is the sum of the schemes induced on its irreducible components, and $\overline{Y}^{\star\circ}=\overline{Y}^{\star}\times_X X^{\circ}$ is the irreducible component of $\overline{Y}^{\circ}$ containing $\overline{y}$. We denote by

$\mathbf{B}_{\pi_1(\overline{Y}^{\star\circ},\overline{y})}$ the classifying topos of the profinite group $\pi_1(\overline{Y}^{\star\circ},\overline{y})$ and by

$$\nu_{\overline{y}}\colon \overline{Y}^{\star\circ}_{\text{fét}} \xrightarrow{\sim} \mathbf{B}_{\pi_1(\overline{Y}^{\star\circ},\overline{y})} \tag{4.4.13.7}$$

the fiber functor at $\overline{y}$ ([3] VI.9.8). We then have (4.3.7.2)

$$\overline{R}^{\overline{y}}_Y = \nu_{\overline{y}}(\overline{\mathscr{B}}_Y|\overline{Y}^{\star\circ}). \tag{4.4.13.8}$$

Since $\nu_{\overline{y}}$ is exact and commutes with direct limits, for every integer $n \geq 0$, we have canonical isomorphisms of $\overline{R}^{\overline{y}}_Y$-modules and $\overline{R}^{\overline{y}}_Y$-algebras, respectively (see 4.4.7),

$$\nu_{\overline{y}}(\mathscr{F}^{(r)}_{Y,n}|\overline{Y}^{\star\circ}) \xrightarrow{\sim} \mathscr{F}^{\overline{y},(r)}_Y/p^n\mathscr{F}^{\overline{y},(r)}_Y, \tag{4.4.13.9}$$

$$\nu_{\overline{y}}(\mathscr{C}^{(r)}_{Y,n}|\overline{Y}^{\star\circ}) \xrightarrow{\sim} \mathscr{C}^{\overline{y},(r)}_Y/p^n\mathscr{C}^{\overline{y},(r)}_Y. \tag{4.4.13.10}$$

4.4.14 Let $(\overline{y} \rightsquigarrow \overline{x})$ be a point of $X_{\text{ét}} \overleftarrow{\times}_{X_{\text{ét}}} \overline{X}^{\circ}_{\text{ét}}$ (4.3.4) such that $\overline{x}$ is above s, $\underline{X}$ the strict localization of X at $\overline{x}$. Recall that giving a neighborhood of the point of $X_{\text{ét}}$ associated with $\overline{x}$ in the site $\mathbf{\acute{E}t}_{/X}$ (resp. $\mathbf{P}$ (4.4.1), resp. $\mathbf{Q}$ (4.4.2)) is equivalent to giving an $\overline{x}$-pointed étale X-scheme (resp. of $\mathbf{P}$, resp. of $\mathbf{Q}$) ([5] IV 6.8.2). These objects naturally form a cofiltered category, that we denote by $\mathfrak{V}_{\overline{x}}$ (resp. $\mathbf{P}_{\overline{x}}$, resp. $\mathbf{Q}_{\overline{x}}$). The categories $\mathbf{P}_{\overline{x}}$ and $\mathbf{Q}_{\overline{x}}$ are $\mathbb{U}$-small, and the canonical injection functors $\mathbf{Q} \to \mathbf{P} \to \mathbf{\acute{E}t}_{/X}$ induce fully faithful and cofinal functors $\mathbf{Q}_{\overline{x}} \to \mathbf{P}_{\overline{x}} \to \mathfrak{V}_{\overline{x}}$.

We take again the notation of 4.3.9, in particular the algebra

$$\overline{R}^{\overline{y}}_{\underline{X}} = \lim_{\substack{\longrightarrow \\ (U,\mathfrak{p})\in\mathfrak{V}^{\circ}_{\overline{x}}}} \overline{R}^{\overline{y}}_U, \tag{4.4.14.1}$$

where $\overline{R}^{\overline{y}}_U$ is the ring defined in (4.4.13.8). We denote by $\widehat{\overline{R}}^{\overline{y}}_{\underline{X}}$ the p-adic Hausdorff completion of $\overline{R}^{\overline{y}}_{\underline{X}}$. We have a canonical isomorphism (4.3.9.4)

$$\nu_{\overline{y}}(\varphi_{\overline{x}}(\overline{\mathscr{B}})) \xrightarrow{\sim} \overline{R}^{\overline{y}}_{\underline{X}}. \tag{4.4.14.2}$$

For any rational number $r \geq 0$, we set

$$\mathscr{F}^{\overline{y},(r)}_{\underline{X}} = \lim_{\substack{\longrightarrow \\ (U,\mathfrak{p})\in\mathbf{Q}^{\circ}_{\overline{x}}}} \mathscr{F}^{\overline{y},(r)}_U \otimes_{\widehat{\overline{R}}^{\overline{y}}_U} \widehat{\overline{R}}^{\overline{y}}_{\underline{X}}, \tag{4.4.14.3}$$

$$\mathscr{C}^{\overline{y},(r)}_{\underline{X}} = \lim_{\substack{\longrightarrow \\ (U,\mathfrak{p})\in\mathbf{Q}^{\circ}_{\overline{x}}}} \mathscr{C}^{\overline{y},(r)}_U \otimes_{\widehat{\overline{R}}^{\overline{y}}_U} \widehat{\overline{R}}^{\overline{y}}_{\underline{X}}, \tag{4.4.14.4}$$

where $\mathscr{F}^{\overline{y},(r)}_U$ is the $\widehat{\overline{R}}^{\overline{y}}_U$-module defined in (4.4.13.1) and $\mathscr{C}^{\overline{y},(r)}_U$ is the $\widehat{\overline{R}}^{\overline{y}}_U$-algebra defined in (4.4.13.2). By (4.4.13.5) and (4.4.13.6), for every object $(U,\mathfrak{p})$ of $\mathbf{Q}_{\overline{x}}$, the canonical morphisms

$$\mathscr{F}_U^{\overline{y},(r)} \otimes_{\widehat{\overline{R}}_U^{\overline{y}}} \widehat{\overline{R}}_{\underline{X}}^{\overline{y}} \to \mathscr{F}_{\underline{X}}^{\overline{y},(r)}, \tag{4.4.14.5}$$

$$\mathscr{C}_U^{\overline{y},(r)} \otimes_{\widehat{\overline{R}}_U^{\overline{y}}} \widehat{\overline{R}}_{\underline{X}}^{\overline{y}} \to \mathscr{C}_{\underline{X}}^{\overline{y},(r)}, \tag{4.4.14.6}$$

are isomorphisms.

Remark 4.4.15 Under the assumptions of 4.4.14, for every integer $n \geq 0$, the canonical morphisms

$$\lim_{\substack{\longrightarrow \\ (U,\mathfrak{p})\in \mathbf{Q}^{\circ}_{\overline{x}}}} \mathscr{F}_U^{\overline{y},(r)}/p^n\mathscr{F}_U^{\overline{y},(r)} \to \mathscr{F}_{\underline{X}}^{\overline{y},(r)}/p^n\mathscr{F}_{\underline{X}}^{\overline{y},(r)}, \tag{4.4.15.1}$$

$$\lim_{\substack{\longrightarrow \\ (U,\mathfrak{p})\in \mathbf{Q}^{\circ}_{\overline{x}}}} \mathscr{C}_U^{\overline{y},(r)}/p^n\mathscr{C}_U^{\overline{y},(r)} \to \mathscr{C}_{\underline{X}}^{\overline{y},(r)}/p^n\mathscr{C}_{\underline{X}}^{\overline{y},(r)}, \tag{4.4.15.2}$$

are isomorphisms.

4.4.16 Let $g : X' \to X$ be an étale morphism of finite type, r a rational number ≥ 0, n an integer ≥ 0. We endow X' with the logarithmic structure $\mathscr{M}_{X'}$ pullback of $\mathscr{M}_X$ and we denote by $f' : (X', \mathscr{M}_{X'}) \to (S, \mathscr{M}_S)$ the morphism induced by f and g. Note that f' is adequate ([3] III.4.7) and that $X'^{\circ} = X^{\circ} \times_X X'$ is the maximal open subscheme of X' where the logarithmic structure $\mathscr{M}_{X'}$ is trivial. We endow $\overline{X}'$ and $\check{\overline{X}}'$ (4.1.1.2) with the logarithmic structures $\mathscr{M}_{\overline{X}'}$ and $\mathscr{M}_{\check{\overline{X}}'}$ pullbacks of $\mathscr{M}_{X'}$. There is essentially a unique étale morphism $\widetilde{g} : \widetilde{X}' \to \widetilde{X}$ that fits into a Cartesian diagram (4.1.5)

$$\begin{array}{ccc} \check{\overline{X}}' & \longrightarrow & \widetilde{X}' \\ {\scriptstyle \check{\overline{g}}}\downarrow & & \downarrow{\scriptstyle \widetilde{g}} \\ \check{\overline{X}} & \longrightarrow & \widetilde{X}. \end{array} \tag{4.4.16.1}$$

We endow $\widetilde{X}'$ with the logarithmic structure $\mathscr{M}_{\widetilde{X}'}$ pullback of $\mathscr{M}_{\widetilde{X}}$, so that $(\widetilde{X}', \mathscr{M}_{\widetilde{X}'})$ is a smooth $(\widetilde{S}, \mathscr{M}_{\widetilde{S}})$-deformation of $(\check{\overline{X}}', \mathscr{M}_{\check{\overline{X}}'})$.

We associate with $(f', \widetilde{X}', \mathscr{M}_{\widetilde{X}'})$ objects analogous to those associated with $(f, \widetilde{X}, \mathscr{M}_{\widetilde{X}})$, denoted by the same symbols equipped with an exponent $'$.

Every étale X'-scheme is naturally an étale X-scheme. We thus define a functor

$$\mathbf{\acute{E}t}_{/X'} \to \mathbf{\acute{E}t}_{/X}, \tag{4.4.16.2}$$

which factors through an equivalence of categories $\mathbf{\acute{E}t}_{/X'} \xrightarrow{\sim} (\mathbf{\acute{E}t}_{/X})_{/X'}$. An object U' of $\mathbf{\acute{E}t}_{/X'}$ is an object of $\mathbf{P}'$ (resp. $\mathbf{Q}'$) if and only if U' is an object of $\mathbf{P}$ (resp. $\mathbf{Q}$).

Every object of E' is naturally an object of E. We thus define a functor

$$\Phi : E' \to E. \tag{4.4.16.3}$$

It factors through an equivalence of categories $E' \xrightarrow{\sim} E_{/(\overline{X}'^{\circ} \to X')}$. We denote by

$$\widehat{\Phi}^*\colon \widehat{E} \to \widehat{E}' \tag{4.4.16.4}$$

the functor defined by composition with Φ.

By ([3] VI.5.38), the covanishing topology of E' is induced by that of E by means of the functor Φ. Hence, Φ is continuous and cocontinuous ([5] III 5.2). It therefore defines a sequence of three adjoint functors:

$$\Phi_!\colon \widetilde{E}' \to \widetilde{E}, \quad \Phi^*\colon \widetilde{E} \to \widetilde{E}', \quad \Phi_*\colon \widetilde{E}' \to \widetilde{E}, \tag{4.4.16.5}$$

in the sense that for two consecutive functors of the sequence, the one on the right is right adjoint to the other. By ([5] III 5.4), the functor $\Phi_!$ factors through an equivalence of categories

$$\widetilde{E}' \xrightarrow{\sim} \widetilde{E}_{/\sigma^*(X')}, \tag{4.4.16.6}$$

where $\sigma^*(X') = (\overline{X}'^{\circ} \to X')^a$ (4.3.2.10). Since $\Phi\colon E' \to E$ is a left adjoint of the functor

$$E \to E', \quad (V \to U) \mapsto (V \times_X X' \to U \times_X X'), \tag{4.4.16.7}$$

the localization morphism $(\Phi^*, \Phi_*)\colon \widetilde{E}' \to \widetilde{E}$ of $\widetilde{E}$ at $\sigma^*(X')$ is identified with the functoriality morphism induced by g ([3] (VI.10.12)), by virtue of ([5] III 2.5).

We have a canonical homomorphism $\Phi^*(\overline{\mathscr{B}}) \to \overline{\mathscr{B}}'$ ([3] (III.8.20.6)), which is an isomorphism by virtue of ([3] III.8.21(i)).

Let n be an integer ≥ 0, Y' an object of $\mathbf{P}'$, $((P', \gamma'), (\mathbb{N}, \iota), \vartheta')$ an adequate chart for the morphism

$$f'|Y'\colon (Y', \mathscr{M}_{X'}|Y') \to (S, \mathscr{M}_S) \tag{4.4.16.8}$$

induced by f'. We can then consider the sheaf $\mathscr{F}'^{(r)}_{Y',n}$ of $\overline{Y}'^{\circ}_{\text{fét}}$ (4.4.7). Since $\mathscr{M}_X|Y' = \mathscr{M}_{X'}|Y'$, considering Y' as object of $\mathbf{P}$, we can also consider the sheaf $\mathscr{F}^{(r)}_{Y',n}$ of $\overline{Y}'^{\circ}_{\text{fét}}$. The morphism $\overline{g}$ being étale, we clearly have $\mathscr{F}^{(r)}_{Y',n} = \mathscr{F}'^{(r)}_{Y',n}$ (see [3] III.14.3).

By 2.10.6(ii), there exists a presheaf $F = \{U \in \mathbf{\acute{E}t}^{\circ}_{/X} \mapsto F_U\}$ on E such that for every $U \in \mathrm{Ob}(\mathbf{Q})$, we have $F_U = \mathscr{F}^{(r)}_{U,n}$, so that we have a canonical isomorphism $\mathscr{F}^{(r)}_n \xrightarrow{\sim} F^a$. Furthermore, we have

$$\widehat{\Phi}^*(F) = \{U' \in \mathbf{\acute{E}t}^{\circ}_{/X'} \mapsto F_{U'}\}. \tag{4.4.16.9}$$

We deduce, in view of 2.10.6(i) and ([5] III 2.3(2)), a canonical $\overline{\mathscr{B}}'_n$-linear isomorphism

$$\Phi^*(\mathscr{F}^{(r)}_n) \xrightarrow{\sim} \mathscr{F}'^{(r)}_n. \tag{4.4.16.10}$$

4.4.17 Let U be an object of $\mathbf{\acute{E}t}_{/X}$. By ([3] VI.10.14) (see also 4.4.16), the topos $\widetilde{E}_{/\sigma^*(U)}$, the localization of $\widetilde{E}$ at $\sigma^*(U)$, is canonically equivalent to the Faltings topos associated with the morphism $\overline{U}^{\circ} \to U$ (4.3.2.10). We denote by

$$j_U\colon \widetilde{E}_{/\sigma^*(U)} \to \widetilde{E} \tag{4.4.17.1}$$

the localization morphism of $\widetilde{E}$ at $\sigma^*(U)$, which is identified with the functoriality morphism induced by the canonical morphism $U \to X$ ([3] (VI.10.12)), and by

$$\beta_U \colon \widetilde{E}_{/\sigma^*(U)} \to \overline{U}^{\circ}_{\text{fét}} \tag{4.4.17.2}$$

the canonical morphism (4.3.2.13).

Let $\mu \colon U' \to U$ be a morphism of $\mathbf{\acute{E}t}_{/X}$. We denote by

$$\Phi_\mu \colon \widetilde{E}_{/\sigma^*(U')} \to \widetilde{E}_{/\sigma^*(U)} \tag{4.4.17.3}$$

the localization morphism associated with the morphism $\sigma^*(U) \to \sigma^*(U)$ ([5] IV 5.5), which is identified with the functoriality morphism induced by μ. The diagrams

$$\begin{array}{ccc} \widetilde{E}_{/\sigma^*(U')} & \xrightarrow{\Phi_\mu} & \widetilde{E}_{/\sigma^*(U)} \\ & \searrow^{j_{U'}} & \downarrow{\scriptstyle j_U} \\ & & \widetilde{E} \end{array} \tag{4.4.17.4}$$

$$\begin{array}{ccc} \widetilde{E}_{/\sigma^*(U')} & \xrightarrow{\Phi_\mu} & \widetilde{E}_{/\sigma^*(U)} \\ {\scriptstyle \beta_{U'}}\downarrow & & \downarrow{\scriptstyle \beta_U} \\ \overline{U}'^{\circ}_{\text{fét}} & \xrightarrow{\overline{\mu}^{\circ}} & \overline{U}^{\circ}_{\text{fét}} \end{array} \tag{4.4.17.5}$$

are commutative up to canonical isomorphisms ([3] (VI.10.12.6)).

Lemma 4.4.18 *Let U be an object of* $\mathbf{P}$, *$((P,\gamma),(\mathbb{N},\iota),\vartheta)$ an adequate chart for the morphism $f|U \colon (U,\mathscr{M}_X|U) \to (S,\mathscr{M}_S)$ induced by f, r a rational number ≥ 0, n an integer ≥ 0. Then, with the notation of* 4.4.17, *we have a canonical morphism*

$$\mathscr{F}^{(r)}_{U,n} \to \beta_{U*}(j^*_U(\mathscr{F}^{(r)}_n)). \tag{4.4.18.1}$$

It is independent of the adequate chart $((P,\gamma),(\mathbb{N},\iota),\vartheta)$ if U is an object of $\mathbf{Q}$.

Let $\mu \colon U' \to U$ be a morphism of $\mathbf{P}$. *We endow the morphism $f|U' \colon (U',\mathscr{M}_X|U') \to (S,\mathscr{M}_S)$ induced by f with the adequate chart induced by $((P,\gamma),(\mathbb{N},\iota),\vartheta)$. Then, the diagram*

$$\begin{array}{ccccc} \beta^*_{U'}(\overline{\mu}^{\circ *}(\mathscr{F}^{(r)}_{U,n})) & \xrightarrow{b} & \Phi^*_\mu(\beta^*_U(\mathscr{F}^{(r)}_{U,n})) & \longrightarrow & \Phi^*_\mu(j^*_U(\mathscr{F}^{(r)}_n)) \\ {\scriptstyle \beta^*_{U'}(a)}\downarrow & & & & \downarrow{\scriptstyle c} \\ \beta^*_{U'}(\mathscr{F}^{(r)}_{U',n}) & & \longrightarrow & & j^*_{U'}(\mathscr{F}^{(r)}_n), \end{array} \tag{4.4.18.2}$$

where $a \colon \overline{\mu}^{\circ *}(\mathscr{F}^{(r)}_{U,n}) \to \mathscr{F}^{(r)}_{U',n}$ (resp. b, resp. c) is the canonical morphism (4.4.8.1) *(resp. the isomorphism underlying* (4.4.17.5), *resp. the isomorphism underlying* (4.4.17.4)), *and the other two arrows are induced by the adjoint of the morphism* (4.4.18.1), *is commutative.*

Indeed, in view of (4.4.16.10), we can reduce to the case where $X = U$. For every object Y of $\mathbf{P}$, we can therefore define the sheaf $\mathscr{F}^{(r)}_{Y,n}$ of $\overline{Y}^{\circ}_{\text{fét}}$ with respect to the adequate chart $((P,\gamma),(\mathbb{N},\iota),\vartheta)$. By virtue of 2.10.6(i), we have a canonical isomorphism

$$\mathscr{F}^{(r)}_n \xrightarrow{\sim} \{Y \in \mathbf{P}^{\circ} \mapsto \mathscr{F}^{(r)}_{Y,n}\}^a. \tag{4.4.18.3}$$

By 2.10.6(ii), there exists a canonical presheaf $F = \{Y \in \mathbf{\acute{E}t}^{\circ}_{/X} \mapsto F_Y\}$ on E such that for every $Y \in \mathrm{Ob}(\mathbf{P})$, we have $F_Y = \mathscr{F}^{(r)}_{Y,n}$, so that we have a canonical isomorphism $\mathscr{F}^{(r)}_n \xrightarrow{\sim} F^a$. The canonical morphism $F \to \mathscr{F}^{(r)}_n$ induces, for every $Y \in \mathrm{Ob}(\mathbf{\acute{E}t}^{\circ}_{/X})$, a morphism

$$F_Y \to \beta_{Y*}(J^*_Y(\mathscr{F}^{(r)}_n)); \tag{4.4.18.4}$$

hence we obtain the morphism (4.4.18.1).

If U is an object of $\mathbf{Q}$, applying the same argument to the presheaf $\{Y \in \mathbf{Q}^{\circ} \mapsto \mathscr{F}^{(r)}_{Y,n}\}$ on $E_{\mathbf{Q}}$, we see that the morphism (4.4.18.1) does not depend on the adequate chart $((P,\gamma),(\mathbb{N},\iota),\vartheta)$.

For every morphism $\mu\colon U' \to U$ of $\mathbf{\acute{E}t}^{\circ}_{/X}$, the diagram

$$\begin{array}{ccccc} F_U & & \xrightarrow{\quad t \quad} & & \overline{\mu}^{\circ}_*(F_{U'}) \\ \downarrow & & & & \downarrow \\ \beta_{U*}(J^*_U(\mathscr{F}^{(r)}_n)) & \xrightarrow{\text{ad}} & \beta_{U*}(\Phi_{\mu*}(\Phi^*_{\mu}(J^*_U(\mathscr{F}^{(r)}_n)))) & \xrightarrow{\lambda} & \overline{\mu}^{\circ}_*(\beta_{U'*}(J^*_{U'}(\mathscr{F}^{(r)}_n))), \end{array} \tag{4.4.18.5}$$

where t is the transition morphism of F, ad is induced by the adjunction morphism $\mathrm{id} \to \Phi_{\mu*}\Phi^*_{\mu}$, λ is induced by the commutative diagrams (4.4.17.4) and (4.4.17.5), and the vertical arrows are induced by (4.4.18.4), is commutative. We immediately deduce that the diagram (4.4.18.2) is commutative.

4.4.19 Let $\overline{x}$ be a geometric point of X above s, $\underline{X}$ the strict localization of X at $\overline{x}$, n an integer ≥ 0. We denote by $\mathfrak{V}_{\overline{x}}$, $\mathbf{P}_{\overline{x}}$ and $\mathbf{Q}_{\overline{x}}$ the categories of the neighborhoods of $\overline{x}$ in the sites $\mathbf{\acute{E}t}_{/X}$, $\mathbf{P}$ and $\mathbf{Q}$, respectively (see 4.4.14). For any object $(U,\iota\colon \overline{x} \to U)$ of $\mathfrak{V}_{\overline{x}}$, we denote also by $\iota\colon \underline{X} \to U$ the X-morphism deduced from ι ([5] VIII 7.3), and we abusively denote by $\overline{\iota}\colon \underline{\overline{X}}^{\circ} \to \overline{U}^{\circ}$ the induced morphism.

We denote by $\underline{\widetilde{E}}$ the Faltings topos associated with the canonical morphism $\underline{\overline{X}}^{\circ} \to \underline{X}$, by

$$\Phi\colon \underline{\widetilde{E}} \to \widetilde{E} \tag{4.4.19.1}$$

the functoriality morphism induced by the canonical morphism $\underline{X} \to X$, by

$$\underline{\beta}\colon \underline{\widetilde{E}} \to \underline{\overline{X}}^{\circ}_{\text{fét}} \tag{4.4.19.2}$$

the canonical morphism (4.3.2.13) and by

$$\theta\colon \underline{\overline{X}}^{\circ}_{\text{fét}} \to \underline{\widetilde{E}} \tag{4.4.19.3}$$

the canonical section of $\underline{\beta}$ defined in ([3] VI.10.23). We denote by

$$\varphi_{\overline{x}}\colon \widetilde{E} \to \overline{X}^{\circ}_{\text{fét}} \tag{4.4.19.4}$$

the composed functor $\theta^* \circ \Phi^*$ (4.3.5.4).

Let $(U, \iota\colon \overline{x} \to U)$ be an object of $\mathfrak{V}_{\overline{x}}$. We take again the notation of 4.4.17. The morphism $\iota\colon \underline{X} \to U$ induces by functoriality a morphism

$$\Phi_\iota\colon \underline{\widetilde{E}} \to \widetilde{E}_{/\sigma^*(U)} \tag{4.4.19.5}$$

that fits into a commutative diagram up to canonical isomorphism

$$\begin{array}{ccc} \underline{\widetilde{E}} & \xrightarrow{\Phi_\iota} & \widetilde{E}_{/\sigma^*(U)} \\ \Big\downarrow{\scriptstyle \underline{\beta}} & & \Big\downarrow{\scriptstyle \beta_U} \\ \underline{\overline{X}}^{\circ}_{\text{fét}} & \xrightarrow{\overline{\iota}} & \overline{U}^{\circ}_{\text{fét}}. \end{array} \tag{4.4.19.6}$$

Let's assume that $(U, \iota\colon \overline{x} \to U)$ is an object of $\mathbf{P}_{\overline{x}}$, and let $((P,\gamma), (\mathbb{N}, \iota), \vartheta)$ be an adequate chart for the morphism $f|U\colon (U, \mathscr{M}_X|U) \to (S, \mathscr{M}_S)$ induced by f. Applying the composed functor $\theta^* \circ \Phi^*_\iota$ to the morphism $\beta^*_U(\mathscr{F}^{(r)}_{U,n}) \to j^*_U(\mathscr{F}^{(r)}_n)$ adjoint of (4.4.18.1) and taking into account the commutative diagram (4.4.19.6), we obtain a morphism

$$\overline{\iota}^*(\mathscr{F}^{(r)}_{U,n}) \to \varphi_{\overline{x}}(\mathscr{F}^{(r)}_n). \tag{4.4.19.7}$$

It is independent of the adequate chart $((P,\gamma), (\mathbb{N}, \iota), \vartheta)$ if U is an object of $\mathbf{Q}$.

The morphism $\beta^*_U(\mathscr{F}^{(r)}_{U,n}) \to j^*_U(\mathscr{F}^{(r)}_n)$ adjoint of (4.4.18.1) is the composition

$$\beta^*_U(\mathscr{F}^{(r)}_{U,n}) \to \beta^*_U(\beta_{U*}(j^*_U(\mathscr{F}^{(r)}_n))) \to j^*_U(\mathscr{F}^{(r)}_n), \tag{4.4.19.8}$$

where the first arrow is the image by the functor β^*_U of (4.4.18.1) and the second arrow is the adjunction morphism. We deduce from this a factorization of the morphism (4.4.19.7) into

$$\overline{\iota}^*(\mathscr{F}^{(r)}_{U,n}) \to \overline{\iota}^*(\beta_{U*}(j^*_U(\mathscr{F}^{(r)}_n))) \to \varphi_{\overline{x}}(\mathscr{F}^{(r)}_n). \tag{4.4.19.9}$$

By ([1] 1.2.4(i)), the second morphism is the composition

$$\overline{\iota}^*(\beta_{U*}(j^*_U(\mathscr{F}^{(r)}_n))) \to \underline{\beta}_*(\Phi^*_\iota(j^*_U(\mathscr{F}^{(r)}_n))) \to \theta^*(\Phi^*(\mathscr{F}^{(r)}_n)), \tag{4.4.19.10}$$

where the first arrow is the base change morphism with respect to the diagram (4.4.19.6) and the second arrow is induced by the base change isomorphism $\underline{\beta}_* \xrightarrow{\sim} \theta^*$ ([3] VI.10.27).

Let $\mu\colon U' \to U$ be a morphism of $\mathbf{P}$, $\iota'\colon \overline{x} \to U'$ an X-morphism, $\iota = \mu \circ \iota'\colon \overline{x} \to U$. Consider first an adequate chart $((P,\gamma), (\mathbb{N}, \iota), \vartheta)$ for the morphism $f|U\colon (U, \mathscr{M}_X|U) \to (S, \mathscr{M}_S)$ induced by f and endow the morphism $f|U'\colon (U', \mathscr{M}_X|U') \to (S, \mathscr{M}_S)$ induced by f with the chart induced by $((P,\gamma), (\mathbb{N}, \iota), \vartheta)$. It then follows from (4.4.18.2) that the diagram

$$\begin{array}{ccc} \overline{\iota}^*(\mathscr{F}^{(r)}_{U,n}) & & \\ \downarrow & \searrow & \\ \overline{\iota}'^*(\mathscr{F}^{(r)}_{U',n}) & \longrightarrow & \varphi_{\overline{x}}(\mathscr{F}^{(r)}_n), \end{array} \qquad (4.4.19.11)$$

where the vertical arrow is induced by the canonical morphism $\overline{\mu}^{\circ *}(\mathscr{F}^{(r)}_{U,n}) \to \mathscr{F}^{(r)}_{U',n}$ (4.4.8.1) and the two other arrows are the morphisms (4.4.19.7), is commutative.

We deduce from the above that if μ is a morphism of **Q**, then the diagram (4.4.19.11) is also commutative.

Proposition 4.4.20 *Let $\overline{x}$ be a geometric point of X, n an integer ≥ 0. We take again the notation of* 4.4.19. *Then,*

(i) *The morphisms* (4.4.19.7) *for* $(U, \iota) \in \mathrm{Ob}(\mathbf{Q}_{\overline{x}})$, *induce isomorphisms*

$$\lim_{\substack{\longrightarrow \\ (U,\iota)\in \mathbf{Q}^{\circ}_{\overline{x}}}} \overline{\iota}^*(\mathscr{F}^{(r)}_{U,n}) \xrightarrow{\sim} \varphi_{\overline{x}}(\mathscr{F}^{(r)}_n), \qquad (4.4.20.1)$$

$$\lim_{\substack{\longrightarrow \\ (U,\iota)\in \mathbf{Q}^{\circ}_{\overline{x}}}} \overline{\iota}^*(\mathscr{C}^{(r)}_{U,n}) \xrightarrow{\sim} \varphi_{\overline{x}}(\mathscr{C}^{(r)}_n). \qquad (4.4.20.2)$$

(ii) *Suppose that the morphism $f : (X, \mathscr{M}_X) \to (S, \mathscr{M}_S)$ admits an adequate chart. Then, the morphisms* (4.4.19.7) *defined with respect to an adequate chart of f, induce isomorphisms*

$$\lim_{\substack{\longrightarrow \\ (U,\iota)\in \mathbf{P}^{\circ}_{\overline{x}}}} \overline{\iota}^*(\mathscr{F}^{(r)}_{U,n}) \xrightarrow{\sim} \varphi_{\overline{x}}(\mathscr{F}^{(r)}_n), \qquad (4.4.20.3)$$

$$\lim_{\substack{\longrightarrow \\ (U,\iota)\in \mathbf{P}^{\circ}_{\overline{x}}}} \overline{\iota}^*(\mathscr{C}^{(r)}_{U,n}) \xrightarrow{\sim} \varphi_{\overline{x}}(\mathscr{C}^{(r)}_n). \qquad (4.4.20.4)$$

(i) This immediately follows from (ii) since the canonical functor $\mathbf{Q}_{\overline{x}} \to \mathbf{P}_{\overline{x}}$ is cofinal.

(ii) By 2.10.6(ii), there exists a canonical presheaf $F = \{U \in \mathbf{\acute{E}t}^{\circ}_{/X} \mapsto F_U\}$ on E such that for every $U \in \mathrm{Ob}(\mathbf{P})$, F_U is the sheaf $\mathscr{F}^{(r)}_{U,n}$ defined with respect to the given adequate chart of f, so that we have a canonical isomorphism $\mathscr{F}^{(r)}_n \xrightarrow{\sim} F^a$. By virtue of ([3] VI.10.37), we have a canonical isomorphism

$$\lim_{\substack{\longrightarrow \\ (U,\iota)\in \mathfrak{V}^{\circ}_{\overline{x}}}} \overline{\iota}^*(F^a_U) \xrightarrow{\sim} \varphi_{\overline{x}}(F^a). \qquad (4.4.20.5)$$

For every $(U, \iota) \in \mathrm{Ob}(\mathbf{P}_{\overline{x}})$, the morphism $\overline{\iota}^*(F^a_U) \to \varphi_{\overline{x}}(F^a)$ is identified with the morphism (4.4.19.7) by the description (4.4.19.9) (see [3] VI.10.34). The isomorphism (4.4.20.3) follows since the canonical functor $\mathbf{P}_{\overline{x}} \to \mathfrak{V}_{\overline{x}}$ is cofinal. It induces the isomorphism (4.4.20.4).

4.4.21 Let r be a rational number ≥ 0. By 4.4.11, we have a canonical $\mathscr{C}_n^{(r)}$-linear isomorphism

$$\Omega^1_{\mathscr{C}_n^{(r)}/\overline{\mathscr{B}}_n} \xrightarrow{\sim} \sigma_n^*(\widetilde{\xi}^{-1}\widetilde{\Omega}^1_{\overline{X}_n/\overline{S}_n}) \otimes_{\overline{\mathscr{B}}_n} \mathscr{C}_n^{(r)}. \tag{4.4.21.1}$$

The universal $\overline{\mathscr{B}}_n$-derivation of $\mathscr{C}_n^{(r)}$ corresponds via this isomorphism to the unique $\overline{\mathscr{B}}_n$-derivation

$$d_{\mathscr{C}_n^{(r)}} : \mathscr{C}_n^{(r)} \to \sigma_n^*(\widetilde{\xi}^{-1}\widetilde{\Omega}^1_{\overline{X}_n/\overline{S}_n}) \otimes_{\overline{\mathscr{B}}_n} \mathscr{C}_n^{(r)} \tag{4.4.21.2}$$

that extends the canonical morphism $\mathscr{F}_n^{(r)} \to \sigma_n^*(\widetilde{\xi}^{-1}\widetilde{\Omega}^1_{\overline{X}_n/\overline{S}_n})$ (4.4.11.1). It follows from 4.4.11(iv) that for all rational numbers $r \geq r' \geq 0$, we have

$$p^{r-r'}(\mathrm{id} \otimes \alpha_n^{r,r'}) \circ d_{\mathscr{C}_n^{(r)}} = d_{\mathscr{C}_n^{(r')}} \circ \alpha_n^{r,r'}. \tag{4.4.21.3}$$

4.4.22 Let r be a rational number ≥ 0. For all integers $m \geq n \geq 1$, we have a canonical $\overline{\mathscr{B}}_m$-linear morphism $\mathscr{F}_m^{(r)} \to \mathscr{F}_n^{(r)}$ compatible with the exact sequence (4.4.11.1) and a canonical homomorphism of $\overline{\mathscr{B}}_m$-algebras $\mathscr{C}_m^{(r)} \to \mathscr{C}_n^{(r)}$ such that the induced morphisms

$$\mathscr{F}_m^{(r)} \otimes_{\overline{\mathscr{B}}_m} \overline{\mathscr{B}}_n \to \mathscr{F}_n^{(r)} \quad \text{and} \quad \mathscr{C}_m^{(r)} \otimes_{\overline{\mathscr{B}}_m} \overline{\mathscr{B}}_n \to \mathscr{C}_n^{(r)} \tag{4.4.22.1}$$

are isomorphisms. These morphisms form compatible systems when m and n vary, so that $(\mathscr{F}_{n+1}^{(r)})_{n\in\mathbb{N}}$ and $(\mathscr{C}_{n+1}^{(r)})_{n\in\mathbb{N}}$ are inverse systems. With the notation of 4.3.11, we call the *Higgs–Tate $\breve{\overline{\mathscr{B}}}$-extension of thickness* r associated with $(f, \widetilde{X}, \mathscr{M}_{\widetilde{X}})$, and denote by $\breve{\mathscr{F}}^{(r)}$, the $\breve{\overline{\mathscr{B}}}$-module $(\mathscr{F}_{n+1}^{(r)})_{n\in\mathbb{N}}$ of $\widetilde{E}_s^{\mathbb{N}^\circ}$. We call the *Higgs–Tate $\breve{\overline{\mathscr{B}}}$-algebra of thickness* r associated with $(f, \widetilde{X}, \mathscr{M}_{\widetilde{X}})$, and denote by $\breve{\mathscr{C}}^{(r)}$, the $\breve{\overline{\mathscr{B}}}$-algebra $(\mathscr{C}_{n+1}^{(r)})_{n\in\mathbb{N}}$ of $\widetilde{E}_s^{\mathbb{N}^\circ}$. These are adic $\breve{\overline{\mathscr{B}}}$-modules ([3] III.7.16). By ([3] III.7.3(i), (III.7.5.4) and (III.7.12.1)), the exact sequence (4.4.11.1) induces an exact sequence of $\breve{\overline{\mathscr{B}}}$-modules

$$0 \to \breve{\overline{\mathscr{B}}} \to \breve{\mathscr{F}}^{(r)} \to \breve{\sigma}^*(\widetilde{\xi}^{-1}\widetilde{\Omega}^1_{\breve{\overline{X}}/\breve{\overline{S}}}) \to 0. \tag{4.4.22.2}$$

Since the $\mathscr{O}_X$-module $\widetilde{\Omega}^1_{X/S}$ is locally free of finite type, the $\breve{\overline{\mathscr{B}}}$-module $\breve{\sigma}^*(\widetilde{\xi}^{-1}\widetilde{\Omega}^1_{\breve{\overline{X}}/\breve{\overline{S}}})$ is locally free of finite type and the sequence (4.4.22.2) is locally split. By ([37] I 4.3.1.7), it induces for every integer $m \geq 1$ an exact sequence of $\breve{\overline{\mathscr{B}}}$-modules

$$0 \to \mathrm{S}^{m-1}_{\breve{\overline{\mathscr{B}}}}(\breve{\mathscr{F}}^{(r)}) \to \mathrm{S}^m_{\breve{\overline{\mathscr{B}}}}(\breve{\mathscr{F}}^{(r)}) \to \breve{\sigma}^*(\mathrm{S}^m_{\mathscr{O}_{\breve{\overline{X}}}}(\widetilde{\xi}^{-1}\widetilde{\Omega}^1_{\breve{\overline{X}}/\breve{\overline{S}}})) \to 0. \tag{4.4.22.3}$$

In particular, the $\breve{\overline{\mathscr{B}}}$-modules $(\mathrm{S}^m_{\breve{\overline{\mathscr{B}}}}(\breve{\mathscr{F}}^{(r)}))_{m\in\mathbb{N}}$ form a filtered direct system. By ([3] III.7.3(i) and (III.7.12.3)), we have a canonical isomorphism of $\breve{\overline{\mathscr{B}}}$-algebras

$$\breve{\mathscr{C}}^{(r)} \xrightarrow{\sim} \varinjlim_{m\geq 0} \mathrm{S}^m_{\breve{\overline{\mathscr{B}}}}(\breve{\mathscr{F}}^{(r)}). \tag{4.4.22.4}$$

Set $\breve{\mathscr{F}} = \breve{\mathscr{F}}^{(0)}$ and $\breve{\mathscr{C}} = \breve{\mathscr{C}}^{(0)}$. We call these the *Higgs–Tate $\breve{\overline{\mathscr{B}}}$-extension* and the *Higgs–Tate $\breve{\overline{\mathscr{B}}}$-algebra*, respectively, associated with $(f, \widetilde{X}, \mathscr{M}_{\widetilde{X}})$. For all rational numbers $r \geq r' \geq 0$, the morphisms $(\mathrm{a}_n^{r,r'})_{n\in\mathbb{N}}$ (4.4.10.6) induce a $\breve{\overline{\mathscr{B}}}$-linear morphism

$$\breve{\mathrm{a}}^{r,r'} : \breve{\mathscr{F}}^{(r)} \to \breve{\mathscr{F}}^{(r')}. \tag{4.4.22.5}$$

The homomorphisms $(\alpha_n^{r,r'})_{n\in\mathbb{N}}$ (4.4.10.7) induce a homomorphism of $\breve{\overline{\mathscr{B}}}$-algebras

$$\breve{\alpha}^{r,r'} : \breve{\mathscr{C}}^{(r)} \to \breve{\mathscr{C}}^{(r')}. \tag{4.4.22.6}$$

For all rational numbers $r \geq r' \geq r'' \geq 0$, we have

$$\breve{\mathrm{a}}^{r,r''} = \breve{\mathrm{a}}^{r',r''} \circ \breve{\mathrm{a}}^{r,r'} \quad \text{and} \quad \breve{\alpha}^{r,r''} = \breve{\alpha}^{r',r''} \circ \breve{\alpha}^{r,r'}. \tag{4.4.22.7}$$

The derivations $(d_{n+1}^{(r)})_{n\in\mathbb{N}}$ (4.4.21.2) define a morphism

$$d_{\breve{\mathscr{C}}^{(r)}} : \breve{\mathscr{C}}^{(r)} \to \breve{\sigma}^*(\widetilde{\xi}^{-1}\widetilde{\Omega}^1_{\breve{\overline{X}}/\breve{\overline{S}}}) \otimes_{\breve{\overline{\mathscr{B}}}} \breve{\mathscr{C}}^{(r)}, \tag{4.4.22.8}$$

which is none other than the universal $\breve{\overline{\mathscr{B}}}$-derivation of $\breve{\mathscr{C}}^{(r)}$. It extends the canonical morphism $\breve{\mathscr{F}}^{(r)} \to \breve{\sigma}^*(\xi^{-1}\widetilde{\Omega}^1_{\breve{\overline{X}}/\breve{\overline{S}}})$. For all rational numbers $r \geq r' \geq 0$, we have

$$p^{r-r'}(\mathrm{id} \otimes \breve{\alpha}^{r,r'}) \circ d_{\breve{\mathscr{C}}^{(r)}} = d_{\breve{\mathscr{C}}^{(r')}} \circ \breve{\alpha}^{r,r'}. \tag{4.4.22.9}$$

Remarks 4.4.23 Let r be a rational number ≥ 0, n an integer ≥ 1.

(i) For every integer $m \geq 0$, the canonical morphisms $\mathrm{S}^m_{\overline{\mathscr{B}}_n}(\mathscr{F}_n^{(r)}) \to \mathscr{C}_n^{(r)}$ and $\mathrm{S}^m_{\breve{\overline{\mathscr{B}}}}(\breve{\mathscr{F}}^{(r)}) \to \breve{\mathscr{C}}^{(r)}$ are injective. Indeed, for every integer $m' \geq m$, the canonical morphism $\mathrm{S}^m_{\overline{\mathscr{B}}_n}(\mathscr{F}_n^{(r)}) \to \mathrm{S}^{m'}_{\overline{\mathscr{B}}_n}(\mathscr{F}_n^{(r)})$ is injective (4.4.22.3). Since filtered injective limits commute with finite inverse limits in $\widetilde{E}_s^{\mathbb{N}^\circ}$, $\mathrm{S}^m_{\overline{\mathscr{B}}_n}(\mathscr{F}_n^{(r)}) \to \mathscr{C}_n^{(r)}$ is injective. The second assertion is deduced from the first by ([3] III.7.3(i)).

(ii) We have $\sigma_n^*(\widetilde{\xi}^{-1}\widetilde{\Omega}^1_{\overline{X}_n/\overline{S}_n}) = d_{\mathscr{C}_n^{(r)}}(\mathscr{F}_n^{(r)}) \subset d_{\mathscr{C}_n^{(r)}}(\mathscr{C}_n^{(r)})$ (4.4.21.2). Consequently, the derivation $d_{\mathscr{C}_n^{(r)}}$ is a Higgs $\overline{\mathscr{B}}_n$-field with coefficients in $\sigma_n^*(\widetilde{\xi}^{-1}\widetilde{\Omega}^1_{\overline{X}_n/\overline{S}_n})$ by 2.5.26(i).

(iii) We have $\breve{\sigma}^*(\widetilde{\xi}^{-1}\widetilde{\Omega}^1_{\breve{\overline{X}}/\breve{\overline{S}}}) = d_{\breve{\mathscr{C}}^{(r)}}(\breve{\mathscr{F}}^{(r)}) \subset d_{\breve{\mathscr{C}}^{(r)}}(\breve{\mathscr{C}}^{(r)})$ (4.4.22.8). Consequently, the derivation $d_{\breve{\mathscr{C}}^{(r)}}$ is a Higgs $\breve{\overline{\mathscr{B}}}$-field with coefficients in $\breve{\sigma}^*(\widetilde{\xi}^{-1}\widetilde{\Omega}^1_{\breve{\overline{X}}/\breve{\overline{S}}})$.

Proposition 4.4.24 *For every rational number $r \geq 0$, the functor*

$$\mathbf{Mod}(\breve{\overline{\mathscr{B}}}) \to \mathbf{Mod}(\breve{\mathscr{C}}^{(r)}), \quad M \mapsto M \otimes_{\breve{\overline{\mathscr{B}}}} \breve{\mathscr{C}}^{(r)} \tag{4.4.24.1}$$

is exact and faithful; in particular, $\breve{\mathscr{C}}^{(r)}$ is $\breve{\overline{\mathscr{B}}}$-flat.

Since the $\mathscr{O}_X$-module $\widetilde{\Omega}^1_{X/S}$ is locally free of finite type, the exact sequence (4.4.22.2) is locally split. A local splitting of this sequence induces, for every integer $m \geq 0$, a local splitting of the exact sequence (4.4.22.3). We deduce from this that $\mathrm{S}^m_{\overline{\breve{\mathscr{B}}}}(\breve{\mathscr{F}}^{(r)})$ is $\overline{\breve{\mathscr{B}}}$-flat and that the canonical homomorphism $\overline{\breve{\mathscr{B}}} \to \breve{\mathscr{C}}^{(r)}$ locally admits sections. The proposition follows in view of (4.4.22.4).

4.4.25 Consider the absolute case, i.e., $(\widetilde{S}, \mathscr{M}_{\widetilde{S}}) = (\mathscr{A}_2(\overline{S}), \mathscr{M}_{\mathscr{A}_2(\overline{S})})$ (3.1.14) and set (4.1.5)

$$(\widetilde{X}', \mathscr{M}_{\widetilde{X}'}) = (\widetilde{X}, \mathscr{M}_{\widetilde{X}}) \times_{(\mathscr{A}_2(\overline{S}), \mathscr{M}_{\mathscr{A}_2(\overline{S})})} (\mathscr{A}_2^*(\overline{S}/S), \mathscr{M}_{\mathscr{A}_2^*(\overline{S}/S)}), \tag{4.4.25.1}$$

where the base change is defined by the morphism pr_2 (3.1.13.6). We assign an exponent $'$ to the objects associated with the $(\mathscr{A}_2^*(\overline{S}/S), \mathscr{M}_{\mathscr{A}_2^*(\overline{S}/S)})$-deformation $(\widetilde{X}', \mathscr{M}_{\widetilde{X}'})$ (4.4.22). We denote by K_0 the field of fractions of W (4.1.1) and by $\mathfrak{d}$ the different of the extension K/K_0 and we set $\rho = v(\pi\mathfrak{d})$ (4.1.1). We denote by

$$\iota\colon \xi\mathscr{O}_C \xrightarrow{\sim} \xi_\pi^*\mathscr{O}_C \tag{4.4.25.2}$$

the $\mathscr{O}_C$-linear isomorphism such that the composition

$$\xi\mathscr{O}_C \xrightarrow{\iota} \xi_\pi^*\mathscr{O}_C \xrightarrow{\cdot p^\rho} p^\rho\xi_\pi^*\mathscr{O}_C \tag{4.4.25.3}$$

coincides with the isomorphism (3.1.10.1). The isomorphism ι induces an $\mathscr{O}_{\breve{\overline{X}}}$-linear isomorphism

$$\nu\colon (\xi_\pi^*)^{-1}\widetilde{\Omega}^1_{\breve{\overline{X}}/\breve{\overline{S}}} \xrightarrow{\sim} \xi^{-1}\widetilde{\Omega}^1_{\breve{\overline{X}}/\breve{\overline{S}}}. \tag{4.4.25.4}$$

It immediately follows from 3.2.21 that for every rational number $r \geq 0$, we have a $\overline{\breve{\mathscr{B}}}$-linear isomorphism

$$\breve{\mathscr{F}}'^{(r)} \xrightarrow{\sim} \breve{\mathscr{F}}^{(r+\rho)} \tag{4.4.25.5}$$

that fits into a commutative diagram

$$\begin{array}{ccccccccc}
0 & \longrightarrow & \overline{\breve{\mathscr{B}}} & \longrightarrow & \breve{\mathscr{F}}'^{(r)} & \longrightarrow & \breve{\sigma}^*((\xi_\pi^*)^{-1}\widetilde{\Omega}^1_{\breve{\overline{X}}/\breve{\overline{S}}}) & \longrightarrow & 0 \\
 & & \| & & \downarrow & & \downarrow{\scriptstyle \nu} & & \\
0 & \longrightarrow & \overline{\breve{\mathscr{B}}} & \longrightarrow & \breve{\mathscr{F}}^{(r+\rho)} & \longrightarrow & \breve{\sigma}^*(\xi^{-1}\widetilde{\Omega}^1_{\breve{\overline{X}}/\breve{\overline{S}}}) & \longrightarrow & 0.
\end{array} \tag{4.4.25.6}$$

We deduce from this an isomorphism of $\overline{\breve{\mathscr{B}}}$-algebras

$$\breve{\mathscr{C}}'^{(r)} \xrightarrow{\sim} \breve{\mathscr{C}}^{(r+\rho)}. \tag{4.4.25.7}$$

By (3.2.21.14), the diagram

$$\begin{array}{ccc} \breve{\mathscr{C}}'^{(r)} & \xrightarrow{d_{\breve{\mathscr{C}}'^{(r)}}} & \breve{\sigma}^*((\xi_\pi^*)^{-1}\widetilde{\Omega}^1_{\breve{\overline{X}}/\breve{\overline{S}}}) \otimes_{\breve{\overline{\mathscr{B}}}} \breve{\mathscr{C}}'^{(r)} \\ \downarrow & & \downarrow \nu \\ \breve{\mathscr{C}}^{(r+\rho)} & \xrightarrow{d_{\breve{\mathscr{C}}^{(r+\rho)}}} & \breve{\sigma}^*(\xi^{-1}\widetilde{\Omega}^1_{\breve{\overline{X}}/\breve{\overline{S}}}) \otimes_{\breve{\overline{\mathscr{B}}}} \breve{\mathscr{C}}^{(r+\rho)} \end{array} \tag{4.4.25.8}$$

is commutative.

4.4.26 Recall that $\mathfrak{X}$ denotes the formal scheme p-adic completion of $\overline{X}$ (4.2.1) and $\widetilde{\xi}^{-1}\widetilde{\Omega}^1_{\mathfrak{X}/\mathscr{S}}$ the p-adic completion of the $\mathscr{O}_{\overline{X}}$-module $\xi^{-1}\widetilde{\Omega}^1_{\overline{X}/\overline{S}}$ (4.2.2.1). By ([2] 2.1.18.6) and (4.3.11.5), we have a canonical $\breve{\overline{\mathscr{B}}}$-linear isomorphism

$$\widehat{\sigma}^*(\widetilde{\xi}^{-1}\widetilde{\Omega}^1_{\mathfrak{X}/\mathscr{S}}) \xrightarrow{\sim} \breve{\sigma}^*(\xi^{-1}\widetilde{\Omega}^1_{\breve{\overline{X}}/\breve{\overline{S}}}), \tag{4.4.26.1}$$

where $\widehat{\sigma}$ is the morphism (4.3.11.4). We denote by

$$\widetilde{\xi}^{-1}\widetilde{\Omega}^1_{\mathfrak{X}/\mathscr{S}} \to \widehat{\sigma}_*(\breve{\sigma}^*(\xi^{-1}\widetilde{\Omega}^1_{\breve{\overline{X}}/\breve{\overline{S}}})) \tag{4.4.26.2}$$

the adjoint morphism and by

$$\delta \colon \widetilde{\xi}^{-1}\widetilde{\Omega}^1_{\mathfrak{X}/\mathscr{S}} \to \mathrm{R}^1\widehat{\sigma}_*(\breve{\overline{\mathscr{B}}}) \tag{4.4.26.3}$$

the composition of (4.4.26.2) and the boundary map of the long exact sequence of cohomology deduced from the exact sequence (4.4.22.2).

Proposition 4.4.27 *There exists one and only one isomorphism of graded $\mathscr{O}_{\mathfrak{X}}[\frac{1}{p}]$-algebras*

$$\wedge(\widetilde{\xi}^{-1}\widetilde{\Omega}^1_{\mathfrak{X}/\mathscr{S}}[\frac{1}{p}]) \xrightarrow{\sim} \oplus_{i\geq 0}\mathrm{R}^i\widehat{\sigma}_*(\breve{\overline{\mathscr{B}}})[\frac{1}{p}] \tag{4.4.27.1}$$

whose component in degree one is the morphism $\delta \otimes_{\mathbb{Z}_p} \mathbb{Q}_p$ (4.4.26.3).

The absolute case (3.1.14) was proved in ([3] III.11.8). For the relative case, the question being local for the Zariski topology of X, we can assume X affine. The proposition then follows from the absolute case, in view of ([41] 3.14) and 4.4.25, especially (4.4.25.6).

Proposition 4.4.28 *Let r, r' be two rational numbers such that $r > r' > 0$. Then,*

(i) *For every integer $n \geq 1$, the canonical homomorphism* (4.4.10.5)

$$\mathscr{O}_{\overline{X}_n} \to \sigma_{n*}(\mathscr{C}_n^{(r)}) \tag{4.4.28.1}$$

is almost injective (4.1.2). *Let $\mathscr{H}_n^{(r)}$ be its cokernel.*

(ii) *There exists a rational number $a > 0$ such that for every integer $n \geq 1$, the morphism*

$$\mathscr{H}_n^{(r)} \to \mathscr{H}_n^{(r')} \tag{4.4.28.2}$$

induced by the homomorphism $\alpha_n^{r,r'} : \mathscr{C}_n^{(r)} \to \mathscr{C}_n^{(r')}$ (4.4.10.7) is annihilated by p^a.

(iii) *There exists a rational number $b > 0$ such that for all integers $n, q \geq 1$, the canonical morphism*

$$\mathrm{R}^q\sigma_{n*}(\mathscr{C}_n^{(r)}) \to \mathrm{R}^q\sigma_{n*}(\mathscr{C}_n^{(r')}) \tag{4.4.28.3}$$

is annihilated by p^b.

The absolute case (3.1.14) was proved in ([3] III.11.13). For the relative case, the question being local for the Zariski topology of X, we can assume X affine. The proposition then follows from the absolute case, by ([41] 3.14) and 4.4.25.

Corollary 4.4.29 *Let r, r' be two rational numbers such that $r > r' > 0$. Then,*

(i) *The canonical homomorphism of $X_{s,\text{ét}}^{\mathbb{N}^\circ}$*

$$\mathscr{O}_{\breve{\overline{X}}} \to \breve{\sigma}_*(\breve{\mathscr{C}}^{(r)}) \tag{4.4.29.1}$$

is almost injective. Let $\breve{\mathscr{H}}^{(r)}$ be its cokernel.

(ii) *There exists a rational number $a > 0$ such that the morphism*

$$\breve{\mathscr{H}}^{(r)} \to \breve{\mathscr{H}}^{(r')} \tag{4.4.29.2}$$

induced by the canonical homomorphism $\breve{\alpha}^{r,r'} : \breve{\mathscr{C}}^{(r)} \to \breve{\mathscr{C}}^{(r')}$ (4.4.22.6) is annihilated by p^a.

(iii) *There exists a rational number $b > 0$ such that for every integer $q \geq 1$, the canonical morphism of $X_{s,\text{ét}}^{\mathbb{N}^\circ}$*

$$\mathrm{R}^q\breve{\sigma}_*(\breve{\mathscr{C}}^{(r)}) \to \mathrm{R}^q\breve{\sigma}_*(\breve{\mathscr{C}}^{(r')}) \tag{4.4.29.3}$$

is annihilated by p^b.

This follows from 4.4.28 and ([3] III.7.3(i) and (III.7.5.5)).

Proposition 4.4.30 *Let r, r' be two rational numbers such that $r > r' > 0$. Then,*

(i) *The canonical homomorphism*

$$\mathscr{O}_{\mathfrak{X}} \to \widehat{\sigma}_*(\breve{\mathscr{C}}^{(r)}) \tag{4.4.30.1}$$

is injective. Let $\mathscr{L}^{(r)}$ be its cokernel.

(ii) *There exists a rational number $a > 0$ such that the morphism*

$$\mathscr{L}^{(r)} \to \mathscr{L}^{(r')} \tag{4.4.30.2}$$

induced by the canonical homomorphism $\breve{\alpha}^{r,r'} : \breve{\mathscr{C}}^{(r)} \to \breve{\mathscr{C}}^{(r')}$ (4.4.22.6) is annihilated by p^a.

(iii) *For every integer $q \geq 1$, there exists a number rational $b > 0$ such that the canonical morphism*

$$\mathrm{R}^q\widehat{\sigma}_*(\breve{\mathscr{C}}^{(r)}) \to \mathrm{R}^q\widehat{\sigma}_*(\breve{\mathscr{C}}^{(r')}) \tag{4.4.30.3}$$

is annihilated by p^b.

The absolute case (3.1.14) was proved in ([3] III.11.16). For the relative case, the question being local for the Zariski topology of X, we can assume X affine. The proposition then follows from the absolute case, in view of ([41] 3.14) and 4.4.25.

Corollary 4.4.31 *Let r, r' be two rational numbers such that $r > r' > 0$. Then,*

(i) *The canonical homomorphism*

$$u^r : \mathscr{O}_{\mathfrak{X}}[\frac{1}{p}] \to \widehat{\sigma}_*(\breve{\mathscr{C}}^{(r)})[\frac{1}{p}] \tag{4.4.31.1}$$

admits (as an $\mathscr{O}_{\mathfrak{X}}[\frac{1}{p}]$-linear morphism) a canonical left inverse

$$v^r : \widehat{\sigma}_*(\breve{\mathscr{C}}^{(r)})[\frac{1}{p}] \to \mathscr{O}_{\mathfrak{X}}[\frac{1}{p}]. \tag{4.4.31.2}$$

(ii) *The composed morphism*

$$\widehat{\sigma}_*(\breve{\mathscr{C}}^{(r)})[\frac{1}{p}] \xrightarrow{v^r} \mathscr{O}_{\mathfrak{X}}[\frac{1}{p}] \xrightarrow{u^{r'}} \widehat{\sigma}_*(\breve{\mathscr{C}}^{(r')})[\frac{1}{p}] \tag{4.4.31.3}$$

is the canonical homomorphism.

(iii) *For every integer $q \geq 1$, the canonical morphism*

$$\mathrm{R}^q\widehat{\sigma}_*(\breve{\mathscr{C}}^{(r)})[\frac{1}{p}] \to \mathrm{R}^q\widehat{\sigma}_*(\breve{\mathscr{C}}^{(r')})[\frac{1}{p}] \tag{4.4.31.4}$$

is zero.

This follows from 4.4.30 (see [3] III.11.17).

Corollary 4.4.32 ([3] III.11.18) *The canonical homomorphism*

$$\mathscr{O}_{\mathfrak{X}}[\frac{1}{p}] \to \varinjlim_{r \in \mathbb{Q}_{>0}} \widehat{\sigma}_*(\breve{\mathscr{C}}^{(r)})[\frac{1}{p}] \tag{4.4.32.1}$$

is an isomorphism, and for every integer $q \geq 1$,

$$\varinjlim_{r \in \mathbb{Q}_{>0}} \mathrm{R}^q\widehat{\sigma}_*(\breve{\mathscr{C}}^{(r)})[\frac{1}{p}] = 0. \tag{4.4.32.2}$$

4.4.33 For any rational number $r \geq 0$, we still denote by

$$d_{\breve{\mathscr{C}}^{(r)}} : \breve{\mathscr{C}}^{(r)} \to \widehat{\sigma}^*(\widetilde{\xi}^{-1}\widetilde{\Omega}^1_{\mathfrak{X}/\mathscr{S}}) \otimes_{\breve{\mathscr{B}}} \breve{\mathscr{C}}^{(r)} \tag{4.4.33.1}$$

the $\overline{\breve{\mathscr{B}}}$-derivation induced by $d_{\breve{\mathscr{C}}^{(r)}}$ (4.4.22.8) and the isomorphism (4.4.26.1), which we identify with the universal $\overline{\breve{\mathscr{B}}}$-derivation of $\breve{\mathscr{C}}^{(r)}$. It is a Higgs $\overline{\breve{\mathscr{B}}}$-field with coefficients in $\widehat{\sigma}^*(\widetilde{\xi}^{-1}\widetilde{\Omega}^1_{\mathfrak{X}/\mathscr{S}})$ (4.4.23). We denote by $\mathbb{K}^\bullet(\breve{\mathscr{C}}^{(r)})$ the Dolbeault complex of the Higgs $\overline{\breve{\mathscr{B}}}$-module $(\breve{\mathscr{C}}^{(r)}, p^r d_{\breve{\mathscr{C}}^{(r)}})$ and by $\widetilde{\mathbb{K}}^\bullet(\breve{\mathscr{C}}^{(r)})$ the augmented Dolbeault complex

$$\overline{\breve{\mathscr{B}}} \to \mathbb{K}^0(\breve{\mathscr{C}}^{(r)}) \to \mathbb{K}^1(\breve{\mathscr{C}}^{(r)}) \to \cdots \to \mathbb{K}^n(\breve{\mathscr{C}}^{(r)}) \to \ldots, \tag{4.4.33.2}$$

where $\overline{\breve{\mathscr{B}}}$ is placed in degree -1 and the differential $\overline{\breve{\mathscr{B}}} \to \breve{\mathscr{C}}^{(r)}$ is the canonical homomorphism.

For all rational numbers $r \geq r' \geq 0$, we have (4.4.22.9)

$$p^r(\mathrm{id} \otimes \breve{\alpha}^{r,r'}) \circ d_{\breve{\mathscr{C}}^{(r)}} = p^{r'} d_{\breve{\mathscr{C}}^{(r')}} \circ \breve{\alpha}^{r,r'}, \tag{4.4.33.3}$$

where $\breve{\alpha}^{r,r'} : \breve{\mathscr{C}}^{(r)} \to \breve{\mathscr{C}}^{(r')}$ is the homomorphism (4.4.22.6). Consequently, $\breve{\alpha}^{r,r'}$ induces a morphism of complexes

$$\breve{\iota}^{r,r'} : \widetilde{\mathbb{K}}^\bullet(\breve{\mathscr{C}}^{(r)}) \to \widetilde{\mathbb{K}}^\bullet(\breve{\mathscr{C}}^{(r')}). \tag{4.4.33.4}$$

We denote by $\mathbb{K}^\bullet_{\mathbb{Q}}(\breve{\mathscr{C}}^{(r)})$ and $\widetilde{\mathbb{K}}^\bullet_{\mathbb{Q}}(\breve{\mathscr{C}}^{(r)})$ the images of $\mathbb{K}^\bullet(\breve{\mathscr{C}}^{(r)})$ and $\widetilde{\mathbb{K}}^\bullet(\breve{\mathscr{C}}^{(r)})$ in $\mathbf{Mod}_{\mathbb{Q}}(\overline{\breve{\mathscr{B}}})$ (4.3.13.1). We will also consider these complexes as complexes of $\mathbf{Ind}\text{-}\mathbf{Mod}(\overline{\breve{\mathscr{B}}})$ via the functor $\alpha_{\overline{\breve{\mathscr{B}}}}$ (4.3.13.4).

Proposition 4.4.34 *For all rational numbers $r > r' > 0$ and every integer q, the canonical morphism* (4.4.33.4)

$$\mathrm{H}^q(\breve{\iota}^{r,r'}_{\mathbb{Q}}) : \mathrm{H}^q(\widetilde{\mathbb{K}}^\bullet_{\mathbb{Q}}(\breve{\mathscr{C}}^{(r)})) \to \mathrm{H}^q(\widetilde{\mathbb{K}}^\bullet_{\mathbb{Q}}(\breve{\mathscr{C}}^{(r')})) \tag{4.4.34.1}$$

is zero.

The absolute case (3.1.14) was proved in ([3] III.11.22). For the relative case, the question being local for the Zariski topology of X ([3] III.6.7), we may assume X affine. The proposition then follows from the absolute case, in view of ([41] 3.14) and 4.4.25, in particular (4.4.25.8).

Corollary 4.4.35 *Let r, r' be two rational numbers such that $r > r' > 0$. Then,*

(i) *The canonical morphism*

$$u^r : \overline{\breve{\mathscr{B}}}_{\mathbb{Q}} \to \mathrm{H}^0(\mathbb{K}^\bullet_{\mathbb{Q}}(\breve{\mathscr{C}}^{(r)})) \tag{4.4.35.1}$$

admits a canonical left inverse

$$v^r : \mathrm{H}^0(\mathbb{K}^\bullet_{\mathbb{Q}}(\breve{\mathscr{C}}^{(r)})) \to \overline{\breve{\mathscr{B}}}_{\mathbb{Q}}. \tag{4.4.35.2}$$

(ii) *The composed morphism*

$$\mathrm{H}^0(\mathbb{K}^\bullet_{\mathbb{Q}}(\breve{\mathscr{C}}^{(r)})) \xrightarrow{v^r} \breve{\overline{\mathscr{B}}}_{\mathbb{Q}} \xrightarrow{u^{r'}} \mathrm{H}^0(\mathbb{K}^\bullet_{\mathbb{Q}}(\breve{\mathscr{C}}^{(r')})) \tag{4.4.35.3}$$

is the canonical morphism.

(iii) *For every integer $q \geq 1$, the canonical morphism*

$$\mathrm{H}^q(\mathbb{K}^\bullet_{\mathbb{Q}}(\breve{\mathscr{C}}^{(r)})) \to \mathrm{H}^q(\mathbb{K}^\bullet_{\mathbb{Q}}(\breve{\mathscr{C}}^{(r')})) \tag{4.4.35.4}$$

is zero.

This follows from 4.4.34 (see [3] III.11.23).

Corollary 4.4.36 ([3] III.11.24) *The canonical morphism of complexes of ind-$\breve{\overline{\mathscr{B}}}$-modules*

$$\breve{\overline{\mathscr{B}}}_{\mathbb{Q}}[0] \to \text{“}\varinjlim_{r\in\mathbb{Q}_{>0}}\text{”}\ \mathbb{K}^\bullet_{\mathbb{Q}}(\breve{\mathscr{C}}^{(r)}) \tag{4.4.36.1}$$

is a quasi-isomorphism.

We first recall that **Ind-Mod**$(\breve{\overline{\mathscr{B}}})$ admits small direct limits and that small filtered direct limits are exact (2.6.7.5). The proposition is therefore equivalent to the fact that the canonical morphism

$$\breve{\overline{\mathscr{B}}}_{\mathbb{Q}} \to \text{“}\varinjlim_{r\in\mathbb{Q}_{>0}}\text{”}\ \mathrm{H}^0(\mathbb{K}^\bullet_{\mathbb{Q}}(\breve{\mathscr{C}}^{(r)})) \tag{4.4.36.2}$$

is an isomorphism, and that for every integer $q \geq 1$,

$$\text{“}\varinjlim_{r\in\mathbb{Q}_{>0}}\text{”}\ \mathrm{H}^q(\mathbb{K}^\bullet_{\mathbb{Q}}(\breve{\mathscr{C}}^{(r)})) = 0. \tag{4.4.36.3}$$

These statements follow immediately from 4.4.35.

Remarks 4.4.37 Although filtered direct limits are not a priori representable in the category $\mathbf{Mod}_{\mathbb{Q}}(\breve{\overline{\mathscr{B}}})$, it follows from 4.4.35 that in this category, the canonical morphism

$$\breve{\overline{\mathscr{B}}}_{\mathbb{Q}} \to \varinjlim_{r\in\mathbb{Q}_{>0}} \mathrm{H}^0(\mathbb{K}^\bullet_{\mathbb{Q}}(\breve{\mathscr{C}}^{(r)})) \tag{4.4.37.1}$$

is an isomorphism, and for every integer $q \geq 1$,

$$\varinjlim_{r\in\mathbb{Q}_{>0}} \mathrm{H}^q(\mathbb{K}^\bullet_{\mathbb{Q}}(\breve{\mathscr{C}}^{(r)})) = 0. \tag{4.4.37.2}$$

4.5 Dolbeault ind-modules

4.5.1 Let r be a rational number ≥ 0. Recall that $\widehat{\sigma}$ denotes the morphism (4.3.11.4) and

$$d_{\breve{\mathscr{C}}^{(r)}} : \breve{\mathscr{C}}^{(r)} \to \widehat{\sigma}^*(\widetilde{\xi}^{-1}\widetilde{\Omega}^1_{\mathfrak{X}/\mathscr{S}}) \otimes_{\overline{\breve{\mathscr{B}}}} \breve{\mathscr{C}}^{(r)} \tag{4.5.1.1}$$

the universal $\overline{\breve{\mathscr{B}}}$-derivation of $\breve{\mathscr{C}}^{(r)}$ (4.4.33.1); the latter is a Higgs $\overline{\breve{\mathscr{B}}}$-field with coefficients in $\widehat{\sigma}^*(\widetilde{\xi}^{-1}\widetilde{\Omega}^1_{\mathfrak{X}/\mathscr{S}})$. We denote by $\mathbf{Ind\text{-}MC}(\breve{\mathscr{C}}^{(r)}/\overline{\breve{\mathscr{B}}})$ the category of ind-$\breve{\mathscr{C}}^{(r)}$-modules with integrable p^r-connection with respect to the extension $\breve{\mathscr{C}}^{(r)}/\overline{\breve{\mathscr{B}}}$ (see 2.8.4 and 2.8.8). Despite the notation, this category is not the category of ind-objects of another category (2.6.3). Each object of $\mathbf{Ind\text{-}MC}(\breve{\mathscr{C}}^{(r)}/\overline{\breve{\mathscr{B}}})$ is a Higgs ind-$\overline{\breve{\mathscr{B}}}$-module with coefficients in $\widehat{\sigma}^*(\widetilde{\xi}^{-1}\widetilde{\Omega}^1_{\mathfrak{X}/\mathscr{S}})$ by 2.8.11(i). Therefore a Dolbeault complex in $\mathbf{Ind\text{-}Mod}(\overline{\breve{\mathscr{B}}})$ can be associated with it (2.8.1.3).

Consider the functors

$$\begin{aligned} \mathrm{I}\mathfrak{S}^{(r)} : \mathbf{Ind\text{-}Mod}(\overline{\breve{\mathscr{B}}}) &\to \mathbf{Ind\text{-}MC}(\breve{\mathscr{C}}^{(r)}/\overline{\breve{\mathscr{B}}}) \\ \mathscr{M} &\mapsto (\breve{\mathscr{C}}^{(r)} \otimes_{\overline{\breve{\mathscr{B}}}} \mathscr{M}, p^r d_{\breve{\mathscr{C}}^{(r)}} \otimes_{\overline{\breve{\mathscr{B}}}} \mathrm{id}_{\mathscr{M}}), \end{aligned} \tag{4.5.1.2}$$

$$\begin{aligned} \mathrm{I}\mathscr{K}^{(r)} : \mathbf{Ind\text{-}MC}(\breve{\mathscr{C}}^{(r)}/\overline{\breve{\mathscr{B}}}) &\to \mathbf{Ind\text{-}Mod}(\overline{\breve{\mathscr{B}}}) \\ (\mathscr{F}, \nabla) &\mapsto \ker(\nabla). \end{aligned} \tag{4.5.1.3}$$

It is clear that $\mathrm{I}\mathfrak{S}^{(r)}$ is a left adjoint of $\mathrm{I}\mathscr{K}^{(r)}$ (2.7.1.7).

We denote by $\mathbf{Ind\text{-}HM}(\mathscr{O}_{\mathfrak{X}}, \widetilde{\xi}^{-1}\widetilde{\Omega}^1_{\mathfrak{X}/\mathscr{S}})$ (resp. $\mathbf{Ind\text{-}HM}(\overline{\breve{\mathscr{B}}}, \widehat{\sigma}^*(\widetilde{\xi}^{-1}\widetilde{\Omega}^1_{\mathfrak{X}/\mathscr{S}}))$) the category of Higgs ind-$\mathscr{O}_{\mathfrak{X}}$-modules (resp. of Higgs ind-$\overline{\breve{\mathscr{B}}}$-modules) with coefficients in $\widetilde{\xi}^{-1}\widetilde{\Omega}^1_{\mathfrak{X}/\mathscr{S}}$ (resp. in $\widehat{\sigma}^*(\widetilde{\xi}^{-1}\widetilde{\Omega}^1_{\mathfrak{X}/\mathscr{S}})$) (2.8.1). The functor $\mathrm{I}\widehat{\sigma}^*$ (4.3.12.3) induces a functor that we denote again by

$$\begin{aligned} \mathrm{I}\widehat{\sigma}^* : \mathbf{Ind\text{-}HM}(\mathscr{O}_{\mathfrak{X}}, \widetilde{\xi}^{-1}\widetilde{\Omega}^1_{\mathfrak{X}/\mathscr{S}}) &\to \mathbf{Ind\text{-}HM}(\overline{\breve{\mathscr{B}}}, \widehat{\sigma}^*(\widetilde{\xi}^{-1}\widetilde{\Omega}^1_{\mathfrak{X}/\mathscr{S}})) \\ (\mathscr{N}, \theta) &\mapsto (\mathrm{I}\widehat{\sigma}^*(\mathscr{N}), \mathrm{I}\widehat{\sigma}^*(\theta)). \end{aligned} \tag{4.5.1.4}$$

In view of 2.7.19 and 2.7.18(i), the functor $\mathrm{I}\widehat{\sigma}_*$ (4.3.12.4) induces a functor that we denote again by

$$\begin{aligned} \mathrm{I}\widehat{\sigma}_* : \mathbf{Ind\text{-}HM}(\overline{\breve{\mathscr{B}}}, \widehat{\sigma}^*(\widetilde{\xi}^{-1}\widetilde{\Omega}^1_{\mathfrak{X}/\mathscr{S}})) &\to \mathbf{Ind\text{-}HM}(\mathscr{O}_{\mathfrak{X}}, \widetilde{\xi}^{-1}\widetilde{\Omega}^1_{\mathfrak{X}/\mathscr{S}}) \\ (\mathscr{M}, \theta) &\mapsto (\mathrm{I}\widehat{\sigma}_*(\mathscr{M}), \mathrm{I}\widehat{\sigma}_*(\theta)). \end{aligned} \tag{4.5.1.5}$$

It follows from 2.7.18(ii) that the functor $\mathrm{I}\widehat{\sigma}^*$ (4.5.1.4) is a left adjoint of $\mathrm{I}\widehat{\sigma}_*$ (4.5.1.5).

In view of 2.8.11(ii), the functor (4.5.1.4) induces a functor

$$\begin{aligned} \mathrm{I}\widehat{\sigma}^{(r)*} : \mathbf{Ind\text{-}HM}(\mathscr{O}_{\mathfrak{X}}, \widetilde{\xi}^{-1}\widetilde{\Omega}^1_{\mathfrak{X}/\mathscr{S}}) &\to \mathbf{Ind\text{-}MC}(\breve{\mathscr{C}}^{(r)}/\overline{\breve{\mathscr{B}}}) \\ (\mathscr{N}, \theta) &\mapsto (\breve{\mathscr{C}}^{(r)} \otimes_{\overline{\breve{\mathscr{B}}}} \mathrm{I}\widehat{\sigma}^*(\mathscr{N}), p^r d_{\breve{\mathscr{C}}^{(r)}} \otimes_{\overline{\breve{\mathscr{B}}}} \mathrm{id} + \mathrm{id} \otimes_{\overline{\breve{\mathscr{B}}}} \mathrm{I}\widehat{\sigma}^*(\theta)). \end{aligned} \tag{4.5.1.6}$$

The functor $\mathrm{I}\widehat{\sigma}_*$ (4.5.1.5) induces a functor

$$\begin{aligned} \mathrm{I}\widehat{\sigma}_*^{(r)}\colon \mathbf{Ind\text{-}MC}(\breve{\mathscr{C}}^{(r)}/\breve{\overline{\mathscr{B}}}) &\to \mathbf{Ind\text{-}HM}(\mathscr{O}_{\mathfrak{X}}, \widetilde{\xi}^{-1}\widetilde{\Omega}^1_{\mathfrak{X}/\mathscr{S}}) \\ (\mathscr{F}, \nabla) &\mapsto (\mathrm{I}\widehat{\sigma}_*(\mathscr{F}), \mathrm{I}\widehat{\sigma}_*(\nabla)). \end{aligned} \tag{4.5.1.7}$$

The functor $\mathrm{I}\widehat{\sigma}^{(r)*}$ (4.5.1.6) is a left adjoint of $\mathrm{I}\widehat{\sigma}_*^{(r)}$ (4.5.1.7).

By (2.7.1.5), the functor $\kappa_{\mathscr{O}_{\mathfrak{X}}}$ (4.3.12.7) induces a functor that we denote again by

$$\kappa_{\mathscr{O}_{\mathfrak{X}}}\colon \mathbf{Ind\text{-}HM}(\mathscr{O}_{\mathfrak{X}}, \widetilde{\xi}^{-1}\widetilde{\Omega}^1_{\mathfrak{X}/\mathscr{S}}) \to \mathbf{HM}(\mathscr{O}_{\mathfrak{X}}, \widetilde{\xi}^{-1}\widetilde{\Omega}^1_{\mathfrak{X}/\mathscr{S}}). \tag{4.5.1.8}$$

We denote by $\mathrm{I}\vec{\sigma}_*^{(r)}$ the composed functor

$$\vec{\sigma}_*^{(r)} = \kappa_{\mathscr{O}_{\mathfrak{X}}} \circ \mathrm{I}\widehat{\sigma}_*^{(r)}\colon \mathbf{Ind\text{-}MC}(\breve{\mathscr{C}}^{(r)}/\breve{\overline{\mathscr{B}}}) \to \mathbf{HM}(\mathscr{O}_{\mathfrak{X}}, \widetilde{\xi}^{-1}\widetilde{\Omega}^1_{\mathfrak{X}/\mathscr{S}}). \tag{4.5.1.9}$$

4.5.2 We take again the notation of 4.2.2. In view of (2.9.7.4), the functor $\alpha_{\mathscr{O}_{\mathfrak{X}}}$ (4.3.13.3) induces a functor

$$\begin{aligned} \mathbf{HI}_{\mathbb{Q}}(\mathscr{O}_{\mathfrak{X}}, \widetilde{\xi}^{-1}\widetilde{\Omega}^1_{\mathfrak{X}/\mathscr{S}}) &\to \mathbf{Ind\text{-}HM}(\mathscr{O}_{\mathfrak{X}}, \widetilde{\xi}^{-1}\widetilde{\Omega}^1_{\mathfrak{X}/\mathscr{S}}) \\ (\mathscr{M}, \mathscr{N}, u, \theta) &\mapsto (\alpha_{\mathscr{O}_{\mathfrak{X}}}(\mathscr{M}_{\mathbb{Q}}), (\mathrm{id} \otimes \alpha_{\mathscr{O}_{\mathfrak{X}}}(u_{\mathbb{Q}})^{-1}) \circ \alpha_{\mathscr{O}_{\mathfrak{X}}}(\theta_{\mathbb{Q}})). \end{aligned} \tag{4.5.2.1}$$

It is fully faithful by (2.6.6.4) and (2.9.10.3). Composing with the functors

$$\mathbf{HM}^{\mathrm{coh}}(\mathscr{O}_{\mathfrak{X}}[\frac{1}{p}], \widetilde{\xi}^{-1}\widetilde{\Omega}^1_{\mathfrak{X}/\mathscr{S}}) \xrightarrow{\sim} \mathbf{HI}^{\mathrm{coh}}_{\mathbb{Q}}(\mathscr{O}_{\mathfrak{X}}, \widetilde{\xi}^{-1}\widetilde{\Omega}^1_{\mathfrak{X}/\mathscr{S}}) \to \mathbf{HI}_{\mathbb{Q}}(\mathscr{O}_{\mathfrak{X}}, \widetilde{\xi}^{-1}\widetilde{\Omega}^1_{\mathfrak{X}/\mathscr{S}}), \tag{4.5.2.2}$$

where the first arrow is a quasi-inverse of the equivalence of categories (4.2.2.4) and the second arrow is the canonical injection, we get a fully faithful functor

$$\mathbf{HM}^{\mathrm{coh}}(\mathscr{O}_{\mathfrak{X}}[\frac{1}{p}], \widetilde{\xi}^{-1}\widetilde{\Omega}^1_{\mathfrak{X}/\mathscr{S}}) \to \mathbf{Ind\text{-}HM}(\mathscr{O}_{\mathfrak{X}}, \widetilde{\xi}^{-1}\widetilde{\Omega}^1_{\mathfrak{X}/\mathscr{S}}). \tag{4.5.2.3}$$

We will identify $\mathbf{HM}^{\mathrm{coh}}(\mathscr{O}_{\mathfrak{X}}[\frac{1}{p}], \widetilde{\xi}^{-1}\widetilde{\Omega}^1_{\mathfrak{X}/\mathscr{S}})$ with a full subcategory of the category $\mathbf{Ind\text{-}HM}(\mathscr{O}_{\mathfrak{X}}, \widetilde{\xi}^{-1}\widetilde{\Omega}^1_{\mathfrak{X}/\mathscr{S}})$ by this functor, which we will omit from the notation. We will therefore consider every coherent Higgs $\mathscr{O}_{\mathfrak{X}}[\frac{1}{p}]$-module with coefficients in $\widetilde{\xi}^{-1}\widetilde{\Omega}^1_{\mathfrak{X}/\mathscr{S}}$ as a Higgs ind-$\mathscr{O}_{\mathfrak{X}}$-module with coefficients in $\widetilde{\xi}^{-1}\widetilde{\Omega}^1_{\mathfrak{X}/\mathscr{S}}$.

For every

$$(\mathscr{N}, \theta) \in \mathrm{Ob}(\mathbf{HM}^{\mathrm{coh}}(\mathscr{O}_{\mathfrak{X}}[\frac{1}{p}], \widetilde{\xi}^{-1}\widetilde{\Omega}^1_{\mathfrak{X}/\mathscr{S}})) \text{and} (\mathscr{F}, \nabla) \in \mathrm{Ob}(\mathbf{Ind\text{-}MC}(\breve{\mathscr{C}}^{(r)}/\breve{\overline{\mathscr{B}}})),$$

we have a bifunctorial canonical map

$$\begin{aligned} &\mathrm{Hom}_{\mathbf{Ind\text{-}MC}(\breve{\mathscr{C}}^{(r)}/\breve{\overline{\mathscr{B}}})}(\mathrm{I}\widehat{\sigma}^{(r)*}(\mathscr{N}, \theta), (\mathscr{F}, \nabla)) \to \\ &\quad \mathrm{Hom}_{\mathbf{HM}^{\mathrm{coh}}(\mathscr{O}_{\mathfrak{X}}, \widetilde{\xi}^{-1}\widetilde{\Omega}^1_{\mathfrak{X}/\mathscr{S}})}((\mathscr{N}, \theta), \vec{\sigma}_*^{(r)}(\mathscr{F}, \nabla)), \end{aligned} \tag{4.5.2.4}$$

induced by the adjunction isomorphism between $\mathrm{I}\widehat{\sigma}^{(r)*}$ and $\mathrm{I}\widehat{\sigma}_*^{(r)}$ and the isomorphism (4.3.16.4). We abusively call *the adjoint* of a morphism $\mathrm{I}\widehat{\sigma}^{(r)*}(\mathscr{N},\theta) \to (\mathscr{F},\nabla)$ of $\mathbf{Ind\text{-}MC}(\breve{\mathscr{C}}^{(r)}/\breve{\overline{\mathscr{B}}})$ its image by the map (4.5.2.4). It immediately follows from the injectivity of the map (4.3.16.5) that the map (4.5.2.4) is injective.

4.5.3 Let r, r' be two rational numbers such that $r \geq r' \geq 0$, $(\mathscr{F},\nabla)$ an ind-$\breve{\mathscr{C}}^{(r)}$-module with integrable p^r-connection with respect to the extension $\breve{\mathscr{C}}^{(r)}/\breve{\overline{\mathscr{B}}}$. By 2.8.9 and (4.4.22.9), the ind-$\breve{\mathscr{C}}^{(r')}$-module $\breve{\mathscr{C}}^{(r')} \otimes_{\breve{\mathscr{C}}^{(r)}} \mathscr{F}$ is then canonically endowed with an integrable $p^{r'}$-connection ∇' with respect to the extension $\breve{\mathscr{C}}^{(r')}/\breve{\overline{\mathscr{B}}}$. We thus define a functor

$$\begin{aligned} \mathrm{I}\varepsilon^{r,r'}: \mathbf{Ind\text{-}MC}(\breve{\mathscr{C}}^{(r)}/\breve{\overline{\mathscr{B}}}) &\to \mathbf{Ind\text{-}MC}(\breve{\mathscr{C}}^{(r')}/\breve{\overline{\mathscr{B}}}) \\ (\mathscr{F},\nabla) &\mapsto (\breve{\mathscr{C}}^{(r')} \otimes_{\breve{\mathscr{C}}^{(r)}} \mathscr{F},\nabla'). \end{aligned} \tag{4.5.3.1}$$

We have a canonical isomorphism of functors from $\mathbf{Ind\text{-}Mod}(\breve{\overline{\mathscr{B}}})$ to the category $\mathbf{Ind\text{-}MC}(\breve{\mathscr{C}}^{(r')}/\breve{\overline{\mathscr{B}}})$

$$\mathrm{I}\varepsilon^{r,r'} \circ \mathrm{I}\mathfrak{S}^{(r)} \xrightarrow{\sim} \mathrm{I}\mathfrak{S}^{(r')}. \tag{4.5.3.2}$$

We have a canonical isomorphism of functors from $\mathbf{Ind\text{-}HM}(\mathscr{O}_{\mathfrak{X}},\widetilde{\xi}^{-1}\widetilde{\Omega}^1_{\mathfrak{X}/\mathscr{S}})$ to $\mathbf{Ind\text{-}MC}(\breve{\mathscr{C}}^{(r')}/\breve{\overline{\mathscr{B}}})$

$$\mathrm{I}\varepsilon^{r,r'} \circ \mathrm{I}\widehat{\sigma}^{(r)*} \xrightarrow{\sim} \mathrm{I}\widehat{\sigma}^{(r')*}. \tag{4.5.3.3}$$

The diagram

$$\begin{array}{ccc} \mathscr{F} & \xrightarrow{\nabla} & \widehat{\sigma}^*(\widetilde{\xi}^{-1}\widetilde{\Omega}^1_{\mathfrak{X}/\mathscr{S}}) \otimes_{\breve{\overline{\mathscr{B}}}} \mathscr{F} \\ \Big\downarrow{\scriptstyle \check{\alpha}^{r,r'} \otimes_{\breve{\mathscr{C}}^{(r)}} \mathrm{id}} & & \Big\downarrow{\scriptstyle \mathrm{id} \otimes_{\breve{\overline{\mathscr{B}}}} \check{\alpha}^{r,r'} \otimes_{\breve{\mathscr{C}}^{(r)}} \mathrm{id}} \\ \breve{\mathscr{C}}^{(r')} \otimes_{\breve{\mathscr{C}}^{(r)}} \mathscr{F} & \xrightarrow{\nabla'} & \widehat{\sigma}^*(\widetilde{\xi}^{-1}\widetilde{\Omega}^1_{\mathfrak{X}/\mathscr{S}}) \otimes_{\breve{\overline{\mathscr{B}}}} \breve{\mathscr{C}}^{(r')} \otimes_{\breve{\mathscr{C}}^{(r)}} \mathscr{F} \end{array} \tag{4.5.3.4}$$

is commutative (2.8.9). We deduce from this a canonical morphism of functors from $\mathbf{Ind\text{-}MC}(\breve{\mathscr{C}}^{(r)}/\breve{\overline{\mathscr{B}}})$ to $\mathbf{Ind\text{-}Mod}(\breve{\overline{\mathscr{B}}})$

$$\mathrm{I}\mathscr{K}^{(r)} \to \mathrm{I}\mathscr{K}^{(r')} \circ \mathrm{I}\varepsilon^{r,r'}, \tag{4.5.3.5}$$

and a canonical morphism of functors from $\mathbf{Ind\text{-}MC}(\breve{\mathscr{C}}^{(r)}/\breve{\overline{\mathscr{B}}})$ to the category $\mathbf{Ind\text{-}HM}(\mathscr{O}_{\mathfrak{X}},\widetilde{\xi}^{-1}\widetilde{\Omega}^1_{\mathfrak{X}/\mathscr{S}})$

$$\mathrm{I}\widehat{\sigma}_*^{(r)} \longrightarrow \mathrm{I}\widehat{\sigma}_*^{(r')} \circ \mathrm{I}\varepsilon^{r,r'}. \tag{4.5.3.6}$$

For every rational number r'' such that $r' \geq r'' \geq 0$, we have a canonical isomorphism of functors from $\mathbf{Ind\text{-}MC}(\breve{\mathscr{C}}^{(r)}/\breve{\overline{\mathscr{B}}})$ to $\mathbf{Ind\text{-}MC}(\breve{\mathscr{C}}^{(r'')}/\breve{\overline{\mathscr{B}}})$

$$\mathrm{I}\varepsilon^{r',r''} \circ \mathrm{I}\varepsilon^{r,r'} \xrightarrow{\sim} \mathrm{I}\varepsilon^{r,r''}. \tag{4.5.3.7}$$

Definition 4.5.4 Let $\mathscr{M}$ be an ind-$\overline{\breve{\mathscr{B}}}$-module, $\mathscr{N}$ a Higgs $\mathscr{O}_{\mathfrak{X}}[\frac{1}{p}]$-bundle with coefficients in $\widetilde{\xi}^{-1}\widetilde{\Omega}^1_{\mathfrak{X}/\mathscr{S}}$ (4.2.3), that we also consider as a Higgs ind-$\mathscr{O}_{\mathfrak{X}}$-module with coefficients in $\widetilde{\xi}^{-1}\widetilde{\Omega}^1_{\mathfrak{X}/\mathscr{S}}$ (4.5.2.3).

(i) Let r be a rational number > 0. We say that $\mathscr{M}$ and $\mathscr{N}$ are *r-associated* if there exists an isomorphism of **Ind-MC**$(\breve{\mathscr{C}}^{(r)}/\overline{\breve{\mathscr{B}}})$

$$\alpha\colon \mathrm{I}\widehat{\sigma}^{(r)*}(\mathscr{N}) \xrightarrow{\sim} \mathrm{I}\mathfrak{S}^{(r)}(\mathscr{M}). \tag{4.5.4.1}$$

We then also say that the triple $(\mathscr{M},\mathscr{N},\alpha)$ is *r-admissible*.

(ii) We say that $\mathscr{M}$ and $\mathscr{N}$ are *associated* if there exists a rational number $r > 0$ such that $\mathscr{M}$ and $\mathscr{N}$ are r-associated.

Note that for all rational numbers $r \geq r' > 0$, if $\mathscr{M}$ and $\mathscr{N}$ are r-associated, they are r'-associated, in view of (4.5.3.2) and (4.5.3.3).

Definition 4.5.5

(i) An ind-$\overline{\breve{\mathscr{B}}}$-module is said to be *Dolbeault* if it is associated with a Higgs $\mathscr{O}_{\mathfrak{X}}[\frac{1}{p}]$-bundle with coefficients in $\widetilde{\xi}^{-1}\widetilde{\Omega}^1_{\mathfrak{X}/\mathscr{S}}$.

(ii) A Higgs $\mathscr{O}_{\mathfrak{X}}[\frac{1}{p}]$-bundle with coefficients in $\widetilde{\xi}^{-1}\widetilde{\Omega}^1_{\mathfrak{X}/\mathscr{S}}$ is said to be *solvable* if it is associated with an ind-$\overline{\breve{\mathscr{B}}}$-module.

These notions depend a priori on the deformation $(\widetilde{X},\mathscr{M}_{\widetilde{X}})$ fixed in 4.1.5. We denote by **Ind-Mod**$^{\mathrm{Dolb}}(\overline{\breve{\mathscr{B}}})$ the full sub-category of **Ind-Mod**$(\overline{\breve{\mathscr{B}}})$ made up of the Dolbeault ind-$\overline{\breve{\mathscr{B}}}$-modules, and by $\mathbf{HM}^{\mathrm{sol}}(\mathscr{O}_{\mathfrak{X}}[\frac{1}{p}],\widetilde{\xi}^{-1}\widetilde{\Omega}^1_{\mathfrak{X}/\mathscr{S}})$ the full subcategory of $\mathbf{HM}(\mathscr{O}_{\mathfrak{X}}[\frac{1}{p}],\widetilde{\xi}^{-1}\widetilde{\Omega}^1_{\mathfrak{X}/\mathscr{S}})$ made up of the solvable Higgs $\mathscr{O}_{\mathfrak{X}}[\frac{1}{p}]$-bundles with coefficients in $\widetilde{\xi}^{-1}\widetilde{\Omega}^1_{\mathfrak{X}/\mathscr{S}}$.

Proposition 4.5.6 *Every Dolbeault ind-$\overline{\breve{\mathscr{B}}}$-module is rational* (4.3.15) *and flat* (2.7.9).

Indeed, let $\mathscr{M}$ be a Dolbeault ind-$\overline{\breve{\mathscr{B}}}$-module, $\mathscr{N}$ a Higgs $\mathscr{O}_{\mathfrak{X}}[\frac{1}{p}]$-bundle with coefficients in $\widetilde{\xi}^{-1}\widetilde{\Omega}^1_{\mathfrak{X}/\mathscr{S}}$, r a rational number > 0 and

$$\mathrm{I}\widehat{\sigma}^{(r)*}(\mathscr{N}) \xrightarrow{\sim} \mathrm{I}\mathfrak{S}^{(r)}(\mathscr{M}) \tag{4.5.6.1}$$

an isomorphism of **Ind-MC**$(\breve{\mathscr{C}}^{(r)}/\overline{\breve{\mathscr{B}}})$.

Let us first show that the multiplication by p on $\mathscr{M}$ is an isomorphism. The question being local for the étale topology of $\mathfrak{X}$ by 2.7.17(iii), we may assume that the canonical homomorphism $\overline{\breve{\mathscr{B}}} \to \breve{\mathscr{C}}^{(r)}$ admits sections (see the proof of 4.4.24). Since the multiplication by p on $\mathscr{M} \otimes_{\overline{\breve{\mathscr{B}}}} \breve{\mathscr{C}}^{(r)}$ is an isomorphism (4.5.6.1), the multiplication by p on $\mathscr{M}$ is also an isomorphism.

We then show that the ind-$\breve{\overline{\mathscr{B}}}$-module $\mathrm{I}\widehat{\sigma}^*(\mathscr{N})$ is flat. In view of 2.7.17(iv), we may assume that $\mathscr{N}$ is free of finite type over $\mathscr{O}_{\mathfrak{X}}[\frac{1}{p}]$. The desired assertion then follows from the fact that $\mathrm{I}\widehat{\sigma}^*(\mathscr{O}_{\mathfrak{X}}[\frac{1}{p}]) = \breve{\overline{\mathscr{B}}}_{\mathbb{Q}}$ (2.9.5.1) is a flat ind-$\breve{\overline{\mathscr{B}}}$-module (2.7.9.1). Consequently, the ind-$\breve{\mathscr{C}}^{(r)}$-module $\mathrm{I}\widehat{\sigma}^*(\mathscr{N}) \otimes_{\breve{\overline{\mathscr{B}}}} \breve{\mathscr{C}}^{(r)}$ is flat (2.7.9.2) and the same is then true for $\mathscr{M} \otimes_{\breve{\overline{\mathscr{B}}}} \breve{\mathscr{C}}^{(r)}$ (4.5.6.1). We deduce from this that the ind-$\breve{\overline{\mathscr{B}}}$-module $\mathscr{M}$ is flat by virtue of 2.7.9.3 and 4.4.24.

4.5.7 For every ind-$\breve{\overline{\mathscr{B}}}$-module $\mathscr{M}$ and all rational numbers $r \geq r' \geq 0$, the morphism (4.5.3.6) and the isomorphism (4.5.3.2) induce a morphism of $\mathbf{Ind\text{-}HM}(\mathscr{O}_{\mathfrak{X}}, \widetilde{\xi}^{-1}\widetilde{\Omega}^1_{\mathfrak{X}/\mathscr{S}})$

$$\mathrm{I}\widehat{\sigma}_*^{(r)}(\mathrm{I}\mathfrak{S}^{(r)}(\mathscr{M})) \to \mathrm{I}\widehat{\sigma}_*^{(r')}(\mathrm{I}\mathfrak{S}^{(r')}(\mathscr{M})). \tag{4.5.7.1}$$

We thus obtain a small filtered direct system $(\mathrm{I}\widehat{\sigma}_*^{(r)}(\mathrm{I}\mathfrak{S}^{(r)}(\mathscr{M})))_{r \in \mathbb{Q}_{\geq 0}}$. We denote by $\mathrm{I}\mathscr{H}$ the functor

$$\mathrm{I}\mathscr{H} : \mathbf{Ind\text{-}Mod}(\breve{\overline{\mathscr{B}}}) \to \mathbf{Ind\text{-}HM}(\mathscr{O}_{\mathfrak{X}}, \widetilde{\xi}^{-1}\widetilde{\Omega}^1_{\mathfrak{X}/\mathscr{S}}), \quad \mathscr{M} \mapsto \text{“}\varinjlim_{r \in \mathbb{Q}_{>0}}\text{”}\, \mathrm{I}\widehat{\sigma}_*^{(r)}(\mathrm{I}\mathfrak{S}^{(r)}(\mathscr{M})). \tag{4.5.7.2}$$

Composing with the functor $\kappa_{\mathscr{O}_{\mathfrak{X}}}$ (4.5.1.8), we obtain the functor

$$\mathscr{H} = \kappa_{\mathscr{O}_{\mathfrak{X}}} \circ \mathrm{I}\mathscr{H} : \mathbf{Ind\text{-}Mod}(\breve{\overline{\mathscr{B}}}) \to \mathbf{HM}(\mathscr{O}_{\mathfrak{X}}, \widetilde{\xi}^{-1}\widetilde{\Omega}^1_{\mathfrak{X}/\mathscr{S}}). \tag{4.5.7.3}$$

Since the functor $\kappa_{\mathscr{O}_{\mathfrak{X}}}$ commutes with small filtered direct limits ([40] 6.3.1), for every ind-$\breve{\overline{\mathscr{B}}}$-module $\mathscr{M}$, we have

$$\mathscr{H}(\mathscr{M}) = \varinjlim_{r \in \mathbb{Q}_{>0}} \vec{\sigma}_*^{(r)}(\mathrm{I}\mathfrak{S}^{(r)}(\mathscr{M})), \tag{4.5.7.4}$$

where the direct system $(\vec{\sigma}_*^{(r)}(\mathrm{I}\mathfrak{S}^{(r)}(\mathscr{M})))_{r \in \mathbb{Q}_{\geq 0}}$ is induced by (4.5.7.1). The functors $\mathrm{I}\mathscr{H}$ and $\mathscr{H}$ apply in particular to $\breve{\overline{\mathscr{B}}}_{\mathbb{Q}}$-modules (4.3.13.4).

For every object $\mathscr{N}$ of $\mathbf{Ind\text{-}HM}(\mathscr{O}_{\mathfrak{X}}, \widetilde{\xi}^{-1}\widetilde{\Omega}^1_{\mathfrak{X}/\mathscr{S}})$ and all rational numbers $r \geq r' \geq 0$, the morphism (4.5.3.5) and the isomorphism (4.5.3.3) induce a morphism of $\mathbf{Ind\text{-}Mod}(\breve{\overline{\mathscr{B}}})$

$$\mathrm{I}\mathscr{K}^{(r)}(\mathrm{I}\widehat{\sigma}^{(r)*}(\mathscr{N})) \to \mathrm{I}\mathscr{K}^{(r')}(\mathrm{I}\widehat{\sigma}^{(r')*}(\mathscr{N})). \tag{4.5.7.5}$$

We thus obtain a small filtered direct system $(\mathrm{I}\mathscr{K}^{(r)}(\mathrm{I}\widehat{\sigma}^{(r)*}(\mathscr{N})))_{r \geq 0}$. We denote by $\mathrm{I}\mathscr{V}$ the functor

$$\mathrm{I}\mathscr{V} : \mathbf{Ind\text{-}HM}(\mathscr{O}_{\mathfrak{X}}, \widetilde{\xi}^{-1}\widetilde{\Omega}^1_{\mathfrak{X}/\mathscr{S}}) \to \mathbf{Ind\text{-}Mod}(\breve{\overline{\mathscr{B}}}), \quad \mathscr{N} \mapsto \text{“}\varinjlim_{r \in \mathbb{Q}_{>0}}\text{”}\, \mathrm{I}\mathscr{K}^{(r)}(\mathrm{I}\widehat{\sigma}^{(r)*}(\mathscr{N})). \tag{4.5.7.6}$$

We denote by $\mathscr{V}$ the composition of the functors $\mathrm{I}\mathscr{V}$ and (4.5.2.3),

$$\mathscr{V}\colon \mathbf{HM}^{\mathrm{coh}}(\mathscr{O}_{\mathfrak{X}}[\frac{1}{p}], \widetilde{\xi}^{-1}\widetilde{\Omega}^1_{\mathfrak{X}/\mathscr{S}}) \to \mathbf{Ind\text{-}HM}(\mathscr{O}_{\mathfrak{X}}, \widetilde{\xi}^{-1}\widetilde{\Omega}^1_{\mathfrak{X}/\mathscr{S}}) \to \mathbf{Ind\text{-}Mod}(\breve{\overline{\mathscr{B}}}). \tag{4.5.7.7}$$

Lemma 4.5.8 *We have a canonical isomorphism of Higgs $\mathscr{O}_{\mathfrak{X}}$-modules with coefficients in $\widetilde{\xi}^{-1}\widetilde{\Omega}^1_{\mathfrak{X}/\mathscr{S}}$,*

$$(\mathscr{O}_{\mathfrak{X}}[\frac{1}{p}], 0) \xrightarrow{\sim} \mathscr{H}(\breve{\overline{\mathscr{B}}}_{\mathbb{Q}}). \tag{4.5.8.1}$$

Indeed, for every rational number $r \geq 0$, we have a canonical isomorphism (4.3.12.9)

$$\vec{\sigma}_*(\breve{\mathscr{C}}^{(r)}_{\mathbb{Q}}) \xrightarrow{\sim} \widehat{\sigma}_*(\breve{\mathscr{C}}^{(r)})[\frac{1}{p}], \tag{4.5.8.2}$$

where $\breve{\mathscr{C}}^{(r)}_{\mathbb{Q}}$ is considered as an ind-$\breve{\overline{\mathscr{B}}}$-module (4.3.13.4). The proposition then follows from 4.4.32.

Lemma 4.5.9 *Let $\mathscr{N}$ be an $\mathscr{O}_{\mathfrak{X}}[\frac{1}{p}]$-bundle with coefficients in $\widetilde{\xi}^{-1}\widetilde{\Omega}^1_{\mathfrak{X}/\mathscr{S}}$ (4.2.3), r a rational number ≥ 0. Then, we have a canonical isomorphism of* **Ind-HM**$(\mathscr{O}_{\mathfrak{X}}, \widetilde{\xi}^{-1}\widetilde{\Omega}^1_{\mathfrak{X}/\mathscr{S}})$

$$\mathscr{N} \otimes_{\mathscr{O}_{\mathfrak{X}}} \mathrm{I}\vec{\sigma}^{(r)}_*(\mathrm{I}\mathfrak{S}^{(r)}(\breve{\overline{\mathscr{B}}})) \xrightarrow{\sim} \mathrm{I}\vec{\sigma}^{(r)}_*(\mathrm{I}\widehat{\sigma}^{(r)*}(\mathscr{N})), \tag{4.5.9.1}$$

where the left-hand side is the tensor product of Higgs ind-modules (2.8.1.7). *Applying the functor $\kappa_{\mathscr{O}_{\mathfrak{X}}}$* (4.5.1.8), *we obtain an isomorphism of* **HM**$(\mathscr{O}_{\mathfrak{X}}, \widetilde{\xi}^{-1}\widetilde{\Omega}^1_{\mathfrak{X}/\mathscr{S}})$

$$\gamma^{(r)}\colon \mathscr{N} \otimes_{\mathscr{O}_{\mathfrak{X}}} \vec{\sigma}^{(r)}_*(\mathrm{I}\mathfrak{S}^{(r)}(\breve{\overline{\mathscr{B}}})) \xrightarrow{\sim} \vec{\sigma}^{(r)}_*(\mathrm{I}\widehat{\sigma}^{(r)*}(\mathscr{N})), \tag{4.5.9.2}$$

where the left-hand side is the tensor product of Higgs ind-modules (2.5.1.7). *Moreover, we have the following properties:*

(i) *The morphism*

$$\mathscr{N} \to \vec{\sigma}^{(r)}_*(\mathrm{I}\widehat{\sigma}^{(r)*}(\mathscr{N})) \tag{4.5.9.3}$$

induced by (4.5.9.2) *and the canonical morphism $\mathscr{O}_{\mathfrak{X}} \to \vec{\sigma}^{(r)}_*(\mathrm{I}\mathfrak{S}^{(r)}(\breve{\overline{\mathscr{B}}}))$ is the adjoint of the identity morphism of $\mathrm{I}\widehat{\sigma}^{(r)*}(\mathscr{N})$* (4.5.2.4).

(ii) *For every rational number r' such that $r \geq r' \geq 0$, the diagram*

$$\begin{array}{ccc} \mathscr{N} \otimes_{\mathscr{O}_{\mathfrak{X}}} \vec{\sigma}^{(r)}_*(\mathrm{I}\mathfrak{S}^{(r)}(\breve{\overline{\mathscr{B}}})) & \xrightarrow{\gamma^{(r)}} & \vec{\sigma}^{(r)}_*(\mathrm{I}\widehat{\sigma}^{(r)*}(\mathscr{N})) \\ \downarrow & & \downarrow \\ \mathscr{N} \otimes_{\mathscr{O}_{\mathfrak{X}}} \vec{\sigma}^{(r')}_*(\mathrm{I}\mathfrak{S}^{(r')}(\breve{\overline{\mathscr{B}}})) & \xrightarrow{\gamma^{(r')}} & \vec{\sigma}^{(r')}_*(\mathrm{I}\widehat{\sigma}^{(r')*}(\mathscr{N})), \end{array} \tag{4.5.9.4}$$

where the vertical arrows are induced by the morphism (4.5.3.6) *and by the isomorphisms* (4.5.3.2) *and* (4.5.3.3), *is commutative.*

Indeed, by 4.3.18, we have canonical isomorphisms of ind-$\mathscr{O}_{\mathfrak{X}}$-modules

$$\mathscr{N} \otimes_{\mathscr{O}_{\mathfrak{X}}} \mathrm{I}\widehat{\sigma}_*(\breve{\mathscr{C}}^{(r)}) \xrightarrow{\sim} \mathrm{I}\widehat{\sigma}_*(\mathrm{I}\widehat{\sigma}^*(\mathscr{N}) \otimes_{\breve{\overline{\mathscr{B}}}} \breve{\mathscr{C}}^{(r)}), \tag{4.5.9.5}$$

$$\widetilde{\xi}^{-1}\widetilde{\Omega}^1_{\mathfrak{X}/\mathscr{S}} \otimes_{\mathscr{O}_{\mathfrak{X}}} \mathrm{I}\widehat{\sigma}_*(\mathrm{I}\widehat{\sigma}^*(\mathscr{N}) \otimes_{\breve{\overline{\mathscr{B}}}} \breve{\mathscr{C}}^{(r)}) \xrightarrow{\sim} \mathrm{I}\widehat{\sigma}_*(\mathrm{I}\widehat{\sigma}^*(\widetilde{\xi}^{-1}\widetilde{\Omega}^1_{\mathfrak{X}/\mathscr{S}} \otimes_{\mathscr{O}_{\mathfrak{X}}} \mathscr{N}) \otimes_{\breve{\overline{\mathscr{B}}}} \breve{\mathscr{C}}^{(r)}), \tag{4.5.9.6}$$

$$\widetilde{\xi}^{-1}\widetilde{\Omega}^1_{\mathfrak{X}/\mathscr{S}} \otimes_{\mathscr{O}_{\mathfrak{X}}} \mathscr{N} \otimes_{\mathscr{O}_{\mathfrak{X}}} \mathrm{I}\widehat{\sigma}_*(\breve{\mathscr{C}}^{(r)}) \xrightarrow{\sim} \mathrm{I}\widehat{\sigma}_*(\mathrm{I}\widehat{\sigma}^*(\widetilde{\xi}^{-1}\widetilde{\Omega}^1_{\mathfrak{X}/\mathscr{S}} \otimes_{\mathscr{O}_{\mathfrak{X}}} \mathscr{N}) \otimes_{\breve{\overline{\mathscr{B}}}} \breve{\mathscr{C}}^{(r)}) \tag{4.5.9.7}$$

where the last isomorphism is induced by the first two, by 2.7.18(i). Moreover, in view of the bifunctoriality of the isomorphism (4.3.18.1), the diagram

$$\begin{array}{ccc} \mathscr{N} \otimes_{\mathscr{O}_{\mathfrak{X}}} \mathrm{I}\widehat{\sigma}_*(\breve{\mathscr{C}}^{(r)}) & \longrightarrow & \mathrm{I}\widehat{\sigma}_*(\mathrm{I}\widehat{\sigma}^*(\mathscr{N}) \otimes_{\breve{\overline{\mathscr{B}}}} \breve{\mathscr{C}}^{(r)}) \\ {\scriptstyle \theta\otimes\mathrm{id}+p^r\mathrm{id}\otimes\mathrm{I}\widehat{\sigma}_*(d_{\breve{\mathscr{C}}^{(r)}})}\downarrow & & \downarrow{\scriptstyle \mathrm{I}\widehat{\sigma}_*(\mathrm{I}\widehat{\sigma}^*(\theta)\otimes\mathrm{id}+p^r\mathrm{id}\otimes d_{\breve{\mathscr{C}}^{(r)}})} \\ \widetilde{\xi}^{-1}\widetilde{\Omega}^1_{\mathfrak{X}/\mathscr{S}} \otimes_{\mathscr{O}_{\mathfrak{X}}} \mathscr{N} \otimes_{\mathscr{O}_{\mathfrak{X}}} \mathrm{I}\widehat{\sigma}_*(\breve{\mathscr{C}}^{(r)}) & \longrightarrow & \mathrm{I}\widehat{\sigma}_*(\mathrm{I}\widehat{\sigma}^*(\widetilde{\xi}^{-1}\widetilde{\Omega}^1_{\mathfrak{X}/\mathscr{S}} \otimes_{\mathscr{O}_{\mathfrak{X}}} \mathscr{N}) \otimes_{\breve{\overline{\mathscr{B}}}} \breve{\mathscr{C}}^{(r)}), \end{array}$$

where θ is the Higgs field of $\mathscr{N}$, is commutative. We then take for morphism (4.5.9.1) the isomorphism (4.5.9.5). In view of (2.7.1.5) and (4.3.16.4), the image of (4.5.9.1) by $\kappa_{\mathscr{O}_{\mathfrak{X}}}$ identifies with the isomorphism (4.5.9.2).

Proposition (i) follows from 2.7.18(ii). Proposition (ii) is a consequence of the bifunctoriality of the isomorphism (4.5.9.2).

4.5.10 Let r be a rational number > 0, $(\mathscr{M}, \mathscr{N}, \alpha)$ an r-admissible triple (4.5.4). For any rational number r' such that $0 < r' \leq r$, we denote by

$$\alpha^{(r')} \colon \mathrm{I}\widehat{\sigma}^{(r')*}(\mathscr{N}) \xrightarrow{\sim} \mathrm{I}\mathfrak{S}^{(r')}(\mathscr{M}) \tag{4.5.10.1}$$

the isomorphism of **Ind-MC**$(\breve{\mathscr{C}}^{(r')}/\breve{\overline{\mathscr{B}}})$ induced by $\mathrm{I}\varepsilon^{r,r'}(\alpha)$ and the isomorphisms (4.5.3.2) and (4.5.3.3), and by

$$\beta^{(r')} \colon \mathscr{N} \to \vec{\sigma}^{(r')}_*(\mathrm{I}\mathfrak{S}^{(r')}(\mathscr{M})) \tag{4.5.10.2}$$

its adjoint (4.5.2.4).

Proposition 4.5.11 *Under the assumptions of* 4.5.10, *let moreover* r', r'' *be two rational numbers such that* $0 < r'' < r' \leq r$. *Then,*

(i) *The composed morphism*

$$\mathscr{N} \xrightarrow{\beta^{(r')}} \vec{\sigma}^{(r')}_*(\mathrm{I}\mathfrak{S}^{(r')}(\mathscr{M})) \longrightarrow \mathscr{H}(\mathscr{M}), \tag{4.5.11.1}$$

where the second arrow is the canonical morphism (4.5.7.4), *is an isomorphism, independent of* r'.

(ii) *The composed morphism*

$$\vec{\sigma}_*^{(r')}(\mathrm{I}\mathfrak{S}^{(r')}(\mathscr{M})) \longrightarrow \mathscr{H}(\mathscr{M}) \xrightarrow{\sim} \mathscr{N} \xrightarrow{\beta^{(r'')}} \vec{\sigma}_*^{(r'')}(\mathrm{I}\mathfrak{S}^{(r'')}(\mathscr{M})), \qquad (4.5.11.2)$$

where the first arrow is the canonical morphism (4.5.7.4) *and the second arrow is the inverse isomorphism of* (4.5.11.1)*, is the canonical morphism deduced from* (4.5.7.1).

(i) For every rational number $0 < t \leq r$, the isomorphism (4.5.9.2) induces an isomorphism of $\mathbf{HM}(\mathscr{O}_{\mathfrak{X}}, \widetilde{\xi}^{-1}\widetilde{\Omega}^1_{\mathfrak{X}/\mathscr{S}})$ that we denote again by

$$\gamma^{(t)} \colon \mathscr{N} \otimes_{\mathscr{O}_{\mathfrak{X}}} \vec{\sigma}_*^{(t)}(\mathrm{I}\mathfrak{S}^{(t)}(\breve{\overline{\mathscr{B}}}_{\mathbb{Q}})) \xrightarrow{\sim} \vec{\sigma}_*^{(t)}(\mathrm{I}\widehat{\sigma}^{(t)*}(\mathscr{N})). \qquad (4.5.11.3)$$

Indeed, for every $\mathscr{O}_{\mathfrak{X}}$-module $\mathscr{M}$, the canonical morphism $\mathscr{N} \otimes_{\mathscr{O}_{\mathfrak{X}}} \mathscr{M} \to \mathscr{N} \otimes_{\mathscr{O}_{\mathfrak{X}}} \mathscr{M}_{\mathbb{Q}}$ is an isomorphism of $\mathbf{Ind}\text{-}\mathbf{Mod}(\mathscr{O}_{\mathfrak{X}})$ (2.6.6.7). We denote by

$$\delta^{(t)} \colon \mathscr{N} \otimes_{\mathscr{O}_{\mathfrak{X}}} \vec{\sigma}_*^{(t)}(\mathrm{I}\mathfrak{S}^{(t)}(\breve{\overline{\mathscr{B}}}_{\mathbb{Q}})) \xrightarrow{\sim} \vec{\sigma}_*^{(t)}(\mathrm{I}\mathfrak{S}^{(t)}(\mathscr{M})) \qquad (4.5.11.4)$$

the composition $\vec{\sigma}_*^{(t)}(\alpha^{(t)}) \circ \gamma^{(t)}$. The diagram

$$\begin{array}{ccc} \mathscr{N} \otimes_{\mathscr{O}_{\mathfrak{X}}} \vec{\sigma}_*^{(r')}(\mathrm{I}\mathfrak{S}^{(r')}(\breve{\overline{\mathscr{B}}}_{\mathbb{Q}})) & \xrightarrow{\delta^{(r')}} & \vec{\sigma}_*^{(r')}(\mathrm{I}\mathfrak{S}^{(r')}(\mathscr{M})) \\ \downarrow & & \downarrow \\ \mathscr{N} \otimes_{\mathscr{O}_{\mathfrak{X}}} \vec{\sigma}_*^{(r'')}(\mathrm{I}\mathfrak{S}^{(r'')}(\breve{\overline{\mathscr{B}}}_{\mathbb{Q}})) & \xrightarrow{\delta^{(r'')}} & \vec{\sigma}_*^{(r'')}(\mathrm{I}\mathfrak{S}^{(r'')}(\mathscr{M})), \end{array} \qquad (4.5.11.5)$$

where the vertical arrows are the canonical morphisms (4.5.7.1), is commutative by virtue of 4.5.9(ii). The direct limit of the isomorphisms $(\delta^{(t)})_{0<t\leq r}$ induces an isomorphism of $\mathbf{HM}(\mathscr{O}_{\mathfrak{X}}, \widetilde{\xi}^{-1}\widetilde{\Omega}^1_{\mathfrak{X}/\mathscr{S}})$

$$\delta \colon \mathscr{N} \otimes_{\mathscr{O}_{\mathfrak{X}}} \mathscr{H}(\breve{\overline{\mathscr{B}}}_{\mathbb{Q}}) \xrightarrow{\sim} \mathscr{H}(\mathscr{M}). \qquad (4.5.11.6)$$

Consider the commutative diagram

$$\begin{array}{ccccc} \mathscr{N} & \xrightarrow{\iota^{(r')}} & \mathscr{N} \otimes_{\mathscr{O}_{\mathfrak{X}}} \vec{\sigma}_*^{(r')}(\mathrm{I}\mathfrak{S}^{(r')}(\breve{\overline{\mathscr{B}}}_{\mathbb{Q}})) & \xrightarrow{\delta^{(r')}} & \vec{\sigma}_*^{(r')}(\mathrm{I}\mathfrak{S}^{(r')}(\mathscr{M})) \\ & \searrow & \downarrow & & \downarrow \\ & & \mathscr{N} \otimes_{\mathscr{O}_{\mathfrak{X}}} \mathscr{H}(\breve{\overline{\mathscr{B}}}_{\mathbb{Q}}) & \xrightarrow{\delta} & \mathscr{H}(\mathscr{M}), \end{array} \qquad (4.5.11.7)$$

where $\iota^{(r')}$ is induced by the canonical morphism $\mathscr{O}_{\mathfrak{X}} \to \vec{\sigma}_*^{(r')}(\mathrm{I}\mathfrak{S}^{(r')}(\breve{\overline{\mathscr{B}}}_{\mathbb{Q}}))$ and the vertical arrows are the canonical morphisms. By 4.5.9(i), we have

$$\delta^{(r')} \circ \iota^{(r')} = \vec{\sigma}_*^{(r')}(\alpha^{(r')}) \circ \gamma^{(r')} \circ \iota^{(r')} = \beta^{(r')}. \qquad (4.5.11.8)$$

The proposition follows by virtue of 4.5.8.

(ii) This follows from (4.5.11.5), (4.5.11.7) and 4.4.31(ii).

Corollary 4.5.12 *For every Dolbeault ind-$\overline{\breve{\mathscr{B}}}$-module $\mathscr{M}$, $\mathscr{H}(\mathscr{M})$ (4.5.7.4) is a solvable Higgs $\mathscr{O}_{\mathfrak{X}}[\frac{1}{p}]$-bundle, associated with $\mathscr{M}$. In particular, $\mathscr{H}$ induces a functor that we denote again by*

$$\begin{array}{rcl}\mathscr{H}\colon \mathbf{Ind\text{-}Mod}^{\mathrm{Dolb}}(\overline{\breve{\mathscr{B}}}) & \to & \mathbf{HM}^{\mathrm{sol}}(\mathscr{O}_{\mathfrak{X}}[\tfrac{1}{p}], \widetilde{\xi}^{-1}\widetilde{\Omega}^1_{\mathfrak{X}/\mathscr{S}}) \\ \mathscr{M} & \mapsto & \mathscr{H}(\mathscr{M}).\end{array} \tag{4.5.12.1}$$

Corollary 4.5.13 *For every Dolbeault ind-$\overline{\breve{\mathscr{B}}}$-module $\mathscr{M}$, there exists a rational number $r > 0$ and an isomorphism of* **Ind-MC**$(\breve{\mathscr{C}}^{(r)}/\overline{\breve{\mathscr{B}}})$,

$$\alpha\colon \mathrm{I}\widehat{\sigma}^{(r)*}(\mathscr{H}(\mathscr{M})) \xrightarrow{\sim} \mathrm{I}\mathfrak{S}^{(r)}(\mathscr{M}), \tag{4.5.13.1}$$

satisfying the following properties. For any rational number r' such that $0 < r' \leq r$, we denote by

$$\alpha^{(r')}\colon \mathrm{I}\widehat{\sigma}^{(r')*}(\mathscr{H}(\mathscr{M})) \xrightarrow{\sim} \mathrm{I}\mathfrak{S}^{(r')}(\mathscr{M}) \tag{4.5.13.2}$$

the isomorphism of **Ind-MC**$(\breve{\mathscr{C}}^{(r')}/\overline{\breve{\mathscr{B}}})$ *induced by $\mathrm{I}\varepsilon^{r,r'}(\alpha)$ and the isomorphisms* (4.5.3.2) *and* (4.5.3.3)*, and by*

$$\beta^{(r')}\colon \mathscr{H}(\mathscr{M}) \to \vec{\sigma}^{(r')}_*(\mathrm{I}\mathfrak{S}^{(r')}(\mathscr{M})) \tag{4.5.13.3}$$

its adjoint (4.5.2.4)*. Then,*

(i) *For every rational number r' such that $0 < r' \leq r$, the morphism $\beta^{(r')}$ is a right inverse of the canonical morphism $\varpi^{(r')}\colon \vec{\sigma}^{(r')}_*(\mathrm{I}\mathfrak{S}^{(r')}(\mathscr{M})) \to \mathscr{H}(\mathscr{M})$.*

(ii) *For all rational numbers r' and r'' such that $0 < r'' < r' \leq r$, the composition*

$$\vec{\sigma}^{(r')}_*(\mathrm{I}\mathfrak{S}^{(r')}(\mathscr{M})) \xrightarrow{\varpi^{(r')}} \mathscr{H}(\mathscr{M}) \xrightarrow{\beta^{(r'')}} \vec{\sigma}^{(r'')}_*(\mathrm{I}\mathfrak{S}^{(r'')}(\mathscr{M})) \tag{4.5.13.4}$$

is the canonical morphism.

Remark 4.5.14 Under the assumptions of 4.5.13, the isomorphism α is a priori not uniquely determined by $(\mathscr{M}, r)$, but for every rational number $0 < r' < r$, the morphism $\alpha^{(r')}$ (4.5.13.2) depends only on $\mathscr{M}$, on which it depends functorially (see the proof of 4.5.20).

4.5.15 Let r be a rational number > 0, $(\mathscr{M}, \mathscr{N}, \alpha)$ an r-admissible triple (4.5.4). To avoid any ambiguity with (4.5.10.1), we denote by

$$\check{\alpha}\colon \mathrm{I}\mathfrak{S}^{(r)}(\mathscr{M}) \to \mathrm{I}\widehat{\sigma}^{(r)*}(\mathscr{N}) \tag{4.5.15.1}$$

the inverse of α in **Ind-MC**$(\breve{\mathscr{C}}^{(r)}/\overline{\breve{\mathscr{B}}})$. For any rational number r' such that $0 < r' \leq r$, we denote by

$$\check{\alpha}^{(r')}\colon \mathrm{I}\mathfrak{S}^{(r')}(\mathscr{M}) \xrightarrow{\sim} \mathrm{I}\widehat{\sigma}^{(r')*}(\mathscr{N}) \tag{4.5.15.2}$$

the isomorphism of **Ind-MC**$(\breve{\mathscr{C}}^{(r')}/\overline{\breve{\mathscr{B}}})$ induced by $\mathrm{I}\varepsilon^{r,r'}(\check{\alpha})$ and the isomorphisms (4.5.3.2) and (4.5.3.2), and by

$$\breve{\beta}^{(r')}\colon \mathscr{M} \to \mathrm{I}\mathscr{K}^{(r')}(\mathrm{I}\widehat{\sigma}^{(r')*}(\mathscr{N})) \tag{4.5.15.3}$$

the adjoint morphism (4.5.1.3).

Proposition 4.5.16 *Under the assumptions of* 4.5.15, *let moreover r', r'' be two rational numbers such that* $0 < r'' < r' \leq r$. *Then,*

(i) *The composed morphism*

$$\mathscr{M} \xrightarrow{\breve{\beta}^{(r')}} \mathrm{I}\mathscr{K}^{(r')}(\mathrm{I}\widehat{\sigma}^{(r')*}(\mathscr{N})) \longrightarrow \mathscr{V}(\mathscr{N})\,, \tag{4.5.16.1}$$

where the second arrow is the canonical morphism (4.5.7.6), *is an isomorphism, independent of r'.*

(ii) *The composed morphism*

$$\mathrm{I}\mathscr{K}^{(r')}(\mathrm{I}\widehat{\sigma}^{(r')*}(\mathscr{N})) \longrightarrow \mathscr{V}(\mathscr{N}) \xrightarrow{\sim} \mathscr{M} \xrightarrow{\breve{\beta}^{(r'')}} \mathrm{I}\mathscr{K}^{(r'')}(\mathrm{I}\widehat{\sigma}^{(r'')*}(\mathscr{N}))\,, \tag{4.5.16.2}$$

where the first arrow is the canonical morphism and the second arrow is the inverse isomorphism of (4.5.16.1), *is the canonical morphism* (4.5.7.5).

(i) Since the ind-$\overline{\breve{\mathscr{B}}}$-module $\mathscr{M}$ is rational and flat by 4.5.6, for every rational number $t \geq 0$, we have canonical isomorphisms of **Ind-Mod**$(\overline{\breve{\mathscr{B}}})$

$$\mathscr{M} \otimes_{\overline{\breve{\mathscr{B}}}} \mathrm{I}\mathscr{K}^{(t)}(\mathrm{I}\mathfrak{S}^{(t)}(\overline{\breve{\mathscr{B}}}_{\mathbb{Q}})) \xrightarrow{\sim} \mathscr{M} \otimes_{\overline{\breve{\mathscr{B}}}} \mathrm{I}\mathscr{K}^{(t)}(\mathrm{I}\mathfrak{S}^{(t)}(\overline{\breve{\mathscr{B}}})) \xrightarrow{\sim} \mathrm{I}\mathscr{K}^{(t)}(\mathrm{I}\mathfrak{S}^{(t)}(\mathscr{M})). \tag{4.5.16.3}$$

We denote by $\gamma^{(t)}$ the composed isomorphism. For any rational number $0 < t \leq r$, we denote by

$$\delta^{(t)}\colon \mathscr{M} \otimes_{\overline{\breve{\mathscr{B}}}} \mathrm{I}\mathscr{K}^{(t)}(\mathrm{I}\mathfrak{S}^{(t)}(\overline{\breve{\mathscr{B}}}_{\mathbb{Q}})) \xrightarrow{\sim} \mathrm{I}\mathscr{K}^{(t)}(\mathrm{I}\widehat{\sigma}^{(t)*}(\mathscr{N})) \tag{4.5.16.4}$$

the isomorphism $\mathscr{K}^{(t)}(\check{\alpha}^{(t)}) \circ \gamma^{(t)}$. The diagram

$$\begin{array}{ccc} \mathscr{M} \otimes_{\overline{\breve{\mathscr{B}}}} \mathrm{I}\mathscr{K}^{(r')}(\mathrm{I}\mathfrak{S}^{(r')}(\overline{\breve{\mathscr{B}}}_{\mathbb{Q}})) & \xrightarrow{\delta^{(r')}} & \mathrm{I}\mathscr{K}^{(r')}(\mathrm{I}\widehat{\sigma}^{(r')*}(\mathscr{N})) \\ \downarrow & & \downarrow \\ \mathscr{M} \otimes_{\overline{\breve{\mathscr{B}}}} \mathrm{I}\mathscr{K}^{(r'')}(\mathrm{I}\mathfrak{S}^{(r'')}(\overline{\breve{\mathscr{B}}}_{\mathbb{Q}})) & \xrightarrow{\delta^{(r'')}} & \mathrm{I}\mathscr{K}^{(r'')}(\mathrm{I}\widehat{\sigma}^{(r'')*}(\mathscr{N})), \end{array} \tag{4.5.16.5}$$

where the vertical arrows are induced by the morphism (4.5.3.5) and the isomorphisms (4.5.3.2) and (4.5.3.3), is clearly commutative.

By 4.4.35, the canonical morphism

$$\mathscr{M} \to \varinjlim_{t \in \mathbb{Q}_{>0}} \mathscr{M} \otimes_{\overline{\breve{\mathscr{B}}}} \mathrm{I}\mathscr{K}^{(t)}(\mathrm{I}\mathfrak{S}^{(t)}(\overline{\breve{\mathscr{B}}}_{\mathbb{Q}})) \tag{4.5.16.6}$$

is an isomorphism. The direct limit of the isomorphisms $(\delta^{(t)})_{0<t\leq r}$ induces an isomorphism

$$\delta\colon \mathscr{M} \xrightarrow{\sim} \mathscr{V}(\mathscr{N}). \tag{4.5.16.7}$$

It immediately follows from the definitions that the diagram

$$\begin{array}{ccc} \mathscr{M} \otimes_{\overline{\breve{\mathscr{B}}}} \mathrm{I}\mathscr{K}^{(r')}(\mathrm{I}\mathfrak{S}^{(r')}(\overline{\breve{\mathscr{B}}}_{\mathbb{Q}})) & \xrightarrow{\delta^{(r')}} & \mathrm{I}\mathscr{K}^{(r')}(\mathrm{I}\widehat{\sigma}^{(r')*}(\mathscr{N})) \\ \uparrow{\scriptstyle \iota^{(r')}} & & \downarrow \\ \mathscr{M} & \xrightarrow{\delta} & \mathscr{V}(\mathscr{N}), \end{array} \tag{4.5.16.8}$$

where $\iota^{(r')}$ is induced by the canonical morphism $\overline{\breve{\mathscr{B}}} \to \mathrm{I}\mathscr{K}^{(r')}(\mathrm{I}\mathfrak{S}^{(r')}(\overline{\breve{\mathscr{B}}}_{\mathbb{Q}}))$ and the unlabeled arrow is the canonical morphism, is commutative. We immediately check that we have

$$\delta^{(r')} \circ \iota^{(r')} = \mathscr{K}^{(r')}(\breve{\alpha}^{(r')}) \circ \gamma^{(r')} \circ \iota^{(r')} = \breve{\beta}^{(r')}. \tag{4.5.16.9}$$

The statement follows.

(ii) This follows from (4.5.16.5) and 4.4.35(ii).

Corollary 4.5.17 *For every Higgs $\mathscr{O}_{\mathfrak{X}}[\frac{1}{p}]$-bundle $\mathscr{N}$ with coefficients in $\widetilde{\xi}^{-1}\widetilde{\Omega}^1_{\mathfrak{X}/\mathscr{S}}$, the ind-$\overline{\breve{\mathscr{B}}}$-module $\mathscr{V}(\mathscr{N})$ (4.5.7.6) is Dolbeault, associated with $\mathscr{N}$. In particular, $\mathscr{V}$ induces a functor that we denote again by*

$$\begin{array}{rccc} \mathscr{V}\colon & \mathbf{HM}^{\mathrm{sol}}(\mathscr{O}_{\mathfrak{X}}[\frac{1}{p}], \widetilde{\xi}^{-1}\widetilde{\Omega}^1_{\mathfrak{X}/\mathscr{S}}) & \to & \mathbf{Ind\text{-}Mod}^{\mathrm{Dolb}}(\overline{\breve{\mathscr{B}}}) \\ & \mathscr{N} & \mapsto & \mathscr{V}(\mathscr{N}). \end{array} \tag{4.5.17.1}$$

Corollary 4.5.18 *For every solvable Higgs $\mathscr{O}_{\mathfrak{X}}[\frac{1}{p}]$-bundle $\mathscr{N}$ with coefficients in $\widetilde{\xi}^{-1}\widetilde{\Omega}^1_{\mathfrak{X}/\mathscr{S}}$, there exist a rational number $r > 0$ and an isomorphism of $\mathbf{Ind\text{-}MC}(\breve{\mathscr{C}}^{(r)}/\overline{\breve{\mathscr{B}}})$*

$$\breve{\alpha}\colon \mathrm{I}\mathfrak{S}^{(r)}(\mathscr{V}(\mathscr{N})) \xrightarrow{\sim} \mathrm{I}\widehat{\sigma}^{(r)*}(\mathscr{N}) \tag{4.5.18.1}$$

satisfying the following properties. For any rational number r' such that $0 < r' \leq r$, we denote by

$$\breve{\alpha}^{(r')}\colon \mathrm{I}\mathfrak{S}^{(r')}(\mathscr{V}(\mathscr{N})) \xrightarrow{\sim} \mathrm{I}\widehat{\sigma}^{(r')*}(\mathscr{N}) \tag{4.5.18.2}$$

the isomorphism of $\mathbf{Ind\text{-}MC}(\breve{\mathscr{C}}^{(r')}/\overline{\breve{\mathscr{B}}})$ induced by $\mathrm{I}\varepsilon^{r,r'}(\breve{\alpha})$ and the isomorphisms (4.5.3.2) and (4.5.3.3), and by

$$\breve{\beta}^{(r')}\colon \mathscr{V}(\mathscr{N}) \to \mathrm{I}\mathscr{K}^{(r')}(\mathrm{I}\widehat{\sigma}^{(r')*}(\mathscr{N})) \tag{4.5.18.3}$$

its adjoint. Then,

(i) *For every rational number r' such that $0 < r' \leq r$, the morphism $\breve{\beta}^{(r')}$ is a right inverse of the canonical morphism $\varpi^{(r')}\colon \mathrm{I}\mathscr{K}^{(r')}(\mathrm{I}\widehat{\sigma}^{(r')*}(\mathscr{N})) \to \mathscr{V}(\mathscr{N})$.*

(ii) *For all rational numbers* r' *and* r'' *such that* $0 < r'' < r' \leq r$, *the composition*

$$\mathrm{I}\mathscr{K}^{(r')}(\widehat{\sigma}^{(r')*}(\mathscr{N})) \xrightarrow{\varpi^{(r')}} \mathscr{V}(\mathscr{N}) \xrightarrow{\check{\beta}^{(r'')}} \mathrm{I}\mathscr{K}^{(r'')}(\widehat{\sigma}^{(r'')*}(\mathscr{N})) \qquad (4.5.18.4)$$

is the canonical morphism.

Remark 4.5.19 Under the assumptions of 4.5.18, the isomorphism $\check{\alpha}$ is a priori not uniquely determined by $(\mathscr{N}, r)$, but for every rational number $0 < r' < r$, the morphism $\check{\alpha}^{(r')}$ (4.5.18.2) depends only on $\mathscr{N}$, on which it depends functorially (see the proof of 4.5.20).

Theorem 4.5.20 *The functors* (4.5.12.1) *and* (4.5.17.1)

$$\mathbf{Ind\text{-}Mod}^{\mathrm{Dolb}}(\overline{\breve{\mathscr{B}}}) \underset{\mathscr{V}}{\overset{\mathscr{H}}{\rightleftarrows}} \mathbf{HM}^{\mathrm{sol}}(\mathscr{O}_{\mathfrak{X}}[\tfrac{1}{p}], \widetilde{\xi}^{-1}\widetilde{\Omega}^{1}_{\mathfrak{X}/\mathscr{S}}) \qquad (4.5.20.1)$$

are equivalences of categories quasi-inverse to each other.

For every object $\mathscr{M}$ of $\mathbf{Ind\text{-}Mod}^{\mathrm{Dolb}}(\overline{\breve{\mathscr{B}}})$, $\mathscr{H}(\mathscr{M})$ is a solvable Higgs $\mathscr{O}_{\mathfrak{X}}[\frac{1}{p}]$-bundle associated with $\mathscr{M}$, by virtue of 4.5.12. We choose a rational number $r_{\mathscr{M}} > 0$ and an isomorphism of $\mathbf{Ind\text{-}MC}(\breve{\mathscr{C}}^{(r_{\mathscr{M}})}/\overline{\breve{\mathscr{B}}})$

$$\alpha_{\mathscr{M}} \colon \mathrm{I}\widehat{\sigma}^{(r_{\mathscr{M}})*}(\mathscr{H}(\mathscr{M})) \xrightarrow{\sim} \mathrm{I}\mathfrak{S}^{(r_{\mathscr{M}})}(\mathscr{M}) \qquad (4.5.20.2)$$

satisfying the properties of 4.5.13. For any rational number r such that $0 < r \leq r_{\mathscr{M}}$, we denote by

$$\alpha^{(r)}_{\mathscr{M}} \colon \mathrm{I}\widehat{\sigma}^{(r)*}(\mathscr{H}(\mathscr{M})) \xrightarrow{\sim} \mathrm{I}\mathfrak{S}^{(r)}(\mathscr{M}) \qquad (4.5.20.3)$$

the isomorphism of $\mathbf{Ind\text{-}MC}(\breve{\mathscr{C}}^{(r)}/\overline{\breve{\mathscr{B}}})$ induced by $\mathrm{I}\varepsilon^{r_{\mathscr{M}},r}(\alpha_{\mathscr{M}})$ and the isomorphisms (4.5.3.2) and (4.5.3.3), by

$$\check{\alpha}_{\mathscr{M}} \colon \mathrm{I}\mathfrak{S}^{(r_{\mathscr{M}})}(\mathscr{M}) \xrightarrow{\sim} \mathrm{I}\widehat{\sigma}^{(r_{\mathscr{M}})*}(\mathscr{H}(\mathscr{M})), \qquad (4.5.20.4)$$

$$\check{\alpha}^{(r)}_{\mathscr{M}} \colon \mathrm{I}\mathfrak{S}^{(r)}(\mathscr{M}) \xrightarrow{\sim} \mathrm{I}\widehat{\sigma}^{(r)*}(\mathscr{H}(\mathscr{M})), \qquad (4.5.20.5)$$

the inverses of $\alpha_{\mathscr{M}}$ and $\alpha^{(r)}_{\mathscr{M}}$, respectively, and by

$$\beta^{(r)}_{\mathscr{M}} \colon \mathscr{H}(\mathscr{M}) \to \vec{\sigma}^{(r)}_{*}(\mathrm{I}\mathfrak{S}^{(r)}(\mathscr{M})), \qquad (4.5.20.6)$$

$$\check{\beta}^{(r)}_{\mathscr{M}} \colon \mathscr{M} \to \mathrm{I}\mathscr{K}^{(r)}(\mathrm{I}\widehat{\sigma}^{(r)*}(\mathscr{H}(\mathscr{M}))), \qquad (4.5.20.7)$$

the adjoint morphisms of $\alpha^{(r)}_{\mathscr{M}}$ and $\check{\alpha}^{(r)}_{\mathscr{M}}$, respectively. Note that $\check{\alpha}^{(r)}_{\mathscr{M}}$ is induced by $\mathrm{I}\varepsilon^{r_{\mathscr{M}},r}(\check{\alpha}_{\mathscr{M}})$ and the isomorphisms (4.5.3.2) and (4.5.3.3). By 4.5.16(i), the composed morphism

$$\mathscr{M} \xrightarrow{\check{\beta}^{(r)}_{\mathscr{M}}} \mathrm{I}\mathscr{K}^{(r)}(\mathrm{I}\widehat{\sigma}^{(r)*}(\mathscr{H}(\mathscr{M}))) \longrightarrow \mathscr{V}(\mathscr{H}(\mathscr{M})), \qquad (4.5.20.8)$$

where the second arrow is the canonical morphism, is an isomorphism, which depends a priori on $\alpha_{\mathscr{M}}$ but not on r. Let us show that this isomorphism depends only on $\mathscr{M}$ (but not on the choice of $\alpha_{\mathscr{M}}$), on which it depends functorially. It suffices to show that for every morphism $u\colon \mathscr{M} \to \mathscr{M}'$ of $\mathbf{Ind\text{-}Mod}^{\mathrm{Dolb}}(\breve{\overline{\mathscr{B}}})$ and every rational number $0 < r < \inf(r_{\mathscr{M}}, r_{\mathscr{M}'})$, the diagram of $\mathbf{Ind\text{-}MC}(\breve{\mathscr{C}}^{(r)}/\breve{\overline{\mathscr{B}}})$

$$\begin{array}{ccc} \mathrm{I}\widehat{\sigma}^{(r)*}(\mathscr{H}(\mathscr{M})) & \xrightarrow{\alpha^{(r)}_{\mathscr{M}}} & \mathrm{I}\mathfrak{S}^{r}(\mathscr{M}) \\ {\scriptstyle \mathrm{I}\widehat{\sigma}^{(r)*}(\mathscr{H}(u))}\downarrow & & \downarrow{\scriptstyle \mathrm{I}\mathfrak{S}^{r}(u)} \\ \mathrm{I}\widehat{\sigma}^{(r)*}(\mathscr{H}(\mathscr{M}')) & \xrightarrow{\alpha^{(r)}_{\mathscr{M}'}} & \mathrm{I}\mathfrak{S}^{r}(\mathscr{M}') \end{array} \tag{4.5.20.9}$$

is commutative. Let r, r' be two rational numbers such that $0 < r < r' < \inf(r_{\mathscr{M}}, r_{\mathscr{M}'})$. Consider the diagram

$$\begin{array}{ccccc} \vec{\sigma}^{(r')}_{*}(\mathrm{I}\mathfrak{S}^{(r')}(\mathscr{M})) & \xrightarrow{\varpi^{(r')}_{\mathscr{M}}} & \mathscr{H}(\mathscr{M}) & \xrightarrow{\beta^{(r)}_{\mathscr{M}}} & \vec{\sigma}^{(r)}_{*}(\mathrm{I}\mathfrak{S}^{(r)}(\mathscr{M})) \\ {\scriptstyle \vec{\sigma}^{(r')}_{*}(\mathrm{I}\mathfrak{S}^{r'}(u))}\downarrow & & \downarrow{\scriptstyle \mathscr{H}(u)} & & \downarrow{\scriptstyle \vec{\sigma}^{(r)}_{*}(\mathrm{I}\mathfrak{S}^{r}(u))} \\ \vec{\sigma}^{(r')}_{*}(\mathrm{I}\mathfrak{S}^{(r')}(\mathscr{M}')) & \xrightarrow{\varpi^{(r')}_{\mathscr{M}'}} & \mathscr{H}(\mathscr{M}') & \xrightarrow{\beta^{(r)}_{\mathscr{M}'}} & \vec{\sigma}^{(r)}_{*}(\mathrm{I}\mathfrak{S}^{r}(\mathscr{M}')), \end{array} \tag{4.5.20.10}$$

where $\varpi^{(r')}_{\mathscr{M}}$ and $\varpi^{(r')}_{\mathscr{M}'}$ are the canonical morphisms. It follows from 4.5.13(ii) that the large rectangle is commutative. Since the left square is commutative and $\varpi^{(r')}_{\mathscr{M}'}$ is surjective by 4.5.13(i), the right square is also commutative. The desired assertion follows in view of the injectivity of (4.5.2.4).

Likewise, for every object $\mathscr{N}$ of $\mathbf{HM}^{\mathrm{sol}}(\mathscr{O}_{\mathfrak{X}}[\frac{1}{p}], \xi^{-1}\widetilde{\Omega}^{1}_{\mathfrak{X}/\mathscr{S}})$, $\mathscr{V}(\mathscr{N})$ is a Dolbeault ind-$\breve{\overline{\mathscr{B}}}$-module associated with $\mathscr{N}$, by virtue of 4.5.17. We choose a rational number $r_{\mathscr{N}} > 0$ and an isomorphism of $\mathbf{Ind\text{-}MC}(\breve{\mathscr{C}}^{(r_{\mathscr{N}})}/\breve{\overline{\mathscr{B}}})$

$$\check{\alpha}_{\mathscr{N}}\colon \mathrm{I}\mathfrak{S}^{(r_{\mathscr{N}})}(\mathscr{V}(\mathscr{N})) \xrightarrow{\sim} \mathrm{I}\widehat{\sigma}^{(r_{\mathscr{N}})*}(\mathscr{N}) \tag{4.5.20.11}$$

satisfying the properties of 4.5.18. For any rational number r such that $0 < r \leq r_{\mathscr{N}}$, we denote by

$$\check{\alpha}^{(r)}_{\mathscr{N}}\colon \mathrm{I}\mathfrak{S}^{(r)}(\mathscr{V}(\mathscr{N})) \xrightarrow{\sim} \mathrm{I}\widehat{\sigma}^{(r)*}(\mathscr{N}) \tag{4.5.20.12}$$

the isomorphism of $\mathbf{Ind\text{-}MC}(\breve{\mathscr{C}}^{(r)}/\breve{\overline{\mathscr{B}}})$ induced by $\mathrm{I}\varepsilon^{r_{\mathscr{N}},r}(\check{\alpha}_{\mathscr{N}})$ and the isomorphisms (4.5.3.2) and (4.5.3.3), by

$$\alpha_{\mathscr{N}}\colon \mathrm{I}\widehat{\sigma}^{(r_{\mathscr{N}})*}(\mathscr{N}) \xrightarrow{\sim} \mathrm{I}\mathfrak{S}^{(r_{\mathscr{N}})}(\mathscr{V}(\mathscr{N})), \tag{4.5.20.13}$$

$$\alpha^{(r)}_{\mathscr{N}}\colon \mathrm{I}\widehat{\sigma}^{(r)*}(\mathscr{N}) \xrightarrow{\sim} \mathrm{I}\mathfrak{S}^{(r)}(\mathscr{V}(\mathscr{N})), \tag{4.5.20.14}$$

the inverses of $\check{\alpha}_{\mathscr{M}}$ and $\check{\alpha}^{(r)}_{\mathscr{N}}$, respectively, and by

$$\check{\beta}^{(r)}_{\mathscr{N}} \colon \mathscr{V}(\mathscr{N}) \to \mathrm{I}\mathscr{K}^{(r)}(\mathrm{I}\widehat{\sigma}^{(r)*}(\mathscr{N})), \qquad (4.5.20.15)$$

$$\beta^{(r)}_{\mathscr{N}} \colon \mathscr{N} \to \mathfrak{F}^{(r)}_{*}(\mathrm{I}\mathfrak{S}^{r}(\mathscr{V}(\mathscr{N}))), \qquad (4.5.20.16)$$

the adjoint morphisms of $\check{\alpha}^{(r)}_{\mathscr{N}}$ and $\alpha^{(r)}_{\mathscr{N}}$, respectively. By 4.5.11(i), the composed morphism

$$\mathscr{N} \xrightarrow{\beta^{(r)}_{\mathscr{N}}} \mathfrak{F}^{(r)}_{*}(\mathrm{I}\mathfrak{S}^{(r)}(\mathscr{V}(\mathscr{N}))) \longrightarrow \mathscr{H}(\mathscr{V}(\mathscr{N})), \qquad (4.5.20.17)$$

where the second arrow is the canonical morphism, is an isomorphism, which depends a priori on $\check{\alpha}_{\mathscr{N}}$ but not on r. Let us show that this isomorphism depends only on $\mathscr{N}$ (but not on the choice of $\check{\alpha}_{\mathscr{N}}$), on which it depends functorially. It suffices to show that for every morphism $v \colon \mathscr{N} \to \mathscr{N}'$ of $\mathbf{HM}^{\mathrm{sol}}(\mathscr{O}_{\mathfrak{X}}[\frac{1}{p}], \xi^{-1}\widetilde{\Omega}^{1}_{\mathfrak{X}/\mathscr{S}})$ and every rational number $0 < r < \inf(r_{\mathscr{N}}, r_{\mathscr{N}'})$, the diagram of $\mathbf{Ind\text{-}MC}(\check{\mathscr{C}}^{(r)}/\overline{\check{\mathscr{B}}})$

$$\begin{array}{ccc}
\mathrm{I}\mathfrak{S}^{(r)}(\mathscr{V}(\mathscr{N})) & \xrightarrow{\check{\alpha}^{(r)}_{\mathscr{N}}} & \mathrm{I}\widehat{\sigma}^{(r)*}(\mathscr{N}) \\
{\scriptstyle \mathrm{I}\mathfrak{S}^{(r)}(\mathscr{V}(v))}\downarrow & & \downarrow{\scriptstyle \mathrm{I}\widehat{\sigma}^{(r)*}(v)} \\
\mathrm{I}\mathfrak{S}^{(r)}(\mathscr{V}(\mathscr{N}')) & \xrightarrow{\check{\alpha}^{(r)}_{\mathscr{N}'}} & \mathrm{I}\widehat{\sigma}^{(r)*}(\mathscr{N}')
\end{array} \qquad (4.5.20.18)$$

is commutative. Let r, r' be two rational numbers such that $0 < r < r' < \inf(r_{\mathscr{N}}, r_{\mathscr{N}'})$. Consider the diagram of $\mathbf{Ind\text{-}Mod}(\overline{\check{\mathscr{B}}})$

$$\begin{array}{ccccc}
\mathrm{I}\mathscr{K}^{(r')}(\mathrm{I}\widehat{\sigma}^{(r')*}(\mathscr{N})) & \xrightarrow{\varpi^{(r')}_{\mathscr{N}}} & \mathscr{V}(\mathscr{N}) & \xrightarrow{\check{\beta}^{(r)}_{\mathscr{M}}} & \mathrm{I}\mathscr{K}^{(r)}(\mathrm{I}\widehat{\sigma}^{(r)*}(\mathscr{N})) \\
{\scriptstyle \mathrm{I}\mathscr{K}^{(r')}(\mathrm{I}\widehat{\sigma}^{(r')*}(v))}\downarrow & & \downarrow{\scriptstyle \mathscr{V}(v)} & & \downarrow{\scriptstyle \mathrm{I}\mathscr{K}^{(r)}(\mathrm{I}\widehat{\sigma}^{(r)*}(v))} \\
\mathrm{I}\mathscr{K}^{(r')}(\mathrm{I}\widehat{\sigma}^{(r')*}(\mathscr{N}')) & \xrightarrow{\varpi^{(r')}_{\mathscr{N}'}} & \mathscr{V}(\mathscr{N}') & \xrightarrow{\check{\beta}^{(r)}_{\mathscr{N}'}} & \mathrm{I}\mathscr{K}^{(r)}(\mathrm{I}\widehat{\sigma}^{(r)*}(\mathscr{N}'))
\end{array} \qquad (4.5.20.19)$$

where $\varpi^{(r')}_{\mathscr{N}}$ and $\varpi^{(r')}_{\mathscr{N}'}$ are the canonical morphisms. It follows from 4.5.18(ii) that the large rectangle is commutative. Since the left square is commutative and $\varpi^{(r')}_{\mathscr{N}}$ is right invertible by 4.5.18(i), the right square is also commutative; the desired assertion follows.

4.5.21 We take again the assumptions and notation of 4.4.25. Recall that we assign an exponent $'$ to the objects associated with the $(\mathscr{A}^{*}_{2}(\overline{S}/S), \mathscr{M}_{\mathscr{A}^{*}_{2}(\overline{S}/S)})$-deformation $(\widetilde{X}', \mathscr{M}_{\widetilde{X}'})$ (4.4.22). We have the functors (4.5.12.1) and (4.5.17.1)

$$\mathbf{Ind\text{-}Mod}^{\mathrm{Dolb}}(\overline{\check{\mathscr{B}}}) \underset{\mathscr{V}}{\overset{\mathscr{H}}{\rightleftarrows}} \mathbf{HM}^{\mathrm{sol}}(\mathscr{O}_{\mathfrak{X}}[\tfrac{1}{p}], \xi^{-1}\widetilde{\Omega}^{1}_{\mathfrak{X}/\mathscr{S}}) \qquad (4.5.21.1)$$

$$\mathbf{Ind\text{-}Mod}^{\mathrm{Dolb}'}(\overline{\check{\mathscr{B}}}) \underset{\mathscr{V}'}{\overset{\mathscr{H}'}{\rightleftarrows}} \mathbf{HM}^{\mathrm{sol}'}(\mathscr{O}_{\mathfrak{X}}[\tfrac{1}{p}], (\xi^{*}_{\pi})^{-1}\widetilde{\Omega}^{1}_{\mathfrak{X}/\mathscr{S}}) \qquad (4.5.21.2)$$

associated with the deformations $(\widetilde{X}, \mathscr{M}_{\widetilde{X}})$ and $(\widetilde{X}', \mathscr{M}_{\widetilde{X}'})$, respectively. The isomorphism ν (4.4.25.4) induces an $\mathscr{O}_{\mathfrak{X}}$-linear isomorphism

$$\lambda_*(\nu)\colon (\xi_\pi^*)^{-1}\widetilde{\Omega}^1_{\mathfrak{X}/\mathscr{S}} \xrightarrow{\sim} \xi^{-1}\widetilde{\Omega}^1_{\mathfrak{X}/\mathscr{S}}. \tag{4.5.21.3}$$

We denote by

$$\tau\colon \mathbf{HM}(\mathscr{O}_{\mathfrak{X}}[\frac{1}{p}], (\xi_\pi^*)^{-1}\widetilde{\Omega}^1_{\mathfrak{X}/\mathscr{S}}) \to \mathbf{HM}(\mathscr{O}_{\mathfrak{X}}[\frac{1}{p}], \xi^{-1}\widetilde{\Omega}^1_{\mathfrak{X}/\mathscr{S}}) \tag{4.5.21.4}$$

the functor induced by $p^\rho\lambda_*(\nu)$.

In view of (4.4.25.8), for every rational number $r \geq 0$, we have canonical isomorphisms

$$\tau \circ \vec{\sigma}_*^{\prime(r)} \circ \mathrm{I}\mathfrak{S}^{\prime(r)} \xrightarrow{\sim} \vec{\sigma}_*^{(r+\rho)} \circ \mathrm{I}\mathfrak{S}^{(r+\rho)}, \tag{4.5.21.5}$$

$$\mathrm{I}\mathscr{K}^{\prime(r)} \circ \mathrm{I}\widehat{\sigma}^{\prime(r)*} \xrightarrow{\sim} \mathrm{I}\mathscr{K}^{(r+\rho)} \circ \mathrm{I}\widehat{\sigma}^{(r+\rho)*} \circ \tau. \tag{4.5.21.6}$$

Proposition 4.5.22 *Under the assumptions of* 4.5.21, *let moreover $\mathscr{M}$ be a Dolbeault ind-$\overline{\mathscr{B}}$-module with respect to the deformation $(\widetilde{X}', \mathscr{M}_{\widetilde{X}'})$, $\mathscr{N}$ a Higgs $\mathscr{O}_{\mathfrak{X}}[\frac{1}{p}]$-bundle with coefficients in $(\xi_\pi^*)^{-1}\widetilde{\Omega}^1_{\mathfrak{X}/\mathscr{S}}$ solvable with respect to the deformation $(\widetilde{X}', \mathscr{M}_{\widetilde{X}'})$. Then,*

(i) *The ind-$\overline{\mathscr{B}}$-module $\mathscr{M}$ is Dolbeault with respect to the deformation $(\widetilde{X}, \mathscr{M}_{\widetilde{X}})$, and the canonical morphisms* (4.5.21.5) *induce a functorial isomorphism*

$$\tau(\mathscr{H}'(\mathscr{M})) \xrightarrow{\sim} \mathscr{H}(\mathscr{M}). \tag{4.5.22.1}$$

(ii) *The Higgs $\mathscr{O}_{\mathfrak{X}}[\frac{1}{p}]$-module $\tau(\mathscr{N})$ with coefficients in $\xi^{-1}\widetilde{\Omega}^1_{\mathfrak{X}/\mathscr{S}}$ is solvable with respect to the deformation $(\widetilde{X}, \mathscr{M}_{\widetilde{X}})$ and the canonical morphisms* (4.5.21.6) *induce a functorial isomorphism*

$$\mathscr{V}'(\mathscr{N}) \xrightarrow{\sim} \mathscr{V}(\tau(\mathscr{N})). \tag{4.5.22.2}$$

(i) By 4.5.13, there exist a rational number $r > 0$ and an isomorphism of $\mathbf{Ind\text{-}MC}(\breve{\mathscr{C}}^{\prime(r)}/\breve{\overline{\mathscr{B}}})$

$$\alpha'\colon \mathrm{I}\widehat{\sigma}^{\prime(r)*}(\mathscr{H}'(\mathscr{M})) \xrightarrow{\sim} \mathrm{I}\mathfrak{S}^{\prime(r)}(\mathscr{M}). \tag{4.5.22.3}$$

By 4.5.14, after decreasing r, if necessary, this isomorphism is canonical and functorial in $\mathscr{M}$. In view of (4.4.25.8), α' induces a functorial isomorphism of $\mathbf{Ind\text{-}MC}(\breve{\mathscr{C}}^{(r+\rho)}/\breve{\overline{\mathscr{B}}})$

$$\alpha\colon \mathrm{I}\widehat{\sigma}^{(r+\rho)*}(\tau(\mathscr{H}'(\mathscr{M}))) \xrightarrow{\sim} \mathrm{I}\mathfrak{S}^{(r+\rho)}(\mathscr{M}). \tag{4.5.22.4}$$

We deduce, by virtue of 4.5.11(i), a functorial isomorphism

$$\tau(\mathscr{H}'(\mathscr{M})) \xrightarrow{\sim} \mathscr{H}(\mathscr{M}). \tag{4.5.22.5}$$

(ii) It suffices to copy the proof of (i) while replacing 4.5.13 and 4.5.11(i) by 4.5.18 and 4.5.16(i).

4.6 Dolbeault $\overline{\breve{\mathscr{B}}}_{\mathbb{Q}}$-modules

4.6.1 Let r be a rational number ≥ 0. We denote simply by $\mathbf{IC}(\breve{\mathscr{C}}^{(r)}/\overline{\breve{\mathscr{B}}})$ the category of integrable p^r-isoconnections with respect to the extension $\breve{\mathscr{C}}^{(r)}/\overline{\breve{\mathscr{B}}}$ (see 2.9.11); we therefore omit the exponent p^r from the notation introduced in 2.9.11 considering that it is redundant with the exponent r of $\breve{\mathscr{C}}^{(r)}$. It is an additive category. We denote by $\mathbf{IC}_{\mathbb{Q}}(\breve{\mathscr{C}}^{(r)}/\overline{\breve{\mathscr{B}}})$ the category of objects of $\mathbf{IC}(\breve{\mathscr{C}}^{(r)}/\overline{\breve{\mathscr{B}}})$ up to isogeny ([3] III.6.1.1). By 2.9.14(i) and 4.4.23(iii), every object of $\mathbf{IC}(\breve{\mathscr{C}}^{(r)}/\overline{\breve{\mathscr{B}}})$ is a Higgs $\overline{\breve{\mathscr{B}}}$-isogeny with coefficients in $\widehat{\sigma}^*(\widetilde{\xi}^{-1}\widetilde{\Omega}^1_{\mathfrak{X}/\mathcal{S}})$ (4.5.1). In particular, one can functorially associate with every object of $\mathbf{IC}_{\mathbb{Q}}(\breve{\mathscr{C}}^{(r)}/\overline{\breve{\mathscr{B}}})$ a Dolbeault complex in $\mathbf{Mod}_{\mathbb{Q}}(\overline{\breve{\mathscr{B}}})$ (see 2.9.9).

Consider the functor

$$\begin{aligned} \mathfrak{S}^{(r)}: \mathbf{Mod}(\overline{\breve{\mathscr{B}}}) &\to \mathbf{IC}(\breve{\mathscr{C}}^{(r)}/\overline{\breve{\mathscr{B}}}) \\ \mathscr{M} &\mapsto (\breve{\mathscr{C}}^{(r)} \otimes_{\overline{\breve{\mathscr{B}}}} \mathscr{M}, \breve{\mathscr{C}}^{(r)} \otimes_{\overline{\breve{\mathscr{B}}}} \mathscr{M}, \mathrm{id}, p^r d_{\breve{\mathscr{C}}^{(r)}} \otimes \mathrm{id}) \end{aligned} \tag{4.6.1.1}$$

and denote again by

$$\mathfrak{S}^{(r)}: \mathbf{Mod}_{\mathbb{Q}}(\overline{\breve{\mathscr{B}}}) \to \mathbf{IC}_{\mathbb{Q}}(\breve{\mathscr{C}}^{(r)}/\overline{\breve{\mathscr{B}}}) \tag{4.6.1.2}$$

the induced functor. Consider also the functor

$$\begin{aligned} \mathscr{K}^{(r)}: \mathbf{IC}(\breve{\mathscr{C}}^{(r)}/\overline{\breve{\mathscr{B}}}) &\to \mathbf{Mod}(\overline{\breve{\mathscr{B}}}) \\ (\mathscr{F}, \mathscr{G}, u, \nabla) &\mapsto \ker(\nabla) \end{aligned} \tag{4.6.1.3}$$

and denote again by

$$\mathscr{K}^{(r)}: \mathbf{IC}_{\mathbb{Q}}(\breve{\mathscr{C}}^{(r)}/\overline{\breve{\mathscr{B}}}) \to \mathbf{Mod}_{\mathbb{Q}}(\overline{\breve{\mathscr{B}}}) \tag{4.6.1.4}$$

the induced functor. It is clear that the functor (4.6.1.1) is a left adjoint of the functor (4.6.1.3). Hence, the functor (4.6.1.2) is a left adjoint of the functor (4.6.1.4).

By 2.9.14(ii), if $(\mathscr{N}, \mathscr{N}', v, \theta)$ is a Higgs $\mathscr{O}_{\mathfrak{X}}$-isogeny with coefficients in $\widetilde{\xi}^{-1}\widetilde{\Omega}^1_{\mathfrak{X}/\mathcal{S}}$ (2.9.9),

$$(\breve{\mathscr{C}}^{(r)} \otimes_{\overline{\breve{\mathscr{B}}}} \widehat{\sigma}^*(\mathscr{N}), \breve{\mathscr{C}}^{(r)} \otimes_{\overline{\breve{\mathscr{B}}}} \widehat{\sigma}^*(\mathscr{N}'), \mathrm{id} \otimes_{\overline{\breve{\mathscr{B}}}} \widehat{\sigma}^*(v), p^r d_{\breve{\mathscr{C}}^{(r)}} \otimes \widehat{\sigma}^*(v) + \mathrm{id} \otimes \widehat{\sigma}^*(\theta)) \tag{4.6.1.5}$$

is an object of $\mathbf{IC}(\breve{\mathscr{C}}^{(r)}/\overline{\breve{\mathscr{B}}})$. We thus obtain a functor (4.2.2)

$$\widehat{\sigma}^{(r)*}: \mathbf{HI}(\mathscr{O}_{\mathfrak{X}}, \widetilde{\xi}^{-1}\widetilde{\Omega}^1_{\mathfrak{X}/\mathcal{S}}) \to \mathbf{IC}(\breve{\mathscr{C}}^{(r)}/\overline{\breve{\mathscr{B}}}). \tag{4.6.1.6}$$

In view of (4.5.2.2), this induces a functor that we denote again by

$$\widehat{\sigma}^{(r)*}\colon \mathbf{HM}^{\mathrm{coh}}(\mathscr{O}_{\mathfrak{X}}[\frac{1}{p}], \widetilde{\xi}^{-1}\widetilde{\Omega}^1_{\mathfrak{X}/\mathscr{S}}) \to \mathbf{IC}_{\mathbb{Q}}(\breve{\mathscr{C}}^{(r)}/\breve{\overline{\mathscr{B}}}). \tag{4.6.1.7}$$

Let $(\mathscr{M}, \mathscr{M}', u, \nabla)$ be an object of $\mathbf{IC}(\breve{\mathscr{C}}^{(r)}/\breve{\overline{\mathscr{B}}})$. In view of ([3] III.12.4(i)), ∇ induces an $\mathscr{O}_{\mathfrak{X}}$-linear morphism

$$\widehat{\sigma}_*(\nabla)\colon \widehat{\sigma}_*(\mathscr{M}) \to \widetilde{\xi}^{-1}\widetilde{\Omega}^1_{\mathfrak{X}/\mathscr{S}} \otimes_{\mathscr{O}_{\mathfrak{X}}} \widehat{\sigma}_*(\mathscr{M}'). \tag{4.6.1.8}$$

One easily checks that $(\widehat{\sigma}_*(\mathscr{M}), \widehat{\sigma}_*(\mathscr{M}'), \widehat{\sigma}_*(u), \widehat{\sigma}_*(\nabla))$ is a Higgs $\mathscr{O}_{\mathfrak{X}}$-isogeny with coefficients in $\widetilde{\xi}^{-1}\widetilde{\Omega}^1_{\mathfrak{X}/\mathscr{S}}$ (see [3] III.12.3(i)). We thus obtain a functor

$$\widehat{\sigma}^{(r)}_*\colon \mathbf{IC}(\breve{\mathscr{C}}^{(r)}/\breve{\overline{\mathscr{B}}}) \to \mathbf{HI}(\mathscr{O}_{\mathfrak{X}}, \widetilde{\xi}^{-1}\widetilde{\Omega}^1_{\mathfrak{X}/\mathscr{S}}). \tag{4.6.1.9}$$

The composition of the functors (4.6.1.9) and (4.2.2.2) induces a functor that we denote again by

$$\widehat{\sigma}^{(r)}_*\colon \mathbf{IC}_{\mathbb{Q}}(\breve{\mathscr{C}}^{(r)}/\breve{\overline{\mathscr{B}}}) \to \mathbf{HM}(\mathscr{O}_{\mathfrak{X}}[\frac{1}{p}], \widetilde{\xi}^{-1}\widetilde{\Omega}^1_{\mathfrak{X}/\mathscr{S}}). \tag{4.6.1.10}$$

It is clear that the functor (4.6.1.6) is a left adjoint of the functor (4.6.1.9). We deduce from this that for all $\mathscr{N} \in \mathrm{Ob}(\mathbf{HM}^{\mathrm{coh}}(\mathscr{O}_{\mathfrak{X}}[\frac{1}{p}], \widetilde{\xi}^{-1}\widetilde{\Omega}^1_{\mathfrak{X}/\mathscr{S}}))$ and $\mathscr{A} \in \mathrm{Ob}(\mathbf{IC}_{\mathbb{Q}}(\breve{\mathscr{C}}^{(r)}/\breve{\overline{\mathscr{B}}}))$, we have a bifunctorial canonical homomorphism

$$\mathrm{Hom}_{\mathbf{IC}_{\mathbb{Q}}(\breve{\mathscr{C}}^{(r)}/\breve{\overline{\mathscr{B}}})}(\widehat{\sigma}^{(r)*}(\mathscr{N}), \mathscr{A}) \to \mathrm{Hom}_{\mathbf{HM}(\mathscr{O}_{\mathfrak{X}}[\frac{1}{p}], \widetilde{\xi}^{-1}\widetilde{\Omega}^1_{\mathfrak{X}/\mathscr{S}})}(\mathscr{N}, \widehat{\sigma}^{(r)}_*(\mathscr{A})), \tag{4.6.1.11}$$

which is injective by ([3] III.6.20 and III.6.21). We abusively call the *adjoint* of a morphism $\widehat{\sigma}^{(r)*}(\mathscr{N}) \to \mathscr{A}$ of $\mathbf{IC}_{\mathbb{Q}}(\breve{\mathscr{C}}^{(r)}/\breve{\overline{\mathscr{B}}})$ its image by the homomorphism (4.6.1.11).

4.6.2 Let r, r' be two rational numbers such that $r \geq r' \geq 0$, $(\mathscr{F}, \mathscr{G}, u, \nabla)$ be a p^r-isoconnection integrable with respect to the extension $\breve{\mathscr{C}}^{(r)}/\breve{\overline{\mathscr{B}}}$. By (4.4.22.9), there exists one and only one $\breve{\overline{\mathscr{B}}}$-linear morphism

$$\nabla'\colon \breve{\mathscr{C}}^{(r')} \otimes_{\breve{\mathscr{C}}^{(r)}} \mathscr{F} \to \widehat{\sigma}^*(\xi^{-1}\widetilde{\Omega}^1_{\mathfrak{X}/\mathscr{S}}) \otimes_{\breve{\overline{\mathscr{B}}}} \breve{\mathscr{C}}^{(r')} \otimes_{\breve{\mathscr{C}}^{(r)}} \mathscr{G} \tag{4.6.2.1}$$

such that for all local sections x' of $\breve{\mathscr{C}}^{(r')}$ and s of $\mathscr{F}$, we have

$$\nabla'(x' \otimes_{\breve{\mathscr{C}}^{(r)}} s) = p^{r'} d_{\breve{\mathscr{C}}^{(r')}}(x') \otimes_{\breve{\mathscr{C}}^{(r)}} u(s) + x' \otimes_{\breve{\mathscr{C}}^{(r)}} \nabla(s). \tag{4.6.2.2}$$

The quadruple $(\breve{\mathscr{C}}^{(r')} \otimes_{\breve{\mathscr{C}}^{(r)}} \mathscr{F}, \breve{\mathscr{C}}^{(r')} \otimes_{\breve{\mathscr{C}}^{(r)}} \mathscr{G}, \mathrm{id} \otimes_{\breve{\mathscr{C}}^{(r)}} u, \nabla')$ is an integrable $p^{r'}$-isoconnection with respect to the extension $\breve{\mathscr{C}}^{(r')}/\breve{\overline{\mathscr{B}}}$. We thus obtain a functor

$$\varepsilon^{r,r'}\colon \mathbf{IC}(\breve{\mathscr{C}}^{(r)}/\breve{\overline{\mathscr{B}}}) \to \mathbf{IC}(\breve{\mathscr{C}}^{(r')}/\breve{\overline{\mathscr{B}}}). \tag{4.6.2.3}$$

This induces a functor that we denote again by

$$\varepsilon^{r,r'} : \mathbf{IC}_{\mathbb{Q}}(\breve{\mathscr{C}}^{(r)}/\overline{\breve{\mathscr{B}}}) \to \mathbf{IC}_{\mathbb{Q}}(\breve{\mathscr{C}}^{(r')}/\overline{\breve{\mathscr{B}}}). \tag{4.6.2.4}$$

We have a canonical isomorphism of functors from $\mathbf{Mod}(\overline{\breve{\mathscr{B}}})$ into $\mathbf{IC}(\breve{\mathscr{C}}^{(r')}/\overline{\breve{\mathscr{B}}})$ (resp. from $\mathbf{Mod}_{\mathbb{Q}}(\overline{\breve{\mathscr{B}}})$ into $\mathbf{IC}_{\mathbb{Q}}(\breve{\mathscr{C}}^{(r')}/\overline{\breve{\mathscr{B}}})$)

$$\varepsilon^{r,r'} \circ \mathfrak{S}^{(r)} \xrightarrow{\sim} \mathfrak{S}^{(r')}, \tag{4.6.2.5}$$

and a canonical isomorphism of functors from $\mathbf{HI}(\mathscr{O}_{\mathfrak{X}}, \xi^{-1}\widetilde{\Omega}^1_{\mathfrak{X}/\mathscr{S}})$ into $\mathbf{IC}(\breve{\mathscr{C}}^{(r')}/\overline{\breve{\mathscr{B}}})$ (resp. from $\mathbf{HM}^{\mathrm{coh}}(\mathscr{O}_{\mathfrak{X}}[\frac{1}{p}], \widetilde{\xi}^{-1}\widetilde{\Omega}^1_{\mathfrak{X}/\mathscr{S}})$ into $\mathbf{IC}_{\mathbb{Q}}(\breve{\mathscr{C}}^{(r')}/\overline{\breve{\mathscr{B}}})$)

$$\varepsilon^{r,r'} \circ \widehat{\sigma}^{(r)*} \xrightarrow{\sim} \widehat{\sigma}^{(r')*}. \tag{4.6.2.6}$$

Moreover, we have a canonical morphism of functors from $\mathbf{IC}(\breve{\mathscr{C}}^{(r)}/\overline{\breve{\mathscr{B}}})$ into $\mathbf{Mod}(\overline{\breve{\mathscr{B}}})$ (resp. from $\mathbf{IC}_{\mathbb{Q}}(\breve{\mathscr{C}}^{(r)}/\overline{\breve{\mathscr{B}}})$ into $\mathbf{Mod}_{\mathbb{Q}}(\overline{\breve{\mathscr{B}}})$)

$$\mathscr{K}^{(r)} \to \mathscr{K}^{(r')} \circ \varepsilon^{r,r'}, \tag{4.6.2.7}$$

and a canonical morphism of functors from $\mathbf{IC}(\breve{\mathscr{C}}^{(r)}/\overline{\breve{\mathscr{B}}})$ into $\mathbf{HI}(\mathscr{O}_{\mathfrak{X}}, \widetilde{\xi}^{-1}\widetilde{\Omega}^1_{\mathfrak{X}/\mathscr{S}})$ (resp. from $\mathbf{IC}_{\mathbb{Q}}(\breve{\mathscr{C}}^{(r)}/\overline{\breve{\mathscr{B}}})$ into $\mathbf{HM}(\mathscr{O}_{\mathfrak{X}}[\frac{1}{p}], \widetilde{\xi}^{-1}\widetilde{\Omega}^1_{\mathfrak{X}/\mathscr{S}})$)

$$\widehat{\sigma}_*^{(r)} \longrightarrow \widehat{\sigma}_*^{(r')} \circ \varepsilon^{r,r'}. \tag{4.6.2.8}$$

For every rational number r'' such that $r' \geq r'' \geq 0$, we have a canonical isomorphism of functors from $\mathbf{IC}(\breve{\mathscr{C}}^{(r)}/\overline{\breve{\mathscr{B}}})$ into $\mathbf{IC}(\breve{\mathscr{C}}^{(r'')}/\overline{\breve{\mathscr{B}}})$ (resp. from $\mathbf{IC}_{\mathbb{Q}}(\breve{\mathscr{C}}^{(r)}/\overline{\breve{\mathscr{B}}})$ into $\mathbf{IC}_{\mathbb{Q}}(\breve{\mathscr{C}}^{(r'')}/\overline{\breve{\mathscr{B}}})$)

$$\varepsilon^{r',r''} \circ \varepsilon^{r,r'} \xrightarrow{\sim} \varepsilon^{r,r''}. \tag{4.6.2.9}$$

4.6.3 Let r be a rational number ≥ 0. We denote by $\mathbf{Mod}_{\mathbb{Q}}(\breve{\mathscr{C}}^{(r)})$ the category of $\breve{\mathscr{C}}^{(r)}$-modules up to isogeny (2.9.7) and by $\mathbf{Ind\text{-}Mod}(\breve{\mathscr{C}}^{(r)})$ the category of ind-$\breve{\mathscr{C}}^{(r)}$-modules (2.7.1). By 2.9.2, we have a fully faithful canonical functor (2.9.2.3)

$$\alpha_{\breve{\mathscr{C}}^{(r)}} : \mathbf{Mod}_{\mathbb{Q}}(\breve{\mathscr{C}}^{(r)}) \to \mathbf{Ind\text{-}Mod}(\breve{\mathscr{C}}^{(r)}), \tag{4.6.3.1}$$

compatible with the functor $\alpha_{\overline{\breve{\mathscr{B}}}}$ (4.3.13.4), via the forgetful functors (2.7.1.6). We immediately check, in view of (2.9.7.4), that for every object $(\mathscr{M}, \mathscr{M}', u, \nabla)$ of $\mathbf{IC}(\breve{\mathscr{C}}^{(r)}/\overline{\breve{\mathscr{B}}})$, the morphism of $\mathbf{Ind\text{-}Mod}(\overline{\breve{\mathscr{B}}})$

$$(\mathrm{id} \otimes \alpha_{\overline{\breve{\mathscr{B}}}}(u_{\mathbb{Q}})^{-1}) \circ \alpha_{\overline{\breve{\mathscr{B}}}}(\nabla_{\mathbb{Q}}) : \alpha_{\breve{\mathscr{C}}^{(r)}}(\mathscr{M}_{\mathbb{Q}}) \to \widehat{\sigma}^*(\widetilde{\xi}^{-1}\widetilde{\Omega}^1_{\mathfrak{X}/\mathscr{S}}) \otimes_{\overline{\breve{\mathscr{B}}}} \alpha_{\breve{\mathscr{C}}^{(r)}}(\mathscr{M}_{\mathbb{Q}}) \tag{4.6.3.2}$$

is an integrable p^r-connection over $\alpha_{\breve{\mathscr{C}}^{(r)}}(\mathscr{M}_{\mathbb{Q}})$ with respect to the extension $\breve{\mathscr{C}}^{(r)}/\overline{\breve{\mathscr{B}}}$ (4.5.1). We thus define a functor

$$\alpha_{\breve{\mathscr{C}}^{(r)}/\breve{\overline{\mathscr{B}}}} \colon \mathbf{IC}_{\mathbb{Q}}(\breve{\mathscr{C}}^{(r)}/\breve{\overline{\mathscr{B}}}) \to \mathbf{Ind\text{-}MC}(\breve{\mathscr{C}}^{(r)}/\breve{\overline{\mathscr{B}}}). \tag{4.6.3.3}$$

It is fully faithful by (2.6.6.4) and (2.9.12.3).

The diagram

$$\begin{array}{ccc} \mathbf{Mod}_{\mathbb{Q}}(\breve{\overline{\mathscr{B}}}) & \xrightarrow{\mathfrak{S}^{(r)}} & \mathbf{IC}_{\mathbb{Q}}(\breve{\mathscr{C}}^{(r)}/\breve{\overline{\mathscr{B}}}) \\ {\scriptstyle \alpha_{\breve{\overline{\mathscr{B}}}}}\downarrow & & \downarrow{\scriptstyle \alpha_{\breve{\mathscr{C}}^{(r)}/\breve{\overline{\mathscr{B}}}}} \\ \mathbf{Ind\text{-}Mod}(\breve{\overline{\mathscr{B}}}) & \xrightarrow{\mathrm{I}\mathfrak{S}^{(r)}} & \mathbf{Ind\text{-}MC}(\breve{\mathscr{C}}^{(r)}/\breve{\overline{\mathscr{B}}}), \end{array} \tag{4.6.3.4}$$

where $\mathfrak{S}^{(r)}$ (resp. $\mathrm{I}\mathfrak{S}^{(r)}$) is the functor (4.6.1.2) (resp. (4.5.1.2)), is commutative up to canonical isomorphism. Likewise, the diagram

$$\begin{array}{ccc} \mathbf{HM}^{\mathrm{coh}}(\mathscr{O}_{\mathfrak{X}}[\frac{1}{p}], \widetilde{\xi}^{-1}\widetilde{\Omega}^1_{\mathfrak{X}/\mathscr{S}}) & \xrightarrow{\widehat{\sigma}^{(r)*}} & \mathbf{IC}_{\mathbb{Q}}(\breve{\mathscr{C}}^{(r)}/\breve{\overline{\mathscr{B}}}) \\ {\scriptstyle u}\downarrow & & \downarrow{\scriptstyle \alpha_{\breve{\mathscr{C}}^{(r)}/\breve{\overline{\mathscr{B}}}}} \\ \mathbf{Ind\text{-}HM}(\mathscr{O}_{\mathfrak{X}}, \widetilde{\xi}^{-1}\widetilde{\Omega}^1_{\mathfrak{X}/\mathscr{S}}) & \xrightarrow{\mathrm{I}\widehat{\sigma}^{(r)*}} & \mathbf{Ind\text{-}MC}(\breve{\mathscr{C}}^{(r)}/\breve{\overline{\mathscr{B}}}), \end{array} \tag{4.6.3.5}$$

where $\widehat{\sigma}^{(r)*}$ (resp. $\mathrm{I}\widehat{\sigma}^{(r)*}$) is the functor (4.6.1.7) (resp. (4.5.1.6)) and u is the functor (4.5.2.3), is commutative up to canonical isomorphism.

Since the functor $\alpha_{\breve{\overline{\mathscr{B}}}}$ is exact by 2.9.3(ii), the diagram

$$\begin{array}{ccc} \mathbf{IC}_{\mathbb{Q}}(\breve{\mathscr{C}}^{(r)}/\breve{\overline{\mathscr{B}}}) & \xrightarrow{\mathscr{K}^{(r)}} & \mathbf{Mod}_{\mathbb{Q}}(\breve{\overline{\mathscr{B}}}) \\ {\scriptstyle \alpha_{\breve{\mathscr{C}}^{(r)}/\breve{\overline{\mathscr{B}}}}}\downarrow & & \downarrow{\scriptstyle \alpha_{\breve{\overline{\mathscr{B}}}}} \\ \mathbf{Ind\text{-}MC}(\breve{\mathscr{C}}^{(r)}/\breve{\overline{\mathscr{B}}}) & \xrightarrow{\mathrm{I}\mathscr{K}^{(r)}} & \mathbf{Ind\text{-}Mod}(\breve{\overline{\mathscr{B}}}), \end{array} \tag{4.6.3.6}$$

where $\mathscr{K}^{(r)}$ (resp. $\mathrm{I}\mathscr{K}^{(r)}$) is the functor (4.6.1.4) (resp. (4.5.1.3)), is commutative up to canonical isomorphism.

By 2.9.5(ii), the diagram

$$\begin{array}{ccc} \mathbf{IC}_{\mathbb{Q}}(\breve{\mathscr{C}}^{(r)}/\breve{\overline{\mathscr{B}}}) & \xrightarrow{(\widehat{\sigma}^{(r)}_*)_{\mathbb{Q}}} & \mathbf{HI}_{\mathbb{Q}}(\mathscr{O}_{\mathfrak{X}}, \widetilde{\xi}^{-1}\widetilde{\Omega}^1_{\mathfrak{X}/\mathscr{S}}) \\ {\scriptstyle \alpha_{\breve{\mathscr{C}}^{(r)}/\breve{\overline{\mathscr{B}}}}}\downarrow & & \downarrow{\scriptstyle v} \\ \mathbf{Ind\text{-}MC}(\breve{\mathscr{C}}^{(r)}/\breve{\overline{\mathscr{B}}}) & \xrightarrow{\mathrm{I}\widehat{\sigma}^{(r)}_*} & \mathbf{Ind\text{-}HM}(\mathscr{O}_{\mathfrak{X}}, \widetilde{\xi}^{-1}\widetilde{\Omega}^1_{\mathfrak{X}/\mathscr{S}}), \end{array} \tag{4.6.3.7}$$

where $\widehat{\sigma}^{(r)}_*$ (resp. $\mathrm{I}\widehat{\sigma}^{(r)}_*$) is the functor (4.6.1.9) (resp. (4.5.1.7)) and v is the functor (4.5.2.1), is commutative up to canonical isomorphism. We deduce from this that the diagram

$$\begin{array}{ccc} \mathbf{IC}_{\mathbb{Q}}(\breve{\mathscr{C}}^{(r)}/\overline{\breve{\mathscr{B}}}) & \xrightarrow{\widehat{\sigma}_*^{(r)}} & \mathbf{HM}(\mathscr{O}_{\mathfrak{X}}[\frac{1}{p}], \widetilde{\xi}^{-1}\widetilde{\Omega}^1_{\mathfrak{X}/\mathscr{S}}) \\ \Big\downarrow {\scriptstyle \alpha_{\breve{\mathscr{C}}^{(r)}/\overline{\breve{\mathscr{B}}}}} & & \Big\downarrow {\scriptstyle w} \\ \mathbf{Ind\text{-}MC}(\breve{\mathscr{C}}^{(r)}/\overline{\breve{\mathscr{B}}}) & \xrightarrow{\vec{\sigma}_*^{(r)}} & \mathbf{HM}(\mathscr{O}_{\mathfrak{X}}, \widetilde{\xi}^{-1}\widetilde{\Omega}^1_{\mathfrak{X}/\mathscr{S}}), \end{array} \tag{4.6.3.8}$$

where $\widehat{\sigma}_*^{(r)}$ (resp. $\vec{\sigma}_*^{(r)}$) is the functor (4.6.1.10) (resp. (4.5.1.9)) and w is the canonical functor, is commutative up to canonical isomorphism.

For all rational numbers $r \geq r' \geq 0$, the diagram

$$\begin{array}{ccc} \mathbf{IC}_{\mathbb{Q}}(\breve{\mathscr{C}}^{(r)}/\overline{\breve{\mathscr{B}}}) & \xrightarrow{\varepsilon^{r,r'}} & \mathbf{IC}_{\mathbb{Q}}(\breve{\mathscr{C}}^{(r')}/\overline{\breve{\mathscr{B}}}) \\ \Big\downarrow {\scriptstyle \alpha_{\breve{\mathscr{C}}^{(r)}/\overline{\breve{\mathscr{B}}}}} & & \Big\downarrow {\scriptstyle \alpha_{\breve{\mathscr{C}}^{(r')}/\overline{\breve{\mathscr{B}}}}} \\ \mathbf{Ind\text{-}MC}(\breve{\mathscr{C}}^{(r)}/\overline{\breve{\mathscr{B}}}) & \xrightarrow{\mathrm{I}\varepsilon^{r,r'}} & \mathbf{Ind\text{-}MC}(\breve{\mathscr{C}}^{(r')}/\overline{\breve{\mathscr{B}}}), \end{array} \tag{4.6.3.9}$$

where $\varepsilon^{r,r'}$ (resp. $\mathrm{I}\varepsilon^{r,r'}$) is the functor (4.6.2.4) (resp. (4.5.3.1)), is commutative up to canonical isomorphism. The isomorphisms (4.6.2.5) and (4.5.3.2) (resp. (4.6.2.6) and (4.5.3.3)) are compatible. The morphisms (4.6.2.7) and (4.5.3.5) (resp. (4.6.2.8) and (4.5.3.6)) are compatible.

Definition 4.6.4 Let $\mathscr{M}$ be a $\overline{\breve{\mathscr{B}}}_{\mathbb{Q}}$-module, $\mathscr{N}$ a Higgs $\mathscr{O}_{\mathfrak{X}}[\frac{1}{p}]$-bundle with coefficients in $\widetilde{\xi}^{-1}\widetilde{\Omega}^1_{\mathfrak{X}/\mathscr{S}}$ (4.2.3).

(i) Let r be a rational number > 0. We say that $\mathscr{M}$ and $\mathscr{N}$ are *r-associated* if there exists an isomorphism of $\mathbf{IC}_{\mathbb{Q}}(\breve{\mathscr{C}}^{(r)}/\overline{\breve{\mathscr{B}}})$

$$\alpha\colon \widehat{\sigma}^{(r)*}(\mathscr{N}) \xrightarrow{\sim} \mathfrak{S}^{(r)}(\mathscr{M}). \tag{4.6.4.1}$$

We then also say that the triple $(\mathscr{M}, \mathscr{N}, \alpha)$ is *r-admissible*.

(ii) We say that $\mathscr{M}$ and $\mathscr{N}$ are *associated* if there exists a rational number $r > 0$ such that $\mathscr{M}$ and $\mathscr{N}$ are r-associated.

Note that for all rational numbers $r \geq r' > 0$, if $\mathscr{M}$ and $\mathscr{N}$ are r-associated, they are r'-associated, in view of (4.6.2.5) and (4.6.2.6).

Lemma 4.6.5 *Let $\mathscr{M}$ be a $\overline{\breve{\mathscr{B}}}_{\mathbb{Q}}$-module, $\mathscr{N}$ a Higgs $\mathscr{O}_{\mathfrak{X}}[\frac{1}{p}]$-bundle with coefficients in $\widetilde{\xi}^{-1}\widetilde{\Omega}^1_{\mathfrak{X}/\mathscr{S}}$, r a rational number > 0. Then, $\mathscr{M}$ and $\mathscr{N}$ are r-associated in the sense of* 4.6.4 *if and only if $\alpha_{\overline{\breve{\mathscr{B}}}}(\mathscr{M})$* (4.3.13.4) *and $\mathscr{N}$ are r-associated in the sense of* 4.5.4.

This immediately follows from (4.6.3.4), (4.6.3.5) and from the full faithfulness of the functor $\alpha_{\breve{\mathscr{C}}^{(r)}/\overline{\breve{\mathscr{B}}}}$ (4.6.3.3).

Definition 4.6.6

(i) A $\overline{\breve{\mathscr{B}}}_{\mathbb{Q}}$-module is said to be *Dolbeault* if it is associated with a Higgs $\mathscr{O}_{\mathfrak{X}}[\frac{1}{p}]$-bundle with coefficients in $\widetilde{\xi}^{-1}\widetilde{\Omega}^1_{\mathfrak{X}/\mathscr{S}}$ (4.6.4).

(ii) A $\overline{\breve{\mathscr{B}}}_{\mathbb{Q}}$-module is said to be *strongly Dolbeault* if it is Dolbeault and adic of finite type (4.3.14).

(iii) A Higgs $\mathscr{O}_{\mathfrak{X}}[\frac{1}{p}]$-bundle with coefficients in $\widetilde{\xi}^{-1}\widetilde{\Omega}^1_{\mathfrak{X}/\mathscr{S}}$ (4.2.3) is said to be *rationally solvable* (resp. *strongly solvable*) if it is associated with a $\overline{\breve{\mathscr{B}}}_{\mathbb{Q}}$-module (resp. adic of finite type).

Observe that these notions depend a priori on the deformation $(\widetilde{X}, \mathscr{M}_{\widetilde{X}})$ fixed in 4.1.5. We denote by $\mathbf{Mod}^{\mathrm{Dolb}}_{\mathbb{Q}}(\overline{\breve{\mathscr{B}}})$ (resp. $\mathbf{Mod}^{\mathrm{sDolb}}_{\mathbb{Q}}(\overline{\breve{\mathscr{B}}})$) the full subcategory of $\mathbf{Mod}_{\mathbb{Q}}(\overline{\breve{\mathscr{B}}})$ made up of the Dolbeault (resp. strongly Dolbeault) $\overline{\breve{\mathscr{B}}}_{\mathbb{Q}}$-modules, and by $\mathbf{HM}^{\mathrm{qsol}}(\mathscr{O}_{\mathfrak{X}}[\frac{1}{p}], \widetilde{\xi}^{-1}\widetilde{\Omega}^1_{\mathfrak{X}/\mathscr{S}})$ (resp. $\mathbf{HM}^{\mathrm{ssol}}(\mathscr{O}_{\mathfrak{X}}[\frac{1}{p}], \widetilde{\xi}^{-1}\widetilde{\Omega}^1_{\mathfrak{X}/\mathscr{S}})$) the full subcategory of $\mathbf{HM}(\mathscr{O}_{\mathfrak{X}}[\frac{1}{p}], \widetilde{\xi}^{-1}\widetilde{\Omega}^1_{\mathfrak{X}/\mathscr{S}})$ made up of the rationally (resp. strongly) solvable Higgs $\mathscr{O}_{\mathfrak{X}}[\frac{1}{p}]$-bundles with coefficients in $\widetilde{\xi}^{-1}\widetilde{\Omega}^1_{\mathfrak{X}/\mathscr{S}}$.

Remark 4.6.7 By 4.6.5, a $\overline{\breve{\mathscr{B}}}_{\mathbb{Q}}$-module $\mathscr{M}$ is Dolbeault if and only if $\alpha_{\overline{\breve{\mathscr{B}}}}(\mathscr{M})$ is Dolbeault (4.5.5).

Remark 4.6.8 Strongly Dolbeault $\overline{\breve{\mathscr{B}}}_{\mathbb{Q}}$-modules (resp. strongly solvable Higgs $\mathscr{O}_{\mathfrak{X}}[\frac{1}{p}]$-bundles) were called *Dolbeault* $\overline{\breve{\mathscr{B}}}_{\mathbb{Q}}$*-module* (resp. *solvable Higgs* $\mathscr{O}_{\mathfrak{X}}[\frac{1}{p}]$*-bundles*) in ([3] III.12.11).

4.6.9 For every $\overline{\breve{\mathscr{B}}}_{\mathbb{Q}}$-module $\mathscr{M}$ and all rational numbers $r \geq r' \geq 0$, the morphism (4.6.2.8) and the isomorphism (4.6.2.5) induce a morphism of $\mathbf{HM}(\mathscr{O}_{\mathfrak{X}}[\frac{1}{p}], \widetilde{\xi}^{-1}\widetilde{\Omega}^1_{\mathfrak{X}/\mathscr{S}})$

$$\widehat{\sigma}^{(r)}_*(\mathfrak{S}^{(r)}(\mathscr{M})) \to \widehat{\sigma}^{(r')}_*(\mathfrak{S}^{(r')}(\mathscr{M})). \tag{4.6.9.1}$$

We thus obtain a filtered direct system $(\widehat{\sigma}^r_+(\mathfrak{S}^{(r)}(\mathscr{M})))_{r\in\mathbb{Q}_{\geq 0}}$. We denote by $\mathscr{H}_{\mathbb{Q}}$ the functor

$$\mathscr{H}_{\mathbb{Q}}\colon \mathbf{Mod}_{\mathbb{Q}}(\overline{\breve{\mathscr{B}}}) \to \mathbf{HM}(\mathscr{O}_{\mathfrak{X}}[\frac{1}{p}], \widetilde{\xi}^{-1}\widetilde{\Omega}^1_{\mathfrak{X}/\mathscr{S}}), \quad \mathscr{M} \mapsto \lim_{\substack{\longrightarrow \\ r\in\mathbb{Q}_{>0}}} \widehat{\sigma}^{(r)}_*(\mathfrak{S}^{(r)}(\mathscr{M})). \tag{4.6.9.2}$$

In view of (4.6.3.4) and (4.6.3.8), the diagram

$$\begin{array}{ccc} \mathbf{Mod}_{\mathbb{Q}}(\overline{\breve{\mathscr{B}}}) & \xrightarrow{\mathscr{H}_{\mathbb{Q}}} & \mathbf{HM}(\mathscr{O}_{\mathfrak{X}}[\frac{1}{p}], \widetilde{\xi}^{-1}\widetilde{\Omega}^1_{\mathfrak{X}/\mathscr{S}}) \\ \downarrow{\scriptstyle \alpha_{\overline{\breve{\mathscr{B}}}}} & & \downarrow{\scriptstyle w} \\ \mathbf{Ind}\text{-}\mathbf{Mod}(\overline{\breve{\mathscr{B}}}) & \xrightarrow{\mathscr{H}} & \mathbf{HM}(\mathscr{O}_{\mathfrak{X}}, \widetilde{\xi}^{-1}\widetilde{\Omega}^1_{\mathfrak{X}/\mathscr{S}}), \end{array} \tag{4.6.9.3}$$

where $\mathscr{H}$ is the functor (4.5.7.3) and w is the canonical functor, is commutative up to canonical isomorphism.

For every object $\mathscr{N}$ of $\mathbf{HM}(\mathscr{O}_{\mathfrak{X}}[\frac{1}{p}], \widetilde{\xi}^{-1}\widetilde{\Omega}^1_{\mathfrak{X}/\mathscr{S}})$ and all rational numbers $r \geq r' \geq 0$, the morphism (4.6.2.7) and the isomorphism (4.6.2.6) induce a morphism of $\mathbf{Mod}_{\mathbb{Q}}(\overline{\breve{\mathscr{B}}})$

$$\mathscr{K}^{(r)}(\widehat{\sigma}^{(r)*}(\mathscr{N})) \to \mathscr{K}^{(r')}(\widehat{\sigma}^{(r')*}(\mathscr{N})). \tag{4.6.9.4}$$

We thus obtain a filtered direct system $(\mathscr{K}^{(r)}(\widehat{\sigma}^{(r)*}(\mathscr{N})))_{r\geq 0}$. Recall (4.4.37) that filtered direct limits are not necessarily representable in the category $\mathbf{Mod}_{\mathbb{Q}}(\breve{\overline{\mathscr{B}}})$.

In the remainder of this section, we establish for the $\breve{\overline{\mathscr{B}}}_{\mathbb{Q}}$-modules statements analogous to those established in 4.5 for the ind-$\breve{\overline{\mathscr{B}}}$-modules, following the point of view adopted in ([3] § III.12).

Lemma 4.6.10 *The $\breve{\overline{\mathscr{B}}}_{\mathbb{Q}}$-module $\breve{\overline{\mathscr{B}}}_{\mathbb{Q}}$ is Dolbeault and we have a canonical isomorphism of Higgs $\mathscr{O}_{\mathfrak{X}}[\frac{1}{p}]$-modules with coefficients in $\widetilde{\xi}^{-1}\widetilde{\Omega}^1_{\mathfrak{X}/\mathscr{S}}$*

$$(\mathscr{O}_{\mathfrak{X}}[\frac{1}{p}],0) \xrightarrow{\sim} \mathscr{H}_{\mathbb{Q}}(\breve{\overline{\mathscr{B}}}_{\mathbb{Q}}). \tag{4.6.10.1}$$

Indeed, the first assertion immediately follows from the definitions and the second from 4.4.32, or equivalently from 4.5.8 and (4.6.9.3).

4.6.11 Let r be a rational number > 0, $\mathscr{M}$ a $\breve{\overline{\mathscr{B}}}_{\mathbb{Q}}$-module, $\mathscr{N}$ a Higgs $\mathscr{O}_{\mathfrak{X}}[\frac{1}{p}]$-bundle with coefficients in $\widetilde{\xi}^{-1}\widetilde{\Omega}^1_{\mathfrak{X}/\mathscr{S}}$ such that the triple $(\mathscr{M},\mathscr{N},\alpha)$ is r-admissible (4.6.4). For any rational number r' such that $0 < r' \leq r$, we denote by

$$\alpha^{(r')}\colon \widehat{\sigma}^{(r')*}(\mathscr{N}) \xrightarrow{\sim} \mathfrak{S}^{(r')}(\mathscr{M}) \tag{4.6.11.1}$$

the isomorphism of $\mathbf{IC}_{\mathbb{Q}}(\breve{\mathscr{C}}^{(r')}/\breve{\overline{\mathscr{B}}})$ induced by $\varepsilon^{r,r'}(\alpha)$ and the isomorphisms (4.6.2.5) and (4.6.2.6), and by

$$\beta^{(r')}\colon \mathscr{N} \to \widehat{\sigma}^{(r')}_*(\mathfrak{S}^{(r')}(\mathscr{M})) \tag{4.6.11.2}$$

its adjoint (4.6.1.11).

Proposition 4.6.12 *Under the assumptions of* 4.6.11, *let moreover r', r'' be two rational numbers such that $0 < r'' < r' \leq r$. Then,*

(i) *The composed morphism*

$$\mathscr{N} \xrightarrow{\beta^{(r')}} \widehat{\sigma}^{(r')}_*(\mathfrak{S}^{(r')}(\mathscr{M})) \longrightarrow \mathscr{H}_{\mathbb{Q}}(\mathscr{M}), \tag{4.6.12.1}$$

where the second arrow is the canonical morphism (4.6.9.2), *is an isomorphism, independent of r'.*

(ii) *The composed morphism*

$$\widehat{\sigma}^{(r')}_*(\mathfrak{S}^{(r')}(\mathscr{M})) \longrightarrow \mathscr{H}_{\mathbb{Q}}(\mathscr{M}) \xrightarrow{\sim} \mathscr{N} \xrightarrow{\beta^{(r'')}} \widehat{\sigma}^{(r'')}_*(\mathfrak{S}^{(r'')}(\mathscr{M})), \tag{4.6.12.2}$$

where the first arrow is the canonical morphism (4.6.9.2) *and the second arrow is the inverse isomorphism of* (4.6.12.1), *is the canonical morphism* (4.6.9.1).

It suffices to copy the proof of ([3] III.12.17) taking into account 4.4.31 and 4.6.10.

Corollary 4.6.13 *For every Dolbeault $\overline{\breve{\mathscr{B}}}_{\mathbb{Q}}$-module $\mathscr{M}$, $\mathscr{H}_{\mathbb{Q}}(\mathscr{M})$ (4.6.9.2) is a rationally solvable Higgs $\mathscr{O}_{\mathfrak{X}}[\frac{1}{p}]$-bundle associated with $\mathscr{M}$. In particular, $\mathscr{H}_{\mathbb{Q}}$ induces a functor that we denote again by*

$$\mathscr{H}_{\mathbb{Q}}\colon \mathbf{Mod}^{\mathrm{Dolb}}_{\mathbb{Q}}(\overline{\breve{\mathscr{B}}}) \to \mathbf{HM}^{\mathrm{qsol}}(\mathscr{O}_{\mathfrak{X}}[\frac{1}{p}], \widetilde{\xi}^{-1}\widetilde{\Omega}^1_{\mathfrak{X}/\mathscr{S}}), \quad \mathscr{M} \mapsto \mathscr{H}_{\mathbb{Q}}(\mathscr{M}). \tag{4.6.13.1}$$

It immediately follows from the definitions that the functor (4.6.13.1) induces a functor that we denote again by

$$\mathscr{H}_{\mathbb{Q}}\colon \mathbf{Mod}^{\mathrm{sDolb}}_{\mathbb{Q}}(\overline{\breve{\mathscr{B}}}) \to \mathbf{HM}^{\mathrm{ssol}}(\mathscr{O}_{\mathfrak{X}}[\frac{1}{p}], \widetilde{\xi}^{-1}\widetilde{\Omega}^1_{\mathfrak{X}/\mathscr{S}}). \tag{4.6.13.2}$$

Corollary 4.6.14 *For every Dolbeault $\overline{\breve{\mathscr{B}}}_{\mathbb{Q}}$-module $\mathscr{M}$, there exist a rational number $r > 0$ and an isomorphism of $\mathbf{IC}_{\mathbb{Q}}(\breve{\mathscr{C}}^{(r)}/\overline{\breve{\mathscr{B}}})$*

$$\alpha\colon \widehat{\sigma}^{(r)*}(\mathscr{H}_{\mathbb{Q}}(\mathscr{M})) \xrightarrow{\sim} \mathfrak{S}^{(r)}(\mathscr{M}) \tag{4.6.14.1}$$

satisfying the following properties. For any rational number r' such that $0 < r' \leq r$, denote by

$$\alpha^{(r')}\colon \widehat{\sigma}^{(r')*}(\mathscr{H}_{\mathbb{Q}}(\mathscr{M})) \xrightarrow{\sim} \mathfrak{S}^{(r')}(\mathscr{M}) \tag{4.6.14.2}$$

the isomorphism of $\mathbf{IC}_{\mathbb{Q}}(\breve{\mathscr{C}}^{(r')}/\overline{\breve{\mathscr{B}}})$ induced by $\varepsilon^{r,r'}(\alpha)$ and the isomorphisms (4.6.2.5) and (4.6.2.6), and by

$$\beta^{(r')}\colon \mathscr{H}_{\mathbb{Q}}(\mathscr{M}) \to \widehat{\sigma}^{(r')}_*(\mathfrak{S}^{(r')}(\mathscr{M})) \tag{4.6.14.3}$$

its adjoint (4.6.1.11). Then,

(i) *For every rational number r' such that $0 < r' \leq r$, the morphism $\beta^{(r')}$ is a right inverse of the canonical morphism $\varpi^{(r')}\colon \widehat{\sigma}^{(r')}_*(\mathfrak{S}^{(r')}(\mathscr{M})) \to \mathscr{H}_{\mathbb{Q}}(\mathscr{M})$.*
(ii) *For all rational numbers r' and r'' such that $0 < r'' < r' \leq r$, the composition*

$$\widehat{\sigma}^{(r')}_*(\mathfrak{S}^{(r')}(\mathscr{M})) \xrightarrow{\varpi^{(r')}} \mathscr{H}_{\mathbb{Q}}(\mathscr{M}) \xrightarrow{\beta^{(r'')}} \widehat{\sigma}^{(r'')}_*(\mathfrak{S}^{(r'')}(\mathscr{M})) \tag{4.6.14.4}$$

is the canonical morphism.

Remark 4.6.15 Under the assumptions of 4.6.14, the isomorphism α is a priori not uniquely determined by $(\mathscr{M}, r)$, but for every rational number $0 < r' < r$, the morphism $\alpha^{r'}$ (4.6.14.2) depends only on $\mathscr{M}$, on which it depends functorially (see the proof of [3] III.12.26).

4.6.16 Let r be a rational number > 0, $\mathscr{M}$ a $\overline{\breve{\mathscr{B}}}_{\mathbb{Q}}$-module, $\mathscr{N}$ a Higgs $\mathscr{O}_{\mathfrak{X}}[\frac{1}{p}]$-bundle with coefficients in $\widetilde{\xi}^{-1}\widetilde{\Omega}^1_{\mathfrak{X}/\mathscr{S}}$ such that the triple $(\mathscr{M}, \mathscr{N}, \alpha)$ is r-admissible (4.6.4). To avoid any ambiguity with (4.6.11.1), denote by

$$\check{\alpha}\colon \mathfrak{S}^{(r)}(\mathscr{M}) \to \widehat{\sigma}^{(r)*}(\mathscr{N}) \tag{4.6.16.1}$$

the inverse of α in $\mathbf{IC}_{\mathbb{Q}}(\breve{\mathscr{C}}^{(r)}/\breve{\overline{\mathscr{B}}})$. For any rational number r' such that $0 < r' \leq r$, we denote by

$$\check{\alpha}^{(r')} \colon \mathfrak{S}^{(r')}(\mathscr{M}) \xrightarrow{\sim} \widehat{\sigma}^{(r')*}(\mathscr{N}) \tag{4.6.16.2}$$

the isomorphism of $\mathbf{IC}_{\mathbb{Q}}(\breve{\mathscr{C}}^{(r')}/\breve{\overline{\mathscr{B}}})$ induced by $\varepsilon^{r,r'}(\check{\alpha})$ and the isomorphisms (4.6.2.5) and (4.6.2.6), and by

$$\check{\beta}^{(r')} \colon \mathscr{M} \to \mathscr{K}^{(r')}(\widehat{\sigma}^{(r')*}(\mathscr{N})) \tag{4.6.16.3}$$

the adjoint morphism.

Proposition 4.6.17 *Under the assumptions* 4.6.16, *let moreover r', r'' be two rational numbers such that $0 < r'' < r' \leq r$. Then,*

(i) *The direct limit $\mathscr{V}_{\mathbb{Q}}(\mathscr{N})$ of the direct system $(\mathscr{K}^{(t)}(\widehat{\sigma}^{(t)*}(\mathscr{N})))_{t \in \mathbb{Q}_{>0}}$ (4.6.9.4) is representable in $\mathbf{Mod}_{\mathbb{Q}}(\breve{\overline{\mathscr{B}}})$.*

(ii) *The composed morphism*

$$\mathscr{M} \xrightarrow{\check{\beta}^{(r')}} \mathscr{K}^{(r')}(\widehat{\sigma}^{(r')*}(\mathscr{N})) \longrightarrow \mathscr{V}_{\mathbb{Q}}(\mathscr{N}), \tag{4.6.17.1}$$

where the second arrow is the canonical morphism, is an isomorphism, independent of r'.

(iii) *The composed morphism*

$$\mathscr{K}^{(r')}(\widehat{\sigma}^{(r')*}(\mathscr{N})) \longrightarrow \mathscr{V}_{\mathbb{Q}}(\mathscr{N}) \xrightarrow{\sim} \mathscr{M} \xrightarrow{\check{\beta}^{(r'')}} \mathscr{K}^{(r'')}(\widehat{\sigma}^{(r'')*}(\mathscr{N})) \tag{4.6.17.2}$$

where the first arrow is the canonical morphism and the second arrow is the inverse isomorphism of (4.6.17.1), *is the canonical morphism* (4.6.9.4).

It suffices to copy the proof of ([3] III.12.22) taking into account 4.4.35.

Corollary 4.6.18 *We have a functor*

$$\mathscr{V}_{\mathbb{Q}} \colon \mathbf{HM}^{\mathrm{qsol}}(\mathscr{O}_{\mathfrak{X}}[\frac{1}{p}], \widetilde{\xi}^{-1}\widetilde{\Omega}^1_{\mathfrak{X}/\mathscr{S}}) \to \mathbf{Mod}^{\mathrm{Dolb}}_{\mathbb{Q}}(\breve{\overline{\mathscr{B}}}), \quad \mathscr{N} \mapsto \varinjlim_{r \in \mathbb{Q}_{>0}} \mathscr{K}^{(r)}(\widehat{\sigma}^{(r)*}(\mathscr{N}))). \tag{4.6.18.1}$$

Moreover, for every object $\mathscr{N}$ of $\mathbf{HM}^{\mathrm{qsol}}(\mathscr{O}_{\mathfrak{X}}[\frac{1}{p}], \widetilde{\xi}^{-1}\widetilde{\Omega}^1_{\mathfrak{X}/\mathscr{S}})$, $\mathscr{V}_{\mathbb{Q}}(\mathscr{N})$ is associated with $\mathscr{N}$.

It immediately follows from the definitions that the functor (4.6.18.1) induces a functor that we denote again by

$$\mathscr{V}_{\mathbb{Q}} \colon \mathbf{HM}^{\mathrm{ssol}}(\mathscr{O}_{\mathfrak{X}}[\frac{1}{p}], \widetilde{\xi}^{-1}\widetilde{\Omega}^1_{\mathfrak{X}/\mathscr{S}}) \to \mathbf{Mod}^{\mathrm{sDolb}}_{\mathbb{Q}}(\breve{\overline{\mathscr{B}}}). \tag{4.6.18.2}$$

Corollary 4.6.19 *For every rationally solvable Higgs $\mathscr{O}_{\mathfrak{X}}[\frac{1}{p}]$-bundle $\mathscr{N}$ with coefficients in $\widetilde{\xi}^{-1}\widetilde{\Omega}^1_{\mathfrak{X}/\mathscr{S}}$, there exist a rational number $r > 0$ and an isomorphism of* $\mathbf{IC}_{\mathbb{Q}}(\breve{\mathscr{C}}^{(r)}/\breve{\overline{\mathscr{B}}})$

$$\check{\alpha} \colon \mathfrak{S}^{(r)}(\mathscr{V}_{\mathbb{Q}}(\mathscr{N})) \xrightarrow{\sim} \widehat{\sigma}^{(r)*}(\mathscr{N}) \tag{4.6.19.1}$$

satisfying the following properties. For any rational number r' such that $0 < r' \leq r$, we denote by

$$\check{\alpha}^{(r')}\colon \mathfrak{S}^{(r')}(\mathscr{V}_{\mathbb{Q}}(\mathscr{N})) \xrightarrow{\sim} \widehat{\sigma}^{(r')*}(\mathscr{N}) \tag{4.6.19.2}$$

the isomorphism of $\mathbf{IC}_{\mathbb{Q}}(\check{\mathscr{C}}^{(r')}/\overline{\check{\mathscr{B}}})$ induced by $\varepsilon^{r,r'}(\check{\alpha})$ and the isomorphisms (4.6.2.5) *and* (4.6.2.6), *and by*

$$\check{\beta}^{(r')}\colon \mathscr{V}_{\mathbb{Q}}(\mathscr{N}) \to \mathscr{K}^{(r')}(\widehat{\sigma}^{(r')*}(\mathscr{N})) \tag{4.6.19.3}$$

its adjoint. Then,

(i) *For every rational number r' such that $0 < r' \leq r$, the morphism $\check{\beta}^{(r')}$ is a right inverse of the canonical morphism $\varpi^{(r')}\colon \mathscr{K}^{(r')}(\widehat{\sigma}^{(r')*}(\mathscr{N})) \to \mathscr{V}_{\mathbb{Q}}(\mathscr{N})$.*
(ii) *For all rational numbers r' and r'' such that $0 < r'' < r' \leq r$, the composition*

$$\mathscr{K}^{(r')}(\widehat{\sigma}^{(r')*}(\mathscr{N})) \xrightarrow{\varpi^{(r')}} \mathscr{V}_{\mathbb{Q}}(\mathscr{N}) \xrightarrow{\check{\beta}^{(r'')}} \mathscr{K}^{(r'')}(\widehat{\sigma}^{(r'')*}(\mathscr{N})) \tag{4.6.19.4}$$

is the canonical morphism.

Remark 4.6.20 Under the assumptions of 4.6.19, the isomorphism $\check{\alpha}$ is a priori not uniquely determined by $(\mathscr{N}, r)$, but for every rational number $0 < r' < r$, the morphism $\check{\alpha}^{r'}$ (4.6.19.2) depends only on $\mathscr{N}$, on which it depends functorially (see the proof of [3] III.12.26).

Lemma 4.6.21 *The diagram*

$$\begin{array}{ccc} \mathbf{HM}^{\mathrm{qsol}}(\mathscr{O}_{\mathfrak{X}}[\frac{1}{p}], \widetilde{\xi}^{-1}\widetilde{\Omega}^1_{\mathfrak{X}/\mathscr{S}}) & \xrightarrow{\mathscr{V}_{\mathbb{Q}}} & \mathbf{Mod}^{\mathrm{Dolb}}_{\mathbb{Q}}(\overline{\check{\mathscr{B}}}) \\ \downarrow & & \downarrow \alpha_{\overline{\check{\mathscr{B}}}} \\ \mathbf{HM}^{\mathrm{sol}}(\mathscr{O}_{\mathfrak{X}}[\frac{1}{p}], \widetilde{\xi}^{-1}\widetilde{\Omega}^1_{\mathfrak{X}/\mathscr{S}}) & \xrightarrow{\mathscr{V}} & \mathbf{Ind}\text{-}\mathbf{Mod}^{\mathrm{Dolb}}(\overline{\check{\mathscr{B}}}), \end{array} \tag{4.6.21.1}$$

where $\mathscr{V}$ is the functor (4.5.17.1) *and the vertical arrows are the canonical functors* (4.6.5), *is commutative up to canonical isomorphism.*

Indeed, let $\mathscr{N}$ be an object of $\mathbf{HM}^{\mathrm{qsol}}(\mathscr{O}_{\mathfrak{X}}[\frac{1}{p}], \widetilde{\xi}^{-1}\widetilde{\Omega}^1_{\mathfrak{X}/\mathscr{S}})$. By virtue of 4.6.19, there exist a rational number $r > 0$ and an isomorphism of $\mathbf{IC}_{\mathbb{Q}}(\check{\mathscr{C}}^{(r)}/\overline{\check{\mathscr{B}}})$

$$\lambda\colon \mathfrak{S}^{(r)}(\mathscr{V}_{\mathbb{Q}}(\mathscr{N})) \xrightarrow{\sim} \widehat{\sigma}^{(r)*}(\mathscr{N}). \tag{4.6.21.2}$$

By 4.6.20, after decreasing r if necessary, this isomorphism is canonical and functorial in $\mathscr{N}$. In view of (4.6.3.4) and (4.6.3.5), $\alpha_{\check{\mathscr{C}}^{(r)}/\overline{\check{\mathscr{B}}}}(\lambda)$ induces an isomorphism of $\mathbf{Ind}\text{-}\mathbf{MC}(\check{\mathscr{C}}^{(r)}/\overline{\check{\mathscr{B}}})$

$$\alpha_{\check{\mathscr{C}}^{(r)}/\overline{\check{\mathscr{B}}}}(\lambda)\colon \mathrm{I}\mathfrak{S}^{(r)}(\alpha_{\overline{\check{\mathscr{B}}}}(\mathscr{V}_{\mathbb{Q}}(\mathscr{N}))) \xrightarrow{\sim} \mathrm{I}\widehat{\sigma}^{(r)*}(\mathscr{N}). \tag{4.6.21.3}$$

By virtue of 4.5.16(i), we deduce from this a functorial isomorphism

$$\alpha_{\breve{\overline{\mathscr{B}}}}(\mathscr{V}_{\mathbb{Q}}(\mathscr{N})) \xrightarrow{\sim} \mathscr{V}(\mathscr{N}). \tag{4.6.21.4}$$

Theorem 4.6.22 *The functors* (4.6.13.1) *and* (4.6.18.1)

$$\mathbf{Mod}^{\mathrm{Dolb}}_{\mathbb{Q}}(\breve{\overline{\mathscr{B}}}) \underset{\mathscr{V}_{\mathbb{Q}}}{\overset{\mathscr{H}_{\mathbb{Q}}}{\rightleftarrows}} \mathbf{HM}^{\mathrm{qsol}}(\mathscr{O}_{\mathfrak{X}}[\tfrac{1}{p}], \widetilde{\xi}^{-1}\widetilde{\Omega}^1_{\mathfrak{X}/\mathscr{S}}) \tag{4.6.22.1}$$

are equivalences of categories quasi-inverse to each other.

It suffices to copy the proof of ([3] III.12.26) taking into account 4.6.12, 4.6.13, 4.6.14, 4.6.17, 4.6.18 and 4.6.19.

Corollary 4.6.23 *The functors* (4.6.13.2) *and* (4.6.18.2)

$$\mathbf{Mod}^{\mathrm{sDolb}}_{\mathbb{Q}}(\breve{\overline{\mathscr{B}}}) \underset{\mathscr{V}_{\mathbb{Q}}}{\overset{\mathscr{H}_{\mathbb{Q}}}{\rightleftarrows}} \mathbf{HM}^{\mathrm{ssol}}(\mathscr{O}_{\mathfrak{X}}[\tfrac{1}{p}], \widetilde{\xi}^{-1}\widetilde{\Omega}^1_{\mathfrak{X}/\mathscr{S}}) \tag{4.6.23.1}$$

are equivalences of categories quasi-inverse to each other.

4.6.24 We take again the assumptions and notation of 4.4.25. Recall that we assign an exponent $'$ to the objects associated with the $(\mathscr{A}_2^*(\overline{S}/S), \mathscr{M}_{\mathscr{A}_2^*(\overline{S}/S)})$-deformation $(\widetilde{X}', \mathscr{M}_{\widetilde{X}'})$ (4.4.22). We have the functors (4.6.13.1) and (4.6.18.1)

$$\mathbf{Mod}^{\mathrm{Dolb}}_{\mathbb{Q}}(\breve{\overline{\mathscr{B}}}) \underset{\mathscr{V}_{\mathbb{Q}}}{\overset{\mathscr{H}_{\mathbb{Q}}}{\rightleftarrows}} \mathbf{HM}^{\mathrm{qsol}}(\mathscr{O}_{\mathfrak{X}}[\tfrac{1}{p}], \xi^{-1}\widetilde{\Omega}^1_{\mathfrak{X}/\mathscr{S}}) \tag{4.6.24.1}$$

$$\mathbf{Mod}^{\mathrm{Dolb}'}_{\mathbb{Q}}(\breve{\overline{\mathscr{B}}}) \underset{\mathscr{V}'_{\mathbb{Q}}}{\overset{\mathscr{H}'_{\mathbb{Q}}}{\rightleftarrows}} \mathbf{HM}^{\mathrm{qsol}'}(\mathscr{O}_{\mathfrak{X}}[\tfrac{1}{p}], (\xi^*_{\pi})^{-1}\widetilde{\Omega}^1_{\mathfrak{X}/\mathscr{S}}) \tag{4.6.24.2}$$

associated with the deformations $(\widetilde{X}, \mathscr{M}_{\widetilde{X}})$ and $(\widetilde{X}', \mathscr{M}_{\widetilde{X}'})$, respectively. The isomorphism ν (4.4.25.4) induces an $\mathscr{O}_{\mathfrak{X}}$-linear isomorphism

$$\lambda_*(\nu)\colon (\xi^*_{\pi})^{-1}\widetilde{\Omega}^1_{\mathfrak{X}/\mathscr{S}} \xrightarrow{\sim} \xi^{-1}\widetilde{\Omega}^1_{\mathfrak{X}/\mathscr{S}}. \tag{4.6.24.3}$$

We denote by

$$\tau\colon \mathbf{HM}(\mathscr{O}_{\mathfrak{X}}[\frac{1}{p}], (\xi^*_{\pi})^{-1}\widetilde{\Omega}^1_{\mathfrak{X}/\mathscr{S}}) \to \mathbf{HM}(\mathscr{O}_{\mathfrak{X}}[\frac{1}{p}], \xi^{-1}\widetilde{\Omega}^1_{\mathfrak{X}/\mathscr{S}}) \tag{4.6.24.4}$$

the functor induced by $p^{\rho}\lambda_*(\nu)$.

In view of (4.4.25.8), for every rational number $r \geq 0$, we have canonical isomorphisms

$$\tau \circ \widehat{\sigma}'^{(r)}_{*} \circ \mathfrak{S}'^{(r)} \xrightarrow{\sim} \widehat{\sigma}^{(r+\rho)}_{*} \circ \mathfrak{S}^{(r+\rho)}, \tag{4.6.24.5}$$

$$\mathscr{K}'^{(r)} \circ \widehat{\sigma}'^{(r)*} \xrightarrow{\sim} \mathscr{K}^{(r+\rho)} \circ \widehat{\sigma}^{(r+\rho)*} \circ \tau. \tag{4.6.24.6}$$

Proposition 4.6.25 *Under the assumptions of* 4.6.24, *let moreover $\mathscr{M}$ be a Dolbeault $\overline{\mathscr{B}}_{\mathbb{Q}}$-module with respect to the deformation $(\widetilde{X}', \mathscr{M}_{\widetilde{X}'})$, $\mathscr{N}$ a Higgs $\mathscr{O}_{\mathfrak{X}}[\frac{1}{p}]$-bundle with coefficients in $(\xi^{*}_{\pi})^{-1}\widetilde{\Omega}^{1}_{\mathfrak{X}/\mathscr{S}}$ rationally solvable with respect to the deformation $(\widetilde{X}', \mathscr{M}_{\widetilde{X}'})$. Then,*

(i) *The $\overline{\mathscr{B}}_{\mathbb{Q}}$-module $\mathscr{M}$ is Dolbeault with respect to the deformation $(\widetilde{X}, \mathscr{M}_{\widetilde{X}})$, and the canonical morphisms* (4.6.24.5) *induce a functorial isomorphism*

$$\tau(\mathscr{H}'_{\mathbb{Q}}(\mathscr{M})) \xrightarrow{\sim} \mathscr{H}_{\mathbb{Q}}(\mathscr{M}). \tag{4.6.25.1}$$

(ii) *The Higgs $\mathscr{O}_{\mathfrak{X}}[\frac{1}{p}]$-module $\tau(\mathscr{N})$ with coefficients in $\xi^{-1}\widetilde{\Omega}^{1}_{\mathfrak{X}/\mathscr{S}}$ is rationally solvable with respect to the deformation $(\widetilde{X}, \mathscr{M}_{\widetilde{X}})$ and the canonical morphisms* (4.6.24.6) *induce a functorial isomorphism*

$$\mathscr{V}'_{\mathbb{Q}}(\mathscr{N}) \xrightarrow{\sim} \mathscr{V}_{\mathbb{Q}}(\tau(\mathscr{N})). \tag{4.6.25.2}$$

(i) By 4.6.14, there exist a rational number $r > 0$ and an isomorphism of $\mathbf{IC}_{\mathbb{Q}}(\breve{\mathscr{C}}'^{(r)}/\breve{\overline{\mathscr{B}}})$

$$\alpha' \colon \widehat{\sigma}'^{(r)*}(\mathscr{H}'_{\mathbb{Q}}(\mathscr{M})) \xrightarrow{\sim} \mathfrak{S}'^{(r)}(\mathscr{M}). \tag{4.6.25.3}$$

In view of 4.6.15, after decreasing r if necessary, this isomorphism is canonical and functorial in $\mathscr{M}$. By (4.4.25.8), α' induces an isomorphism of $\mathbf{IC}(\breve{\mathscr{C}}^{(r+\rho)}/\breve{\overline{\mathscr{B}}})$

$$\alpha \colon \widehat{\sigma}^{(r+\rho)*}(\tau(\mathscr{H}'_{\mathbb{Q}}(\mathscr{M}))) \xrightarrow{\sim} \mathfrak{S}^{(r+\rho)}(\mathscr{M}). \tag{4.6.25.4}$$

We deduce, by virtue of 4.6.12(i), a functorial isomorphism

$$\tau(\mathscr{H}'_{\mathbb{Q}}(\mathscr{M})) \xrightarrow{\sim} \mathscr{H}_{\mathbb{Q}}(\mathscr{M}). \tag{4.6.25.5}$$

(ii) It suffices to copy the proof of (i) while replacing 4.6.12(i) by 4.6.17(ii).

Proposition 4.6.26 ([3] III.15.8) *Assume that the scheme X is affine and the $\mathscr{O}_X$-module $\widetilde{\Omega}^{1}_{X/S}$ is free. Then, every strongly solvable Higgs $\mathscr{O}_{\mathfrak{X}}[\frac{1}{p}]$-bundle $(\mathscr{N}, \theta)$ with coefficients in $\xi^{-1}\widetilde{\Omega}^{1}_{\mathfrak{X}/\mathscr{S}}$ is small* (4.2.11).

In fact, the proposition ([3] III.15.8) is formulated in the absolute case (3.1.14), but the proof applies mutatis mutandis to the relative case.

Remark 4.6.27 The proof of proposition 4.6.26 in ([3] III.15.8) crucially uses the condition that $(\mathscr{N}, \theta)$ is *strongly solvable*. Indeed, this proof relies on ([3] III.12.31) which requires the $\breve{\overline{\mathscr{B}}}_{\mathbb{Q}}$-module $\mathscr{M}$ to be adic of finite type, a necessary condition to have the isomorphism ([3] (III.12.29.9)).

4.7 Cohomology of Dolbeault ind-modules

Lemma 4.7.1 *For every ind-$\breve{\overline{\mathscr{B}}}$-module $\mathscr{M}$ that is rational* (4.3.15) *and flat* (2.7.9) *and every integer $q \geq 0$, the canonical morphisms, the first one of ind-$\mathscr{O}_{\mathfrak{X}}$-modules and the second one of $\mathscr{O}_{\mathfrak{X}}$-modules,*

$$\mathrm{R}^q\mathrm{I}\widehat{\sigma}_*(\mathscr{M}) \rightarrow \text{“}\varinjlim_{r\in\mathbb{Q}_{>0}}\text{”}\, \mathrm{R}^q\mathrm{I}\widehat{\sigma}_*(\mathscr{M}\otimes_{\breve{\overline{\mathscr{B}}}}\mathbb{K}^\bullet(\breve{\mathscr{C}}^{(r)})), \tag{4.7.1.1}$$

$$\mathrm{R}^q\vec{\sigma}_*(\mathscr{M}) \rightarrow \varinjlim_{r\in\mathbb{Q}_{>0}} \mathrm{R}^q\vec{\sigma}_*(\mathscr{M}\otimes_{\breve{\overline{\mathscr{B}}}}\mathbb{K}^\bullet(\breve{\mathscr{C}}^{(r)})), \tag{4.7.1.2}$$

where $\mathbb{K}^\bullet(\breve{\mathscr{C}}^{(r)})$ is the Dolbeault complex of the Higgs $\breve{\overline{\mathscr{B}}}$-module $(\breve{\mathscr{C}}^{(r)}, p^r d_{\breve{\mathscr{C}}^{(r)}})$ (4.4.33) *and $\vec{\sigma}_*$ is the functor* (4.3.12.9), *are isomorphisms.*

Indeed, by virtue of 4.4.36, the canonical morphism of complexes of ind-$\breve{\overline{\mathscr{B}}}$-modules

$$\breve{\overline{\mathscr{B}}}_{\mathbb{Q}}[0] \rightarrow \text{“}\varinjlim_{r\in\mathbb{Q}_{>0}}\text{”}\, \mathbb{K}^\bullet_{\mathbb{Q}}(\breve{\mathscr{C}}^{(r)}) \tag{4.7.1.3}$$

is a quasi-isomorphism. Moreover, $\mathscr{M}$ being rational, for every $\breve{\overline{\mathscr{B}}}$-module $\mathscr{F}$, the canonical morphism $\mathscr{M}\otimes_{\breve{\overline{\mathscr{B}}}}\mathscr{F} \rightarrow \mathscr{M}\otimes_{\breve{\overline{\mathscr{B}}}}\mathscr{F}_{\mathbb{Q}}$ (2.6.6.7) is an isomorphism. Since $\mathscr{M}$ is flat, we deduce from this that the canonical morphism of complexes of ind-$\breve{\overline{\mathscr{B}}}$-modules

$$\mathscr{M}[0] \rightarrow \text{“}\varinjlim_{r\in\mathbb{Q}_{>0}}\text{”}\, \mathscr{M}\otimes_{\breve{\overline{\mathscr{B}}}}\mathbb{K}^\bullet(\breve{\mathscr{C}}^{(r)}) \tag{4.7.1.4}$$

is a quasi-isomorphism. Since $\mathrm{R}^q\mathrm{I}\widehat{\sigma}_*$ commutes with small filtered direct limits (2.6.9.4), we deduce first that (4.7.1.1) is an isomorphism, and then that (4.7.1.2) is an isomorphism (2.6.5.2).

Lemma 4.7.2 *Let $\mathscr{N}$ be a Higgs $\mathscr{O}_{\mathfrak{X}}[\frac{1}{p}]$-bundle with coefficients in $\widetilde{\xi}^{-1}\widetilde{\Omega}^1_{\mathfrak{X}/\mathscr{S}}$. We denote by $\mathbb{K}^\bullet(\mathscr{N})$ the Dolbeault complex of $\mathscr{N}$* (2.5.1.2) *and for any rational number $r \geq 0$, by $\widehat{\sigma}^{(r)*}(\mathscr{N})$ the object of $\mathbf{IC}_{\mathbb{Q}}(\breve{\mathscr{C}}^{(r)}/\breve{\overline{\mathscr{B}}})$ associated with $\mathscr{N}$* (4.6.1.7) *and by $\mathbb{K}^\bullet(\widehat{\sigma}^{(r)*}(\mathscr{N}))$ its Dolbeault complex in $\mathbf{Mod}_{\mathbb{Q}}(\breve{\overline{\mathscr{B}}})$* (4.6.1). *Then, we have a functorial canonical morphism of $\mathbf{D}^+(\mathbf{Mod}(\mathscr{O}_{\mathfrak{X}}[\frac{1}{p}]))$*

$$\mathbb{K}^\bullet(\mathscr{N}) \rightarrow \mathrm{R}\widehat{\sigma}_{\mathbb{Q}*}(\mathbb{K}^\bullet(\widehat{\sigma}^{(r)*}(\mathscr{N}))), \tag{4.7.2.1}$$

where $\widehat{\sigma}_{\mathbb{Q}*}$ is the functor (4.3.13.11). *It is in fact a morphism of direct systems indexed by $r \in \mathbb{Q}_{>0}$, where the transition morphisms on the right-hand side are induced by the homomorphisms $\breve{\alpha}^{r,r'}$* (4.4.22.6) *for $r \geq r' > 0$. Moreover, for every integer $q \geq 0$, the induced morphism*

$$\mathrm{H}^q(\mathbb{K}^\bullet(\mathscr{N})) \to \varinjlim_{t\in\mathbb{Q}_{>r}} \mathrm{R}^q\widehat{\sigma}_{\mathbb{Q}*}(\mathbb{K}^\bullet(\widehat{\sigma}^{(r)*}(\mathscr{N}))) \tag{4.7.2.2}$$

is an isomorphism.

Let r be a rational number > 0. We have a canonical morphism of complexes of $\mathscr{O}_{\mathfrak{X}}[\frac{1}{p}]$-modules

$$\mathbb{K}^\bullet(\mathscr{N}) \to \widehat{\sigma}_{\mathbb{Q}*}(\mathbb{K}^\bullet(\widehat{\sigma}^{(r)*}(\mathscr{N}))). \tag{4.7.2.3}$$

Since the abelian category $\mathbf{Mod}_{\mathbb{Q}}(\breve{\overline{\mathscr{B}}})$ has enough injectives (2.9.6), there exist a complex of injective $\breve{\overline{\mathscr{B}}}_{\mathbb{Q}}$-modules bounded from below $\mathscr{L}^\bullet$ and a quasi-isomorphism $u\colon \mathbb{K}^\bullet(\underline{\widehat{\sigma}}'^{(r)*}(\mathscr{N})) \to \mathscr{L}^\bullet$ ([59] 013K). The latter induces a morphism of $\mathbf{D}^+(\mathbf{Mod}(\mathscr{O}_{\mathfrak{X}}[\frac{1}{p}]))$

$$\widehat{\sigma}_{\mathbb{Q}*}(\mathbb{K}^\bullet(\widehat{\sigma}^{(r)*}(\mathscr{N}))) \to \mathrm{R}\widehat{\sigma}_{\mathbb{Q}*}(\mathbb{K}^\bullet(\widehat{\sigma}^{(r)*}(\mathscr{N}))), \tag{4.7.2.4}$$

which depends only on $\mathbb{K}^\bullet(\widehat{\sigma}^{(r)*}(\mathscr{N}))$, but not on u ([59] 05TG). We take for morphism (4.7.2.1) the composition of (4.7.2.3) and (4.7.2.4).

In view of ([3] III.12.4(i) and III.12.2(i)), for every integer $j \geq 0$, $d_{\breve{\mathscr{C}}^{(r)}}$ (4.5.1.1) induces an $\mathscr{O}_{\mathfrak{X}}$-linear morphism

$$\delta^{j,(r)}\colon \mathrm{R}^j\widehat{\sigma}_*(\breve{\mathscr{C}}^{(r)}) \to \widetilde{\xi}^{-1}\widetilde{\Omega}^1_{\mathfrak{X}/\mathscr{S}} \otimes_{\mathscr{O}_{\mathfrak{X}}} \mathrm{R}^j\widehat{\sigma}_*(\breve{\mathscr{C}}^{(r)}), \tag{4.7.2.5}$$

which is a Higgs $\mathscr{O}_{\mathfrak{X}}$-field on $\mathrm{R}^j\widehat{\sigma}_*(\breve{\mathscr{C}}^{(r)})$ with coefficients in $\widetilde{\xi}^{-1}\widetilde{\Omega}^1_{\mathfrak{X}/\mathscr{S}}$. We denote by θ the Higgs $\mathscr{O}_{\mathfrak{X}}[\frac{1}{p}]$-field on $\mathscr{N}$ and by $\vartheta^{j,(r)}_{\mathrm{tot}} = \theta \otimes \mathrm{id} + p^r \mathrm{id} \otimes \delta^{j,(r)}$ the total $\mathscr{O}_{\mathfrak{X}}[\frac{1}{p}]$-Higgs field on $\mathscr{N} \otimes_{\mathscr{O}_{\mathfrak{X}}} \mathrm{R}^j\widehat{\sigma}_*(\widehat{\mathscr{C}}^{(r)})$ (2.5.1.7). By ([3] III.12.4(ii)), for every integer $i \geq 0$, we have a canonical $\mathscr{O}_{\mathfrak{X}}[\frac{1}{p}]$-linear isomorphism

$$\mathrm{R}^j\widehat{\sigma}_{\mathbb{Q}*}(\mathbb{K}^i(\widehat{\sigma}^{(r)*}(\mathscr{N}))) \xrightarrow{\sim} \mathbb{K}^i(\mathscr{N} \otimes_{\mathscr{O}_{\mathfrak{X}}} \mathrm{R}^j\widehat{\sigma}_*(\breve{\mathscr{C}}^{(r)}), \vartheta^{j,(r)}_{\mathrm{tot}}), \tag{4.7.2.6}$$

compatible with the morphisms induced by the differentials of the two Dolbeault complexes.

Moreover, the abelian category $\mathbf{Mod}_{\mathbb{Q}}(\breve{\overline{\mathscr{B}}})$ having enough injectives, we have a canonical functorial spectral sequence

$${}^r\mathrm{E}_1^{i,j} = \mathrm{R}^j\widehat{\sigma}_{\mathbb{Q}*}(\mathbb{K}^i(\widehat{\sigma}^{(r)*}(\mathscr{N}))) \Rightarrow \mathrm{R}^{i+j}\widehat{\sigma}_{\mathbb{Q}*}(\mathbb{K}^\bullet(\widehat{\sigma}^{(r)*}(\mathscr{N}))). \tag{4.7.2.7}$$

This induces, for every integer $i \geq 0$, a canonical morphism

$$\mathrm{H}^i({}^t\mathrm{E}_1^{\bullet,0}) \to \mathrm{R}^i\widehat{\sigma}_{\mathbb{Q}*}(\underline{\mathbb{K}}^\bullet(\underline{\widehat{\sigma}}^{(r)*}(\mathscr{N}))), \tag{4.7.2.8}$$

which is none other than the morphism induced by (4.7.2.4) by ([29] (0.11.3.4.2)).

By virtue of 4.4.32 and (4.7.2.6), for every $i \geq 0$, we have a canonical isomorphism

$$\varinjlim_{r\in\mathbb{Q}_{>0}} {}^r\mathrm{E}_1^{i,0} \xrightarrow{\sim} \mathbb{K}^i(\mathscr{N}, \theta), \tag{4.7.2.9}$$

and for every $j \geq 1$, we have

$$\varinjlim_{r \in \mathbb{Q}_{>0}} {}^{r}\mathrm{E}_1^{i,j} = 0. \tag{4.7.2.10}$$

Moreover, the isomorphisms (4.7.2.9) (for $i \in \mathbb{N}$) form an isomorphism of complexes. Since small filtered direct limits exist and are exact in $\mathbf{Mod}(\mathscr{O}_{\mathfrak{X}}[\frac{1}{p}])$ ([5] II 4.3), we deduce that the morphisms (4.7.2.2) are isomorphisms.

Lemma 4.7.3 *Let $\mathscr{N}$ be a Higgs $\mathscr{O}_{\mathfrak{X}}[\frac{1}{p}]$-bundle with coefficients in $\widetilde{\xi}^{-1}\widetilde{\Omega}^1_{\mathfrak{X}/\mathscr{S}}$. We denote by $\mathbb{K}^\bullet(\mathscr{N})$ the Dolbeault complex of $\mathscr{N}$ and for any rational number $r \geq 0$, by $\mathrm{I}\widehat{\sigma}^{(r)*}(\mathscr{N})$ the object of* **Ind-MC**$(\breve{\mathscr{C}}^{(r)}/\overline{\breve{\mathscr{B}}})$ *associated with $\mathscr{N}$* (4.5.1.6) *and by $\mathbb{K}^\bullet(\mathrm{I}\widehat{\sigma}^{(r)*}(\mathscr{N}))$ its Dolbeault complex* (2.8.1.3). *Then, we have a canonical functorial morphism of* $\mathbf{D}^+(\mathbf{Mod}(\mathscr{O}_{\mathfrak{X}}))$

$$\mathbb{K}^\bullet(\mathscr{N}) \to \mathrm{R}\vec{\sigma}_*(\mathbb{K}^\bullet(\mathrm{I}\widehat{\sigma}^{(r)*}(\mathscr{N}))), \tag{4.7.3.1}$$

where $\vec{\sigma}_*$ is the functor (4.3.12.9). *It is in fact a morphism of direct systems indexed by $r \in \mathbb{Q}_{>0}$, where the transition morphisms on the right-hand side are induced by the homomorphisms $\check{\alpha}^{r,r'}$* (4.4.22.6) *for $r \geq r' > 0$. Moreover, the induced morphism of* $\mathbf{D}^+(\mathbf{Mod}(\mathscr{O}_{\mathfrak{X}}))$

$$\mathbb{K}^\bullet(\mathscr{N}) \to \mathrm{R}\vec{\sigma}_*(\text{“}\varinjlim_{t \in \mathbb{Q}_{>r}}\text{”}\, \mathbb{K}^\bullet(\mathrm{I}\widehat{\sigma}^{(r)*}(\mathscr{N}))) \tag{4.7.3.2}$$

is an isomorphism.

In view of (4.6.3.5) and (4.3.13.15), we take for morphism (4.7.3.1) the canonical image of the morphism (4.7.2.1). By 2.7.2, for every integer $q \geq 0$, we have $\mathrm{R}^q\vec{\sigma}_* = \kappa_{\mathscr{O}_{\mathfrak{X}}} \circ \mathrm{R}^q\mathrm{I}\widehat{\sigma}_*$ (4.3.12.9) and this functor commutes with small filtered direct limits (2.6.9.4). It then follows from 4.7.2 that (4.7.3.2) is an isomorphism.

Theorem 4.7.4 *Let $\mathscr{M}$ be a Dolbeault ind-$\overline{\breve{\mathscr{B}}}$-module* (4.5.5), *$q$ an integer ≥ 0. We denote by $\mathbb{K}^\bullet(\mathscr{H}(\mathscr{M}))$ the Dolbeault complex of the Higgs $\mathscr{O}_{\mathfrak{X}}[\frac{1}{p}]$-bundle $\mathscr{H}(\mathscr{M})$* (4.5.12). *Then, we have a canonical functorial isomorphism of* $\mathbf{D}^+(\mathbf{Mod}(\mathscr{O}_{\mathfrak{X}}))$

$$\mathrm{R}\vec{\sigma}_*(\mathscr{M}) \xrightarrow{\sim} \mathbb{K}^\bullet(\mathscr{H}(\mathscr{M})), \tag{4.7.4.1}$$

where $\vec{\sigma}_*$ is the functor (4.3.12.9).

Indeed, by virtue of 4.5.13, there exist a rational number $r_{\mathscr{M}} > 0$ and an isomorphism of **Ind-MC**$(\breve{\mathscr{C}}^{(r_{\mathscr{M}})}/\overline{\breve{\mathscr{B}}})$,

$$\alpha_{\mathscr{M}}\colon \mathrm{I}\widehat{\sigma}^{(r_{\mathscr{M}})*}(\mathscr{H}(\mathscr{M})) \xrightarrow{\sim} \mathrm{I}\mathfrak{S}^{(r_{\mathscr{M}})}(\mathscr{M}), \tag{4.7.4.2}$$

satisfying the properties 4.5.13(i)-(ii). For any rational number r such that $0 < r < r_{\mathscr{M}}$, we denote by

$$\alpha^{(r)}_{\mathscr{M}}\colon \mathrm{I}\widehat{\sigma}^{(r)*}(\mathscr{H}(\mathscr{M})) \xrightarrow{\sim} \mathrm{I}\mathfrak{S}^{(r)}(\mathscr{M}) \tag{4.7.4.3}$$

the isomorphism of **Ind-MC**$(\breve{\mathscr{C}}^{(r)}/\overline{\breve{\mathscr{B}}})$ induced by $\mathrm{I}\varepsilon^{r_{\mathscr{M}},r}(\alpha_{\mathscr{M}})$ and the isomorphisms (4.5.3.2) and (4.5.3.3). By the proof of 4.5.20, $\alpha^{(r)}_{\mathscr{M}}$, depends only on $\mathscr{M}$ (but not on $\alpha_{\mathscr{M}}$) on which it depends functorially. We denote by $\mathbb{K}^\bullet(\mathrm{I}\widehat{\sigma}^{(r)*}(\mathscr{H}(\mathscr{M})))$ the Dolbeault complex of $\mathrm{I}\widehat{\sigma}^{(r)*}(\mathscr{H}(\mathscr{M}))$ (2.8.1.3). The isomorphism $\alpha^{(r)}_{\mathscr{M}}$ induces an isomorphism (4.4.33)

$$\mathbb{K}^\bullet(\mathrm{I}\widehat{\sigma}^{(r)*}(\mathscr{H}(\mathscr{M}))) \xrightarrow{\sim} \mathscr{M} \otimes_{\breve{\overline{\mathscr{B}}}} \mathbb{K}^\bullet(\breve{\mathscr{C}}^{(r)}), \tag{4.7.4.4}$$

where $\mathbb{K}^\bullet(\breve{\mathscr{C}}^{(r)})$ is the Dolbeault complex of the Higgs $\breve{\overline{\mathscr{B}}}$-module $(\breve{\mathscr{C}}^{(r)}, p^r d_{\breve{\mathscr{C}}^{(r)}})$ (4.4.33). These isomorphisms form an isomorphism of direct systems (for $0 < r < r_{\mathscr{M}}$). We deduce from this a canonical functorial isomorphism of complexes of ind-$\breve{\overline{\mathscr{B}}}$-modules

$$\text{“}\varinjlim_{r\in\mathbb{Q}_{>0}}\text{”}\, \mathbb{K}^\bullet(\mathrm{I}\widehat{\sigma}^{(r)*}(\mathscr{H}(\mathscr{M}))) \xrightarrow{\sim} \text{“}\varinjlim_{r\in\mathbb{Q}_{>0}}\text{”}\, \mathscr{M} \otimes_{\breve{\overline{\mathscr{B}}}} \mathbb{K}^\bullet(\breve{\mathscr{C}}^{(r)}). \tag{4.7.4.5}$$

Since the ind-$\breve{\overline{\mathscr{B}}}$-module $\mathscr{M}$ is rational and flat by 4.5.6, the theorem follows in view of (4.7.1.4) and 4.7.3.

Corollary 4.7.5 *Let $\mathscr{M}$ be a Dolbeault $\breve{\overline{\mathscr{B}}}_{\mathbb{Q}}$-module* (4.6.6), $\mathbb{K}^\bullet(\mathscr{H}_{\mathbb{Q}}(\mathscr{M}))$ *the Dolbeault complex of the Higgs $\mathscr{O}_{\mathfrak{X}}[\frac{1}{p}]$-bundle $\mathscr{H}_{\mathbb{Q}}(\mathscr{M})$* (4.6.9.2). *Then, we have a canonical functorial isomorphism of* $\mathbf{D}^+(\mathbf{Mod}(\mathscr{O}_{\mathfrak{X}}))$

$$\mathrm{R}\widehat{\sigma}_{\mathbb{Q}*}(\mathscr{M}) \xrightarrow{\sim} \mathbb{K}^\bullet(\mathscr{H}_{\mathbb{Q}}(\mathscr{M})), \tag{4.7.5.1}$$

where $\mathrm{R}\widehat{\sigma}_{\mathbb{Q}}$ denotes again the composition of the functor (4.3.13.14) *and the canonical functor from* $\mathbf{D}^+(\mathbf{Mod}(\mathscr{O}_{\mathfrak{X}}[\frac{1}{p}]))$ *to* $\mathbf{D}^+(\mathbf{Mod}(\mathscr{O}_{\mathfrak{X}}))$. *In particular, for every integer $q \geq 0$, we have a canonical isomorphism of $\mathscr{O}_{\mathfrak{X}}[\frac{1}{p}]$-modules*

$$\mathrm{R}^q\widehat{\sigma}_{\mathbb{Q}*}(\mathscr{M}) \xrightarrow{\sim} \mathrm{H}^q(\mathbb{K}^\bullet(\mathscr{H}_{\mathbb{Q}}(\mathscr{M}))). \tag{4.7.5.2}$$

This follows from 4.7.4, (4.6.9.3) and (4.3.13.15)

Remark 4.7.6 The isomorphisms (4.7.5.2) have been proved in ([3] III.12.34) in the absolute case (3.1.14). The proof is essentially the same as that of 4.7.4. Passing to ind-modules makes it possible to obtain the isomorphism (4.7.5.1).

4.8 Dolbeault Modules Over a Small Affine Scheme

4.8.1 We assume in this section that X is an object of $\mathbf{P}$ (4.4.1) and that X_s is non-empty. We fix an adequate chart $((P,\gamma),(\mathbb{N},\iota),\vartheta)$ for f (3.1.15). We set $R = \Gamma(X, \mathscr{O}_X)$, $R_1 = R \otimes_{\mathscr{O}_K} \mathscr{O}_{\overline{K}}$ and

$$\widetilde{\Omega}^1_{R/\mathscr{O}_K} = \Gamma(X, \widetilde{\Omega}^1_{X/S}). \tag{4.8.1.1}$$

If A is a ring and M an A-module, we denote again by A (resp. M) the constant sheaf with value A (resp. M) of $\overline{X}^\circ_{\text{fét}}$ or $(\overline{X}^\circ_{\text{fét}})^{\mathbb{N}^\circ}$, depending on the context.

We denote by $\breve{\overline{\mathscr{B}}}_X$ the ring $(\overline{\mathscr{B}}_{X,n+1})_{n\in\mathbb{N}}$ of $(\overline{X}^{\circ}_{\text{fét}})^{\mathbb{N}^{\circ}}$ (4.3.10.2), by $\mathbf{Mod}(\breve{\overline{\mathscr{B}}}_X)$ the category of $\breve{\overline{\mathscr{B}}}_X$-modules of $(\overline{X}^{\circ}_{\text{fét}})^{\mathbb{N}^{\circ}}$ and by $\mathbf{Mod}_{\mathbb{Q}}(\breve{\overline{\mathscr{B}}}_X)$ the category of objects of $\mathbf{Mod}(\breve{\overline{\mathscr{B}}}_X)$ up to isogeny (2.9.1.1). We denote by

$$\mathbf{Mod}(\breve{\overline{\mathscr{B}}}_X) \to \mathbf{Mod}_{\mathbb{Q}}(\breve{\overline{\mathscr{B}}}_X), \quad \mathscr{M} \mapsto \mathscr{M}_{\mathbb{Q}} \tag{4.8.1.2}$$

the canonical functor. The category $\mathbf{Mod}_{\mathbb{Q}}(\breve{\overline{\mathscr{B}}}_X)$ is abelian symmetric monoidal, having $\breve{\overline{\mathscr{B}}}_{X,\mathbb{Q}}$ as unit object. Objects of $\mathbf{Mod}_{\mathbb{Q}}(\breve{\overline{\mathscr{B}}}_X)$ will also be called $\breve{\overline{\mathscr{B}}}_{X,\mathbb{Q}}$*-modules*.

We denote by

$$\breve{\beta}\colon (\widetilde{E}_s^{\mathbb{N}^{\circ}}, \breve{\overline{\mathscr{B}}}) \to ((\overline{X}^{\circ}_{\text{fét}})^{\mathbb{N}^{\circ}}, \breve{\overline{\mathscr{B}}}_X) \tag{4.8.1.3}$$

the morphism of ringed topos induced by the $(\beta_{n+1})_{n\in\mathbb{N}}$ (4.3.10.4). For modules, we use the notation $\breve{\beta}^{-1}$ to denote the pullback in the sense of abelian sheaves, and we keep the notation $\breve{\beta}^*$ for the pullback in the sense of modules.

The functor $\breve{\beta}_*$ induces an additive and left-exact functor

$$\breve{\beta}_{\mathbb{Q}*}\colon \mathbf{Mod}_{\mathbb{Q}}(\breve{\overline{\mathscr{B}}}) \to \mathbf{Mod}_{\mathbb{Q}}(\breve{\overline{\mathscr{B}}}_X). \tag{4.8.1.4}$$

The functor $\breve{\beta}^*$ induces an additive functor

$$\breve{\beta}^*_{\mathbb{Q}}\colon \mathbf{Mod}_{\mathbb{Q}}(\breve{\overline{\mathscr{B}}}_X) \to \mathbf{Mod}_{\mathbb{Q}}(\breve{\overline{\mathscr{B}}}). \tag{4.8.1.5}$$

Definition 4.8.2 We say that a $\breve{\overline{\mathscr{B}}}_{X,\mathbb{Q}}$-module is *adic of finite type* if it is isomorphic to $\mathscr{M}_{\mathbb{Q}}$ (4.8.1.2), where $\mathscr{M}$ is an adic $\breve{\overline{\mathscr{B}}}_X$-module of finite type ([3] III.7.16).

We denote by $\mathbf{Mod}^{\text{aft}}(\breve{\overline{\mathscr{B}}}_X)$ the full subcategory of $\mathbf{Mod}(\breve{\overline{\mathscr{B}}}_X)$ made up of the adic $\breve{\overline{\mathscr{B}}}_X$-modules of finite type and by $\mathbf{Mod}^{\text{aft}}_{\mathbb{Q}}(\breve{\overline{\mathscr{B}}}_X)$ the category of objects of $\mathbf{Mod}^{\text{aft}}(\breve{\overline{\mathscr{B}}}_X)$ up to isogeny (2.9.1). The canonical functor

$$\mathbf{Mod}^{\text{aft}}_{\mathbb{Q}}(\breve{\overline{\mathscr{B}}}_X) \to \mathbf{Mod}_{\mathbb{Q}}(\breve{\overline{\mathscr{B}}}_X) \tag{4.8.2.1}$$

being fully faithful, a $\breve{\overline{\mathscr{B}}}_{X,\mathbb{Q}}$-module is adic of finite type if and only if it is in the essential image of this functor.

Proposition 4.8.3 ([3] III.13.2) *For every coherent $\mathscr{O}_{\mathfrak{X}}$-module $\mathscr{N}$, we have a canonical and functorial $\breve{\overline{\mathscr{B}}}$-linear isomorphism*

$$\breve{\beta}^*(\mathscr{N}(\mathfrak{X}) \otimes_{\widehat{R_1}} \breve{\overline{\mathscr{B}}}_X) \xrightarrow{\sim} \widehat{\sigma}^*(\mathscr{N}), \tag{4.8.3.1}$$

where $\widehat{\sigma}$ is the morphism of ringed topos (4.3.11.4).

4.8.4 Let r be a rational number ≥ 0. With the notation of 4.4.7, we denote by $\breve{\mathscr{F}}^{(r)}_X$ the $\breve{\overline{\mathscr{B}}}_X$-module $(\mathscr{F}^{(r)}_{X,n+1})_{n\in\mathbb{N}}$ and by $\breve{\mathscr{C}}^{(r)}_X$ the $\breve{\overline{\mathscr{B}}}_X$-algebra $(\mathscr{C}^{(r)}_{X,n+1})_{n\in\mathbb{N}}$. By ([3] III.7.3(i) and (III.7.12.1)), we have an exact sequence of $\breve{\overline{\mathscr{B}}}_X$-modules

$$0 \to \breve{\overline{\mathscr{B}}}_X \to \breve{\mathscr{F}}_X^{(r)} \to \xi^{-1}\widetilde{\Omega}^1_{R/\mathscr{O}_K} \otimes_R \breve{\overline{\mathscr{B}}}_X \to 0. \tag{4.8.4.1}$$

In view of ([3] III.7.3(i) and (III.7.12.3)), we have a canonical isomorphism of $\breve{\overline{\mathscr{B}}}_X$-algebras

$$\breve{\mathscr{C}}_X^{(r)} \xrightarrow{\sim} \varinjlim_{m\geq 0} \mathrm{S}^m_{\breve{\overline{\mathscr{B}}}_X}(\breve{\mathscr{F}}_X^{(r)}). \tag{4.8.4.2}$$

For all rational numbers $r \geq r' \geq 0$, the morphisms $(\mathrm{a}^{r,r'}_{X,n+1})_{n\in\mathbb{N}}$ (4.4.7.3) induce a $\breve{\overline{\mathscr{B}}}_X$-linear morphism

$$\breve{\mathrm{a}}^{r,r'}_X : \breve{\mathscr{F}}_X^{(r)} \to \breve{\mathscr{F}}_X^{(r')}. \tag{4.8.4.3}$$

The homomorphisms $(\alpha^{r,r'}_{X,n+1})_{n\in\mathbb{N}}$ (4.4.7.4) induce a homomorphism of $\breve{\overline{\mathscr{B}}}_X$-algebras

$$\breve{\alpha}^{r,r'}_X : \breve{\mathscr{C}}_X^{(r)} \to \breve{\mathscr{C}}_X^{(r')}. \tag{4.8.4.4}$$

For all rational numbers $r \geq r' \geq r'' \geq 0$, we have

$$\breve{\mathrm{a}}^{r,r''}_X = \breve{\mathrm{a}}^{r',r''}_X \circ \breve{\mathrm{a}}^{r,r'}_X \quad \text{and} \quad \breve{\alpha}^{r,r''}_X = \breve{\alpha}^{r',r''}_X \circ \breve{\alpha}^{r,r'}_X. \tag{4.8.4.5}$$

We have a canonical $\breve{\mathscr{C}}_X^{(r)}$-linear isomorphism

$$\Omega^1_{\breve{\mathscr{C}}_X^{(r)}/\breve{\overline{\mathscr{B}}}_X} \xrightarrow{\sim} \xi^{-1}\widetilde{\Omega}^1_{R/\mathscr{O}_K} \otimes_R \breve{\mathscr{C}}_X^{(r)}. \tag{4.8.4.6}$$

The universal $\breve{\overline{\mathscr{B}}}_X$-derivation of $\breve{\mathscr{C}}_X^{(r)}$ corresponds via this isomorphism to the unique $\breve{\overline{\mathscr{B}}}_X$-derivation

$$d_{\breve{\mathscr{C}}_X^{(r)}} : \breve{\mathscr{C}}_X^{(r)} \to \xi^{-1}\widetilde{\Omega}^1_{R/\mathscr{O}_K} \otimes_R \breve{\mathscr{C}}_X^{(r)} \tag{4.8.4.7}$$

which extends the canonical morphism $\breve{\mathscr{F}}_X^{(r)} \to \xi^{-1}\widetilde{\Omega}^1_{R/\mathscr{O}_K} \otimes_R \breve{\overline{\mathscr{B}}}_X$ (4.8.4.1). Since

$$\xi^{-1}\widetilde{\Omega}^1_{R/\mathscr{O}_K} \otimes_R \breve{\overline{\mathscr{B}}}_X = d_{\breve{\mathscr{C}}_X^{(r)}}(\breve{\mathscr{F}}_X^{(r)}) \subset d_{\breve{\mathscr{C}}_X^{(r)}}(\breve{\mathscr{C}}_X^{(r)}), \tag{4.8.4.8}$$

the derivation $d_{\breve{\mathscr{C}}_X^{(r)}}$ is a Higgs $\breve{\overline{\mathscr{B}}}_X$-field with coefficients in $\xi^{-1}\widetilde{\Omega}^1_{R/\mathscr{O}_K}$ by 2.5.26(i). For all rational numbers $r \geq r' \geq 0$, we have

$$p^{r-r'}(\mathrm{id} \otimes \breve{\alpha}^{r,r'}_X) \circ d_{\breve{\mathscr{C}}_X^{(r)}} = d_{\breve{\mathscr{C}}_X^{(r')}} \circ \breve{\alpha}^{r,r'}_X. \tag{4.8.4.9}$$

Proposition 4.8.5 *For every rational number $r \geq 0$, the canonical morphisms*

$$\breve{\beta}^*(\breve{\mathscr{F}}_X^{(r)}) \to \breve{\mathscr{F}}^{(r)}, \tag{4.8.5.1}$$

$$\breve{\beta}^*(\breve{\mathscr{C}}_X^{(r)}) \to \breve{\mathscr{C}}^{(r)}, \tag{4.8.5.2}$$

are isomorphisms. Moreover, for all rational numbers $r \geq r' \geq 0$, the morphisms $\breve{\beta}^*(\breve{\mathrm{a}}^{r,r'}_X)$ and $\breve{\beta}^*(\breve{\alpha}^{r,r'}_X)$ identify with the morphisms $\breve{\mathrm{a}}^{r,r'}$ (4.4.22.5) *and $\breve{\alpha}^{r,r'}$* (4.4.22.6), *respectively.*

The absolute case (3.1.14) where X is an object of $\mathbf{Q}$ (4.4.2) was proved in ([3] III.13.4). The condition that X is an object of $\mathbf{Q}$ is in fact unnecessary (see 4.4.12). The relative case is treated in the same way.

4.8.6 Let r be a rational number ≥ 0. We denote simply by $\mathbf{IC}(\breve{\mathscr{C}}_X^{(r)}/\breve{\overline{\mathscr{B}}}_X)$ the category of integrable p^r-isoconnections with respect to the extension $\breve{\mathscr{C}}_X^{(r)}/\breve{\overline{\mathscr{B}}}_X$ (see 2.9.11); we therefore omit the exponent p^r from the notation introduced in 2.9.11 considering that it is redundant with the exponent r of $\breve{\mathscr{C}}_X^{(r)}$. It is an additive category. We denote by $\mathbf{IC}_{\mathbb{Q}}(\breve{\mathscr{C}}_X^{(r)}/\breve{\overline{\mathscr{B}}}_X)$ the category of objects of $\mathbf{IC}(\breve{\mathscr{C}}_X^{(r)}/\breve{\overline{\mathscr{B}}}_X)$ up to isogeny.

Consider the functor

$$\begin{aligned} \mathfrak{S}_X^{(r)}\colon \mathbf{Mod}(\breve{\overline{\mathscr{B}}}_X) &\to \mathbf{IC}(\breve{\mathscr{C}}_X^{(r)}/\breve{\overline{\mathscr{B}}}_X) \\ \mathscr{M} &\mapsto (\breve{\mathscr{C}}_X^{(r)} \otimes_{\breve{\overline{\mathscr{B}}}_X} \mathscr{M}, \breve{\mathscr{C}}_X^{(r)} \otimes_{\breve{\overline{\mathscr{B}}}_X} \mathscr{M}, \mathrm{id}, p^r d_{\breve{\mathscr{C}}_X^{(r)}} \otimes \mathrm{id}) \end{aligned} \tag{4.8.6.1}$$

and denote again by

$$\mathfrak{S}_X^{(r)}\colon \mathbf{Mod}_{\mathbb{Q}}(\breve{\overline{\mathscr{B}}}_X) \to \mathbf{IC}_{\mathbb{Q}}(\breve{\mathscr{C}}_X^{(r)}/\breve{\overline{\mathscr{B}}}_X) \tag{4.8.6.2}$$

the induced functor.

We take again the notation of 4.2.2. By 2.9.14, if $(\mathscr{N}, \mathscr{N}', v, \theta)$ is a Higgs $\mathscr{O}_{\mathfrak{X}}$-isogeny with coefficients in $\widetilde{\xi}^{-1}\widetilde{\Omega}^1_{\mathfrak{X}/\mathscr{S}}$,

$$(\breve{\mathscr{C}}_X^{(r)} \otimes_{\overline{R_1}} \mathscr{N}(\mathfrak{X}), \breve{\mathscr{C}}_X^{(r)} \otimes_{\overline{R_1}} \mathscr{N}'(\mathfrak{X}), \mathrm{id} \otimes_{\overline{R_1}} v, p^r d_{\breve{\mathscr{C}}_X^{(r)}} \otimes v + \mathrm{id} \otimes \theta) \tag{4.8.6.3}$$

is an object of $\mathbf{IC}(\breve{\mathscr{C}}_X^{(r)}/\breve{\overline{\mathscr{B}}}_X)$. We thus obtain a functor

$$\widehat{\sigma}_X^{(r)*}\colon \mathbf{HI}(\mathscr{O}_{\mathfrak{X}}, \widetilde{\xi}^{-1}\widetilde{\Omega}^1_{\mathfrak{X}/\mathscr{S}}) \to \mathbf{IC}(\breve{\mathscr{C}}_X^{(r)}/\breve{\overline{\mathscr{B}}}_X). \tag{4.8.6.4}$$

By (4.2.2.4), the latter induces a functor that we denote again by

$$\widehat{\sigma}_X^{(r)*}\colon \mathbf{HM}^{\mathrm{coh}}(\mathscr{O}_{\mathfrak{X}}[\tfrac{1}{p}], \widetilde{\xi}^{-1}\widetilde{\Omega}^1_{\mathfrak{X}/\mathscr{S}}) \to \mathbf{IC}_{\mathbb{Q}}(\breve{\mathscr{C}}_X^{(r)}/\breve{\overline{\mathscr{B}}}_X). \tag{4.8.6.5}$$

For all rational numbers $r \geq r' \geq 0$, we have a canonical functor

$$\varepsilon_X^{r,r'}\colon \mathbf{IC}(\breve{\mathscr{C}}_X^{(r)}/\breve{\overline{\mathscr{B}}}_X) \to \mathbf{IC}(\breve{\mathscr{C}}_X^{(r')}/\breve{\overline{\mathscr{B}}}_X) \tag{4.8.6.6}$$

analogous to the functor (4.6.2.3). It induces a functor that we denote again by

$$\varepsilon_X^{r,r'}\colon \mathbf{IC}_{\mathbb{Q}}(\breve{\mathscr{C}}_X^{(r)}/\breve{\overline{\mathscr{B}}}_X) \to \mathbf{IC}_{\mathbb{Q}}(\breve{\mathscr{C}}_X^{(r')}/\breve{\overline{\mathscr{B}}}_X). \tag{4.8.6.7}$$

We have a canonical isomorphism of functors from $\mathbf{Mod}(\breve{\overline{\mathscr{B}}}_X)$ to $\mathbf{IC}(\breve{\mathscr{C}}_X^{(r')}/\breve{\overline{\mathscr{B}}}_X)$ (resp. from $\mathbf{Mod}_{\mathbb{Q}}(\breve{\overline{\mathscr{B}}}_X)$ to $\mathbf{IC}_{\mathbb{Q}}(\breve{\mathscr{C}}_X^{(r')}/\breve{\overline{\mathscr{B}}}_X)$)

$$\varepsilon_X^{r,r'} \circ \mathfrak{S}_X^{(r)} \xrightarrow{\sim} \mathfrak{S}_X^{(r')}, \tag{4.8.6.8}$$

and a canonical isomorphism of functors from $\mathbf{HI}(\mathscr{O}_{\mathfrak{X}}, \xi^{-1}\widetilde{\Omega}^1_{\mathfrak{X}/\mathscr{S}})$ to $\mathbf{IC}(\breve{\mathscr{C}}^{(r')}_X/\breve{\overline{\mathscr{B}}}_X)$ (resp. from $\mathbf{HM}^{\mathrm{coh}}(\mathscr{O}_{\mathfrak{X}}[\frac{1}{p}], \widetilde{\xi}^{-1}\widetilde{\Omega}^1_{\mathfrak{X}/\mathscr{S}})$ to $\mathbf{IC}_{\mathbb{Q}}(\breve{\mathscr{C}}^{(r')}_X/\breve{\overline{\mathscr{B}}}_X)$)

$$\varepsilon^{r,r'}_X \circ \widehat{\sigma}^{(r)*}_X \xrightarrow{\sim} \widehat{\sigma}^{(r')*}_X. \tag{4.8.6.9}$$

Proposition 4.8.7 ([3] III.13.7) *For every rational number $r \geq 0$, the diagrams of functors*

$$\begin{array}{ccc} \mathbf{Mod}(\breve{\overline{\mathscr{B}}}_X) & \xrightarrow{\mathfrak{S}^{(r)}_X} & \mathbf{IC}(\breve{\mathscr{C}}^{(r)}_X/\breve{\overline{\mathscr{B}}}_X) \\ \downarrow{\scriptstyle \breve{\beta}^*} & & \downarrow{\scriptstyle \breve{\beta}^*} \\ \mathbf{Mod}(\breve{\overline{\mathscr{B}}}) & \xrightarrow{\mathfrak{S}^{(r)}} & \mathbf{IC}(\breve{\mathscr{C}}^{(r)}/\breve{\overline{\mathscr{B}}}) \end{array} \tag{4.8.7.1}$$

$$\begin{array}{ccc} \mathbf{HI}^{\mathrm{coh}}(\mathscr{O}_{\mathfrak{X}}, \widetilde{\xi}^{-1}\widetilde{\Omega}^1_{\mathfrak{X}/\mathscr{S}}) & \xrightarrow{\widehat{\sigma}^{(r)*}_X} & \mathbf{IC}(\breve{\mathscr{C}}^{(r)}_X/\breve{\overline{\mathscr{B}}}_X) \\ & \searrow{\scriptstyle \widehat{\sigma}^{(r)*}} & \downarrow{\scriptstyle \breve{\beta}^*} \\ & & \mathbf{IC}(\breve{\mathscr{C}}^{(r)}/\breve{\overline{\mathscr{B}}}), \end{array} \tag{4.8.7.2}$$

where the pullback $\breve{\beta}^*$ for p^r-isoconnections is defined in 2.9.13, *are commutative up to canonical isomorphisms.*

Definition 4.8.8 Let $\mathscr{M}$ be a $\breve{\overline{\mathscr{B}}}_{X,\mathbb{Q}}$-module (4.8.1), $\mathscr{N}$ a Higgs $\mathscr{O}_{\mathfrak{X}}[\frac{1}{p}]$-bundle with coefficients in $\widetilde{\xi}^{-1}\widetilde{\Omega}^1_{\mathfrak{X}/\mathscr{S}}$ (4.2.3).

(i) Let r be a rational number > 0. We say that $\mathscr{M}$ and $\mathscr{N}$ are *r-associated* if there exists an isomorphism of $\mathbf{IC}_{\mathbb{Q}}(\breve{\mathscr{C}}^{(r)}_X/\breve{\overline{\mathscr{B}}}_X)$

$$\alpha\colon \widehat{\sigma}^{(r)*}_X(\mathscr{N}) \xrightarrow{\sim} \mathfrak{S}^{(r)}_X(\mathscr{M}). \tag{4.8.8.1}$$

We then also say that the triple $(\mathscr{M}, \mathscr{N}, \alpha)$ is *r-admissible*.

(ii) We say that $\mathscr{M}$ and $\mathscr{N}$ are *associated* if there exists a rational number $r > 0$ such that $\mathscr{M}$ and $\mathscr{N}$ are r-associated.

Note that for all rational numbers $r \geq r' > 0$, if $\mathscr{M}$ and $\mathscr{N}$ are r-associated, they are r'-associated, in view of (4.8.6.8) and (4.8.6.9).

Remark 4.8.9 Note that every $\mathscr{O}_{\mathfrak{X}}[\frac{1}{p}]$-module locally projective of finite type (2.1.11) is in fact projective of finite type, i.e., is a direct factor of a free $\mathscr{O}_{\mathfrak{X}}[\frac{1}{p}]$-module, by ([3] III.6.17).

Lemma 4.8.10 *Let $\mathscr{M}$ be a $\breve{\overline{\mathscr{B}}}_{X,\mathbb{Q}}$-module, $\mathscr{N}$ a Higgs $\mathscr{O}_{\mathfrak{X}}[\frac{1}{p}]$-bundle with coefficients in $\widetilde{\xi}^{-1}\widetilde{\Omega}^1_{\mathfrak{X}/\mathscr{S}}$, r a number rational > 0. If $\mathscr{M}$ and $\mathscr{N}$ are r-associated, then the $\breve{\overline{\mathscr{B}}}_{\mathbb{Q}}$-module $\breve{\beta}^*_{\mathbb{Q}}(\mathscr{M})$ and the Higgs $\mathscr{O}_{\mathfrak{X}}[\frac{1}{p}]$-bundle $\mathscr{N}$ are r-associated in the sense of* 4.6.4.

This immediately follows from 4.8.7.

4.8.11 The scheme $\overline{X}$ being locally irreducible by ([2] 4.2.7(iii)), it is the sum of the schemes induced on its irreducible components that we denote by $\overline{X}_1, \ldots, \overline{X}_c$. For any $1 \leq i \leq c$, we set $\mathscr{R}_i = \Gamma(\overline{X}_i, \mathscr{O}_{\overline{X}})$ and denote by $\widehat{\mathscr{R}}_i$ the p-adic Hausdorff completion of $\mathscr{R}_i$. We fix a geometric point $\overline{y}_i$ of $\overline{X}_i^\circ = \overline{X}_i \times_X X^\circ$ and set $\Delta_i = \pi_1(\overline{X}_i^\circ, \overline{y}_i)$. We denote by $\mathbf{B}_{\Delta_i}$ the classifying topos of Δ_i, by

$$\nu_i \colon \overline{X}_{i,\text{fét}}^\circ \xrightarrow{\sim} \mathbf{B}_{\Delta_i} \tag{4.8.11.1}$$

the fiber functor of $\overline{X}_{i,\text{fét}}^\circ$ at $\overline{y}_i$ ([3] (VI.9.8.4)) and by

$$\mu_i \colon \mathbf{B}_{\Delta_i} \to \overline{X}_{i,\text{fét}}^\circ \tag{4.8.11.2}$$

the quasi-inverse functor defined in ([3] (VI.9.8.3)).

Note that the ring $R_1 = R \otimes_{\mathscr{O}_K} \mathscr{O}_{\overline{K}}$ is identified with the product of the rings $\mathscr{R}_i$ for $1 \leq i \leq c$.

4.8.12 We keep the notation of 4.8.11 and take again those of 3.2. For any integer $n \geq 1$, we consider the logarithmic scheme $(X^{(n)}, \mathscr{M}_{X^{(n)}})$ defined by the formula (3.2.2.3). For every $1 \leq i \leq c$, we get a X-morphism

$$\overline{y}_i \to \varprojlim_{n \geq 1} X^{(n)}. \tag{4.8.12.1}$$

We can then apply the constructions of 3.4, in particular those of 3.4.5.

Let $\mathscr{N}$ be a coherent $\mathscr{O}_{\mathfrak{X}}$-module that is $\mathscr{S}$-flat, θ a *quasi-small* Higgs $\mathscr{O}_{\mathfrak{X}}$-field on $\mathscr{N}$ with coefficients in $\widetilde{\xi}^{-1}\widetilde{\Omega}^1_{\mathfrak{X}/\mathscr{S}}$ (4.2.9). For any $1 \leq i \leq c$, we set $N_i = \Gamma(\overline{X}_{i,s}, \mathscr{N})$ and we denote by

$$\theta_i \colon N_i \to \widetilde{\xi}^{-1}\widetilde{\Omega}^1_{R/\mathscr{O}_K} \otimes_R N_i \tag{4.8.12.2}$$

the Higgs $\widehat{\mathscr{R}}_i$-field induced by θ. The Higgs $\widehat{\mathscr{R}}_i$-module (N_i, θ_i) is quasi-small (3.4.3), and N_i is $\mathscr{O}_C$-flat. We denote by φ_i the quasi-small $\widehat{\mathscr{R}}_i$-representation of Δ_i on N_i associated with (N_i, θ_i) by the functor (3.4.5.8). For any integer $n \geq 1$, we denote by $\mathscr{P}_{n,i}$ the $\mathscr{R}_i$-module $\mu_i(N_i/p^n N_i, \varphi_i)$ of $\overline{X}_{i,\text{fét}}^\circ$ (4.8.11.2). There then exists an R_1-module $\mathscr{P}_n$ of $\overline{X}_{\text{fét}}^\circ$, unique up to unique isomorphism, such that for every $1 \leq i \leq c$, we have $\mathscr{P}_n|\overline{X}_i^\circ = \mathscr{P}_{n,i}$. By ([3] III.2.11), $\mathscr{P}_n$ is of finite type on R_1. The $(\mathscr{P}_{n+1})_{n \in \mathbb{N}}$ naturally form an inverse system. For all integers $n \geq m \geq 1$, the morphism $\mathscr{P}_n/p^m\mathscr{P}_n \to \mathscr{P}_m$ induced by the transition morphism $\mathscr{P}_n \to \mathscr{P}_m$ is an isomorphism. The R_1-module $\breve{\mathscr{P}} = (\mathscr{P}_{n+1})_{n \in \mathbb{N}}$ of $(\overline{X}_{\text{fét}}^\circ)^{\mathbb{N}^\circ}$ is adic of finite type by ([3] III.7.14). We thus obtain a functor (4.2.9)

$$\begin{aligned} \mathbf{HM}^{\text{qsf}}(\mathscr{O}_{\mathfrak{X}}, \widetilde{\xi}^{-1}\widetilde{\Omega}^1_{\mathfrak{X}/\mathscr{S}}) &\to \mathbf{Mod}^{\text{aft}}((\overline{X}_{\text{fét}}^\circ)^{\mathbb{N}^\circ}, R_1) \\ (\mathscr{N}, \theta) &\mapsto \breve{\mathscr{P}} = (\mathscr{P}_{n+1})_{n \in \mathbb{N}}. \end{aligned} \tag{4.8.12.3}$$

Proposition 4.8.13 *Let r, ε be two rational numbers such that $r \geq 0$ and $\varepsilon > r + \frac{1}{p-1}$, $\mathscr{N}$ a coherent $\mathscr{O}_{\mathfrak{X}}$-module that is $\mathscr{S}$-flat, θ an ε-quasi-small Higgs $\mathscr{O}_{\mathfrak{X}}$-field over $\mathscr{N}$ with coefficients in $\widetilde{\xi}^{-1}\widetilde{\Omega}^1_{\mathfrak{X}/\mathscr{S}}$* (4.2.9). *Let $\breve{\mathscr{P}}$ be the R_1-module of $(\overline{X}^{\circ}_{\text{fét}})^{\mathbb{N}^{\circ}}$ associated with $(\mathscr{N}, \theta)$ by the functor* (4.8.12.3) *and denote again by $\mathscr{N}$ the Higgs $\mathscr{O}_{\mathfrak{X}}$-isogeny $(\mathscr{N}, \mathscr{N}, \mathrm{id}, \theta)$ with coefficients in $\widetilde{\xi}^{-1}\widetilde{\Omega}^1_{\mathfrak{X}/\mathscr{S}}$* (4.2.2). *Then, there exists a functorial isomorphism of* $\mathbf{IC}(\breve{\mathscr{C}}^{(r)}_X/\breve{\overline{\mathscr{B}}}_X)$ (4.8.6)

$$\widehat{\sigma}^{(r)*}_X(\mathscr{N}) \xrightarrow{\sim} \mathfrak{S}^{(r)}_X(\breve{\mathscr{P}} \otimes_{R_1} \breve{\overline{\mathscr{B}}}_X). \qquad (4.8.13.1)$$

It suffices to copy the proof of ([3] III.13.8), which corresponds to the absolute case (3.1.14), using in the general case 3.4.18 instead of ([3] II. 13.17).

Corollary 4.8.14 *We keep the assumptions of* 4.8.13, *and suppose, moreover, that the $\mathscr{O}_{\mathfrak{X}}[\frac{1}{p}]$-module $\mathscr{N}[\frac{1}{p}]$ is projective of finite type* (4.8.9). *Then,*

(i) *The $\breve{\overline{\mathscr{B}}}_{X,\mathbb{Q}}$-module $(\breve{\mathscr{P}} \otimes_{R_1} \breve{\overline{\mathscr{B}}}_X)_{\mathbb{Q}}$ and the Higgs $\mathscr{O}_{\mathfrak{X}}[\frac{1}{p}]$-bundle $\mathscr{N}[\frac{1}{p}]$ are associated in the sense of* 4.8.8.
(ii) *The Higgs $\mathscr{O}_{\mathfrak{X}}[\frac{1}{p}]$-bundle $\mathscr{N}[\frac{1}{p}]$ is strongly solvable, the $\breve{\overline{\mathscr{B}}}_{\mathbb{Q}}$-module $(\breve{\beta}^*(\breve{\mathscr{P}} \otimes_{R_1} \breve{\overline{\mathscr{B}}}_X))_{\mathbb{Q}}$ is strongly Dolbeault, and we have a functorial isomorphism of $\breve{\overline{\mathscr{B}}}_{\mathbb{Q}}$-modules*

$$\mathscr{V}_{\mathbb{Q}}(\mathscr{N}[\frac{1}{p}]) \xrightarrow{\sim} \breve{\beta}^*(\breve{\mathscr{P}} \otimes_{R_1} \breve{\overline{\mathscr{B}}}_X)_{\mathbb{Q}}, \qquad (4.8.14.1)$$

where $\mathscr{V}_{\mathbb{Q}}$ is the functor (4.6.18.1).

(i) This follows from 4.8.13.

(ii) This follows from (i), 4.8.10 and 4.6.17(ii). Note that the $\breve{\overline{\mathscr{B}}}$-module $\breve{\beta}^*(\breve{\mathscr{P}} \otimes_{R_1} \breve{\overline{\mathscr{B}}}_X)$ is adic of finite type by (4.8.12.3) and ([3] III.7.5).

Corollary 4.8.15 *A Higgs $\mathscr{O}_{\mathfrak{X}}[\frac{1}{p}]$-bundle with coefficients in $\widetilde{\xi}^{-1}\widetilde{\Omega}^1_{\mathfrak{X}/\mathscr{S}}$ is small* (4.2.11) *if and only if it is strongly solvable* (4.6.6).

This follows from 4.6.26 and 4.8.14.

Proposition 4.8.16 *We keep the assumptions of* 4.8.12 *and suppose, moreover, that the $\mathscr{O}_{\mathfrak{X}}[\frac{1}{p}]$-module $\mathscr{N}[\frac{1}{p}]$ is projective of finite type* (4.8.9). *Then,*

(i) *The adjunction morphism*

$$\breve{\mathscr{P}} \otimes_{R_1} \breve{\overline{\mathscr{B}}}_X \to \breve{\beta}_*(\breve{\beta}^*(\breve{\mathscr{P}} \otimes_{R_1} \breve{\overline{\mathscr{B}}}_X)), \qquad (4.8.16.1)$$

where $\breve{\beta}$ denotes the morphism of ringed topos (4.8.1.3), *is an isogeny.*
(ii) *For every integer $q \geq 1$, $\mathrm{R}^q\breve{\beta}_*(\breve{\beta}^*(\breve{\mathscr{P}} \otimes_{R_1} \breve{\overline{\mathscr{B}}}_X))$ is of finite exponent* (2.9.1).

Indeed, by virtue of ([3] III.6.17 and [1] 1.10.2(iii)), there exists a coherent $\mathscr{O}_{\mathfrak{X}}$-module $\mathscr{N}'$, two integers $t, m \geq 0$ and an $\mathscr{O}_{\mathfrak{X}}$-linear p^t-isomorphism $u\colon \mathscr{N} \oplus \mathscr{N}' \to \mathscr{O}^m_{\mathfrak{X}}$ (2.1.14).

(i) In view of ([2] 2.6.3 and [3] III.7.5), it suffices to show that for every integer $n \geq 0$, the canonical morphism

$$a\colon \mathscr{P}_n \otimes_{R_1} \overline{\mathscr{B}}_{X,n} \to \beta_*(\beta^{-1}(\mathscr{P}_n) \otimes_{R_1} \overline{\mathscr{B}}_n), \tag{4.8.16.2}$$

where β is the morphism of topos (4.3.2.13), is a p^{3t}-isomorphism, or that for all $1 \leq i \leq c$, the fiber $\nu_i(a|\overline{X}_i^\circ)$ (4.8.11.1) is a p^{3t}-isomorphism

For every object V of $\mathbf{\acute{E}t}_{\mathrm{f}/\overline{X}_i^\circ}$ such that the sheaf $\mathscr{P}_n|V$ is constant, the canonical morphism

$$a_V\colon (\mathscr{P}_n \otimes_{R_1} \overline{\mathscr{B}}_{X,n})(V) \to (\beta^{-1}(\mathscr{P}_n) \otimes_{R_1} \overline{\mathscr{B}}_n)(V \to X) \tag{4.8.16.3}$$

is a p^{3t}-isomorphism. Indeed, using u, we are reduced to the case where $\mathscr{P}_n$ is the constant sheaf with value $\mathscr{R}_i/p^n\mathscr{R}_i$ (4.8.12), i.e., corresponding to the trivial representation of Δ_i on this R_1-module, in which case a_V is an α-isomorphism by virtue of ([2] 4.6.29).

The morphism of R_1-modules underlying $\nu_i(a|\overline{X}_i^\circ)$ is the direct limit of the morphisms a_{V_j}, where $(V_j)_{j\in J_i}$ is the normalized universal cover of $\overline{X}_i^\circ$ in $\overline{y}_i$ ([3] VI.9.8). The representation $\nu_i(\mathscr{P}_n)$ being discrete, the subset J_i' of J_i made up of the elements j such that $\mathscr{P}_n|V_j$ is constant is cofinal in J_i, and the desired assertion follows.

(ii) By virtue of ([2] 4.6.30(ii)), for all integers $1 \leq i \leq c$ and $n \geq 0$, the direct limit

$$\varinjlim_{j\in J_i} \mathrm{H}^q((V_j \to X), \overline{\mathscr{B}}_n), \tag{4.8.16.4}$$

where $(V_j)_{j\in J_i}$ is defined above, is α-zero. Using u, we deduce from this that

$$p^{3t}\varinjlim_{j\in J_i'} \mathrm{H}^q((V_j \to X), \beta^{-1}(\mathscr{P}_n) \otimes_{R_1} \overline{\mathscr{B}}_n)) = 0, \tag{4.8.16.5}$$

where J_i' is the ordered set defined above. Consequently, $\nu_i(\mathrm{R}^q\beta_*(\beta^{-1}(\mathscr{P}_n) \otimes_{R_1} \overline{\mathscr{B}}_n)|\overline{X}_i^\circ)$ vanishes, in view of ([5] V 5.1 and IV (6.3.3)). The proposition follows ([3] III.7.5).

Corollary 4.8.17 *Under the assumptions of* 4.8.14, *we have a functorial isomorphism of* $\breve{\overline{\mathscr{B}}}_{X,\mathbb{Q}}$*-modules*

$$\breve{\beta}_{\mathbb{Q}*}(\mathscr{V}_{\mathbb{Q}}(\mathscr{N}[\tfrac{1}{p}])) \xrightarrow{\sim} (\breve{\mathscr{P}} \otimes_{R_1} \breve{\overline{\mathscr{B}}}_X)_{\mathbb{Q}}, \tag{4.8.17.1}$$

where $\breve{\beta}_{\mathbb{Q}}$ *denotes the morphism of ringed topos* (4.8.1.4).

This follows from 4.8.14(ii) and 4.8.16(i).

Proposition 4.8.18 *Let* $\mathscr{M}$ *be a strongly Dolbeault* $\breve{\overline{\mathscr{B}}}_{\mathbb{Q}}$*-module,* $(\mathscr{N}, \theta)$ *a quasi-small Higgs sub-*$\mathscr{O}_{\mathfrak{X}}$*-module with coefficients in* $\widetilde{\xi}^{-1}\widetilde{\Omega}^1_{\mathfrak{X}/\mathscr{S}}$ *of* $\mathscr{H}_{\mathbb{Q}}(\mathscr{M})$ (4.2.9) *such that the* $\mathscr{O}_{\mathfrak{X}}$*-module* $\mathscr{N}$ *is coherent and* $\mathscr{S}$*-flat and it generates* $\mathscr{H}_{\mathbb{Q}}(\mathscr{M})$ *over* $\mathscr{O}_{\mathfrak{X}}[\frac{1}{p}]$*,* $\breve{\mathscr{P}}$ *the adic* R_1*-module of finite type of* $(\overline{X}^\circ_{\mathrm{f\acute{e}t}})^{\mathbb{N}^\circ}$ *associated with* $(\mathscr{N}, \theta)$ *by the functor* (4.8.12.3).

Then, there exist adjoint isomorphisms

$$\mathscr{M} \xrightarrow{\sim} (\breve{\beta}^*(\breve{\mathscr{P}} \otimes_{R_1} \overline{\breve{\mathscr{B}}}_X))_{\mathbb{Q}}, \tag{4.8.18.1}$$

$$\breve{\beta}_{\mathbb{Q}*}(\mathscr{M}) \xrightarrow{\sim} (\breve{\mathscr{P}} \otimes_{R_1} \overline{\breve{\mathscr{B}}}_X)_{\mathbb{Q}}. \tag{4.8.18.2}$$

Indeed, by 4.6.22 and 4.8.14(ii), we have an isomorphism of $\overline{\breve{\mathscr{B}}}_{\mathbb{Q}}$-modules

$$\mathscr{M} \xrightarrow{\sim} (\breve{\beta}^*(\breve{\mathscr{P}} \otimes_{R_1} \overline{\breve{\mathscr{B}}}_X))_{\mathbb{Q}}. \tag{4.8.18.3}$$

By virtue of 4.8.16(i), we deduce by adjunction an isomorphism

$$\breve{\beta}_{\mathbb{Q}*}(\mathscr{M}) \xrightarrow{\sim} (\breve{\mathscr{P}} \otimes_{R_1} \overline{\breve{\mathscr{B}}}_X)_{\mathbb{Q}}. \tag{4.8.18.4}$$

Remark 4.8.19 By 4.6.13 and 4.8.15, for every strongly Dolbeault $\overline{\breve{\mathscr{B}}}_{\mathbb{Q}}$-module $\mathscr{M}$, there exists a quasi-small Higgs sub-$\mathscr{O}_{\mathfrak{X}}$-module $(\mathscr{N}, \theta)$ of $\mathscr{H}_{\mathbb{Q}}(\mathscr{M})$ satisfying the required properties in 4.8.18.

Corollary 4.8.20 *Let $\mathscr{M}$ be a strongly Dolbeault $\overline{\breve{\mathscr{B}}}_{\mathbb{Q}}$-module. Then,*

(i) *The $\overline{\breve{\mathscr{B}}}_{X,\mathbb{Q}}$-module $\breve{\beta}_{\mathbb{Q}*}(\mathscr{M})$ is adic of type* (4.8.2) *and it is associated with the Higgs $\mathscr{O}_{\mathfrak{X}}[\frac{1}{p}]$-bundle $\mathscr{H}_{\mathbb{Q}}(\mathscr{M})$ in the sense of* 4.8.8.
(ii) *The adjunction morphism $\breve{\beta}^*_{\mathbb{Q}}(\breve{\beta}_{\mathbb{Q}*}(\mathscr{M})) \to \mathscr{M}$ is an isomorphism.*
(iii) *The adjunction morphism $\breve{\beta}_{\mathbb{Q}*}(\mathscr{M}) \to \breve{\beta}_{\mathbb{Q}*}(\breve{\beta}^*_{\mathbb{Q}}(\breve{\beta}_{\mathbb{Q}*}(\mathscr{M})))$ is an isomorphism.*

Assertions (i) and (ii) follow from 4.8.18, 4.8.19 and 4.8.14(i). Assertion (iii) follows from (ii) in view of the commutative diagram

$$\begin{array}{ccc} & & \breve{\beta}_* \\ & \overset{\text{id}}{\nearrow} & \uparrow{\scriptstyle \breve{\beta}_*(\text{adj})} \\ \breve{\beta}_* & \xrightarrow[\text{adj}]{} & \breve{\beta}_*\breve{\beta}^*\breve{\beta}_* \end{array} \tag{4.8.20.1}$$

where the arrows denoted by "adj" are the adjunction morphisms.

Definition 4.8.21 We say that a $\overline{\breve{\mathscr{B}}}_{X,\mathbb{Q}}$-module is *strongly Dolbeault* if it is isomorphic to $\breve{\beta}_{\mathbb{Q}*}(\mathscr{M})$ (4.8.1.4) for a strongly Dolbeault $\overline{\breve{\mathscr{B}}}_{\mathbb{Q}}$-module $\mathscr{M}$.

It follows from 4.8.18 that every strongly Dolbeault $\overline{\breve{\mathscr{B}}}_{X,\mathbb{Q}}$-module is adic of finite type and that it is associated with a strongly solvable Higgs $\mathscr{O}_{\mathfrak{X}}[\frac{1}{p}]$-bundle, in the sense of 4.8.8.

We denote by $\mathbf{Mod}^{\text{sDolb}}_{\mathbb{Q}}(\overline{\breve{\mathscr{B}}}_X)$ the full subcategory of $\mathbf{Mod}_{\mathbb{Q}}(\overline{\breve{\mathscr{B}}}_X)$ made up of strongly Dolbeault $\overline{\breve{\mathscr{B}}}_{X,\mathbb{Q}}$-modules.

Proposition 4.8.22 *The functors* (4.8.1.4) *and* (4.8.1.5)

$$\mathbf{Mod}^{\mathrm{sDolb}}_{\mathbb{Q}}(\breve{\overline{\mathscr{B}}}_X) \underset{\breve{\beta}_{\mathbb{Q}*}}{\overset{\breve{\beta}^*_{\mathbb{Q}}}{\rightleftarrows}} \mathbf{Mod}^{\mathrm{sDolb}}_{\mathbb{Q}}(\breve{\overline{\mathscr{B}}}) \tag{4.8.22.1}$$

induce equivalences of categories quasi-inverse to each other between the category of strongly Dolbeault $\breve{\overline{\mathscr{B}}}_{X,\mathbb{Q}}$-modules and that of strongly Dolbeault $\breve{\overline{\mathscr{B}}}_{\mathbb{Q}}$-modules.

This follows from 4.8.18.

4.8.23 We take again the notation of 4.8.11. For any integer $1 \leq i \leq c$, we set

$$\overline{\mathscr{R}}_i = \nu_i(\overline{\mathscr{B}}_X|\overline{X}^{\circ}_i), \tag{4.8.23.1}$$

and denote by $\widehat{\overline{\mathscr{R}}}_i$ its p-adic Hausdorff completion. We endow the rings $\widehat{\overline{\mathscr{R}}}_i$ and $\widehat{\overline{\mathscr{R}}}_i[\frac{1}{p}]$ with the p-adic topologies (2.1.1). We say that a continuous $\widehat{\overline{\mathscr{R}}}_i$-representation of Δ_i (2.1.2) is *p-adic of finite type* if the underlying $\widehat{\overline{\mathscr{R}}}_i$-module is endowed with the p-adic topology and if it is separated of finite type. Observe that every $\widehat{\overline{\mathscr{R}}}_i$-module of finite type is complete for the p-adic topology ([10] chap. III § 2.11 cor. 1 of prop. 16). We denote by $\mathbf{Rep}^{\mathrm{disc}}_{\overline{\mathscr{R}}_i}(\Delta_i)$ the category of discrete $\overline{\mathscr{R}}_i$-representations of Δ_i, by $\mathbf{Rep}^{p\text{-atf}}_{\widehat{\overline{\mathscr{R}}}_i}(\Delta_i)$ the full subcategory of $\mathbf{Rep}_{\widehat{\overline{\mathscr{R}}}_i}(\Delta_i)$ made up of the p-adic $\widehat{\overline{\mathscr{R}}}_i$-representations of finite type of Δ_i and by $\mathbf{Rep}^{\mathrm{Dolb}}_{\widehat{\overline{\mathscr{R}}}_i[\frac{1}{p}]}(\Delta_i)$ the full subcategory of $\mathbf{Rep}_{\widehat{\overline{\mathscr{R}}}_i[\frac{1}{p}]}(\Delta_i)$ made up of the Dolbeault $\widehat{\overline{\mathscr{R}}}_i[\frac{1}{p}]$-representations of Δ_i (3.3.9).

The functor ν_i (4.8.11.1) induces a functor

$$\nu_i\colon \mathbf{Mod}(\overline{\mathscr{B}}_X) \to \mathbf{Rep}^{\mathrm{disc}}_{\overline{\mathscr{R}}_i}(\Delta_i). \tag{4.8.23.2}$$

We deduce by taking the inverse limit a functor

$$\widehat{\nu}_i\colon \mathbf{Mod}(\breve{\overline{\mathscr{B}}}_X) \to \mathbf{Rep}_{\widehat{\overline{\mathscr{R}}}_i}(\Delta_i). \tag{4.8.23.3}$$

This induces a functor that we denote again by

$$\widehat{\nu}_i\colon \mathbf{Mod}_{\mathbb{Q}}(\breve{\overline{\mathscr{B}}}_X) \to \mathbf{Rep}_{\widehat{\overline{\mathscr{R}}}_i[\frac{1}{p}]}(\Delta_i). \tag{4.8.23.4}$$

Lemma 4.8.24 *With the notation of* 4.8.23, *the functors* $(\widehat{\nu}_i)_{1\leq i\leq c}$ (4.8.23.3) *induce an equivalence of categories*

$$\mathbf{Mod}^{\mathrm{aft}}(\breve{\overline{\mathscr{B}}}_X) \xrightarrow{\sim} \prod_{1\leq i\leq c} \mathbf{Rep}^{p\text{-atf}}_{\widehat{\overline{\mathscr{R}}}_i}(\Delta_i). \tag{4.8.24.1}$$

Indeed, denoting by

$$\lambda\colon (\overline{X}^{\circ}_{\text{fét}})^{\mathbb{N}^{\circ}} \to \overline{X}^{\circ}_{\text{fét}} \tag{4.8.24.2}$$

the canonical morphism (2.1.9.1), the $\lambda^*(\overline{X}^{\circ}_i)$ $(1 \leq i \leq c)$ form a disjoint covering of the final object of $(\overline{X}^{\circ}_{\text{fét}})^{\mathbb{N}^{\circ}}$. The proposition therefore follows from ([3] III.7.6 and III.7.21).

Lemma 4.8.25 *Let N be a coherent $\widehat{R_1}$-module such that the $\widehat{R_1}[\frac{1}{p}]$-module $N[\frac{1}{p}]$ is projective, A an $\widehat{R_1}$-algebra, $\widehat{A}$ its p-adic Hausdorff completion. Then, the canonical $\widehat{A}$-linear morphism*

$$N \otimes_{\widehat{R_1}} \widehat{A} \to N \widehat{\otimes}_{\widehat{R_1}} A, \tag{4.8.25.1}$$

where the tensor product $\widehat{\otimes}$ is completed for the p-adic topology, is an isogeny.

Indeed, by ([1] 1.10.2(iii)), there exists a coherent $\widehat{R_1}$-module N', two integers $t, n \geq 0$ and a p^t-isomorphism $u\colon N \oplus N' \to \widehat{R_1}^{\,n}$. Consider the canonical commutative diagram

$$\begin{array}{ccc} (N \oplus N') \otimes_{\widehat{R_1}} \widehat{A} & \xrightarrow{\ v\ } & (N \oplus N') \widehat{\otimes}_{\widehat{R_1}} A \\ {\scriptstyle u \otimes \mathrm{id}_{\widehat{A}}} \downarrow & & \downarrow {\scriptstyle u \widehat{\otimes} \mathrm{id}_A} \\ \widehat{A}^n & = & \widehat{A}^n. \end{array} \tag{4.8.25.2}$$

Since $u \otimes \mathrm{id}_{\widehat{A}}$ and $u \widehat{\otimes} \mathrm{id}_A$ are p^t-isomorphisms, v is a p^{2t}-isomorphism; the proposition follows ([2] 2.6.3).

Lemma 4.8.26 *We take again the notation of* 4.8.23. *Let i be an integer such that $1 \leq i \leq c$, (N, θ), (N', θ') two quasi-small Higgs $\widehat{\mathscr{R}}_i$-modules with coefficients in $\widetilde{\xi}^{-1}\Omega^1_{R/\mathscr{O}_K}$* (3.4.3) *such that the $\widehat{\mathscr{R}}_i$-modules N and N' are coherent and $\mathscr{O}_C$-flat and the $\widehat{\mathscr{R}}_i[\frac{1}{p}]$-modules $N[\frac{1}{p}]$ and $N'[\frac{1}{p}]$ are projective. We denote by φ (resp. φ') the $\widehat{\mathscr{R}}_i$-representation of Δ_i on N (resp. N') associated with (N, θ) (resp. (N', θ')) by the functor* (3.4.5.8). *Then, the canonical morphism*

$$\mathrm{Hom}_{\widehat{\overline{\mathscr{R}}}_i[\Delta_i]}((N, \varphi)\widehat{\otimes}_{\widehat{\mathscr{R}}_i} \widehat{\overline{\mathscr{R}}}_i, (N', \varphi')\widehat{\otimes}_{\widehat{\mathscr{R}}_i} \widehat{\overline{\mathscr{R}}}_i)[\frac{1}{p}] \tag{4.8.26.1}$$

$$\to \mathrm{Hom}_{\widehat{\overline{\mathscr{R}}}_i[\frac{1}{p}][\Delta_i]}((N, \varphi)\widehat{\otimes}_{\widehat{\mathscr{R}}_i} \widehat{\overline{\mathscr{R}}}_i[\frac{1}{p}], (N', \varphi')\widehat{\otimes}_{\widehat{\mathscr{R}}_i} \widehat{\overline{\mathscr{R}}}_i[\frac{1}{p}])$$

is an isomorphism.

Observe first that the $\widehat{\mathscr{R}}_i$-module N being of finite presentation, the canonical morphism

$$\mathrm{Hom}_{\widehat{\overline{\mathscr{R}}}_i}(N \otimes_{\widehat{\mathscr{R}}_i} \widehat{\overline{\mathscr{R}}}_i, N' \otimes_{\widehat{\mathscr{R}}_i} \widehat{\overline{\mathscr{R}}}_i)[\frac{1}{p}] \to \mathrm{Hom}_{\widehat{\overline{\mathscr{R}}}_i[\frac{1}{p}]}(N \otimes_{\widehat{\mathscr{R}}_i} \widehat{\overline{\mathscr{R}}}_i[\frac{1}{p}], N' \otimes_{\widehat{\mathscr{R}}_i} \widehat{\overline{\mathscr{R}}}_i[\frac{1}{p}]) \tag{4.8.26.2}$$

is an isomorphism. Therefore, the canonical morphism

$$j\colon \mathrm{Hom}_{\widehat{\overline{\mathscr{R}}}_i[\Delta_i]}((N,\varphi)\otimes_{\widehat{\mathscr{R}}_i}\widehat{\overline{\mathscr{R}}}_i,(N',\varphi')\otimes_{\widehat{\mathscr{R}}_i}\widehat{\overline{\mathscr{R}}}_i)[\frac{1}{p}] \tag{4.8.26.3}$$
$$\to \mathrm{Hom}_{\widehat{\overline{\mathscr{R}}}_i[\frac{1}{p}][\Delta_i]}((N,\varphi)\otimes_{\widehat{\mathscr{R}}_i}\widehat{\overline{\mathscr{R}}}_i[\frac{1}{p}],(N',\varphi')\otimes_{\widehat{\mathscr{R}}_i}\widehat{\overline{\mathscr{R}}}_i[\frac{1}{p}])$$

is injective.

Similarly, the canonical morphism

$$\mathrm{Hom}_{\widehat{\mathscr{R}}_i}(N,N')[\frac{1}{p}]\to \mathrm{Hom}_{\widehat{\mathscr{R}}_i[\frac{1}{p}]}(N[\frac{1}{p}],N'[\frac{1}{p}]) \tag{4.8.26.4}$$

is an isomorphism. Since N' is $\mathscr{O}_C$-flat, we deduce from this that the canonical morphism

$$i\colon \mathrm{Hom}_{\mathbf{HM}(\widehat{\mathscr{R}}_i)}((N,\theta),(N',\theta'))[\frac{1}{p}]\to \mathrm{Hom}_{\mathbf{HM}(\widehat{\mathscr{R}}_i[\frac{1}{p}])}((N[\frac{1}{p}],\theta),(N'[\frac{1}{p}],\theta')), \tag{4.8.26.5}$$

where we abusively denoted by θ (resp. θ') the Higgs field induced by θ (resp. θ') on $N[\frac{1}{p}]$ (resp. $N'[\frac{1}{p}]$), is an isomorphism.

Consider the commutative diagram

$$\begin{array}{ccc}
\mathrm{Hom}_{\mathbf{HM}(\widehat{\mathscr{R}}_i)}((N,\theta),(N',\theta'))[\frac{1}{p}] & \xrightarrow{\ u\ } & \mathrm{Hom}_{\widehat{\overline{\mathscr{R}}}_i[\Delta_i]}((N,\varphi)\otimes_{\widehat{\mathscr{R}}_i}\widehat{\overline{\mathscr{R}}}_i,(N',\varphi')\otimes_{\widehat{\mathscr{R}}_i}\widehat{\overline{\mathscr{R}}}_i)[\frac{1}{p}] \\
\Big\downarrow{\scriptstyle i} & & \Big\downarrow{\scriptstyle j} \\
\mathrm{Hom}_{\mathbf{HM}(\widehat{\mathscr{R}}_i[\frac{1}{p}])}((N[\frac{1}{p}],\theta),(N'[\frac{1}{p}],\theta')) & \xrightarrow{\ v\ } & \mathrm{Hom}_{\widehat{\overline{\mathscr{R}}}_i[\frac{1}{p}][\Delta_i]}((N,\varphi)\otimes_{\widehat{\mathscr{R}}_i}\widehat{\overline{\mathscr{R}}}_i[\frac{1}{p}],(N',\varphi')\otimes_{\widehat{\mathscr{R}}_i}\widehat{\overline{\mathscr{R}}}_i[\frac{1}{p}])
\end{array}$$

where u is induced by the functor (3.4.5.8) and v by the functor (3.4.27.2). By virtue of 3.4.28, the Higgs $\widehat{\mathscr{R}}_i[\frac{1}{p}]$-modules $(N[\frac{1}{p}],\theta)$ and $(N'[\frac{1}{p}],\theta')$ are solvable and the morphism v is identified with the morphism

$$\mathrm{Hom}_{\mathbf{HM}(\widehat{\mathscr{R}}_i[\frac{1}{p}])}((N[\frac{1}{p}],\theta),(N'[\frac{1}{p}],\theta'))\to \mathrm{Hom}_{\widehat{\overline{\mathscr{R}}}_i[\frac{1}{p}][\Delta_i]}(\mathbb{V}(N[\frac{1}{p}],\theta),\mathbb{V}(N[\frac{1}{p}],\theta))$$

induced by the functor $\mathbb{V}$ (3.3.7.2). By 3.3.16, v is therefore an isomorphism. Since i is an isomorphism and j is injective, we deduce that j is bijective. To conclude the proof of the proposition, it suffices to observe that the canonical morphism of $\widehat{\overline{\mathscr{R}}}_i$-representations of Δ_i

$$(N,\varphi)\otimes_{\widehat{\mathscr{R}}_i}\overline{\mathscr{R}}_i\to (N,\varphi)\widehat{\otimes}_{\widehat{\mathscr{R}}_i}\overline{\mathscr{R}}_i \tag{4.8.26.6}$$

is an isogeny by 4.8.25 and ([2] 2.6.3), and the same for (N',φ').

Lemma 4.8.27 *With the notation of* 4.8.23, *for every strongly Dolbeault* $\check{\overline{\mathscr{B}}}_{X,\mathbb{Q}}$-*module* $\mathscr{M}$ (4.8.21) *and every integer* $1\le i\le c$, $\widehat{v}_i(\mathscr{M})$ *is a Dolbeault* $\widehat{\overline{\mathscr{R}}}_i[\frac{1}{p}]$-*representation of* Δ_i (3.3.9).

Indeed, by 4.8.18 and 4.8.19, there exist a coherent $\mathscr{O}_{\mathfrak{X}}$-module $\mathscr{N}$ such that $\mathscr{N}$ is $\mathscr{S}$-flat and the $\mathscr{O}_{\mathfrak{X}}[\frac{1}{p}]$-module $\mathscr{N}[\frac{1}{p}]$ is projective of finite type (4.8.9), a quasi-small Higgs $\mathscr{O}_{\mathfrak{X}}$-field θ on $\mathscr{N}$ with coefficients in $\widetilde{\xi}^{-1}\widetilde{\Omega}^1_{\mathfrak{X}/\mathscr{S}}$ (4.2.9) and an isomorphism

$$\mathscr{M} \xrightarrow{\sim} (\check{\mathscr{P}} \otimes_{R_1} \overline{\check{\mathscr{B}}}_X)_{\mathbb{Q}}, \tag{4.8.27.1}$$

where $\check{\mathscr{P}}$ is the adic R_1-module of finite type of $(\overline{X}^{\circ}_{\text{fét}})^{\mathbb{N}^{\circ}}$ associated with $(\mathscr{N}, \theta)$ by the functor (4.8.12.3). We set $N_i = \Gamma(\overline{X}_{i,s}, \mathscr{N})$, which is a coherent $\widehat{\mathscr{R}}_i$-module, and denote by φ_i the quasi-small $\widehat{\mathscr{R}}_i$-representation of Δ_i on N_i, associated in 4.8.12 with θ. By 4.8.24, since the $\widehat{\mathscr{R}}_i[\frac{1}{p}]$-module $N_i[\frac{1}{p}]$ is projective, the isomorphism (4.8.27.1) then induces an isomorphism of $\widehat{\overline{\mathscr{R}}}_i[\frac{1}{p}]$-representations of Δ_i

$$\widehat{\nu}_i(\mathscr{M}) \xrightarrow{\sim} (N_i, \varphi_i) \otimes_{\widehat{\mathscr{R}}_i} \widehat{\overline{\mathscr{R}}}_i[\frac{1}{p}]. \tag{4.8.27.2}$$

We deduce from this that the $\widehat{\overline{\mathscr{R}}}_i[\frac{1}{p}]$-representation $\widehat{\nu}_i(\mathscr{M})$ of Δ_i is Dolbeault by virtue of 3.4.28.

Lemma 4.8.28 *With the notation of* 4.8.23, *for every integer* $1 \leq i \leq c$ *and every Dolbeault* $\widehat{\overline{\mathscr{R}}}_i[\frac{1}{p}]$*- representation* M *of* Δ_i, *there exists a strongly Dolbeault* $\overline{\check{\mathscr{B}}}_{X,\mathbb{Q}}$*-module* $\mathscr{M}$ *such that* $\widehat{\nu}_i(\mathscr{M}) = M$ *and* $\widehat{\nu}_j(\mathscr{M}) = 0$ *for all* $1 \leq j \leq c$ *with* $j \neq i$.

Indeed, by 3.3.16 and 3.4.30, the Higgs $\widehat{\mathscr{R}}_i[\frac{1}{p}]$-module $\mathbb{H}(M)$ is small. Therefore, there exists a coherent sub-$\widehat{\mathscr{R}}_i$-module N of $\mathbb{H}(M)$ which generates it over $\widehat{\mathscr{R}}_i[\frac{1}{p}]$, such that the Higgs $\widehat{\mathscr{R}}_i[\frac{1}{p}]$-field of $\mathbb{H}(M)$ induces on N a quasi-small Higgs $\widehat{\mathscr{R}}_i$-field θ with coefficients in $\widetilde{\xi}^{-1}\widetilde{\Omega}^1_{R/\mathscr{O}_K}$. We denote by φ the quasi-small $\widehat{\mathscr{R}}_i$-representation of Δ_i on N associated with (N, θ) by the functor (3.4.5.8). By virtue of 3.3.16 and 3.4.28(ii), we have an $\widehat{\overline{\mathscr{R}}}_i$-linear and Δ_i-equivariant isomorphism

$$M \xrightarrow{\sim} (N, \varphi) \otimes_{\widehat{\mathscr{R}}_i} \widehat{\overline{\mathscr{R}}}_i[\frac{1}{p}]. \tag{4.8.28.1}$$

Let $\mathscr{N}$ be the coherent $\mathscr{O}_{\mathfrak{X}}$-module such that $\Gamma(\overline{X}_{i,s}, \mathscr{N}) = N$ and $\Gamma(\overline{X}_{j,s}, \mathscr{N}) = 0$ for all $1 \leq j \leq c$ with $j \neq i$. We denote by θ the quasi-small Higgs $\mathscr{O}_{\mathfrak{X}}$-field induced on $\mathscr{N}$ by the Higgs $\widehat{\mathscr{R}}_i$-field θ on N. We denote by $\check{\mathscr{P}}$ the adic R_1-module of finite type of $(\overline{X}^{\circ}_{\text{fét}})^{\mathbb{N}^{\circ}}$ associated with $(\mathscr{N}, \theta)$ by the functor (4.8.12.3). It follows from 4.8.14 and 4.8.17 that the $\overline{\check{\mathscr{B}}}_{X,\mathbb{Q}}$-module $\mathscr{M} = (\check{\mathscr{P}} \otimes_{R_1} \overline{\check{\mathscr{B}}}_X)_{\mathbb{Q}}$ is strongly Dolbeault. By the proof of 4.8.27, we have an isomorphism of $\widehat{\overline{\mathscr{R}}}_i[\frac{1}{p}]$-representations of Δ_i

$$\widehat{\nu}_i(\mathscr{M}) \xrightarrow{\sim} (N, \varphi) \otimes_{\widehat{\mathscr{R}}_i} \widehat{\overline{\mathscr{R}}}_i[\frac{1}{p}], \tag{4.8.28.2}$$

and $\widehat{\nu}_j(\mathscr{M}) = 0$ for all $1 \leq j \leq c$ with $j \neq i$; the proposition follows.

Proposition 4.8.29 *With the notation of* 4.8.23, *the functors* $(\widehat{\nu}_i)_{1\le i\le c}$ (4.8.23.4) *induce an equivalence of categories*

$$\mathbf{Mod}^{\mathrm{sDolb}}_{\mathbb{Q}}(\breve{\overline{\mathscr{B}}}_X) \xrightarrow{\sim} \prod_{1\le i\le c} \mathbf{Rep}^{\mathrm{Dolb}}_{\widehat{\overline{\mathscr{R}}}_i[\frac{1}{p}]}(\Delta_i). \tag{4.8.29.1}$$

Indeed, the functor (4.8.29.1) is well defined by virtue of 4.8.27. It is fully faithful by 4.8.18, 4.8.24 and 4.8.26, and it is essentially surjective by 4.8.28.

4.8.30 We take again the notation of 4.8.11. For any integer $1 \le i \le c$, we denote by β_i the composed functor

$$\beta_i \colon \widetilde{E} \to \mathbf{B}_{\Delta_i}, \quad F \mapsto \nu_i \circ (\beta_*(F)|\overline{X}_i^\circ), \tag{4.8.30.1}$$

where β is the morphism of topos (4.3.2.13). We thus define a functor from the category of abelian sheaves of $\widetilde{E}$ to that of $\mathbb{Z}[\Delta_i]$-modules. The latter being left-exact, we denote by $\mathrm{R}^q\beta_i$ $(q \ge 0)$ its right derived functors. For every abelian sheaf F of $\widetilde{E}$ and every integer $q \ge 0$, we have a canonical functorial isomorphism

$$\mathrm{R}^q\beta_i(F) \xrightarrow{\sim} \nu_i \circ (\mathrm{R}^q\beta_*(F)|\overline{X}_i^\circ). \tag{4.8.30.2}$$

Forgetting the action of Δ_i, we have a canonical functorial isomorphism

$$\mathrm{R}^q\beta_i(F) \xrightarrow{\sim} \varinjlim_{j\in J_i} \mathrm{H}^q((V_j \to X), F), \tag{4.8.30.3}$$

where $(V_j)_{j\in J_i}$ is the normalized universal cover of $\overline{X}_i^\circ$ at $\overline{y}_i$ ([2] 2.1.20).

For any abelian sheaf $F = (F_n)_{n\ge 0}$ of $\widetilde{E}^{\mathbb{N}^\circ}$, we denote by $\widehat{\beta}_i(F)$ the inverse limit of the $\beta_i(F_n)$'s,

$$\widehat{\beta}_i(F) = \varprojlim_{n\ge 0} \beta_i(F_n). \tag{4.8.30.4}$$

We thus define a functor from the category of abelian sheaves of $\widetilde{E}^{\mathbb{N}^\circ}$ to that of $\mathbb{Z}[\Delta_i]$-modules. The latter being left-exact, we abusively denote by $\mathrm{R}^q\widehat{\beta}_i(F)$ $(q \ge 0)$ its right derived functors. By ([39] 1.6), we have a canonical exact sequence

$$0 \to \mathrm{R}^1\varprojlim_{n\ge 0} \mathrm{R}^{q-1}\beta_i(F_n) \to \mathrm{R}^q\widehat{\beta}_i(F) \to \varprojlim_{n\ge 0} \mathrm{R}^q\beta_i(F_n) \to 0, \tag{4.8.30.5}$$

where we set $\mathrm{R}^{-1}\beta_i(F_n) = 0$ for all $n \ge 0$.

For every integer $q \ge 0$, the functor $\mathrm{R}^q\widehat{\beta}_i$ induces a functor that we also denote by

$$\mathrm{R}^q\widehat{\beta}_i \colon \mathbf{Mod}(\breve{\overline{\mathscr{B}}}) \to \mathbf{Rep}_{\widehat{\overline{\mathscr{R}}}_i}(\Delta_i), \tag{4.8.30.6}$$

to the category of $\widehat{\overline{\mathscr{R}}}_i$-representations of Δ_i, where $\widehat{\overline{\mathscr{R}}}_i$ is the p-adic Hausdorff completion of the ring $\overline{\mathscr{R}}_i$ (4.8.23.1). The latter induces a functor that we also denote by

$$\mathrm{R}^q\widehat{\beta}_i\colon \mathbf{Mod}_{\mathbb{Q}}(\breve{\overline{\mathscr{B}}}) \to \mathbf{Rep}_{\widehat{\overline{\mathscr{R}}}_i[\frac{1}{p}]}(\Delta_i). \tag{4.8.30.7}$$

It easily follows from (4.8.30.4), (4.8.30.5) and ([5] V 3.5) that the functors $\mathrm{R}^q\widehat{\beta}_i$ $(q \geq 0)$ (4.8.30.6) are the right derived functors of the functor $\widehat{\beta}_i = \mathrm{R}^0\widehat{\beta}_i$ on the category of $\breve{\overline{\mathscr{B}}}$-modules. Consequently, the functors $\mathrm{R}^q\widehat{\beta}_i$ $(q \geq 0)$ (4.8.30.7) are the right derived functors of the functor $\widehat{\beta}_i = \mathrm{R}^0\widehat{\beta}_i$ on the category of $\breve{\overline{\mathscr{B}}}_{\mathbb{Q}}$-modules (2.9.6).

Theorem 4.8.31 *We keep the notation of* 4.8.30. *Then,*

(i) *The functors* $(\widehat{\beta}_i)_{1\leq i\leq c}$ (4.8.30.7) *induce an equivalence of categories*

$$\mathbf{Mod}^{\mathrm{sDolb}}_{\mathbb{Q}}(\breve{\overline{\mathscr{B}}}) \xrightarrow{\sim} \prod_{1\leq i\leq c} \mathbf{Rep}^{\mathrm{Dolb}}_{\widehat{\overline{\mathscr{R}}}_i[\frac{1}{p}]}(\Delta_i). \tag{4.8.31.1}$$

(ii) *For every strongly Dolbeault $\breve{\overline{\mathscr{B}}}_{\mathbb{Q}}$-module $\mathscr{M}$ and every integer $q \geq 1$, we have*

$$\mathrm{R}^q\widehat{\beta}_i(\mathscr{M}) = 0. \tag{4.8.31.2}$$

(i) Indeed, for all $1 \leq i \leq c$, we have $\widehat{\beta}_i = \widehat{\nu}_i \circ \breve{\beta}_*$, where $\widehat{\beta}_i = \mathrm{R}^0\widehat{\beta}_i$ is the functor (4.8.30.6), $\widehat{\nu}_i$ is the functor (4.8.23.3) and $\breve{\beta}$ is the morphism of ringed topos (4.8.1.3). The proposition then follows from 4.8.22 and 4.8.29.

(ii) Indeed, by 4.8.18, there exist an adic $\breve{\overline{\mathscr{B}}}_X$-module of finite type $\breve{\mathscr{F}}$ and two adjoint isomorphisms

$$\mathscr{M} \xrightarrow{\sim} (\breve{\beta}^*(\breve{\mathscr{F}}))_{\mathbb{Q}}, \tag{4.8.31.3}$$

$$\breve{\beta}_{\mathbb{Q}*}(\mathscr{M}) \xrightarrow{\sim} \breve{\mathscr{F}}_{\mathbb{Q}}. \tag{4.8.31.4}$$

Moreover, by virtue of 4.8.16(ii), the $\breve{\overline{\mathscr{B}}}_X$-module $\mathrm{R}^q\breve{\beta}_*(\breve{\beta}^*(\breve{\mathscr{F}}))$ is of finite exponent. The proposition then follows from (4.8.30.5) applied to the sheaf $\breve{\beta}^*(\breve{\mathscr{F}})$, (4.8.30.2), ([3] III.7.5) and ([39] 1.15).

Proposition 4.8.32 *We keep the notation of* 4.8.30. *Then,*

(i) *For every strongly solvable $\mathscr{O}_{\mathfrak{X}}[\frac{1}{p}]$-bundle $\mathscr{N}$ with coefficients in $\widetilde{\xi}^{-1}\widetilde{\Omega}^1_{\mathfrak{X}/\mathscr{S}}$* (4.6.6), *and every $1 \leq i \leq c$, the Higgs $\widehat{\overline{\mathscr{R}}}_i[\frac{1}{p}]$-module $N_i = \Gamma(\overline{X}_{i,s}, \mathscr{N})$ with coefficients in $\widetilde{\xi}^{-1}\widetilde{\Omega}^1_{\mathfrak{X}/\mathscr{S}}(\overline{X}_{i,s})$ is solvable in the sense of* 3.3.10.

(ii) *For every integer $1 \leq i \leq c$, the diagram of functors*

$$\begin{array}{ccc} \mathbf{HM}^{\mathrm{ssol}}(\mathscr{O}_{\mathfrak{X}}[\frac{1}{p}], \widetilde{\xi}^{-1}\widetilde{\Omega}^1_{\mathfrak{X}/\mathscr{S}}) & \xrightarrow{\mathscr{V}_{\mathbb{Q}}} & \mathbf{Mod}^{\mathrm{sDolb}}_{\mathbb{Q}}(\breve{\overline{\mathscr{B}}}) \\ \downarrow{\scriptstyle \Gamma(\overline{X}_{i,s},-)} & & \downarrow{\scriptstyle \widehat{\beta}_i} \\ \mathbf{HM}^{\mathrm{sol}}(\widehat{\overline{\mathscr{R}}}_i[\frac{1}{p}], \widetilde{\xi}^{-1}\widetilde{\Omega}^1_{\mathfrak{X}/\mathscr{S}}(\overline{X}_{i,s})) & \xrightarrow{\mathbb{V}} & \mathbf{Rep}^{\mathrm{Dolb}}_{\widehat{\overline{\mathscr{R}}}_i[\frac{1}{p}]}(\Delta_i), \end{array} \tag{4.8.32.1}$$

where $\mathscr{V}_{\mathbb{Q}}$ is the functor (4.6.18.2) *and* $\mathbb{V}$ *is the functor* (3.3.7.2), *is commutative up to canonical isomorphism.*

(iii) *For every integer* $1 \leq i \leq c$, *the diagram of functors*

$$\begin{array}{ccc} \mathbf{Mod}^{\mathrm{sDolb}}_{\mathbb{Q}}(\breve{\overline{\mathscr{B}}}) & \xrightarrow{\mathscr{H}_{\mathbb{Q}}} & \mathbf{HM}^{\mathrm{ssol}}(\mathscr{O}_{\mathfrak{X}}[\frac{1}{p}], \widetilde{\xi}^{-1}\widetilde{\Omega}^1_{\mathfrak{X}/\mathscr{S}}) \\ \Big\downarrow{\scriptstyle \widehat{\beta}_i} & & \Big\downarrow{\scriptstyle \Gamma(\overline{X}_{i,s},-)} \\ \mathbf{Rep}^{\mathrm{Dolb}}_{\widehat{\overline{\mathscr{R}}}_i[\frac{1}{p}]}(\Delta_i) & \xrightarrow{\mathbb{H}} & \mathbf{HM}^{\mathrm{sol}}(\widehat{\mathscr{R}}_i[\frac{1}{p}], \widetilde{\xi}^{-1}\widetilde{\Omega}^1_{\mathfrak{X}/\mathscr{S}}(\overline{X}_{i,s})), \end{array} \tag{4.8.32.2}$$

where $\mathscr{H}_{\mathbb{Q}}$ is the functor (4.6.13.2) *and* $\mathbb{H}$ *is the functor* (3.3.6.2), *is commutative up to canonical isomorphism.*

(i) Indeed, since $\mathscr{N}$ is small by virtue of 4.6.26, N_i is small by 4.2.12 and therefore solvable by virtue of 3.4.30.

(ii) This follows from 3.4.28(ii), (4.8.14.1) and (4.8.18.2), taking into account 4.8.15 and 4.2.16.

(iii) This follows from (ii), 3.3.16 and 4.6.23.

4.9 Pullback of a Dolbeault ind-module by an Étale Morphism

4.9.1 Let $g \colon X' \to X$ be an étale morphism of finite type. We endow X' with the logarithmic structure $\mathscr{M}_{X'}$ pullback of $\mathscr{M}_X$ and we denote by $f' \colon (X', \mathscr{M}_{X'}) \to (S, \mathscr{M}_S)$ the morphism induced by f and g. Observe that f' is adequate ([3] III.4.7) and that $X'^{\circ} = X^{\circ} \times_X X'$ is the maximal open subscheme of X' where the logarithmic structure $\mathscr{M}_{X'}$ is trivial. We endow $\overline{X}'$ and $\check{\overline{X}}'$ (4.1.1.2) with logarithmic structures $\mathscr{M}_{\overline{X}'}$ and $\mathscr{M}_{\check{\overline{X}}'}$ pullbacks of $\mathscr{M}_{X'}$. There exists essentially a unique étale morphism $\widetilde{g} \colon \widetilde{X}' \to \widetilde{X}$ which fits into a Cartesian diagram (4.1.5)

$$\begin{array}{ccc} \check{\overline{X}}' & \longrightarrow & \widetilde{X}' \\ \Big\downarrow{\scriptstyle \check{\overline{g}}} & & \Big\downarrow{\scriptstyle \widetilde{g}} \\ \check{\overline{X}} & \longrightarrow & \widetilde{X}. \end{array} \tag{4.9.1.1}$$

We endow $\widetilde{X}'$ with the logarithmic structure $\mathscr{M}_{\widetilde{X}'}$ pullback of $\mathscr{M}_{\widetilde{X}}$, so that $(\widetilde{X}', \mathscr{M}_{\widetilde{X}'})$ is a smooth $(\widetilde{S}, \mathscr{M}_{\widetilde{S}})$-deformation of $(\check{\overline{X}}', \mathscr{M}_{\check{\overline{X}}'})$.

We associate with $(f', \widetilde{X}', \mathscr{M}_{\widetilde{X}'})$ objects analogous to those defined in 4.3–4.6 for $(f, \widetilde{X}, \mathscr{M}_{\widetilde{X}})$, denoted by the same symbols equipped with an exponent $'$. We denote by

$$\Phi \colon \widetilde{E}' \to \widetilde{E}, \tag{4.9.1.2}$$

$$\Phi_s \colon \widetilde{E}'_s \to \widetilde{E}_s, \tag{4.9.1.3}$$

the morphisms of topos induced by functoriality by g ([3] (III.8.5.3) and (III.9.118)). By ([3] VI.10.14), Φ is identified with the localization morphism of $\widetilde{E}$ at $\sigma^*(X')$. Moreover, we have a canonical homomorphism $\Phi^*(\overline{\mathscr{B}}) \to \overline{\mathscr{B}}'$ ([3] (III.8.20.6)), which is an isomorphism by virtue of ([3] III.8.21(i)). For every integer $n \geq 1$, Φ_s is underlying a canonical morphism of ringed topos ([3] (III.9.11.11))

$$\Phi_n \colon (\widetilde{E}'_s, \overline{\mathscr{B}}'_n) \to (\widetilde{E}_s, \overline{\mathscr{B}}_n). \tag{4.9.1.4}$$

The homomorphism $\Phi_s^*(\overline{\mathscr{B}}_n) \to \overline{\mathscr{B}}'_n$ being an isomorphism by ([3] III.9.13), there is no difference for $\overline{\mathscr{B}}_n$-modules between the pullback by Φ_s in the sense of abelian sheaves and the pullback by Φ_n in the sense of modules. The diagram of morphisms of ringed topos ([3] (III.9.11.12))

$$\begin{array}{ccc} (\widetilde{E}'_s, \overline{\mathscr{B}}'_n) & \xrightarrow{\Phi_n} & (\widetilde{E}_s, \overline{\mathscr{B}}_n) \\ {\scriptstyle \sigma'_n}\big\downarrow & & \big\downarrow{\scriptstyle \sigma_n} \\ (X'_{s,\text{ét}}, \mathscr{O}_{\overline{X}'_n}) & \xrightarrow{\overline{g}_n} & (X_{s,\text{ét}}, \mathscr{O}_{\overline{X}_n}), \end{array} \tag{4.9.1.5}$$

where $\overline{g}_n$ is the morphism induced by g, is commutative up to canonical isomorphism.

For every rational number $r \geq 0$, we have a canonical $\overline{\mathscr{B}}'_n$-linear isomorphism (4.4.16.10)

$$\nu_n^{(r)} \colon \Phi_n^*(\mathscr{F}_n^{(r)}) \xrightarrow{\sim} \mathscr{F}_n'^{(r)}, \tag{4.9.1.6}$$

where $\mathscr{F}_n^{(r)}$ and $\mathscr{F}_n'^{(r)}$ are the modules defined in (4.4.10.4). We have a canonical isomorphism (4.3.3.4)

$$\sigma_n^*(\widetilde{\xi}^{-1}\widetilde{\Omega}^1_{\overline{X}_n/\overline{S}_n}) \xrightarrow{\sim} \sigma^{-1}(\widetilde{\xi}^{-1}\widetilde{\Omega}^1_{X/S}) \otimes_{\sigma^{-1}(\mathscr{O}_X)} \overline{\mathscr{B}}_n. \tag{4.9.1.7}$$

Then, by virtue of ([3] VI.5.34(ii), VI.8.9 and VI.5.17), $\sigma_n^*(\xi^{-1}\widetilde{\Omega}^1_{\overline{X}_n/\overline{S}_n})$ is the sheaf of $\widetilde{E}$ associated with the presheaf on E defined by the correspondence

$$\{U \in \mathbf{\acute{E}t}^{\circ}_{/X} \mapsto \xi^{-1}\widetilde{\Omega}^1_{X/S}(U) \otimes_{\mathscr{O}_X(U)} \overline{\mathscr{B}}_{U,n}\}. \tag{4.9.1.8}$$

We deduce from this that the diagram

$$\begin{array}{ccccccccc} 0 & \longrightarrow & \overline{\mathscr{B}}'_n & \longrightarrow & \Phi_n^*(\mathscr{F}_n^{(r)}) & \longrightarrow & \Phi_n^*(\sigma_n^*(\widetilde{\xi}^{-1}\widetilde{\Omega}^1_{\overline{X}_n/\overline{S}_n})) & \longrightarrow & 0 \\ & & \big\| & & \big\downarrow{\scriptstyle \nu_n^{(r)}} & & \big\downarrow & & \\ 0 & \longrightarrow & \overline{\mathscr{B}}'_n & \longrightarrow & \mathscr{F}_n'^{(r)} & \longrightarrow & \sigma_n'^*(\widetilde{\xi}^{-1}\widetilde{\Omega}^1_{\overline{X}'_n/\overline{S}_n}) & \longrightarrow & 0 \end{array} \tag{4.9.1.9}$$

where the horizontal lines are induced by the exact sequences (4.4.11.1) and the right vertical arrow is the isomorphism induced by the canonical isomorphism

$$\overline{g}_n^*(\widetilde{\xi}^{-1}\widetilde{\Omega}^1_{\overline{X}_n/\overline{S}_n}) \xrightarrow{\sim} \widetilde{\xi}^{-1}\widetilde{\Omega}^1_{\overline{X}'_n/\overline{S}_n} \tag{4.9.1.10}$$

and the commutative diagram (4.9.1.5), is commutative.

4.9.2 We denote by $\mathfrak{X}'$ the formal scheme p-adic completion of $\overline{X}'$, by

$$\mathfrak{g}\colon \mathfrak{X}' \to \mathfrak{X} \tag{4.9.2.1}$$

the extension of $\overline{g}\colon \overline{X}' \to \overline{X}$ to the completions and by

$$\breve{\Phi}\colon (\widetilde{E}_s'^{\mathbb{N}^\circ}, \breve{\overline{\mathscr{B}}}') \to (\widetilde{E}_s^{\mathbb{N}^\circ}, \breve{\overline{\mathscr{B}}}) \tag{4.9.2.2}$$

the morphism of ringed topos induced by the morphisms $(\Phi_n)_{n\geq 1}$ (4.9.1.4) (see [3] III.7.5). By ([3] III.9.14), $\breve{\Phi}$ is canonically isomorphic to the localization morphism of the ringed topos $(\widetilde{E}_s^{\mathbb{N}^\circ}, \breve{\overline{\mathscr{B}}})$ at $\lambda^*(\sigma_s^*(X_s'))$, where $\lambda\colon \widetilde{E}_s^{\mathbb{N}^\circ} \to \widetilde{E}_s$ is the canonical morphism of topos defined in (2.1.9.1). Consequently, there is no difference for the $\breve{\overline{\mathscr{B}}}$-modules between the pullback by $\breve{\Phi}$ in the sense of abelian sheaves and the pullback in the sense of modules.

It immediately follows from (4.9.1.5) that the diagram of morphisms of ringed topos

$$\begin{array}{ccc} (\widetilde{E}_s'^{\mathbb{N}^\circ}, \breve{\overline{\mathscr{B}}}') & \xrightarrow{\breve{\Phi}} & (\widetilde{E}_s^{\mathbb{N}^\circ}, \breve{\overline{\mathscr{B}}}) \\ \big\downarrow{\scriptstyle \widehat{\sigma}'} & & \big\downarrow{\scriptstyle \widehat{\sigma}} \\ (X'_{s,\mathrm{zar}}, \mathscr{O}_{\mathfrak{X}'}) & \xrightarrow{\mathfrak{g}} & (X_{s,\mathrm{zar}}, \mathscr{O}_{\mathfrak{X}}), \end{array} \tag{4.9.2.3}$$

where $\widehat{\sigma}$ and $\widehat{\sigma}'$ are the morphisms defined in (4.3.11.4), is commutative up to canonical isomorphism.

For every rational number $r \geq 0$, the isomorphisms (4.9.1.6) induce an isomorphism

$$\breve{\nu}^{(r)}\colon \breve{\Phi}^*(\breve{\mathscr{F}}^{(r)}) \xrightarrow{\sim} \breve{\mathscr{F}}'^{(r)}, \tag{4.9.2.4}$$

where $\breve{\mathscr{F}}^{(r)}$ and $\breve{\mathscr{F}}'^{(r)}$ are the modules defined in (4.4.22.2). The commutative diagrams (4.9.1.9) induce a commutative diagram

$$\begin{array}{ccccccccc} 0 & \to & \breve{\Phi}^*(\breve{\overline{\mathscr{B}}}) & \to & \breve{\Phi}^*(\breve{\mathscr{F}}^{(r)}) & \to & \breve{\Phi}^*(\widehat{\sigma}^*(\widetilde{\xi}^{-1}\widetilde{\Omega}^1_{\overline{X}/\overline{S}})) & \to & 0 \\ & & \Big\| & & \big\downarrow{\scriptstyle \breve{\nu}^{(r)}} & & \big\downarrow{\scriptstyle \delta} & & \\ 0 & \to & \breve{\overline{\mathscr{B}}}' & \to & \breve{\mathscr{F}}'^{(r)} & \to & \widehat{\sigma}'^*(\widetilde{\xi}^{-1}\widetilde{\Omega}^1_{\overline{X}'/\overline{S}}) & \to & 0 \end{array} \tag{4.9.2.5}$$

where δ is the isomorphism induced by the canonical isomorphism

$$\mathfrak{g}^*(\widetilde{\xi}^{-1}\widetilde{\Omega}^1_{\mathfrak{X}/\mathscr{S}}) \xrightarrow{\sim} \widetilde{\xi}^{-1}\widetilde{\Omega}^1_{\mathfrak{X}'/\mathscr{S}} \tag{4.9.2.6}$$

and the commutative diagram (4.9.2.3).

In view of (4.4.22.4), the isomorphism (4.9.2.4) induces an isomorphism of $\overline{\breve{\mathscr{B}}}'$-algebras

$$\breve{\mu}^{(r)} \colon \breve{\Phi}^*(\breve{\mathscr{C}}^{(r)}) \xrightarrow{\sim} \breve{\mathscr{C}}'^{(r)}. \tag{4.9.2.7}$$

It immediately follows from (4.9.2.5) that the diagram

$$\begin{array}{ccc} \breve{\Phi}^*(\breve{\mathscr{C}}^{(r)}) & \xrightarrow{\breve{\mu}^{(r)}} & \breve{\mathscr{C}}'^{(r)} \\ \Big\downarrow{\scriptstyle \breve{\Phi}^*(d_{\breve{\mathscr{C}}^{(r)}})} & & \Big\downarrow{\scriptstyle d_{\breve{\mathscr{C}}'^{(r)}}} \\ \breve{\Phi}^*(\widehat{\sigma}^*(\widetilde{\xi}^{-1}\widetilde{\Omega}^1_{\mathfrak{X}/\mathscr{S}}) \otimes_{\overline{\breve{\mathscr{B}}}} \breve{\mathscr{C}}^{(r)}) & \xrightarrow{\delta\otimes\breve{\nu}^{(r)}} & \widehat{\sigma}'^*(\widetilde{\xi}^{-1}\widetilde{\Omega}^1_{\mathfrak{X}'/\mathscr{S}}) \otimes_{\overline{\breve{\mathscr{B}}}'} \breve{\mathscr{C}}'^{(r)}, \end{array} \tag{4.9.2.8}$$

where $d_{\breve{\mathscr{C}}^{(r)}}$ and $d_{\breve{\mathscr{C}}'^{(r)}}$ are the derivations (4.5.1.1), is commutative. For all rational numbers $r \geq r' \geq 0$, the diagram

$$\begin{array}{ccc} \breve{\Phi}^*(\breve{\mathscr{C}}^{(r)}) & \xrightarrow{\breve{\nu}^{(r)}} & \breve{\mathscr{C}}'^{(r)} \\ \Big\downarrow{\scriptstyle \breve{\Phi}^*(\breve{\alpha}^{r,r'})} & & \Big\downarrow{\scriptstyle \breve{\alpha}'^{r,r'}} \\ \breve{\Phi}^*(\breve{\mathscr{C}}^{(r')}) & \xrightarrow{\breve{\nu}^{(r')}} & \breve{\mathscr{C}}'^{(r')}, \end{array} \tag{4.9.2.9}$$

where $\breve{\alpha}^{r,r'}$ and $\breve{\alpha}'^{r,r'}$ are the canonical homomorphisms (4.4.22.6), is commutative.

4.9.3 In view of the isomorphism (4.9.2.6), the functor $\mathfrak{g}^*$ induces functors

$$\mathfrak{g}^* \colon \mathbf{HM}(\mathscr{O}_{\mathfrak{X}}, \widetilde{\xi}^{-1}\widetilde{\Omega}^1_{\mathfrak{X}/\mathscr{S}}) \to \mathbf{HM}(\mathscr{O}_{\mathfrak{X}'}, \widetilde{\xi}^{-1}\widetilde{\Omega}^1_{\mathfrak{X}'/\mathscr{S}}), \tag{4.9.3.1}$$

$$\mathfrak{g}^* \colon \mathbf{HI}(\mathscr{O}_{\mathfrak{X}}, \widetilde{\xi}^{-1}\widetilde{\Omega}^1_{\mathfrak{X}/\mathscr{S}}) \to \mathbf{HI}(\mathscr{O}_{\mathfrak{X}'}, \widetilde{\xi}^{-1}\widetilde{\Omega}^1_{\mathfrak{X}'/\mathscr{S}}), \tag{4.9.3.2}$$

$$\mathrm{I}\mathfrak{g}^* \colon \mathbf{Ind\text{-}HM}(\mathscr{O}_{\mathfrak{X}}, \widetilde{\xi}^{-1}\widetilde{\Omega}^1_{\mathfrak{X}/\mathscr{S}}) \to \mathbf{Ind\text{-}HM}(\mathscr{O}_{\mathfrak{X}'}, \widetilde{\xi}^{-1}\widetilde{\Omega}^1_{\mathfrak{X}'/\mathscr{S}}). \tag{4.9.3.3}$$

The functor $\breve{\Phi}^*$ (4.9.2.2) induces a functor (2.7.10.1)

$$\mathrm{I}\breve{\Phi}^* \colon \mathbf{Ind\text{-}Mod}(\overline{\breve{\mathscr{B}}}) \to \mathbf{Ind\text{-}Mod}(\overline{\breve{\mathscr{B}}}'). \tag{4.9.3.4}$$

Let r be a rational number ≥ 0. By 2.8.10, the functor $\breve{\Phi}^*$ (4.9.2.2) and the isomorphism (4.9.2.7) induce a functor that we also denote by (4.5.1)

$$\mathrm{I}\breve{\Phi}^* \colon \mathbf{Ind\text{-}MC}(\breve{\mathscr{C}}^{(r)}/\overline{\breve{\mathscr{B}}}) \to \mathbf{Ind\text{-}MC}(\breve{\mathscr{C}}'^{(r)}/\overline{\breve{\mathscr{B}}}'). \tag{4.9.3.5}$$

The diagrams of functors

$$\begin{array}{ccc} \mathbf{Ind\text{-}Mod}(\overline{\breve{\mathscr{B}}}) & \xrightarrow{\mathrm{I}\mathfrak{S}^{(r)}} & \mathbf{Ind\text{-}MC}(\breve{\mathscr{C}}^{(r)}/\overline{\breve{\mathscr{B}}}) \\ \Big\downarrow{\scriptstyle \mathrm{I}\breve{\Phi}^*} & & \Big\downarrow{\scriptstyle \mathrm{I}\breve{\Phi}^*} \\ \mathbf{Ind\text{-}Mod}(\overline{\breve{\mathscr{B}}}') & \xrightarrow{\mathrm{I}\mathfrak{S}'^{(r)}} & \mathbf{Ind\text{-}MC}(\breve{\mathscr{C}}'^{(r)}/\overline{\breve{\mathscr{B}}}'), \end{array} \tag{4.9.3.6}$$

where the horizontal arrows are the functors (4.5.1.2), and

$$\begin{array}{ccc}
\mathbf{Ind\text{-}HM}(\mathscr{O}_{\mathfrak{X}}, \widetilde{\xi}^{-1}\widetilde{\Omega}^1_{\mathfrak{X}/\mathscr{S}}) & \xrightarrow{\mathrm{I}\widehat{\sigma}^{(r)*}} & \mathbf{Ind\text{-}MC}(\breve{\mathscr{C}}^{(r)}/\overline{\breve{\mathscr{B}}}) \\
\Big\downarrow{\scriptstyle \mathrm{I}\mathfrak{g}^*} & & \Big\downarrow{\scriptstyle \mathrm{I}\breve{\Phi}^*} \\
\mathbf{Ind\text{-}HM}(\mathscr{O}_{\mathfrak{X}'}, \widetilde{\xi}^{-1}\widetilde{\Omega}^1_{\mathfrak{X}'/\mathscr{S}}) & \xrightarrow{\mathrm{I}\widehat{\sigma}'^{(r)*}} & \mathbf{Ind\text{-}MC}(\breve{\mathscr{C}}'^{(r)}/\overline{\breve{\mathscr{B}}}'),
\end{array} \tag{4.9.3.7}$$

where the horizontal arrows are the functors (4.5.1.6), are clearly commutative up to canonical isomorphisms.

By ([3] III.9.14) and 2.7.15, the diagram of functors

$$\begin{array}{ccc}
\mathbf{Ind\text{-}MC}(\breve{\mathscr{C}}^{(r)}/\overline{\breve{\mathscr{B}}}) & \xrightarrow{\mathrm{I}\mathscr{K}^{(r)}} & \mathbf{Mod}(\overline{\breve{\mathscr{B}}}) \\
\Big\downarrow{\scriptstyle \mathrm{I}\breve{\Phi}^*} & & \Big\downarrow{\scriptstyle \mathrm{I}\breve{\Phi}^*} \\
\mathbf{Ind\text{-}MC}(\breve{\mathscr{C}}'^{(r)}/\overline{\breve{\mathscr{B}}}') & \xrightarrow{\mathrm{I}\mathscr{K}'^{(r)}} & \mathbf{Mod}(\overline{\breve{\mathscr{B}}}'),
\end{array} \tag{4.9.3.8}$$

where the horizontal arrows are the functors (4.5.1.3), is commutative up to canonical isomorphism.

The base change morphism (2.7.13.4) with respect to the diagram (4.9.2.3) induces a morphism of functors from $\mathbf{Ind\text{-}MC}(\breve{\mathscr{C}}^{(r)}/\overline{\breve{\mathscr{B}}})$ to $\mathbf{Ind\text{-}HM}(\mathscr{O}_{\mathfrak{X}'}, \widetilde{\xi}^{-1}\widetilde{\Omega}^1_{\mathfrak{X}'/\mathscr{S}})$

$$\mathrm{I}\mathfrak{g}^* \circ \mathrm{I}\widehat{\sigma}^{(r)}_* \to \mathrm{I}\widehat{\sigma}'^{(r)}_* \circ \mathrm{I}\breve{\Phi}^*, \tag{4.9.3.9}$$

where $\mathrm{I}\widehat{\sigma}^{(r)}_*$ and $\mathrm{I}\widehat{\sigma}'^{(r)}_*$ are the functors (4.5.1.7). Since the functor $\mathfrak{g}^*$ commutes with direct limits, the functor $\mathrm{I}\mathfrak{g}^*$ commutes with the functor $\kappa_{\mathscr{O}_{\mathfrak{X}}}$ (4.5.1.8). We deduce a morphism of functors from $\mathbf{Ind\text{-}MC}(\breve{\mathscr{C}}^{(r)}/\overline{\breve{\mathscr{B}}})$ to $\mathbf{HM}(\mathscr{O}_{\mathfrak{X}'}, \widetilde{\xi}^{-1}\widetilde{\Omega}^1_{\mathfrak{X}'/\mathscr{S}})$

$$\mathfrak{g}^* \circ \vec{\sigma}^{(r)}_* \to \vec{\sigma}'^{(r)}_* \circ \mathrm{I}\breve{\Phi}^*, \tag{4.9.3.10}$$

where $\vec{\sigma}^{(r)}_*$ and $\vec{\sigma}'^{(r)}_*$ are the functors (4.5.1.9).

By ([5] XVII 2.1.3), the morphism (4.9.3.9) is the adjoint of the composed morphism

$$\mathrm{I}\widehat{\sigma}'^{(r)*} \circ \mathrm{I}\mathfrak{g}^* \circ \mathrm{I}\widehat{\sigma}^{(r)}_* \xrightarrow{\sim} \mathrm{I}\breve{\Phi}^* \circ \mathrm{I}\widehat{\sigma}^{(r)*} \circ \mathrm{I}\widehat{\sigma}^{(r)}_* \to \mathrm{I}\breve{\Phi}^*, \tag{4.9.3.11}$$

where the first arrow is the isomorphism underlying the diagram (4.9.3.7) and the second arrow is the adjunction morphism. Consequently, for every object $\mathscr{N}$ of $\mathbf{Ind\text{-}HM}(\mathscr{O}_{\mathfrak{X}}, \widetilde{\xi}^{-1}\widetilde{\Omega}^1_{\mathfrak{X}/\mathscr{S}})$ and every object $\mathscr{F}$ of $\mathbf{Ind\text{-}MC}(\breve{\mathscr{C}}^{(r)}/\overline{\breve{\mathscr{B}}})$, the diagram of maps of sets

$$\begin{array}{ccc}
\mathrm{Hom}_{\mathbf{Ind\text{-}MC}(\breve{\mathscr{C}}^{(r)}/\breve{\overline{\mathscr{B}}})}(\mathrm{I}\widehat{\sigma}^{(r)*}(\mathscr{N}),\mathscr{F}) & \xrightarrow{\ a\ } & \mathrm{Hom}_{\mathbf{Ind\text{-}MC}(\breve{\mathscr{C}}'^{(r)}/\breve{\overline{\mathscr{B}}}')}(\mathrm{I}\widehat{\sigma}'^{(r)*}(\mathrm{Ig}^*(\mathscr{N})),\mathrm{I}\check{\Phi}^*(\mathscr{F})) \\
\downarrow & & \downarrow \\
\mathrm{Hom}_{\mathbf{Ind\text{-}HM}(\mathscr{O}_{\mathfrak{X}},\widetilde{\xi}^{-1}\widetilde{\Omega}^1_{\mathfrak{X}/\mathscr{S}})}(\mathscr{N},\mathrm{I}\widehat{\sigma}^{(r)}_*(\mathscr{F})) & \xrightarrow{\ b\ } & \mathrm{Hom}_{\mathbf{Ind\text{-}HM}(\mathscr{O}_{\mathfrak{X}'},\widetilde{\xi}^{-1}\widetilde{\Omega}^1_{\mathfrak{X}'/\mathscr{S}})}(\mathrm{Ig}^*(\mathscr{N}),\mathrm{I}\widehat{\sigma}'^{(r)}_*(\mathrm{I}\check{\Phi}^*(\mathscr{F}))),
\end{array}\tag{4.9.3.12}$$

where the vertical arrows are the adjunction isomorphisms, a is induced by the functor $\mathrm{I}\check{\Phi}^*$ and the isomorphism underlying the diagram (4.9.3.7), and b is induced by the functor Ig^* and the morphism (4.9.3.9), is commutative.

For all rational numbers $r \geq r' \geq 0$, the diagram of functors

$$\begin{array}{ccc}
\mathbf{Ind\text{-}MC}(\breve{\mathscr{C}}^{(r)}/\breve{\overline{\mathscr{B}}}) & \xrightarrow{\mathrm{I}\varepsilon^{r,r'}} & \mathbf{Ind\text{-}MC}(\breve{\mathscr{C}}^{(r')}/\breve{\overline{\mathscr{B}}}) \\
{\scriptstyle \mathrm{I}\check{\Phi}^*}\downarrow & & \downarrow{\scriptstyle \mathrm{I}\check{\Phi}^*} \\
\mathbf{Ind\text{-}MC}(\breve{\mathscr{C}}'^{(r)}/\breve{\overline{\mathscr{B}}}') & \xrightarrow{\mathrm{I}\varepsilon'^{r,r'}} & \mathbf{Ind\text{-}MC}(\breve{\mathscr{C}}'^{(r')}/\breve{\overline{\mathscr{B}}}'),
\end{array}\tag{4.9.3.13}$$

where the horizontal arrows are the functors (4.5.3.1), is commutative up to canonical isomorphism. It immediately follows from (4.9.2.9) that the diagram of morphisms of functors

$$\begin{array}{ccc}
\mathrm{Ig}^* \circ \mathrm{I}\widehat{\sigma}^{(r)}_* & \longrightarrow & \mathrm{Ig}^* \circ \mathrm{I}\widehat{\sigma}^{(r')}_* \circ \mathrm{I}\varepsilon^{r,r'} \\
\downarrow & & \downarrow \\
\mathrm{I}\widehat{\sigma}'^{(r)}_* \circ \mathrm{I}\check{\Phi}^* \longrightarrow \mathrm{I}\widehat{\sigma}'^{(r')}_* \circ \mathrm{I}\varepsilon'^{r,r'} \circ \mathrm{I}\check{\Phi}^* & = & \mathrm{I}\widehat{\sigma}'^{(r')}_* \circ \mathrm{I}\check{\Phi}^* \circ \mathrm{I}\varepsilon^{r,r'},
\end{array}\tag{4.9.3.14}$$

where the horizontal arrows are induced by the morphism (4.5.3.6), the vertical arrows are induced by the morphism (4.9.3.9) and the identification denoted by a symbol = comes from the diagram (4.9.3.13), is commutative. Consequently, the composed morphism

$$\mathrm{Ig}^* \circ \mathrm{I}\widehat{\sigma}^{(r)}_* \circ \mathrm{I}\mathfrak{S}^{(r)} \to \mathrm{I}\widehat{\sigma}'^{(r)}_* \circ \mathrm{I}\check{\Phi}^* \circ \mathrm{I}\mathfrak{S}^{(r)} \xrightarrow{\sim} \mathrm{I}\widehat{\sigma}'^{(r)}_* \circ \mathrm{I}\mathfrak{S}'^{(r)} \circ \mathrm{I}\check{\Phi}^*, \tag{4.9.3.15}$$

where the first arrow is induced by (4.9.3.9) and the second arrow is the isomorphism underlying the diagram (4.9.3.6), induces by taking the direct limit, for $r \in \mathbb{Q}_{>0}$, a morphism of functors from $\mathbf{Ind\text{-}Mod}(\breve{\overline{\mathscr{B}}})$ to $\mathbf{Ind\text{-}HM}(\mathscr{O}_{\mathfrak{X}'},\widetilde{\xi}^{-1}\widetilde{\Omega}^1_{\mathfrak{X}'/\mathscr{S}})$

$$\mathrm{Ig}^* \circ \mathrm{I}\mathscr{H} \to \mathrm{I}\mathscr{H}' \circ \mathrm{I}\check{\Phi}^*, \tag{4.9.3.16}$$

where $\mathscr{H}$ and $\mathscr{H}'$ are the functors (4.5.7.2). Since the functor $\mathfrak{g}^*$ commutes with direct limits, the functor Ig^* commutes with the functor $\kappa_{\mathscr{O}_{\mathfrak{X}}}$ (4.5.1.8). We deduce a morphism of functors from $\mathbf{Ind\text{-}Mod}(\breve{\overline{\mathscr{B}}})$ to $\mathbf{HM}(\mathscr{O}_{\mathfrak{X}'},\widetilde{\xi}^{-1}\widetilde{\Omega}^1_{\mathfrak{X}'/\mathscr{S}})$

$$\mathfrak{g}^* \circ \mathscr{H} \to \mathscr{H}' \circ \mathrm{I}\check{\Phi}^*, \tag{4.9.3.17}$$

where $\mathscr{H}$ and $\mathscr{H}'$ are the functors (4.5.7.3).

Proposition 4.9.4 *Let g be an open immersion. Then,*

(i) *For every rational number $r \geq 0$, the morphism* (4.9.3.9) *is an isomorphism. It makes commutative the diagram of functors*

$$\begin{array}{ccc} \mathbf{Ind\text{-}MC}(\breve{\mathscr{C}}^{(r)}/\overline{\breve{\mathscr{B}}}) & \xrightarrow{\mathrm{I}\widehat{\sigma}_*^{(r)}} & \mathbf{Ind\text{-}HM}(\mathscr{O}_{\mathfrak{X}}, \xi^{-1}\widetilde{\Omega}^1_{\mathfrak{X}/\mathscr{S}}) \\ \downarrow{\scriptstyle \mathrm{I}\breve{\Phi}^*} & & \downarrow{\scriptstyle \mathrm{I}\mathfrak{g}^*} \\ \mathbf{Ind\text{-}MC}(\breve{\mathscr{C}}'^{(r)}/\overline{\breve{\mathscr{B}}}') & \xrightarrow{\mathrm{I}\widehat{\sigma}_*'^{(r)}} & \mathbf{Ind\text{-}HM}(\mathscr{O}_{\mathfrak{X}'}, \xi^{-1}\widetilde{\Omega}^1_{\mathfrak{X}'/\mathscr{S}}). \end{array} \tag{4.9.4.1}$$

(ii) *The morphism* (4.9.3.17) *is an isomorphism. It makes commutative the diagram of functors*

$$\begin{array}{ccc} \mathbf{Ind\text{-}Mod}(\overline{\breve{\mathscr{B}}}) & \xrightarrow{\mathscr{H}} & \mathbf{HM}(\mathscr{O}_{\mathfrak{X}}, \widetilde{\xi}^{-1}\widetilde{\Omega}^1_{\mathfrak{X}/\mathscr{S}}) \\ \downarrow{\scriptstyle \mathrm{I}\breve{\Phi}^*} & & \downarrow{\scriptstyle \mathfrak{g}^*} \\ \mathbf{Ind\text{-}Mod}(\overline{\breve{\mathscr{B}}}') & \xrightarrow{\mathscr{H}'} & \mathbf{HM}(\mathscr{O}_{\mathfrak{X}'}, \widetilde{\xi}^{-1}\widetilde{\Omega}^1_{\mathfrak{X}'/\mathscr{S}}). \end{array} \tag{4.9.4.2}$$

(i) This follows from ([3] III.9.15) and 2.7.13.
(ii) This follows from (i) and the definitions.

Proposition 4.9.5 *Let $\mathscr{M}$ be a Dolbeault ind-$\overline{\breve{\mathscr{B}}}$-module, $\mathscr{N}$ a solvable Higgs $\mathscr{O}_{\mathfrak{X}}[\frac{1}{p}]$-bundle with coefficients in $\widetilde{\xi}^{-1}\widetilde{\Omega}^1_{\mathfrak{X}/\mathscr{S}}$. Then, $\mathrm{I}\breve{\Phi}^*(\mathscr{M})$ is a Dolbeault ind-$\overline{\breve{\mathscr{B}}}'$-module and $\mathfrak{g}^*(\mathscr{N})$ is a solvable Higgs $\mathscr{O}_{\mathfrak{X}'}[\frac{1}{p}]$-bundle with coefficients in $\widetilde{\xi}^{-1}\widetilde{\Omega}^1_{\mathfrak{X}'/\mathscr{S}}$. If, moreover, $\mathscr{M}$ and $\mathscr{N}$ are associated, $\mathrm{I}\breve{\Phi}^*(\mathscr{M})$ and $\mathfrak{g}^*(\mathscr{N})$ are associated.*

Suppose there exists a rational number $r > 0$ and an isomorphism of $\mathbf{Ind\text{-}MC}(\breve{\mathscr{C}}^{(r)}/\overline{\breve{\mathscr{B}}})$

$$\alpha\colon \mathrm{I}\widehat{\sigma}^{(r)*}(\mathscr{N}) \xrightarrow{\sim} \mathrm{I}\mathfrak{S}^{(r)}(\mathscr{M}). \tag{4.9.5.1}$$

By (4.9.3.6), (4.9.3.7) and (2.9.5.1), $\mathrm{I}\breve{\Phi}^*(\alpha)$ induces an isomorphism of $\mathbf{Ind\text{-}MC}(\breve{\mathscr{C}}'^{(r)}/\overline{\breve{\mathscr{B}}}')$

$$\alpha'\colon \mathrm{I}\widehat{\sigma}'^{(r)*}(\mathfrak{g}^*(\mathscr{N})) \xrightarrow{\sim} \mathrm{I}\mathfrak{S}'^{(r)}(\mathrm{I}\breve{\Phi}^*(\mathscr{M})); \tag{4.9.5.2}$$

the statement follows.

4.9.6 By 4.9.5, $\mathrm{I}\breve{\Phi}^*$ induces a functor

$$\mathrm{I}\breve{\Phi}^*\colon \mathbf{Ind\text{-}Mod}^{\mathrm{Dolb}}(\overline{\breve{\mathscr{B}}}) \to \mathbf{Ind\text{-}Mod}^{\mathrm{Dolb}}(\overline{\breve{\mathscr{B}}}'), \tag{4.9.6.1}$$

and $\mathfrak{g}^*$ induces a functor

$$\mathfrak{g}^*\colon \mathbf{HM}^{\mathrm{sol}}(\mathscr{O}_{\mathfrak{X}}[\frac{1}{p}], \widetilde{\xi}^{-1}\widetilde{\Omega}^1_{\mathfrak{X}/\mathscr{S}}) \to \mathbf{HM}^{\mathrm{sol}}(\mathscr{O}_{\mathfrak{X}'}[\frac{1}{p}], \widetilde{\xi}^{-1}\widetilde{\Omega}^1_{\mathfrak{X}'/\mathscr{S}}). \tag{4.9.6.2}$$

Proposition 4.9.7

(i) *The diagram of functors*

$$\begin{array}{ccc} \mathbf{Ind\text{-}Mod}^{\mathrm{Dolb}}(\overline{\check{\mathscr{B}}}) & \xrightarrow{\ \mathscr{H}\ } & \mathbf{HM}^{\mathrm{sol}}(\mathscr{O}_{\mathfrak{X}}[\frac{1}{p}], \widetilde{\xi}^{-1}\widetilde{\Omega}^1_{\mathfrak{X}/\mathscr{S}}) \\ \Big\downarrow{\scriptstyle \mathrm{I}\check{\Phi}^*} & & \Big\downarrow{\scriptstyle \mathfrak{g}^*} \\ \mathbf{Ind\text{-}Mod}^{\mathrm{Dolb}}(\overline{\check{\mathscr{B}}}') & \xrightarrow{\ \mathscr{H}'\ } & \mathbf{HM}^{\mathrm{sol}}(\mathscr{O}_{\mathfrak{X}'}[\frac{1}{p}], \widetilde{\xi}^{-1}\widetilde{\Omega}^1_{\mathfrak{X}'/\mathscr{S}}), \end{array} \tag{4.9.7.1}$$

where $\mathscr{H}$ and $\mathscr{H}'$ are the functors (4.5.12), *is commutative up to canonical isomorphism.*

(ii) *The diagram of functors*

$$\begin{array}{ccc} \mathbf{HM}^{\mathrm{sol}}(\mathscr{O}_{\mathfrak{X}}[\frac{1}{p}], \widetilde{\xi}^{-1}\widetilde{\Omega}^1_{\mathfrak{X}/\mathscr{S}}) & \xrightarrow{\ \mathscr{V}\ } & \mathbf{Ind\text{-}Mod}^{\mathrm{Dolb}}(\overline{\check{\mathscr{B}}}) \\ \Big\downarrow{\scriptstyle \mathfrak{g}^*} & & \Big\downarrow{\scriptstyle \mathrm{I}\check{\Phi}^*} \\ \mathbf{HM}^{\mathrm{sol}}(\mathscr{O}_{\mathfrak{X}'}[\frac{1}{p}], \widetilde{\xi}^{-1}\widetilde{\Omega}^1_{\mathfrak{X}'/\mathscr{S}}) & \xrightarrow{\ \mathscr{V}'\ } & \mathbf{Ind\text{-}Mod}^{\mathrm{Dolb}}(\overline{\check{\mathscr{B}}}'), \end{array} \tag{4.9.7.2}$$

where $\mathscr{V}$ and $\mathscr{V}'$ are the functors (4.5.17.1), *is commutative up to canonical isomorphism.*

(i) By 4.5.13, there exist a rational number $r_{\mathscr{M}} > 0$ and an isomorphism of $\mathbf{Ind\text{-}MC}(\check{\mathscr{C}}^{(r_{\mathscr{M}})}/\overline{\check{\mathscr{B}}})$

$$\alpha_{\mathscr{M}} : \mathrm{I}\widehat{\sigma}^{(r_{\mathscr{M}})*}(\mathscr{H}(\mathscr{M})) \xrightarrow{\sim} \mathrm{I}\mathfrak{S}^{(r_{\mathscr{M}})}(\mathscr{M}) \tag{4.9.7.3}$$

satisfying properties (i) and (ii) of *loc. cit.* For any rational number r such that $0 < r \leq r_{\mathscr{M}}$, we denote by

$$\alpha^{(r)}_{\mathscr{M}} : \mathrm{I}\widehat{\sigma}^{(r)*}(\mathscr{H}(\mathscr{M})) \xrightarrow{\sim} \mathrm{I}\mathfrak{S}^{(r)}(\mathscr{M}) \tag{4.9.7.4}$$

the isomorphism of $\mathbf{Ind\text{-}MC}(\check{\mathscr{C}}^{(r)}/\overline{\check{\mathscr{B}}})$ induced by $\mathrm{I}\varepsilon^{r_{\mathscr{M}},r}(\alpha_{\mathscr{M}})$ (4.5.3.1) and the isomorphisms (4.5.3.2) and (4.5.3.3). In view of (4.9.3.6), (4.9.3.7) and (2.9.5.1), $\mathrm{I}\check{\Phi}^*(\alpha_{\mathscr{M}})$ induces an isomorphism of $\mathbf{Ind\text{-}MC}(\check{\mathscr{C}}'^{(r_{\mathscr{M}})}/\overline{\check{\mathscr{B}}}')$

$$\alpha'_{\mathscr{M}} : \mathrm{I}\widehat{\sigma}'^{(r_{\mathscr{M}})*}(\mathfrak{g}^*(\mathscr{H}(\mathscr{M}))) \xrightarrow{\sim} \mathrm{I}\mathfrak{S}'^{(r_{\mathscr{M}})}(\mathrm{I}\check{\Phi}^*(\mathscr{M})). \tag{4.9.7.5}$$

Similarly, $\check{\Phi}^*(\alpha^{(r)}_{\mathscr{M}})$ induces an isomorphism of $\mathbf{Ind\text{-}MC}(\check{\mathscr{C}}'^{(r)}/\overline{\check{\mathscr{B}}}')$

$$\alpha'^{(r)}_{\mathscr{M}} : \mathrm{I}\widehat{\sigma}'^{(r)*}(\mathfrak{g}^*(\mathscr{H}(\mathscr{M}))) \xrightarrow{\sim} \mathrm{I}\mathfrak{S}'^{(r)}(\mathrm{I}\check{\Phi}^*(\mathscr{M})), \tag{4.9.7.6}$$

which can also be deduced from $\mathrm{I}\varepsilon'^{r_{\mathscr{M}},r}(\alpha'_{\mathscr{M}})$ by (4.9.3.13). We denote by

$$\beta'^{(r)}_{\mathscr{M}} : \mathfrak{g}^*(\mathscr{H}(\mathscr{M})) \to \vec{\sigma}'^{(r)}_*(\mathrm{I}\mathfrak{S}'^{(r)}(\mathrm{I}\check{\Phi}^*(\mathscr{M})) \tag{4.9.7.7}$$

its adjoint (4.5.2.4). By 4.9.5, $\mathrm{I}\breve{\Phi}^*(\mathscr{M})$ is a Dolbeault ind-$\breve{\overline{\mathscr{B}}}'$-module and $\mathfrak{g}^*(\mathscr{H}(\mathscr{M}))$ is a solvable Higgs $\mathscr{O}_{\mathfrak{X}'}[\frac{1}{p}]$-bundle with coefficients in $\widetilde{\xi}^{-1}\widetilde{\Omega}^1_{\mathfrak{X}'/\mathscr{S}}$, associated with $\mathrm{I}\breve{\Phi}^*(\mathscr{M})$. Consequently, by virtue of 4.5.11(i), the composed morphism

$$\mathfrak{g}^*(\mathscr{H}(\mathscr{M})) \xrightarrow{\beta'^{(r)}_{\mathscr{M}}} \vec{\sigma}'^{(r)}_*(\mathrm{I}\mathfrak{S}'^{(r)}(\mathrm{I}\breve{\Phi}^*(\mathscr{M})) \longrightarrow \mathscr{H}'(\mathrm{I}\breve{\Phi}^*(\mathscr{M})), \tag{4.9.7.8}$$

where the second arrow is the canonical morphism (4.5.7.4), is an isomorphism which depends a priori on $\alpha_{\mathscr{M}}$ but not on r. By the proof of 4.5.20, for every morphism $u\colon \mathscr{M} \to \mathscr{M}'$ of $\mathbf{Ind\text{-}Mod}^{\mathrm{Dolb}}(\breve{\overline{\mathscr{B}}})$ and every rational number r such that $0 < r < \inf(r_{\mathscr{M}}, r_{\mathscr{M}'})$, the diagram of $\mathbf{Ind\text{-}MC}(\breve{\mathscr{C}}^{(r)}/\breve{\overline{\mathscr{B}}})$

$$\begin{array}{ccc} \mathrm{I}\widehat{\sigma}^{(r)*}(\mathscr{H}(\mathscr{M})) & \xrightarrow{\alpha^{(r)}_{\mathscr{M}}} & \mathrm{I}\mathfrak{S}^{(r)}(\mathscr{M}) \\ \Big\downarrow{\scriptstyle \mathrm{I}\widehat{\sigma}^{(r)*}(\mathscr{H}(u))} & & \Big\downarrow{\scriptstyle \mathrm{I}\mathfrak{S}^{(r)}(u)} \\ \mathrm{I}\widehat{\sigma}^{(r)*}(\mathscr{H}(\mathscr{M}')) & \xrightarrow{\alpha^{(r)}_{\mathscr{M}'}} & \mathrm{I}\mathfrak{S}^{(r)}(\mathscr{M}') \end{array} \tag{4.9.7.9}$$

is commutative. We deduce from this that the composed isomorphism (4.9.7.8)

$$\mathfrak{g}^*(\mathscr{H}(\mathscr{M})) \xrightarrow{\sim} \mathscr{H}'(\mathrm{I}\breve{\Phi}^*(\mathscr{M})) \tag{4.9.7.10}$$

depends only on $\mathscr{M}$ (but not on the choice of $\alpha_{\mathscr{M}}$), on which it depends functorially; the statement follows.

(ii) The proof is similar to that of (i) and is left to the reader.

Remarks 4.9.8 Let $\mathscr{M}$ be a Dolbeault ind-$\breve{\overline{\mathscr{B}}}$-module.

(i) The canonical morphism (4.9.3.17)

$$\mathfrak{g}^*(\mathscr{H}(\mathscr{M})) \to \mathscr{H}'(\breve{\Phi}^*(\mathscr{M})) \tag{4.9.8.1}$$

is an isomorphism; it is the isomorphism underlying the commutative diagram (4.9.7.1). Indeed, take again the notation of the proof of 4.9.7(i) and denote, moreover, by

$$\beta^{(r)}_{\mathscr{M}}\colon \mathscr{H}(\mathscr{M}) \to \vec{\sigma}^{(r)}_*(\mathfrak{S}^r(\mathscr{M})) \tag{4.9.8.2}$$

the adjoint morphism of $\alpha^{(r)}_{\mathscr{M}}$. It follows from (4.9.3.12) that the morphism $\beta'^{(r)}_{\mathscr{M}}$ (4.9.7.7) is equal to the composition

$$\begin{aligned} \mathfrak{g}^*(\mathscr{H}(\mathscr{M})) &\xrightarrow{\mathfrak{g}^*(\beta^{(r)}_{\mathscr{M}})} \mathfrak{g}^*(\vec{\sigma}^{(r)}_*(\mathrm{I}\mathfrak{S}^{(r)}(\mathscr{M}))) \longrightarrow \vec{\sigma}'^{(r)}_*(\mathrm{I}\breve{\Phi}^*(\mathrm{I}\mathfrak{S}^{(r)}(\mathscr{M}))) \\ &\xrightarrow{\sim} \vec{\sigma}'^{(r)}_*(\mathfrak{S}'^{(r)}(\mathrm{I}\breve{\Phi}^*(\mathscr{M}))), \end{aligned} \tag{4.9.8.3}$$

where the second arrow is the morphism (4.9.3.10) and the last arrow is the isomorphism underlying the commutative diagram (4.9.3.6). Moreover, the direct limit of the morphisms $\beta^r_{\mathscr{M}}$, for $r \in \mathbb{Q}_{>0}$, is the identity, and the direct limit of the morphisms $\beta'^{(r)}_{\mathscr{M}}$, for $r \in \mathbb{Q}_{>0}$, is equal to the composed isomorphism (4.9.7.8), underlying the commutative diagram (4.9.7.1).

(ii) Let r be a rational number > 0,

$$\alpha\colon \mathrm{I}\widehat{\sigma}^{(r)*}(\mathscr{H}(\mathscr{M})) \xrightarrow{\sim} \mathrm{I}\mathfrak{S}^{(r)}(\mathscr{M}) \tag{4.9.8.4}$$

an isomorphism of **Ind-MC**$(\breve{\mathscr{C}}^{(r)}/\breve{\overline{\mathscr{B}}})$ satisfying the properties of 4.5.13. In view of (i), (4.9.3.6) and (4.9.3.7), we can identify $\mathrm{I}\check{\Phi}^*(\alpha)$ with an isomorphism

$$\alpha'\colon \mathrm{I}\widehat{\sigma}'^{(r)*}(\mathscr{H}'(\check{\Phi}^*(\mathscr{M}))) \xrightarrow{\sim} \mathrm{I}\mathfrak{S}'^{(r)}(\mathrm{I}\check{\Phi}^*(\mathscr{M})). \tag{4.9.8.5}$$

Moreover, $\mathrm{I}\check{\Phi}^*(\mathscr{M})$ is a Dolbeault ind-$\breve{\overline{\mathscr{B}}}'$-module by 4.9.5. It immediately follows from 4.5.11 that α' satisfies the properties of 4.5.13.

4.9.9 The functor $\check{\Phi}^*$ (4.9.2.2) induces a functor (2.9.8.1)

$$\check{\Phi}^*_{\mathbb{Q}}\colon \mathbf{Mod}_{\mathbb{Q}}(\breve{\overline{\mathscr{B}}}) \to \mathbf{Mod}_{\mathbb{Q}}(\breve{\overline{\mathscr{B}}}'). \tag{4.9.9.1}$$

We develop in ([3] III.14.7 and III.14.8) the compatibilities for $\breve{\overline{\mathscr{B}}}_{\mathbb{Q}}$-modules, analogous to those developed in 4.9.3 and 4.9.4 for ind-$\breve{\overline{\mathscr{B}}}$-modules. We deduce from this the following proposition.

Proposition 4.9.10 ([3] III.14.9) *Let $\mathscr{M}$ be a Dolbeault (resp. strongly Dolbeault) $\breve{\overline{\mathscr{B}}}_{\mathbb{Q}}$-module, $\mathscr{N}$ a rationally solvable (resp. strongly solvable) Higgs $\mathscr{O}_{\mathfrak{X}}[\frac{1}{p}]$-bundle with coefficients in $\widetilde{\xi}^{-1}\widetilde{\Omega}^1_{\mathfrak{X}/\mathscr{S}}$. Then, $\check{\Phi}^*_{\mathbb{Q}}(\mathscr{M})$ is a Dolbeault (resp. strongly Dolbeault) $\breve{\overline{\mathscr{B}}}'_{\mathbb{Q}}$-module and $\mathfrak{g}^*(\mathscr{N})$ is a rationally solvable (resp. strongly solvable) Higgs $\mathscr{O}_{\mathfrak{X}'}[\frac{1}{p}]$-bundle with coefficients in $\widetilde{\xi}^{-1}\widetilde{\Omega}^1_{\mathfrak{X}'/\mathscr{S}}$. If, moreover, $\mathscr{M}$ and $\mathscr{N}$ are associated, $\check{\Phi}^*_{\mathbb{Q}}(\mathscr{M})$ and $\mathfrak{g}^*(\mathscr{N})$ are associated.*

Proposition 4.9.11 *Every strongly solvable Higgs $\mathscr{O}_{\mathfrak{X}}[\frac{1}{p}]$-bundle $(\mathscr{N}, \theta)$ with coefficients in $\widetilde{\xi}^{-1}\widetilde{\Omega}^1_{\mathfrak{X}/\mathscr{S}}$ (4.6.6) is locally small (4.2.11).*

This follows from 4.6.26 and 4.9.10.

4.9.12 By 4.9.10, $\check{\Phi}^*$ induces a functor

$$\check{\Phi}^*\colon \mathbf{Mod}^{\mathrm{Dolb}}_{\mathbb{Q}}(\breve{\overline{\mathscr{B}}}) \to \mathbf{Mod}^{\mathrm{Dolb}}_{\mathbb{Q}}(\breve{\overline{\mathscr{B}}}'), \tag{4.9.12.1}$$

and $\mathfrak{g}^*$ induces a functor

$$\mathfrak{g}^*\colon \mathbf{HM}^{\mathrm{qsol}}(\mathscr{O}_{\mathfrak{X}}[\frac{1}{p}], \widetilde{\xi}^{-1}\widetilde{\Omega}^1_{\mathfrak{X}/\mathscr{S}}) \to \mathbf{HM}^{\mathrm{qsol}}(\mathscr{O}_{\mathfrak{X}'}[\frac{1}{p}], \widetilde{\xi}^{-1}\widetilde{\Omega}^1_{\mathfrak{X}'/\mathscr{S}}). \tag{4.9.12.2}$$

Proposition 4.9.13 ([3] III.14.11)

(i) *The diagram of functors*

$$\begin{array}{ccc} \mathbf{Mod}^{\mathrm{Dolb}}_{\mathbb{Q}}(\breve{\overline{\mathscr{B}}}) & \xrightarrow{\mathscr{H}_{\mathbb{Q}}} & \mathbf{HM}^{\mathrm{qsol}}(\mathscr{O}_{\mathfrak{X}}[\frac{1}{p}], \xi^{-1}\widetilde{\Omega}^1_{\mathfrak{X}/\mathscr{S}}) \\ \downarrow{\scriptstyle \check{\Phi}^*} & & \downarrow{\scriptstyle \mathfrak{g}^*} \\ \mathbf{Mod}^{\mathrm{Dolb}}_{\mathbb{Q}}(\breve{\overline{\mathscr{B}}}') & \xrightarrow{\mathscr{H}'_{\mathbb{Q}}} & \mathbf{HM}^{\mathrm{qsol}}(\mathscr{O}_{\mathfrak{X}'}[\frac{1}{p}], \xi^{-1}\widetilde{\Omega}^1_{\mathfrak{X}'/\mathscr{S}}), \end{array} \qquad (4.9.13.1)$$

where $\mathscr{H}_{\mathbb{Q}}$ and $\mathscr{H}'_{\mathbb{Q}}$ are the functors (4.6.13.1), *is commutative up to canonical isomorphism.*

(ii) *The diagram of functors*

$$\begin{array}{ccc} \mathbf{HM}^{\mathrm{qsol}}(\mathscr{O}_{\mathfrak{X}}[\frac{1}{p}], \xi^{-1}\widetilde{\Omega}^1_{\mathfrak{X}/\mathscr{S}}) & \xrightarrow{\mathscr{V}_{\mathbb{Q}}} & \mathbf{Mod}^{\mathrm{Dolb}}_{\mathbb{Q}}(\breve{\overline{\mathscr{B}}}) \\ \downarrow{\scriptstyle \mathfrak{g}^*} & & \downarrow{\scriptstyle \check{\Phi}^*} \\ \mathbf{HM}^{\mathrm{qsol}}(\mathscr{O}_{\mathfrak{X}'}[\frac{1}{p}], \xi^{-1}\widetilde{\Omega}^1_{\mathfrak{X}'/\mathscr{S}}) & \xrightarrow{\mathscr{V}'_{\mathbb{Q}}} & \mathbf{Mod}^{\mathrm{Dolb}}_{\mathbb{Q}}(\breve{\overline{\mathscr{B}}}'), \end{array} \qquad (4.9.13.2)$$

where $\mathscr{V}_{\mathbb{Q}}$ and $\mathscr{V}'_{\mathbb{Q}}$ are the functors (4.6.18.1), *is commutative up to canonical isomorphism.*

The proof is identical to that of 4.9.7.

4.10 Stacky Properties of Dolbeault Modules

4.10.1 We denote by ψ the composed morphism

$$\psi\colon \widetilde{E}_s^{\mathbb{N}^\circ} \xrightarrow{\lambda} \widetilde{E}_s \xrightarrow{\sigma_s} X_{s,\mathrm{\acute{e}t}} \xrightarrow{a_{\mathrm{\acute{e}t}}} X_{\mathrm{\acute{e}t}}, \qquad (4.10.1.1)$$

where λ is the canonical morphism of topos defined in (2.1.9.1), σ_s the canonical morphism of topos (4.3.3.3) and $a\colon X_s \to X$ the canonical injection (4.1.4.1). For any object U of $\mathbf{\acute{E}t}_{/X}$, we denote by $f_U\colon (U, \mathscr{M}_X|U) \to (S, \mathscr{M}_S)$ the morphism induced by f, and by $\widetilde{U} \to \widetilde{X}$ the unique étale morphism lifting $\check{\overline{U}} \to \check{\overline{X}}$, so that $(\widetilde{U}, \mathscr{M}_{\widetilde{X}}|\widetilde{U})$ is an $(\widetilde{S}, \mathscr{M}_{\widetilde{S}})$-smooth deformation of $(\check{\overline{U}}, \mathscr{M}_{\check{\overline{X}}}|\check{\overline{U}})$. The localization of the ringed topos $(\widetilde{E}_s^{\mathbb{N}^\circ}, \breve{\overline{\mathscr{B}}})$ at $\psi^*(U)$ is canonically equivalent to the analogous ringed topos associated with f_U by virtue of ([3] III.9.14). For every rational number $r \geq 0$, $\check{\mathscr{C}}^{(r)}|\psi^*(U)$ is identified with the Higgs–Tate $(\breve{\overline{\mathscr{B}}}|\psi^*(U))$-algebra of thickness r associated with the deformation $(\widetilde{U}, \mathscr{M}_{\widetilde{X}}|\widetilde{U})$ by (4.9.2.7). We denote by $\underline{\mathbf{Ind\text{-}Mod}}_U(\breve{\overline{\mathscr{B}}})$ the category of ind-$(\breve{\overline{\mathscr{B}}}|\psi^*(U))$-modules, by $\underline{\mathbf{Ind\text{-}Mod}}_U^{\mathrm{Dolb}}(\breve{\overline{\mathscr{B}}})$ the subcategory of Dolbeault ind-$(\breve{\overline{\mathscr{B}}}|\psi^*(U))$-modules relative to the deformation $(\widetilde{U}, \mathscr{M}_{\widetilde{X}}|\widetilde{U})$, and by

$\underline{\mathbf{Ind\text{-}MC}}_U(\breve{\mathscr{C}}^{(r)}/\breve{\overline{\mathscr{B}}})$ the category of ind-$(\breve{\mathscr{C}}^{(r)}|\psi^*(U))$-modules with integrable p^r-connection with respect to extension $(\breve{\mathscr{C}}^{(r)}|\psi^*(U))/(\breve{\overline{\mathscr{B}}}|\psi^*(U))$ (4.5.1).

For any morphism $g\colon U' \to U$ of $\mathbf{\acute{E}t}_{/X}$, we denote by

$$\begin{array}{ccc}\underline{\mathbf{Ind\text{-}Mod}}_U(\breve{\overline{\mathscr{B}}}) & \to & \underline{\mathbf{Ind\text{-}Mod}}_{U'}(\breve{\overline{\mathscr{B}}}) \\ \mathscr{M} & \mapsto & \mathscr{M}|\psi^*(U')\end{array} \tag{4.10.1.2}$$

the restriction functor (2.7.15). This induces, by 4.9.5, a functor

$$\begin{array}{ccc}\underline{\mathbf{Ind\text{-}Mod}}_U^{\mathrm{Dolb}}(\breve{\overline{\mathscr{B}}}) & \to & \underline{\mathbf{Ind\text{-}Mod}}_{U'}^{\mathrm{Dolb}}(\breve{\overline{\mathscr{B}}}) \\ \mathscr{M} & \mapsto & \mathscr{M}|\psi^*(U').\end{array} \tag{4.10.1.3}$$

For any rational number $r \geq 0$, we denote by

$$\begin{array}{ccc}\underline{\mathbf{Ind\text{-}MC}}_U(\breve{\mathscr{C}}^{(r)}/\breve{\overline{\mathscr{B}}}) & \to & \underline{\mathbf{Ind\text{-}MC}}_{U'}(\breve{\mathscr{C}}^{(r)}/\breve{\overline{\mathscr{B}}}) \\ \mathscr{F} & \mapsto & \mathscr{F}|\psi^*(U')\end{array} \tag{4.10.1.4}$$

the restriction functor.

We denote by $\mathbf{\acute{E}t}_{\mathrm{coh}/X}$ the full subcategory of $\mathbf{\acute{E}t}_{/X}$ made up of the étale schemes of finite presentation over X (2.1.12) and by

$$\underline{\mathbf{Ind\text{-}Mod}}^{\mathrm{Dolb}}(\breve{\overline{\mathscr{B}}}) \to \mathbf{\acute{E}t}_{\mathrm{coh}/X} \tag{4.10.1.5}$$

the cleaved and normalized fibered category whose fiber above an object U of $\mathbf{\acute{E}t}_{\mathrm{coh}/X}$ is the category $\underline{\mathbf{Ind\text{-}Mod}}_U^{\mathrm{Dolb}}(\breve{\overline{\mathscr{B}}})$ and the inverse image functor by a morphism $U' \to U$ of $\mathbf{\acute{E}t}_{\mathrm{coh}/X}$ is the restriction functor (4.10.1.3).

Lemma 4.10.2 *Let r be a rational number ≥ 0, $\mathscr{F}, \mathscr{F}'$ two objects of $\mathbf{Ind\text{-}MC}(\breve{\mathscr{C}}^{(r)}/\breve{\overline{\mathscr{B}}})$, $(U_i)_{i\in I}$ an étale covering of X. For any $(i,j) \in I^2$, we set $U_{ij} = U_i \times_X U_j$. Then, the diagram of maps of sets*

$$\mathrm{Hom}_{\underline{\mathbf{Ind\text{-}MC}}_X(\breve{\mathscr{C}}^{(r)}/\breve{\overline{\mathscr{B}}})}(\mathscr{F},\mathscr{F}') \to \prod_{i\in I} \mathrm{Hom}_{\underline{\mathbf{Ind\text{-}MC}}_{U_i}(\breve{\mathscr{C}}^{(r)}/\breve{\overline{\mathscr{B}}})}(\mathscr{F}|\psi^*(U_i),\mathscr{F}'|\psi^*(U_i))$$
$$\rightrightarrows \prod_{(i,j)\in I^2} \mathrm{Hom}_{\underline{\mathbf{Ind\text{-}MC}}_{U_{ij}}(\breve{\mathscr{C}}^{(r)}/\breve{\overline{\mathscr{B}}})}(\mathscr{F}|\psi^*(U_{ij}),\mathscr{F}'|\psi^*(U_{ij})) \tag{4.10.2.1}$$

is exact.

Indeed, since X is quasi-compact, we may assume that I is finite, in which case the assertion easily follows from 2.7.17.

Proposition 4.10.3 *Let $\mathscr{M}$ be an ind-$\breve{\overline{\mathscr{B}}}$-module, $(U_i)_{i\in I}$ a covering of $\mathbf{\acute{E}t}_{\mathrm{coh}/X}$. Then $\mathscr{M}$ is Dolbeault if and only if for all $i \in I$, the ind-$(\breve{\overline{\mathscr{B}}}|\psi^*(U_i))$-module $\mathscr{M}|\psi^*(U_i)$ is Dolbeault.*

Indeed, the condition is necessary by virtue of 4.9.5. Suppose that for all $i \in I$, $\mathscr{M}|\psi^*(U_i)$ is Dolbeault and let us show that $\mathscr{M}$ is Dolbeault. Since X is quasi-compact, we may assume that I is finite. For any $i \in I$, we denote by $\mathfrak{X}_i$ the formal scheme p-adic completion of $\overline{U}_i$. For any $(i,j) \in I^2$, we set $U_{ij} = U_i \times_X U_j$ and denote by $\mathfrak{X}_{ij}$ the formal scheme p-adic completion of $\overline{U}_{ij}$. By virtue of 4.9.7(i), we have a canonical isomorphism of Higgs $\mathscr{O}_{\mathfrak{X}_{ij}}[\frac{1}{p}]$-modules with coefficients in $\xi^{-1}\widetilde{\Omega}^1_{\mathfrak{X}_{ij}/\mathscr{S}}$

$$\mathscr{H}_i(\mathscr{M}|\psi^*(U_i)) \otimes_{\mathscr{O}_{\mathfrak{X}_i}} \mathscr{O}_{\mathfrak{X}_{ij}} \xrightarrow{\sim} \mathscr{H}_{ij}(\mathscr{M}|\psi^*(U_{ij})), \tag{4.10.3.1}$$

where $\mathscr{H}_i$ and $\mathscr{H}_{ij}$ are the functors (4.5.12.1) associated respectively with $(f_{U_i}, (\widetilde{U}_i, \mathscr{M}_{\widetilde{X}}|\widetilde{U}_i))$ and $(f_{U_{ij}}, (\widetilde{U}_{ij}, \mathscr{M}_{\widetilde{X}}|\widetilde{U}_{ij}))$. We deduce from this a descent datum δ on the Higgs modules $(\mathscr{H}_i(\mathscr{M}|\psi^*(U_i)))_{i\in I}$ with respect to the étale covering $(\mathfrak{X}_i \to \mathfrak{X})_{i\in I}$. Since the latter is effective by ([3] III.6.22), there exist a Higgs $\mathscr{O}_{\mathfrak{X}}[\frac{1}{p}]$-bundle $\mathscr{N}$ with coefficients in $\xi^{-1}\widetilde{\Omega}^1_{\mathfrak{X}/\mathscr{S}}$ and for every $i \in I$, an isomorphism of Higgs $\mathscr{O}_{\mathfrak{X}_i}[\frac{1}{p}]$-modules

$$\mathscr{N} \otimes_{\mathscr{O}_{\mathfrak{X}}} \mathscr{O}_{\mathfrak{X}_i} \xrightarrow{\sim} \mathscr{H}_i(\mathscr{M}|\psi^*(U_i)), \tag{4.10.3.2}$$

that induce the descent datum δ.

For any $(i,j) \in I^2$ and any rational number $r > 0$, we denote by $\mathrm{I}\widehat{\sigma}_i^{(r)*}$ and $\mathrm{I}\mathfrak{S}_i^{(r)}$ (resp. $\mathrm{I}\widehat{\sigma}_{ij}^{(r)*}$ and $\mathrm{I}\mathfrak{S}_{ij}^{(r)}$) the functors (4.5.1.4) and (4.5.1.2) associated with $(f_{U_i}, (\widetilde{U}_i, \mathscr{M}_{\widetilde{X}}|\widetilde{U}_i))$ (resp. $(f_{U_{ij}}, (\widetilde{U}_{ij}, \mathscr{M}_{\widetilde{X}}|\widetilde{U}_{ij}))$). By 4.5.13, for every $i \in I$, there exist a rational number $r_i > 0$ and an isomorphism of $\underline{\mathbf{Ind}\text{-}\mathbf{MC}}_{U_i}(\breve{\mathscr{C}}^{(r)}/\breve{\overline{\mathscr{B}}})$

$$\alpha_i \colon \mathrm{I}\widehat{\sigma}_i^{(r_i)*}(\mathscr{H}_i(\mathscr{M}|\psi^*(U_i))) \xrightarrow{\sim} \mathrm{I}\mathfrak{S}_i^{(r_i)}(\mathscr{M}|\psi^*(U_i)) \tag{4.10.3.3}$$

satisfying properties (i) and (ii) of *loc. cit.* For every $(i,j) \in I^2$, $\mathscr{M}|\psi^*(U_{ij})$ is Dolbeault by virtue of 4.9.5. By (4.9.3.6), (4.9.3.7) and 4.9.7(i), $\alpha_i|\psi^*(U_{ij})$ identifies with an isomorphism

$$\alpha_i|\psi^*(U_{ij}) \colon \mathrm{I}\widehat{\sigma}_{ij}^{(r_i)*}(\mathscr{H}_{ij}(\mathscr{M}|\psi^*(U_{ij}))) \xrightarrow{\sim} \mathrm{I}\mathfrak{S}_{ij}^{(r_i)}(\mathscr{M}|\psi^*(U_{ij})). \tag{4.10.3.4}$$

This satisfies the properties of 4.5.13, in view of 4.9.8. For any rational number r such that $0 < r \leq r_i$, we denote by $\mathrm{I}\varepsilon_i^{r_i,r} \colon \underline{\mathbf{Ind}\text{-}\mathbf{MC}}_{U_i}(\breve{\mathscr{C}}^{(r_i)}/\breve{\overline{\mathscr{B}}}) \to \underline{\mathbf{Ind}\text{-}\mathbf{MC}}_{U_i}(\breve{\mathscr{C}}^{(r)}/\breve{\overline{\mathscr{B}}})$ the functor (4.5.3.1) associated with $(f_{U_i}, \widetilde{U}_i, \mathscr{M}_{\widetilde{X}}|\widetilde{U}_i)$ and by

$$\alpha_i^{(r)} \colon \mathrm{I}\widehat{\sigma}_i^{(r)*}(\mathscr{H}_i(\mathscr{M}|\psi^*(U_i))) \xrightarrow{\sim} \mathrm{I}\mathfrak{S}_i^{(r)}(\mathscr{M}|\psi^*(U_i)) \tag{4.10.3.5}$$

the isomorphism of $\underline{\mathbf{Ind}\text{-}\mathbf{MC}}_{U_i}(\breve{\mathscr{C}}^{(r)}/\breve{\overline{\mathscr{B}}})$ induced by $\mathrm{I}\varepsilon_i^{r_i,r}(\alpha_i)$ and the isomorphisms (4.5.3.2) and (4.5.3.3). By (4.9.3.6) and (4.9.3.7), we can identify $\alpha_i^{(r)}$ with an isomorphism

$$\alpha_i^{(r)} \colon \mathrm{I}\widehat{\sigma}^{(r)*}(\mathscr{N})|\psi^*(U_i) \xrightarrow{\sim} \mathrm{I}\mathfrak{S}^{(r)}(\mathscr{M})|\psi^*(U_i). \tag{4.10.3.6}$$

It follows from the proof of 4.5.20 that for every rational number $0 < r < \inf(r_i, r_j)$, we have in $\underline{\mathbf{Ind\text{-}MC}}_{U_{ij}}(\breve{\mathscr{C}}^{(r)}/\overline{\breve{\mathscr{B}}})$

$$\alpha_i^{(r)}|\psi^*(U_{ij}) = \alpha_j^{(r)}|\psi^*(U_{ij}). \tag{4.10.3.7}$$

By virtue of 4.10.2, for every rational number $0 < r < \inf(r_i, i \in I)$, the isomorphisms $(\alpha_i^{(r)})_{i\in I}$ glue to an isomorphism of $\mathbf{Ind\text{-}MC}(\breve{\mathscr{C}}^{(r)}/\overline{\breve{\mathscr{B}}})$

$$\alpha^{(r)} \colon \mathrm{I}\widehat{\sigma}^{(r)*}(\mathscr{N}) \xrightarrow{\sim} \mathrm{I}\mathfrak{S}^{(r)}(\mathscr{M}). \tag{4.10.3.8}$$

Consequently, $\mathscr{M}$ is Dolbeault.

Corollary 4.10.4 *The property for an ind-$\overline{\breve{\mathscr{B}}}$-module to be Dolbeault does not depend on the choice of the deformation $(\widetilde{X}, \mathscr{M}_{\widetilde{X}})$* (4.1.5) *provided that we remain in the same setting, absolute or relative* (3.1.14).

Indeed, the question is local by 4.10.3, and if X is affine, all smooth $(\widetilde{S}, \mathscr{M}_{\widetilde{S}})$-deformations of $(\check{\overline{X}}, \mathscr{M}_{\check{\overline{X}}})$ are isomorphic by virtue of ([41] 3.14).

Proposition 4.10.5 *The following conditions are equivalent:*

(i) *The fibered category* (4.10.1.5)

$$\underline{\mathbf{Ind\text{-}Mod}}^{\mathrm{Dolb}}(\overline{\breve{\mathscr{B}}}) \to \mathbf{\acute{E}t}_{\mathrm{coh}/X} \tag{4.10.5.1}$$

is a stack ([23] II 1.2.1).

(ii) *For every covering $(U_i \to U)_{i\in I}$ of $\mathbf{\acute{E}t}_{\mathrm{coh}/X}$, denoting by $\mathscr{U}$ (resp. for any $i \in I$, by $\mathscr{U}_i$) the formal scheme p-adic completion of $\overline{U}$ (resp. $\overline{U}_i$), a Higgs $\mathscr{O}_{\mathscr{U}}[\frac{1}{p}]$-bundle $\mathscr{N}$ with coefficients in $\widetilde{\xi}^{-1}\widetilde{\Omega}^1_{\mathscr{U}/\mathscr{S}}$ is solvable if and only if, for every $i \in I$, the Higgs $\mathscr{O}_{\mathscr{U}_i}[\frac{1}{p}]$-bundle $\mathscr{N} \otimes_{\mathscr{O}_{\mathscr{U}}} \mathscr{O}_{\mathscr{U}_i}$ with coefficients in $\widetilde{\xi}^{-1}\widetilde{\Omega}^1_{\mathscr{U}_i/\mathscr{S}}$ is solvable.*

Let $(U_i \to U)_{i\in I}$ be a covering of $\mathbf{\acute{E}t}_{\mathrm{coh}/X}$. For any $(i, j) \in I^2$, we set $U_{ij} = U_i \times_X U_j$. We denote by $\mathscr{U}$ the formal scheme p-adic completion of $\overline{U}$ and by $\mathscr{H}^\star$ and $\mathscr{V}^\star$ the functors (4.5.12.1) and (4.5.17.1) associated with $(f_U, \widetilde{\overline{U}}, \mathscr{M}_{\widetilde{X}}|\widetilde{\overline{U}})$. For any $i \in I$, we denote by $\mathscr{U}_i$ the formal scheme p-adic completion of $\overline{U}_i$ and by $\mathscr{H}_i$ and $\mathscr{V}_i$ the functors (4.5.12.1) and (4.5.17.1) associated with $(f_{U_i}, \widetilde{\overline{U}}_i, \mathscr{M}_{\widetilde{X}}|\widetilde{\overline{U}}_i)$.

Let us first show (i)⇒(ii). Let $\mathscr{N}$ be a Higgs $\mathscr{O}_{\mathscr{U}}[\frac{1}{p}]$-bundle with coefficients in $\widetilde{\xi}^{-1}\widetilde{\Omega}^1_{\mathscr{U}/\mathscr{S}}$. If $\mathscr{N}$ is solvable, for every $i \in I$, $\mathscr{N} \otimes_{\mathscr{O}_{\mathscr{U}}} \mathscr{O}_{\mathscr{U}_i}$ is solvable by 4.9.5. Conversely, suppose that for every $i \in I$, $\mathscr{N} \otimes_{\mathscr{O}_{\mathscr{U}}} \mathscr{O}_{\mathscr{U}_i}$ is solvable and let us show that $\mathscr{N}$ is solvable. For every $i \in I$, $\mathscr{M}_i = \mathscr{V}_i(\mathscr{N} \otimes_{\mathscr{O}_{\mathscr{U}}} \mathscr{O}_{\mathscr{U}_i})$ is a Dolbeault ind-$\overline{\breve{\mathscr{B}}}|\psi^*(U_i)$-module. By 4.9.7(ii), the canonical descent datum on the Higgs bundles $(\mathscr{N} \otimes_{\mathscr{O}_{\mathscr{U}}} \mathscr{O}_{\mathscr{U}_i})_{i\in I}$ with respect to the étale covering $(\mathscr{U}_i \to \mathscr{U})_{i\in I}$ induces a descent datum δ on the Dolbeault modules $(\mathscr{M}_i)_{i\in I}$ with respect to the covering $(U_i \to U)_{i\in I}$. The latter being effective by (i), there exist a Dolbeault ind-$\overline{\breve{\mathscr{B}}}|\psi^*(U)$-module $\mathscr{M}$ and for every $i \in I$, an isomorphism of ind-$\overline{\breve{\mathscr{B}}}|\psi^*(U_i)$-modules

$$\mathscr{M}|\psi^*(U_i) \xrightarrow{\sim} \mathscr{M}_i \tag{4.10.5.2}$$

that induce the descent datum δ. By virtue of 4.5.20 and 4.9.7(i), we have a canonical isomorphism of Higgs $\mathscr{O}_{\mathscr{U}}[\frac{1}{p}]$-bundles $\mathscr{H}^\star(\mathscr{M}) \xrightarrow{\sim} \mathscr{N}$. Consequently, $\mathscr{N}$ is solvable.

Next, let us show (ii)$\Rightarrow$(i). For all ind-$\breve{\overline{\mathscr{B}}}|\psi^*(U)$-modules $\mathscr{M}$ and $\mathscr{M}'$, the map of sets

$$\begin{aligned} &\mathrm{Hom}_{\underline{\mathbf{Ind\text{-}Mod}}_U(\breve{\overline{\mathscr{B}}})}(\mathscr{M},\mathscr{M}') \to \prod_{i\in I} \mathrm{Hom}_{\underline{\mathbf{Ind\text{-}Mod}}_{U_i}(\breve{\overline{\mathscr{B}}})}(\mathscr{M}|\psi^*(U_i),\mathscr{M}'|\psi^*(U_i)) \\ &\rightrightarrows \prod_{(i,j)\in I^2} \mathrm{Hom}_{\underline{\mathbf{Ind\text{-}Mod}}_{U_{ij}}(\breve{\overline{\mathscr{B}}}))}(\mathscr{M}|\psi^*(U_{ij}),\mathscr{M}'|\psi^*(U_{ij})) \end{aligned} \tag{4.10.5.3}$$

is exact. Indeed, since U is quasi-compact, we may assume that I is finite, in which case the assertion follows from 2.7.17.

For every $i \in I$, let $\mathscr{M}_i$ be a Dolbeault ind-$\breve{\overline{\mathscr{B}}}_{\mathbb{Q}}|\psi^*(U_i)$-module and let δ be a descent datum on $(\mathscr{M}_i)_{i\in I}$ with respect to the covering $(U_i \to U)_{i\in I}$. Let us show that δ is effective. By assumption, for every $i \in I$, $\mathscr{N}_i = \mathscr{H}_i(\mathscr{M}_i)$ is a solvable Higgs $\mathscr{O}_{\mathscr{U}_i}[\frac{1}{p}]$-bundle with coefficients in $\widetilde{\xi}^{-1}\widetilde{\Omega}^1_{\mathscr{U}_i/\mathscr{S}}$. In view of 4.9.7(i), δ induces a descent datum γ on the Higgs bundles $(\mathscr{N}_i)_{i\in I}$ with respect to the étale covering $(\mathscr{U}_i \to \mathscr{U})_{\in I}$. This being effective by ([3] III.6.22), there exist a Higgs $\mathscr{O}_{\mathscr{U}}[\frac{1}{p}]$-bundle $\mathscr{N}$ and for every $i \in I$, an isomorphism of Higgs $\mathscr{O}_{\mathscr{U}_i}[\frac{1}{p}]$-modules

$$\mathscr{N} \otimes_{\mathscr{O}_{\mathscr{U}}} \mathscr{O}_{\mathscr{U}_i} \xrightarrow{\sim} \mathscr{N}_i, \tag{4.10.5.4}$$

that induce the descent datum γ. By (ii), $\mathscr{N}$ is solvable. Consequently, $\mathscr{M} = \mathscr{V}^\star(\mathscr{N})$ is a Dolbeault ind-$\breve{\overline{\mathscr{B}}}_{\mathbb{Q}}|\psi^*(U)$-module. By 4.5.20 and 4.9.7(ii), for every $i \in I$, we have a canonical isomorphism of ind-$\breve{\overline{\mathscr{B}}}_{\mathbb{Q}}|\psi^*(U_i)$-modules $\mathscr{M}|\psi^*(U_i) \xrightarrow{\sim} \mathscr{M}_i$, that induces the descent datum δ, proving the assertion.

4.10.6 For any $U \in \mathrm{Ob}(\mathbf{\acute{E}t}_{/X})$, we denote by $\underline{\mathbf{Mod}}_U(\breve{\overline{\mathscr{B}}})$ the category of $(\breve{\overline{\mathscr{B}}}|\psi^*(U))$-modules, by $\underline{\mathbf{Mod}}_{\mathbb{Q},U}(\breve{\overline{\mathscr{B}}})$ the category of $(\breve{\overline{\mathscr{B}}}|\psi^*(U))_{\mathbb{Q}}$-modules and by $\underline{\mathbf{Mod}}^{\mathrm{Dolb}}_{\mathbb{Q},U}(\breve{\overline{\mathscr{B}}})$ (resp. $\underline{\mathbf{Mod}}^{\mathrm{sDolb}}_{\mathbb{Q},U}(\breve{\overline{\mathscr{B}}})$) the subcategory of Dolbeault (resp. strongly Dolbeault) $(\breve{\overline{\mathscr{B}}}|\psi^*(U))_{\mathbb{Q}}$-modules relative to the deformation $(\widetilde{U}, \mathscr{M}_{\widetilde{X}}|\widetilde{U})$ (4.10.1). For every morphism $g\colon U' \to U$ of $\mathbf{\acute{E}t}_{/X}$, the restriction functor

$$\begin{aligned} \underline{\mathbf{Mod}}_U(\breve{\overline{\mathscr{B}}}) &\to \underline{\mathbf{Mod}}_{U'}(\breve{\overline{\mathscr{B}}}) \\ \mathscr{M} &\mapsto \mathscr{M}|\psi^*(U') \end{aligned} \tag{4.10.6.1}$$

induces a functor

$$\begin{aligned} \underline{\mathbf{Mod}}_{\mathbb{Q},U}(\breve{\overline{\mathscr{B}}}) &\to \underline{\mathbf{Mod}}_{\mathbb{Q},U'}(\breve{\overline{\mathscr{B}}}) \\ \mathscr{M} &\mapsto \mathscr{M}|\psi^*(U'). \end{aligned} \tag{4.10.6.2}$$

By 4.9.10, the latter induces two functors

$$\begin{array}{ccc} \underline{\mathbf{Mod}}^{\mathrm{Dolb}}_{\mathbb{Q},U}(\overline{\breve{\mathscr{B}}}) & \to & \underline{\mathbf{Mod}}^{\mathrm{Dolb}}_{\mathbb{Q},U'}(\overline{\breve{\mathscr{B}}}) \\ \mathscr{M} & \mapsto & \mathscr{M}|\psi^*(U'), \end{array} \tag{4.10.6.3}$$

$$\begin{array}{ccc} \underline{\mathbf{Mod}}^{\mathrm{sDolb}}_{\mathbb{Q},U}(\overline{\breve{\mathscr{B}}}) & \to & \underline{\mathbf{Mod}}^{\mathrm{sDolb}}_{\mathbb{Q},U'}(\overline{\breve{\mathscr{B}}}) \\ \mathscr{M} & \mapsto & \mathscr{M}|\psi^*(U'). \end{array} \tag{4.10.6.4}$$

We denote by $\mathbf{MOD}(\overline{\breve{\mathscr{B}}})$ the fibered $(\widetilde{E}_s^{\mathbb{N}^\circ})$-category of $\overline{\breve{\mathscr{B}}}$-modules over $\widetilde{E}_s^{\mathbb{N}^\circ}$ ([23] II 3.4.1), which is in fact split ([24] VI § 9). It is a stack over $\widetilde{E}_s^{\mathbb{N}^\circ}$ by ([23] II3.4.4). We denote by

$$\underline{\mathbf{Mod}}(\overline{\breve{\mathscr{B}}}) \to \mathbf{\acute{E}t}_{\mathrm{coh}/X} \tag{4.10.6.5}$$

the base change of $\mathbf{MOD}(\overline{\breve{\mathscr{B}}})$ ([24] VI § 3) by $\psi^* \circ \varepsilon$, where ψ is the morphism (4.10.1.1) and $\varepsilon\colon \mathbf{\acute{E}t}_{\mathrm{coh}/X} \to X_{\mathrm{\acute{e}t}}$ is the canonical functor. It is also a stack by ([23] II 3.1.1). We deduce from this a fibered category

$$\underline{\mathbf{Mod}}_{\mathbb{Q}}(\overline{\breve{\mathscr{B}}}) \to \mathbf{\acute{E}t}_{\mathrm{coh}/X}, \tag{4.10.6.6}$$

whose fiber over an object U of $\mathbf{\acute{E}t}_{\mathrm{coh}/X}$ is the category $\underline{\mathbf{Mod}}_{\mathbb{Q},U}(\overline{\breve{\mathscr{B}}})$ and the inverse image functor by a morphism $U' \to U$ of $\mathbf{\acute{E}t}_{\mathrm{coh}/X}$ is the restriction functor (4.10.6.2). It is not a priori a stack. It induces two fibered categories

$$\underline{\mathbf{Mod}}^{\mathrm{Dolb}}_{\mathbb{Q}}(\overline{\breve{\mathscr{B}}}) \to \mathbf{\acute{E}t}_{\mathrm{coh}/X}, \tag{4.10.6.7}$$

$$\underline{\mathbf{Mod}}^{\mathrm{sDolb}}_{\mathbb{Q}}(\overline{\breve{\mathscr{B}}}) \to \mathbf{\acute{E}t}_{\mathrm{coh}/X}, \tag{4.10.6.8}$$

whose fibers above an object U of $\mathbf{\acute{E}t}_{\mathrm{coh}/X}$ are the categories $\underline{\mathbf{Mod}}^{\mathrm{Dolb}}_{\mathbb{Q},U}(\overline{\breve{\mathscr{B}}})$ and $\underline{\mathbf{Mod}}^{\mathrm{sDolb}}_{\mathbb{Q},U}(\overline{\breve{\mathscr{B}}})$, respectively, and the inverse image functors by a morphism $U' \to U$ of $\mathbf{\acute{E}t}_{\mathrm{coh}/X}$ are the restriction functors (4.10.6.3) and (4.10.6.4).

Proposition 4.10.7 *Let $\mathscr{M}$ be a $\overline{\breve{\mathscr{B}}}_{\mathbb{Q}}$-module (resp. an adic $\overline{\breve{\mathscr{B}}}_{\mathbb{Q}}$-module of finite type), $(U_i)_{i\in I}$ a covering of $\mathbf{\acute{E}t}_{\mathrm{coh}/X}$. Then $\mathscr{M}$ is Dolbeault (resp. strongly Dolbeault) if and only if for every $i \in I$, the $(\overline{\breve{\mathscr{B}}}|\psi^*(U_i))_{\mathbb{Q}}$-module $\mathscr{M}|\psi^*(U_i)$ is Dolbeault (resp. strongly Dolbeault)* (4.10.6.2).

It suffices to copy the proof of ([3] III.15.4).

Corollary 4.10.8 *The property for a $\overline{\breve{\mathscr{B}}}_{\mathbb{Q}}$-module to be Dolbeault (resp. strongly Dolbeault) does not depend on the choice of the deformation $(\widetilde{X}, \mathscr{M}_{\widetilde{X}})$* (4.1.5) *provided that we keep the same setting, absolute or relative* (3.1.14).

Indeed, the fact that a $\overline{\breve{\mathscr{B}}}_{\mathbb{Q}}$-module is Dolbeault (resp. strongly Dolbeault) is a local question by 4.10.7, and if X is affine, all smooth $(\widetilde{S}, \mathscr{M}_{\widetilde{S}})$-deformations of $(\check{\overline{X}}, \mathscr{M}_{\check{\overline{X}}})$ are isomorphic by virtue of ([41] 3.14).

Proposition 4.10.9 *The following conditions are equivalent:*

(i) *The fibered category* (4.10.6.7)

$$\underline{\mathbf{Mod}}^{\mathrm{Dolb}}_{\mathbb{Q}}(\check{\overline{\mathscr{B}}}) \to \mathbf{\acute{E}t}_{\mathrm{coh}/X} \tag{4.10.9.1}$$

is a stack ([23] II 1.2.1).

(ii) *For every covering $(U_i \to U)_{i\in I}$ of $\mathbf{\acute{E}t}_{\mathrm{coh}/X}$, denoting by $\mathscr{U}$ (resp. for any $i \in I$, by $\mathscr{U}_i$) the formal scheme p-adic completion of $\overline{U}$ (resp. $\overline{U}_i$), a Higgs $\mathscr{O}_{\mathscr{U}}[\frac{1}{p}]$-bundle $\mathscr{N}$ with coefficients in $\widetilde{\xi}^{-1}\widetilde{\Omega}^1_{\mathscr{U}/\mathscr{S}}$ is rationally solvable* (4.6.6) *if and only if for every $i \in I$, the Higgs $\mathscr{O}_{\mathscr{U}_i}[\frac{1}{p}]$-bundle $\mathscr{N} \otimes_{\mathscr{O}_{\mathscr{U}}} \mathscr{O}_{\mathscr{U}_i}$ with coefficients in $\widetilde{\xi}^{-1}\widetilde{\Omega}^1_{\mathscr{U}_i/\mathscr{S}}$ is rationally solvable.*

It suffices to copy the proof of ([3] III.15.5).

Proposition 4.10.10 *The following conditions are equivalent:*

(i) *The fibered category* (4.10.6.8)

$$\underline{\mathbf{Mod}}^{\mathrm{sDolb}}_{\mathbb{Q}}(\check{\overline{\mathscr{B}}}) \to \mathbf{\acute{E}t}_{\mathrm{coh}/X} \tag{4.10.10.1}$$

is a stack.

(ii) *For every covering $(U_i \to U)_{i\in I}$ of $\mathbf{\acute{E}t}_{\mathrm{coh}/X}$, denoting by $\mathscr{U}$ (resp. for any $i \in I$, by $\mathscr{U}_i$) the formal scheme p-adic completion of $\overline{U}$ (resp. $\overline{U}_i$), a Higgs $\mathscr{O}_{\mathscr{U}}[\frac{1}{p}]$-bundle $\mathscr{N}$ with coefficients in $\widetilde{\xi}^{-1}\widetilde{\Omega}^1_{\mathscr{U}/\mathscr{S}}$ is strongly solvable* (4.6.6) *if and only if for every $i \in I$, the Higgs $\mathscr{O}_{\mathscr{U}_i}[\frac{1}{p}]$-bundle $\mathscr{N} \otimes_{\mathscr{O}_{\mathscr{U}}} \mathscr{O}_{\mathscr{U}_i}$ with coefficients in $\widetilde{\xi}^{-1}\widetilde{\Omega}^1_{\mathscr{U}_i/\mathscr{S}}$ is strongly solvable.*

It suffices to copy the proof of ([3] III.15.5), which corresponds to the absolute case (3.1.14).

Proposition 4.10.11 *Under the conditions of* 4.10.10, *a Higgs $\mathscr{O}_{\mathfrak{X}}[\frac{1}{p}]$-bundle with coefficients in $\widetilde{\xi}^{-1}\widetilde{\Omega}^1_{\mathfrak{X}/\mathscr{S}}$ is locally small* (4.2.11) *if and only if it is strongly solvable* (4.6.6).

This follows from 4.8.15.

Remark 4.10.12 Under the conditions of 4.10.5 (resp. 4.10.9, resp. 4.10.10), the property for a Higgs $\mathscr{O}_{\mathfrak{X}}[\frac{1}{p}]$-bundle with coefficients in $\widetilde{\xi}^{-1}\widetilde{\Omega}^1_{\mathfrak{X}/\mathscr{S}}$ to be solvable (resp. rationally solvable, resp. strongly solvable) does not depend on the choice of the deformation $(\widetilde{X}, \mathscr{M}_{\widetilde{X}})$ (4.1.5) provided that we keep the same setting, absolute or relative (3.1.14). Indeed, the question is local by assumption, and if X is affine, all smooth $(\widetilde{S}, \mathscr{M}_{\widetilde{S}})$-deformations of $(\check{\overline{X}}, \mathscr{M}_{\check{\overline{X}}})$ are isomorphic.

4.11 Hodge–Tate Modules

Definition 4.11.1 We call a *Hodge–Tate $\breve{\overline{\mathscr{B}}}_{\mathbb{Q}}$-module* any Dolbeault $\breve{\overline{\mathscr{B}}}_{\mathbb{Q}}$-module $\mathscr{M}$ (4.6.6) whose associated Higgs $\mathscr{O}_{\mathfrak{X}}[\frac{1}{p}]$-bundle $\mathscr{H}_{\mathbb{Q}}(\mathscr{M})$ (4.6.13.1) is nilpotent (4.2.5).

We denote by $\mathbf{Mod}^{\mathrm{HT}}_{\mathbb{Q}}(\breve{\overline{\mathscr{B}}})$ the full sub-category of $\mathbf{Mod}_{\mathbb{Q}}(\breve{\overline{\mathscr{B}}})$ made up of the Hodge–Tate $\breve{\overline{\mathscr{B}}}_{\mathbb{Q}}$-modules, and by $\mathbf{HM}^{\mathrm{qsolnilp}}(\mathscr{O}_{\mathfrak{X}}[\frac{1}{p}], \widetilde{\xi}^{-1}\widetilde{\Omega}^1_{\mathfrak{X}/\mathscr{S}})$ the full subcategory of $\mathbf{HM}(\mathscr{O}_{\mathfrak{X}}[\frac{1}{p}], \widetilde{\xi}^{-1}\widetilde{\Omega}^1_{\mathfrak{X}/\mathscr{S}})$ made up of the rationally solvable and nilpotent Higgs $\mathscr{O}_{\mathfrak{X}}[\frac{1}{p}]$-bundles with coefficients in $\widetilde{\xi}^{-1}\widetilde{\Omega}^1_{\mathfrak{X}/\mathscr{S}}$ (4.6.6).

Proposition 4.11.2 *The functors* $\mathscr{H}_{\mathbb{Q}}$ (4.6.13.1) *and* $\mathscr{V}_{\mathbb{Q}}$ (4.6.18.1) *induce equivalences of categories quasi-inverse to each other.*

$$\mathbf{Mod}^{\mathrm{HT}}_{\mathbb{Q}}(\breve{\overline{\mathscr{B}}}) \underset{\mathscr{V}_{\mathbb{Q}}}{\overset{\mathscr{H}_{\mathbb{Q}}}{\rightleftarrows}} \mathbf{HM}^{\mathrm{qsolnilp}}(\mathscr{O}_{\mathfrak{X}}[\tfrac{1}{p}], \widetilde{\xi}^{-1}\widetilde{\Omega}^1_{\mathfrak{X}/\mathscr{S}}). \tag{4.11.2.1}$$

This immediately follows from 4.6.22.

Proposition 4.11.3 *Suppose that X is an object of* $\mathbf{P}$ (4.4.1). *Then, every nilpotent Higgs $\mathscr{O}_{\mathfrak{X}}[\frac{1}{p}]$-bundle with coefficients in $\widetilde{\xi}^{-1}\widetilde{\Omega}^1_{\mathfrak{X}/\mathscr{S}}$ is strongly solvable* (4.6.6).

Indeed, $(\mathscr{N}, \theta)$ is small by 4.2.13. It is therefore strongly solvable by virtue of 4.8.15.

Corollary 4.11.4 *Suppose that X is an object of* $\mathbf{P}$ (4.4.1). *Then, every Hodge–Tate $\breve{\overline{\mathscr{B}}}_{\mathbb{Q}}$-module is strongly Dolbeault* (4.6.6), *and is in particular adic of finite type* (4.3.14).

This follows from 4.11.3 and 4.6.22.

Corollary 4.11.5 *Under the conditions of* 4.10.10, *every nilpotent Higgs $\mathscr{O}_{\mathfrak{X}}[\frac{1}{p}]$-bundle with coefficients in $\widetilde{\xi}^{-1}\widetilde{\Omega}^1_{\mathfrak{X}/\mathscr{S}}$ is strongly solvable, and every Hodge–Tate $\breve{\overline{\mathscr{B}}}_{\mathbb{Q}}$-module is strongly Dolbeault* (4.6.6), *and is in particular adic of finite type.*

This follows from 4.11.3 and 4.6.22.

Theorem 4.11.6 *Suppose that X is an object of* $\mathbf{P}$ (4.4.1); *we take again the notation of* 4.8.30. *For any integer $1 \leq i \leq c$, let $\mathbf{Rep}^{\mathrm{HT}}_{\widehat{\overline{\mathscr{R}}}_i[\frac{1}{p}]}(\Delta_i)$ be the subcategory of $\mathbf{Rep}_{\widehat{\overline{\mathscr{R}}}_i[\frac{1}{p}]}(\Delta_i)$ made up of the Hodge–Tate $\widehat{\overline{\mathscr{R}}}_i[\frac{1}{p}]$-representations of Δ_i* (3.5.6). *Then, the functors $(\widehat{\beta}_i)_{1\leq i\leq c}$* (4.8.30.7) *induce an equivalence of categories*

$$\mathbf{Mod}^{\mathrm{HT}}_{\mathbb{Q}}(\breve{\overline{\mathscr{B}}}) \xrightarrow{\sim} \prod_{1\leq i\leq c} \mathbf{Rep}^{\mathrm{HT}}_{\widehat{\overline{\mathscr{R}}}_i[\frac{1}{p}]}(\Delta_i). \tag{4.11.6.1}$$

This follows from 4.8.31(i), 4.8.32(iii) and 4.2.6.

4.11.7 We take again the notation of 4.10.6. For any object U of $\mathbf{\acute{E}t}_{/X}$, we denote by

$$\underline{\mathbf{Mod}}^{\mathrm{HT}}_{U,\mathbb{Q}}(\breve{\overline{\mathscr{B}}}) \tag{4.11.7.1}$$

the category of Hodge–Tate $(\breve{\overline{\mathscr{B}}}|\psi^*(U))_{\mathbb{Q}}$-modules with respect to the deformation $(\widetilde{U}, \mathscr{M}_{\widetilde{X}}|\widetilde{U})$. By 4.9.10, for every morphism $g\colon U' \to U$ of $\mathbf{\acute{E}t}_{/X}$, the restriction functor (4.10.6.2) induces a functor

$$\begin{array}{rcl}\underline{\mathbf{Mod}}^{\mathrm{HT}}_{U,\mathbb{Q}}(\breve{\overline{\mathscr{B}}}) & \to & \underline{\mathbf{Mod}}^{\mathrm{HT}}_{U',\mathbb{Q}}(\breve{\overline{\mathscr{B}}}) \\ \mathscr{M} & \mapsto & \mathscr{M}|\psi^*(U_i).\end{array} \tag{4.11.7.2}$$

The fibered category (4.10.6.7) induces a fibered category

$$\underline{\mathbf{Mod}}^{\mathrm{HT}}_{\mathbb{Q}}(\breve{\overline{\mathscr{B}}}) \to \mathbf{\acute{E}t}_{\mathrm{coh}/X} \tag{4.11.7.3}$$

whose fiber above an object U of $\mathbf{\acute{E}t}_{\mathrm{coh}/X}$ is the category $\underline{\mathbf{Mod}}^{\mathrm{HT}}_{U,\mathbb{Q}}(\breve{\overline{\mathscr{B}}})$ and the inverse image functor by a morphism $U' \to U$ of $\mathbf{\acute{E}t}_{\mathrm{coh}/X}$ is the restriction functor (4.11.7.2).

Proposition 4.11.8 *Let $\mathscr{M}$ be a $\breve{\overline{\mathscr{B}}}_{\mathbb{Q}}$-module, $(U_i)_{i\in I}$ a covering of $\mathbf{\acute{E}t}_{\mathrm{coh}/X}$. Then, $\mathscr{M}$ is Hodge–Tate if and only if for every $i \in I$, the $(\breve{\overline{\mathscr{B}}}|\psi^*(U_i))_{\mathbb{Q}}$-module $\mathscr{M}|\psi^*(U_i)$ is Hodge–Tate.*

This follows from 4.10.7 and 4.9.13(i).

Corollary 4.11.9 *The property for a $\breve{\overline{\mathscr{B}}}_{\mathbb{Q}}$-module to be Hodge–Tate does not depend on the choice of the deformation $(\widetilde{X}, \mathscr{M}_{\widetilde{X}})$ (4.1.5) nor even on the absolute or relative setting* (3.1.14).

Indeed, the property for a $\breve{\overline{\mathscr{B}}}_{\mathbb{Q}}$-module to be Hodge–Tate being a local question by 4.11.8, we may assume that X is affine satisfying the assumptions of 4.8.1. All smooth $(\widetilde{S}, \mathscr{M}_{\widetilde{S}})$-deformations of $(\check{\overline{X}}, \mathscr{M}_{\check{\overline{X}}})$ are then isomorphic by virtue of ([41] 3.14). Therefore, the property for a $\breve{\overline{\mathscr{B}}}_{\mathbb{Q}}$-module to be Hodge–Tate does not depend on the choice of the $(\widetilde{S}, \mathscr{M}_{\widetilde{S}})$-deformation $(\widetilde{X}, \mathscr{M}_{\widetilde{X}})$ provided that we keep the same setting, absolute or relative.

To show that this property does not depend on the absolute or relative setting, we take again the assumptions and notation of 4.4.25 and 4.6.24. In view of 4.6.25, it suffices to show that if $\mathscr{M}$ is a Hodge–Tate $\overline{\mathscr{B}}_{\mathbb{Q}}$-module with respect to the deformation $(\widetilde{X}, \mathscr{M}_{\widetilde{X}})$, it is Hodge–Tate with respect to the deformation $(\widetilde{X}', \mathscr{M}_{\widetilde{X}'})$. We suppose therefore that the $\overline{\mathscr{B}}_{\mathbb{Q}}$-module $\mathscr{M}$ is Hodge–Tate with respect to the deformation $(\widetilde{X}, \mathscr{M}_{\widetilde{X}})$, and denote by θ the canonical Higgs field on $\mathscr{H}_{\mathbb{Q}}(\mathscr{M})$ with coefficients in $\xi^{-1}\widetilde{\Omega}^1_{\mathfrak{X}/\mathscr{S}}$ (4.6.24.1). By the proofs of 4.2.13 and 3.5.2(i), for every $\varepsilon \geq 0$, there exists a coherent sub-$\mathscr{O}_{\mathfrak{X}}$-module $\mathscr{N}_\varepsilon$ of $\mathscr{H}_{\mathbb{Q}}(\mathscr{M})$ that generates it over $\mathscr{O}_{\mathfrak{X}}[\frac{1}{p}]$ such that

$$\theta(\mathscr{N}_\varepsilon) \subset p^\varepsilon \xi^{-1}\widetilde{\Omega}^1_{\mathfrak{X}/\mathscr{S}} \otimes_{\mathscr{O}_{\mathfrak{X}}} \mathscr{N}_\varepsilon. \tag{4.11.9.1}$$

Taking $\varepsilon > \rho + \frac{1}{p-1}$ (4.4.25), we then have (3.1.10.1)

$$\theta(\mathscr{N}_\varepsilon) \subset p^{\varepsilon-\rho}(\xi_\pi^*)^{-1}\widetilde{\Omega}^1_{\mathfrak{X}/\mathscr{S}} \otimes_{\mathscr{O}_\mathfrak{X}} \mathscr{N}_\varepsilon. \tag{4.11.9.2}$$

Consequently, θ induces on $\mathscr{N}_\varepsilon$ a Higgs field with coefficients in $(\xi_\pi^*)^{-1}\widetilde{\Omega}^1_{\mathfrak{X}/\mathscr{S}}$ that we denote by θ' to avoid confusion. We take again the assumptions and notation of 4.8.12 and denote by $\breve{\mathscr{P}}$ the R_1-module of $(\overline{X}^\circ_{\text{fét}})^{\mathbb{N}^\circ}$ associated with $(\mathscr{N}_\varepsilon, \theta)$ by the functor

$$\mathbf{HM}^{\text{qsf}}(\mathscr{O}_\mathfrak{X}, \xi^{-1}\widetilde{\Omega}^1_{\mathfrak{X}/\mathscr{S}}) \to \mathbf{Mod}((\overline{X}^\circ_{\text{fét}})^{\mathbb{N}^\circ}, R_1) \tag{4.11.9.3}$$

defined in (4.8.12.3). It immediately follows from 3.4.5 that $\breve{\mathscr{P}}$ is also the R_1-module of $(\overline{X}^\circ_{\text{fét}})^{\mathbb{N}^\circ}$ associated with $(\mathscr{N}_\varepsilon, \theta')$ by the analogous functor

$$\mathbf{HM}^{\text{qsf}}(\mathscr{O}_\mathfrak{X}, (\xi_\pi^*)^{-1}\widetilde{\Omega}^1_{\mathfrak{X}/\mathscr{S}}) \to \mathbf{Mod}((\overline{X}^\circ_{\text{fét}})^{\mathbb{N}^\circ}, R_1). \tag{4.11.9.4}$$

By virtue of 4.6.22 and 4.8.14 applied to $(\mathscr{N}_\varepsilon, \theta)$, we have an isomorphism

$$\mathscr{M} \xrightarrow{\sim} \breve{\beta}^*(\breve{\mathscr{P}} \otimes_{R_1} \breve{\overline{\mathscr{B}}}_X)_\mathbb{Q}. \tag{4.11.9.5}$$

It then follows from 4.8.14 applied to $(\mathscr{N}_\varepsilon, \theta')$ that $\mathscr{M}$ is strongly Dolbeault with respect to the deformation $(\widetilde{X}', \mathscr{M}_{\widetilde{X}'})$ and we have an isomorphism

$$\mathscr{H}'_\mathbb{Q}(\mathscr{M}) \xrightarrow{\sim} (\mathscr{N}_\varepsilon \otimes_{\mathbb{Z}_p} \mathbb{Q}_p, \theta'). \tag{4.11.9.6}$$

Consequently, $\mathscr{M}$ is Hodge–Tate with respect to the deformation $(\widetilde{X}', \mathscr{M}_{\widetilde{X}'})$. The proposition follows.

Proposition 4.11.10 *The following conditions are equivalent:*

(i) *The fibered category* (4.11.7.3)

$$\underline{\mathbf{Mod}}^{\text{HT}}_\mathbb{Q}(\breve{\overline{\mathscr{B}}}) \to \mathbf{\acute{E}t}_{\text{coh}/X} \tag{4.11.10.1}$$

is a stack ([23] II 1.2.1).

(ii) *Every $\mathscr{O}_\mathfrak{X}[\frac{1}{p}]$-nilpotent Higgs bundle with coefficients in $\widetilde{\xi}^{-1}\widetilde{\Omega}^1_{\mathfrak{X}/\mathscr{S}}$ is rationally solvable.*

Moreover, these conditions are satisfied if the conditions of 4.10.9 *are.*

Indeed, adapting the proof of ([3] III.15.5), we show that condition (i) is equivalent to the following condition:

(ii') For every covering $(U_i \to U)_{i\in I}$ of $\mathbf{\acute{E}t}_{\text{coh}/X}$, denoting by $\mathscr{U}$ (resp. for any $i \in I$, by $\mathscr{U}_i$) the formal scheme p-adic completion of $\overline{U}$ (resp. $\overline{U}_i$), a *nilpotent* Higgs $\mathscr{O}_\mathscr{U}[\frac{1}{p}]$-bundle $\mathscr{N}$ with coefficients in $\widetilde{\xi}^{-1}\widetilde{\Omega}^1_{\mathscr{U}/\mathscr{S}}$ is rationally solvable if and only if for every $i \in I$, the Higgs $\mathscr{O}_{\mathscr{U}_i}[\frac{1}{p}]$-bundle $\mathscr{N} \otimes_{\mathscr{O}_\mathscr{U}} \mathscr{O}_{\mathscr{U}_i}$ with coefficients in $\widetilde{\xi}^{-1}\widetilde{\Omega}^1_{\mathscr{U}_i/\mathscr{S}}$ is rationally solvable.

We clearly have (ii)⇒(ii'), and we have (ii')⇒(ii) by virtue of 4.11.3. Finally, condition 4.10.9(ii) clearly implies (ii').

4.12 Dolbeault and Hodge–Tate Local Systems

4.12.1 For any $\mathbb{U}$-topos T, we denote by $\breve{\mathbb{Z}}_p$ the $\mathbb{Z}_p$-algebra $(\mathbb{Z}/p^n\mathbb{Z})_{n\geq 1}$ of $T^{\mathbb{N}^\circ}$ (2.1.9) and by

$$Q_{\breve{\mathbb{Z}}_p}: \mathbf{Mod}(\breve{\mathbb{Z}}_p, T^{\mathbb{N}^\circ}) \to \mathbf{Mod}_{\mathbb{Q}}(\breve{\mathbb{Z}}_p, T^{\mathbb{N}^\circ}), \quad M \mapsto M_{\mathbb{Q}}, \tag{4.12.1.1}$$

the canonical functor (2.9.7.1). Recall that the objects of $\mathbf{Mod}_{\mathbb{Q}}(\breve{\mathbb{Z}}_p, T^{\mathbb{N}^\circ})$ are also called $\breve{\mathbb{Z}}_{p,\mathbb{Q}}$*-modules*. We denote by

$$\alpha_{\breve{\mathbb{Z}}_p}: \mathbf{Mod}_{\mathbb{Q}}(\breve{\mathbb{Z}}_p, T^{\mathbb{N}^\circ}) \to \mathbf{Ind\text{-}Mod}(\breve{\mathbb{Z}}_p, T^{\mathbb{N}^\circ}) \tag{4.12.1.2}$$

the canonical functor (2.9.7.3) .

4.12.2 We denote by

$$\breve{\psi}: (\overline{X}^\circ_{\text{ét}})^{\mathbb{N}^\circ} \to \widetilde{E}^{\mathbb{N}^\circ} \tag{4.12.2.1}$$

the morphism of topos induced by ψ (4.3.2.15), that we will naturally consider as a morphism of topos ringed by the rings $\breve{\mathbb{Z}}_p$. It follows from ([3] III.7.5 and VI.10.9(iii)) that the canonical homomorphism $\breve{\mathbb{Z}}_p \to \breve{\psi}_*(\breve{\mathbb{Z}}_p)$ is an isomorphism. We will use for $\breve{\psi}$ the notation introduced in 2.7.10 and 2.9.8.

For every integer $n \geq 0$, every $\overline{\mathscr{B}}_n$-module $\mathscr{M}_n$ of $\widetilde{E}$ is naturally an object of $\widetilde{E}_s$ ([3] III.9.7). Moreover, the functor δ_* (4.3.3.2) being exact, we have a canonical isomorphism

$$\mathrm{R}\Gamma(\widetilde{E}, \mathscr{M}_n) \xrightarrow{\sim} \mathrm{R}\Gamma(\widetilde{E}_s, \mathscr{M}_n). \tag{4.12.2.2}$$

Consequently, taking into account ([3] III.7.5 and III.7.11), every $\breve{\overline{\mathscr{B}}}$-module $\mathscr{M}$ of $\widetilde{E}^{\mathbb{N}^\circ}$ is naturally an object of $\widetilde{E}_s^{\mathbb{N}^\circ}$, and we have a canonical isomorphism

$$\mathrm{R}\Gamma(\widetilde{E}^{\mathbb{N}^\circ}, \mathscr{M}) \xrightarrow{\sim} \mathrm{R}\Gamma(\widetilde{E}_s^{\mathbb{N}^\circ}, \mathscr{M}). \tag{4.12.2.3}$$

It follows that every ind-$\breve{\overline{\mathscr{B}}}$-module of $\widetilde{E}^{\mathbb{N}^\circ}$ is naturally an ind-$\breve{\overline{\mathscr{B}}}$-module of $\widetilde{E}_s^{\mathbb{N}^\circ}$ (2.6.4.4).

Definition 4.12.3

(i) We say that a $\breve{\mathbb{Z}}_p$-module $M = (M_n)_{n\in\mathbb{N}}$ of $(\overline{X}^\circ_{\text{ét}})^{\mathbb{N}^\circ}$ is a *local system* (or that M is a $\breve{\mathbb{Z}}_p$*-local system*) if the following two conditions are satisfied:

(a) M is p-adic, in other words, for all integers $n \geq m \geq 0$, the morphism $M_n/p^m M_n \to M_m$ deduced from the transition morphism $M_n \to M_m$ is an isomorphism;

(b) for every integer $n \geq 0$, the $\mathbb{Z}/p^n\mathbb{Z}$-module M_n of $\overline{X}^\circ_{\text{ét}}$ is locally constant constructible.

(ii) A $\breve{\mathbb{Z}}_{p,\mathbb{Q}}$-module of $(\overline{X}^\circ_{\text{ét}})^{\mathbb{N}^\circ}$ is said to be a *local system* if it is isomorphic to $M_{\mathbb{Q}}$ for a $\breve{\mathbb{Z}}_p$-local system M (4.12.1.1).

Definition 4.12.4

(i) We say that an ind-$\breve{\mathbb{Z}}_p$-module M of $(\overline{X}^{\circ}_{\text{ét}})^{\mathbb{N}^\circ}$ is *Dolbeault* if the ind-$\breve{\overline{\mathscr{B}}}$-module $\mathrm{I}\breve{\psi}_*(M) \otimes_{\breve{\mathbb{Z}}_p} \breve{\overline{\mathscr{B}}}$ is Dolbeault (4.5.5).

(ii) A $\breve{\mathbb{Z}}_{p,\mathbb{Q}}$-local system M of $(\overline{X}^{\circ}_{\text{ét}})^{\mathbb{N}^\circ}$ is said to be *Dolbeault* (resp. *Hodge–Tate*) if the $\breve{\overline{\mathscr{B}}}_{\mathbb{Q}}$-module $\breve{\psi}_{\mathbb{Q}*}(M) \otimes_{\breve{\mathbb{Z}}_{p,\mathbb{Q}}} \breve{\overline{\mathscr{B}}}_{\mathbb{Q}}$ is Dolbeault (4.6.6) (resp. Hodge–Tate (4.11.1)).

By (2.9.5.1), (2.9.7.4) and 4.6.7, a $\breve{\mathbb{Z}}_{p,\mathbb{Q}}$-local system M of $(\overline{X}^{\circ}_{\text{ét}})^{\mathbb{N}^\circ}$ is Dolbeault if and only if the ind-$\breve{\mathbb{Z}}_p$-module $\alpha_{\breve{\mathbb{Z}}_p}(M)$ is Dolbeault.

Remark 4.12.5 Every Dolbeault $\breve{\mathbb{Z}}_{p,\mathbb{Q}}$-local system M of $(\overline{X}^{\circ}_{\text{ét}})^{\mathbb{N}^\circ}$ is strongly Dolbeault (4.6.6). Indeed, the $\breve{\overline{\mathscr{B}}}_{\mathbb{Q}}$-module $\breve{\psi}_{\mathbb{Q}*}(M) \otimes_{\breve{\mathbb{Z}}_{p,\mathbb{Q}}} \breve{\overline{\mathscr{B}}}_{\mathbb{Q}}$ is adic of finite type (4.3.14) by ([3] III.7.5, VI.9.20 and VI.10.9(iii)).

Theorem 4.12.6 *Let $M = (M_n)_{n\geq 0}$ be a $\breve{\mathbb{Z}}_p$-local system of $(\overline{X}^{\circ}_{\text{ét}})^{\mathbb{N}^\circ}$, $\mathscr{M} = \breve{\psi}_*(M) \otimes_{\breve{\mathbb{Z}}_p} \breve{\overline{\mathscr{B}}}$, $\mathbb{K}^\bullet$ the Dolbeault complex of the Higgs $\mathscr{O}_{\mathfrak{X}}[\frac{1}{p}]$-module $\mathscr{H}_{\mathbb{Q}}(\mathscr{M}_{\mathbb{Q}})$ (4.6.9.2). Assume that the morphism $f\colon X \to S$ is proper and that the $\breve{\mathbb{Z}}_{p,\mathbb{Q}}$-local system $M_{\mathbb{Q}}$ is Dolbeault, i.e., that the $\breve{\overline{\mathscr{B}}}_{\mathbb{Q}}$-module $\mathscr{M}_{\mathbb{Q}}$ is Dolbeault. Then, there exists a canonical spectral sequence*

$$\mathrm{E}_2^{i,j} = \mathrm{H}^i(X_s, \mathrm{H}^j(\mathbb{K}^\bullet)) \Rightarrow \mathrm{H}^{i+j}((\overline{X}^{\circ}_{\text{ét}})^{\mathbb{N}^\circ}, M) \otimes_{\mathbb{Z}_p} C. \tag{4.12.6.1}$$

Indeed, by virtue of 4.7.5 and (4.12.2.3), the Cartan–Leray spectral sequence

$$\mathrm{E}_2^{i,j} = \mathrm{H}^i(X_s, \mathrm{R}^j\widehat{\sigma}_*(\mathscr{M})) \Rightarrow \mathrm{H}^{i+j}(\widetilde{E}_s^{\mathbb{N}^\circ}, \mathscr{M}) \tag{4.12.6.2}$$

induces a spectral sequence

$$\mathrm{E}_2^{i,j} = \mathrm{H}^i(X_s, \mathrm{H}^j(\mathbb{K}^\bullet)) \Rightarrow \mathrm{H}^{i+j}(\widetilde{E}^{\mathbb{N}^\circ}, \mathscr{M}) \otimes_{\mathbb{Z}} \mathbb{Q}. \tag{4.12.6.3}$$

For any $n \geq 0$, let $\mathscr{M}_n = \psi_*(M_n) \otimes_{\mathbb{Z}_p} \overline{\mathscr{B}}$, so that $\mathscr{M} = (\mathscr{M}_n)_{n\geq 0}$ ([3] VI.7.5). By ([3] VI.7.10), for every integer $q \geq 0$, we have an exact sequence

$$0 \to \mathrm{R}^1\lim_{\substack{\longleftarrow \\ n\geq 1}} \mathrm{H}^{q-1}(\widetilde{E}, \mathscr{M}_n) \to \mathrm{H}^q(\widetilde{E}^{\mathbb{N}^\circ}, \mathscr{M}) \to \lim_{\substack{\longleftarrow \\ n\geq 1}} \mathrm{H}^q(\widetilde{E}, \mathscr{M}_n) \to 0. \tag{4.12.6.4}$$

By virtue of ([2] 4.8.13), for every $n \geq 0$, we have a canonical morphism

$$u_n^q\colon \mathrm{H}^q(\overline{X}^{\circ}_{\text{ét}}, M_n) \otimes_{\mathbb{Z}_p} \mathscr{O}_{\overline{K}} \to \mathrm{H}^q(\widetilde{E}, \mathscr{M}_n), \tag{4.12.6.5}$$

which is an α-isomorphism. The morphisms

$$\lim_{\substack{\longleftarrow \\ n\geq 1}} u_n^q \quad \text{and} \quad \mathrm{R}^1\lim_{\substack{\longleftarrow \\ n\geq 1}} u_n^q \tag{4.12.6.6}$$

are thus α-isomorphisms ([21] 2.4.2(ii)).

Similarly, we have an exact sequence

$$0 \to \mathrm{R}^1 \lim_{\substack{\longleftarrow \\ n \geq 0}} \mathrm{H}^{q-1}(\overline{X}^{\circ}_{\text{ét}}, M_n) \to \mathrm{H}^q((\overline{X}^{\circ}_{\text{ét}})^{\mathbb{N}^{\circ}}, M) \to \lim_{\substack{\longleftarrow \\ n \geq 0}} \mathrm{H}^q(\overline{X}^{\circ}_{\text{ét}}, M_n) \to 0. \tag{4.12.6.7}$$

The groups $\mathrm{H}^q(\overline{X}^{\circ}_{\text{ét}}, M_n)$ being finite by virtue of ([15] Th.finitude 1.1), the inverse system $(\mathrm{H}^q(\overline{X}^{\circ}_{\text{ét}}, M_n))_{n \geq 1}$ satisfies the Mittag–Leffler condition. The canonical morphism

$$\mathrm{H}^q((\overline{X}^{\circ}_{\text{ét}})^{\mathbb{N}^{\circ}}, M) \to \lim_{\substack{\longleftarrow \\ n \geq 0}} \mathrm{H}^q(\overline{X}^{\circ}_{\text{ét}}, M_n) \tag{4.12.6.8}$$

is therefore an isomorphism by ([39] 1.15). We also deduce from this that the canonical morphism

$$\mathrm{H}^q(\widetilde{E}_s^{\mathbb{N}^{\circ}}, \mathscr{M}) \to \lim_{\substack{\longleftarrow \\ n \geq 1}} \mathrm{H}^q(\widetilde{E}_s, \mathscr{M}_n) \tag{4.12.6.9}$$

is an α-isomorphism in view of ([48] theo. 1).

By ([25] VI 2.2.2 and the remark after 2.2.3), the inverse system $\mathrm{H}^q = (\mathrm{H}^q(\overline{X}^{\circ}_{\text{ét}}, M_n))_{n \geq 0}$ is AR-p-adic constructible, in other words, there exist a Noetherian p-adic inverse system of abelian groups $A^q = (A^q_n)_{n \geq 0}$ and an AR-isomorphism $A^q \to \mathrm{H}^q$ ([25] V 3.2.2). We deduce from this an isomorphism

$$\lim_{\substack{\longleftarrow \\ n \geq 0}} A^q_n \xrightarrow{\sim} \lim_{\substack{\longleftarrow \\ n \geq 0}} \mathrm{H}^q(\overline{X}^{\circ}_{\text{ét}}, M_n). \tag{4.12.6.10}$$

In particular, the $\mathbb{Z}_p$-module $\mathrm{H}^q((\overline{X}^{\circ}_{\text{ét}})^{\mathbb{N}^{\circ}}, M)$ is of finite type. The morphism of inverse systems $(A^q_n \otimes_{\mathbb{Z}_p} \mathscr{O}_C)_{n \geq 0} \to (\mathrm{H}^q(\overline{X}^{\circ}_{\text{ét}}, M_n) \otimes_{\mathbb{Z}_p} \mathscr{O}_C)_{n \geq 0}$ is also an AR-isomorphism. Since the source is a p-adic system, we deduce from this that the canonical morphism

$$\mathrm{H}^q((\overline{X}^{\circ}_{\text{ét}})^{\mathbb{N}^{\circ}}, M) \otimes_{\mathbb{Z}_p} \mathscr{O}_C \to \lim_{\substack{\longleftarrow \\ n \geq 0}} \mathrm{H}^q(\overline{X}^{\circ}_{\text{ét}}, M_n) \otimes_{\mathbb{Z}_p} \mathscr{O}_C \tag{4.12.6.11}$$

is an isomorphism. The proposition then follows from (4.12.6.3), (4.12.6.5), (4.12.6.8), (4.12.6.9) and (4.12.6.11).

Remarks 4.12.7 In 4.12.6, if we take $M = \breve{\mathbb{Z}}_p$, then $\mathscr{M} = \breve{\overline{\mathscr{B}}}$, the $\breve{\overline{\mathscr{B}}}_{\mathbb{Q}}$-module $\breve{\overline{\mathscr{B}}}_{\mathbb{Q}}$ is Dolbeault and $\mathscr{H}(\breve{\overline{\mathscr{B}}}_{\mathbb{Q}})$ is equal to $\mathscr{O}_{\mathfrak{X}}[\frac{1}{p}]$ endowed with the zero Higgs field (4.6.10). The spectral sequence (1.4.15.1) is none other than the Hodge–Tate spectral sequence ([2] 6.4.6). Note that the construction 4.12.6 of this spectral sequence shows directly that it degenerates at E_2 and that the abutment filtration is split without using Tate's theorem on the Galois cohomology of $C(j)$. This construction applies in particular by taking for $(\widetilde{X}, \mathscr{M}_{\widetilde{X}})$ in the relative case (3.1.14) the trivial deformation (4.1.6).

Chapter 5
Relative Cohomologies of Higgs–Tate Algebras. Local Study

5.1 Assumptions and Notation. Review of Relative Galois Cohomology

5.1.1 In this chapter, K denotes a complete discrete valuation field of characteristic 0, with *algebraically closed* residue field k of characteristic $p > 0$, $\mathscr{O}_K$ the valuation ring of K, $\overline{K}$ an algebraic closure of K, $\mathscr{O}_{\overline{K}}$ the integral closure of $\mathscr{O}_K$ in $\overline{K}$, $\mathfrak{m}_{\overline{K}}$ the maximal ideal of $\mathscr{O}_{\overline{K}}$ and G_K the Galois group of $\overline{K}$ over K. We denote by $\mathscr{O}_C$ the p-adic Hausdorff completion of $\mathscr{O}_{\overline{K}}$, by $\mathfrak{m}_C$ its maximal ideal, by C its field of fractions and by v its valuation, normalized by $v(p) = 1$. Let $\widehat{\mathbb{Z}}(1)$, $\mathbb{Z}_p(1)$ and $\mu_{p^\infty}(\mathscr{O}_{\overline{K}})$ be the $\mathbb{Z}[G_K]$-modules

$$\widehat{\mathbb{Z}}(1) = \varprojlim_{n\geq 1} \mu_n(\mathscr{O}_{\overline{K}}), \tag{5.1.1.1}$$

$$\mathbb{Z}_p(1) = \varprojlim_{n\geq 0} \mu_{p^n}(\mathscr{O}_{\overline{K}}), \tag{5.1.1.2}$$

$$\mu_{p^\infty}(\mathscr{O}_{\overline{K}}) = \varinjlim_{n\geq 0} \mu_{p^n}(\mathscr{O}_{\overline{K}}), \tag{5.1.1.3}$$

where $\mu_n(\mathscr{O}_{\overline{K}})$ denotes the subgroup of nth roots of unity in $\mathscr{O}_{\overline{K}}$. For any $\mathbb{Z}_p[G_K]$-module M and any integer n, we set $M(n) = M \otimes_{\mathbb{Z}_p} \mathbb{Z}_p(1)^{\otimes n}$.

For any $\mathbb{Z}_p$-module A, we denote by $\widehat{A}$ its p-adic Hausdorff completion.

We set $S = \mathrm{Spec}(\mathscr{O}_K)$, $\overline{S} = \mathrm{Spec}(\mathscr{O}_{\overline{K}})$ and $\check{\overline{S}} = \mathrm{Spec}(\mathscr{O}_C)$. We denote by s (resp. η, resp. $\overline{\eta}$) the closed point of S (resp. generic point of S, resp. generic point of $\overline{S}$). For any integer $n \geq 1$, we set $S_n = \mathrm{Spec}(\mathscr{O}_K/p^n\mathscr{O}_K)$. For any S-scheme X, we set

$$\overline{X} = X \times_S \overline{S}, \quad \check{\overline{X}} = X \times_S \check{\overline{S}} \quad \text{and} \quad X_n = X \times_S S_n. \tag{5.1.1.4}$$

We endow S with the logarithmic structure $\mathscr{M}_S$ defined by its closed point, and $\overline{S}$ and $\check{\overline{S}}$ with the logarithmic structures $\mathscr{M}_{\overline{S}}$ and $\mathscr{M}_{\check{\overline{S}}}$ pullbacks of $\mathscr{M}_S$.

A. Abbes, M. Gros, *The p-adic Simpson Correspondence and Hodge-Tate Local Systems*, Lecture Notes in Mathematics 2345, https://doi.org/10.1007/978-3-031-55914-3_5

5.1.2 Since $\mathscr{O}_{\overline{K}}$ is a non-discrete valuation ring of height 1, we can develop the α-algebra (or almost algebra) on this ring ([2] 2.10.1) (see [2] 2.6-2.10). We choose a compatible system $(\beta_n)_{n>0}$ of nth roots of p in $\mathscr{O}_{\overline{K}}$. For any rational number $\varepsilon > 0$, we set $p^{\varepsilon} = (\beta_n)^{\varepsilon n}$, where n is an integer > 0 such that εn is an integer.

5.1.3 In this chapter, $f\colon (X, \mathscr{M}_X) \to (S, \mathscr{M}_S)$ and $f'\colon (X', \mathscr{M}_{X'}) \to (S, \mathscr{M}_S)$ denote *adequate* morphisms of logarithmic schemes ([3] III.4.7) and

$$g\colon (X', \mathscr{M}_{X'}) \to (X, \mathscr{M}_X) \tag{5.1.3.1}$$

a smooth and saturated $(S, \mathscr{M}_S)$-morphism. We assume that the schemes $X = \mathrm{Spec}(R)$ and $X' = \mathrm{Spec}(R')$ are affine and that X'_s is non-empty. Moreover, except in 5.2.12, we assume that the morphism g admits a relative adequate chart ([2] 5.1.11)

$$((P', \gamma'), (P, \gamma), (\mathbb{N}, \iota), \vartheta\colon \mathbb{N} \to P, h\colon P \to P'), \tag{5.1.3.2}$$

that we fix. We set $\pi = \iota(1)$, which is a uniformizer of $\mathscr{O}_K$.

We denote by X° the maximal open subscheme of X where the logarithmic structure $\mathscr{M}_X$ is trivial; it is an open subscheme of X_η. The immersion $j\colon X^\circ \to X$ is schematically dominant ([3] III.4.2(iv)). For any X-scheme U, we set

$$U^\circ = U \times_X X^\circ. \tag{5.1.3.3}$$

We denote by $X'^{\rhd}$ the maximal open subscheme of X' where the logarithmic structure $\mathscr{M}_{X'}$ is trivial; it is an open subscheme of $X'^\circ = X' \times_X X^\circ$. The immersion $j'\colon X'^{\rhd} \to X'$ is schematically dominant. For any X'-scheme U', we set

$$U'^{\rhd} = U' \times_{X'} X'^{\rhd}. \tag{5.1.3.4}$$

5.1.4 We set

$$\widetilde{\Omega}^1_{R/\mathscr{O}_K} = \Gamma(X, \Omega^1_{(X,\mathscr{M}_X)/(S,\mathscr{M}_S)}), \tag{5.1.4.1}$$

which is a free R-module of finite type ([2] (4.5.9.2)). Likewise, we set

$$\widetilde{\Omega}^1_{R'/\mathscr{O}_K} = \Gamma(X', \Omega^1_{(X',\mathscr{M}_{X'})/(S,\mathscr{M}_S)}), \tag{5.1.4.2}$$

$$\widetilde{\Omega}^1_{R'/R} = \Gamma(X', \Omega^1_{(X',\mathscr{M}_{X'})/(X,\mathscr{M}_X)}), \tag{5.1.4.3}$$

which are free R'-modules of finite type. We then have a split exact sequence of R'-modules

$$0 \to \widetilde{\Omega}^1_{R/\mathscr{O}_K} \otimes_R R' \to \widetilde{\Omega}^1_{R'/\mathscr{O}_K} \to \widetilde{\Omega}^1_{R'/R} \to 0. \tag{5.1.4.4}$$

5.1.5 For any integer $n \geq 1$, we set

$$\mathscr{O}_{K_n} = \mathscr{O}_K[\zeta]/(\zeta^n - \pi), \tag{5.1.5.1}$$

which is a discrete valuation ring (5.1.3). We denote by K_n the field of fractions of $\mathscr{O}_{K_n}$ and by π_n the residue class of ζ in $\mathscr{O}_{K_n}$, which is a uniformizer of $\mathscr{O}_{K_n}$. We set $S^{(n)} = \mathrm{Spec}(\mathscr{O}_{K_n})$, which we endow with the logarithmic structure $\mathscr{M}_{S^{(n)}}$ defined by its closed point. We denote by $\iota_n \colon \mathbb{N} \to \Gamma(S^{(n)}, \mathscr{M}_{S^{(n)}})$ the homomorphism defined by $\iota_n(1) = \pi_n$; this is a chart for $(S^{(n)}, \mathscr{M}_{S^{(n)}})$.

Consider the direct system of monoids $(\mathbb{N}^{(n)})_{n\geq 1}$, indexed by the set $\mathbb{Z}_{\geq 1}$ ordered by the divisibility relation, defined by $\mathbb{N}^{(n)} = \mathbb{N}$ for every $n \geq 1$ and whose transition homomorphism $\mathbb{N}^{(n)} \to \mathbb{N}^{(mn)}$ (for $m, n \geq 1$) is the Frobenius endomorphism of order m of $\mathbb{N}$ (i.e., the mth power). We denote $\mathbb{N}^{(1)}$ simply by $\mathbb{N}$. The logarithmic schemes $(S^{(n)}, \mathscr{M}_{S^{(n)}})_{n\geq 1}$ naturally form an inverse system. For all integers $m, n \geq 1$, with the notation of 2.1.4, we have a Cartesian diagram of morphisms of logarithmic schemes

$$\begin{array}{ccc} (S^{(mn)}, \mathscr{M}_{S^{(mn)}}) & \xrightarrow{\iota^a_{mn}} & \mathbf{A}_{\mathbb{N}^{(mn)}} \\ \downarrow & & \downarrow \\ (S^{(n)}, \mathscr{M}_{S^{(n)}}) & \xrightarrow{\iota^a_n} & \mathbf{A}_{\mathbb{N}^{(n)}} \end{array} \tag{5.1.5.2}$$

where ι^a_n (resp. ι^a_{mn}) is the morphism associated with ι_n (resp. ι_{mn}) and $\mathbf{A}_{\mathbb{N}^{(n)}}$ is defined in 2.1.4 (see [3] II.5.13).

5.1.6 We denote by $(P^{(n)})_{n\geq 1}$ the direct system of monoids, indexed by the set $\mathbb{Z}_{\geq 1}$ ordered by the divisibility relation, defined by $P^{(n)} = P$ for any $n \geq 1$ and whose transition homomorphism $i_{n,mn} \colon P^{(n)} \to P^{(mn)}$ (for $m, n \geq 1$) is the Frobenius endomorphism of order m of P (i.e., the mth power). We denote $P^{(1)}$ simply by P.

For any $n \geq 1$, we set (with the notation of 2.1.4)

$$(X^{(n)}, \mathscr{M}_{X^{(n)}}) = (X, \mathscr{M}_X) \times_{\mathbf{A}_P} \mathbf{A}_{P^{(n)}}. \tag{5.1.6.1}$$

We recall (3.2.2.5) that there exists a unique morphism

$$f^{(n)} \colon (X^{(n)}, \mathscr{M}_{X^{(n)}}) \to (S^{(n)}, \mathscr{M}_{S^{(n)}}) \tag{5.1.6.2}$$

that fits into the commutative diagram

$$\begin{array}{ccccc} (X^{(n)}, \mathscr{M}_{X^{(n)}}) & & \longrightarrow & & \mathbf{A}_{P^{(n)}} \\ \downarrow & \searrow^{f^{(n)}} & & \swarrow^{\mathbf{A}_\vartheta} & \downarrow \\ & (S^{(n)}, \mathscr{M}_{S^{(n)}}) & \xrightarrow{\iota^a_n} & \mathbf{A}_{\mathbb{N}^{(n)}} & \\ & \downarrow & & \downarrow & \\ & (S, \mathscr{M}_S) & \xrightarrow{\iota^a} & \mathbf{A}_{\mathbb{N}} & \\ \downarrow & \nearrow^{f} & & \nwarrow^{\mathbf{A}_\vartheta} & \downarrow \\ (X, \mathscr{M}_X) & & \longrightarrow & & \mathbf{A}_P. \end{array} \tag{5.1.6.3}$$

5.1.7 We denote by $(P'^{[n]})_{n\geq 1}$ the direct system of monoids indexed by the set $\mathbb{Z}_{\geq 1}$ ordered by the divisibility relation, defined by $P'^{[n]} = P'$ for any $n \geq 1$ and whose transition homomorphism $P'^{[n]} \to P'^{[mn]}$ (for $m, n \geq 1$) is the Frobenius endomorphism of order m of P'. We denote $P'^{[1]}$ simply by P'.

For any $n \geq 1$, we set (with the notation of 2.1.4)

$$(X'^{[n]}, \mathscr{M}_{X'^{[n]}}) = (X', \mathscr{M}_{X'}) \times_{\mathbf{A}_{P'}} \mathbf{A}_{P'^{[n]}}. \tag{5.1.7.1}$$

Then there exists a unique morphism

$$f'^{[n]} \colon (X'^{[n]}, \mathscr{M}_{X'^{[n]}}) \to (S^{(n)}, \mathscr{M}_{S^{(n)}}) \tag{5.1.7.2}$$

that fits into the commutative diagram

$$\begin{array}{ccccc}
(X'^{[n]}, \mathscr{M}_{X'^{[n]}}) & & \longrightarrow & & \mathbf{A}_{P'^{[n]}} \\
\downarrow & \searrow^{f'^{[n]}} & & \swarrow^{\mathbf{A}_{h\circ\vartheta}} & \downarrow \\
 & (S^{(n)}, \mathscr{M}_{S^{(n)}}) & \longrightarrow & \mathbf{A}_{\mathbb{N}^{(n)}} & \\
 & \downarrow & & \downarrow & \\
 & (S, \mathscr{M}_S) & \longrightarrow & \mathbf{A}_{\mathbb{N}} & \\
\downarrow & \nearrow^{f'} & & \nwarrow^{\mathbf{A}_{h\circ\vartheta}} & \downarrow \\
(X', \mathscr{M}_{X'}) & & \longrightarrow & & \mathbf{A}_{P'}.
\end{array} \tag{5.1.7.3}$$

5.1.8 For any $n \geq 1$, we set

$$(X'^{(n)}, \mathscr{M}_{X'^{(n)}}) = (X^{(n)}, \mathscr{M}_{X^{(n)}}) \times_{(X, \mathscr{M}_X)} (X', \mathscr{M}_{X'}), \tag{5.1.8.1}$$

the product being indifferently taken in the category of logarithmic schemes or in that of saturated logarithmic schemes ([3] II.5.20). We denote by

$$f'^{(n)} \colon (X'^{(n)}, \mathscr{M}_{X'^{(n)}}) \to (S^{(n)}, \mathscr{M}_{S^{(n)}}) \tag{5.1.8.2}$$

the morphism deduced from $f^{(n)}$ (5.1.6.2).

There is a unique $(S^{(n)}, \mathscr{M}_{S^{(n)}})$-morphism

$$g^{(n)} \colon (X'^{[n]}, \mathscr{M}_{X'^{[n]}}) \to (X^{(n)}, \mathscr{M}_{X^{(n)}}) \tag{5.1.8.3}$$

above g and the morphism $\mathbf{A}_h \colon \mathbf{A}_{P'^{[n]}} \to \mathbf{A}_{P^{(n)}}$. This induces a canonical morphism

$$(X'^{[n]}, \mathscr{M}_{X'^{[n]}}) \to (X'^{(n)}, \mathscr{M}_{X'^{(n)}}) \tag{5.1.8.4}$$

above $(X', \mathscr{M}_{X'})$ and $(X^{(n)}, \mathscr{M}_{X^{(n)}})$.

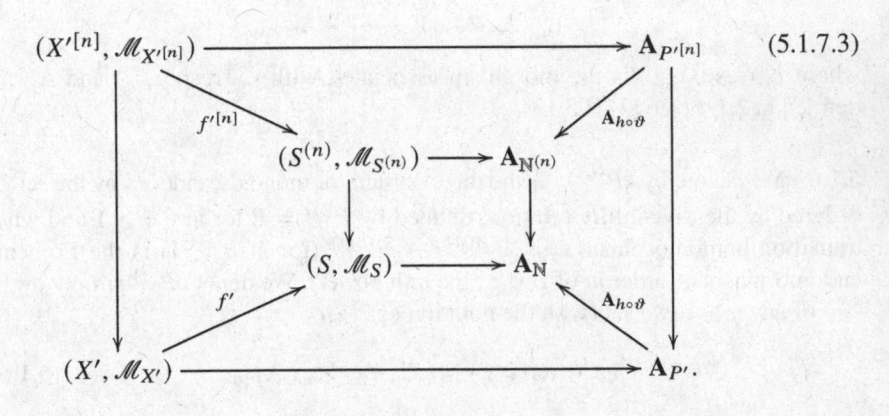

5.1.9 Let $\overline{y}'$ be a geometric point of $\overline{X}'^{\triangleright}$ and $\overline{y}$ its image in $\overline{X}^{\circ}$ (see 5.1.1 and 5.1.3). The schemes $\overline{X}$ and $\overline{X}'$ being locally irreducible by ([2] 4.2.7(iii)), they are the sums of the schemes induced on their irreducible components. We denote by $\overline{X}^{\star}$ (resp. $\overline{X}'^{\star}$) the irreducible component of $\overline{X}$ (resp. $\overline{X}'$) containing $\overline{y}$ (resp. $\overline{y}'$). Similarly, $\overline{X}^{\circ}$ (resp. $\overline{X}'^{\triangleright}$) is the sum of the schemes induced on its irreducible components and $\overline{X}^{\star\circ} = \overline{X}^{\star} \times_X X^{\circ}$ (resp. $\overline{X}'^{\star\triangleright} = \overline{X}'^{\star} \times_{X'} X'^{\triangleright}$) is the irreducible component of $\overline{X}^{\circ}$ (resp. $\overline{X}'^{\triangleright}$) containing $\overline{y}$ (resp. $\overline{y}'$).

We denote by Δ the profinite group $\pi_1(\overline{X}^{\star\circ}, \overline{y})$ and by $(V_i)_{i\in I}$ the normalized universal cover of $\overline{X}^{\star\circ}$ at $\overline{y}$ ([3] VI.9.8). For any $i \in I$, we denote by $\overline{X}^{V_i}$ the integral closure of $\overline{X}$ in V_i. We set

$$\overline{R} = \lim_{\substack{\longrightarrow \\ i\in I}} \Gamma(\overline{X}^{V_i}, \mathscr{O}_{\overline{X}^{V_i}}), \tag{5.1.9.1}$$

which is naturally endowed with a discrete action of Δ by ring automorphisms. Observe that $\overline{R}$ is the discrete representation $\overline{R}_X^{\overline{y}}$ of Δ defined in (4.3.7.2) ([2] (4.1.9.3)).

We denote by Δ' the profinite group $\pi_1(\overline{X}'^{\star\triangleright}, \overline{y}')$ and by $(W_j)_{j\in J}$ the normalized universal cover of $\overline{X}'^{\star\triangleright}$ at $\overline{y}'$. For any $j \in J$, we denote by $\overline{X}'^{W_j}$ the integral closure of $\overline{X}'$ in W_j. We set

$$\overline{R}' = \lim_{\substack{\longrightarrow \\ j\in J}} \Gamma(\overline{X}'^{W_j}, \mathscr{O}_{\overline{X}'^{W_j}}), \tag{5.1.9.2}$$

which is naturally endowed with a discrete action of Δ' by ring automorphisms. Observe that $\overline{R}'$ is the discrete representation $\overline{R}_{X'}^{\overline{y}'}$ of Δ' defined in (4.3.7.2).

For every $i \in I$, we have a canonical $\overline{X}^{\circ}$-morphism $\overline{y} \to V_i$. We deduce from this an $\overline{X}'^{\triangleright}$-morphism $\overline{y}' \to V_i \times_{\overline{X}^{\circ}} \overline{X}'^{\triangleright}$. The scheme $V_i \times_{\overline{X}^{\circ}} \overline{X}'^{\triangleright}$ being locally irreducible, it is the sum of the schemes induced on its irreducible components. We denote by V_i' the irreducible component of $V_i \times_{\overline{X}^{\circ}} \overline{X}'^{\triangleright}$ containing the image of $\overline{y}'$ and by $\overline{X}'^{V_i'}$ the integral closure of $\overline{X}'$ in V_i'. The schemes $(V_i')_{i\in I}$ naturally form an inverse system of $\overline{y}'$-pointed connected finite étale covers of $\overline{X}'^{\star\triangleright}$. For any $i \in I$, we denote by Π_i the open subgroup of Δ' corresponding to V_i', i.e., the kernel of the canonical action of Δ' on the fiber of V_i' above $\overline{y}'$ ([2] (2.1.20.2)). We denote by Π the closed subgroup of Δ' defined by

$$\Pi = \cap_{i\in I} \Pi_i. \tag{5.1.9.3}$$

Setting $\Delta^{\circ} = \Delta'/\Pi$, we have a canonical homomorphism $\Delta' \to \Delta$ which factors through an injective homomorphism $\Delta^{\circ} \to \Delta$.

Consider the ring

$$\overline{R}^{\circ} = \lim_{\substack{\longrightarrow \\ i\in I}} \Gamma(\overline{X}'^{V_i'}, \mathscr{O}_{\overline{X}'^{V_i'}}), \tag{5.1.9.4}$$

which is naturally endowed with a discrete action of Δ° by ring automorphisms. We have a Δ°-equivariant canonical homomorphism $\overline{R} \to \overline{R}^{\circ}$.

For all $i \in I$ and $j \in J$, there exists at most one morphism of pointed $\overline{X}'^{\star\rhd}$-schemes $W_j \to V'_i$. Moreover, for every $i \in I$, there exist $j \in J$ and a morphism of pointed $\overline{X}'^{\star\rhd}$-schemes $W_j \to V'_i$. We therefore have a canonical homomorphism

$$\varinjlim_{i\in I} \Gamma(\overline{X}'^{V'_i}, \mathscr{O}_{\overline{X}'^{V'_i}}) \to \varinjlim_{j\in J} \Gamma(\overline{X}'^{W_j}, \mathscr{O}_{\overline{X}'^{W_j}}). \tag{5.1.9.5}$$

We deduce from this a Δ'-equivariant canonical homomorphism of R'_1-algebras

$$\overline{R}^{\circ} \to \overline{R}'. \tag{5.1.9.6}$$

5.1.10 By (5.1.8.4), we have canonical morphisms of inverse systems of logarithmic schemes, indexed by the set $\mathbb{Z}_{\geq 1}$ ordered by the divisibility relation,

$$(X'^{[n]}, \mathscr{M}_{X'^{[n]}}) \to (X'^{(n)}, \mathscr{M}_{X'^{(n)}}) \to (X^{(n)}, \mathscr{M}_{X^{(n)}}) \to (S^{(n)}, \mathscr{M}_{S^{(n)}}). \tag{5.1.10.1}$$

For all integers $m, n \geq 1$, the canonical morphism $X'^{[mn]} \to X'^{[n]}$ is finite and surjective. Then, there exists an X'-morphism ([30] 8.3.8(i))

$$\overline{y}' \to \varprojlim_{n\geq 1} X'^{[n]}. \tag{5.1.10.2}$$

We fix such a morphism in the remainder of this section. This induces an S-morphism

$$\overline{S} \to \varprojlim_{n\geq 1} S^{(n)}. \tag{5.1.10.3}$$

For any integer $n \geq 1$, we set

$$\overline{X}'^{[n]} = X'^{[n]} \times_{S^{(n)}} \overline{S}, \quad \overline{X}'^{[n]\rhd} = \overline{X}'^{[n]} \times_{X'} X'^{\rhd}, \tag{5.1.10.4}$$

$$\overline{X}'^{(n)} = X'^{(n)} \times_{S^{(n)}} \overline{S}, \quad \overline{X}'^{(n)\rhd} = \overline{X}'^{(n)} \times_{X'} X'^{\rhd}, \tag{5.1.10.5}$$

$$\overline{X}^{(n)} = X^{(n)} \times_{S^{(n)}} \overline{S}, \quad \overline{X}^{(n)\circ} = \overline{X}^{(n)} \times_{X} X^{\circ}. \tag{5.1.10.6}$$

The morphisms (5.1.10.1) induce morphisms of inverse systems of $\overline{S}$-schemes

$$\overline{X}'^{[n]} \to \overline{X}'^{(n)} \to \overline{X}^{(n)}. \tag{5.1.10.7}$$

We deduce from (5.1.10.1) and (5.1.10.2) an $\overline{X}'$-morphism

$$\overline{y}' \to \varprojlim_{n\geq 1} \overline{X}'^{[n]}. \tag{5.1.10.8}$$

5.1.11 For every integer $n \geq 1$, the scheme $\overline{X}'^{[n]}$ is normal and locally irreducible by ([2] 4.2.7(iii)). It is therefore the sum of the schemes induced on its irreducible components. We denote by $\overline{X}'^{[n]\star}$ the irreducible component of $\overline{X}'^{[n]}$ containing the image of $\overline{y}'$ (5.1.10.8). Similarly, $\overline{X}'^{[n]\rhd}$ is the sum of the schemes induced on its irreducible components and $\overline{X}'^{[n]\star\triangleright} = \overline{X}'^{[n]\star} \times_{X'} X'^{\triangleright}$ is the irreducible component of $\overline{X}'^{[n]\triangleright}$ containing the image of $\overline{y}'$. We set

$$R'_n = \Gamma(\overline{X}'^{[n]\star}, \mathscr{O}_{\overline{X}'^{[n]}}). \tag{5.1.11.1}$$

The rings $(R'_n)_{n\geq 1}$ naturally form a direct system. We set

$$R'_\infty = \varinjlim_{n\geq 1} R'_n, \tag{5.1.11.2}$$

$$R'_{p^\infty} = \varinjlim_{n\geq 0} R'_{p^n}. \tag{5.1.11.3}$$

The morphism (5.1.10.8) induces injective homomorphisms of R'_1-algebras

$$R'_{p^\infty} \to R'_\infty \to \overline{R}'. \tag{5.1.11.4}$$

Similarly, for every integer $n \geq 1$, the scheme $\overline{X}^{(n)}$ is the sum of the schemes induced on its irreducible components. We denote by $\overline{X}^{(n)\star}$ the irreducible component of $\overline{X}^{(n)}$ containing the image of $\overline{y}'$ (5.1.10.8). The scheme $\overline{X}^{(n)\circ}$ is the sum of the schemes induced on its irreducible components and $\overline{X}^{(n)\star\circ} = \overline{X}^{(n)\star} \times_X X^\circ$ is the irreducible component of $\overline{X}^{(n)\circ}$ containing the image of $\overline{y}'$. We set

$$R_n = \Gamma(\overline{X}^{(n)\star}, \mathscr{O}_{\overline{X}^{(n)}}). \tag{5.1.11.5}$$

The rings $(R_n)_{n\geq 1}$ naturally form a direct system. We set

$$R_\infty = \varinjlim_{n\geq 1} R_n, \tag{5.1.11.6}$$

$$R_{p^\infty} = \varinjlim_{n\geq 0} R_{p^n}. \tag{5.1.11.7}$$

The morphisms (5.1.10.7) and (5.1.10.8) induce injective homomorphisms of R_1-algebras

$$R_{p^\infty} \to R_\infty \to \overline{R}. \tag{5.1.11.8}$$

For every integer $n \geq 1$, the scheme $\overline{X}'^{(n)}$ is normal and locally irreducible by (5.1.8.1) and ([3] III.4.2(iii)). It is therefore the sum of the schemes induced on its irreducible components. Moreover, the immersion $\overline{X}'^{(n)\triangleright} \to \overline{X}'^{(n)}$ is schematically dominant ([30] 11.10.5). We denote by $\overline{X}'^{(n)\star}$ the irreducible component of $\overline{X}'^{(n)}$ containing the image of $\overline{y}'$ (5.1.10.8). The scheme $\overline{X}'^{(n)\triangleright}$ is also the sum of the schemes induced on its irreducible components and $\overline{X}'^{(n)\star\triangleright} = \overline{X}'^{(n)\star} \times_{X'} X'^{\triangleright}$ is the irreducible component of $\overline{X}'^{(n)\triangleright}$ containing the image of $\overline{y}'$. We set

$$R^\circ_n = \Gamma(\overline{X}'^{(n)\star}, \mathscr{O}_{\overline{X}'^{(n)}}). \tag{5.1.11.9}$$

The rings $(R^\diamond_n)_{n\geq 1}$ naturally form a direct system. We set

$$R^\diamond_\infty = \lim_{\substack{\longrightarrow \\ n\geq 1}} R^\diamond_n, \tag{5.1.11.10}$$

$$R^\diamond_{p^\infty} = \lim_{\substack{\longrightarrow \\ n\geq 0}} R^\diamond_{p^n}. \tag{5.1.11.11}$$

They are normal integral domains by ([27] 0.6.1.6(i) and 0.6.5.12(ii)). In view of the definition (5.1.9.4), the morphisms (5.1.10.7) and (5.1.10.8) induce injective homomorphisms of $R^\diamond_1$-algebras

$$R^\diamond_{p^\infty} \to R^\diamond_\infty \to \overline{R}^\diamond. \tag{5.1.11.12}$$

The morphisms of (5.1.10.7) induce homomorphisms of direct systems of algebras

$$(R_n)_{n\geq 1} \to (R^\diamond_n)_{n\geq 1} \to (R'_n)_{n\geq 1}. \tag{5.1.11.13}$$

The homomorphism $R^\diamond_1 \to R'_1$ is an isomorphism.

5.1.12 Let n be an integer ≥ 1. It follows from ([2] 4.2.7(v)) and from the proof of ([3] II.6.8(iv)) that the morphism $\overline{X}'^{[n]\star\rhd} \to \overline{X}'^{\star\rhd}$ is a Galois finite étale cover of group Δ'_n, a subgroup of $\mathrm{Hom}_{\mathbb{Z}}(P'^{\mathrm{gp}}/h^{\mathrm{gp}}(\vartheta^{\mathrm{gp}}(\mathbb{Z})), \mu_n(\overline{K}))$. The group Δ'_n acts naturally on R'_n. If n is a power of p, the canonical morphism

$$\overline{X}'^{[n]\star} \to \overline{X}'^{[n]} \times_{\overline{X}'} \overline{X}'^{\star} \tag{5.1.12.1}$$

is an isomorphism by virtue of ([3] II.6.6(v)), and we therefore have

$$\Delta'_n \xrightarrow{\sim} \mathrm{Hom}_{\mathbb{Z}}(P'^{\mathrm{gp}}/h^{\mathrm{gp}}(\vartheta^{\mathrm{gp}}(\mathbb{Z})), \mu_n(\overline{K})). \tag{5.1.12.2}$$

The groups $(\Delta'_n)_{n\geq 1}$ naturally form an inverse system. We set

$$\Delta'_\infty = \lim_{\substack{\longleftarrow \\ n\geq 1}} \Delta'_n, \tag{5.1.12.3}$$

$$\Delta'_{p^\infty} = \lim_{\substack{\longleftarrow \\ n\geq 0}} \Delta'_{p^n}. \tag{5.1.12.4}$$

We identify Δ'_∞ (resp. Δ'_{p^∞}) with the Galois group of the extension of the fields of fractions of R'_∞ over R'_1 (resp. R'_{p^∞} on R'_1). We have canonical homomorphisms (5.1.1.1)

$$\begin{array}{ccc} \Delta'_\infty & \hookrightarrow & \mathrm{Hom}(P'^{\mathrm{gp}}/h^{\mathrm{gp}}(\vartheta^{\mathrm{gp}}(\mathbb{Z})), \widehat{\mathbb{Z}}(1)) \\ \downarrow & & \downarrow \\ \Delta'_{p^\infty} & \xrightarrow{\sim} & \mathrm{Hom}(P'^{\mathrm{gp}}/h^{\mathrm{gp}}(\vartheta^{\mathrm{gp}}(\mathbb{Z})), \mathbb{Z}_p(1)). \end{array} \tag{5.1.12.5}$$

The kernel Σ'_0 of the canonical homomorphism $\Delta'_\infty \to \Delta'_{p^\infty}$ is a profinite group of order prime to p. Moreover, the morphism (5.1.10.8) determines a surjective homomorphism $\Delta' \to \Delta'_\infty$ (5.1.9). We denote by Σ' its kernel. The homomorphisms (5.1.11.4) are Δ'-equivariant.

Similarly, the morphism $\overline{X}^{(n)\star\circ} \to \overline{X}^{\star\circ}$ is a Galois finite étale cover of group Δ_n, a subgroup of $\mathrm{Hom}_{\mathbb{Z}}(P^{\mathrm{gp}}/\vartheta^{\mathrm{gp}}(\mathbb{Z}), \mu_n(\overline{K}))$. The group Δ_n acts naturally on R_n. If n is a power of p, the canonical morphism

$$\overline{X}^{(n)\star} \to \overline{X}^{(n)} \times_{\overline{X}} \overline{X}^{\star} \tag{5.1.12.6}$$

is an isomorphism, and we therefore have $\Delta_n \xrightarrow{\sim} \mathrm{Hom}_{\mathbb{Z}}(P^{\mathrm{gp}}/\vartheta^{\mathrm{gp}}(\mathbb{Z}), \mu_n(\overline{K}))$.

The groups $(\Delta_n)_{n\geq 1}$ naturally form an inverse system. We set

$$\Delta_\infty = \lim_{\substack{\longleftarrow \\ n\geq 1}} \Delta_n, \tag{5.1.12.7}$$

$$\Delta_{p^\infty} = \lim_{\substack{\longleftarrow \\ n\geq 0}} \Delta_{p^n}. \tag{5.1.12.8}$$

We identify Δ_∞ (resp. Δ_{p^∞}) with the Galois group of the extension of the fields of fractions of R_∞ over R_1 (resp. R_{p^∞} on R_1). We have canonical homomorphisms (5.1.1.1)

$$\begin{array}{ccc} \Delta_\infty & \hookrightarrow & \mathrm{Hom}(P^{\mathrm{gp}}/\vartheta^{\mathrm{gp}}(\mathbb{Z}), \widehat{\mathbb{Z}}(1)) \\ \downarrow & & \downarrow \\ \Delta_{p^\infty} & \xrightarrow{\sim} & \mathrm{Hom}(P^{\mathrm{gp}}/\vartheta^{\mathrm{gp}}(\mathbb{Z}), \mathbb{Z}_p(1)). \end{array} \tag{5.1.12.9}$$

The kernel Σ_0 of the canonical homomorphism $\Delta_\infty \to \Delta_{p^\infty}$ is a profinite group of order prime to p. Moreover, the morphisms (5.1.10.7) and (5.1.10.8) determine a surjective homomorphism $\Delta \to \Delta_\infty$ (5.1.9). We denote by Σ its kernel. The homomorphisms (5.1.11.8) are Δ-equivariant.

It follows from ([2] 5.2.8(iii)) that the morphism $\overline{X}'^{[n]\star\triangleright} \to \overline{X}'^{(n)\star\triangleright}$ is a Galois finite étale cover of group $\mathfrak{S}_n$, a subgroup of $\mathrm{Hom}_{\mathbb{Z}}(P'^{\mathrm{gp}}/h^{\mathrm{gp}}(P^{\mathrm{gp}}), \mu_n(\overline{K}))$. It follows from 3.2.4 and (5.1.8.1) that the morphism $\overline{X}'^{(n)\star\triangleright} \to \overline{X}'^{\star\triangleright}$ is a Galois finite étale cover of group Δ_n°, a subgroup of Δ_n. The group Δ_n° acts naturally on R_n°. We have a canonical exact sequence

$$0 \to \mathfrak{S}_n \to \Delta'_n \to \Delta_n^\circ \to 0. \tag{5.1.12.10}$$

Moreover, the diagram

$$\begin{array}{ccc} \mathfrak{S}_n & \longrightarrow & \mathrm{Hom}_{\mathbb{Z}}(P'^{\mathrm{gp}}/h^{\mathrm{gp}}(P^{\mathrm{gp}}), \mu_n(\overline{K})) \\ \downarrow & & \downarrow \\ \Delta'_n & \longrightarrow & \mathrm{Hom}_{\mathbb{Z}}(P'^{\mathrm{gp}}/h^{\mathrm{gp}}(\vartheta^{\mathrm{gp}}(\mathbb{Z})), \mu_n(\overline{K})), \end{array} \tag{5.1.12.11}$$

where the right vertical arrow is the canonical homomorphism, is commutative. By virtue of ([2] 5.2.14(iv)), if n is a power of p, the canonical morphism

$$\overline{X}'^{(n)\star} \to \overline{X}^{(n)} \times_{\overline{X}} \overline{X}'^{\star} \tag{5.1.12.12}$$

is an isomorphism, and we have

$$\mathfrak{S}_n = \mathrm{Hom}(P'^{\mathrm{gp}}/h^{\mathrm{gp}}(P^{\mathrm{gp}}), \mu_n(\overline{K})), \tag{5.1.12.13}$$
$$\Delta_n^{\diamond} = \mathrm{Hom}(P^{\mathrm{gp}}/\vartheta^{\mathrm{gp}}(\mathbb{Z}), \mu_n(\overline{K})). \tag{5.1.12.14}$$

The groups $(\mathfrak{S}_n)_{n\geq 1}$ and $(\Delta_n^{\diamond})_{n\geq 1}$ naturally form inverse systems. We set

$$\mathfrak{S}_\infty = \lim_{\substack{\longleftarrow \\ n\geq 1}} \mathfrak{S}_n, \tag{5.1.12.15}$$
$$\mathfrak{S}_{p^\infty} = \lim_{\substack{\longleftarrow \\ n\geq 0}} \mathfrak{S}_{p^n}, \tag{5.1.12.16}$$
$$\Delta_\infty^{\diamond} = \lim_{\substack{\longleftarrow \\ n\geq 1}} \Delta_n^{\diamond}, \tag{5.1.12.17}$$
$$\Delta_{p^\infty}^{\diamond} = \lim_{\substack{\longleftarrow \\ n\geq 0}} \Delta_{p^n}^{\diamond}. \tag{5.1.12.18}$$

We identify $\Delta_\infty^{\diamond}$ (resp. $\Delta_{p^\infty}^{\diamond}$) with the Galois group of the extension of the fields of fractions of $R_\infty^{\diamond}$ (resp. $R_{p^\infty}^{\diamond}$) over R_1'. By Galois theory, the sequences

$$0 \to \mathfrak{S}_\infty \to \Delta_\infty' \to \Delta_\infty^{\diamond} \to 0, \tag{5.1.12.19}$$

$$0 \to \mathfrak{S}_{p^\infty} \to \Delta_{p^\infty}' \to \Delta_{p^\infty}^{\diamond} \to 0, \tag{5.1.12.20}$$

deduced from (5.1.12.10) are exact. The group $\mathfrak{S}_\infty$ (resp. $\mathfrak{S}_{p^\infty}$) therefore identifies with the Galois group of the extension of the fields of fractions of R_∞' over $R_\infty^{\diamond}$ (resp. R_{p^∞}' over $R_{p^\infty}^{\diamond}$). Since the kernel of the canonical homomorphism $\Delta_\infty^{\diamond} \to \Delta_{p^\infty}^{\diamond}$ is a profinite group of order prime to p, we deduce from this that the canonical homomorphism $\mathfrak{S}_\infty \to \mathfrak{S}_{p^\infty}$ is surjective. We denote by $\mathfrak{N}$ its kernel, which is a profinite group of order prime to p, so that we have the exact sequence

$$0 \to \mathfrak{N} \to \mathfrak{S}_\infty \to \mathfrak{S}_{p^\infty} \to 0. \tag{5.1.12.21}$$

We denote by $\mathfrak{S}$ the kernel of the canonical homomorphism $\Delta' \to \Delta_\infty^{\diamond}$. We then have a canonical isomorphism

$$\mathfrak{S} \xrightarrow{\sim} \lim_{\substack{\longleftarrow \\ n\geq 1}} \pi_1(\overline{X}'^{(n)\star\triangleright}, \overline{y}'). \tag{5.1.12.22}$$

5.1.13 We can summarize the main constructions of this section in the following commutative diagram of extensions of normal integral domains

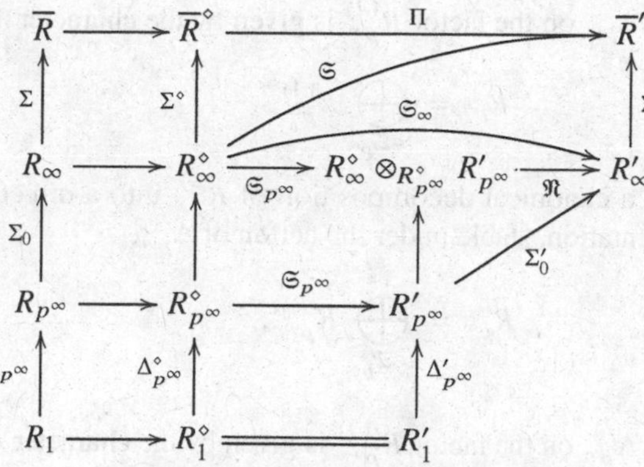

$$\begin{array}{ccccc}
\overline{R} & \longrightarrow & \overline{R}^{\circ} & \xrightarrow{\quad\Pi\quad} & \overline{R}' \\
\Sigma\uparrow & & \Sigma^{\circ}\uparrow & & \uparrow\Sigma' \\
R_{\infty} & \longrightarrow & R_{\infty}^{\circ} & \xrightarrow{\mathfrak{S}_{p^{\infty}}} R_{\infty}^{\circ}\otimes_{R_{p^{\infty}}^{\circ}}R_{p^{\infty}}' \xrightarrow{\mathfrak{N}} & R_{\infty}' \\
\Sigma_0\uparrow & & \uparrow & & \uparrow \\
R_{p^{\infty}} & \longrightarrow & R_{p^{\infty}}^{\circ} & \xrightarrow{\mathfrak{S}_{p^{\infty}}} R_{p^{\infty}}' & \\
\Delta_{p^{\infty}}\uparrow & & \Delta_{p^{\infty}}^{\circ}\uparrow & & \uparrow\Delta_{p^{\infty}}' \\
R_1 & \longrightarrow & R_1^{\circ} & ======= & R_1'
\end{array}$$
(5.1.13.1)

where for some integral extensions, we have denoted the Galois group of the associated extension of the fields of fractions. Observe that the homomorphism $R_1^{\circ} \to \overline{R}^{\circ}$ induces a Galois extension of the fields of fractions, and hence so is the homomorphism $R_{\infty}^{\circ} \to \overline{R}^{\circ}$. We denote by $\overline{F}^{\circ}$ (resp. F_{∞}°) the field of fractions of $\overline{R}^{\circ}$ (resp. R_{∞}°) and by Σ° the Galois group of the extension $\overline{F}^{\circ}/F_{\infty}^{\circ}$.

Proposition 5.1.14 ([2] 5.2.15)

(i) *The ring $R_{\infty}^{\circ} \otimes_{R_{p^{\infty}}^{\circ}} R_{p^{\infty}}'$ is a normal integral domain, and we have a canonical isomorphism* (5.1.12.21)

$$R_{\infty}^{\circ} \otimes_{R_{p^{\infty}}^{\circ}} R_{p^{\infty}}' \xrightarrow{\sim} (R_{\infty}')^{\mathfrak{N}}. \tag{5.1.14.1}$$

(ii) *For every $a \in \mathscr{O}_{\overline{K}}$, the canonical homomorphism*

$$(R_{\infty}^{\circ}/aR_{\infty}^{\circ}) \otimes_{R_{p^{\infty}}^{\circ}} R_{p^{\infty}}' \to (R_{\infty}'/aR_{\infty}')^{\mathfrak{N}} \tag{5.1.14.2}$$

is an isomorphism.

Proposition 5.1.15 *The R_{∞}°-algebra $\overline{R}^{\circ}$ is α-faithfully flat.*

This follows from ([2] 2.9.12 and 5.2.15) and ([3] V.12.4 and V.12.9).

5.1.16 We set

$$\Xi_{p^n} = \mathrm{Hom}(\Delta_{p^{\infty}}, \mu_{p^n}(\mathscr{O}_{\overline{K}})), \tag{5.1.16.1}$$

$$\Xi_{p^{\infty}} = \mathrm{Hom}(\Delta_{p^{\infty}}, \mu_{p^{\infty}}(\mathscr{O}_{\overline{K}})), \tag{5.1.16.2}$$

$$\Xi_{p^n}' = \mathrm{Hom}(\Delta_{p^{\infty}}', \mu_{p^n}(\mathscr{O}_{\overline{K}})), \tag{5.1.16.3}$$

$$\Xi_{p^{\infty}}' = \mathrm{Hom}(\Delta_{p^{\infty}}', \mu_{p^{\infty}}(\mathscr{O}_{\overline{K}})). \tag{5.1.16.4}$$

We identify Ξ_{p^n} (resp. Ξ_{p^n}') with a subgroup of $\Xi_{p^{\infty}}$ (resp. $\Xi_{p^{\infty}}'$). By ([3] II.8.9), there exists a canonical decomposition of $R_{p^{\infty}}$ into a direct sum of R_1-modules of finite presentation, stable under the action of $\Delta_{p^{\infty}}$,

$$R_{p^{\infty}} = \bigoplus_{\lambda \in \Xi_{p^{\infty}}} R_{p^{\infty}}^{(\lambda)}, \tag{5.1.16.5}$$

such that the action of Δ_{p^∞} on the factor $R^{(\lambda)}_{p^\infty}$ is given by the character λ. Moreover, for every $n \geq 0$, we have

$$R_{p^n} = \bigoplus_{\lambda \in \Xi_{p^n}} R^{(\lambda)}_{p^\infty}. \tag{5.1.16.6}$$

Similarly, there exists a canonical decomposition of R'_{p^∞} into a direct sum of R'_1-modules of finite presentation, stable under the action of Δ'_{p^∞},

$$R'_{p^\infty} = \bigoplus_{\lambda \in \Xi'_{p^\infty}} R'^{(\lambda)}_{p^\infty}, \tag{5.1.16.7}$$

such that the action of Δ'_{p^∞} on the factor $R'^{(\lambda)}_{p^\infty}$ is given by the character λ. Moreover, for every $n \geq 0$, we have

$$R'_{p^n} = \bigoplus_{\lambda \in \Xi'_{p^n}} R'^{(\lambda)}_{p^\infty}. \tag{5.1.16.8}$$

We set

$$\chi_{p^n} = \mathrm{Hom}(\mathfrak{S}_{p^\infty}, \mu_{p^n}(\mathscr{O}_{\overline{K}})), \tag{5.1.16.9}$$
$$\chi_{p^\infty} = \mathrm{Hom}(\mathfrak{S}_{p^\infty}, \mu_{p^\infty}(\mathscr{O}_{\overline{K}})). \tag{5.1.16.10}$$

We identify χ_{p^n} with a subgroup of χ_{p^∞}. By virtue of (5.1.12.20) and ([2] 5.2.14(iv)), the canonical sequence of $\mathbb{Z}_p$-modules

$$0 \to \mathfrak{S}_{p^\infty} \to \Delta'_{p^\infty} \to \Delta_{p^\infty} \to 0 \tag{5.1.16.11}$$

is exact and split. We deduce from this an exact sequence

$$0 \to \Xi_{p^\infty} \to \Xi'_{p^\infty} \to \chi_{p^\infty} \to 0. \tag{5.1.16.12}$$

Lemma 5.1.17 ([2] 5.2.24) *The canonical homomorphism $R^\diamond_{p^\infty} \to R'_{p^\infty}$ induces an isomorphism*

$$R^\diamond_{p^\infty} \xrightarrow{\sim} \bigoplus_{\lambda \in \Xi_{p^\infty}} R'^{(\lambda)}_{p^\infty}. \tag{5.1.17.1}$$

5.1.18 For each $\nu \in \chi_{p^\infty}$, fix a lifting $\widetilde{\nu} \in \Xi'_{p^\infty}$ (5.1.16.12) and set

$$R'^{(\nu)}_{p^\infty} = \bigoplus_{\lambda \in \Xi_{p^\infty}} R'^{(\widetilde{\nu}+\lambda)}_{p^\infty}, \tag{5.1.18.1}$$

which is naturally an $R^\diamond_{p^\infty}$-module (5.1.17). The decomposition (5.1.16.7) then induces a decomposition into $R^\diamond_{p^\infty}$-modules

$$R'_{p^\infty} = \bigoplus_{\nu \in \chi_{p^\infty}} R'^{(\nu)}_{p^\infty}, \tag{5.1.18.2}$$

such that the action of $\mathfrak{S}_{p^\infty}$ on the factor $R'^{(\nu)}_{p^\infty}$ is given by the character ν.

Lemma 5.1.19 ([2] 5.2.26) *For every integer $n \geq 0$, the canonical homomorphism $R^{\diamond}_{p^\infty} \otimes_{R^{\diamond}_{p^n}} R'_{p^n} \to R'_{p^\infty}$ induces an isomorphism*

$$R^{\diamond}_{p^\infty} \otimes_{R^{\diamond}_{p^n}} R'_{p^n} \xrightarrow{\sim} \bigoplus_{\nu \in \chi_{p^n}} R'^{(\nu)}_{p^\infty}. \qquad (5.1.19.1)$$

In particular, for every $\nu \in \chi_{p^n}$, the $R^{\diamond}_{p^\infty}$-module $R'^{(\nu)}_{p^\infty}$ is of finite presentation.

5.1.20 Let a be a nonzero element of $\mathscr{O}_{\overline{K}}$. Since the torsion subgroup of $P'^{\rm gp}/h^{\rm gp}(P^{\rm gp})$ is of order prime to p, we have, by ([2] (5.2.12.3)), a canonical isomorphism

$$(P'^{\rm gp}/h^{\rm gp}(P^{\rm gp})) \otimes_{\mathbb{Z}} \mathbb{Z}_p(-1) \xrightarrow{\sim} \mathrm{Hom}_{\mathbb{Z}_p}(\mathfrak{S}_{p^\infty}, \mathbb{Z}_p). \qquad (5.1.20.1)$$

We deduce, taking into account ([45] IV 1.1.4), an $R^{\diamond}_{\infty}$-linear isomorphism (5.1.4.3)

$$\widetilde{\Omega}^1_{R'/R} \otimes_{R'} (R^{\diamond}_{\infty}/aR^{\diamond}_{\infty})(-1) \xrightarrow{\sim} \mathrm{Hom}_{\mathbb{Z}}(\mathfrak{S}_{p^\infty}, R^{\diamond}_{\infty}/aR^{\diamond}_{\infty}). \qquad (5.1.20.2)$$

In the following, we interpret the target of this morphism as a cohomology group. We set

$$\delta = \dim_{\mathbb{Q}}((P'^{\rm gp}/h^{\rm gp}(P^{\rm gp})) \otimes_{\mathbb{Z}} \mathbb{Q}). \qquad (5.1.20.3)$$

Proposition 5.1.21 ([2] 5.2.32) *Let a be a nonzero element of $\mathscr{O}_{\overline{K}}$.*

(i) *There exists one and only one homomorphism of $\overline{R}^{\diamond}$-graded algebras*

$$\wedge(\widetilde{\Omega}^1_{R'/R} \otimes_{R'} (\overline{R}^{\diamond}/a\overline{R}^{\diamond})(-1)) \to \mathrm{H}^*(\Pi, \overline{R}'/a\overline{R}') \qquad (5.1.21.1)$$

whose component in degree one is induced by (5.1.20.2) (see 5.1.13). It is α-injective and its cokernel is annihilated by $p^{\frac{1}{p-1}}\mathfrak{m}_{\overline{K}}$.

(ii) *The $\overline{R}^{\diamond}$-module $\mathrm{H}^i(\Pi, \overline{R}'/a\overline{R}')$ is of α-finite presentation for every $i \geq 0$, and is α-zero for every $i \geq \delta + 1$ (5.1.20.3).*

(iii) *For any integers $r' \geq r \geq 0$, we denote by*

$$\hbar_{r,r'}: \mathrm{H}^*(\Pi, \overline{R}'/p^{r'}\overline{R}') \to \mathrm{H}^*(\Pi, \overline{R}'/p^r\overline{R}') \qquad (5.1.21.2)$$

the canonical morphism. Then, for every integer $r \geq 1$, there exists an integer $r' \geq r$, depending only on δ but not on the other data in 5.1.3, such that for every integer $r'' \geq r'$, the images of $\hbar_{r,r'}$ and $\hbar_{r,r''}$ are α-isomorphic.

5.2 Relative Galois Cohomologies of Higgs–Tate Algebras

5.2.1 The assumptions and notation of 5.1 are in effect in this section, in particular those of 5.1.3. Moreover, we take again the notation introduced in 3.1.14. We endow $\check{\overline{X}} = X \times_S \check{\overline{S}}$ and $\check{\overline{X}}' = X' \times_S \check{\overline{S}}$ (5.1.1.4) with the logarithmic structures $\mathscr{M}_{\check{\overline{X}}}$ and $\mathscr{M}_{\check{\overline{X}}'}$ pullbacks respectively of $\mathscr{M}_X$ and $\mathscr{M}_{X'}$. We assume that there exist smooth

$(\widetilde{S}, \mathscr{M}_{\widetilde{S}})$-deformations $(\widetilde{X}, \mathscr{M}_{\widetilde{X}})$ of $(\check{\overline{X}}, \mathscr{M}_{\check{\overline{X}}})$ and $(\widetilde{X}', \mathscr{M}_{\widetilde{X}'})$ of $(\check{\overline{X}}', \mathscr{M}_{\check{\overline{X}}'})$ (4.1.5.3) and an $(\widetilde{S}, \mathscr{M}_{\widetilde{S}})$-morphism

$$\widetilde{g} \colon (\widetilde{X}', \mathscr{M}_{\widetilde{X}'}) \to (\widetilde{X}, \mathscr{M}_{\widetilde{X}}) \tag{5.2.1.1}$$

that fits into a commutative diagram (with Cartesian squares)

$$\begin{array}{ccc} (\check{\overline{X}}', \mathscr{M}_{\check{\overline{X}}'}) & \longrightarrow & (\widetilde{X}', \mathscr{M}_{\widetilde{X}'}) \\ \downarrow & \square & \downarrow{\scriptstyle \widetilde{g}} \\ (\check{\overline{X}}, \mathscr{M}_{\check{\overline{X}}}) & \longrightarrow & (\widetilde{X}, \mathscr{M}_{\widetilde{X}}) \\ \downarrow & \square & \downarrow \\ (\check{\overline{S}}, \mathscr{M}_{\check{\overline{S}}}) & \longrightarrow & (\widetilde{S}, \mathscr{M}_{\widetilde{S}}). \end{array} \tag{5.2.1.2}$$

Observe that the squares are Cartesian both in the category of logarithmic schemes and in that of fine logarithmic schemes. *We fix in the remainder of this chapter the deformations and the $(\widetilde{S}, \mathscr{M}_{\widetilde{S}})$-morphism $\widetilde{g}$ (5.2.1.1).*

Lemma 5.2.2 *The morphism $\widetilde{g}$ (5.2.1.1) is smooth.*

Indeed, by virtue of ([41] 3.12 or [45] IV 3.2.3), it suffices to show that the canonical morphism

$$\widetilde{g}^*(\Omega^1_{(\widetilde{X}, \mathscr{M}_{\widetilde{X}})/(\widetilde{S}, \mathscr{M}_{\widetilde{S}})}) \to \Omega^1_{(\widetilde{X}', \mathscr{M}_{\widetilde{X}'})/(\widetilde{S}, \mathscr{M}_{\widetilde{S}})} \tag{5.2.2.1}$$

is locally left invertible, in other words that it locally induces an isomorphism from the left-hand side to a direct factor of the right-hand side, or equivalently that for every point x' of $\widetilde{X}'$ (or what amounts to the same of $\check{\overline{X}}'$), of image x of $\widetilde{X}$, the canonical morphism

$$\widetilde{g}^*(\Omega^1_{(\widetilde{X}, \mathscr{M}_{\widetilde{X}})/(\widetilde{S}, \mathscr{M}_{\widetilde{S}})})_{x'} \to \Omega^1_{(\widetilde{X}', \mathscr{M}_{\widetilde{X}'})/(\widetilde{S}, \mathscr{M}_{\widetilde{S}}), x'} \tag{5.2.2.2}$$

is left invertible. Since the left and right-hand sides of this morphism are free $\mathscr{O}_{\widetilde{X}', x'}$-modules of finite type, this last condition is equivalent to the fact that the canonical morphism

$$\Omega^1_{(\widetilde{X}, \mathscr{M}_{\widetilde{X}})/(\widetilde{S}, \mathscr{M}_{\widetilde{S}}), x} \otimes_{\mathscr{O}_{\widetilde{X}, x}} \kappa(x') \to \Omega^1_{(\widetilde{X}', \mathscr{M}_{\widetilde{X}'})/(\widetilde{S}, \mathscr{M}_{\widetilde{S}}), x'} \otimes_{\mathscr{O}_{\widetilde{X}', x'}} \kappa(x') \tag{5.2.2.3}$$

is injective by virtue of ([30] 0.19.1.12). The latter identifies with the morphism

$$\Omega^1_{(\check{\overline{X}}, \mathscr{M}_{\check{\overline{X}}})/(\check{\overline{S}}, \mathscr{M}_{\check{\overline{S}}}), x} \otimes_{\mathscr{O}_{\check{\overline{X}}, x}} \kappa(x') \to \Omega^1_{(\check{\overline{X}}', \mathscr{M}_{\check{\overline{X}}'})/(\check{\overline{S}}, \mathscr{M}_{\check{\overline{S}}}), x'} \otimes_{\mathscr{O}_{\check{\overline{X}}', x'}} \kappa(x') \tag{5.2.2.4}$$

induced by the smooth morphism $g \times_S \check{\overline{S}}$ (5.1.3.1). It is therefore injective, which proves the statement.

5.2.3 We set

$$\mathbb{X} = \mathrm{Spec}(\overline{R}) \quad \text{and} \quad \widehat{\mathbb{X}} = \mathrm{Spec}(\widehat{\overline{R}}), \tag{5.2.3.1}$$

that we endow with the logarithmic structures pullbacks of $\mathscr{M}_X$, denoted respectively by $\mathscr{M}_{\mathbb{X}}$ and $\mathscr{M}_{\widehat{\mathbb{X}}}$. Taking again the notation of 3.2.11, we denote by $(\widetilde{\mathbb{X}}, \mathscr{M}_{\widetilde{\mathbb{X}}})$ one of the two logarithmic schemes

$$(\mathscr{A}_2(\mathbb{X}), \mathscr{M}_{\mathscr{A}_2(\mathbb{X})}) \quad \text{or} \quad (\mathscr{A}_2^*(\mathbb{X}/S), \mathscr{M}_{\mathscr{A}_2^*(\mathbb{X}/S)}), \tag{5.2.3.2}$$

depending on whether we are in the absolute or relative case (3.1.14), and by

$$i_X \colon (\widehat{\mathbb{X}}, \mathscr{M}_{\widehat{\mathbb{X}}}) \to (\widetilde{\mathbb{X}}, \mathscr{M}_{\widetilde{\mathbb{X}}}) \tag{5.2.3.3}$$

the canonical exact closed immersion.

Likewise, we put

$$\mathbb{X}' = \mathrm{Spec}(\overline{R}') \quad \text{and} \quad \widehat{\mathbb{X}}' = \mathrm{Spec}(\widehat{\overline{R}}'), \tag{5.2.3.4}$$

that we endow with the logarithmic structures pullbacks of $\mathscr{M}_{X'}$, denoted respectively by $\mathscr{M}_{\mathbb{X}'}$ and $\mathscr{M}_{\widehat{\mathbb{X}}'}$. We denote by $(\widetilde{\mathbb{X}}', \mathscr{M}_{\widetilde{\mathbb{X}}'})$ one of the two logarithmic schemes

$$(\mathscr{A}_2(\mathbb{X}'), \mathscr{M}_{\mathscr{A}_2(\mathbb{X}')}) \quad \text{or} \quad (\mathscr{A}_2^*(\mathbb{X}'/S), \mathscr{M}_{\mathscr{A}_2^*(\mathbb{X}'/S)}), \tag{5.2.3.5}$$

depending on whether we are in the absolute or relative case (3.1.14), and by

$$i_{X'} \colon (\widehat{\mathbb{X}}', \mathscr{M}_{\widehat{\mathbb{X}}'}) \to (\widetilde{\mathbb{X}}', \mathscr{M}_{\widetilde{\mathbb{X}}'}) \tag{5.2.3.6}$$

the canonical exact closed immersion (see 3.2.11).

We have a canonical morphism of algebras $\overline{R} \to \overline{R}'$ (5.1.9) and a canonical homomorphism of monoids $Q_X \to Q_{X'}$, where Q_X and $Q_{X'}$ are the monoids defined in 3.2.5 with respect to the logarithmic schemes $(X, \mathscr{M}_X)$ and $(X', \mathscr{M}_{X'})$. The diagram of homomorphisms of monoids

$$\begin{array}{ccccc} P & \xrightarrow{\nu} & Q_X & \xrightarrow{\tau_X} & \mathrm{W}(\overline{R}^{\flat}) \\ \downarrow{\scriptstyle h} & & \downarrow & & \downarrow \\ P' & \xrightarrow{\nu'} & Q_{X'} & \xrightarrow{\tau_{X'}} & \mathrm{W}(\overline{R}'^{\flat}), \end{array} \tag{5.2.3.7}$$

where τ_X and $\tau_{X'}$ are the homomorphisms (3.2.5.2) and ν and ν' are the homomorphisms (3.2.6.5), is commutative.

We deduce from the above a morphism of $(\check{\overline{S}}, \mathscr{M}_{\check{\overline{S}}})$-logarithmic schemes

$$\widehat{\mathrm{g}} \colon (\widehat{\mathbb{X}}', \mathscr{M}_{\widehat{\mathbb{X}}'}) \to (\widehat{\mathbb{X}}, \mathscr{M}_{\widehat{\mathbb{X}}}), \tag{5.2.3.8}$$

and a morphism of $(\widetilde{S}, \mathscr{M}_{\widetilde{S}})$-logarithmic schemes

$$\widetilde{\mathrm{g}} \colon (\widetilde{\mathbb{X}}', \mathscr{M}_{\widetilde{\mathbb{X}}'}) \to (\widetilde{\mathbb{X}}, \mathscr{M}_{\widetilde{\mathbb{X}}}), \tag{5.2.3.9}$$

compatible with $\widehat{\mathfrak{g}}$.

Let U be a Zariski open subscheme of $\widehat{\mathbb{X}}$ and U' a Zariski open subscheme of $\widehat{\mathbb{X}}'$ such that $\widehat{\mathfrak{g}}(U') \subset U$. We denote by $\widetilde{U}$ the open subscheme of $\widetilde{\mathbb{X}}$ corresponding to U and by $\widetilde{U}'$ the open subscheme of $\widetilde{\mathbb{X}}'$ corresponding to U'.

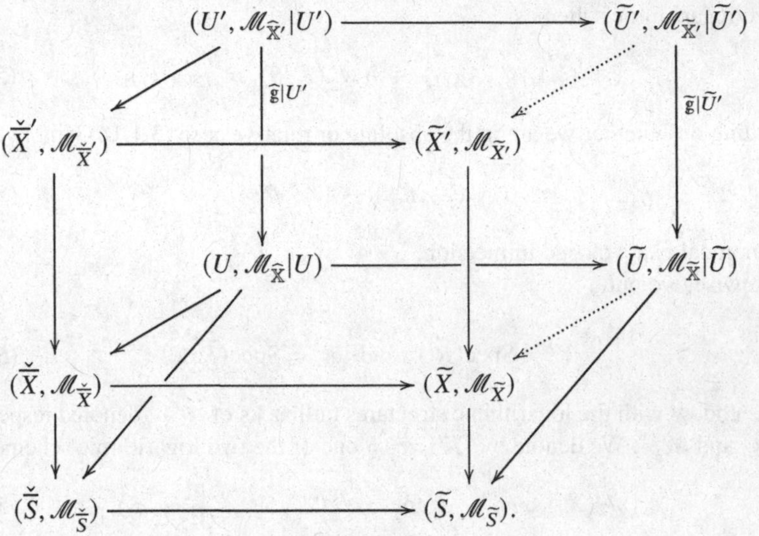

(5.2.3.10)

5.2.4 We set (3.1.15.3)

$$\mathrm{T} = \mathrm{Hom}_{\widehat{\overline{R}}}(\widetilde{\Omega}^1_{R/\mathscr{O}_K} \otimes_R \widehat{\overline{R}}, \xi\widehat{\overline{R}}). \tag{5.2.4.1}$$

Its dual $\widehat{\overline{R}}$-module identifies with $\widetilde{\xi}^{-1}\widetilde{\Omega}^1_{R/\mathscr{O}_K} \otimes_R \widehat{\overline{R}}$ (see 3.1.14); we denote by $\mathscr{G}$ the associated symmetric $\widehat{\overline{R}}$-algebra (2.1.10)

$$\mathscr{G} = \mathrm{S}_{\widehat{\overline{R}}}(\widetilde{\xi}^{-1}\widetilde{\Omega}^1_{R/\mathscr{O}_K} \otimes_R \widehat{\overline{R}}). \tag{5.2.4.2}$$

We denote by $\widehat{\mathbb{X}}_{\mathrm{zar}}$ the Zariski topos of $\widehat{\mathbb{X}}$, by $\widetilde{\mathrm{T}}$ the $\mathscr{O}_{\widehat{\mathbb{X}}}$-module associated with T and by $\mathbf{T}$ the vector bundle associated with its dual, i.e.,

$$\mathbf{T} = \mathrm{Spec}(\mathscr{G}). \tag{5.2.4.3}$$

We set, likewise,

$$\mathrm{T}' = \mathrm{Hom}_{\widehat{\overline{R}}'}(\widetilde{\Omega}^1_{R'/\mathscr{O}_K} \otimes_{R'} \widehat{\overline{R}}', \xi\widehat{\overline{R}}'). \tag{5.2.4.4}$$

Its dual $\widehat{\overline{R}}'$-module identifies with $\widetilde{\xi}^{-1}\widetilde{\Omega}^1_{R'/\mathscr{O}_K} \otimes_{R'} \widehat{\overline{R}}'$; we denote by $\mathscr{G}'$ the associated symmetric $\widehat{\overline{R}}'$-algebra

$$\mathscr{G}' = \mathrm{S}_{\overline{\widehat{R}}'}(\widetilde{\xi}^{-1}\widetilde{\Omega}^1_{R'/\mathscr{O}_K} \otimes_{R'} \overline{\widehat{R}}'). \tag{5.2.4.5}$$

We denote by $\widehat{\mathbb{X}}'_{\text{zar}}$ the Zariski topos of $\widehat{\mathbb{X}}'$, by $\widetilde{\mathrm{T}}'$ the $\mathscr{O}_{\widehat{\mathbb{X}}'}$-module associated with T′ and by $\mathbf{T}'$ the $\widehat{\mathbb{X}}'$-vector bundle associated with its dual, i.e.,

$$\mathbf{T}' = \mathrm{Spec}(\mathscr{G}'). \tag{5.2.4.6}$$

Finally, we set

$$\mathrm{T}_{R'/R} = \mathrm{Hom}_{\overline{\widehat{R}}'}(\widetilde{\Omega}^1_{R'/R} \otimes_{R'} \overline{\widehat{R}}', \widetilde{\xi}\overline{\widehat{R}}'). \tag{5.2.4.7}$$

Its dual $\overline{\widehat{R}}'$-module identifies with $\widetilde{\xi}^{-1}\widetilde{\Omega}^1_{R'/R} \otimes_{R'} \overline{\widehat{R}}'$; we denote by $\mathscr{G}_{R'/R}$ the associated symmetric $\overline{\widehat{R}}'$-algebra (2.1.10)

$$\mathscr{G}_{R'/R} = \mathrm{S}_{\overline{\widehat{R}}'}(\widetilde{\xi}^{-1}\widetilde{\Omega}^1_{R'/R} \otimes_{R'} \overline{\widehat{R}}'). \tag{5.2.4.8}$$

We denote by $\widetilde{\mathrm{T}}_{X'/X}$ the $\mathscr{O}_{\widehat{\mathbb{X}}'}$-module associated with $\mathrm{T}_{R'/R}$ and by $\mathbf{T}_{X'/X}$ the $\widehat{\mathbb{X}}'$-vector bundle associated with its dual, i.e.,

$$\mathbf{T}_{X'/X} = \mathrm{Spec}(\mathscr{G}_{R'/R}). \tag{5.2.4.9}$$

The split exact sequence (5.1.4.4) induces an exact sequence of $\mathscr{O}_{\widehat{\mathbb{X}}'}$-modules

$$0 \longrightarrow \widetilde{\mathrm{T}}_{X'/X} \xrightarrow{\iota} \widetilde{\mathrm{T}}' \xrightarrow{u} \widehat{\mathfrak{g}}^*(\widetilde{\mathrm{T}}) \longrightarrow 0. \tag{5.2.4.10}$$

We deduce from this an exact sequence of $\widehat{\mathbb{X}}'$-vector bundles

$$0 \longrightarrow \mathbf{T}_{X'/X} \longrightarrow \mathbf{T}' \longrightarrow \mathbf{T} \times_{\widehat{\mathbb{X}}} \widehat{\mathbb{X}}' \longrightarrow 0. \tag{5.2.4.11}$$

5.2.5 Let U be a Zariski open subscheme of $\widehat{\mathbb{X}}$, $\widetilde{U}$ the corresponding open subscheme of $\widetilde{\mathbb{X}}$ (5.2.3). We denote by $\mathscr{L}(U)$ the set of $\widetilde{S}$-morphisms represented by dotted arrows that complete the lower face of the parallelepiped of the diagram (5.2.3.10) in such a way that it remains commutative. By 3.2.14, the functor $U \mapsto \mathscr{L}(U)$ is a $\widetilde{\mathrm{T}}$-torsor of $\widehat{\mathbb{X}}_{\text{zar}}$; it is the Higgs–Tate torsor associated with $(\widetilde{X}, \mathscr{M}_{\widetilde{X}})$. The $\widetilde{\mathrm{T}}$-torsor $\mathscr{L}$ is naturally endowed with a Δ-equivariant structure (see 3.2.15 and 5.1.9). We denote by $\mathscr{F}$ the $\widehat{\overline{R}}$-module of affine functions on $\mathscr{L}$ (see [3] II.4.9). This fits into a canonical exact sequence

$$0 \to \widehat{\overline{R}} \to \mathscr{F} \to \widetilde{\xi}^{-1}\widetilde{\Omega}^1_{R/\mathscr{O}_K} \otimes_R \widehat{\overline{R}} \to 0. \tag{5.2.5.1}$$

The $\widehat{\overline{R}}$-module $\mathscr{F}$ is naturally endowed with an $\widehat{\overline{R}}$-semi-linear action of Δ such that the morphisms of the sequence (5.2.5.1) are Δ-equivariant (see 3.2.15). By ([37] I 4.3.1.7), this sequence induces for every integer $n \geq 1$ an exact sequence (2.1.10)

$$0 \to \mathrm{S}^{n-1}_{\widehat{\overline{R}}}(\mathscr{F}) \to \mathrm{S}^n_{\widehat{\overline{R}}}(\mathscr{F}) \to \mathrm{S}^n_{\widehat{\overline{R}}}(\widetilde{\xi}^{-1}\widetilde{\Omega}^1_{R/\mathscr{O}_K} \otimes_R \widehat{\overline{R}}) \to 0. \tag{5.2.5.2}$$

The $\widehat{\overline{R}}$-modules $(\mathrm{S}^n_{\widehat{\overline{R}}}(\mathscr{F}))_{n\in\mathbb{N}}$ therefore form a filtered direct system, whose direct limit

$$\mathscr{C} = \lim_{\substack{\longrightarrow \\ n\geq 0}} \mathrm{S}^n_{\widehat{\overline{R}}}(\mathscr{F}) \tag{5.2.5.3}$$

is naturally endowed with the structure of an $\widehat{\overline{R}}$-algebra and an action of Δ by ring automorphisms, compatible with its action on $\widehat{\overline{R}}$; it is the Higgs–Tate algebra associated with $(\widetilde{X}, \mathscr{M}_{\widetilde{X}})$ (3.2.16). By ([3] II.4.10), the $\widehat{\mathbb{X}}$-scheme

$$\mathbf{L} = \mathrm{Spec}(\mathscr{C}) \tag{5.2.5.4}$$

is naturally a homogeneous principal $\mathbf{T}$-bundle over $\widehat{\mathbb{X}}$ that canonically represents $\mathscr{L}$. It is endowed with a Δ-equivariant structure (see 3.2.15). This structure determines a left action of Δ on $\mathbf{L}$ compatible with its action on $\widehat{\mathbb{X}}$.

5.2.6 Let U' be a Zariski open subscheme of $\widehat{\mathbb{X}}'$, $\widetilde{U}'$ the corresponding open subscheme of $\widetilde{\mathbb{X}}'$. We denote by $\mathscr{L}'(U')$ the set of $\widetilde{S}$-morphisms represented by the dotted arrows that complete the upper face of the parallelepiped of the diagram (5.2.3.10) in such a way that it remains commutative. The functor $U' \mapsto \mathscr{L}'(U')$ is a $\widetilde{\mathrm{T}}'$-torsor of $\widehat{\mathbb{X}}'_{\mathrm{zar}}$; it is the Higgs–Tate torsor associated with $(\widetilde{X}', \mathscr{M}_{\widetilde{X}'})$. The $\widetilde{\mathrm{T}}'$-torsor $\mathscr{L}'$ is naturally endowed with a Δ'-equivariant structure (see 3.2.15 and 5.1.9). We denote by $\mathscr{F}'$ the $\widehat{\overline{R}'}$-module of affine functions on $\mathscr{L}'$. This fits into a canonical exact sequence

$$0 \to \widehat{\overline{R}'} \to \mathscr{F}' \to \widetilde{\xi}^{-1}\widetilde{\Omega}^1_{R'/\mathscr{O}_K} \otimes_{R'} \widehat{\overline{R}'} \to 0. \tag{5.2.6.1}$$

The $\widehat{\overline{R}'}$-module $\mathscr{F}'$ is naturally endowed with an $\widehat{\overline{R}'}$-semi-linear action of Δ' such that the morphisms of the sequence (5.2.6.1) are Δ'-equivariant (see 3.2.15). This sequence induces for every integer $n \geq 1$ an exact sequence (2.1.10)

$$0 \to \mathrm{S}^{n-1}_{\widehat{\overline{R}'}}(\mathscr{F}') \to \mathrm{S}^n_{\widehat{\overline{R}'}}(\mathscr{F}') \to \mathrm{S}^n_{\widehat{\overline{R}'}}(\widetilde{\xi}^{-1}\widetilde{\Omega}^1_{R'/\mathscr{O}_K} \otimes_{R'} \widehat{\overline{R}'}) \to 0. \tag{5.2.6.2}$$

The $\widehat{\overline{R}}$-modules $(\mathrm{S}^n_{\widehat{\overline{R}'}}(\mathscr{F}'))_{n\in\mathbb{N}}$ therefore form a filtered direct system, whose direct limit

$$\mathscr{C}' = \lim_{\substack{\longrightarrow \\ n\geq 0}} \mathrm{S}^n_{\widehat{\overline{R}'}}(\mathscr{F}') \tag{5.2.6.3}$$

is naturally endowed with the structure of an $\widehat{\overline{R}'}$-algebra and an action of Δ' by ring automorphisms, compatible with its action on $\widehat{\overline{R}'}$; it is the Higgs–Tate algebra associated with $(\widetilde{X}', \mathscr{M}_{\widetilde{X}'})$ (3.2.16). The $\widehat{\mathbb{X}}'$-scheme

$$\mathbf{L}' = \mathrm{Spec}(\mathscr{C}') \tag{5.2.6.4}$$

is naturally a homogeneous principal $\mathbf{T}'$-bundle over $\widehat{\mathbb{X}}'$ that canonically represents $\mathscr{L}'$. Then, we have a canonical $\widehat{\mathbb{X}}'$-morphism

$$\mathbf{T}' \times_{\widehat{\mathbb{X}}'} \mathbf{L}' \to \mathbf{L}'. \tag{5.2.6.5}$$

The principal homogeneous $\mathbf{T}'$-bundle $\mathbf{L}'$ is endowed with a Δ'-equivariant structure (see 3.2.15). This structure determines an action on the left of Δ' on $\mathbf{L}'$ compatible with its action on $\widehat{\mathbb{X}}'$.

5.2.7 We denote by $\mathscr{L}^+$ the affine pullback $\widehat{\mathfrak{g}}^+(\mathscr{L})$ of the $\widetilde{\mathrm{T}}$-torsor $\mathscr{L}$ by the morphism $\widehat{\mathfrak{g}}\colon \widehat{\mathbb{X}}' \to \widehat{\mathbb{X}}$ ([3] II.4.5), i.e., the $\widehat{\mathfrak{g}}^*(\widetilde{\mathrm{T}})$-torsor of $\widehat{\mathbb{X}}'_{\mathrm{zar}}$ deduced from the $\widehat{\mathfrak{g}}^{-1}(\widetilde{\mathrm{T}})$-torsor $\widehat{\mathfrak{g}}^*(\mathscr{L})$ by extension of its structural group by the canonical homomorphism $\widehat{\mathfrak{g}}^{-1}(\widetilde{\mathrm{T}}) \to \widehat{\mathfrak{g}}^*(\widetilde{\mathrm{T}})$:

$$\widehat{\mathfrak{g}}^+(\mathscr{L}) = \widehat{\mathfrak{g}}^*(\mathscr{L}) \wedge^{\widehat{\mathfrak{g}}^{-1}(\widetilde{\mathrm{T}})} \widehat{\mathfrak{g}}^*(\widetilde{\mathrm{T}}). \tag{5.2.7.1}$$

Since $\widehat{\mathbb{X}}$ is affine, the $\widetilde{\mathrm{T}}$-torsor $\mathscr{L}$ is trivial; let $\sigma \in \mathscr{L}(\widehat{\mathbb{X}})$. On the other hand, the morphism $(\widetilde{X}', \mathscr{M}_{\widetilde{X}'}) \to (\widetilde{X}, \mathscr{M}_{\widetilde{X}})$ being smooth and $\widehat{\mathbb{X}}'$ being affine, there exists a $\sigma' \in \mathscr{L}'(\widehat{\mathbb{X}}')$ that completes the right side face of the parallelepiped of the diagram (5.2.3.10) (for $U = \widehat{\mathbb{X}}$ and $U' = \widehat{\mathbb{X}}'$) in such a way that it remains commutative. In the following, we will say that σ' is a lifting of σ, or that σ and σ' are *compatible*. Then, there exists a unique u-equivariant morphism (5.2.4.10)

$$\nu\colon \mathscr{L}' \to \mathscr{L}^+ \tag{5.2.7.2}$$

that sends σ' to the canonical image of σ in $\mathscr{L}^+(\widehat{\mathbb{X}}')$. It immediately follows from 5.2.8 below that the morphism ν does not depend on the choice of the pair of compatible sections (σ, σ').

By ([3] II.4.12 and II.4.13), the morphism ν induces an $\widehat{\overline{R}'}$-linear morphism

$$\mathscr{F} \otimes_{\widehat{\overline{R}}} \widehat{\overline{R}'} \to \mathscr{F}' \tag{5.2.7.3}$$

that fits into a commutative diagram

$$\begin{array}{ccccccccc} 0 & \longrightarrow & \widehat{\overline{R}'} & \longrightarrow & \mathscr{F} \otimes_{\widehat{\overline{R}}} \widehat{\overline{R}'} & \longrightarrow & \xi^{-1}\widetilde{\Omega}^1_{R/\mathscr{O}_K} \otimes_R \widehat{\overline{R}'} & \longrightarrow & 0 \\ & & \| & & \downarrow & & \downarrow & & \\ 0 & \longrightarrow & \widehat{\overline{R}'} & \longrightarrow & \mathscr{F}' & \longrightarrow & \xi^{-1}\widetilde{\Omega}^1_{R'/\mathscr{O}_K} \otimes_{R'} \widehat{\overline{R}'} & \longrightarrow & 0 \end{array} \tag{5.2.7.4}$$

where the rows are induced by the exact sequences (5.2.5.1) and (5.2.6.1) and the third vertical arrow is induced by the canonical morphism (5.1.4.4).

By ([3] II.4.6), the morphism ν (5.2.7.2) induces a $\mathbf{T}'$-equivariant morphism (5.2.4.11)

$$\nu\colon \mathbf{L}' \to \mathbf{L} \times_{\widehat{\mathbb{X}}} \widehat{\mathbb{X}}', \tag{5.2.7.5}$$

and therefore a homomorphism of $\widehat{\overline{R}}$-algebras

$$\mathscr{C} \to \mathscr{C}'. \tag{5.2.7.6}$$

The latter extends the $\widehat{\overline{R}}$-linear morphism $\mathscr{F} \to \mathscr{F}'$ induced by (5.2.7.3) and can be deduced from it in view of (5.2.7.4) and the definitions (5.2.5.3) and (5.2.6.3).

The morphism $\widehat{\mathrm{g}}$ being Δ'-equivariant (5.1.9), the principal homogeneous $(\mathbf{T}\times_{\widehat{\mathbb{X}}}\widehat{\mathbb{X}}')$-bundle $\mathbf{L} \times_{\widehat{\mathbb{X}}} \widehat{\mathbb{X}}'$ over $\widehat{\mathbb{X}}'$ is naturally endowed with a Δ'-equivariant structure.

Lemma 5.2.8 *Let* $(\sigma, \sigma'), (\sigma_1, \sigma_1') \in \mathscr{L}(\widehat{\mathbb{X}}) \times \mathscr{L}'(\widehat{\mathbb{X}}')$ *be two pairs of compatible sections* (5.2.7). *Then, we have*

$$u(\sigma_1' - \sigma') = \sigma_1 - \sigma \in \mathrm{T} \otimes_{\widehat{\overline{R}}} \widehat{\overline{R}'}, \tag{5.2.8.1}$$

where u is the morphism defined in (5.2.4.10). *In particular, the morphism* ν (5.2.7.2) *does not depend on the choice of the pair of compatible sections which is used to define it.*

This immediately follows by functoriality from the diagram (5.2.3.10).

Lemma 5.2.9 *The morphism* $\mathbf{T}_{X'/X} \times_{\widehat{\mathbb{X}}'} \mathbf{L}' \to \mathbf{L}$ *induced by the morphisms* (5.2.6.5) *and* $\mathbf{T}_{X'/X} \to \mathbf{T}'$ (5.2.4.11) *define on* $\mathbf{L}'$ *the structure of a principal homogeneous* $\mathbf{L} \times_{\widehat{\mathbb{X}}} \mathbf{T}_{X'/X}$*-bundle over* $\mathbf{L} \times_{\widehat{\mathbb{X}}} \widehat{\mathbb{X}}'$ (5.2.7.5).

This immediately follows from the exact sequence of $\widehat{\mathbb{X}}'$-vector bundles (5.2.4.11) and from the fact that the morphism $\nu\colon \mathbf{L}' \to \mathbf{L} \times_{\widehat{\mathbb{X}}} \widehat{\mathbb{X}}'$ (5.2.7.5) is $\mathbf{T}'$-equivariant.

Lemma 5.2.10 *The morphism* $\nu\colon \mathbf{L}' \to \mathbf{L} \times_{\widehat{\mathbb{X}}} \widehat{\mathbb{X}}'$ (5.2.7.5) *is* Δ'*-equivariant. In particular, the homomorphism* $\mathscr{C} \to \mathscr{C}'$ (5.2.7.6) *is* Δ'*-equivariant.*

Let $(\sigma, \sigma') \in \mathscr{L}(\widehat{\mathbb{X}}) \times \mathscr{L}'(\widehat{\mathbb{X}}')$ be a pair of compatible sections (5.2.7), $c \in \Delta'$. We denote by ${}^c\sigma$ the section of $\mathscr{L}(\widehat{\mathbb{X}})$ defined by the composed morphism (3.2.15.13)

$$(\widetilde{\mathbb{X}}, \mathscr{M}_{\widetilde{\mathbb{X}}}) \xrightarrow{c^{-1}} (\widetilde{\mathbb{X}}, \mathscr{M}_{\widetilde{\mathbb{X}}}) \xrightarrow{\sigma} (\widetilde{X}, \mathscr{M}_{\widetilde{X}}) \tag{5.2.10.1}$$

and by ${}^c\sigma'$ the section of $\mathscr{L}'(\widehat{\mathbb{X}}')$ defined by the composed morphism

$$(\widetilde{\mathbb{X}}', \mathscr{M}_{\widetilde{\mathbb{X}}'}) \xrightarrow{c^{-1}} (\widetilde{\mathbb{X}}', \mathscr{M}_{\widetilde{\mathbb{X}}'}) \xrightarrow{\sigma'} (\widetilde{X}', \mathscr{M}_{\widetilde{X}'}). \tag{5.2.10.2}$$

Since the morphism $\widetilde{\mathrm{g}}\colon (\widetilde{\mathbb{X}}', \mathscr{M}_{\widetilde{\mathbb{X}}'}) \to (\widetilde{\mathbb{X}}, \mathscr{M}_{\widetilde{\mathbb{X}}})$ is Δ'-equivariant, the sections ${}^c\sigma$ and ${}^c\sigma'$ are compatible (5.2.3.10). Consequently, $\nu({}^c\sigma') = {}^c\sigma$ and therefore $\nu(c(\sigma')) = c(\sigma)$ by virtue of (3.2.15.11). Since the canonical homomorphism $\mu\colon \mathbf{T}' \to \mathbf{T} \times_{\widehat{\mathbb{X}}} \widehat{\mathbb{X}}'$ (5.2.4.11) is Δ'-equivariant and ν is μ-equivariant, we deduce from this that ν is Δ'-equivariant.

5.2.11 Consider smooth $(\widetilde{S}, \mathscr{M}_{\widetilde{S}})$-deformations $(\widetilde{X}^{\natural}, \mathscr{M}_{\widetilde{X}^{\natural}})$ of $(\check{\overline{X}}, \mathscr{M}_{\check{\overline{X}}})$ and $(\widetilde{X}'^{\natural}, \mathscr{M}_{\widetilde{X}'^{\natural}})$ of $(\check{\overline{X}}', \mathscr{M}_{\check{\overline{X}}'})$ (4.1.5) and a smooth $(\widetilde{S}, \mathscr{M}_{\widetilde{S}})$-morphism

$$g^{\natural}\colon (\widetilde{X}'^{\natural}, \mathscr{M}_{\widetilde{X}'^{\natural}}) \to (\widetilde{X}^{\natural}, \mathscr{M}_{\widetilde{X}^{\natural}}) \tag{5.2.11.1}$$

that fits into a commutative diagram

$$\begin{array}{ccc} (\check{\check{X}}', \mathscr{M}_{\check{\check{X}}'}) & \longrightarrow & (\widetilde{X}'^{\natural}, \mathscr{M}_{\widetilde{X}'^{\natural}}) \\ \downarrow & & \downarrow {\scriptstyle \widetilde{g}^{\natural}} \\ (\check{\check{X}}, \mathscr{M}_{\check{\check{X}}}) & \longrightarrow & (\widetilde{X}^{\natural}, \mathscr{M}_{\widetilde{X}^{\natural}}) \\ \downarrow & & \downarrow \\ (\check{\check{S}}, \mathscr{M}_{\check{\check{S}}}) & \longrightarrow & (\widetilde{S}, \mathscr{M}_{\widetilde{S}}). \end{array} \tag{5.2.11.2}$$

By ([41] 3.14), there exists an isomorphism of $(\widetilde{S}, \mathscr{M}_{\widetilde{S}})$-deformations

$$h\colon (\widetilde{X}, \mathscr{M}_{\widetilde{X}}) \xrightarrow{\sim} (\widetilde{X}^{\natural}, \mathscr{M}_{\widetilde{X}^{\natural}}). \tag{5.2.11.3}$$

Consider $(\widetilde{X}', \mathscr{M}_{\widetilde{X}'})$ as a smooth $(\widetilde{X}, \mathscr{M}_{\widetilde{X}})$-deformation of $(\check{\check{X}}', \mathscr{M}_{\check{\check{X}}'})$ and $(\widetilde{X}'^{\natural}, \mathscr{M}_{\widetilde{X}'^{\natural}})$ as a smooth $(\widetilde{X}^{\natural}, \mathscr{M}_{\widetilde{X}^{\natural}})$-deformation of $(\check{\check{X}}', \mathscr{M}_{\check{\check{X}}'})$. By ([41] 3.14), there exists an isomorphism of $(\widetilde{S}, \mathscr{M}_{\widetilde{S}})$-deformations

$$h'\colon (\widetilde{X}', \mathscr{M}_{\widetilde{X}'}) \xrightarrow{\sim} (\widetilde{X}'^{\natural}, \mathscr{M}_{\widetilde{X}'^{\natural}}) \tag{5.2.11.4}$$

that fits into a commutative diagram

$$\begin{array}{ccc} (\widetilde{X}', \mathscr{M}_{\widetilde{X}'}) & \xrightarrow{h'} & (\widetilde{X}'^{\natural}, \mathscr{M}_{\widetilde{X}'^{\natural}}) \\ {\scriptstyle \widetilde{g}} \downarrow & & \downarrow {\scriptstyle \widetilde{g}^{\natural}} \\ (\widetilde{X}, \mathscr{M}_{\widetilde{X}}) & \xrightarrow{h} & (\widetilde{X}^{\natural}, \mathscr{M}_{\widetilde{X}^{\natural}}). \end{array} \tag{5.2.11.5}$$

We assign an exponent $\natural$ to the objects associated with these deformations (see 5.2.5 and 5.2.6). The isomorphism of $\widetilde{\mathrm{T}}$-torsors $\mathscr{L} \xrightarrow{\sim} \mathscr{L}^{\natural}$, $\psi \mapsto h \circ \psi$ (5.2.3.10) induces an $\widehat{\overline{R}}$-linear and Δ-equivariant isomorphism

$$\mathscr{F}^{\natural} \xrightarrow{\sim} \mathscr{F}, \tag{5.2.11.6}$$

that fits into a commutative diagram (5.2.5.1)

$$\begin{array}{ccccccccc} 0 & \longrightarrow & \widehat{\overline{R}} & \longrightarrow & \mathscr{F}^{\natural} & \longrightarrow & \widetilde{\xi}^{-1}\widetilde{\Omega}^1_{R/\mathscr{O}_K} \otimes_R \widehat{\overline{R}} & \longrightarrow & 0 \\ & & \| & & \downarrow & & \| & & \\ 0 & \longrightarrow & \widehat{\overline{R}} & \longrightarrow & \mathscr{F} & \longrightarrow & \widetilde{\xi}^{-1}\widetilde{\Omega}^1_{R/\mathscr{O}_K} \otimes_R \widehat{\overline{R}} & \longrightarrow & 0. \end{array} \tag{5.2.11.7}$$

We deduce from this a Δ-equivariant $\overline{\widehat{R}}$-isomorphism

$$\mathscr{C}^{\natural} \xrightarrow{\sim} \mathscr{C}. \tag{5.2.11.8}$$

The isomorphism of $\widetilde{\mathrm{T}}'$-torsors $\mathscr{L}' \xrightarrow{\sim} \mathscr{L}'^{\natural}$, $\psi' \mapsto h' \circ \psi'$ (5.2.3.10) induces an $\widehat{\overline{R}'}$-linear and Δ'-equivariant isomorphism

$$\mathscr{F}'^{\natural} \xrightarrow{\sim} \mathscr{F}', \tag{5.2.11.9}$$

that fits into a commutative diagram (5.2.6.1)

$$\begin{array}{ccccccccc} 0 & \longrightarrow & \widehat{\overline{R}'} & \longrightarrow & \mathscr{F}'^{\natural} & \longrightarrow & \xi^{-1}\widetilde{\Omega}^1_{R'/\mathscr{O}_K} \otimes_{R'} \widehat{\overline{R}'} & \longrightarrow & 0 \\ & & \| & & \downarrow & & \| & & \\ 0 & \longrightarrow & \widehat{\overline{R}'} & \longrightarrow & \mathscr{F}' & \longrightarrow & \xi^{-1}\widetilde{\Omega}^1_{R'/\mathscr{O}_K} \otimes_{R'} \widehat{\overline{R}'} & \longrightarrow & 0. \end{array} \tag{5.2.11.10}$$

We deduce from this a Δ'-equivariant $\widehat{\overline{R}'}$-isomorphism

$$\mathscr{C}'^{\natural} \xrightarrow{\sim} \mathscr{C}'. \tag{5.2.11.11}$$

We immediately check (5.2.7) that the diagram

$$\begin{array}{ccc} \mathscr{C} & \longrightarrow & \mathscr{C}' \\ \downarrow & & \downarrow \\ \mathscr{C}^{\natural} & \longrightarrow & \mathscr{C}'^{\natural}, \end{array} \tag{5.2.11.12}$$

where the vertical arrows are the isomorphisms (5.2.11.8) and (5.2.11.11) and the horizontal arrows are the homomorphisms (5.2.7.6), is commutative.

5.2.12 In this paragraph, we exceptionally replace the assumption that g admits a relative adequate chart (5.1.3.2) by the following assumptions: the morphisms f and f' admit adequate charts and the logarithmic scheme $(X', \mathscr{M}_{X'})$ admits a fine and saturated chart $M' \to \Gamma(X', \mathscr{M}_{X'})$ inducing an isomorphism

$$M' \xrightarrow{\sim} \Gamma(X', \mathscr{M}_{X'})/(X', \mathscr{O}^{\times}_{X'}). \tag{5.2.12.1}$$

These three charts are a priori independent of each other. Moreover, we fix an adequate chart $((P,\gamma), (\mathbb{N}, \iota), \vartheta)$ for f (3.1.15). We can then consider the commutative diagram

$$\begin{array}{ccc} (\widehat{\mathbb{X}}', \mathscr{M}_{\widehat{\mathbb{X}}'}) & \xrightarrow{i'_{X'}} & (\widetilde{\mathbb{X}}', \mathscr{M}'_{\widetilde{\mathbb{X}}'}) \\ {\scriptstyle \widehat{\mathfrak{g}}}\downarrow & & \downarrow \\ (\widehat{\mathbb{X}}, \mathscr{M}_{\widehat{\mathbb{X}}}) & \xrightarrow{i_X} & (\widetilde{\mathbb{X}}, \mathscr{M}_{\widetilde{\mathbb{X}}}), \end{array} \tag{5.2.12.2}$$

where $\widehat{\mathfrak{g}}$ is the morphism defined in (5.2.3.8), $\mathscr{M}'_{\widetilde{\mathbb{X}}'}$ (resp. $\mathscr{M}_{\widetilde{\mathbb{X}}}$) is the logarithmic structure on $\widetilde{\mathbb{X}}'$ (resp. $\widetilde{\mathbb{X}}$) defined in 3.2.11 and $i'_{X'}$ (resp. i_X) is the exact closed immersion (3.2.11.4) (resp. (3.2.11.5)). Observe here the difference between $\mathscr{M}'_{\widetilde{\mathbb{X}}'}$ and $\mathscr{M}_{\widetilde{\mathbb{X}}}$ (see 3.2.13). We then have the representations $\mathscr{F}$ and $\mathscr{C}$ of Δ defined in 5.2.5 with respect to i_X. Adapting 5.2.6, we define the representations $\mathscr{F}'$ and $\mathscr{C}'$ of Δ' with respect to $i'_{X'}$. We immediately see that the results 5.2.7–5.2.10 still hold. In particular, there exists a canonical $\widehat{\overline{R}}$-linear and Δ'-equivariant morphism

$$\mathscr{F} \to \mathscr{F}', \tag{5.2.12.3}$$

and a Δ'-equivariant homomorphism of $\widehat{\overline{R}}$-algebras

$$\mathscr{C} \to \mathscr{C}' \tag{5.2.12.4}$$

that extends it.

5.2.13 For any rational number $r \geq 0$, we denote by $\mathscr{F}^{(r)}$ the $\widehat{\overline{R}}$-representation of Δ deduced from $\mathscr{F}$ (5.2.5.1) by pullback by the multiplication by p^r on $\widetilde{\xi}^{-1}\widetilde{\Omega}^1_{R/\mathscr{O}_K} \otimes_R \widehat{\overline{R}}$, so that we have a locally split exact sequence of $\widehat{\overline{R}}$-modules

$$0 \to \widehat{\overline{R}} \longrightarrow \mathscr{F}^{(r)} \to \widetilde{\xi}^{-1}\widetilde{\Omega}^1_{R/\mathscr{O}_K} \otimes_R \widehat{\overline{R}} \to 0. \tag{5.2.13.1}$$

This induces for every integer $n \geq 1$ an exact sequence

$$0 \to \mathrm{S}^{n-1}_{\widehat{\overline{R}}}(\mathscr{F}^{(r)}) \to \mathrm{S}^{n}_{\widehat{\overline{R}}}(\mathscr{F}^{(r)}) \to \mathrm{S}^{n}_{\widehat{\overline{R}}}(\widetilde{\xi}^{-1}\widetilde{\Omega}^1_{R/\mathscr{O}_K} \otimes_R \widehat{\overline{R}}) \to 0. \tag{5.2.13.2}$$

The $\widehat{\overline{R}}$-modules $(\mathrm{S}^{n}_{\widehat{\overline{R}}}(\mathscr{F}^{(r)}))_{n\in\mathbb{N}}$ thus form a filtered direct system, whose direct limit

$$\mathscr{C}^{(r)} = \varinjlim_{n\geq 0} \mathrm{S}^{n}_{\widehat{\overline{R}}}(\mathscr{F}^{(r)}) \tag{5.2.13.3}$$

is naturally endowed with the structure of an $\widehat{\overline{R}}$-algebra and an action of Δ by ring automorphisms, compatible with its action on $\widehat{\overline{R}}$; it is the Higgs–Tate algebra of thickness r associated to $(\widetilde{X}, \mathscr{M}_{\widetilde{X}})$ (3.2.17). We denote by $\widehat{\mathscr{C}}^{(r)}$ the p-adic Hausdorff completion of $\mathscr{C}^{(r)}$ that we endow with the p-adic topology. We endow $\widehat{\mathscr{C}}^{(r)}[\frac{1}{p}]$ with the p-adic topology (2.1.1).

For all rational numbers $r' \geq r \geq 0$, we have a canonical injective and Δ-equivariant $\widehat{\overline{R}}$-homomorphism $\alpha^{r,r'} : \mathscr{C}^{(r')} \to \mathscr{C}^{(r)}$.

5.2.14 For any rational number $r \geq 0$, we denote by $\mathscr{F}'^{(r)}$ the $\widehat{\overline{R}}'$-representation of Δ' deduced from $\mathscr{F}'$ (5.2.6.1) by pullback by the multiplication by p^r on $\widetilde{\xi}^{-1}\widetilde{\Omega}^1_{R'/\mathscr{O}_K} \otimes_{R'} \widehat{\overline{R}}'$, so that we have a locally split exact sequence of $\widehat{\overline{R}}'$-modules

$$0 \to \widehat{\overline{R}'} \longrightarrow \mathscr{F}'^{(r)} \to \widetilde{\xi}^{-1}\widetilde{\Omega}^1_{R'/\mathscr{O}_K} \otimes_{R'} \widehat{\overline{R}'} \to 0. \tag{5.2.14.1}$$

This induces for every integer $n \geq 1$ an exact sequence

$$0 \to \mathrm{S}^{n-1}_{\widehat{\overline{R}'}}(\mathscr{F}'^{(r)}) \to \mathrm{S}^{n}_{\widehat{\overline{R}'}}(\mathscr{F}'^{(r)}) \to \mathrm{S}^{n}_{\widehat{\overline{R}'}}(\widetilde{\xi}^{-1}\widetilde{\Omega}^1_{R'/\mathscr{O}_K} \otimes_{R'} \widehat{\overline{R}'}) \to 0. \tag{5.2.14.2}$$

The $\widehat{\overline{R}'}$-modules $(\mathrm{S}^{n}_{\widehat{\overline{R}'}}(\mathscr{F}'^{(r)}))_{n\in\mathbb{N}}$ thus form a filtered direct system, whose direct limit

$$\mathscr{C}'^{(r)} = \varinjlim_{n\geq 0} \mathrm{S}^{n}_{\widehat{\overline{R}'}}(\mathscr{F}'^{(r)}) \tag{5.2.14.3}$$

is naturally endowed with the structure of an $\widehat{\overline{R}'}$-algebra and an action of Δ' by ring automorphisms, compatible with its action on $\widehat{\overline{R}'}$; it is the Higgs–Tate algebra of thickness r associated to $(\widetilde{X}', \mathscr{M}_{\widetilde{X}'})$ (3.2.17).

For all rational numbers $r' \geq r \geq 0$, we have a canonical injective and Δ'-equivariant homomorphism $\alpha'^{r,r'} \colon \mathscr{C}'^{(r')} \to \mathscr{C}'^{(r)}$.

The morphism (5.2.7.3) induces a Δ'-equivariant $\widehat{\overline{R}'}$-linear morphism $\mathscr{F}^{(r)} \otimes_{\widehat{\overline{R}}} \widehat{\overline{R}'} \to \mathscr{F}'^{(r)}$ and hence a Δ'-equivariant homomorphism of $\widehat{\overline{R}}$-algebras

$$\mathscr{C}^{(r)} \to \mathscr{C}'^{(r)}. \tag{5.2.14.4}$$

For any rational number $t \geq r$, we consider the $\mathscr{C}^{(r)}$-algebra

$$\mathscr{C}'^{(t,r)} = \mathscr{C}'^{(t)} \otimes_{\mathscr{C}^{(t)}} \mathscr{C}^{(r)} \tag{5.2.14.5}$$

deduced from $\mathscr{C}'^{(t)}$ by base change by the homomorphism $\alpha^{t,r} \colon \mathscr{C}^{(t)} \to \mathscr{C}^{(r)}$ (5.2.13). We denote by $\widehat{\mathscr{C}}'^{(t,r)}$ the p-adic Hausdorff completion of $\mathscr{C}'^{(t,r)}$ that we endow with the p-adic topology. We endow $\widehat{\mathscr{C}}'^{(t,r)}[\frac{1}{p}]$ with the p-adic topology (2.1.1). The actions of Δ' on the rings $\mathscr{C}'^{(t)}$ and $\mathscr{C}^{(r)}$ induce an action on $\mathscr{C}'^{(t,r)}$ by ring automorphisms, compatible with its action on $\mathscr{C}^{(r)}$.

For all rational numbers t, t', r' such that $t' \geq t \geq r$ and $t' \geq r' \geq r$, the diagram

$$\begin{array}{ccccc} \mathscr{C}^{(r')} & \xleftarrow{\alpha^{t',r'}} & \mathscr{C}^{(t')} & \longrightarrow & \mathscr{C}'^{(t')} \\ \downarrow{\scriptstyle \alpha^{r',r}} & & \downarrow{\scriptstyle \alpha^{t',t}} & & \downarrow{\scriptstyle \alpha'^{t',t}} \\ \mathscr{C}^{(r)} & \xleftarrow[\alpha^{t,r}]{} & \mathscr{C}^{(t)} & \longrightarrow & \mathscr{C}'^{(t)}, \end{array} \tag{5.2.14.6}$$

where the unlabeled arrows are the homomorphisms (5.2.14.4), is commutative. We deduce from this a canonical homomorphism of $\mathscr{C}^{(r')}$-algebras

$$\mathscr{C}'^{(t',r')} \to \mathscr{C}'^{(t,r)}. \tag{5.2.14.7}$$

Proposition 5.2.15 *Let r, t, t' be three rational numbers such that $t' > t > r \geq 0$, $\overline{R}^\diamond$ the ring defined in* (5.1.9.4), Π *the group defined in* (5.1.9.3). *Then,*

(i) *For every integer $n \geq 1$, the canonical homomorphism*

$$(\mathscr{C}^{(r)}/p^n\mathscr{C}^{(r)}) \otimes_{\overline{R}} \overline{R}^\diamond \to (\mathscr{C}'^{(t,r)}/p^n\mathscr{C}'^{(t,r)})^\Pi \qquad (5.2.15.1)$$

is α-injective (5.1.2). *Let $\mathscr{H}_n^{(t,r)}$ be its cokernel.*

(ii) *There exists an integer $a \geq 0$, depending on t, t' and $\ell = \dim(X'/X)$, but not on the morphisms f, f' and g satisfying the conditions of* 5.1.3 *and* 5.2.1, *such that for every integer $n \geq 1$, the canonical morphism $\mathscr{H}_n^{(t',r)} \to \mathscr{H}_n^{(t,r)}$ is annihilated by p^a.*

(iii) *There exists an integer $b \geq 0$, depending on t, t' and ℓ, but not on the morphisms f, f' and g satisfying the conditions of* 5.1.3 *and* 5.2.1, *such that for all integers $n, q \geq 1$, the canonical morphism*

$$\mathrm{H}^q(\Pi, \mathscr{C}'^{(t',r)}/p^n\mathscr{C}'^{(t',r)}) \to \mathrm{H}^q(\Pi, \mathscr{C}'^{(t,r)}/p^n\mathscr{C}'^{(t,r)}) \qquad (5.2.15.2)$$

is annihilated by p^b.

The remainder of this section is devoted to the proof of this statement, which will be deduced in 5.2.35 from 5.2.34.

Corollary 5.2.16 *Let r be a rational number ≥ 0.*

(i) *The canonical morphism*

$$(\mathscr{C}^{(r)}\widehat{\otimes}_{\overline{R}}\overline{R}^\diamond)[\frac{1}{p}] \to \varinjlim_{t\in\mathbb{Q}_{>r}} (\widehat{\mathscr{C}}'^{(t,r)}[\frac{1}{p}])^\Pi, \qquad (5.2.16.1)$$

where the tensor product $\widehat{\otimes}$ is completed for the p-adic topology, is an isomorphism.

(ii) *For every integer $i \geq 1$, we have*

$$\varinjlim_{t\in\mathbb{Q}_{>r}} \mathrm{H}^i_{\mathrm{cont}}(\Pi, \widehat{\mathscr{C}}'^{(t,r)}[\frac{1}{p}]) = 0. \qquad (5.2.16.2)$$

We take again the notation of 5.2.15. It follows from 5.2.15(i) and ([21] 2.4.2(ii)) that for every rational number $t > r$, the canonical sequence

$$0 \to \mathscr{C}^{(r)}\widehat{\otimes}_{\overline{R}}\overline{R}^\diamond \to (\widehat{\mathscr{C}}'^{(t,r)})^\Pi \to \varprojlim_{n\geq 0} \mathscr{H}_n^{(t,r)} \qquad (5.2.16.3)$$

is α-exact. Proposition (i) follows by inverting p and then taking the direct limit over the rational numbers $t > r$, in view of 5.2.15(ii).

Since we have ([3] (II.3.10.2))

$$\mathrm{R}^1\varprojlim_n (\mathscr{C}^{(r)}/p^n\mathscr{C}^{(r)}) \otimes_{\overline{R}} \overline{R}^\diamond = 0, \qquad (5.2.16.4)$$

it follows from 5.2.15(i) and ([21] 2.4.2(ii)) that the canonical morphism

$$\mathrm{R}^1 \lim_{\substack{\longleftarrow \\ n \geq 0}} (\mathscr{C}'^{(t,r)}/p^n \mathscr{C}'^{(t,r)})^{\Pi} \to \mathrm{R}^1 \lim_{\substack{\longleftarrow \\ n \geq 0}} \mathscr{H}_n^{(t,r)} \tag{5.2.16.5}$$

is an α-isomorphism.

By ([3] (II.3.10.4) and (II.3.10.5)), for every $i \geq 1$, we have a canonical exact sequence

$$0 \to \mathrm{R}^1 \lim_{\substack{\longleftarrow \\ n}} \mathrm{H}^{i-1}(\Pi, \mathscr{C}'^{(t,r)}/p^n \mathscr{C}'^{(t,r)}) \to \mathrm{H}^i_{\mathrm{cont}}(\Pi, \widehat{\mathscr{C}}'^{(t,r)})$$
$$\to \lim_{\substack{\longleftarrow \\ n}} \mathrm{H}^i(\Pi, \mathscr{C}'^{(t,r)}/p^n \mathscr{C}'^{(t,r)}) \to 0.$$

We deduce, in view of (5.2.16.5) and 5.2.15(ii)-(iii), that for all integers $t' > t > r$, there exists a rational number $\gamma \geq 0$ such that the canonical morphism

$$\mathrm{H}^i_{\mathrm{cont}}(\Pi, \widehat{\mathscr{C}}'^{(t',r)}) \to \mathrm{H}^i_{\mathrm{cont}}(\Pi, \widehat{\mathscr{C}}'^{(t,r)}) \tag{5.2.16.6}$$

is annihilated by p^γ. Proposition (ii) follows by taking the direct limit over the rational numbers $t > r$.

5.2.17 For any rational number $r \geq 0$, we denote by $\mathscr{G}^{(r)}$ the sub-$\widehat{\overline{R}}$-algebra of $\mathscr{G}$ (5.2.4.2) defined by (2.1.10)

$$\mathscr{G}^{(r)} = \mathrm{S}_{\widehat{\overline{R}}}(p^r \widetilde{\xi}^{-1} \widetilde{\Omega}^1_{R/\mathscr{O}_K} \otimes_R \widehat{\overline{R}}) \tag{5.2.17.1}$$

and by $\widehat{\mathscr{G}}^{(r)}$ its p-adic Hausdorff completion that we endow with the p-adic topology. For all rational numbers $r' \geq r \geq 0$, we have a canonical injective homomorphism $a^{r,r'} : \mathscr{G}^{(r')} \to \mathscr{G}^{(r)}$.

5.2.18 Let $(\widetilde{X}_0, \mathscr{M}_{\widetilde{X}_0})$ be the smooth $(\mathscr{A}_2(\overline{S}), \mathscr{M}_{\mathscr{A}_2(\overline{S})})$-deformation of $(\check{\overline{X}}, \mathscr{M}_{\check{\overline{X}}})$ defined by the chart (P, γ) (5.1.3.2) (see 3.4.6), $\mathscr{L}_0$ the associated Higgs–Tate torsor, $\psi_0 \in \mathscr{L}_0(\widehat{\mathbb{X}})$ the section defined by the same chart (3.4.6.4). We denote by $\mathbf{L}_0$ the principal homogeneous $\mathbf{T}$-bundle over $\widehat{\mathbb{X}}$ associated with $\mathscr{L}_0$ (3.2.14) and by $\mathtt{t}$ the isomorphism of principal homogeneous $\mathbf{T}$-bundles over $\widehat{\mathbb{X}}$

$$\mathtt{t} \colon \mathbf{T} \xrightarrow{\sim} \mathbf{L}_0, v \mapsto v + \psi_0. \tag{5.2.18.1}$$

The canonical structure of a Δ-equivariant principal homogeneous $\mathbf{T}$-bundle on $\mathbf{L}_0$ (3.2.15) induces via $\mathtt{t}$ the structure of a Δ-equivariant principal homogeneous $\mathbf{T}$-bundle on $\mathbf{T}$. The latter determines a left action of Δ on $\mathbf{T}$ compatible with its action on $\widehat{\mathbb{X}}$. We deduce from this an action

$$\varphi \colon \Delta \to \mathrm{Aut}_{\overline{R_1}}(\mathscr{G}) \tag{5.2.18.2}$$

of Δ on $\mathscr{G}$ (5.2.4.2) by ring automorphisms, compatible with its action on $\widehat{\overline{R}}$; for every $c \in \Delta$, $\varphi(c)$ is induced by the automorphism of $\mathbf{T}$ defined by c^{-1}.

By 3.4.12, for every rational number $r \geq 0$, the sub-$\widehat{\overline{R}}$-algebra $\mathscr{G}^{(r)}$ (5.2.17.1) of $\mathscr{G}$ is stable by the action φ of Δ on $\mathscr{G}$, and the induced actions of Δ on $\mathscr{G}^{(r)}$ and $\widehat{\mathscr{G}}^{(r)}$ are continuous for the p-adic topologies. Unless expressly stated otherwise, we endow $\mathscr{G}^{(r)}$ and $\widehat{\mathscr{G}}^{(r)}$ with the actions of Δ induced by φ.

5.2.19 We take again the notation of 5.1.11. We denote by $\mathfrak{G}$ the sub-$\widehat{R_1}$-algebra of $\mathscr{G}$ (5.2.4.2) defined by (2.1.10)

$$\mathfrak{G} = \mathrm{S}_{\widehat{R_1}}(\widetilde{\xi}^{-1}\widetilde{\Omega}^1_{R/\mathscr{O}_K} \otimes_R \widehat{R_1}). \tag{5.2.19.1}$$

By 3.4.10, the action φ of Δ on $\mathscr{G}$ preserves $\mathfrak{G}$, and the induced action on $\mathfrak{G}$ factors through Δ_{p^∞} (5.1.12.8). We denote also by φ the action of Δ_{p^∞} on $\mathfrak{G}$ thus defined. By adapting the proof of 3.4.12, we show that it is continuous for the p-adic topology on $\mathfrak{G}$. We set

$$\mathscr{G}_\infty = \mathfrak{G} \otimes_{\widehat{R_1}} \widehat{R_\infty}, \tag{5.2.19.2}$$

$$\mathscr{G}_{p^\infty} = \mathfrak{G} \otimes_{\widehat{R_1}} \widehat{R_{p^\infty}}. \tag{5.2.19.3}$$

The action φ of Δ_{p^∞} on $\mathfrak{G}$ induces actions of Δ_{p^∞} on $\mathscr{G}_{p^\infty}$ and of Δ_∞ on $\mathscr{G}_\infty$.

For any rational number $r \geq 0$, we denote by $\mathscr{G}_\infty^{(r)}$ the sub-$\widehat{R_\infty}$-algebra of $\mathscr{G}_\infty$ defined by

$$\mathscr{G}_\infty^{(r)} = \mathrm{S}_{\widehat{R_\infty}}(p^r\widetilde{\xi}^{-1}\widetilde{\Omega}^1_{R/\mathscr{O}_K} \otimes_R \widehat{R_\infty}), \tag{5.2.19.4}$$

and by $\widehat{\mathscr{G}}_\infty^{(r)}$ its p-adic Hausdorff completion. For every rational number $r' \geq r$, we have a canonical injective homomorphism $\mathscr{G}_\infty^{(r')} \to \mathscr{G}_\infty^{(r)}$. By the proof of 3.4.12, $\mathscr{G}_\infty^{(r)}$ is stable under the action of Δ_∞ on $\mathscr{G}_\infty = \mathscr{G}_\infty^{(0)}$, and the induced actions of Δ_∞ on $\mathscr{G}_\infty^{(r)}$ and $\widehat{\mathscr{G}}_\infty^{(r)}$ are continuous for the p-adic topologies. Unless expressly stated otherwise, we endow $\mathscr{G}_\infty^{(r)}$ and $\widehat{\mathscr{G}}_\infty^{(r)}$ with these actions and with the p-adic topologies.

We denote by $\mathscr{G}_{p^\infty}^{(r)}$ the sub-$\widehat{R_{p^\infty}}$-algebra of $\mathscr{G}_{p^\infty}$ defined by

$$\mathscr{G}_{p^\infty}^{(r)} = \mathrm{S}_{\widehat{R_{p^\infty}}}(p^r\widetilde{\xi}^{-1}\widetilde{\Omega}^1_{R/\mathscr{O}_K} \otimes_R \widehat{R_{p^\infty}}), \tag{5.2.19.5}$$

and by $\widehat{\mathscr{G}}_{p^\infty}^{(r)}$ its p-adic Hausdorff completion. The algebra $\mathscr{G}_{p^\infty}^{(r)}$ satisfies properties analogous to those satisfied by $\mathscr{G}_\infty^{(r)}$. In particular, $\mathscr{G}_{p^\infty}^{(r)}$ is stable by the action of Δ_{p^∞} on $\mathscr{G}_{p^\infty} = \mathscr{G}_{p^\infty}^{(0)}$, and the induced actions of Δ_{p^∞} on $\mathscr{G}_{p^\infty}^{(r)}$ and $\widehat{\mathscr{G}}_{p^\infty}^{(r)}$ are continuous for the p-adic topologies. Unless expressly stated otherwise, we endow $\mathscr{G}_{p^\infty}^{(r)}$ and $\widehat{\mathscr{G}}_{p^\infty}^{(r)}$ with these actions and with the p-adic topologies.

5.2.20 For any rational number $r \geq 0$, we denote by $\mathscr{G}'^{(r)}$ the sub-$\widehat{\overline{R}'}$-algebra of $\mathscr{G}'$ (5.2.4.5) defined by (2.1.10)

$$\mathscr{G}'^{(r)} = \mathrm{S}_{\widehat{\overline{R}'}}(p^r\widetilde{\xi}^{-1}\widetilde{\Omega}^1_{R'/\mathscr{O}_K} \otimes_{R'} \widehat{\overline{R}'}). \tag{5.2.20.1}$$

For all rational numbers $r' \geq r \geq 0$, we have a canonical injective homomorphism $a'^{r,r'} : \mathscr{G}'^{(r')} \to \mathscr{G}'^{(r)}$.

The canonical morphism $\widetilde{\Omega}^1_{R/\mathscr{O}_K} \otimes_R R' \to \widetilde{\Omega}^1_{R'/\mathscr{O}_K}$ induces a homomorphism of $\widehat{\overline{R}}$-algebras $\mathscr{G}^{(r)} \to \mathscr{G}'^{(r)}$ (5.2.17.1). For any rational number $t \geq r$, we consider the $\mathscr{G}^{(r)}$-algebra

$$\mathscr{G}'^{(t,r)} = \mathscr{G}'^{(t)} \otimes_{\mathscr{G}^{(t)}} \mathscr{G}^{(r)} \qquad (5.2.20.2)$$

deduced from $\mathscr{G}'^{(t)}$ by base change by the canonical homomorphism $a^{t,r} : \mathscr{G}^{(t)} \to \mathscr{G}^{(r)}$. We denote by $\widehat{\mathscr{G}}'^{(t,r)}$ the p-adic Hausdorff completion of $\mathscr{G}'^{(t,r)}$ that we endow with the p-adic topology.

For all rational numbers t, t', r' such that $t' \geq t \geq r$ and $t' \geq r' \geq r$, the diagram

$$\begin{array}{ccccc} \mathscr{G}^{(r')} & \xleftarrow{a^{t',r'}} & \mathscr{G}^{(t')} & \longrightarrow & \mathscr{G}'^{(t')} \\ \downarrow{\scriptstyle a^{r',r}} & & \downarrow{\scriptstyle a^{t',t}} & & \downarrow{\scriptstyle a'^{t',t}} \\ \mathscr{G}^{(r)} & \xleftarrow[a^{t,r}]{} & \mathscr{G}^{(t)} & \longrightarrow & \mathscr{G}'^{(t)} \end{array} \qquad (5.2.20.3)$$

is commutative. We deduce from this a canonical homomorphism of $\mathscr{G}^{(r')}$-algebras

$$a'^{t',t,r',r} : \mathscr{G}'^{(t',r')} \to \mathscr{G}'^{(t,r)}. \qquad (5.2.20.4)$$

5.2.21 Let $(\widetilde{X}'_0, \mathscr{M}_{\widetilde{X}'_0})$ be the $(\mathscr{A}_2(\overline{S}), \mathscr{M}_{\mathscr{A}_2(\overline{S})})$-smooth deformation of $(\check{\overline{X}}', \mathscr{M}_{\check{\overline{X}}'})$ defined by the chart (P', γ') (5.1.3.2) (see 3.4.6), $\mathscr{L}'_0$ the associated Higgs–Tate torsor, $\psi'_0 \in \mathscr{L}'_0(\widehat{\mathbb{X}}')$ the section defined by the same chart (3.4.6.4). We denote by $\mathbf{L}'_0$ the principal homogeneous $\mathbf{T}'$-bundle over $\widehat{\mathbb{X}}'$ associated with $\mathscr{L}'_0$ (3.2.14) and by $\mathtt{t}'$ the isomorphism of principal homogeneous $\mathbf{T}'$-bundles over $\widehat{\mathbb{X}}'$

$$\mathtt{t}' : \mathbf{T}' \xrightarrow{\sim} \mathbf{L}'_0, \quad v \mapsto v + \psi'_0. \qquad (5.2.21.1)$$

The canonical structure of a Δ'-equivariant principal homogeneous $\mathbf{T}'$-bundle over $\mathbf{L}'_0$ (3.2.15) induces via $\mathtt{t}'$ the structure of a Δ'-equivariant principal homogeneous $\mathbf{T}'$-bundle over $\mathbf{T}'$. The latter determines a left action of Δ' on $\mathbf{T}'$ compatible with its action on $\widehat{\mathbb{X}}'$. We deduce from this an action

$$\varphi' : \Delta' \to \mathrm{Aut}_{\overline{R_1}'}(\mathscr{G}') \qquad (5.2.21.2)$$

of Δ' on $\mathscr{G}'$ (5.2.4.2) by ring automorphisms, compatible with its action on $\widehat{\overline{R}}'$; for all $\gamma \in \Delta'$, $\varphi'(\gamma)$ is induced by the automorphism of $\mathbf{T}'$ defined by γ^{-1}.

By 3.4.12, for every rational number $r \geq 0$, the sub-$\widehat{\overline{R}}'$-algebra $\mathscr{G}'^{(r)}$ (5.2.20.1) of $\mathscr{G}'$ is stable by the action φ' of Δ' on $\mathscr{G}'$, and the induced action of Δ' on $\mathscr{G}'^{(r)}$ is continuous for the p-adic topology. Unless expressly stated otherwise, we endow $\mathscr{G}'^{(r)}$ with the action of Δ' induced by φ'.

5.2.22 By (5.1.3.2) and ([2] 5.1.11), with the notation of 2.1.4, we have a canonical strict étale morphism

$$(\check{\overline{X}}', \mathscr{M}_{\check{\overline{X}}'}) \to (\check{\overline{X}}, \mathscr{M}_{\check{\overline{X}}}) \times_{\mathbf{A}_P} \mathbf{A}_{P'}, \tag{5.2.22.1}$$

the product being taken indifferently in the category of logarithmic schemes or in that of fine logarithmic schemes ([3] I.5.19). Moreover, with the notation of 5.2.18 and 5.2.21, we have two canonical strict étale $(\widetilde{S}, \mathscr{M}_{\widetilde{S}})$-morphisms (3.4.6.1)

$$(\widetilde{X}'_0, \mathscr{M}_{\widetilde{X}'_0}) \to (\widetilde{S}, \mathscr{M}_{\widetilde{S}}) \times_{\mathbf{A}_{\mathbb{N}}} \mathbf{A}_{P'} \quad \text{and} \quad (\widetilde{X}_0, \mathscr{M}_{\widetilde{X}_0}) \times_{\mathbf{A}_P} \mathbf{A}_{P'} \to (\widetilde{S}, \mathscr{M}_{\widetilde{S}}) \times_{\mathbf{A}_{\mathbb{N}}} \mathbf{A}_{P'}, \tag{5.2.22.2}$$

that induce by reduction over $(\check{\overline{S}}, \mathscr{M}_{\check{\overline{S}}})$ the canonical strict étale morphisms

$$(\check{\overline{X}}', \mathscr{M}_{\check{\overline{X}}'}) \to (\check{\overline{S}}, \mathscr{M}_{\check{\overline{S}}}) \times_{\mathbf{A}_{\mathbb{N}}} \mathbf{A}_{P'} \quad \text{and} \quad (\check{\overline{X}}, \mathscr{M}_{\check{\overline{X}}}) \times_{\mathbf{A}_P} \mathbf{A}_{P'} \to (\check{\overline{S}}, \mathscr{M}_{\check{\overline{S}}}) \times_{\mathbf{A}_{\mathbb{N}}} \mathbf{A}_{P'}. \tag{5.2.22.3}$$

The morphism (5.2.22.1) therefore extends into a unique strict étale morphism

$$(\widetilde{X}'_0, \mathscr{M}_{\widetilde{X}'_0}) \to (\widetilde{X}_0, \mathscr{M}_{\widetilde{X}_0}) \times_{\mathbf{A}_P} \mathbf{A}_{P'} \tag{5.2.22.4}$$

over $(\widetilde{S}, \mathscr{M}_{\widetilde{S}}) \times_{\mathbf{A}_{\mathbb{N}}} \mathbf{A}_{P'}$. We denote by $\widetilde{g}_0 \colon (\widetilde{X}'_0, \mathscr{M}_{\widetilde{X}'_0}) \to (\widetilde{X}_0, \mathscr{M}_{\widetilde{X}_0})$ the induced $(\widetilde{S}, \mathscr{M}_{\widetilde{S}})$-morphism. This induces a $\mathbf{T}'$-equivariant morphism (5.2.7.5)

$$\mathbf{L}'_0 \to \mathbf{L}_0 \times_{\widehat{\mathbb{X}}} \widehat{\mathbb{X}}'. \tag{5.2.22.5}$$

The diagram

$$\begin{array}{ccc} (\widetilde{\mathbb{X}}', \mathscr{M}_{\widetilde{\mathbb{X}}'}) & \xrightarrow{\ \widetilde{\mathfrak{g}}\ } & (\widetilde{\mathbb{X}}, \mathscr{M}_{\widetilde{\mathbb{X}}}) \\ {\scriptstyle \phi'_0}\downarrow & & \downarrow{\scriptstyle \phi_0} \\ (\widetilde{S}, \mathscr{M}_{\widetilde{S}}) \times_{\mathbf{A}_{\mathbb{N}}} \mathbf{A}_{P'} & \longrightarrow & (\widetilde{S}, \mathscr{M}_{\widetilde{S}}) \times_{\mathbf{A}_{\mathbb{N}}} \mathbf{A}_{P}, \end{array} \tag{5.2.22.6}$$

where ϕ_0 and ϕ'_0 are the morphisms defined by the horizontal arrows of (5.2.3.7), is commutative. Moreover, the diagram of canonical morphisms

$$\begin{array}{ccc} (\widehat{\mathbb{X}}', \mathscr{M}_{\widehat{\mathbb{X}}'}) & \xrightarrow{\ \widehat{\mathfrak{g}}\ } & (\widehat{\mathbb{X}}, \mathscr{M}_{\widehat{\mathbb{X}}}) \\ \downarrow & & \downarrow \\ (\check{\overline{X}}', \mathscr{M}_{\check{\overline{X}}'}) & \longrightarrow & (\check{\overline{X}}, \mathscr{M}_{\check{\overline{X}}}) \\ \downarrow & & \downarrow \\ (\check{\overline{S}}, \mathscr{M}_{\check{\overline{S}}}) \times_{\mathbf{A}_{\mathbb{N}}} \mathbf{A}_{P'} & \longrightarrow & (\check{\overline{S}}, \mathscr{M}_{\check{\overline{S}}}) \times_{\mathbf{A}_{\mathbb{N}}} \mathbf{A}_{P} \end{array} \tag{5.2.22.7}$$

is commutative. We deduce from this that the diagram

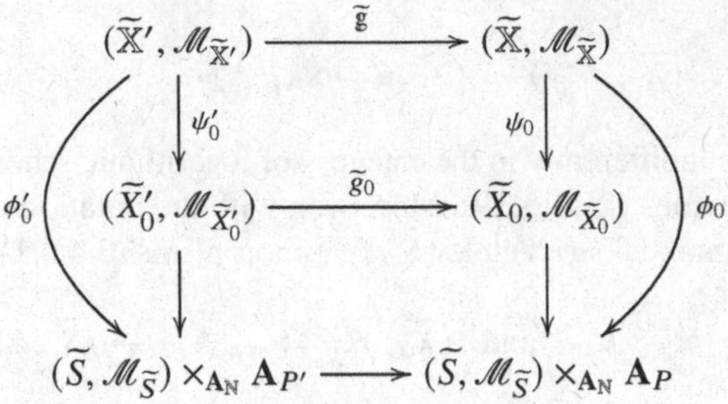
$(\widetilde{X}', \mathscr{M}_{\widetilde{X}'})$
$\widetilde{g}$
$(\widetilde{X}, \mathscr{M}_{\widetilde{X}})$
ψ'_0
ψ_0
ϕ'_0
$(\widetilde{X}'_0, \mathscr{M}_{\widetilde{X}'_0})$
$\widetilde{g}_0$
$(\widetilde{X}_0, \mathscr{M}_{\widetilde{X}_0})$
ϕ_0
$(\widetilde{S}, \mathscr{M}_{\widetilde{S}}) \times_{\mathbf{A}_{\mathbb{N}}} \mathbf{A}_{P'}$
$(\widetilde{S}, \mathscr{M}_{\widetilde{S}}) \times_{\mathbf{A}_{\mathbb{N}}} \mathbf{A}_{P}$

(5.2.22.8)

is commutative (3.4.6.4). The sections ψ_0 and ψ'_0 are therefore (5.2.7) compatible. In view of 5.2.10, we deduce from this that the canonical $\widehat{\overline{R}}$-homomorphism $\mathscr{G} \to \mathscr{G}'$ is Δ'-equivariant for the actions φ of Δ on $\mathscr{G}$ (5.2.18.2) and φ' of Δ' on $\mathscr{G}'$ (5.2.21.2). Consequently, for every rational number $r \geq 0$, the canonical homomorphism $\mathscr{G}^{(r)} \to \mathscr{G}'^{(r)}$ is Δ'-equivariant (5.2.21). We deduce, for all rational numbers $t \geq r \geq 0$, an action of Δ' on $\mathscr{G}'^{(t,r)}$ (5.2.20.2) by ring automorphisms, compatible with the action of Δ on $\mathscr{G}^{(r)}$. This action as well as the induced action of Δ' on $\widehat{\mathscr{G}}'^{(t,r)}$ are continuous for the p-adic topologies.

5.2.23 We take again the notation of 5.1.11. We denote by $\mathfrak{G}'$ the sub-$\widehat{R'_1}$-algebra of $\mathscr{G}'$ (5.2.4.5) defined by (2.1.10)

$$\mathfrak{G}' = \mathrm{S}_{\widehat{R'_1}}(\widetilde{\xi}^{-1}\widetilde{\Omega}^1_{R'/\mathscr{O}_K} \otimes_{R'} \widehat{R'_1}). \tag{5.2.23.1}$$

By 3.4.10, the action φ' of Δ' on $\mathscr{G}'$ preserves $\mathfrak{G}'$, and the induced action on $\mathfrak{G}'$ factors through Δ'_{p^∞} (5.1.12.4). We denote also by φ' the action of Δ'_{p^∞} on $\mathfrak{G}'$ thus defined. It is continuous for the p-adic topology on $\mathfrak{G}'$ (see [3] II.11.9). We set

$$\mathscr{G}'_\infty = \mathfrak{G}' \otimes_{\widehat{R'_1}} \widehat{R'_\infty}, \tag{5.2.23.2}$$

$$\mathscr{G}'_{p^\infty} = \mathfrak{G}' \otimes_{\widehat{R'_1}} \widehat{R'_{p^\infty}}. \tag{5.2.23.3}$$

The action φ' of Δ'_{p^∞} on $\mathfrak{G}'$ induces actions of Δ'_{p^∞} on $\mathscr{G}'_{p^\infty}$ and of Δ'_∞ on $\mathscr{G}'_\infty$.

Let r be a rational number ≥ 0. We denote by $\mathscr{G}'^{(r)}_\infty$ the sub-$\widehat{R'_\infty}$-algebra of $\mathscr{G}'_\infty$ defined by

$$\mathscr{G}'^{(r)}_\infty = \mathrm{S}_{\widehat{R'_\infty}}(p^r\widetilde{\xi}^{-1}\widetilde{\Omega}^1_{R'/\mathscr{O}_K} \otimes_{R'} \widehat{R'_\infty}). \tag{5.2.23.4}$$

For every rational number $r' \geq r$, we have a canonical injective homomorphism $\mathscr{G}'^{(r')}_\infty \to \mathscr{G}'^{(r)}_\infty$. By the proof of 3.4.12, $\mathscr{G}'^{(r)}_\infty$ is stable by the action of Δ'_∞ on $\mathscr{G}'_\infty = \mathscr{G}'^{(0)}_\infty$, and the induced action of Δ'_∞ on $\mathscr{G}'^{(r)}_\infty$ is continuous for the p-adic topology.

The canonical morphism $\widetilde{\Omega}^1_{R/\mathscr{O}_K} \otimes_R R' \to \widetilde{\Omega}^1_{R'/\mathscr{O}_K}$ induces a Δ'_∞-equivariant homomorphism of $\widehat{R_\infty}$-algebras $\mathscr{G}^{(r)}_\infty \to \mathscr{G}'^{(r)}_\infty$ (5.2.19.4). For any rational number $t \geq r$, we consider the $\mathscr{G}^{(r)}_\infty$-algebra

$$\mathscr{G}'^{(t,r)}_{\infty} = \mathscr{G}'^{(t)}_{\infty} \otimes_{\mathscr{G}^{(t)}_{\infty}} \mathscr{G}^{(r)}_{\infty} \tag{5.2.23.5}$$

deduced from $\mathscr{G}'^{(t)}_{\infty}$ by base change by the canonical homomorphism $\mathscr{G}^{(t)}_{\infty} \to \mathscr{G}^{(r)}_{\infty}$. We denote by $\widehat{\mathscr{G}}'^{(t,r)}_{\infty}$ the p-adic Hausdorff completion of $\mathscr{G}'^{(t,r)}_{\infty}$, which we always assume endowed with the p-adic topology. The action of Δ'_{∞} on $\mathscr{G}'^{(t)}_{\infty}$ induces an action of Δ'_{∞} on $\mathscr{G}'^{(t,r)}_{\infty}$ by ring automorphisms, compatible with the action of Δ_{∞} on $\mathscr{G}^{(r)}_{\infty}$. This action as well as the induced action of Δ'_{∞} on $\widehat{\mathscr{G}}'^{(t,r)}_{\infty}$ are continuous for the p-adic topologies.

We denote by $\mathscr{G}'^{(r)}_{p^\infty}$ the sub-$\widehat{R'_{p^\infty}}$-algebra of $\mathscr{G}'_{p^\infty}$ defined by

$$\mathscr{G}'^{(r)}_{p^\infty} = \mathrm{S}_{\widehat{R'_{p^\infty}}}(p^r \widetilde{\xi}^{-1} \widetilde{\Omega}^1_{R'/\mathscr{O}_K} \otimes_{R'} \widehat{R'_{p^\infty}}). \tag{5.2.23.6}$$

The algebra $\mathscr{G}'^{(r)}_{p^\infty}$ satisfies properties analogous to those satisfied by $\mathscr{G}'^{(r)}_{\infty}$. In particular, $\mathscr{G}'^{(r)}_{p^\infty}$ is stable by the action of Δ'_{p^∞} on $\mathscr{G}'_{p^\infty} = \mathscr{G}'^{(0)}_{p^\infty}$, and the induced action of Δ'_{p^∞} on $\mathscr{G}'^{(r)}_{p^\infty}$ is continuous for the p-adic topology.

The canonical morphism $\widetilde{\Omega}^1_{R/\mathscr{O}_K} \otimes_R R' \to \widetilde{\Omega}^1_{R'/\mathscr{O}_K}$ induces a Δ'_{p^∞}-equivariant homomorphism of $\widehat{R_{p^\infty}}$-algebras $\mathscr{G}^{(r)}_{p^\infty} \to \mathscr{G}'^{(r)}_{p^\infty}$ (5.2.19.5). For any rational number $t \geq r$, we consider the $\mathscr{G}^{(r)}_{p^\infty}$-algebra

$$\mathscr{G}'^{(t,r)}_{p^\infty} = \mathscr{G}'^{(t)}_{p^\infty} \otimes_{\mathscr{G}^{(t)}_{p^\infty}} \mathscr{G}^{(r)}_{p^\infty} \tag{5.2.23.7}$$

deduced from $\mathscr{G}'^{(t)}_{p^\infty}$ by base change by the canonical homomorphism $\mathscr{G}^{(t)}_{p^\infty} \to \mathscr{G}^{(r)}_{p^\infty}$. We denote by $\widehat{\mathscr{G}}'^{(t,r)}_{p^\infty}$ the p-adic Hausdorff completion of $\mathscr{G}'^{(t,r)}_{p^\infty}$ that we endow with the p-adic topology. The action of Δ'_{p^∞} on $\mathscr{G}'^{(t)}_{p^\infty}$ induces an action of Δ'_{p^∞} on $\mathscr{G}'^{(t,r)}_{p^\infty}$ by ring automorphisms, compatible with the action of Δ_{p^∞} on $\mathscr{G}^{(r)}_{p^\infty}$. This action as well as the induced action of Δ'_{p^∞} on $\widehat{\mathscr{G}}'^{(t,r)}_{p^\infty}$ are continuous for the p-adic topologies.

We mainly consider the actions of the subgroups $\mathfrak{S} \subset \Delta'$ (5.1.12.22), $\mathfrak{S}_\infty \subset \Delta'_\infty$ (5.1.12.15) and $\mathfrak{S}_{p^\infty} \subset \Delta'_{p^\infty}$ (5.1.12.16) on the algebras introduced above (see 5.1.13).

Lemma 5.2.24 *Let r, t be two rational numbers such that $t \geq r \geq 0$, i an integer ≥ 0. Then,*

(i) *The canonical morphism*

$$\mathrm{H}^i_{\mathrm{cont}}(\mathfrak{S}_{p^\infty}, \mathscr{G}'^{(t,r)}_{p^\infty} \widehat{\otimes}_{R^\circ_{p^\infty}} R^\circ_\infty) \to \mathrm{H}^i_{\mathrm{cont}}(\mathfrak{S}_\infty, \widehat{\mathscr{G}}'^{(t,r)}_\infty), \tag{5.2.24.1}$$

where the rings R°_∞ and $R^\circ_{p^\infty}$ are defined in (5.1.11.10) *and* (5.1.11.11) *and the tensor product $\widehat{\otimes}$ is completed for the p-adic topology, is an isomorphism.*

(ii) *The canonical morphism*

$$\mathrm{H}^i_{\mathrm{cont}}(\mathfrak{S}_\infty, \widehat{\mathscr{G}}'^{(t,r)}_\infty) \to \mathrm{H}^i_{\mathrm{cont}}(\mathfrak{S}, \widehat{\mathscr{G}}'^{(t,r)}) \tag{5.2.24.2}$$

is an α-isomorphism.

(i) Let n be an integer ≥ 0. The canonical homomorphism

$$\mathscr{G}'^{(t,r)}_{p^\infty} \otimes_{R^\diamond_{p^\infty}} (R^\diamond_\infty / p^n R^\diamond_\infty) \to (\mathscr{G}'^{(t,r)}_\infty / p^n \mathscr{G}'^{(t,r)}_\infty)^{\mathfrak{N}} \tag{5.2.24.3}$$

is an isomorphism by virtue of 5.1.14(ii). Since the cohomological p-dimension of $\mathfrak{N}$ vanishes (5.1.12.21) ([49] I cor. 2 of prop. 14), the canonical morphism

$$\mathrm{H}^i(\mathfrak{S}_{p^\infty}, \mathscr{G}'^{(t,r)}_{p^\infty} \otimes_{R^\diamond_{p^\infty}} (R^\diamond_\infty / p^n R^\diamond_\infty)) \to \mathrm{H}^i(\mathfrak{S}_\infty, \mathscr{G}'^{(t,r)}_\infty / p^n \mathscr{G}'^{(t,r)}_\infty) \tag{5.2.24.4}$$

is an isomorphism. We deduce by ([3] (II.3.10.4) and (II.3.10.5)) that the morphism (5.2.24.1) is an isomorphism.

(ii) Let n be an integer ≥ 0. The canonical homomorphism

$$\mathscr{G}'^{(t,r)}_\infty / p^n \mathscr{G}'^{(t,r)}_\infty \to (\mathscr{G}'^{(t,r)} / p^n \mathscr{G}'^{(t,r)})^{\Sigma'} \tag{5.2.24.5}$$

is an α-isomorphism by virtue of ([3] II.6.22). We deduce by ([3] II.6.20) that the canonical morphism

$$\psi_n : \mathrm{H}^i(\mathfrak{S}_\infty, \mathscr{G}'^{(t,r)}_\infty / p^n \mathscr{G}'^{(t,r)}_\infty) \to \mathrm{H}^i(\mathfrak{S}, \mathscr{G}'^{(t,r)} / p^n \mathscr{G}'^{(t,r)}) \tag{5.2.24.6}$$

is an α-isomorphism. We denote by A_n (resp. C_n) the kernel (resp. cokernel) of ψ_n. Then the $\mathscr{O}_{\overline{K}}$-modules

$$\varprojlim_{n\geq 0} A_n, \quad \varprojlim_{n\geq 0} C_n, \quad \mathrm{R}^1\varprojlim_{n\geq 0} A_n, \quad \mathrm{R}^1\varprojlim_{n\geq 0} C_n$$

are α-zero by virtue of ([21] 2.4.2(ii)). Consequently, the morphisms

$$\varprojlim_{n\geq 0} \psi_n \quad \text{and} \quad \mathrm{R}^1\varprojlim_{n\geq 0} \psi_n$$

are α-isomorphisms. We deduce by ([3] (II.3.10.4) and (II.3.10.5)) that the morphism (5.2.24.2) is an α-isomorphism.

5.2.25 We set $\lambda = \vartheta(1)$ (5.1.3.2). The homomorphism $h^{\mathrm{gp}} : P^{\mathrm{gp}} \to P'^{\mathrm{gp}}$ is injective, and the torsion subgroup of $P'^{\mathrm{gp}}/h^{\mathrm{gp}}(P^{\mathrm{gp}})$ is prime to p ([2] 5.1.11). Consider the canonical exact sequence

$$0 \longrightarrow P^{\mathrm{gp}}/\lambda\mathbb{Z} \xrightarrow{h^{\mathrm{gp}}} P'^{\mathrm{gp}}/h^{\mathrm{gp}}(\lambda)\mathbb{Z} \longrightarrow P'^{\mathrm{gp}}/h^{\mathrm{gp}}(P^{\mathrm{gp}}) \longrightarrow 0. \tag{5.2.25.1}$$

The $\mathbb{Z}$-modules $P^{\mathrm{gp}}/\mathbb{Z}\lambda$ and $P'^{\mathrm{gp}}/h^{\mathrm{gp}}(\lambda)\mathbb{Z}$ are free of finite type by ([2] 4.2.2); let d and d' be their respective ranks and set $\ell = d' - d$. By virtue of ([2] (5.2.12.3)), we have a canonical isomorphism

$$\mathfrak{S}_{p^\infty} \xrightarrow{\sim} \mathrm{Hom}(P'^{\mathrm{gp}}/h^{\mathrm{gp}}(P^{\mathrm{gp}}), \mathbb{Z}_p(1)). \tag{5.2.25.2}$$

We deduce from this an isomorphism

$$(P'^{\mathrm{gp}}/h^{\mathrm{gp}}(P^{\mathrm{gp}})) \otimes_{\mathbb{Z}} \mathbb{Z}_p \xrightarrow{\sim} \mathrm{Hom}(\mathfrak{S}_{p^\infty}, \mathbb{Z}_p(1)). \tag{5.2.25.3}$$

By ([45] IV 1.1.4), we have canonical linear isomorphisms (5.1.4)

$$\widetilde{\Omega}^1_{R'/R} \xrightarrow{\sim} (P'^{\mathrm{gp}}/h^{\mathrm{gp}}(P^{\mathrm{gp}})) \otimes_{\mathbb{Z}} R', \tag{5.2.25.4}$$

$$\widetilde{\Omega}^1_{R'/\mathscr{O}_K} \xrightarrow{\sim} (P'^{\mathrm{gp}}/h^{\mathrm{gp}}(\lambda)\mathbb{Z}) \otimes_{\mathbb{Z}} R', \tag{5.2.25.5}$$

$$\widetilde{\Omega}^1_{R/\mathscr{O}_K} \xrightarrow{\sim} (P^{\mathrm{gp}}/\lambda\mathbb{Z}) \otimes_{\mathbb{Z}} R. \tag{5.2.25.6}$$

Let $(t_i)_{1\le i\le \ell}$ be elements of P'^{gp} and let $(t_i)_{\ell+1\le i\le d'}$ be elements of P^{gp}, such that the images of $(t_i)_{1\le i\le \ell}$ in $P'^{\mathrm{gp}}/h^{\mathrm{gp}}(P^{\mathrm{gp}}) \otimes_{\mathbb{Z}} \mathbb{Z}_p$ form a $\mathbb{Z}_p$-basis, the images of $(t_i)_{\ell+1\le i\le d'}$ in $P^{\mathrm{gp}}/\lambda\mathbb{Z}$ form a $\mathbb{Z}$-basis, and the images of $t_1,\ldots,t_\ell, h^{\mathrm{gp}}(t_{\ell+1}),\ldots, h^{\mathrm{gp}}(t_{d'})$ in $P'^{\mathrm{gp}}/h^{\mathrm{gp}}(\lambda)\mathbb{Z}$ form a $\mathbb{Z}$-basis. To lighten the notation, we will abusively denote $(h^{\mathrm{gp}}(t_i))_{\ell+1\le i\le d'}$ simply by $(t_i)_{\ell+1\le i\le d'}$, which does not induce any ambiguity since h^{gp} is injective.

The $(d\log(t_i))_{1\le i\le d'}$ form an R'-basis of $\widetilde{\Omega}^1_{R'/\mathscr{O}_K}$ (5.2.25.5). For any $1 \le i \le d'$ and any $\underline{n} = (n_1,\ldots,n_{d'}) \in \mathbb{N}^{d'}$, we set $y_i = \widetilde{\xi}^{-1} d\log(t_i) \in \widetilde{\xi}^{-1}\widetilde{\Omega}^1_{R'/\mathscr{O}_K} \subset \mathscr{G}'_{p^\infty}$ (5.2.23.3), $|\underline{n}| = \sum_{i=1}^{d'} n_i$ and $\underline{y}^{\underline{n}} = \prod_{i=1}^{d'} y_i^{n_i} \in \mathscr{G}'_{p^\infty}$.

We denote by W the ring of Witt vectors with coefficients in k with respect to p, by K_0 the field of fractions of W and by $\mathfrak{d}$ the different of the extension K/K_0. We set $\rho = 0$ in the absolute case (3.1.14) and $\rho = v(\pi\mathfrak{d})$ in the relative case. By (3.1.10.1) and (3.1.5.3), we have a canonical $\mathscr{O}_C$-linear isomorphism

$$\mathscr{O}_C(1) \xrightarrow{\sim} p^{\rho+\frac{1}{p-1}}\widetilde{\xi}\mathscr{O}_C. \tag{5.2.25.7}$$

For any $1 \le i \le d'$, we denote by χ_{t_i} the image of t_i by the homomorphism (5.2.25.3), and by χ_i the composed homomorphism

$$\mathfrak{S}_{p^\infty} \xrightarrow{\chi_{t_i}} \mathbb{Z}_p(1) \xrightarrow{\log([\,])} p^{\rho+\frac{1}{p-1}}\widetilde{\xi}\mathscr{O}_C, \tag{5.2.25.8}$$

where the second arrow is induced by the isomorphism (5.2.25.7). Observe that $\chi_{t_i} = 1$ for all $\ell+1 \le i \le d'$.

By 3.4.10, for every $\gamma \in \mathfrak{S}_{p^\infty}$ and every $1 \le i \le d'$, we have

$$\varphi'(\gamma)(y_i) = y_i - \widetilde{\xi}^{-1}\chi_i(\gamma). \tag{5.2.25.9}$$

Let ζ be a $\mathbb{Z}_p$-basis of $\mathbb{Z}_p(1)$. Since the $(\chi_{t_i})_{1\le i\le \ell}$ form a $\mathbb{Z}_p$-basis of $\mathrm{Hom}(\mathfrak{S}_{p^\infty}, \mathbb{Z}_p(1))$, there exists a unique $\mathbb{Z}_p$-basis $(\gamma_i)_{1\le i\le \ell}$ of $\mathfrak{S}_{p^\infty}$ such that $\chi_{t_i}(\gamma_j) = \delta_{ij}\zeta$ for all $1 \le i,j \le \ell$. For any integer $1 \le i \le \ell$, we denote by $\chi^{>i}_{p^\infty}$ the subgroup of (5.1.16.10)

$$\chi_{p^\infty} = \mathrm{Hom}(\mathfrak{S}_{p^\infty}, \mu_{p^\infty}(\mathscr{O}_{\overline{K}})) \tag{5.2.25.10}$$

made up of the homomorphisms $\nu\colon \mathfrak{S}_{p^\infty} \to \mu_{p^\infty}(\mathscr{O}_{\overline{K}})$ such that $\nu(\gamma_j) = 1$ for all $1 \le j \le i$. Then, we have $\chi^{>0}_{p^\infty} = \chi_{p^\infty}$ and $\chi^{>\ell}_{p^\infty} = 0$.

We observe that R'_{p^∞} is separated for the p-adic topology and therefore identifies with a subring of $\widehat{R'_{p^\infty}}$; this follows for example from the proof of ([3] II.8.9), more precisely, with the notation of *loc. cit.*, from ([3] (II.8.9.8) and (II.8.9.18)) and from the fact that $\mathrm{Spec}(R_1)$ is an open set of $\mathrm{Spec}(C_1)$.

Let t, r be two rational numbers such that $t \geq r \geq 0$. For any element $\underline{n} = (n_1, \ldots, n_{d'})$ of $\mathbb{N}^{d'}$, we set

$$|\underline{n}|^{(t,r)} = t \sum_{i=1}^{\ell} n_i + r \sum_{i=\ell+1}^{d'} n_i. \tag{5.2.25.11}$$

For any integer $0 \leq i \leq \ell$ and any $\nu \in \chi_{p^\infty}^{>i}$, taking into account (5.1.18.2) and with the same notation, we denote by $\mathscr{G}'^{(t,r),>i}_{p^\infty}(\nu)$ and $\mathscr{G}'^{(t,r),>i}_{p^\infty}$ the sub-$R^\circ_{p^\infty}$-modules of $\mathscr{G}'^{(t,r)}_{p^\infty}$ defined by

$$\mathscr{G}'^{(t,r),>i}_{p^\infty}(\nu) = \bigoplus_{\underline{n} \in J_i} p^{|\underline{n}|^{(t,r)}} R'^{(\nu)}_{p^\infty} \underline{y}^{\underline{n}}, \tag{5.2.25.12}$$

$$\mathscr{G}'^{(t,r),>i}_{p^\infty} = \bigoplus_{\nu' \in \chi_{p^\infty}^{>i}} \mathscr{G}'^{(t,r),>i}_{p^\infty}(\nu'), \tag{5.2.25.13}$$

where J_i is the subset of $\mathbb{N}^{d'}$ made up of the elements $\underline{n} = (n_1, \ldots, n_{d'})$ such that $n_1 = \cdots = n_i = 0$. We denote by $\widehat{\mathscr{G}}'^{(t,r),>i}_{p^\infty}$ the p-adic Hausdorff completion of $\mathscr{G}'^{(t,r),>i}_{p^\infty}$. Since R'_{p^∞} is separated for the p-adic topology, so is $R'^{(\nu)}_{p^\infty}$, $\mathscr{G}'^{(t,r),>i}_{p^\infty}(\nu)$ and $\mathscr{G}'^{(t,r),>i}_{p^\infty}$. Consequently, $\mathscr{G}'^{(t,r),>i}_{p^\infty}$ identifies with a sub-$R^\circ_{p^\infty}$-module of $\widehat{\mathscr{G}}'^{(t,r),>i}_{p^\infty}$.

The p-adic topology of R'_{p^∞} being induced by the p-adic topology of $\widehat{R'_{p^\infty}}$, we deduce easily from (5.1.18.2) that the p-adic topology of $\mathscr{G}'^{(t,r),>i}_{p^\infty}$ is induced by the p-adic topology of $\mathscr{G}'^{(t,r)}_{p^\infty}$. Hence, $\widehat{\mathscr{G}}'^{(t,r),>i}_{p^\infty}$ is the closure of $\mathscr{G}'^{(t,r),>i}_{p^\infty}$ in $\widehat{\mathscr{G}}'^{(t,r)}_{p^\infty}$.

It follows from (5.1.18.2) and (5.2.25.9) that for every $1 \leq j \leq \ell$ and every $\nu \in \chi_{p^\infty}^{>i}$, γ_j preserves $\mathscr{G}'^{(t,r),>i}_{p^\infty}(\nu)$ and thus also $\mathscr{G}'^{(t,r),>i}_{p^\infty}$. If $1 \leq j \leq i$, γ_j fixes $\mathscr{G}'^{(t,r),>i}_{p^\infty}(\nu)$ and $\mathscr{G}'^{(t,r),>i}_{p^\infty}$.

Proposition 5.2.26 *We keep the assumptions of* 5.2.25, *let i be an integer such that $1 \leq i \leq \ell = d' - d$, and let t, t', r be three rational numbers such that $t' > t > r \geq 0$. Then,*

(i) *We have $\widehat{\mathscr{G}}'^{(t,r),>0}_{p^\infty} = \widehat{\mathscr{G}}'^{(t,r)}_{p^\infty}$ and $\widehat{\mathscr{G}}'^{(t,r),>\ell}_{p^\infty} = \mathscr{G}^{(r)}_{p^\infty} \widehat{\otimes}_{R_{p^\infty}} R^\circ_{p^\infty}$, where the tensor product $\widehat{\otimes}$ is completed for the p-adic topology.*

(ii) *The sequence*

$$0 \longrightarrow \widehat{\mathscr{G}}'^{(t,r),>i}_{p^\infty} \xrightarrow{\ u\ } \widehat{\mathscr{G}}'^{(t,r),>i-1}_{p^\infty} \xrightarrow{\gamma_i - \mathrm{id}} \widehat{\mathscr{G}}'^{(t,r),>i-1}_{p^\infty}, \tag{5.2.26.1}$$

where u is the canonical morphism, is exact.

(iii) *There exists an integer $\alpha \geq 0$, depending on t and t' but not on the morphisms f, f' and g satisfying the conditions of* 5.1.3 *and* 5.2.1, *such that we have*

$$p^{\alpha} \cdot \widehat{\mathscr{G}}_{p^\infty}^{\prime(t',r),>i-1} \subset (\gamma_i - \mathrm{id})(\widehat{\mathscr{G}}_{p^\infty}^{\prime(t,r),>i-1}). \tag{5.2.26.2}$$

(i) Denoting by e the trivial character of χ_{p^∞}, we have $R_{p^\infty}^{\prime(e)} = R_{p^\infty}^{\diamond}$ by 5.1.17. We deduce from this that $\widehat{\mathscr{G}}_{p^\infty}^{\prime(t,r),>\ell} = \mathscr{G}_{p^\infty}^{(r)} \widehat{\otimes}_{R_{p^\infty}} R_{p^\infty}^{\diamond}$. We set

$$\mathrm{S}^{\prime(r)} = \mathrm{S}_{R'_{p^\infty}}(p^r \widetilde{\xi}^{-1} \widetilde{\Omega}^1_{R'/\mathscr{O}_K} \otimes_{R'} R'_{p^\infty}), \tag{5.2.26.3}$$

$$\mathrm{S}^{(r)} = \mathrm{S}_{R'_{p^\infty}}(p^r \widetilde{\xi}^{-1} \widetilde{\Omega}^1_{R/\mathscr{O}_K} \otimes_{R'} R'_{p^\infty}). \tag{5.2.26.4}$$

In view of (5.1.18.2), we have

$$\mathscr{G}_{p^\infty}^{\prime(t,r),>0} = \mathrm{S}^{\prime(t)} \otimes_{\mathrm{S}^{(t)}} \mathrm{S}^{(r)}, \tag{5.2.26.5}$$

which implies that $\widehat{\mathscr{G}}_{p^\infty}^{\prime(t,r),>0} = \widehat{\mathscr{G}}_{p^\infty}^{\prime(t,r)}$.

(ii) Since for every integer $n \geq 0$, we clearly have

$$p^n \cdot \mathscr{G}_{p^\infty}^{\prime(t,r),>i} = \mathscr{G}_{p^\infty}^{\prime(t,r),>i} \cap (p^n \cdot \mathscr{G}_{p^\infty}^{\prime(t,r),>i-1}), \tag{5.2.26.6}$$

the canonical morphism $u \colon \widehat{\mathscr{G}}_{p^\infty}^{\prime(t,r),>i} \to \widehat{\mathscr{G}}_{p^\infty}^{\prime(t,r),>i-1}$ is injective.

For any $\nu \in \chi_{p^\infty}^{>i-1}$, denote by $(R_{p^\infty}^{\prime(\nu)})^{\wedge}$ and $(\mathscr{G}_{p^\infty}^{\prime(t,r),>i-1}(\nu))^{\wedge}$ the p-adic Hausdorff completions of $R_{p^\infty}^{\prime(\nu)}$ and $\mathscr{G}_{p^\infty}^{\prime(t,r),>i-1}(\nu)$, respectively. Observe that $(\mathscr{G}_{p^\infty}^{\prime(t,r),>i-1}(\nu))^{\wedge}$ identifies with a sub-$R_{p^\infty}^{\diamond}$-module of $\widehat{\mathscr{G}}_{p^\infty}^{\prime(t,r),>i-1}$, stable by γ_i. In view of (5.2.25.13), every element x of $\widehat{\mathscr{G}}_{p^\infty}^{\prime(t,r),>i-1}$ can be written as the sum of a series

$$\sum_{\nu \in \chi_{p^\infty}^{>i-1}} x_\nu, \tag{5.2.26.7}$$

where $x_\nu \in (\mathscr{G}_{p^\infty}^{\prime(t,r),>i-1}(\nu))^{\wedge}$ and, for every integer $n \geq 0$, except for a finite number of $\nu \in \chi_{p^\infty}^{>i-1}$, $x_\nu \in p^n(\mathscr{G}_{p^\infty}^{\prime(t,r),>i-1}(\nu))^{\wedge}$. Such an element x vanishes if and only if x_ν vanishes for every $\nu \in \chi_{p^\infty}^{>i-1}$.

Let $\nu \in \chi_{p^\infty}^{>i-1}$. For every

$$z = \sum_{\underline{n} \in J_{i-1}} p^{|\underline{n}|^{(t,r)}} a_{\underline{n}} y^{\underline{n}} \in \mathscr{G}_{p^\infty}^{\prime(t,r),>i-1}(\nu), \tag{5.2.26.8}$$

we have (5.2.25.9)

$$(\gamma_i - \mathrm{id})(z) = \sum_{\underline{n} \in J_{i-1}} p^{|\underline{n}|^{(t,r)}} b_{\underline{n}} y^{\underline{n}} \in \mathscr{G}_{p^\infty}^{\prime(t,r),>i-1}(\nu), \tag{5.2.26.9}$$

where for every $\underline{n} = (n_1, \ldots, n_{d'}) \in J_{i-1}$,

$$b_{\underline{n}} = (\nu(\gamma_i) - 1)a_{\underline{n}} + \sum_{\underline{m}=(m_1,\dots,m_{d'})\in J_{i-1}(\underline{n})} p^{t(m_i-n_i)}\binom{m_i}{n_i}\nu(\gamma_i)a_{\underline{m}}w^{m_i-n_i}, \tag{5.2.26.10}$$

$J_{i-1}(\underline{n})$ denotes the subset of J_{i-1} made up of the elements $\underline{m} = (m_1, \dots, m_{d'})$ such that $m_j = n_j$ for $j \neq i$ and $m_i > n_i$, and $w = -\widetilde{\xi}^{-1}\log([\zeta])$ is an element of valuation $\rho + \frac{1}{p-1}$ of $\mathscr{O}_C$ (5.2.25.7).

Let z be an element of $(\mathscr{G}'^{(t,r),>i-1}_{p^\infty}(\nu))^\wedge$. Then z can be written as the sum of a series

$$z = \sum_{\underline{n}\in J_{i-1}} p^{|\underline{n}|^{(t,r)}} a_{\underline{n}} y^{\underline{n}}, \tag{5.2.26.11}$$

where $a_{\underline{n}} \in (R'^{(\nu)}_{p^\infty})^\wedge$ and $a_{\underline{n}}$ tends to 0 when $|\underline{n}|$ tends to infinity. Since $R'^{(\nu)}_{p^\infty}$ is $\mathscr{O}_C$-flat, so is $(R'^{(\nu)}_{p^\infty})^\wedge$ (see the proof of [3] II.6.14). Consequently, z vanishes if and only if $a_{\underline{n}}$ vanishes for every $\underline{n} \in J_{i-1}$. We immediately see that $(\gamma_i - \mathrm{id})(z)$ is still given by the formula (5.2.26.9).

Suppose that $\gamma_i(z) = z$ and $\nu(\gamma_i) \neq 1$. Since $(R'^{(\nu)}_{p^\infty})^\wedge$ is $\mathscr{O}_C$-flat and $v(\nu(\gamma_i) - 1) \leq \frac{1}{p-1}$, for every $\underline{n} = (n_1, \dots, n_{d'}) \in J_{i-1}$, we have

$$a_{\underline{n}} = -(\nu(\gamma_i) - 1)^{-1} \sum_{\underline{m}=(m_1,\dots,m_{d'})\in J_{i-1}(\underline{n})} p^{t(m_i-n_i)}\binom{m_i}{n_i}\nu(\gamma_i)a_{\underline{m}}w^{m_i-n_i}.$$

We deduce from this that for every $\alpha \in \mathbb{N}$ and every $\underline{n} \in J_{i-1}$, we have $a_{\underline{n}} \in p^{t\alpha}(R'^{(\nu)}_{p^\infty})^\wedge$ (we prove it by induction on α); hence $z = 0$ since $(R'^{(\nu)}_{p^\infty})^\wedge$ is separated for the p-adic topology. Hence, $\gamma_i - \mathrm{id}$ is injective on $(\mathscr{G}'^{(t,r),>i-1}_{p^\infty}(\nu))^\wedge$.

Suppose that $\gamma_i(z) = z$ and $\nu(\gamma_i) = 1$, so that $\nu \in \chi^{>i}_{p^\infty}$. Then, for every $\underline{n} = (n_1, \dots, n_{d'}) \in J_{i-1}$, if we set $\underline{n}' = (n'_1, \dots, n'_{d'}) \in J_{i-1}(\underline{n})$ with $n'_i = n_i + 1$, we have

$$(n_i + 1)!a_{\underline{n}'} = - \sum_{\underline{m}=(m_1,\dots,m_{d'})\in J_{i-1}(\underline{n}')} p^{t(m_i-n_i-1)} m_i! a_{\underline{m}} \frac{w^{m_i-n_i-1}}{(m_i-n_i)!}.$$

We have $w^{m-1}/m! \in \mathscr{O}_C$ for every integer $m \geq 1$. We deduced that for every $\alpha \in \mathbb{N}$ and every $\underline{n} = (n_1, \dots, n_{d'}) \in J_{i-1}$ such that $n_i \geq 1$, we have $n_i!a_{\underline{n}} \in p^{t\alpha}(R'^{(\nu)}_{p^\infty})^\wedge$ (we prove it by induction on α); so $a_{\underline{n}} = 0$. Hence $z \in (\mathscr{G}'^{(t,r),>i}_{p^\infty}(\nu))^\wedge$. We deduce from this that the sequence

$$0 \longrightarrow \widehat{\mathscr{G}}'^{(t,r),>i}_{p^\infty}(\nu) \xrightarrow{u_\nu} \widehat{\mathscr{G}}'^{(t,r),>i-1}_{p^\infty}(\nu) \xrightarrow{\gamma_i - \mathrm{id}} \widehat{\mathscr{G}}'^{(t,r),>i-1}_{p^\infty}(\nu), \tag{5.2.26.12}$$

where u_ν is the canonical morphism, is exact.

It follows from the above that the sequence (5.2.26.1) is exact.

(iii) In view of (5.2.26.7), it suffices to show that there exists an integer $\alpha \geq 0$ depending only on t and t' such that for every $\nu \in \chi^{>i-1}_{p^\infty}$, we have

$$p^\alpha \cdot (\mathscr{G}'^{(t',r),>i-1}_{p^\infty}(\nu))^\wedge \subset (\gamma_i - \mathrm{id})((\mathscr{G}'^{(t,r),>i-1}_{p^\infty}(\nu))^\wedge). \tag{5.2.26.13}$$

Assume $\nu(\gamma_i) \neq 1$. By (5.2.26.9) applied to the elements of $(\mathscr{G}'^{(t,r),>i-1}_{p^\infty}(\nu))^\wedge$, we have

$$(\nu(\gamma_i)-1)(\mathscr{G}'^{(t,r),>i-1}_{p^\infty}(\nu))^\wedge \subset (\gamma_i - \mathrm{id})((\mathscr{G}'^{(t,r),>i-1}_{p^\infty}(\nu))^\wedge) + (\nu(\gamma_i)-1)p^t(\mathscr{G}'^{(t,r),>i-1}_{p^\infty}(\nu))^\wedge.$$

We deduce from this that

$$(\nu(\gamma_i)-1)(\mathscr{G}'^{(t,r),>i-1}_{p^\infty}(\nu))^\wedge \subset (\gamma_i - \mathrm{id})((\mathscr{G}'^{(t,r),>i-1}_{p^\infty}(\nu))^\wedge). \qquad (5.2.26.14)$$

We obtain the desired inclusion (5.2.26.13) with $\alpha = 1$ because $v(\nu(\gamma_i)-1) \leq \frac{1}{p-1}$.

Suppose that $\nu(\gamma_i) = 1$, so $\nu \in \chi^{>i}_{p^\infty}$. In view of (5.2.26.9), to establish the desired inclusion (5.2.26.13), it suffices to show that there exists an integer $\alpha \geq 0$ depending only on t and t' such that for every element

$$\sum_{\underline{n}\in J_{i-1}} p^{|\underline{n}|(t',r)} b_{\underline{n}} \underline{y}^{\underline{n}} \in \widehat{\mathscr{G}}'^{(t',r),>i-1}_{p^\infty}(\nu), \qquad (5.2.26.15)$$

where $b_{\underline{n}} \in (R'^{(\nu)}_{p^\infty})^\wedge$ and $b_{\underline{n}}$ tends to 0 when $|\underline{n}|$ tends to infinity, the system of linear equations defined, for $\underline{n} = (n_1, \dots, n_{d'}) \in J_{i-1}$, by

$$p^\alpha p^{(t'-t)\sum_{j=1}^{\ell} n_j} n_i! b_{\underline{n}} = \sum_{\underline{m}=(m_1,\dots,m_{d'})\in J_{i-1}(\underline{n})} p^{t(m_i-n_i)} m_i! a_{\underline{m}} \frac{w^{m_i-n_i}}{(m_i-n_i)!}, \qquad (5.2.26.16)$$

admits a solution $a_{\underline{m}} \in (R'^{(\nu)}_{p^\infty})^\wedge$ for $\underline{m} \in J_{i-1}$ such that $a_{\underline{m}}$ tends to 0 when $|\underline{m}|$ tends to infinity. For $\underline{n} = (n_1, \dots, n_{d'}) \in J_{i-1}$, we set

$$a'_{\underline{n}} = n_i! p^{-\frac{n_i}{p-1}} a_{\underline{n}}, \qquad (5.2.26.17)$$

$$b'_{\underline{n}} = p^\alpha p^{(t'-t)\sum_{j=1}^{\ell} n_j} n_i! p^{-\frac{n_i}{p-1}} b_{\underline{n}}, \qquad (5.2.26.18)$$

so that the equation (5.2.26.16) becomes

$$b'_{\underline{n}} = p^{t+\frac{1}{p-1}} w \sum_{\underline{m}=(m_1,\dots,m_{d'})\in J_{i-1}(\underline{n})} p^{(t+\frac{1}{p-1})(m_i-n_i-1)} a'_{\underline{m}} \frac{w^{(m_i-n_i-1)}}{(m_i-n_i)!}. \qquad (5.2.26.19)$$

Consider the $R^\circ_{p^\infty}$-linear endomorphism Φ of $(\oplus_{\underline{n}\in J_{i-1}} R'^{(\nu)}_{p^\infty})^\wedge$ defined, for a sequence $(x_{\underline{n}})_{\underline{n}\in J_{i-1}}$ of elements of $(R'^{(\nu)}_{p^\infty})^\wedge$ tending to 0 when $|\underline{n}|$ tends to infinity, by

$$\Phi(\sum_{\underline{n}\in J_{i-1}} x_{\underline{n}}) = \sum_{\underline{n}\in J_{i-1}} z_{\underline{n}}, \qquad (5.2.26.20)$$

where for $\underline{n} = (n_1, \dots, n_{d'}) \in J_{i-1}$,

$$z_{\underline{n}} = \sum_{\underline{m}=(m_1,\dots,m_{d'})\in\{\underline{n}\}\cup J_{i-1}(\underline{n})} p^{(t+\frac{1}{p-1})(m_i-n_i)} x_{\underline{m}} \frac{w^{(m_i-n_i)}}{(m_i-n_i+1)!}. \qquad (5.2.26.21)$$

Since Φ is congruent to the identity modulo $p^{t+\frac{1}{p-1}}$, it is surjective by virtue of ([1] 1.8.5). Hence, for every sequence $b'_{\underline{n}} \in p^{t+\frac{1}{p-1}} w(R'^{(\nu)}_{p^\infty})^\wedge$ for $\underline{n} \in J_{i-1}$ tending to 0 when $|\underline{n}|$ tends to infinity, the equation (5.2.26.19) admits a solution $a'_{\underline{m}} \in (R'^{(\nu)}_{p^\infty})^\wedge$ for $\underline{m} \in J_{i-1}$ tending to 0 when $|\underline{m}|$ tends to infinity; just take $a'_{\underline{n}} = x_{\underline{n}}$ a preimage by the morphism Φ (5.2.26.20) of the sequence $z_{\underline{n}} = p^{-t-\frac{1}{p-1}} w^{-1} b'_{\underline{n}}$. On the other hand, $v(w) = \rho + \frac{1}{p-1}$ and there exists an integer $\alpha \geq 0$ such that for every $n \in \mathbb{N}$, we have

$$(t' - t)n + v(n!) - \frac{n}{p-1} + \alpha \geq t + \rho + \frac{2}{p-1}. \tag{5.2.26.22}$$

The desired assertion follows by taking for $b'_{\underline{n}}$ for $\underline{n} \in J_{i-1}$ the elements defined by (5.2.26.18). Observe that $v(n!) \leq \frac{n}{p-1}$ for every integer $n \geq 0$ (5.2.26.17).

5.2.27 We take again the notation of 5.2.25. Let t, r be two rational numbers such that $t \geq r \geq 0$, i an integer such that $0 \leq i \leq \ell$. To lighten the notation, we set (5.2.25.13)

$$G'^{(t,r)} = \mathscr{G}'^{(t,r),>0}_{p^\infty}, \tag{5.2.27.1}$$

which is an R'_{p^∞}-algebra by (5.1.18.2). We have $\mathscr{G}'^{(t,r)}_{p^\infty} = G'^{(t,r)} \otimes_{R'_{p^\infty}} \widehat{R'_{p^\infty}}$ (5.2.23.3). Since the ring R'_{p^∞} is separated for the p-adic topology (5.2.25), $G'^{(t,r)}$ identifies with a sub-R'_{p^∞}-algebra of $\widehat{\mathscr{G}}'^{(t,r)}_{p^\infty}$, stable by the action of $\mathfrak{S}_{p^\infty}$ by virtue of (5.2.25.9). The p-adic Hausdorff completion of $G'^{(t,r)}$ identifies with $\widehat{\mathscr{G}}'^{(t,r)}_{p^\infty}$. Moreover, with the notation of (5.1.13.1), the canonical morphism

$$G'^{(t,r)} \widehat{\otimes}_{R^\diamond_{p^\infty}} R^\diamond_\infty \to \mathscr{G}'^{(t,r)}_{p^\infty} \widehat{\otimes}_{R^\diamond_{p^\infty}} R^\diamond_\infty, \tag{5.2.27.2}$$

where the tensor product $\widehat{\otimes}$ is completed for the p-adic topology, is an isomorphism. For every integer $0 \leq i \leq \ell$, the $R^\diamond_\infty$-module $\mathscr{G}'^{(t,r),>i}_{p^\infty} \otimes_{R^\diamond_{p^\infty}} R^\diamond_\infty$ being a direct factor of $G'^{(t,r)} \otimes_{R^\diamond_{p^\infty}} R^\diamond_\infty$, $\mathscr{G}'^{(t,r),>i}_{p^\infty} \widehat{\otimes}_{R^\diamond_{p^\infty}} R^\diamond_\infty$ is a sub-$\widehat{R^\diamond_\infty}$-module of $\mathscr{G}'^{(t,r)}_{p^\infty} \widehat{\otimes}_{R^\diamond_{p^\infty}} R^\diamond_\infty$.

Proposition 5.2.28 *We keep the assumptions of* 5.2.27, *let i be an integer such that $1 \leq i \leq \ell = d' - d$, and let t, t', r be three rational numbers such that $t' > t > r \geq 0$. Then,*

(i) *We have $\mathscr{G}'^{(t,r),>0}_{p^\infty} \widehat{\otimes}_{R^\diamond_{p^\infty}} R^\diamond_\infty = \mathscr{G}'^{(t,r)}_{p^\infty} \widehat{\otimes}_{R^\diamond_{p^\infty}} R^\diamond_\infty$ and $\mathscr{G}'^{(t,r),>\ell}_{p^\infty} \widehat{\otimes}_{R^\diamond_{p^\infty}} R^\diamond_\infty = \mathscr{G}^{(t,r)}_{p^\infty} \widehat{\otimes}_{R_{p^\infty}} R^\diamond_\infty$, where the tensor product $\widehat{\otimes}$ is completed for the p-adic topology.*

(ii) *The sequence*

$$0 \longrightarrow \mathscr{G}'^{(t,r),>i}_{p^\infty} \widehat{\otimes}_{R^\diamond_{p^\infty}} R^\diamond_\infty \xrightarrow{u} \mathscr{G}'^{(t,r),>i-1}_{p^\infty} \widehat{\otimes}_{R^\diamond_{p^\infty}} R^\diamond_\infty \xrightarrow{\gamma_i - \mathrm{id}} \mathscr{G}'^{(t,r),>i-1}_{p^\infty} \widehat{\otimes}_{R^\diamond_{p^\infty}} R^\diamond_\infty, \tag{5.2.28.1}$$

where u is the canonical morphism, is exact.

(iii) *There exists an integer $\alpha \geq 0$, depending on t and t' but not on the morphisms f, f' and g satisfying the conditions of* 5.1.3 *and* 5.2.1*, such that we have*

$$p^{\alpha}(\mathscr{G}'^{(t',r),>i-1}_{p^\infty}\widehat{\otimes}_{R^\diamond_{p^\infty}} R^\diamond_\infty) \subset (\gamma_i - \mathrm{id})(\mathscr{G}'^{(t,r),>i-1}_{p^\infty}\widehat{\otimes}_{R^\diamond_{p^\infty}} R^\diamond_\infty). \qquad (5.2.28.2)$$

(i) This immediately follows from 5.2.26(i).

(ii) Since for every integer $n \geq 0$, we clearly have

$$p^n(\mathscr{G}'^{(t,r),>i}_{p^\infty} \otimes_{R^\diamond_{p^\infty}} R^\diamond_\infty) = (\mathscr{G}'^{(t,r),>i}_{p^\infty} \otimes_{R^\diamond_{p^\infty}} R^\diamond_\infty) \cap p^n(\mathscr{G}'^{(t,r),>i-1}_{p^\infty} \otimes_{R^\diamond_{p^\infty}} R^\diamond_\infty), \qquad (5.2.28.3)$$

the canonical morphism $u\colon \mathscr{G}'^{(t,r),>i}_{p^\infty}\widehat{\otimes}_{R^\diamond_{p^\infty}} R^\diamond_\infty \to \mathscr{G}'^{(t,r),>i-1}_{p^\infty}\widehat{\otimes}_{R^\diamond_{p^\infty}} R^\diamond_\infty$ is injective.

For any $\nu \in \chi^{>i-1}_{p^\infty}$, we denote by $R'^{(\nu)}_{p^\infty}\widehat{\otimes}_{R^\diamond_{p^\infty}} R^\diamond_\infty$ and $\mathscr{G}'^{(t,r),>i-1}_{p^\infty}(\nu)\widehat{\otimes}_{R^\diamond_{p^\infty}} R^\diamond_\infty$ the p-adic Hausdorff completions of $R'^{(\nu)}_{p^\infty} \otimes_{R^\diamond_{p^\infty}} R^\diamond_\infty$ and $\mathscr{G}'^{(t,r),>i-1}_{p^\infty}(\nu) \otimes_{R^\diamond_{p^\infty}} R^\diamond_\infty$, respectively. Observe that $\mathscr{G}'^{(t,r),>i-1}_{p^\infty}(\nu)\widehat{\otimes}_{R^\diamond_{p^\infty}} R^\diamond_\infty$ identifies with a sub-$\widehat{R^\diamond_{p^\infty}}$-module of $\mathscr{G}'^{(t,r),>i-1}_{p^\infty}\widehat{\otimes}_{R^\diamond_{p^\infty}} R^\diamond_\infty$, stable by γ_i. In view of (5.2.25.13), every element x of $\mathscr{G}'^{(t,r),>i-1}_{p^\infty}\widehat{\otimes}_{R^\diamond_{p^\infty}} R^\diamond_\infty$ can be written as the sum of a series

$$\sum_{\nu \in \chi^{>i-1}_{p^\infty}} x_\nu, \qquad (5.2.28.4)$$

where $x_\nu \in \mathscr{G}'^{(t,r),>i-1}_{p^\infty}(\nu)\widehat{\otimes}_{R^\diamond_{p^\infty}} R^\diamond_\infty$ and, for every integer $n \geq 0$, except for a finite number of $\nu \in \chi^{>i-1}_{p^\infty}$, $x_\nu \in p^n(\mathscr{G}'^{(t,r),>i-1}_{p^\infty}(\nu)\widehat{\otimes}_{R^\diamond_{p^\infty}} R^\diamond_\infty)$. Such an element x vanishes if and only if x_ν vanishes for every $\nu \in \chi^{>i-1}_{p^\infty}$.

Let $\nu \in \chi^{>i-1}_{p^\infty}$. By (5.2.26.9) and with the same notation, for every

$$z = \sum_{\underline{n} \in J_{i-1}} p^{|\underline{n}|^{(t,r)}} a_{\underline{n}} \underline{y}^{\underline{n}} \in \mathscr{G}'^{(t,r),>i-1}_{p^\infty}(\nu) \otimes_{R^\diamond_{p^\infty}} R^\diamond_\infty, \qquad (5.2.28.5)$$

where $a_{\underline{n}} \in R'^{(\nu)}_{p^\infty} \otimes_{R^\diamond_{p^\infty}} R^\diamond_\infty$, we have (5.2.25.9)

$$(\gamma_i - \mathrm{id})(z) = \sum_{\underline{n} \in J_{i-1}} p^{|\underline{n}|^{(t,r)}} b_{\underline{n}} \underline{y}^{\underline{n}} \in \mathscr{G}'^{(t,r),>i-1}_{p^\infty}(\nu) \otimes_{R^\diamond_{p^\infty}} R^\diamond_\infty, \qquad (5.2.28.6)$$

where for every $\underline{n} = (n_1, \ldots, n_{d'}) \in J_{i-1}$,

$$b_{\underline{n}} = (\nu(\gamma_i) - 1)a_{\underline{n}} + \sum_{\underline{m}=(m_1,\ldots,m_{d'}) \in J_{i-1}(\underline{n})} p^{t(m_i - n_i)} \binom{m_i}{n_i} \nu(\gamma_i) a_{\underline{m}} w^{m_i - n_i}. \qquad (5.2.28.7)$$

Let z be an element of $\mathscr{G}'^{(t,r),>i-1}_{p^\infty}(\nu)\widehat{\otimes}_{R^\diamond_{p^\infty}} R^\diamond_\infty$. Then z can be written as the sum of a series

$$z = \sum_{\underline{n}\in J_{i-1}} p^{|\underline{n}|^{(t,r)}} a_{\underline{n}} \underline{y}^{\underline{n}}, \tag{5.2.28.8}$$

where $a_{\underline{n}} \in R'^{(\nu)}_{p^\infty}\widehat{\otimes}_{R^\diamond_{p^\infty}} R^\diamond_\infty$ and $a_{\underline{n}}$ tends to 0 when $|\underline{n}|$ tends to infinity. As the module $R'^{(\nu)}_{p^\infty} \otimes_{R^\diamond_{p^\infty}} R^\diamond_\infty$ is a direct factor of $R'_{p^\infty} \otimes_{R^\diamond_{p^\infty}} R^\diamond_\infty$ (5.1.18.2) and $R'_{p^\infty} \otimes_{R^\diamond_{p^\infty}} R^\diamond_\infty$ is a subring of R'_∞ (5.1.14.1), we see that $R'^{(\nu)}_{p^\infty} \otimes_{R^\diamond_{p^\infty}} R^\diamond_\infty$ is $\mathscr{O}_C$-flat. Therefore so is $R'^{(\nu)}_{p^\infty}\widehat{\otimes}_{R^\diamond_{p^\infty}} R^\diamond_\infty$ (see the proof of [3] II.6.14). Consequently, z vanishes if and only if $a_{\underline{n}}$ vanishes for every $\underline{n} \in J_{i-1}$. We immediately see that $(\gamma_i - \mathrm{id})(z)$ is still given by the formula (5.2.28.6).

Suppose that $\gamma_i(z) = z$ and $\nu(\gamma_i) \neq 1$. Since $R'^{(\nu)}_{p^\infty}\widehat{\otimes}_{R^\diamond_{p^\infty}} R^\diamond_\infty$ is $\mathscr{O}_C$-flat and $v(\nu(\gamma_i) - 1) \leq \frac{1}{p-1}$, for every $\underline{n} = (n_1, \dots, n_{d'}) \in J_{i-1}$, we have

$$a_{\underline{n}} = -(\nu(\gamma_i) - 1)^{-1} \sum_{\underline{m}=(m_1,\dots,m_{d'})\in J_{i-1}(\underline{n})} p^{t(m_i-n_i)}\binom{m_i}{n_i}\nu(\gamma_i)a_{\underline{m}}w^{m_i-n_i}.$$

We deduce from this that for every $\alpha \in \mathbb{N}$ and every $\underline{n} \in J_{i-1}$, we have $a_{\underline{n}} \in p^{t\alpha}R'^{(\nu)}_{p^\infty}\widehat{\otimes}_{R^\diamond_{p^\infty}} R^\diamond_\infty$ (we prove it by induction on α); so $z = 0$. Hence, $\gamma_i - \mathrm{id}$ is injective on $\mathscr{G}'^{(t,r),>i-1}_{p^\infty}(\nu)\widehat{\otimes}_{R^\diamond_{p^\infty}} R^\diamond_\infty$.

Suppose that $\gamma_i(z) = z$ and $\nu(\gamma_i) = 1$, so that $\nu \in \chi^{>i}_{p^\infty}$. Then, for every $\underline{n} = (n_1, \dots, n_{d'}) \in J_{i-1}$, if we set $\underline{n}' = (n'_1, \dots, n'_{d'}) \in J_{i-1}(\underline{n})$ with $n'_i = n_i + 1$, we have

$$(n_i + 1)!a_{\underline{n}'} = - \sum_{\underline{m}=(m_1,\dots;m_{d'})\in J_{i-1}(\underline{n}')} p^{t(m_i-n_i-1)}m_i!a_{\underline{m}}\frac{w^{m_i-n_i-1}}{(m_i-n_i)!}.$$

We have $w^{m-1}/m! \in \mathscr{O}_C$ for every integer $m \geq 1$. We deduce from this that for every $\alpha \in \mathbb{N}$ and every $\underline{n} = (n_1, \dots, n_{d'}) \in J_{i-1}$ such that $n_i \geq 1$, we have $n_i!a_{\underline{n}} \in p^{t\alpha}R'^{(\nu)}_{p^\infty}\widehat{\otimes}_{R^\diamond_{p^\infty}} R^\diamond_\infty$ (we prove it by induction on α); so $a_{\underline{n}} = 0$. Hence $z \in \mathscr{G}'^{(t,r),>i}_{p^\infty}(\nu)\widehat{\otimes}_{R^\diamond_{p^\infty}} R^\diamond_\infty$. We deduce from this that the sequence

$$0 \longrightarrow \mathscr{G}'^{(t,r),>i}_{p^\infty}(\nu)\widehat{\otimes}_{R^\diamond_{p^\infty}} R^\diamond_\infty \xrightarrow{u_\nu} \mathscr{G}'^{(t,r),>i-1}_{p^\infty}(\nu)\widehat{\otimes}_{R^\diamond_{p^\infty}} R^\diamond_\infty \xrightarrow{\gamma_i-\mathrm{id}} \mathscr{G}'^{(t,r),>i-1}_{p^\infty}(\nu)\widehat{\otimes}_{R^\diamond_{p^\infty}} R^\diamond_\infty, \tag{5.2.28.9}$$

where u_ν is the canonical morphism, is exact.

It follows from the above that the sequence (5.2.28.1) is exact.

(iii) By virtue of 5.2.26(iii), there exists an integer $\alpha \geq 0$, depending on t and t' but not on the data in 5.1.3, such that

$$p^\alpha(\widehat{\mathscr{G}}'^{(t',r),>i-1}_{p^\infty} \otimes_{R^\diamond_{p^\infty}} R^\diamond_\infty) \subset (\gamma_i - \mathrm{id})(\widehat{\mathscr{G}}'^{(t,r),>i-1}_{p^\infty} \otimes_{R^\diamond_{p^\infty}} R^\diamond_\infty). \tag{5.2.28.10}$$

Setting $M = (\gamma_i-\mathrm{id})(\widehat{\mathscr{G}}'^{(t,r),>i-1}_{p^\infty}\otimes_{R^\diamond_{p^\infty}} R^\diamond_\infty)$, we deduce from this by p-adic completion a commutative diagram

$$\begin{array}{ccc} p^{\alpha}(\mathscr{G}'^{(t',r),>i-1}_{p^\infty}\widehat{\otimes}_{R^\circ_{p^\infty}} R^\circ_\infty) \longrightarrow & \widehat{M} & \longrightarrow \mathscr{G}'^{(t,r),>i-1}_{p^\infty}\widehat{\otimes}_{R^\circ_{p^\infty}} R^\circ_\infty \\ & \uparrow & \nearrow_{\gamma_i-\mathrm{id}} \\ & \mathscr{G}'^{(t,r),>i-1}_{p^\infty}\widehat{\otimes}_{R^\circ_{p^\infty}} R^\circ_\infty & \end{array} \tag{5.2.28.11}$$

where the vertical arrow is surjective by virtue of ([1] 1.8.5). The required proposition follows.

5.2.29 We keep the assumptions of 5.2.27. For any rational numbers $t \geq r \geq 0$, we define by induction, for any integer $0 \leq i \leq \ell$, a complex $\mathbb{K}_i^{(t,r),\bullet}$ of continuous $\widehat{R^\circ_\infty}$-representations of $\mathfrak{S}_{p^\infty}$ by setting $\mathbb{K}_0^{(t,r),\bullet} = \mathscr{G}'^{(t,r)}_{p^\infty}\widehat{\otimes}_{R^\circ_{p^\infty}} R^\circ_\infty[0]$ and for any $1 \leq i \leq \ell$, $\mathbb{K}_i^{(t,r),\bullet}$ is the homotopy fiber of the morphism

$$\gamma_i - \mathrm{id}\colon \mathbb{K}_{i-1}^{(t,r),\bullet} \to \mathbb{K}_{i-1}^{(t,r),\bullet}. \tag{5.2.29.1}$$

It follows from ([3] II.3.25 and (II.2.7.9)) that we have a canonical isomorphism in $\mathbf{D}^+(\mathbf{Mod}(\widehat{R^\circ_\infty}))$

$$\mathrm{C}^\bullet_{\mathrm{cont}}(\mathfrak{S}_{p^\infty}, \mathscr{G}'^{(t,r)}_{p^\infty}\widehat{\otimes}_{R^\circ_{p^\infty}} R^\circ_\infty) \xrightarrow{\sim} \mathbb{K}_\ell^{(t,r),\bullet}. \tag{5.2.29.2}$$

For all rational numbers $t' \geq t \geq r \geq 0$, the canonical homomorphism $\mathscr{G}'^{(t',r)}_{p^\infty} \to \mathscr{G}'^{(t,r)}_{p^\infty}$ induces for every integer $0 \leq i \leq \ell$ a morphism $\mathbb{K}_i^{(t',r),\bullet} \to \mathbb{K}_i^{(t,r),\bullet}$ of complexes of continuous $\widehat{R^\circ_\infty}$-representations of $\mathfrak{S}_{p^\infty}$.

Proposition 5.2.30 *We keep the assumptions of* 5.2.27*, let i be an integer such that* $0 \leq i \leq \ell = d' - d$*, and let* t, t', r *be three rational numbers such that* $t' > t > r \geq 0$*. Then,*

(i) *We have a canonical and* $\mathfrak{S}_{p^\infty}$*-equivariant* $\widehat{R^\circ_\infty}$*-linear isomorphism*

$$\mathscr{G}'^{(t,r),>i}_{p^\infty}\widehat{\otimes}_{R^\circ_{p^\infty}} R^\circ_\infty \xrightarrow{\sim} \mathrm{H}^0(\mathbb{K}_i^{(t,r),\bullet}). \tag{5.2.30.1}$$

(ii) *There exists an integer* $\alpha_i \geq 0$*, depending on* t, t' *and* i*, but not on the morphisms* f, f' *and* g *satisfying the conditions of* 5.1.3 *and* 5.2.1*, such that for every integer* $j \geq 1$*, the canonical morphism*

$$\mathrm{H}^j(\mathbb{K}_i^{(t',r),\bullet}) \to \mathrm{H}^j(\mathbb{K}_i^{(t,r),\bullet}) \tag{5.2.30.2}$$

is annihilated by p^{α_i}.

We proceed by induction on i. The proposition is immediate for $i = 0$ by 5.2.28(i). Suppose $i \geq 1$ and the proposition proved for $i - 1$. The distinguished triangle

$$\mathbb{K}_i^{(t,r),\bullet} \longrightarrow \mathbb{K}_{i-1}^{(t,r),\bullet} \xrightarrow{\gamma_i - \mathrm{id}} \mathbb{K}_{i-1}^{(t,r),\bullet} \xrightarrow{+1} \tag{5.2.30.3}$$

and the induction hypothesis induce an exact sequence

$$0 \longrightarrow \mathrm{H}^0(\mathbb{K}_i^{(t,r),\bullet}) \longrightarrow \mathscr{G}_{p^\infty}^{\prime(t,r),>i-1}\widehat{\otimes}_{R^\circ_{p^\infty}} R^\circ_\infty \xrightarrow{\gamma_i-\mathrm{id}} \mathscr{G}_{p^\infty}^{\prime(t,r),>i-1}\widehat{\otimes}_{R^\circ_{p^\infty}} R^\circ_\infty, \tag{5.2.30.4}$$

which implies proposition (i) by virtue of 5.2.28(ii).

For every integer $j \geq 1$, the distinguished triangle (5.2.30.3) induces an exact sequence of $\widehat{R^\circ_\infty}$-modules

$$0 \to C_j^{(t,r)} \to \mathrm{H}^j(\mathbb{K}_i^{(t,r),\bullet}) \to D_j^{(t,r)} \to 0, \tag{5.2.30.5}$$

where $C_j^{(t,r)}$ is a quotient of $\mathrm{H}^{j-1}(\mathbb{K}_{i-1}^{(t,r),\bullet})$ and $D_j^{(t,r)}$ is a submodule of $\mathrm{H}^j(\mathbb{K}_{i-1}^{(t,r),\bullet})$. Moreover, by the induction hypothesis, we have a canonical isomorphism

$$C_1^{(t,r)} \xrightarrow{\sim} (\mathscr{G}_{p^\infty}^{\prime(t,r),>i-1}\widehat{\otimes}_{R^\circ_{p^\infty}} R^\circ_\infty)/(\gamma_i - 1)(\mathscr{G}_{p^\infty}^{\prime(t,r),>i-1}\widehat{\otimes}_{R^\circ_{p^\infty}} R^\circ_\infty). \tag{5.2.30.6}$$

The canonical morphisms $\mathbb{K}_{i-1}^{(t',r),\bullet} \to \mathbb{K}_{i-1}^{(t,r),\bullet}$ and $\mathbb{K}_i^{(t',r),\bullet} \to \mathbb{K}_i^{(t,r),\bullet}$ induce morphisms $C_j^{(t',r)} \to C_j^{(t,r)}$ and $D_j^{(t',r)} \to D_j^{(t,r)}$ that fit into a commutative diagram

$$\begin{array}{ccccccccc} 0 & \longrightarrow & C_j^{(t',r)} & \longrightarrow & \mathrm{H}^j(\mathbb{K}_i^{(t',r),\bullet}) & \longrightarrow & D_j^{(t',r)} & \longrightarrow & 0 \\ & & \downarrow & & \downarrow & & \downarrow & & \\ 0 & \longrightarrow & C_j^{(t,r)} & \longrightarrow & \mathrm{H}^j(\mathbb{K}_i^{(t,r),\bullet}) & \longrightarrow & D_j^{(t,r)} & \longrightarrow & 0 \end{array} \tag{5.2.30.7}$$

where the middle vertical arrow is the canonical morphism.

We set $t'' = (t + t')/2$. By the induction hypothesis, there exists an integer $\alpha_{i-1} \geq 0$, depending only on t, t' and $i - 1$, such that for every integer $j \geq 1$, the morphism $D_j^{(t',r)} \to D_j^{(t'',r)}$ is annihilated by $p^{\alpha_{i-1}}$. On the other hand, in view of the induction hypothesis and by virtue of (5.2.30.6) and 5.2.28(iii), there exists an integer $\alpha'_{i-1} \geq 0$, depending only on t, t' and $i - 1$, such that for every integer $j \geq 1$, the morphism $C_j^{(t'',r)} \to C_j^{(t,r)}$ is annihilated by $p^{\alpha'_{i-1}}$. Proposition (ii) follows by taking $\alpha_i = \alpha_{i-1} + \alpha'_{i-1}$.

Corollary 5.2.31 *Let t, t', r be three rational numbers such that $t' > t > r \geq 0$. Then,*

(i) *The canonical homomorphism*

$$\mathscr{G}_{p^\infty}^{(r)}\widehat{\otimes}_{R_{p^\infty}} R^\circ_\infty \to (\mathscr{G}_{p^\infty}^{\prime(t,r)}\widehat{\otimes}_{R^\circ_{p^\infty}} R^\circ_\infty)^{\mathfrak{S}_{p^\infty}} \tag{5.2.31.1}$$

is an isomorphism.

(ii) *There exists an integer $\alpha \geq 0$, depending on t, t' and $\ell = d' - d$, but not on the morphisms f, f' and g satisfying the conditions of* 5.1.3 *and* 5.2.1, *such that for every integer $j \geq 1$, the canonical morphism*

$$\mathrm{H}^j_{\mathrm{cont}}(\mathfrak{S}_{p^\infty}, \mathscr{G}_{p^\infty}^{\prime(t',r)}\widehat{\otimes}_{R^\circ_{p^\infty}} R^\circ_\infty) \to \mathrm{H}^j_{\mathrm{cont}}(\mathfrak{S}_{p^\infty}, \mathscr{G}_{p^\infty}^{\prime(t,r)}\widehat{\otimes}_{R^\circ_{p^\infty}} R^\circ_\infty) \tag{5.2.31.2}$$

is annihilated by p^α.

This follows from (5.2.29.2) and 5.2.30.

Corollary 5.2.32 *Let t, t', r be three rational numbers such that $t' > t > r \geq 0$, $\mathfrak{S}$ the group defined in* (5.1.12.22)*. Then,*

(i) *The canonical homomorphism*

$$\mathscr{G}^{(r)}_{p^\infty} \widehat{\otimes}_{R_{p^\infty}} R^\circ_\infty \to (\widehat{\mathscr{G}}'^{(t,r)})^{\mathfrak{S}} \qquad (5.2.32.1)$$

is an α-isomorphism.

(ii) *There exists an integer $\alpha \geq 0$, depending on t, t' and $\ell = d' - d$, but not on the morphisms f, f' and g satisfying the conditions of* 5.1.3 *and* 5.2.1*, such that for every integer $j \geq 1$, the canonical morphism*

$$\mathrm{H}^j_{\mathrm{cont}}(\mathfrak{S}, \widehat{\mathscr{G}}'^{(t',r)}) \to \mathrm{H}^j_{\mathrm{cont}}(\mathfrak{S}, \widehat{\mathscr{G}}'^{(t,r)}) \qquad (5.2.32.2)$$

is annihilated by p^α.

This follows from 5.2.24 and 5.2.31 (see 5.1.13).

Proposition 5.2.33 *Let t, t', r be three rational numbers such that $t' > t > r \geq 0$. Then,*

(i) *For every integer $n \geq 1$, the canonical homomorphism*

$$(\mathscr{G}^{(r)}_{p^\infty}/p^n\mathscr{G}^{(r)}_{p^\infty}) \otimes_{R_{p^\infty}} R^\circ_\infty \to (\mathscr{G}'^{(t,r)}/p^n\mathscr{G}'^{(r)})^{\mathfrak{S}} \qquad (5.2.33.1)$$

is α-injective. Let $\mathscr{K}^{(r)}_n$ be its cokernel.

(ii) *There exists an integer $\alpha \geq 0$, depending on t, t' and $\ell = d' - d$, but not on the morphisms f, f' and g satisfying the conditions of* 5.1.3 *and* 5.2.1*, such that for every integer $n \geq 1$, the canonical morphism $\mathscr{K}^{(t',r)}_n \to \mathscr{K}^{(t,r)}_n$ is annihilated by p^α.*

(iii) *There exists an integer $\gamma \geq 0$, depending on t, t' and ℓ, but not on the morphisms f, f' and g satisfying the conditions of* 5.1.3 *and* 5.2.1*, such that for all integers $n, q \geq 1$, the canonical morphism*

$$\mathrm{H}^q(\mathfrak{S}, \mathscr{G}'^{(t',r)}/p^n\mathscr{G}'^{(t',r)}) \to \mathrm{H}^q(\mathfrak{S}, \mathscr{G}'^{(t,r)}/p^n\mathscr{G}'^{(t,r)}) \qquad (5.2.33.2)$$

is annihilated by p^γ.

(i) This follows from 5.2.32(i) and from the long exact sequence of cohomology associated with the short exact sequence of $\mathbb{Z}_p$-representations of $\mathfrak{S}$

$$0 \longrightarrow \widehat{\mathscr{G}}'^{(t,r)} \xrightarrow{\cdot p^n} \widehat{\mathscr{G}}'^{(t,r)} \longrightarrow \mathscr{G}'^{(r)}/p^n\mathscr{G}'^{(t,r)} \longrightarrow 0. \qquad (5.2.33.3)$$

We also deduce an α-injective $\widehat{R^\circ_\infty}$-linear morphism

$$\mathscr{K}^{(t,r)}_n \to \mathrm{H}^1_{\mathrm{cont}}(\mathfrak{S}, \widehat{\mathscr{G}}'^{(t,r)}). \qquad (5.2.33.4)$$

(ii) This follows from (5.2.33.4) and 5.2.32(ii).

(iii) For all integers $n, q \geq 1$, the long exact sequence of cohomology deduced from (5.2.33.3) provides an exact sequence of $\widehat{R_1}$-modules

$$0 \to \mathrm{H}^q_{\mathrm{cont}}(\mathfrak{S}, \widehat{\mathscr{G}}'^{(t,r)})/p^n\mathrm{H}^q_{\mathrm{cont}}(\mathfrak{S}, \widehat{\mathscr{G}}'^{(t,r)}) \to \mathrm{H}^q(\mathfrak{S}, \mathscr{G}'^{(t,r)}/p^n\mathscr{G}'^{(t,r)}) \to T_n^{(t,r),q} \to 0, \tag{5.2.33.5}$$

where $T_n^{(t,r),q}$ is a p^n-torsion submodule of $\mathrm{H}^{q+1}_{\mathrm{cont}}(\mathfrak{S}, \widehat{\mathscr{G}}'^{(t,r)})$. We set $t'' = (t+t')/2$. By 5.2.32(ii), there exists an integer $\beta' > 0$, depending only of t, t' and ℓ, such that for every integer $q \geq 1$, the canonical morphisms

$$\mathrm{H}^q_{\mathrm{cont}}(\mathfrak{S}, \widehat{\mathscr{G}}'^{(t',r)}) \to \mathrm{H}^q_{\mathrm{cont}}(\mathfrak{S}, \widehat{\mathscr{G}}'^{(t'',r)}) \text{ and } \mathrm{H}^q_{\mathrm{cont}}(\mathfrak{S}, \widehat{\mathscr{G}}'^{(t'',r)}) \to \mathrm{H}^q_{\mathrm{cont}}(\mathfrak{S}, \widehat{\mathscr{G}}'^{(t,r)})$$

are annihilated by $p^{\beta'}$. The proposition follows by taking $\beta = 2\beta'$.

Corollary 5.2.34 *Let t, t', r be three rational numbers such that $t' > t > r \geq 0$, Π the group defined in (5.1.9.3). Then,*

(i) *For every integer $n \geq 1$, the canonical homomorphism*

$$(\mathscr{G}^{(r)}/p^n\mathscr{G}^{(r)}) \otimes_{\overline{R}} \overline{R}^\diamond \to (\mathscr{G}'^{(t,r)}/p^n\mathscr{G}'^{(t,r)})^\Pi \tag{5.2.34.1}$$

is α-injective. Let $\mathscr{H}_n^{(t,r)}$ be its cokernel.

(ii) *There exists an integer $\alpha \geq 0$, depending on t, t' and $\ell = d' - d$, but not on the morphisms f, f' and g satisfying the conditions of* 5.1.3 *and* 5.2.1, *such that for every integer $n \geq 1$, the canonical morphism $\mathscr{H}_n^{(t',r)} \to \mathscr{H}_n^{(t,r)}$ is annihilated by p^α.*

(iii) *There exists an integer $\gamma \geq 0$, depending on t, t' and ℓ, but not on the morphisms f, f' and g satisfying the conditions of* 5.1.3 *and* 5.2.1, *such that for all integers $n, q \geq 1$, the canonical morphism*

$$\mathrm{H}^q(\Pi, \mathscr{G}'^{(t',r)}/p^n\mathscr{G}'^{(t',r)}) \to \mathrm{H}^q(\Pi, \mathscr{G}'^{(t,r)}/p^n\mathscr{G}'^{(t,r)}) \tag{5.2.34.2}$$

is annihilated by p^γ.

We use the notation of 5.1.13. It follows from ([2] 5.2.17) and from the spectral sequence

$$\mathrm{E}_2^{ij} = \mathrm{H}^i(\Sigma^\circ, \mathrm{H}^j(\Pi, \widehat{\mathscr{G}}'^{(t,r)}/p^n\widehat{\mathscr{G}}'^{(t,r)})) \Rightarrow \mathrm{H}^{i+j}(\mathfrak{S}, \widehat{\mathscr{G}}'^{(t,r)}/p^n\widehat{\mathscr{G}}'^{(t,r)}) \tag{5.2.34.3}$$

that for every integer $i \geq 0$, the canonical morphism

$$\mathrm{H}^i(\mathfrak{S}, \mathscr{G}'^{(t,r)}/p^n\mathscr{G}'^{(t,r)}) \to \mathrm{H}^i(\Pi, \mathscr{G}'^{(t,r)}/p^n\mathscr{G}'^{(t,r)})^{\Sigma^\circ} \tag{5.2.34.4}$$

is an α-isomorphism, and the canonical $\overline{R}^\diamond$-linear morphism

$$\mathrm{H}^i(\Pi, \mathscr{G}'^{(t,r)}/p^n\mathscr{G}'^{(t,r)})^{\Sigma^\circ} \otimes_{R^\circ_\infty} \overline{R}^\diamond \to \mathrm{H}^i(\Pi, \mathscr{G}'^{(t,r)}/p^n\mathscr{G}'^{(t,r)}) \tag{5.2.34.5}$$

is an α-isomorphism. Note that the $\overline{R}^\diamond$-representations $\mathrm{H}^j(\Pi, \widehat{\mathscr{G}}'^{(t,r)}/p^n\widehat{\mathscr{G}}'^{(t,r)})$ of Σ° are discrete ([47] § 7.2 page 257). Hence, for every integer $i \geq 0$, the canonical $\overline{R}^\diamond$-linear morphism

$$\mathrm{H}^i(\mathfrak{S}, \mathscr{G}'^{(t,r)}/p^n\mathscr{G}'^{(t,r)}) \otimes_{R^\circ_\infty} \overline{R}^\diamond \to \mathrm{H}^i(\Pi, \mathscr{G}'^{(t,r)}/p^n\mathscr{G}'^{(t,r)}) \tag{5.2.34.6}$$

is an α-isomorphism. Moreover, the $R^{\diamond}_{\infty}$-algebra $\overline{R}^{\diamond}$ is α-flat by 5.1.15. The proposition then follows from 5.2.33.

5.2.35 We can now prove the proposition 5.2.15. In view of 5.2.11, 5.2.18 and 5.2.21, we are reduced to the case where $\widetilde{g}$ (5.2.1.1) is the morphism $\widetilde{g}_0$ considered in 5.2.22. The sections $(\psi_0, \psi'_0) \in \mathscr{L}_0(\widehat{\mathbb{X}}) \times \mathscr{L}'_0(\widehat{\mathbb{X}}')$ are then compatible (5.2.7). They define a Δ-equivariant isomorphism of $\widehat{\overline{R}}$-algebras $\mathscr{G} \xrightarrow{\sim} \mathscr{C}$ (5.2.4.2) and a Δ'-equivariant isomorphism of $\widehat{\overline{R}'}$-algebras $\mathscr{G}' \xrightarrow{\sim} \mathscr{C}'$ (5.2.4.5) that fit into a commutative diagram

$$\begin{array}{ccc} \mathscr{G} & \longrightarrow & \mathscr{C} \\ \downarrow & & \downarrow \\ \mathscr{G}' & \longrightarrow & \mathscr{C}' \end{array} \tag{5.2.35.1}$$

where the vertical arrows are the canonical homomorphisms (5.2.7.6). By ([3] II.12.1), they also define a Δ-equivariant isomorphism of $\widehat{\overline{R}}$-algebras $\mathscr{G}^{(r)} \xrightarrow{\sim} \mathscr{C}^{(r)}$ (5.2.17.1) and a Δ'-equivariant isomorphism of $\widehat{\overline{R}'}$-algebras $\mathscr{G}'^{(r)} \xrightarrow{\sim} \mathscr{C}'^{(r)}$ (5.2.20.1) which fit into a commutative diagram

$$\begin{array}{ccc} \mathscr{G}^{(r)} & \longrightarrow & \mathscr{C}^{(r)} \\ \downarrow & & \downarrow \\ \mathscr{G}'^{(r)} & \longrightarrow & \mathscr{C}'^{(r)} \end{array} \tag{5.2.35.2}$$

where the vertical arrows are the canonical homomorphisms (5.2.14.4). We deduce from this a Δ'-equivariant isomorphism of $\widehat{\overline{R}'}$-algebras $\mathscr{G}'^{(t,r)} \xrightarrow{\sim} \mathscr{C}'^{(t,r)}$ (5.2.20.2) compatible with the isomorphism $\mathscr{G}^{(r)} \xrightarrow{\sim} \mathscr{C}^{(r)}$. The proposition 5.2.15 then follows from 5.2.34.

5.3 Relative Dolbeault Cohomology of Higgs–Tate Algebras

5.3.1 We keep the assumptions and notation of 5.1 and 5.2 in this section. For any rational number $r \geq 0$, we denote by

$$d_{\mathscr{C}'^{(r)}} : \mathscr{C}'^{(r)} \to \widetilde{\xi}^{-1}\widetilde{\Omega}^1_{R'/\mathscr{O}_K} \otimes_{R'} \mathscr{C}'^{(r)} \tag{5.3.1.1}$$

the universal $\widehat{\overline{R}'}$-derivation of the algebra $\mathscr{C}'^{(r)}$ defined in (5.2.14.3) (see 3.2.19). The morphism

$$\underline{d}_{\mathscr{C}'^{(r)}} : \mathscr{C}'^{(r)} \to \widetilde{\xi}^{-1}\widetilde{\Omega}^1_{R'/R} \otimes_{R'} \mathscr{C}'^{(r)} \tag{5.3.1.2}$$

induced by $d_{\mathscr{C}'^{(r)}}$ identifies with the universal $(\mathscr{C}^{(r)} \otimes_{\widehat{\overline{R}}} \widehat{\overline{R}'})$-derivation of $\mathscr{C}'^{(r)}$. This follows from 5.2.9 when $r = 0$ and the proof in the general case is similar. Since

$\widetilde{\xi}^{-1}\widetilde{\Omega}^1_{R'/R}\otimes_{R'}\widehat{\overline{R}'} = \underline{d}_{\mathscr{C}'^{(r)}}(\mathscr{F}'^{(r)}) \subset \underline{d}_{\mathscr{C}'^{(r)}}(\mathscr{C}'^{(r)})$, $\underline{d}_{\mathscr{C}'^{(r)}}$ is a Higgs $(\mathscr{C}^{(r)}\otimes_{\widehat{\overline{R}}}\widehat{\overline{R}'})$-field with coefficients in $\widetilde{\xi}^{-1}\widetilde{\Omega}^1_{R'/R}$ by 2.5.26(i).

For all rational numbers $r' \geq r \geq 0$, we have

$$p^{r'}(\mathrm{id}\times\alpha'^{r,r'})\circ d_{\mathscr{C}'^{(r')}} = p^r d_{\mathscr{C}'^{(r)}}\circ\alpha'^{r,r'}, \tag{5.3.1.3}$$

where $\alpha'^{r,r'}\colon \mathscr{C}'^{(r')}\to\mathscr{C}'^{(r)}$ is the canonical homomorphism.

For all rational numbers $t \geq r \geq 0$, the morphism

$$\underline{d}_{\mathscr{C}'^{(t,r)}}\colon \mathscr{C}'^{(t,r)} \to \widetilde{\xi}^{-1}\widetilde{\Omega}^1_{R'/R}\otimes_{R'}\mathscr{C}'^{(t,r)} \tag{5.3.1.4}$$

induced by $\underline{d}_{\mathscr{C}'^{(t)}}$ identifies with the universal $(\mathscr{C}^{(r)}\otimes_{\widehat{\overline{R}}}\widehat{\overline{R}'})$-derivation of the algebra $\mathscr{C}'^{(t,r)}$ defined in (5.2.14.5). We denote by

$$\underline{d}_{\widehat{\mathscr{C}}'^{(t,r)}}\colon \widehat{\mathscr{C}}'^{(t,r)} \to \widetilde{\xi}^{-1}\widetilde{\Omega}^1_{R'/R}\otimes_{R'}\widehat{\mathscr{C}}'^{(t,r)} \tag{5.3.1.5}$$

its extension to the completions (note that the R'-module $\widetilde{\Omega}^1_{R'/R}$ is free of finite type). The derivations $\underline{d}_{\mathscr{C}'^{(t,r)}}$ and $\underline{d}_{\widehat{\mathscr{C}}'^{(t,r)}}$ are Higgs $(\mathscr{C}^{(r)}\otimes_{\widehat{\overline{R}}}\widehat{\overline{R}'})$-fields with coefficients in $\widetilde{\xi}^{-1}\widetilde{\Omega}^1_{R'/R}$. We denote by $\underline{\mathbb{K}}^\bullet(\widehat{\mathscr{C}}'^{(t,r)})$ the Dolbeault complex of $(\widehat{\mathscr{C}}'^{(t,r)}, p^t\underline{d}_{\widehat{\mathscr{C}}'^{(t,r)}})$ and by $\widetilde{\underline{\mathbb{K}}}^\bullet(\widehat{\mathscr{C}}'^{(t,r)})$ the augmented Dolbeault complex

$$\mathscr{C}^{(r)}\widehat{\otimes}_{\overline{R}}\overline{R}' \to \underline{\mathbb{K}}^0(\widehat{\mathscr{C}}'^{(t,r)}) \to \underline{\mathbb{K}}^1(\widehat{\mathscr{C}}'^{(t,r)}) \to \cdots \to \underline{\mathbb{K}}^n(\widehat{\mathscr{C}}'^{(t,r)}) \to \ldots, \tag{5.3.1.6}$$

where $\mathscr{C}^{(r)}\widehat{\otimes}_{\overline{R}}\overline{R}'$ is placed in degree -1, the tensor product $\widehat{\otimes}$ is completed for the p-adic topology and the differential $\mathscr{C}^{(r)}\widehat{\otimes}_{\overline{R}}\overline{R}' \to \widehat{\mathscr{C}}'^{(t,r)}$ is the canonical homomorphism.

In view of (5.3.1.3), for all rational numbers $t' \geq t \geq r \geq 0$, the morphism $\widehat{\alpha}'^{t,t'}$ induces a morphism of complexes

$$\iota^{t',t,r}\colon \widetilde{\underline{\mathbb{K}}}^\bullet(\widehat{\mathscr{C}}'^{(t',r)}) \to \widetilde{\underline{\mathbb{K}}}^\bullet(\widehat{\mathscr{C}}'^{(t,r)}). \tag{5.3.1.7}$$

Proposition 5.3.2 *Let t', t, r be three rational numbers such that $t' > t > r \geq 0$. Then,*

(i) *There exists a rational number $\alpha \geq 0$ depending on t and t' but not on the morphisms f, f' and g satisfying the conditions of* 5.1.3 *and* 5.2.1, *such that*

$$p^\alpha\iota^{t,t',r}\colon \widetilde{\underline{\mathbb{K}}}^\bullet(\widehat{\mathscr{C}}'^{(t',r)}) \to \widetilde{\underline{\mathbb{K}}}^\bullet(\widehat{\mathscr{C}}'^{(t,r)}), \tag{5.3.2.1}$$

where $\iota^{t,t',r}$ is the morphism (5.3.1.7), *is homotopic to* 0 *by an $\widehat{\overline{R}'}$-linear homotopy.*

(ii) *The canonical morphism*

$$\iota^{t,t',r}\otimes_{\mathbb{Z}_p}\mathbb{Q}_p\colon \widetilde{\underline{\mathbb{K}}}^\bullet(\widehat{\mathscr{C}}'^{(t',r)})\otimes_{\mathbb{Z}_p}\mathbb{Q}_p \to \widetilde{\underline{\mathbb{K}}}^\bullet(\widehat{\mathscr{C}}'^{(t,r)})\otimes_{\mathbb{Z}_p}\mathbb{Q}_p \tag{5.3.2.2}$$

is homotopic to 0 *by a continuous homotopy.*

Let $(\sigma, \sigma') \in \mathscr{L}(\widehat{\mathbb{X}}) \times \mathscr{L}'(\widehat{\mathbb{X}}')$ be a pair of compatible sections (5.2.7). By ([3] II.12.1), it defines an isomorphism of $\widehat{\overline{R}}$-algebras $\mathscr{G}^{(r)} \xrightarrow{\sim} \mathscr{C}^{(r)}$ (5.2.17.1) and an isomorphism of $\widehat{\overline{R}'}$-algebras $\mathscr{G}'^{(r)} \xrightarrow{\sim} \mathscr{C}'^{(r)}$ (5.2.20.1) that fit into a commutative diagram

$$\begin{array}{ccc} \mathscr{G}^{(r)} & \longrightarrow & \mathscr{C}^{(r)} \\ \downarrow & & \downarrow \\ \mathscr{G}'^{(r)} & \longrightarrow & \mathscr{C}'^{(r)} \end{array} \tag{5.3.2.3}$$

where the vertical arrows are the canonical homomorphisms (5.2.14.4). We deduce from this an isomorphism of $\widehat{\overline{R}'}$-algebras $\mathscr{G}'^{(t,r)} \xrightarrow{\sim} \mathscr{C}'^{(t,r)}$ (5.2.20.2) compatible with the isomorphism $\mathscr{G}^{(r)} \xrightarrow{\sim} \mathscr{C}^{(r)}$. We identify the $\widehat{\overline{R}}$-algebras $\mathscr{G}^{(r)}$ and $\mathscr{C}^{(r)}$ and the $\widehat{\overline{R}}$-algebras $\mathscr{G}'^{(t,r)}$ and $\mathscr{C}'^{(t,r)}$ by these isomorphisms, and we adapt the notation of 5.3.1 accordingly.

Consider the canonical exact sequence (5.1.3.2)

$$0 \longrightarrow P^{\mathrm{gp}}/\lambda\mathbb{Z} \xrightarrow{h^{\mathrm{gp}}} P'^{\mathrm{gp}}/h^{\mathrm{gp}}(\lambda)\mathbb{Z} \longrightarrow P'^{\mathrm{gp}}/h^{\mathrm{gp}}(P^{\mathrm{gp}}) \longrightarrow 0. \tag{5.3.2.4}$$

The $\mathbb{Z}$-modules $P^{\mathrm{gp}}/\mathbb{Z}\lambda$ and $P'^{\mathrm{gp}}/h^{\mathrm{gp}}(\lambda)\mathbb{Z}$ are free of finite type by ([2] 4.2.2); let d and d' be their respective ranks and set $\ell = d' - d$. Let $(t_i)_{1 \le i \le \ell}$ be elements of P'^{gp} and let $(t_i)_{\ell+1 \le i \le d'}$ be elements of P^{gp}, such that the images of $(t_i)_{1 \le i \le \ell}$ in $P'^{\mathrm{gp}}/h^{\mathrm{gp}}(P^{\mathrm{gp}}) \otimes_{\mathbb{Z}} \mathbb{Z}_p$ form a $\mathbb{Z}_p$-basis, the images of $(t_i)_{\ell+1 \le i \le d'}$ in $P^{\mathrm{gp}}/\lambda\mathbb{Z}$ form a $\mathbb{Z}$-basis, and the images of the $t_1, \dots, t_\ell, h^{\mathrm{gp}}(t_{\ell+1}), \dots, h^{\mathrm{gp}}(t_{d'})$ in $P'^{\mathrm{gp}}/h^{\mathrm{gp}}(\lambda)\mathbb{Z}$ form a $\mathbb{Z}$-basis. To lighten the notation, we will abusively denote $(h^{\mathrm{gp}}(t_i))_{\ell+1 \le i \le d'}$ simply by $(t_i)_{\ell+1 \le i \le d'}$, which does not induce any ambiguity since h^{gp} is injective. The $(d\log(t_i))_{1 \le i \le d'}$ form an R'-basis of $\widetilde{\Omega}^1_{R'/\mathscr{O}_K}$ (5.2.25.5). For any $1 \le i \le d'$, set $y_i = \widetilde{\xi}^{-1} d\log(t_i) \in \widetilde{\xi}^{-1}\widetilde{\Omega}^1_{R'/\mathscr{O}_K} \subset \mathscr{G}'$. For any $1 \le i \le \ell$, we denote by J_i the subset of $\mathbb{N}^{d'}$ made up of the elements $\underline{n} = (n_1, \dots, n_{d'})$ such that $n_1 = \cdots = n_i = 0$.

We denote by

$$\tau^{-1} \colon \widehat{\mathscr{G}}'^{(t',r)} \otimes_{\mathbb{Z}_p} \mathbb{Q}_p \to (\mathscr{G}^{(r)} \widehat{\otimes}_{\overline{R}} \overline{R}') \otimes_{\mathbb{Z}_p} \mathbb{Q}_p \tag{5.3.2.5}$$

the $\widehat{\overline{R}'}$-linear morphism defined by

$$\tau^{-1}\Big(\sum_{\underline{n}=(n_1,\dots,n_{d'}) \in \mathbb{N}^{d'}} a_{\underline{n}} \prod_{1 \le i \le d'} y_i^{n_i} \Big) = \sum_{\underline{n}=(n_1,\dots,n_{d'}) \in J_\ell} a_{\underline{n}} \prod_{\ell+1 \le i \le d'} y_i^{n_i}.$$

For every integer $m \ge 0$, there exists one and only one $\widehat{\overline{R}'}$-linear morphism

$$\tau^m \colon \widetilde{\xi}^{-m-1}\widetilde{\Omega}^{m+1}_{R'/R} \otimes_{R'} \widehat{\mathscr{G}}'^{(t',r)} \otimes_{\mathbb{Z}_p} \mathbb{Q}_p \to \widetilde{\xi}^{-m}\widetilde{\Omega}^m_{R'/R} \otimes_{R'} \widehat{\mathscr{G}}'^{(t,r)} \otimes_{\mathbb{Z}_p} \mathbb{Q}_p \tag{5.3.2.6}$$

such that for all $1 \le i_1 < \cdots < i_{m+1} \le \ell$, we have

$$\tau^m\Big(\sum_{\underline{n}=(n_1,\dots,n_{d'})\in\mathbb{N}^{d'}} a_{\underline{n}} \prod_{1\le i\le d'} y_i^{n_i} \otimes \widetilde{\xi}^{-1} d\log(t_{i_1}) \wedge\cdots\wedge \widetilde{\xi}^{-1} d\log(t_{i_{m+1}})\Big)$$
$$= \sum_{\underline{n}=(n_1,\dots,n_{d'})\in J_{i_1-1}} \frac{a_{\underline{n}}}{n_{i_1}+1} \prod_{1\le i\le d'} y_i^{n_i+\delta_{ii_1}} \otimes \widetilde{\xi}^{-1} d\log(t_{i_2}) \wedge\cdots\wedge \widetilde{\xi}^{-1} d\log(t_{i_{m+1}}).$$

Let α be a rational number such that

$$\alpha \ge \sup_{x\in\mathbb{Q}_{\ge 0}} (\log_p(x+1)+(x+1)t-xt'), \tag{5.3.2.7}$$

where $\log_p$ is the logarithm function of base p. For every integer $m\ge 0$, we clearly have

$$p^\alpha\tau^m(\widetilde{\xi}^{-m-1}\widetilde{\Omega}^{m+1}_{R'/R}\otimes_{R'}\widehat{\mathscr{G}}'^{(t',r)}) \subset \widetilde{\xi}^{-m}\widetilde{\Omega}^{m}_{R'/R}\otimes_{R'}\widehat{\mathscr{G}}'^{(t,r)}. \tag{5.3.2.8}$$

We check immediately that the morphisms $(p^\alpha\tau^m)_{m\ge -1}$ define a homotopy linking 0 to the morphism $p^\alpha\iota^{t,t',r}$. The proposition follows.

Corollary 5.3.3 *For every rational number $r \ge 0$, the canonical morphism of complexes*

$$(\mathscr{C}^{(r)}\widehat{\otimes}_{\overline{R}}\overline{R}')\otimes_{\mathbb{Z}_p}\mathbb{Q}_p[0] \to \varinjlim_{t\in\mathbb{Q}_{>r}} \underline{\mathbb{K}}^\bullet(\widehat{\mathscr{C}}'^{(t,r)})\otimes_{\mathbb{Z}_p}\mathbb{Q}_p \tag{5.3.3.1}$$

is a quasi-isomorphism.

Chapter 6
Relative Cohomology of Dolbeault Modules

6.1 Assumptions and Notation. Relative Faltings Topos

6.1.1 In this chapter, K denotes a complete discrete valuation field of characteristic 0, with *algebraically closed* residue field k of characteristic $p > 0$, $\mathscr{O}_K$ the valuation ring of K, $\overline{K}$ an algebraic closure of K, $\mathscr{O}_{\overline{K}}$ the integral closure of $\mathscr{O}_K$ in $\overline{K}$, $\mathfrak{m}_{\overline{K}}$ the maximal ideal of $\mathscr{O}_{\overline{K}}$ and G_K the Galois group of $\overline{K}$ over K. We denote by $\mathscr{O}_C$ the p-adic Hausdorff completion of $\mathscr{O}_{\overline{K}}$, $\mathfrak{m}_C$ its maximal ideal, C its field of fractions and v its valuation, normalized by $v(p) = 1$. We denote by $\mathbb{Z}_p(1)$ the $\mathbb{Z}[G_K]$-module

$$\mathbb{Z}_p(1) = \lim_{\substack{\longleftarrow \\ n\geq 0}} \mu_{p^n}(\mathscr{O}_{\overline{K}}), \tag{6.1.1.1}$$

where $\mu_{p^n}(\mathscr{O}_{\overline{K}})$ denotes the subgroup of p^nth roots of unity in $\mathscr{O}_{\overline{K}}$. For any $\mathbb{Z}_p[G_K]$-module M and any integer n, we set $M(n) = M \otimes_{\mathbb{Z}_p} \mathbb{Z}_p(1)^{\otimes n}$.

For any abelian group A, we denote by $\widehat{A}$ its p-adic Hausdorff completion.

We set $S = \mathrm{Spec}(\mathscr{O}_K)$, $\overline{S} = \mathrm{Spec}(\mathscr{O}_{\overline{K}})$ and $\check{\overline{S}} = \mathrm{Spec}(\mathscr{O}_C)$. We denote by s (resp. η, resp. $\overline{\eta}$) the closed point of S (resp. generic point of S, resp. generic point of $\overline{S}$). For any integer $n \geq 1$, we set $S_n = \mathrm{Spec}(\mathscr{O}_K/p^n\mathscr{O}_K)$. For any S-scheme X, we set

$$\overline{X} = X \times_S \overline{S}, \quad \check{\overline{X}} = X \times_S \check{\overline{S}} \quad \text{and} \quad X_n = X \times_S S_n. \tag{6.1.1.2}$$

We endow S with the logarithmic structure $\mathscr{M}_S$ defined by its closed point, and $\overline{S}$ and $\check{\overline{S}}$ with the logarithmic structures $\mathscr{M}_{\overline{S}}$ and $\mathscr{M}_{\check{\overline{S}}}$ pullbacks of $\mathscr{M}_S$.

6.1.2 Since $\mathscr{O}_{\overline{K}}$ is a non-discrete valuation ring of height 1, it is possible to develop the α-algebra (or almost algebra) on this ring ([2] 2.10.1) (see [2] 2.6-2.10). We choose a compatible system $(\beta_n)_{n>0}$ of nth roots of p in $\mathscr{O}_{\overline{K}}$. For any rational number $\varepsilon > 0$, we set $p^\varepsilon = (\beta_n)^{\varepsilon n}$, where n is an integer > 0 such that εn is an integer.

A. Abbes, M. Gros, *The p-adic Simpson Correspondence and Hodge-Tate Local Systems*, Lecture Notes in Mathematics 2345, https://doi.org/10.1007/978-3-031-55914-3_6

6.1.3 In this chapter, $f\colon (X, \mathcal{M}_X) \to (S, \mathcal{M}_S)$ and $f'\colon (X', \mathcal{M}_{X'}) \to (S, \mathcal{M}_S)$ denote *adequate* morphisms of logarithmic schemes ([3] III.4.7) and

$$g\colon (X', \mathcal{M}_{X'}) \to (X, \mathcal{M}_X) \tag{6.1.3.1}$$

a smooth and saturated $(S, \mathcal{M}_S)$-morphism. We denote by X° the maximal open subscheme of X where the logarithmic structure $\mathcal{M}_X$ is trivial; it is an open subscheme of X_η. We denote by $j\colon X^\circ \to X$ the canonical injection. For any X-scheme U, we set

$$U^\circ = U \times_X X^\circ. \tag{6.1.3.2}$$

We denote by $\hbar\colon \overline{X} \to X$ and $h\colon \overline{X}^\circ \to X$ the canonical morphisms (6.1.1.2), so that we have $h = \hbar \circ j_{\overline{X}}$.

We denote by $X'^{\rhd}$ the maximal open subscheme of X' where the logarithmic structure $\mathcal{M}_{X'}$ is trivial; it is an open subscheme of $X'^\circ = X' \times_X X^\circ$. We denote by $j'\colon X'^{\rhd} \to X'$ the canonical injection. For any X'-scheme U', we set

$$U'^{\rhd} = U' \times_{X'} X'^{\rhd}. \tag{6.1.3.3}$$

We denote by $\hbar'\colon \overline{X}' \to X'$ and $h'\colon \overline{X}'^{\rhd} \to X'$ the canonical morphisms, so that we have $h' = \hbar' \circ j'_{\overline{X}'}$.

6.1.4 We take again the notation introduced in 3.1.14. We endow $\check{\overline{X}} = X \times_S \check{\overline{S}}$ and $\check{\overline{X}}' = X' \times_S \check{\overline{S}}$ (6.1.1.2) with the logarithmic structures $\mathcal{M}_{\check{\overline{X}}}$ and $\mathcal{M}_{\check{\overline{X}}'}$ pullbacks respectively of $\mathcal{M}_X$ and $\mathcal{M}_{X'}$. We assume that there exist smooth $(\widetilde{S}, \mathcal{M}_{\widetilde{S}})$-deformations $(\widetilde{X}, \mathcal{M}_{\widetilde{X}})$ of $(\check{\overline{X}}, \mathcal{M}_{\check{\overline{X}}})$ and $(\widetilde{X}', \mathcal{M}_{\widetilde{X}'})$ of $(\check{\overline{X}}', \mathcal{M}_{\check{\overline{X}}'})$ (4.1.5) and a $(\widetilde{S}, \mathcal{M}_{\widetilde{S}})$-morphism

$$\widetilde{g}\colon (\widetilde{X}', \mathcal{M}_{\widetilde{X}'}) \to (\widetilde{X}, \mathcal{M}_{\widetilde{X}}) \tag{6.1.4.1}$$

that fits into a commutative diagram (with Cartesian squares)

$$\begin{array}{ccc}
(\check{\overline{X}}', \mathcal{M}_{\check{\overline{X}}'}) & \longrightarrow & (\widetilde{X}', \mathcal{M}_{\widetilde{X}'}) \\
\downarrow{\scriptstyle \check{\overline{g}}} & \Box & \downarrow{\scriptstyle \widetilde{g}} \\
(\check{\overline{X}}, \mathcal{M}_{\check{\overline{X}}}) & \longrightarrow & (\widetilde{X}, \mathcal{M}_{\widetilde{X}}) \\
\downarrow & \Box & \downarrow \\
(\check{\overline{S}}, \mathcal{M}_{\check{\overline{S}}}) & \longrightarrow & (\widetilde{S}, \mathcal{M}_{\widetilde{S}}).
\end{array} \tag{6.1.4.2}$$

Note that the squares are Cartesian both in the category of logarithmic schemes and in that of fine logarithmic schemes. *We fix in this chapter the deformations and the $(\widetilde{S}, \mathcal{M}_{\widetilde{S}})$-morphism $\widetilde{g}$ (6.1.4.1).*

Note that $\widetilde{g}$ is smooth by virtue of 5.2.2.

Remark 6.1.5 In the relative case (3.1.14), there exists an $(\widetilde{S}, \mathscr{M}_{\widetilde{S}})$-canonical smooth deformation $\widetilde{g}$ of g, namely

$$\widetilde{g} = g \times_{(S,\mathscr{M}_S)} (\widetilde{S}, \mathscr{M}_{\widetilde{S}}) : (X', \mathscr{M}_{X'}) \times_{(S,\mathscr{M}_S)} (\widetilde{S}, \mathscr{M}_{\widetilde{S}}) \to (X, \mathscr{M}_X) \times_{(S,\mathscr{M}_S)} (\widetilde{S}, \mathscr{M}_{\widetilde{S}}), \tag{6.1.5.1}$$

where we consider $(\widetilde{S}, \mathscr{M}_{\widetilde{S}})$ as a logarithmic scheme above $(S, \mathscr{M}_S)$ via pr_1 (3.1.13.5), the product being indifferently taken in the category of logarithmic schemes or in that of fine logarithmic schemes.

6.1.6 For any integer $n \geq 1$, we denote by $a: X_s \to X$, $a_n: X_s \to X_n$, $a': X'_s \to X'$ and $a'_n: X'_s \to X'_n$ the canonical injections (6.1.1.2). The residue field of $\mathscr{O}_K$ being algebraically closed, there exists a unique S-morphism $s \to \overline{S}$. This induces closed immersions $\overline{a}: X_s \to \overline{X}$, $\overline{a}_n: X_s \to \overline{X}_n$, $\overline{a}': X'_s \to \overline{X}'$ and $\overline{a}'_n: X'_s \to \overline{X}'_n$ lifting a, a_n, a' and a'_n, respectively. Since $\overline{a}_n$ (resp. $\overline{a}'_n$) is a universal homeomorphism, we can consider $\mathscr{O}_{\overline{X}_n}$ (resp. $\mathscr{O}_{\overline{X}'_n}$) as a sheaf of $X_{s,\mathrm{zar}}$ or $X_{s,\mathrm{\acute{e}t}}$ (resp. $X'_{s,\mathrm{zar}}$ or $X'_{s,\mathrm{\acute{e}t}}$), depending on the context (see 2.1.13).

6.1.7 To lighten the notation, we set

$$\widetilde{\Omega}^1_{X/S} = \Omega^1_{(X,\mathscr{M}_X)/(S,\mathscr{M}_S)}, \tag{6.1.7.1}$$

that we consider as a sheaf of X_{zar} or $X_{\mathrm{\acute{e}t}}$, depending on the context (2.1.13). It is a locally free $\mathscr{O}_X$-module of finite type. Likewise, we put

$$\widetilde{\Omega}^1_{X'/S} = \Omega^1_{(X',\mathscr{M}_{X'})/(S,\mathscr{M}_S)} \quad \text{and} \quad \widetilde{\Omega}^1_{X'/X} = \Omega^1_{(X',\mathscr{M}_{X'})/(X,\mathscr{M}_X)}, \tag{6.1.7.2}$$

that we consider as a sheaf of X'_{zar} or $X'_{\mathrm{\acute{e}t}}$, depending on the context. They are locally free $\mathscr{O}_{X'}$-modules of finite type. Moreover, we have a canonical locally split exact sequence of $\mathscr{O}_{X'}$-modules

$$0 \to g^*(\widetilde{\Omega}^1_{X/S}) \to \widetilde{\Omega}^1_{X'/S} \to \widetilde{\Omega}^1_{X'/X} \to 0. \tag{6.1.7.3}$$

Following the conventions of 6.1.6, for any integer $n \geq 1$, we set

$$\widetilde{\Omega}^1_{\overline{X}_n/\overline{S}_n} = \widetilde{\Omega}^1_{X/S} \otimes_{\mathscr{O}_X} \mathscr{O}_{\overline{X}_n}, \tag{6.1.7.4}$$

that we consider as a sheaf of $X_{s,\mathrm{zar}}$ or $X_{s,\mathrm{\acute{e}t}}$, depending on the context, and

$$\widetilde{\Omega}^1_{\overline{X}'_n/\overline{S}_n} = \widetilde{\Omega}^1_{X'/S} \otimes_{\mathscr{O}_{X'}} \mathscr{O}_{\overline{X}'_n} \quad (\text{resp. } \widetilde{\Omega}^1_{\overline{X}'_n/\overline{X}_n} = \widetilde{\Omega}^1_{X'/X} \otimes_{\mathscr{O}_{X'}} \mathscr{O}_{\overline{X}'_n}), \tag{6.1.7.5}$$

that we consider as a sheaf of $X'_{s,\mathrm{zar}}$ or $X'_{s,\mathrm{\acute{e}t}}$, depending on the context.

6.1.8 We denote by

$$\pi\colon E \to \mathbf{\acute{E}t}_{/X} \tag{6.1.8.1}$$

the fibered Faltings $\mathbb{U}$-site associated with the morphism $h\colon \overline{X}^{\circ} \to X$ (6.1.3.2) (see 4.3.2). We endow E with the covanishing topology defined by π (4.3.2) and we denote by $\widetilde{E}$ the topos of sheaves of $\mathbb{U}$-sets on E, called the Faltings topos associated with h.

We denote by $\mathbf{\acute{E}t}_{\mathrm{coh}/X}$ the full subcategory of $\mathbf{\acute{E}t}_{/X}$ made up of the étale schemes of finite presentation over X, endowed with the topology induced by that of $\mathbf{\acute{E}t}_{/X}$ (2.1.12). Since X is Noetherian and therefore quasi-separated, $\mathbf{\acute{E}t}_{\mathrm{coh}/X}$ is a $\mathbb{U}$-small family, topologically generating the site $\mathbf{\acute{E}t}_{/X}$ and is stable by fiber products. We denote by

$$\pi_{\mathrm{coh}}\colon E_{\mathrm{coh}} \to \mathbf{\acute{E}t}_{\mathrm{coh}/X} \tag{6.1.8.2}$$

the fibered site deduced from π by base change by the canonical injection functor $\mathbf{\acute{E}t}_{\mathrm{coh}/X} \to \mathbf{\acute{E}t}_{/X}$. We endow E_{coh} with the covanishing topology defined by π_{coh}. By ([3] VI.10.4), the canonical projection $E_{\mathrm{coh}} \to E$ induces by restriction an equivalence between the topos $\widetilde{E}$ and the topos of sheaves of $\mathbb{U}$-sets on E_{coh}. Moreover, the covanishing topology of E_{coh} is induced by that of E.

We denote by

$$\sigma\colon \widetilde{E} \to X_{\mathrm{\acute{e}t}}, \tag{6.1.8.3}$$

$$\beta\colon \widetilde{E} \to \overline{X}^{\circ}_{\mathrm{f\acute{e}t}}, \tag{6.1.8.4}$$

$$\psi\colon \overline{X}^{\circ}_{\mathrm{\acute{e}t}} \to \widetilde{E}, \tag{6.1.8.5}$$

the canonical morphisms (4.3.2.12), (4.3.2.13) and (4.3.2.15).

We denote by $X_{\mathrm{\acute{e}t}} \overleftarrow{\times}_{X_{\mathrm{\acute{e}t}}} \overline{X}^{\circ}_{\mathrm{\acute{e}t}}$ the covanishing topos of the morphism $h_{\mathrm{\acute{e}t}}\colon \overline{X}^{\circ}_{\mathrm{\acute{e}t}} \to X_{\mathrm{\acute{e}t}}$ ([3] VI.3.12) and by

$$\rho\colon X_{\mathrm{\acute{e}t}} \overleftarrow{\times}_{X_{\mathrm{\acute{e}t}}} \overline{X}^{\circ}_{\mathrm{\acute{e}t}} \to \widetilde{E} \tag{6.1.8.6}$$

the canonical morphism ([3] VI.10.15).

We denote by $\widetilde{E}_s$ the closed subtopos of $\widetilde{E}$ complement of the open object $\sigma^*(X_\eta)$ (4.3.3),

$$\delta\colon \widetilde{E}_s \to \widetilde{E} \tag{6.1.8.7}$$

the canonical embedding and

$$\sigma_s\colon \widetilde{E}_s \to X_{s,\mathrm{\acute{e}t}} \tag{6.1.8.8}$$

the morphism of topos induced by σ (4.3.3.3).

We also consider analogous objects and notation for f', that we endow with a $'$.

6.1.9 For any $(V \to U) \in \mathrm{Ob}(E)$, we denote by $\overline{U}^V$ the integral closure of $\overline{U}$ in V. We denote by $\overline{\mathscr{B}}$ the presheaf on E defined for any $(V \to U) \in \mathrm{Ob}(E)$, by

$$\overline{\mathscr{B}}((V \to U)) = \Gamma(\overline{U}^V, \mathscr{O}_{\overline{U}^V}). \tag{6.1.9.1}$$

It is a ring of $\widetilde{E}$ ([3] III.8.16). By ([3] III.8.17), we have a canonical homomorphism

$$\sigma^*(\hbar_*(\mathscr{O}_{\overline{X}})) \to \overline{\mathscr{B}}. \tag{6.1.9.2}$$

Unless explicitly stated otherwise, we consider σ (6.1.8.3) as a morphism of ringed topos

$$\sigma\colon (\widetilde{E}, \overline{\mathscr{B}}) \to (X_{\text{ét}}, \hbar_*(\mathscr{O}_{\overline{X}})). \tag{6.1.9.3}$$

For any integer $n \geq 1$, we set

$$\overline{\mathscr{B}}_n = \overline{\mathscr{B}}/p^n\overline{\mathscr{B}}. \tag{6.1.9.4}$$

It is a ring of $\widetilde{E}_s$ ([3] III.9.7). We denote by

$$\sigma_n\colon (\widetilde{E}_s, \overline{\mathscr{B}}_n) \to (X_{s,\text{ét}}, \mathscr{O}_{\overline{X}_n}) \tag{6.1.9.5}$$

the morphism of ringed topos induced by σ (6.1.9.3) (see [3] (III.9.9.4)).

We also consider analogous objects and notation for f', that we endow with a $'$.

6.1.10 The functor

$$\Theta^+\colon E \to E', \quad (V \to U) \mapsto (V \times_{X^\circ} X'^{\rhd} \to U \times_X X') \tag{6.1.10.1}$$

is continuous and left exact ([3] VI.10.12). It therefore defines a morphism of topos

$$\Theta\colon \widetilde{E}' \to \widetilde{E}. \tag{6.1.10.2}$$

It immediately follows from the definitions that the squares of the diagram

$$\begin{array}{ccccc} X'_{\text{ét}} & \xleftarrow{\sigma'} & \widetilde{E}' & \xrightarrow{\beta'} & \overline{X}'^{\rhd}_{\text{fét}} \\ \downarrow g & & \downarrow \Theta & & \downarrow \gamma \\ X_{\text{ét}} & \xleftarrow{\sigma} & \widetilde{E} & \xrightarrow{\beta} & \overline{X}^{\circ}_{\text{fét}}, \end{array} \tag{6.1.10.3}$$

where $\gamma\colon \overline{X}'^{\rhd} \to \overline{X}^{\circ}$ is the morphism induced by g, are commutative up to canonical isomorphisms.

We have a canonical isomorphism $\Theta^*(\sigma^*(X_\eta)) \simeq \sigma'^*(X'_\eta)$ (6.1.10.3). Hence, by virtue of ([5] IV 9.4.3), there exists a morphism of topos

$$\theta\colon \widetilde{E}'_s \to \widetilde{E}_s \tag{6.1.10.4}$$

unique up to canonical isomorphism such that the diagram

$$\begin{array}{ccc} \widetilde{E}'_s & \xrightarrow{\theta} & \widetilde{E}_s \\ \downarrow \delta' & & \downarrow \delta \\ \widetilde{E}' & \xrightarrow{\Theta} & \widetilde{E} \end{array} \tag{6.1.10.5}$$

is commutative up to isomorphism, and even 2-Cartesian. It follows from (6.1.10.3) and ([5] IV 9.4.3) that the diagram of morphism of topos

$$\begin{array}{ccc} \widetilde{E}'_s & \xrightarrow{\theta} & \widetilde{E}_s \\ {\scriptstyle \sigma'_s}\downarrow & & \downarrow{\scriptstyle \sigma_s} \\ X'_{s,\text{ét}} & \xrightarrow{g_s} & X_{s,\text{ét}} \end{array} \tag{6.1.10.6}$$

is commutative up to canonical isomorphism.

6.1.11 For any $(V \to U) \in \mathrm{Ob}(E)$, let $(V' \to U') = \Theta^+(V \to U)$ so that we have a commutative diagram

$$\begin{array}{ccccccc} \overline{X}'^{\rhd} & \longleftarrow & V' & \longrightarrow & \overline{U}' & \longrightarrow & \overline{X}' \\ \downarrow & \square & \downarrow & & \downarrow & \square & \downarrow \\ \overline{X}^{\circ} & \longleftarrow & V & \longrightarrow & \overline{U} & \longrightarrow & \overline{X}. \end{array} \tag{6.1.11.1}$$

We deduce from this a morphism

$$\overline{U}'^{V'} \to \overline{U}^{V}, \tag{6.1.11.2}$$

and hence a ring homomorphism of $\widetilde{E}$

$$\overline{\mathscr{B}} \to \Theta_*(\overline{\mathscr{B}}'). \tag{6.1.11.3}$$

We consider in the following Θ (6.1.10.2) as a morphism of ringed topos (respectively by $\overline{\mathscr{B}}'$ and $\overline{\mathscr{B}}$). For modules, we use the notation Θ^{-1} to denote the pullback in the sense of abelian sheaves and we keep the notation Θ^* for the pullback in the sense of modules.

For every integer $n \geq 1$, the canonical homomorphism $\Theta^{-1}(\overline{\mathscr{B}}) \to \overline{\mathscr{B}}'$ induces a homomorphism $\theta^*(\overline{\mathscr{B}}_n) \to \overline{\mathscr{B}}'_n$. The morphism θ is therefore underlying a morphism of ringed topos, which we denote by

$$\theta_n \colon (\widetilde{E}'_s, \overline{\mathscr{B}}'_n) \to (\widetilde{E}_s, \overline{\mathscr{B}}_n). \tag{6.1.11.4}$$

In view of (6.1.10.6), we immediately verify that the diagram of morphism of topos

$$\begin{array}{ccc} (\widetilde{E}'_s, \overline{\mathscr{B}}'_n) & \xrightarrow{\theta_n} & (\widetilde{E}_s, \overline{\mathscr{B}}_n) \\ {\scriptstyle \sigma'_n}\downarrow & & \downarrow{\scriptstyle \sigma_n} \\ (X'_{s,\text{ét}}, \mathscr{O}_{\overline{X}'_n}) & \xrightarrow{\overline{g}_n} & (X_{s,\text{ét}}, \mathscr{O}_{\overline{X}_n}) \end{array} \tag{6.1.11.5}$$

is commutative up to canonical isomorphism.

6.1.12 Let U be an object of $\mathbf{\acute{E}t}_{/X}$. By ([3] VI.10.14), the topos $\widetilde{E}_{/\sigma^*(U)}$, localization of $\widetilde{E}$ at $\sigma^*(U)$, is canonically equivalent to the Faltings topos associated with the morphism $\overline{U}^{\circ} \to U$ (4.3.2.10). We denote by

$$j_U\colon \widetilde{E}_{/\sigma^*(U)} \to \widetilde{E} \tag{6.1.12.1}$$

the localization morphism of $\widetilde{E}$ at $\sigma^*(U)$, which identifies with the functoriality morphism induced by the canonical morphism $U \to X$, and by

$$\beta_U\colon \widetilde{E}_{/\sigma^*(U)} \to \overline{U}^{\circ}_{\text{fét}} \tag{6.1.12.2}$$

the canonical morphism (4.3.2.13).

We consider the analogous notation for the topos $\widetilde{E}'$ and objects of $\mathbf{\acute{E}t}_{/X'}$, that we endow with an exponent $'$.

Let $\mu\colon U' \to U$ be a morphism above $g\colon X' \to X$ such that $U' \to X'$ and $U \to X$ are étale. We abusively denote by $\overline{\mu}\colon \overline{U}'^{\triangleright} \to \overline{U}^{\circ}$ the morphism induced by μ, and we denote by

$$\Theta_{\mu}\colon \widetilde{E}'_{/\sigma'^*(U')} \to \widetilde{E}_{/\sigma^*(U)} \tag{6.1.12.3}$$

the functoriality morphism induced by μ (6.1.10). The diagrams

$$\begin{array}{ccc} \widetilde{E}'_{/\sigma'^*(U')} & \xrightarrow{\Theta_{\mu}} & \widetilde{E}_{/\sigma^*(U)} \\ {\scriptstyle j'_{U'}}\downarrow & & \downarrow{\scriptstyle j_U} \\ \widetilde{E}' & \xrightarrow{\Theta} & \widetilde{E} \end{array} \tag{6.1.12.4}$$

$$\begin{array}{ccc} \widetilde{E}'_{/\sigma'^*(U')} & \xrightarrow{\Theta_{\mu}} & \widetilde{E}_{/\sigma^*(U)} \\ {\scriptstyle \beta'_{U'}}\downarrow & & \downarrow{\scriptstyle \beta_U} \\ \overline{U}'^{\triangleright}_{\text{fét}} & \xrightarrow{\overline{\mu}} & \overline{U}^{\circ}_{\text{fét}} \end{array} \tag{6.1.12.5}$$

are commutative up to canonical isomorphisms ([3] (VI.10.12.6)).

6.1.13 Let $\overline{x}'$ be a geometric point of X', $\overline{x} = g(\overline{x}')$, $\underline{X}$ (resp. $\underline{X}'$) the strict localization of X at $\overline{x}$ (resp. X' at $\overline{x}'$), $\underline{g}\colon \underline{X}' \to \underline{X}$ and $\gamma\colon \overline{\underline{X}}'^{\triangleright} \to \overline{\underline{X}}^{\circ}$ the morphisms induced by g. We denote by $\underline{\widetilde{E}}$ (resp. $\underline{\widetilde{E}}'$) the Faltings topos associated with the canonical morphism $\overline{\underline{X}}^{\circ} \to X$ (resp. $\overline{\underline{X}}'^{\triangleright} \to X'$), by

$$\underline{\Theta}\colon \underline{\widetilde{E}}' \to \underline{\widetilde{E}} \tag{6.1.13.1}$$

the functoriality morphism induced by the morphism $\underline{g}$, and by

$$\underline{\beta}\colon \underline{\widetilde{E}} \to \overline{\underline{X}}^{\circ}_{\text{fét}}, \tag{6.1.13.2}$$

$$\underline{\beta}'\colon \underline{\widetilde{E}}' \to \overline{\underline{X}}'^{\triangleright}_{\text{fét}}, \tag{6.1.13.3}$$

the canonical morphisms (4.3.2.13). By ([3] (VI.10.12.6)), the diagram

$$\begin{array}{ccc} \underline{\widetilde{E}}' & \xrightarrow{\underline{\Theta}} & \underline{\widetilde{E}} \\ {\scriptstyle \underline{\beta}'}\downarrow & & \downarrow{\scriptstyle \underline{\beta}} \\ \underline{\overline{X}}'^{\rhd}_{\text{fét}} & \xrightarrow{\underline{\gamma}} & \underline{\overline{X}}^{\circ}_{\text{fét}} \end{array} \tag{6.1.13.4}$$

is commutative up to canonical isomorphism. We denote by

$$\theta\colon \underline{\overline{X}}^{\circ}_{\text{fét}} \to \underline{\widetilde{E}}, \tag{6.1.13.5}$$

$$\theta'\colon \underline{\overline{X}}'^{\rhd}_{\text{fét}} \to \underline{\widetilde{E}}', \tag{6.1.13.6}$$

the canonical sections of $\underline{\beta}$ and $\underline{\beta}'$ ([3] VI.10.23). The diagram

$$\begin{array}{ccc} \underline{\overline{X}}'^{\rhd}_{\text{fét}} & \xrightarrow{\underline{\gamma}} & \underline{\overline{X}}^{\circ}_{\text{fét}} \\ {\scriptstyle \theta'}\downarrow & & \downarrow{\scriptstyle \theta} \\ \underline{\widetilde{E}}' & \xrightarrow{\underline{\Theta}} & \underline{\widetilde{E}} \end{array} \tag{6.1.13.7}$$

is commutative up to canonical isomorphism. Indeed, for every $\underline{X}$-scheme U that is étale, separated and of finite presentation, if we denote by U^{f} its $\underline{X}$-finite part (i.e., the disjoint sum of the strict localizations of U at the points of $U_{\overline{x}}$), then $U^{\text{f}} \times_{\underline{X}} \underline{X}'$ is the $\underline{X}'$-finite part of $U \times_{\underline{X}} \underline{X}'$ (see [3] VI.10.22).

Moreover, the diagram

$$\begin{array}{ccc} \underline{\widetilde{E}}' & \xrightarrow{\underline{\Theta}} & \underline{\widetilde{E}} \\ {\scriptstyle \Phi'}\downarrow & & \downarrow{\scriptstyle \Phi} \\ \widetilde{E}' & \xrightarrow{\Theta} & \widetilde{E}, \end{array} \tag{6.1.13.8}$$

where the vertical arrows are the functoriality morphisms induced by the canonical morphisms $\underline{X} \to X$ and $\underline{X}' \to X'$, is commutative up to canonical isomorphism. We denote by

$$\varphi_{\overline{x}}\colon \widetilde{E} \to \underline{\overline{X}}^{\circ}_{\text{fét}}, \tag{6.1.13.9}$$

$$\varphi'_{\overline{x}'}\colon \widetilde{E}' \to \underline{\overline{X}}'^{\rhd}_{\text{fét}}, \tag{6.1.13.10}$$

the composed functors $\theta^* \circ \Phi^*$ and $\theta'^* \circ \Phi'^*$ (see 4.3.5). We therefore have a canonical isomorphism of functors

$$\underline{\gamma}^* \circ \varphi_{\overline{x}} \xrightarrow{\sim} \varphi'_{\overline{x}'} \circ \Theta^*. \tag{6.1.13.11}$$

6.1.14 We denote by G the *relative Faltings* $\mathbb{U}$*-site* associated with the pair of morphisms $(h\colon \overline{X}^{\circ} \to X, g\colon X' \to X)$ ([2] 3.4.1). Objects of the category underlying G are the triple $(U, U' \to U, V \to U)$ made up of an X-scheme U and of two morphisms $U' \to U$ and $V \to U$ above g and h respectively, i.e., commutative diagrams of morphisms of schemes

$$\begin{array}{ccccc} U' & \longrightarrow & U & \longleftarrow & V \\ \downarrow & & \downarrow & & \downarrow \\ X' & \xrightarrow{g} & X & \xleftarrow{h} & \overline{X}^{\circ} \end{array} \tag{6.1.14.1}$$

such that the morphisms $U \to X$ and $U' \to X'$ are étale and the morphism $V \to \overline{U}^{\circ}$ finite étale; such an object will be denoted by $(U' \to U \leftarrow V)$. Let $(U' \to U \leftarrow V)$ and $(U'_1 \to U_1 \leftarrow V_1)$ be two objects of G. A morphism from $(U'_1 \to U_1 \leftarrow V_1)$ to $(U' \to U \leftarrow V)$ consists of three morphisms $U'_1 \to U'$, $U_1 \to U$ and $V_1 \to V$ above X', X and $\overline{X}^{\circ}$ respectively, that makes the diagram

$$\begin{array}{ccccc} U'_1 & \longrightarrow & U_1 & \longleftarrow & V_1 \\ \downarrow & & \downarrow & & \downarrow \\ U' & \longrightarrow & U & \longleftarrow & V \end{array} \tag{6.1.14.2}$$

commutative.

We endow G with the *covanishing* topology ([2] 3.4.1), i.e., the topology generated by the coverings

$$\{(U'_i \to U_i \leftarrow V_i) \to (U' \to U \leftarrow V)\}_{i \in I}$$

of the following three types:

(a) $V_i = V$, $U_i = U$ for all $i \in I$, and $(U'_i \to U')_{i \in I}$ is a covering family.
(b) $U'_i = U'$, $U_i = U$ for all $i \in I$, and $(V_i \to V)_{i \in I}$ is a covering family.
(c) $I = \{1\}$, $U'_1 = U'$ and the morphism $V_1 \to V \times_U U_1$ is an isomorphism (there is no condition on the morphism $U_1 \to U$).

We denote by $\widetilde{G}$ the *relative Faltings* $\mathbb{U}$*-topos* associated with (h, g), that is, the topos of sheaves of $\mathbb{U}$-sets on G.

We denote by

$$\pi\colon \widetilde{G} \to X'_{\text{ét}}, \tag{6.1.14.3}$$

$$\lambda\colon \widetilde{G} \to \overline{X}^{\circ}_{\text{fét}}, \tag{6.1.14.4}$$

the canonical morphisms ([2] 3.4.4). By ([2] 3.4.18), the canonical commutative diagram

$$\begin{array}{ccc} X' & \xleftarrow{h'} & \overline{X}'^{\triangleright} \\ {\scriptstyle g}\downarrow & & \downarrow{\scriptstyle \gamma} \\ X & \xleftarrow{h} & \overline{X}^{\circ} \end{array} \tag{6.1.14.5}$$

induces morphisms of topos

$$\widetilde{E}' \xrightarrow{\tau} \widetilde{G} \xrightarrow{\mathrm{g}} \widetilde{E} \tag{6.1.14.6}$$

whose composition is the morphism $\Theta\colon \widetilde{E}' \to \widetilde{E}$ (6.1.10.2). The triangles and squares of the diagram of morphism of topos

$$\begin{array}{ccccc} & & \widetilde{E}' & \xrightarrow{\beta'} & \overline{X}'^{\rhd}_{\text{fét}} \\ & \swarrow^{\sigma'} & \downarrow{\scriptstyle \tau} & & \downarrow{\scriptstyle \gamma_{\text{fét}}} \\ X'_{\text{ét}} & \xleftarrow{\pi} & \widetilde{G} & \xrightarrow{\lambda} & \overline{X}^{\circ}_{\text{fét}} \\ \downarrow{\scriptstyle g_{\text{ét}}} & & \downarrow{\scriptstyle \mathfrak{g}} & \swarrow_{\beta} & \\ X_{\text{ét}} & \xleftarrow{\sigma} & \widetilde{E} & & \end{array} \tag{6.1.14.7}$$

are commutative up to canonical isomorphisms ([2] (3.4.18.4)).

We denote by $X'_{\text{ét}} \overleftarrow{\times}_{X_{\text{ét}}} \overline{X}^{\circ}_{\text{ét}}$ the oriented product of the morphisms $g_{\text{ét}}\colon X'_{\text{ét}} \to X_{\text{ét}}$ and $f_{\text{ét}}\colon \overline{X}^{\circ}_{\text{ét}} \to X_{\text{ét}}$ ([3] VI.3.10) and by

$$\varrho\colon X'_{\text{ét}} \overleftarrow{\times}_{X_{\text{ét}}} \overline{X}^{\circ}_{\text{ét}} \to \widetilde{G} \tag{6.1.14.8}$$

the canonical morphism of topos ([2] (3.4.9.2)). We check immediately that the squares of the diagram of morphisms of topos

$$\begin{array}{ccc} X'_{\text{ét}} \overleftarrow{\times}_{X'_{\text{ét}}} \overline{X}'^{\rhd}_{\text{ét}} & \xrightarrow{\rho'} & \widetilde{E}' \\ \downarrow & & \downarrow{\scriptstyle \tau} \\ X'_{\text{ét}} \overleftarrow{\times}_{X_{\text{ét}}} \overline{X}^{\circ}_{\text{ét}} & \xrightarrow{\varrho} & \widetilde{G} \\ \downarrow & & \downarrow{\scriptstyle \mathfrak{g}} \\ X_{\text{ét}} \overleftarrow{\times}_{X_{\text{ét}}} \overline{X}^{\circ}_{\text{ét}} & \xrightarrow{\rho} & \widetilde{E}, \end{array} \tag{6.1.14.9}$$

where ρ and ρ' are the canonical morphisms (6.1.8.6), are commutative up to canonical isomorphisms ([3] VI.4.10 and [2] 3.4.17).

6.1.15 Consider a commutative diagram of morphisms of schemes

$$\begin{array}{ccc} U' & \longrightarrow & U \\ \downarrow & & \downarrow \\ X' & \longrightarrow & X \end{array} \tag{6.1.15.1}$$

such that the vertical arrows are étale morphisms. For any presheaf F on G, we define the presheaf $F_{U'\to U}$ on $\mathbf{\acute{E}t}_{\mathrm{f}/\overline{U}^{\circ}}$ by setting for any $V \in \mathrm{Ob}(\mathbf{\acute{E}t}_{\mathrm{f}/\overline{U}^{\circ}})$,

$$F_{U'\to U}(V) = F(U' \to U \leftarrow V). \tag{6.1.15.2}$$

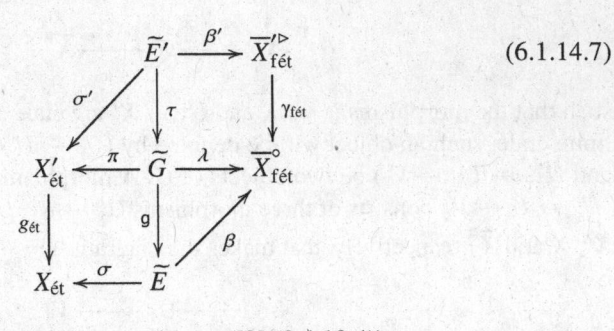

If F is a sheaf of $\widetilde{G}$, then $F_{U' \to U}$ is a sheaf of $\overline{U}^{\circ}_{\text{fét}}$.

6.1.16 To any geometric point $\overline{x}'$ of X', we associate a category $\mathfrak{C}_{\overline{x}'}$ in the following way. The objects of $\mathfrak{C}_{\overline{x}'}$ are the commutative diagrams of morphisms of schemes

$$\begin{array}{ccccc} & & U' & \longrightarrow & U \\ & \nearrow & \downarrow & & \downarrow \\ \overline{x}' & \longrightarrow & X' & \longrightarrow & X \end{array} \qquad (6.1.16.1)$$

such that the morphisms $U' \to X'$ and $U \to X$ are étale. Such an object will be denoted by $(\overline{x}' \to U' \to U)$. Let $(\overline{x}' \to U' \to U)$, $(\overline{x}' \to U'_1 \to U_1)$ be two objects of $\mathfrak{C}_{\overline{x}'}$. A morphism from $(\overline{x}' \to U'_1 \to U_1)$ to $(\overline{x}' \to U' \to U)$ consists of an X'-morphism $U'_1 \to U'$ and an X-morphism $U_1 \to U$ such that the diagram

$$\begin{array}{ccccc} & & U'_1 & \longrightarrow & U_1 \\ & \nearrow & \downarrow & & \downarrow \\ \overline{x}' & \longrightarrow & U' & \longrightarrow & U \end{array} \qquad (6.1.16.2)$$

is commutative. Observe that fiber products are representable in $\mathfrak{C}_{\overline{x}'}$ (see the proof of [3] VI.10.3). Finite inverse limits are therefore representable in $\mathfrak{C}_{\overline{x}'}$ (see [5] I 2.3). Consequently, the category $\mathfrak{C}_{\overline{x}'}$ is cofiltered ([5] I 2.7.1).

6.1.17 Let $(\overline{y} \rightsquigarrow \overline{x}')$ be a point of $X'_{\text{ét}} \overleftarrow{\times}_{X_{\text{ét}}} \overline{X}^{\circ}_{\text{ét}}$ ([2] 3.4.23), $\underline{X}'$ the strict localization of X' at $\overline{x}'$, $\underline{X}$ the strict localization of X at $g(\overline{x}')$, $\mathfrak{C}_{\overline{x}'}$ the category associated with $\overline{x}'$ in 6.1.16. We denote by $u\colon \overline{y} \to \underline{X}$ the X-morphism that defines the point $(\overline{y} \rightsquigarrow \overline{x}')$, and by $v\colon \overline{y} \to \overline{\underline{X}}^{\circ}$ the induced $\overline{X}^{\circ}$-morphism (6.1.3.2). For every object $(\overline{x}' \to U' \to U)$ of $\mathfrak{C}_{\overline{x}'}$, we have a canonical X'-morphism $\underline{X}' \to U'$ and a canonical X-morphism $\underline{X} \to U$ that fit into a commutative diagram

$$\begin{array}{ccccc} & & \underline{X}' & \xrightarrow{g} & \underline{X} \\ & \nearrow & \downarrow & & \downarrow \\ \overline{x}' & \longrightarrow & U' & \longrightarrow & U \end{array} \qquad (6.1.17.1)$$

where we have again denoted by g the morphism induced by g. We deduce from this a morphism $\overline{\underline{X}} \to \overline{U}$. The morphism $v\colon \overline{y} \to \overline{\underline{X}}^{\circ}$ then induces a geometric point of $\overline{U}^{\circ}$ that we denote again by $\overline{y}$. The diagram

$$\begin{array}{ccc} \underline{X} & \xleftarrow{u} & \overline{y} \\ \downarrow & & \downarrow \\ U & \longleftarrow & \overline{U}^{\circ} \end{array} \qquad (6.1.17.2)$$

is commutative.

We denote by $\varrho(\overline{y} \rightsquigarrow \overline{x}')$ the image of $(\overline{y} \rightsquigarrow \overline{x}')$ by the morphism ϱ (6.1.14.8), which is therefore a point of $\widetilde{G}$. By ([2] (6.5.9.5)), for every presheaf F on G, we have a canonical isomorphism

$$F^a_{\varrho(\overline{y}\rightsquigarrow\overline{x}')} \xrightarrow{\sim} \lim_{\substack{\longrightarrow \\ (\overline{x}'\to U'\to U)\in\mathfrak{C}^{\circ}_{\overline{x}'}}} (F^a_{U'\to U})_{\rho_{\overline{U}^{\circ}}(\overline{y})}, \tag{6.1.17.3}$$

where $F_{U'\to U}$ is the presheaf on $\mathbf{\acute{E}t}_{\mathrm{f}/\overline{U}^{\circ}}$ defined in (6.1.15.2), the exponent ^{a} denotes the associated sheaves and $\rho_{\overline{U}^{\circ}}\colon \overline{U}^{\circ}_{\text{ét}} \to \overline{U}^{\circ}_{\text{fét}}$ is the canonical morphism (2.1.12.1).

6.1.18 Let $\overline{x}'$ be a geometric point of X', $\underline{X}'$ the strict localization of X' at $\overline{x}'$, $\underline{X}$ the strict localization of X at $g(\overline{x}')$. We denote by $\underline{G}$ (resp. $\underline{\widetilde{G}}$) the relative Faltings site (resp. topos) associated with the pair of morphisms $(\underline{h}\colon \underline{\overline{X}}^{\circ} \to \underline{X}, \underline{g}\colon \underline{X}' \to \underline{X})$ induced by h and g (6.1.3), by

$$\Phi\colon \underline{\widetilde{G}} \to \widetilde{G} \tag{6.1.18.1}$$

the functoriality morphism ([2] (3.4.10.3)) and by

$$\vartheta\colon \underline{\overline{X}}^{\circ}_{\text{fét}} \to \underline{\widetilde{G}} \tag{6.1.18.2}$$

the morphism defined in ([2] (3.4.26.9)). We set

$$\phi_{\overline{x}'} = \vartheta^* \circ \Phi^*\colon \widetilde{G} \to \underline{\overline{X}}^{\circ}_{\text{fét}}. \tag{6.1.18.3}$$

By ([2] 3.4.34), for every abelian group F of $\widetilde{G}$ and every $q \geq 0$, we have a canonical and functorial isomorphism

$$\mathrm{R}^q\pi_*(F)_{\overline{x}'} \xrightarrow{\sim} \mathrm{H}^q(\underline{\overline{X}}^{\circ}_{\text{fét}}, \phi_{\overline{x}'}(F)). \tag{6.1.18.4}$$

We denote by

$$\varphi'_{\overline{x}'}\colon \widetilde{E}' \to \underline{\overline{X}}'^{\triangleright}_{\text{fét}} \tag{6.1.18.5}$$

the canonical functor (6.1.13.10) and by

$$\underline{\gamma}\colon \underline{\overline{X}}'^{\triangleright} \to \underline{\overline{X}}^{\circ} \tag{6.1.18.6}$$

the morphism induced by $\underline{g}$.

Proposition 6.1.19 ([2] 6.5.16) *We keep the assumptions and notation of* 6.1.18.

(i) *For every abelian group F of $\widetilde{E}'$ and every integer $q \geq 0$, we have a canonical functorial isomorphism*

$$\phi_{\overline{x}'}(\mathrm{R}^q\tau_*(F)) \xrightarrow{\sim} \mathrm{R}^q\underline{\gamma}_{\text{fét}*}(\varphi'_{\overline{x}'}(F)). \tag{6.1.19.1}$$

(ii) *For every exact sequence of abelian sheaves* $0 \to F' \to F \to F'' \to 0$ *of* $\widetilde{E}'$ *and every* $q \geq 0$, *the diagram*

$$\begin{array}{ccc} \phi_{\overline{x}'}(\mathrm{R}^q\tau_*(F'')) & \longrightarrow & \phi_{\overline{x}'}(\mathrm{R}^{q+1}\tau_*(F')) \\ \downarrow & & \downarrow \\ \mathrm{R}^q\underline{\gamma}_{\mathrm{f\acute{e}t}*}(\varphi_{\overline{x}'}(F'')) & \longrightarrow & \mathrm{R}^{q+1}\underline{\gamma}_{\mathrm{f\acute{e}t}*}(\varphi_{\overline{x}'}(F')), \end{array} \tag{6.1.19.2}$$

where the vertical arrows are the canonical isomorphisms (6.1.19.1) *and the horizontal arrows are the boundaries of the long exact sequences of cohomology, is commutative.*

6.1.20 We denote by $\widetilde{G}_s$ the closed subtopos of $\widetilde{G}$ complement of the open object $\pi^*(X'_\eta)$ ([5] IV 9.3.5), and by

$$\kappa\colon \widetilde{G}_s \to \widetilde{G} \tag{6.1.20.1}$$

the canonical embedding ([5] IV 9.3.5) (see [2] 6.5.2). By virtue of ([5] IV 9.4.3), there exists a morphism of topos

$$\pi_s\colon \widetilde{G}_s \to X'_{s,\mathrm{\acute{e}t}} \tag{6.1.20.2}$$

unique up to canonical isomorphism such that the diagram

$$\begin{array}{ccc} \widetilde{G}_s & \xrightarrow{\pi_s} & X'_{s,\mathrm{\acute{e}t}} \\ {\scriptstyle\kappa}\downarrow & & \downarrow{\scriptstyle a'} \\ \widetilde{G} & \xrightarrow{\pi} & X'_{\mathrm{\acute{e}t}} \end{array} \tag{6.1.20.3}$$

where a' is the canonical injection, is commutative up to isomorphism, and even 2-Cartesian.

We have a canonical isomorphism $\tau^*(\pi^*(X'_\eta)) \simeq \sigma'^*(X'_\eta)$ (6.1.14.7). Hence, by virtue of ([5] IV 9.4.3), there exists a morphism of topos

$$\tau_s\colon \widetilde{E}'_s \to \widetilde{G}_s \tag{6.1.20.4}$$

unique up to canonical isomorphism such that the diagram

$$\begin{array}{ccc} \widetilde{E}'_s & \xrightarrow{\tau_s} & \widetilde{G}_s \\ {\scriptstyle\delta'}\downarrow & & \downarrow{\scriptstyle\kappa} \\ \widetilde{E}' & \xrightarrow{\tau} & \widetilde{G} \end{array} \tag{6.1.20.5}$$

is commutative up to isomorphism (see 6.1.8).

The functors δ'_* and κ_* being exact, for every abelian group F of $\widetilde{E}'_s$ and every integer $i \geq 0$, we have a canonical isomorphism

$$\kappa_*(\mathrm{R}^i\tau_{s*}(F)) \xrightarrow{\sim} \mathrm{R}^i\tau_*(\delta'_*F). \tag{6.1.20.6}$$

It follows from (6.1.14.7) and ([5] IV 9.4.3) that the diagram of morphisms of topos

$$\begin{array}{ccc} \widetilde{E}'_s & \xrightarrow{\tau_s} & \widetilde{G}_s \\ {\scriptstyle \sigma'_s}\downarrow & \swarrow{\scriptstyle \pi_s} & \\ X'_{s,\text{ét}} & & \end{array} \tag{6.1.20.7}$$

is commutative up to canonical isomorphism.

We have a canonical isomorphism $\mathfrak{g}^*(\sigma^*(X_\eta)) \simeq \pi^*(X'_\eta)$ (6.1.14.7). Hence, by virtue of ([5] IV 9.4.3), there exists a morphism of topos

$$\mathfrak{g}_s\colon \widetilde{G}_s \to \widetilde{E}_s \tag{6.1.20.8}$$

unique up to canonical isomorphism such that the diagram

$$\begin{array}{ccc} \widetilde{G}_s & \xrightarrow{\mathfrak{g}_s} & \widetilde{E}_s \\ {\scriptstyle \kappa}\downarrow & & \downarrow{\scriptstyle \delta} \\ \widetilde{G} & \xrightarrow{\mathfrak{g}} & \widetilde{E} \end{array} \tag{6.1.20.9}$$

is commutative up to isomorphism (see 6.1.8).

The functors κ_* and δ_* being exact, for every abelian group F of $\widetilde{G}_s$ and every integer $i \geq 0$, we have a canonical isomorphism

$$\delta_*(\mathrm{R}^i\mathfrak{g}_{s*}(F)) \xrightarrow{\sim} \mathrm{R}^i\mathfrak{g}_*(\kappa_*F). \tag{6.1.20.10}$$

It follows from (6.1.14.7) and ([5] IV 9.4.3) that the diagram of morphisms of topos

$$\begin{array}{ccc} \widetilde{G}_s & \xrightarrow{\mathfrak{g}_s} & \widetilde{E}_s \\ {\scriptstyle \pi_s}\downarrow & & \downarrow{\scriptstyle \sigma_s} \\ X'_{s,\text{ét}} & \xrightarrow{g_{s,\text{ét}}} & X_{s,\text{ét}} \end{array} \tag{6.1.20.11}$$

is commutative up to canonical isomorphism.

It follows again from ([5] IV 9.4.3) that the composition $\mathfrak{g}_s \circ \tau_s$ is the morphism (6.1.10.4)

$$\theta\colon \widetilde{E}'_s \to \widetilde{E}_s. \tag{6.1.20.12}$$

6.1.21 For any object $(U' \to U \leftarrow V)$ of G, let $\overline{U}'^V$ be the integral closure of $\overline{U}'$ in $U' \times_U V$. We denote by $\overline{\mathscr{B}}^!$ the presheaf on G defined for any $(U' \to U \leftarrow V) \in \mathrm{Ob}(G)$ by

$$\overline{\mathscr{B}}^!(U' \to U \leftarrow V) = \Gamma(\overline{U}'^V, \mathscr{O}_{\overline{U}'^V}). \tag{6.1.21.1}$$

Since $\overline{X}'$ is normal and locally irreducible ([3] III.4.2(iii)), $\overline{\mathscr{B}}^!$ is a sheaf for the covanishing topology of G by ([2] 3.6.4).

By ([2] 6.5.18), we have canonical homomorphisms

$$\overline{\mathscr{B}}^! \to \tau_*(\overline{\mathscr{B}}'), \tag{6.1.21.2}$$

$$\overline{\mathscr{B}} \to \mathrm{g}_*(\overline{\mathscr{B}}^!), \tag{6.1.21.3}$$

the first of which is an isomorphism by virtue of ([2] 6.5.19). We also have a canonical homomorphism

$$\hbar'_*(\mathscr{O}_{\overline{X}'}) \to \pi_*(\overline{\mathscr{B}}^!). \tag{6.1.21.4}$$

For any integer $n \geq 1$, we set

$$\overline{\mathscr{B}}^!_n = \overline{\mathscr{B}}^!/p^n\overline{\mathscr{B}}^!. \tag{6.1.21.5}$$

It is a ring of $\widetilde{G}_s$ ([2] 6.5.27). The canonical homomorphism $\pi^*(\hbar_*(\mathscr{O}_{\overline{X}'})) \to \overline{\mathscr{B}}^!$ (6.1.21.4) induces a homomorphism $\pi_s^*(\mathscr{O}_{\overline{X}'_n}) \to \overline{\mathscr{B}}^!_n$ of $\widetilde{G}_s$ ([2] 6.5.28). The morphism π_s (6.1.20.2) is therefore underlying a morphism of ringed topos, that we denote by

$$\pi_n \colon (\widetilde{G}_s, \overline{\mathscr{B}}^!_n) \to (\overline{X}'_{s,\text{ét}}, \mathscr{O}_{\overline{X}'_n}). \tag{6.1.21.6}$$

The canonical homomorphism $\tau^*(\overline{\mathscr{B}}^!) \to \overline{\mathscr{B}}'$ (6.1.21.2) induces a homomorphism $\tau_s^*(\overline{\mathscr{B}}^!_n) \to \overline{\mathscr{B}}'_n$ of $\widetilde{E}'_s$. The morphism τ_s (6.1.20.4) is therefore underlying a morphism of ringed topos, that we denote by

$$\tau_n \colon (\widetilde{E}'_s, \overline{\mathscr{B}}'_n) \to (\widetilde{G}_s, \overline{\mathscr{B}}^!_n). \tag{6.1.21.7}$$

The canonical homomorphism $\mathrm{g}^{-1}(\overline{\mathscr{B}}) \to \overline{\mathscr{B}}^!$ (6.1.21.3) induces a homomorphism $\mathrm{g}_s^*(\overline{\mathscr{B}}_n) \to \overline{\mathscr{B}}^!_n$ of $\widetilde{G}_s$. The morphism g_s (6.1.20.8) is therefore underlying a morphism of ringed topos, that we denote by

$$\mathrm{g}_n \colon (\widetilde{G}_s, \overline{\mathscr{B}}^!_n) \to (\widetilde{E}_s, \overline{\mathscr{B}}_n). \tag{6.1.21.8}$$

We check immediately that the composition $\mathrm{g}_n \circ \tau_n$ is the morphism (6.1.11.4)

$$\theta_n \colon (\widetilde{E}'_s, \overline{\mathscr{B}}'_n) \to (\widetilde{E}_s, \overline{\mathscr{B}}_n). \tag{6.1.21.9}$$

By ([2] 6.5.28.10), the triangle and the square of the diagram of morphisms of ringed topos

$$\begin{array}{ccccc} (\widetilde{E}'_s, \overline{\mathscr{B}}'_n) & \xrightarrow{\tau_n} & (\widetilde{G}_s, \overline{\mathscr{B}}^!_n) & \xrightarrow{\mathrm{g}_n} & (\widetilde{E}_s, \overline{\mathscr{B}}_n) \\ & \searrow^{\sigma'_n} & \downarrow{\scriptstyle \pi_n} & & \downarrow{\scriptstyle \sigma_n} \\ & & (X'_{s,\text{ét}}, \mathscr{O}_{\overline{X}'_n}) & \xrightarrow{\overline{g}_n} & (X_{s,\text{ét}}, \mathscr{O}_{\overline{X}_n}) \end{array} \tag{6.1.21.10}$$

are commutative up to canonical isomorphisms ([1] 1.2.3).

6.1.22 We take again the notation of 2.1.9. We denote by $\breve{\overline{\mathscr{B}}}$ the ring $(\overline{\mathscr{B}}_{n+1})_{n\in\mathbb{N}}$ of $\widetilde{E}_s^{\mathbb{N}^\circ}$ (6.1.9.4), by $\breve{\overline{\mathscr{B}}}'$ the ring $(\overline{\mathscr{B}}'_{n+1})_{n\in\mathbb{N}}$ of $\widetilde{E}_s'^{\mathbb{N}^\circ}$ and by $\breve{\overline{\mathscr{B}}}^!$ the ring $(\overline{\mathscr{B}}^!_{n+1})_{n\in\mathbb{N}}$ of $\widetilde{G}_s^{\mathbb{N}^\circ}$ (6.1.21.5) . We denote by $\mathscr{O}_{\breve{\overline{X}}}$ the ring $(\mathscr{O}_{\overline{X}_{n+1}})_{n\in\mathbb{N}}$ of $X^{\mathbb{N}^\circ}_{s,\text{ét}}$ or $X^{\mathbb{N}^\circ}_{s,\text{zar}}$, depending on the context (2.1.13), and by $\widetilde{\xi}^{-1}\widetilde{\Omega}^1_{\breve{\overline{X}}/\breve{\overline{S}}}$ the $\mathscr{O}_{\breve{\overline{X}}}$-module $(\widetilde{\xi}^{-1}\widetilde{\Omega}^1_{\overline{X}_{n+1}/\overline{S}_{n+1}})_{n\in\mathbb{N}}$ (6.1.7.4). We denote by $\mathscr{O}_{\breve{\overline{X}}'}$ the ring $(\mathscr{O}_{\overline{X}'_{n+1}})_{n\in\mathbb{N}}$ of $X'^{\mathbb{N}^\circ}_{s,\text{ét}}$ or $X'^{\mathbb{N}^\circ}_{s,\text{zar}}$, depending on the context, and by $\widetilde{\xi}^{-1}\widetilde{\Omega}^1_{\breve{\overline{X}}'/\breve{\overline{S}}}$ (resp. $\widetilde{\xi}^{-1}\widetilde{\Omega}^1_{\breve{\overline{X}}'/\breve{\overline{X}}}$) the $\mathscr{O}_{\breve{\overline{X}}'}$-module $(\widetilde{\xi}^{-1}\widetilde{\Omega}^1_{\overline{X}'_{n+1}/\overline{S}_{n+1}})_{n\in\mathbb{N}}$ (resp. $(\widetilde{\xi}^{-1}\widetilde{\Omega}^1_{\overline{X}'_{n+1}/\overline{X}_{n+1}})_{n\in\mathbb{N}}$) (6.1.7.5). The reader should not confuse $\mathscr{O}_{\breve{\overline{X}}}$ and $\mathscr{O}_{\overline{X}}$ (resp. $\mathscr{O}_{\breve{\overline{X}}'}$ and $\mathscr{O}_{\overline{X}'}$) (6.1.1.2). Consider the diagram of morphisms of ringed topos

$$\begin{array}{ccc} (\widetilde{E}_s'^{\mathbb{N}^\circ}, \breve{\overline{\mathscr{B}}}') & & \\ \downarrow{\scriptstyle \breve{\tau}} & \searrow^{\breve{\sigma}'} & \\ (\widetilde{G}_s^{\mathbb{N}^\circ}, \breve{\overline{\mathscr{B}}}^!) & \xrightarrow{\breve{\pi}} & (X'^{\mathbb{N}^\circ}_{s,\text{ét}}, \mathscr{O}_{\breve{\overline{X}}'}) \\ \downarrow{\scriptstyle \breve{\mathrm{g}}} & & \downarrow{\scriptstyle \breve{\overline{g}}} \\ (\widetilde{E}_s^{\mathbb{N}^\circ}, \breve{\overline{\mathscr{B}}}) & \xrightarrow{\breve{\sigma}} & (X^{\mathbb{N}^\circ}_{s,\text{ét}}, \mathscr{O}_{\breve{\overline{X}}}) \end{array} \qquad (\breve{\theta}\colon (\widetilde{E}_s'^{\mathbb{N}^\circ}, \breve{\overline{\mathscr{B}}}') \to (\widetilde{E}_s^{\mathbb{N}^\circ}, \breve{\overline{\mathscr{B}}})) \tag{6.1.22.1}$$

induced by $(\theta_{n+1})_{n\in\mathbb{N}}$ (6.1.11.4), $(\tau_{n+1})_{n\in\mathbb{N}}$ (6.1.21.7), $(\mathrm{g}_{n+1})_{n\in\mathbb{N}}$ (6.1.21.8), $(\sigma_{n+1})_{n\in\mathbb{N}}$ (6.1.9.5), $(\sigma'_{n+1})_{n\in\mathbb{N}}$, $(\pi_{n+1})_{n\in\mathbb{N}}$ (6.1.21.6) and $\overline{g}$ (see [3] III.7.5). The two triangles and the square are commutative up to canonical isomorphisms (6.1.21.10).

6.1.23 We set $\mathscr{S} = \mathrm{Spf}(\mathscr{O}_C)$ and we denote by $\mathfrak{X}$ (resp. $\mathfrak{X}'$) the formal scheme p-adic completion of $\overline{X}$ (resp. $\overline{X}'$), and by $\mathfrak{g}\colon \mathfrak{X}' \to \mathfrak{X}$ the morphism induced by g (6.1.3.1). For any integer $n \geq 1$, we denote by

$$u_n\colon (X_{s,\text{ét}}, \mathscr{O}_{\overline{X}_n}) \to (X_{s,\text{zar}}, \mathscr{O}_{\overline{X}_n}) \tag{6.1.23.1}$$

the canonical morphism (2.1.13.5). We denote by

$$\breve{u}\colon (X^{\mathbb{N}^\circ}_{s,\text{ét}}, \mathscr{O}_{\breve{\overline{X}}}) \to (X^{\mathbb{N}^\circ}_{s,\text{zar}}, \mathscr{O}_{\breve{\overline{X}}}) \tag{6.1.23.2}$$

the morphism of ringed topos defined by $(u_{n+1})_{n\in\mathbb{N}}$ and by

$$\lambda\colon (X^{\mathbb{N}^\circ}_{s,\mathrm{zar}}, \mathscr{O}_{\breve{\mathfrak{X}}}) \to (X_{s,\mathrm{zar}}, \mathscr{O}_{\mathfrak{X}}) \tag{6.1.23.3}$$

the morphism of ringed topos for which the functor λ_* is the inverse limit functor (2.1.9.1). We denote by

$$\widehat{\sigma}\colon (\widetilde{E}^{\mathbb{N}^\circ}_s, \breve{\overline{\mathscr{B}}}) \to (X_{s,\mathrm{zar}}, \mathscr{O}_{\mathfrak{X}}) \tag{6.1.23.4}$$

the composed morphism $\lambda \circ \breve{u} \circ \breve{\sigma}$. We also consider the analogous notation for f', that we endow with an exponent $'$. We denote by

$$\widehat{\pi}\colon (\widetilde{G}^{\mathbb{N}^\circ}_s, \breve{\overline{\mathscr{B}}}^!) \to (X'_{s,\mathrm{zar}}, \mathscr{O}_{\mathfrak{X}'}) \tag{6.1.23.5}$$

the composed morphism $\lambda' \circ \breve{u}' \circ \breve{\pi}$. By (6.1.22.1), the diagram of morphisms of ringed topos

$$\begin{array}{ccccc} & & \overset{\check{\theta}}{\frown\!\!\!\frown\!\!\!\frown\!\!\!\rightarrow} & & \\ (\widetilde{E}'^{\mathbb{N}^\circ}_s, \breve{\overline{\mathscr{B}}}') & \xrightarrow{\check{\tau}} & (\widetilde{G}^{\mathbb{N}^\circ}_s, \breve{\overline{\mathscr{B}}}^!) & \xrightarrow{\breve{\mathfrak{g}}} & (\widetilde{E}^{\mathbb{N}^\circ}_s, \breve{\overline{\mathscr{B}}}) \\ & \searrow^{\widehat{\sigma}'} & \Big\downarrow{\scriptstyle\widehat{\pi}} & & \Big\downarrow{\scriptstyle\widehat{\sigma}} \\ & & (X'_{s,\mathrm{zar}}, \mathscr{O}_{\mathfrak{X}'}) & \xrightarrow{\mathfrak{g}} & (X_{s,\mathrm{zar}}, \mathscr{O}_{\mathfrak{X}}) \end{array} \tag{6.1.23.6}$$

is commutative up to canonical isomorphism ([1] 1.2.3).

For every $\mathscr{O}_{\mathfrak{X}}$-module $\mathscr{F}$ of $X_{s,\mathrm{zar}}$, we have a canonical isomorphism

$$\widehat{\sigma}^*(\mathscr{F}) \xrightarrow{\sim} (\sigma^*_n(u^*_n(\mathscr{F}/p^{n+1}\mathscr{F})))_{n\in\mathbb{N}}. \tag{6.1.23.7}$$

Similarly, for every $\mathscr{O}_{\mathfrak{X}'}$-module $\mathscr{F}'$ of $X'_{s,\mathrm{zar}}$, we have a canonical isomorphism

$$\widehat{\pi}^*(\mathscr{F}') \xrightarrow{\sim} (\pi^*_n(u'^*_n(\mathscr{F}'/p^{n+1}\mathscr{F}')))_{n\in\mathbb{N}}. \tag{6.1.23.8}$$

In particular, $\widehat{\sigma}^*(\mathscr{F})$ and $\widehat{\pi}^*(\mathscr{F}')$ are adic ([3] III.7.18).

We use the notation introduced in 2.7.10 and 2.9.8 for morphisms of ringed topos.

6.1.24 We take again the notation of 4.3.12, 4.3.13 and 4.3.16 for the morphisms f and f'. We have a canonical fully faithful functor (4.3.16.3)

$$\mathbf{Mod}^{\mathrm{coh}}(\mathscr{O}_{\mathfrak{X}}[\frac{1}{p}]) \to \mathbf{Ind}\text{-}\mathbf{Mod}(\mathscr{O}_{\mathfrak{X}}). \tag{6.1.24.1}$$

We identify $\mathbf{Mod}^{\mathrm{coh}}(\mathscr{O}_{\mathfrak{X}}[\frac{1}{p}])$ with a full subcategory of $\mathbf{Ind}\text{-}\mathbf{Mod}(\mathscr{O}_{\mathfrak{X}})$ by this functor. We therefore consider any coherent $\mathscr{O}_{\mathfrak{X}}[\frac{1}{p}]$-module also as an ind-$\mathscr{O}_{\mathfrak{X}}$-module. The functor $\mathrm{I}\widehat{\sigma}^*$ (4.3.12.3) therefore induces a functor that we denote again by

$$\mathrm{I}\widehat{\sigma}^* : \mathbf{Mod}^{\mathrm{coh}}(\mathscr{O}_{\mathfrak{X}}[\frac{1}{p}]) \to \mathbf{Ind\text{-}Mod}(\overline{\breve{\mathscr{B}}}). \tag{6.1.24.2}$$

Moreover, the functor

$$\widehat{\sigma}^*_{\mathbb{Q}} : \mathbf{Mod}_{\mathbb{Q}}(\mathscr{O}_{\mathfrak{X}}) \to \mathbf{Mod}_{\mathbb{Q}}(\overline{\breve{\mathscr{B}}}) \tag{6.1.24.3}$$

induces, in view of (4.3.16.2), a functor that we denote by

$$\widehat{\sigma}^*_{\mathbb{Q}} : \mathbf{Mod}^{\mathrm{coh}}(\mathscr{O}_{\mathfrak{X}}[\frac{1}{p}]) \to \mathbf{Mod}_{\mathbb{Q}}(\overline{\breve{\mathscr{B}}}). \tag{6.1.24.4}$$

We adopt the same conventions for coherent $\mathscr{O}_{\mathfrak{X}'}[\frac{1}{p}]$-modules.

We denote by $\mathbf{Mod}(\overline{\breve{\mathscr{B}}}^!)$ the category of $\overline{\breve{\mathscr{B}}}^!$-modules of $\widetilde{G}_s^{\mathbb{N}^\circ}$, by $\mathbf{Mod}_{\mathbb{Q}}(\overline{\breve{\mathscr{B}}}^!)$ the category of $\overline{\breve{\mathscr{B}}}^!$-modules of $\widetilde{G}_s^{\mathbb{N}^\circ}$ up to isogeny (2.9.2) and by $\mathbf{Ind\text{-}Mod}(\overline{\breve{\mathscr{B}}}^!)$ the category of ind-$\overline{\breve{\mathscr{B}}}^!$-modules of $\widetilde{G}_s^{\mathbb{N}^\circ}$ (2.7.1). We denote by

$$\iota_{\overline{\breve{\mathscr{B}}}^!} : \mathbf{Mod}(\overline{\breve{\mathscr{B}}}^!) \to \mathbf{Ind\text{-}Mod}(\overline{\breve{\mathscr{B}}}^!) \tag{6.1.24.5}$$

the canonical functor, which is exact and fully faithful (2.7.1). We will identify $\mathbf{Mod}(\overline{\breve{\mathscr{B}}}^!)$ with a full subcategory of $\mathbf{Ind\text{-}Mod}(\overline{\breve{\mathscr{B}}}^!)$ by this functor, which we will omit from the notation.

By 2.7.10, the morphism of ringed topos $\widehat{\pi}$ (6.1.23.5) induces two adjoint additive functors

$$\mathrm{I}\widehat{\pi}^* : \mathbf{Ind\text{-}Mod}(\mathscr{O}_{\mathfrak{X}'}) \to \mathbf{Ind\text{-}Mod}(\overline{\breve{\mathscr{B}}}^!), \tag{6.1.24.6}$$

$$\mathrm{I}\widehat{\pi}_* : \mathbf{Ind\text{-}Mod}(\overline{\breve{\mathscr{B}}}^!) \to \mathbf{Ind\text{-}Mod}(\mathscr{O}_{\mathfrak{X}'}). \tag{6.1.24.7}$$

The functor $\mathrm{I}\widehat{\pi}^*$ (resp. $\mathrm{I}\widehat{\pi}_*$) is right-exact (resp. left-exact). In view of (4.3.16.3), the functor $\mathrm{I}\widehat{\pi}^*$ induces a functor that we denote again by

$$\mathrm{I}\widehat{\pi}^* : \mathbf{Mod}^{\mathrm{coh}}(\mathscr{O}_{\mathfrak{X}'}[\frac{1}{p}]) \to \mathbf{Ind\text{-}Mod}(\overline{\breve{\mathscr{B}}}^!). \tag{6.1.24.8}$$

The functor $\mathrm{I}\widehat{\pi}_*$ admits a right derived functor

$$\mathrm{RI}\widehat{\pi}_* : \mathbf{D}^+(\mathbf{Ind\text{-}Mod}(\overline{\breve{\mathscr{B}}}^!)) \to \mathbf{D}^+(\mathbf{Ind\text{-}Mod}(\mathscr{O}_{\mathfrak{X}'})). \tag{6.1.24.9}$$

We denote by $\vec{\pi}_*$ the composed functor (4.3.12.7)

$$\vec{\pi}_* = \kappa_{\mathscr{O}_{\mathfrak{X}'}} \circ \mathrm{I}\widehat{\pi}_* : \mathbf{Ind\text{-}Mod}(\overline{\breve{\mathscr{B}}}^!) \to \mathbf{Mod}(\mathscr{O}_{\mathfrak{X}'}). \tag{6.1.24.10}$$

This one is left exact. It admits a right derived functor

$$\mathrm{R}\vec{\pi}_* : \mathbf{D}^+(\mathbf{Ind\text{-}Mod}(\overline{\breve{\mathscr{B}}}^!)) \to \mathbf{D}^+(\mathbf{Mod}(\mathscr{O}_{\mathfrak{X}'})), \tag{6.1.24.11}$$

canonically isomorphic to $\kappa_{\mathscr{O}_{\mathfrak{X}'}} \circ \mathrm{RI}\widehat{\pi}_*$ ([40] 13.3.13).

By 2.9.2, we have a canonical functor (2.9.2.3)

$$\alpha_{\overline{\breve{\mathscr{B}}}^{!}} \colon \mathbf{Mod}_{\mathbb{Q}}(\overline{\breve{\mathscr{B}}}^{!}) \to \mathbf{Ind\text{-}Mod}(\overline{\breve{\mathscr{B}}}^{!}), \tag{6.1.24.12}$$

which is fully faithful (2.6.6.5) and exact (2.9.3). We will identify $\mathbf{Mod}_{\mathbb{Q}}(\overline{\breve{\mathscr{B}}}^{!})$ with a full subcategory of $\mathbf{Ind\text{-}Mod}(\overline{\breve{\mathscr{B}}}^{!})$ by this functor, which we will omit from the notation. We will therefore consider any $\overline{\breve{\mathscr{B}}}^{!}_{\mathbb{Q}}$-module as an ind-$\overline{\breve{\mathscr{B}}}^{!}$-module.

By 2.9.7, the morphism of ringed topos $\widehat{\pi}$ (6.1.23.5) induces two adjoint additive functors

$$\widehat{\pi}^*_{\mathbb{Q}} \colon \mathbf{Mod}_{\mathbb{Q}}(\mathscr{O}_{\mathfrak{X}'}) \to \mathbf{Mod}_{\mathbb{Q}}(\overline{\breve{\mathscr{B}}}^{!}), \tag{6.1.24.13}$$

$$\widehat{\pi}_{\mathbb{Q}*} \colon \mathbf{Mod}_{\mathbb{Q}}(\overline{\breve{\mathscr{B}}}^{!}) \to \mathbf{Mod}_{\mathbb{Q}}(\mathscr{O}_{\mathfrak{X}'}). \tag{6.1.24.14}$$

The functor $\widehat{\pi}^*_{\mathbb{Q}}$ (resp. $\widehat{\pi}_{\mathbb{Q}*}$) is right-exact (resp. left-exact). Taking into account (4.3.16.2), the functor $\widehat{\pi}^*_{\mathbb{Q}}$ induces a functor that we denote again by

$$\widehat{\pi}^*_{\mathbb{Q}} \colon \mathbf{Mod}^{\mathrm{coh}}(\mathscr{O}_{\mathfrak{X}'}[\frac{1}{p}]) \to \mathbf{Mod}_{\mathbb{Q}}(\overline{\breve{\mathscr{B}}}^{!}). \tag{6.1.24.15}$$

We denote again by

$$\widehat{\pi}_{\mathbb{Q}*} \colon \mathbf{Mod}_{\mathbb{Q}}(\overline{\breve{\mathscr{B}}}^{!}) \to \mathbf{Mod}(\mathscr{O}_{\mathfrak{X}'}[\frac{1}{p}]) \tag{6.1.24.16}$$

the composition of the functor $\widehat{\pi}_{\mathbb{Q}*}$ (6.1.24.14) and the canonical exact functor (4.3.13.10)

$$\mathbf{Mod}_{\mathbb{Q}}(\mathscr{O}_{\mathfrak{X}'}) \to \mathbf{Mod}(\mathscr{O}_{\mathfrak{X}'}[\frac{1}{p}]). \tag{6.1.24.17}$$

These abuses of notation do not lead to any confusion.

By 2.9.6, the functor $\widehat{\pi}_{\mathbb{Q}*}$ (6.1.24.14) admits a right derived functor

$$\mathrm{R}\widehat{\pi}_{\mathbb{Q}*} \colon \mathbf{D}^+(\mathbf{Mod}_{\mathbb{Q}}(\overline{\breve{\mathscr{B}}}^{!})) \to \mathbf{D}^+(\mathbf{Mod}_{\mathbb{Q}}(\mathscr{O}_{\mathfrak{X}'})). \tag{6.1.24.18}$$

By virtue of (2.9.6.7), the diagram

$$\begin{array}{ccc} \mathbf{D}^+(\mathbf{Mod}_{\mathbb{Q}}(\overline{\breve{\mathscr{B}}}^{!})) & \xrightarrow{\mathrm{R}\widehat{\pi}_{\mathbb{Q}*}} & \mathbf{D}^+(\mathbf{Mod}_{\mathbb{Q}}(\mathscr{O}_{\mathfrak{X}'})) \\ \downarrow{\scriptstyle \alpha_{\overline{\breve{\mathscr{B}}}^{!}}} & & \downarrow{\scriptstyle \alpha_{\mathscr{O}_{\mathfrak{X}'}}} \\ \mathbf{D}^+(\mathbf{Ind\text{-}Mod}(\overline{\breve{\mathscr{B}}}^{!})) & \xrightarrow{\mathrm{RI}\widehat{\pi}_*} & \mathbf{D}^+(\mathbf{Ind\text{-}Mod}(\mathscr{O}_{\mathfrak{X}'})) \end{array} \tag{6.1.24.19}$$

is commutative up to a canonical isomorphism.

The functor $\widehat{\pi}_{\mathbb{Q}*}$ (6.1.24.16) admits a right derived functor

$$\mathrm{R}\widehat{\pi}_{\mathbb{Q}*}\colon \mathbf{D}^+(\mathbf{Mod}_{\mathbb{Q}}(\overline{\check{\mathscr{B}}}^{!})) \to \mathbf{D}^+(\mathbf{Mod}(\mathscr{O}_{\mathfrak{X}'}[\frac{1}{p}])) \tag{6.1.24.20}$$

which is none other than the composition of the functor $\mathrm{R}\widehat{\pi}_{\mathbb{Q}*}$ (6.1.24.18) and the functor (6.1.24.17). It follows from (4.3.13.13) that the diagram

$$\begin{array}{ccc} \mathbf{D}^+(\mathbf{Mod}_{\mathbb{Q}}(\overline{\check{\mathscr{B}}}^{!})) & \xrightarrow{\mathrm{R}\widehat{\pi}_{\mathbb{Q}*}} & \mathbf{D}^+(\mathbf{Mod}(\mathscr{O}_{\mathfrak{X}'}[\frac{1}{p}])) \\ \Big\downarrow{\scriptstyle \alpha_{\overline{\check{\mathscr{B}}}^{!}}} & & \Big\downarrow \\ \mathbf{D}^+(\mathbf{Ind}\text{-}\mathbf{Mod}(\overline{\check{\mathscr{B}}}^{!})) & \xrightarrow{\mathrm{R}\vec{\pi}_*} & \mathbf{D}^+(\mathbf{Mod}(\mathscr{O}_{\mathfrak{X}'})), \end{array} \tag{6.1.24.21}$$

where $\mathrm{R}\vec{\pi}_*$ is the functor (6.1.24.11) and the unlabeled arrow is the canonical functor, is commutative up to a canonical isomorphism.

6.1.25 We take again the notation of 4.2 for the formal $\mathscr{S}$-schemes $\mathfrak{X}$ and $\mathfrak{X}'$, in particular, $\widetilde{\xi}^{-1}\widetilde{\Omega}^1_{\mathfrak{X}/\mathscr{S}}$ denotes the p-adic completion of the $\mathscr{O}_{\overline{X}}$-module ([1] 2.5.1)

$$\widetilde{\xi}^{-1}\widetilde{\Omega}^1_{\check{\overline{X}}/\check{\overline{S}}} = \widetilde{\xi}^{-1}\widetilde{\Omega}^1_{X/S} \otimes_{\mathscr{O}_X} \mathscr{O}_{\check{\overline{X}}}, \tag{6.1.25.1}$$

and $\widetilde{\xi}^{-1}\widetilde{\Omega}^1_{\mathfrak{X}'/\mathscr{S}}$ the p-adic completion of the $\mathscr{O}_{\overline{X}'}$-module

$$\widetilde{\xi}^{-1}\widetilde{\Omega}^1_{\check{\overline{X}}'/\check{\overline{S}}} = \widetilde{\xi}^{-1}\widetilde{\Omega}^1_{X'/S} \otimes_{\mathscr{O}_{X'}} \mathscr{O}_{\check{\overline{X}}'}. \tag{6.1.25.2}$$

We denote moreover by $\widetilde{\xi}^{-1}\widetilde{\Omega}^1_{\mathfrak{X}'/\mathfrak{X}}$ the p-adic completion of the $\mathscr{O}_{\overline{X}'}$-module

$$\widetilde{\xi}^{-1}\widetilde{\Omega}^1_{\check{\overline{X}}'/\check{\overline{X}}} = \widetilde{\xi}^{-1}\widetilde{\Omega}^1_{X'/X} \otimes_{\mathscr{O}_{X'}} \mathscr{O}_{\check{\overline{X}}'}. \tag{6.1.25.3}$$

The locally split exact sequence (6.1.7.3) induces an exact sequence

$$0 \to \mathfrak{g}^*(\widetilde{\Omega}^1_{\mathfrak{X}/\mathscr{S}}) \to \widetilde{\Omega}^1_{\mathfrak{X}'/\mathscr{S}} \to \widetilde{\Omega}^1_{\mathfrak{X}'/\mathfrak{X}} \to 0. \tag{6.1.25.4}$$

The Koszul filtration $\mathrm{W}^\bullet\widetilde{\Omega}^\bullet_{\mathfrak{X}'/\mathscr{S}}$ of the exterior $\mathscr{O}_{\mathfrak{X}'}$-algebra $\widetilde{\Omega}^\bullet_{\mathfrak{X}'/\mathscr{S}}$ associated with the exact sequence (6.1.25.4), defined in (2.5.11.2), induces for any integer $j \geq 0$ an exact sequence (2.5.11.5)

$$0 \to \mathfrak{g}^*(\widetilde{\Omega}^1_{\mathfrak{X}/\mathscr{S}}) \otimes_{\mathscr{O}_{\mathfrak{X}'}} \widetilde{\Omega}^{j-1}_{\mathfrak{X}'/\mathfrak{X}} \to \mathrm{W}^0(\widetilde{\Omega}^\bullet_{\mathfrak{X}'/\mathscr{S}})/\mathrm{W}^2(\widetilde{\Omega}^\bullet_{\mathfrak{X}'/\mathscr{S}}) \to \widetilde{\Omega}^j_{\mathfrak{X}'/\mathfrak{X}} \to 0. \tag{6.1.25.5}$$

In view of the projection formula ([59] 0B54), we deduce from this a morphism of $\mathbf{D}^+(\mathbf{Mod}(\mathscr{O}_{\mathfrak{X}}))$

$$\mathrm{R}\mathfrak{g}_*(\widetilde{\Omega}^j_{\mathfrak{X}'/\mathfrak{X}}) \to \widetilde{\Omega}^1_{\mathfrak{X}/\mathscr{S}} \otimes_{\mathscr{O}_{\mathfrak{X}}} \mathrm{R}\mathfrak{g}_*(\widetilde{\Omega}^{j-1}_{\mathfrak{X}'/\mathfrak{X}})[+1], \tag{6.1.25.6}$$

that we call the *Kodaira–Spencer map of* $\mathfrak{g}$ (see [42] 1.2).

6.1.26 We denote by $\mathbf{HM}(\mathscr{O}_{\mathfrak{X}'}, \widetilde{\xi}^{-1}\widetilde{\Omega}^1_{\mathfrak{X}'/\mathfrak{X}})$ the category of Higgs $\mathscr{O}_{\mathfrak{X}'}$-modules with coefficients in $\widetilde{\xi}^{-1}\widetilde{\Omega}^1_{\mathfrak{X}'/\mathfrak{X}}$ (2.5.1). Such a Higgs module is said to be *coherent* if the underlying $\mathscr{O}_{\mathfrak{X}'}$-module is coherent. We denote by $\mathbf{HM}^{\mathrm{coh}}(\mathscr{O}_{\mathfrak{X}'}, \widetilde{\xi}^{-1}\widetilde{\Omega}^1_{\mathfrak{X}'/\mathfrak{X}})$ the full subcategory of $\mathbf{HM}(\mathscr{O}_{\mathfrak{X}'}, \widetilde{\xi}^{-1}\widetilde{\Omega}^1_{\mathfrak{X}'/\mathfrak{X}})$ made up of coherent Higgs modules. By a *Higgs* $\mathscr{O}_{\mathfrak{X}'}[\frac{1}{p}]$*-module with coefficients in* $\widetilde{\xi}^{-1}\widetilde{\Omega}^1_{\mathfrak{X}'/\mathfrak{X}}$, we mean a Higgs $\mathscr{O}_{\mathfrak{X}'}[\frac{1}{p}]$-module with coefficients in $\widetilde{\xi}^{-1}\widetilde{\Omega}^1_{\mathfrak{X}'/\mathfrak{X}}[\frac{1}{p}]$. Such a Higgs module is said to be *coherent* if the underlying $\mathscr{O}_{\mathfrak{X}'}[\frac{1}{p}]$-module is coherent. We denote by $\mathbf{HM}(\mathscr{O}_{\mathfrak{X}'}[\frac{1}{p}], \widetilde{\xi}^{-1}\widetilde{\Omega}^1_{\mathfrak{X}'/\mathfrak{X}})$ the category of Higgs $\mathscr{O}_{\mathfrak{X}'}[\frac{1}{p}]$-modules with coefficients in $\widetilde{\xi}^{-1}\widetilde{\Omega}^1_{\mathfrak{X}'/\mathfrak{X}}$ and by $\mathbf{HM}^{\mathrm{coh}}(\mathscr{O}_{\mathfrak{X}'}[\frac{1}{p}], \widetilde{\xi}^{-1}\widetilde{\Omega}^1_{\mathfrak{X}'/\mathfrak{X}})$ the full subcategory made up of coherent Higgs modules. We omit the Higgs field from the notation of a Higgs module when we do not explicitly need it.

We denote by $\mathbf{HI}(\mathscr{O}_{\mathfrak{X}'}, \widetilde{\xi}^{-1}\widetilde{\Omega}^1_{\mathfrak{X}'/\mathfrak{X}})$ the category of Higgs $\mathscr{O}_{\mathfrak{X}'}$-isogenies with coefficients in $\widetilde{\xi}^{-1}\widetilde{\Omega}^1_{\mathfrak{X}'/\mathfrak{X}}$ (2.9.9), by $\mathbf{HI}^{\mathrm{coh}}(\mathscr{O}_{\mathfrak{X}'}, \widetilde{\xi}^{-1}\widetilde{\Omega}^1_{\mathfrak{X}'/\mathfrak{X}})$ the full subcategory made up of the quadruples $(\mathscr{M}, \mathscr{N}, u, \theta)$ such that $\mathscr{M}$ and $\mathscr{N}$ are coherent $\mathscr{O}_{\mathfrak{X}'}$-modules and by $\mathbf{HI}_{\mathbb{Q}}(\mathscr{O}_{\mathfrak{X}'}, \widetilde{\xi}^{-1}\widetilde{\Omega}^1_{\mathfrak{X}'/\mathfrak{X}})$ (resp. $\mathbf{HI}^{\mathrm{coh}}_{\mathbb{Q}}(\mathscr{O}_{\mathfrak{X}'}, \widetilde{\xi}^{-1}\widetilde{\Omega}^1_{\mathfrak{X}'/\mathfrak{X}})$) the category of objects of $\mathbf{HI}(\mathscr{O}_{\mathfrak{X}'}, \widetilde{\xi}^{-1}\widetilde{\Omega}^1_{\mathfrak{X}'/\mathfrak{X}})$ (resp. $\mathbf{HI}^{\mathrm{coh}}(\mathscr{O}_{\mathfrak{X}'}, \widetilde{\xi}^{-1}\widetilde{\Omega}^1_{\mathfrak{X}'/\mathfrak{X}})$) up to isogeny ([3] III.6.1.1). By ([3] III.6.20), we have a canonical equivalence of categories

$$\mathbf{HI}^{\mathrm{coh}}_{\mathbb{Q}}(\mathscr{O}_{\mathfrak{X}'}, \widetilde{\xi}^{-1}\widetilde{\Omega}^1_{\mathfrak{X}'/\mathfrak{X}}) \xrightarrow{\sim} \mathbf{HM}^{\mathrm{coh}}(\mathscr{O}_{\mathfrak{X}'}[\frac{1}{p}], \widetilde{\xi}^{-1}\widetilde{\Omega}^1_{\mathfrak{X}'/\mathfrak{X}}). \tag{6.1.26.1}$$

We denote by $\mathbf{Ind\text{-}HM}(\mathscr{O}_{\mathfrak{X}'}, \widetilde{\xi}^{-1}\widetilde{\Omega}^1_{\mathfrak{X}'/\mathscr{S}})$ (resp. $\mathbf{Ind\text{-}HM}(\mathscr{O}_{\mathfrak{X}'}, \widetilde{\xi}^{-1}\widetilde{\Omega}^1_{\mathfrak{X}'/\mathfrak{X}})$) the category of Higgs ind-$\mathscr{O}_{\mathfrak{X}'}$-modules with coefficients in $\widetilde{\xi}^{-1}\widetilde{\Omega}^1_{\mathfrak{X}'/\mathscr{S}}$ (resp. $\widetilde{\xi}^{-1}\widetilde{\Omega}^1_{\mathfrak{X}'/\mathfrak{X}}$) (2.8.1). Recall that we have a fully faithful functor (4.5.2.3)

$$\mathbf{HM}^{\mathrm{coh}}(\mathscr{O}_{\mathfrak{X}'}[\frac{1}{p}], \widetilde{\xi}^{-1}\widetilde{\Omega}^1_{\mathfrak{X}'/\mathscr{S}}) \to \mathbf{Ind\text{-}HM}(\mathscr{O}_{\mathfrak{X}'}, \widetilde{\xi}^{-1}\widetilde{\Omega}^1_{\mathfrak{X}'/\mathscr{S}}). \tag{6.1.26.2}$$

Similarly, we have a fully faithful functor

$$\mathbf{HM}^{\mathrm{coh}}(\mathscr{O}_{\mathfrak{X}'}[\frac{1}{p}], \widetilde{\xi}^{-1}\widetilde{\Omega}^1_{\mathfrak{X}'/\mathfrak{X}}) \to \mathbf{Ind\text{-}HM}(\mathscr{O}_{\mathfrak{X}'}, \widetilde{\xi}^{-1}\widetilde{\Omega}^1_{\mathfrak{X}'/\mathfrak{X}}). \tag{6.1.26.3}$$

We identify $\mathbf{HM}^{\mathrm{coh}}(\mathscr{O}_{\mathfrak{X}'}[\frac{1}{p}], \widetilde{\xi}^{-1}\widetilde{\Omega}^1_{\mathfrak{X}'/\mathscr{S}})$ (resp. $\mathbf{HM}^{\mathrm{coh}}(\mathscr{O}_{\mathfrak{X}'}[\frac{1}{p}], \widetilde{\xi}^{-1}\widetilde{\Omega}^1_{\mathfrak{X}'/\mathfrak{X}})$) with a full subcategory of $\mathbf{Ind\text{-}HM}(\mathscr{O}_{\mathfrak{X}'}, \widetilde{\xi}^{-1}\widetilde{\Omega}^1_{\mathfrak{X}'/\mathscr{S}})$ (resp. $\mathbf{Ind\text{-}HM}(\mathscr{O}_{\mathfrak{X}'}, \widetilde{\xi}^{-1}\widetilde{\Omega}^1_{\mathfrak{X}'/\mathfrak{X}})$) by these functors, which we omit from the notation.

Definition 6.1.27 We call a *Higgs* $\mathscr{O}_{\mathfrak{X}'}[\frac{1}{p}]$*-bundle with coefficients in* $\widetilde{\xi}^{-1}\widetilde{\Omega}^1_{\mathfrak{X}'/\mathfrak{X}}$ any Higgs $\mathscr{O}_{\mathfrak{X}'}[\frac{1}{p}]$-module with coefficients in $\widetilde{\xi}^{-1}\widetilde{\Omega}^1_{\mathfrak{X}'/\mathfrak{X}}$ whose underlying $\mathscr{O}_{\mathfrak{X}'}[\frac{1}{p}]$-module is locally projective of finite type (2.1.11).

Proposition 6.1.28 *If the morphism $g\colon X' \to X$ is proper, for all integers $i, j \geq 0$, we have the following properties.*

(i) *The $\mathscr{O}_{X_\eta}$-module $\mathrm{R}^i g_*(\widetilde{\Omega}^j_{X'/X})|X_\eta$ is locally free of finite type.*

(ii) *We have a canonical isomorphism*

$$\mathrm{R}^i \mathfrak{g}_*(\widetilde{\Omega}^j_{\mathfrak{X}'/\mathfrak{X}}) \xrightarrow{\sim} a^{-1}(\mathrm{R}^i g_*(\widetilde{\Omega}^j_{X'/X})) \otimes_{a^{-1}(\mathscr{O}_X)} \mathscr{O}_{\mathfrak{X}}, \qquad (6.1.28.1)$$

where $a\colon X_s \to X$ is the canonical injection.

(iii) *The $\mathscr{O}_{\mathfrak{X}}[\frac{1}{p}]$-module $\mathrm{R}^i \mathfrak{g}_*(\widetilde{\Omega}^j_{\mathfrak{X}'/\mathfrak{X}}) \otimes_{\mathscr{O}_{\mathfrak{X}}} \mathscr{O}_{\mathfrak{X}}[\frac{1}{p}]$ is locally projective of finite type* (2.1.11).

(i) Observe first that the morphism g being saturated, is exact ([45] III 2.5.2), and that the stalks of the monoid $(\mathscr{M}_X/\mathscr{O}_X^{\times})|X_\eta$ are free since $\mathscr{M}_X|X_\eta$ is defined by a divisor with normal crossings ([3] III.4.7). The proposition then follows from ([38] 7.2) (see [12] 5.5 for the smooth case without logarithmic structures).

(ii) This follows from ([1] 2.5.5(ii) and 2.12.2).

(iii) The question being local, we may suppose X affine. We set $\mathscr{F} = \mathrm{R}^i g_*(\widetilde{\Omega}^j_{X'/X})$. By (i), there exists a coherent $\mathscr{O}_X$-module $\mathscr{G}$, an integer $n \geq 1$ and a $\mathscr{O}_X$-linear morphism $u\colon \mathscr{F} \oplus \mathscr{G} \to \mathscr{O}_X^n$ inducing an isomorphism on X_η. There exists an integer $m \geq 0$ such that p^m annihilates the kernel and the cokernel of u. Then, by ([2] 2.6.3), there exists an $\mathscr{O}_X$-linear morphism $v\colon \mathscr{O}_X^n \to \mathscr{F} \oplus \mathscr{G}$ such that $u \circ v = p^{2m}\mathrm{id}_{\mathscr{O}_X^n}$ and $v \circ u = p^{2m}\mathrm{id}_{\mathscr{F}\oplus\mathscr{G}}$, i.e., u is an isogeny (2.9.1). The proposition follows in view of (ii).

6.2 Base Change

6.2.1 Let $\mathscr{M}$ be an $\mathscr{O}_{\mathfrak{X}'}$-module, $\mathscr{N}$ a $\overline{\breve{\mathscr{B}}}'$-module, q an integer ≥ 0. The adjunction morphism $\widehat{\pi}^*(\mathscr{M}) \to \check{\tau}_*(\widehat{\sigma}'^*(\mathscr{M}))$ (6.1.23.6) and the cup-product induce a bifunctorial morphism

$$\widehat{\pi}^*(\mathscr{M}) \otimes_{\overline{\breve{\mathscr{B}}}^!} \mathrm{R}^q\check{\tau}_*(\mathscr{N}) \to \mathrm{R}^q\check{\tau}_*(\widehat{\sigma}'^*(\mathscr{M}) \otimes_{\overline{\breve{\mathscr{B}}}'} \mathscr{N}). \qquad (6.2.1.1)$$

This induces for all objects $\mathscr{F}$ of $\mathbf{Mod}_{\mathbb{Q}}(\mathscr{O}_{\mathfrak{X}'})$ and $\mathscr{G}$ of $\mathbf{Mod}_{\mathbb{Q}}(\overline{\breve{\mathscr{B}}}')$ a bifunctorial morphism of $\mathbf{Mod}_{\mathbb{Q}}(\overline{\breve{\mathscr{B}}}^!)$

$$\widehat{\pi}^*_{\mathbb{Q}}(\mathscr{F}) \otimes_{\overline{\breve{\mathscr{B}}}^!_{\mathbb{Q}}} \mathrm{R}^q\check{\tau}_{\mathbb{Q}*}(\mathscr{G}) \to \mathrm{R}^q\check{\tau}_{\mathbb{Q}*}(\widehat{\sigma}'^*_{\mathbb{Q}}(\mathscr{F}) \otimes_{\overline{\breve{\mathscr{B}}}'_{\mathbb{Q}}} \mathscr{G}). \qquad (6.2.1.2)$$

In view of (2.7.10.5), the morphism (6.2.1.1) also induces for all objects $\mathscr{F}$ of $\mathbf{Ind\text{-}Mod}(\mathscr{O}_{\mathfrak{X}'})$ and $\mathscr{G}$ of $\mathbf{Ind\text{-}Mod}(\overline{\breve{\mathscr{B}}}')$ a bifunctorial morphism of $\mathbf{Ind\text{-}Mod}(\overline{\breve{\mathscr{B}}}^!)$

$$\mathrm{I}\widehat{\pi}^*(\mathscr{F}) \otimes_{\overline{\breve{\mathscr{B}}}^!} \mathrm{R}^q\mathrm{I}\check{\tau}_*(\mathscr{H}) \to \mathrm{R}^q\mathrm{I}\check{\tau}_*(\mathrm{I}\widehat{\sigma}'^*(\mathscr{F}) \otimes_{\overline{\breve{\mathscr{B}}}'} \mathscr{H}). \qquad (6.2.1.3)$$

In view of (4.2.1.2) (resp. (6.1.24.1)), the morphism (6.2.1.2) (resp. (6.2.1.3)) exists when $\mathscr{F}$ is a $\mathscr{O}_{\mathfrak{X}'}[\frac{1}{p}]$-coherent module.

Lemma 6.2.2 *Let q be an integer ≥ 0. Then,*

(i) *For every locally free $\mathscr{O}_{\mathfrak{X}'}$-module of finite type $\mathscr{M}$ and every $\overline{\breve{\mathscr{B}}}'$-module $\mathscr{N}$, the canonical morphism* (6.2.1.1)

$$\widehat{\pi}^*(\mathscr{M}) \otimes_{\overline{\breve{\mathscr{B}}}'} \mathrm{R}^q\breve{\tau}_*(\mathscr{N}) \to \mathrm{R}^q\breve{\tau}_*(\widehat{\sigma}'^*(\mathscr{M}) \otimes_{\overline{\breve{\mathscr{B}}}'} \mathscr{N}) \tag{6.2.2.1}$$

is an isomorphism.

(ii) *For every locally projective $\mathscr{O}_{\mathfrak{X}'}[\frac{1}{p}]$-module of finite type $\mathscr{F}$* (2.1.11) *and every $\overline{\breve{\mathscr{B}}}'_{\mathbb{Q}}$-module $\mathscr{G}$, the canonical morphism* (6.2.1.2)

$$\widehat{\pi}^*_{\mathbb{Q}}(\mathscr{F}) \otimes_{\overline{\breve{\mathscr{B}}}'_{\mathbb{Q}}} \mathrm{R}^q\breve{\tau}_{\mathbb{Q}*}(\mathscr{G}) \to \mathrm{R}^q\breve{\tau}_{\mathbb{Q}*}(\widehat{\sigma}'^*_{\mathbb{Q}}(\mathscr{F}) \otimes_{\overline{\breve{\mathscr{B}}}'_{\mathbb{Q}}} \mathscr{G}) \tag{6.2.2.2}$$

is an isomorphism.

(iii) *For every locally free $\mathscr{O}_{\mathfrak{X}'}$-module of finite type $\mathscr{F}$ and every ind-$\overline{\breve{\mathscr{B}}}'$-module $\mathscr{G}$, the canonical morphism* (6.2.1.3)

$$\widehat{\pi}^*(\mathscr{F}) \otimes_{\overline{\breve{\mathscr{B}}}'} \mathrm{R}^q\mathrm{I}\breve{\tau}_*(\mathscr{G}) \to \mathrm{R}^q\mathrm{I}\breve{\tau}_*(\widehat{\sigma}'^*(\mathscr{F}) \otimes_{\overline{\breve{\mathscr{B}}}'} \mathscr{G}) \tag{6.2.2.3}$$

is an isomorphism.

(iv) *For every locally projective $\mathscr{O}_{\mathfrak{X}'}[\frac{1}{p}]$-module of finite type $\mathscr{F}$* (2.1.11) *and every ind-$\overline{\breve{\mathscr{B}}}'$-module $\mathscr{G}$, the canonical morphism* (6.2.1.3)

$$\mathrm{I}\widehat{\pi}^*(\mathscr{F}) \otimes_{\overline{\breve{\mathscr{B}}}'} \mathrm{R}^q\mathrm{I}\breve{\tau}_*(\mathscr{G}) \to \mathrm{R}^q\mathrm{I}\breve{\tau}_*(\mathrm{I}\widehat{\sigma}'^*(\mathscr{F}) \otimes_{\overline{\breve{\mathscr{B}}}'} \mathscr{G}) \tag{6.2.2.4}$$

is an isomorphism.

(i) This is immediate.

(ii) There exists a finite covering $(U'_i)_{i\in I}$ by Zariski open subschemes of X' such that for every $i \in I$, the restriction of $\mathscr{F}$ to $(U'_i)_s$ is a direct factor of a free $(\mathscr{O}_{\mathfrak{X}'}|(U'_i)_s)[\frac{1}{p}]$-module of finite type. For any $i \in I$, we denote by $\breve{\pi}^*((U'_i)_s)$ the constant object of $\widetilde{G}_s^{\mathbb{N}^\circ}$ with value $\pi_s^*((U'_i)_s)$ (6.1.20.2). The objects $(\breve{\pi}^*((U'_i)_s))_{i\in I}$ then form a covering of the final object of $\widetilde{G}_s^{\mathbb{N}^\circ}$ (see [3] III.7.4 and III.7.5). We leave to the reader to write the analogous statement for $\widetilde{E}_s'^{\mathbb{N}^\circ}$. In view of ([3] III.6.7(ii)), we may then reduce to the case where $\mathscr{F}$ is a direct factor of a free $\mathscr{O}_{\mathfrak{X}'}[\frac{1}{p}]$-module of finite type, and even to the case where $\mathscr{F}$ is a free $\mathscr{O}_{\mathfrak{X}'}[\frac{1}{p}]$-module of finite type, in which case the assertion is obvious.

(iii) This follows from 2.7.14. It can also be deduced from (i), in view of (2.7.10.5) and the fact that the tensor product and the functor $\mathrm{R}^q\mathrm{I}\breve{\tau}_*$ commute with small filtered direct limits by 2.7.3 and 2.6.4(ii).

(iv) In view of 2.7.17(iii), we can reduce to the case where $\mathscr{F}$ is a free $\mathscr{O}_{\mathfrak{X}'}[\frac{1}{p}]$-module of finite type, see the proof of (ii). The assertion then follows from the fact that the tensor product and the functor $\mathrm{R}^q\mathrm{I}\breve{\tau}_*$ commute with small filtered direct limits by 2.7.3, (2.7.10.5) and 2.6.4(ii).

Lemma 6.2.3 *There exists an integer $N \geq 0$ such that for every integer $n \geq 0$, the following properties hold.*

(i) *For every exact sequence of $\mathscr{O}_{\overline{X}_n}$-modules of $X_{s,\text{ét}}$,*

$$0 \to \mathscr{F}' \to \mathscr{F} \to \mathscr{F}'' \to 0,$$

the sequence

$$0 \to \sigma_n^*(\mathscr{F}') \to \sigma_n^*(\mathscr{F}) \to \sigma_n^*(\mathscr{F}'') \to 0 \tag{6.2.3.1}$$

is p^N-exact (2.1.14).

(ii) *For every exact sequence of $\mathscr{O}_{\overline{X}'_n}$-modules of $X'_{s,\text{ét}}$,*

$$0 \to \mathscr{F}' \to \mathscr{F} \to \mathscr{F}'' \to 0,$$

the sequence

$$0 \to \pi_n^*(\mathscr{F}') \to \pi_n^*(\mathscr{F}) \to \pi_n^*(\mathscr{F}'') \to 0 \tag{6.2.3.2}$$

is p^N-exact.

Since X is quasi-compact, we may assume that the morphism f (6.1.3) admits an adequate chart ([3] III.4.7). Let us show that if N denotes the integer given by the proposition ([2] 5.3.9), the integer $N+1$ is suitable.

(i) By ([2] 5.3.9), for every point $(\overline{y} \rightsquigarrow \overline{x})$ of $X_{\text{ét}} \overleftarrow{\times}_{X_{\text{ét}}} \overline{X}^\circ_{\text{ét}}$ (4.3.4) such that $\overline{x}$ is above s, and every exact sequence of $\mathscr{O}_{\overline{X},\overline{x}}$-modules $0 \to M' \to M \to M'' \to 0$, the sequence

$$0 \to M' \otimes_{\mathscr{O}_{\overline{X},\overline{x}}} \overline{\mathscr{B}}_{\rho(\overline{y} \rightsquigarrow \overline{x})} \to M \otimes_{\mathscr{O}_{\overline{X},\overline{x}}} \overline{\mathscr{B}}_{\rho(\overline{y} \rightsquigarrow \overline{x})} \to M'' \otimes_{\mathscr{O}_{\overline{X},\overline{x}}} \overline{\mathscr{B}}_{\rho(\overline{y} \rightsquigarrow \overline{x})} \to 0 \tag{6.2.3.3}$$

is p^N-exact. Note that k being algebraically closed (6.1.1), $\overline{x}$ is naturally a geometric point of $\overline{X}$. Property (i) follows in view of ([3] (VI.10.18.1) and III.9.5).

(ii) Let $(\overline{y} \rightsquigarrow \overline{x}')$ be a point of $X'_{\text{ét}} \overleftarrow{\times}_{X_{\text{ét}}} \overline{X}^\circ_{\text{ét}}$ ([2] 3.4.23) such that $\overline{x}'$ is above s. We set $\overline{x} = g(\overline{x}')$ and denote by $(\overline{y} \rightsquigarrow \overline{x})$ the image of $(\overline{y} \rightsquigarrow \overline{x}')$ by the canonical morphism

$$X'_{\text{ét}} \overleftarrow{\times}_{X_{\text{ét}}} \overline{X}^\circ_{\text{ét}} \to X_{\text{ét}} \overleftarrow{\times}_{X_{\text{ét}}} \overline{X}^\circ_{\text{ét}}. \tag{6.2.3.4}$$

By ([2] 6.5.29), for every integer $n \geq 0$, the canonical homomorphism

$$(\overline{\mathscr{B}}_{\rho(\overline{y} \rightsquigarrow \overline{x})}/p^n \overline{\mathscr{B}}_{\rho(\overline{y} \rightsquigarrow \overline{x})}) \otimes_{\mathscr{O}_{\overline{X},\overline{x}}} \mathscr{O}_{\overline{X}',\overline{x}'} \to \overline{\mathscr{B}}^!_{\varrho(\overline{y} \rightsquigarrow \overline{x}')}/p^n \overline{\mathscr{B}}^!_{\varrho(\overline{y} \rightsquigarrow \overline{x}')} \tag{6.2.3.5}$$

is an α-isomorphism (6.1.14.9). It then follows from ([2] 5.3.9) that for every exact sequence of $(\mathscr{O}_{\overline{X}',\overline{x}'}/p^n \mathscr{O}_{\overline{X}',\overline{x}'})$-modules $0 \to M' \to M \to M'' \to 0$, the sequence

$$0 \to M' \otimes_{\mathscr{O}_{\overline{X}',\overline{x}'}} \overline{\mathscr{B}}^!_{\varrho(\overline{y} \rightsquigarrow \overline{x}')} \to M \otimes_{\mathscr{O}_{\overline{X}',\overline{x}'}} \overline{\mathscr{B}}^!_{\varrho(\overline{y} \rightsquigarrow \overline{x}')} \to M'' \otimes_{\mathscr{O}_{\overline{X}',\overline{x}'}} \overline{\mathscr{B}}^!_{\varrho(\overline{y} \rightsquigarrow \overline{x}')} \to 0 \tag{6.2.3.6}$$

is p^{N+1}-exact. Property (ii) follows in view of ([2] (3.4.23.1) and 6.5.3).

Lemma 6.2.4 *There is an integer $N \geq 0$ satisfying the following properties:*

(i) *For every exact sequence of $\mathscr{O}_{\mathfrak{X}}$-modules $0 \to \mathscr{F}' \to \mathscr{F} \to \mathscr{F}'' \to 0$ where $\mathscr{F}''$ is $\mathscr{S}$-flat, the sequence of $\check{\overline{\mathscr{B}}}$-modules*

$$0 \to \widehat{\sigma}^*(\mathscr{F}') \to \widehat{\sigma}^*(\mathscr{F}) \to \widehat{\sigma}^*(\mathscr{F}'') \to 0 \tag{6.2.4.1}$$

is p^N-exact (2.1.14).

(ii) *For every exact sequence of $\mathscr{O}_{\mathfrak{X}'}$-modules $0 \to \mathscr{F}' \to \mathscr{F} \to \mathscr{F}'' \to 0$ where $\mathscr{F}''$ is $\mathscr{S}$-flat, the sequence*

$$0 \to \widehat{\pi}^*(\mathscr{F}') \to \widehat{\pi}^*(\mathscr{F}) \to \widehat{\pi}^*(\mathscr{F}'') \to 0 \tag{6.2.4.2}$$

is p^N-exact.

Let us show that the integer N provided by the proposition 6.2.3 is suitable. Let $0 \to \mathscr{F}' \to \mathscr{F} \to \mathscr{F}'' \to 0$ be an exact sequence of $\mathscr{O}_{\mathfrak{X}}$-modules such that $\mathscr{F}''$ is $\mathscr{S}$-flat. For every integer $n \geq 0$, the sequence of $\mathscr{O}_{\overline{X}_n}$-modules of $X_{s,\mathrm{zar}}$

$$0 \to \mathscr{F}'/p^n\mathscr{F}' \to \mathscr{F}/p^n\mathscr{F} \to \mathscr{F}''/p^n\mathscr{F}'' \to 0 \tag{6.2.4.3}$$

is exact. Since u_n is flat (6.1.23.1), the sequence of $\overline{\mathscr{B}}_n$-modules

$$0 \to \sigma_n^*(u_n^*(\mathscr{F}'/p^n\mathscr{F}')) \to \sigma_n^*(u_n^*(\mathscr{F}/p^n\mathscr{F})) \to \sigma_n^*(u_n^*(\mathscr{F}''/p^n\mathscr{F}'')) \to 0 \tag{6.2.4.4}$$

is p^N-exact by virtue of 6.2.3. Property (i) follows in view of (6.1.23.7) and ([3] III.7.3(i)). Property (ii) is proved in the same way.

Proposition 6.2.5 *The functors* (6.1.24.4) *and* (6.1.24.15)

$$\widehat{\sigma}_{\mathbb{Q}}^*\colon \mathbf{Mod}^{\mathrm{coh}}(\mathscr{O}_{\mathfrak{X}}[\frac{1}{p}]) \to \mathbf{Mod}_{\mathbb{Q}}(\check{\overline{\mathscr{B}}}), \tag{6.2.5.1}$$

$$\widehat{\sigma}_{\mathbb{Q}}^{\prime *}\colon \mathbf{Mod}^{\mathrm{coh}}(\mathscr{O}_{\mathfrak{X}'}[\frac{1}{p}]) \to \mathbf{Mod}_{\mathbb{Q}}(\check{\overline{\mathscr{B}}}'), \tag{6.2.5.2}$$

$$\widehat{\pi}_{\mathbb{Q}}^*\colon \mathbf{Mod}^{\mathrm{coh}}(\mathscr{O}_{\mathfrak{X}'}[\frac{1}{p}]) \to \mathbf{Mod}_{\mathbb{Q}}(\check{\overline{\mathscr{B}}}^!), \tag{6.2.5.3}$$

are exact.

Indeed, by ([3] III.6.16 and III.6.1.4), every exact sequence of coherent $\mathscr{O}_{\mathfrak{X}}[\frac{1}{p}]$-modules $0 \to \mathscr{G}' \to \mathscr{G} \to \mathscr{G}'' \to 0$ is obtained from an exact sequence of coherent $\mathscr{O}_{\mathfrak{X}}$-modules $0 \to \mathscr{F}' \to \mathscr{F} \to \mathscr{F}'' \to 0$ by inverting p. Since the kernel $\mathscr{F}''_{\mathrm{tor}}$ of the canonical morphism $\mathscr{F}'' \to \mathscr{F}''[\frac{1}{p}]$ is a coherent $\mathscr{O}_{\mathfrak{X}}$-module ([1] 2.10.14), replacing $\mathscr{F}''$ by $\mathscr{F}''/\mathscr{F}''_{\mathrm{tor}}$, we reduce to the case where $\mathscr{F}''$ is $\mathscr{S}$-flat. It then follows from 6.2.4 that the sequence

$$0 \to \widehat{\sigma}_{\mathbb{Q}}^*(\mathscr{G}') \to \widehat{\sigma}_{\mathbb{Q}}^*(\mathscr{G}) \to \widehat{\sigma}_{\mathbb{Q}}^*(\mathscr{G}'') \to 0 \tag{6.2.5.4}$$

is exact. We prove similarly that the functors (6.2.5.2) and (6.2.5.3) are exact.

Corollary 6.2.6 *The functors* (6.1.24.2) *and* (6.1.24.8)

$$\mathrm{I}\widehat{\sigma}^*\colon \mathbf{Mod}^{\mathrm{coh}}(\mathscr{O}_{\mathfrak{X}}[\frac{1}{p}]) \to \mathbf{Ind\text{-}Mod}(\overline{\breve{\mathscr{B}}}), \tag{6.2.6.1}$$

$$\mathrm{I}\widehat{\sigma}'^*\colon \mathbf{Mod}^{\mathrm{coh}}(\mathscr{O}_{\mathfrak{X}'}[\frac{1}{p}]) \to \mathbf{Ind\text{-}Mod}(\overline{\breve{\mathscr{B}}}'), \tag{6.2.6.2}$$

$$\mathrm{I}\widehat{\pi}^*\colon \mathbf{Mod}^{\mathrm{coh}}(\mathscr{O}_{\mathfrak{X}'}[\frac{1}{p}]) \to \mathbf{Ind\text{-}Mod}(\overline{\breve{\mathscr{B}}}^!), \tag{6.2.6.3}$$

are exact.

This follows from 2.9.3(ii) and 6.2.5.

6.2.7 For every integer $n \geq 0$, the diagram of morphisms of ringed topos (6.1.21.10)

$$\begin{array}{ccc} (\widetilde{G}_s, \overline{\mathscr{B}}^!_n) & \xrightarrow{\pi_n} & (X'_{s,\text{ét}}, \mathscr{O}_{\overline{X}'_n}) \\ \downarrow{\scriptstyle \mathfrak{g}_n} & & \downarrow{\scriptstyle \overline{g}_n} \\ (\widetilde{E}_s, \overline{\mathscr{B}}_n) & \xrightarrow{\sigma_n} & (X_{s,\text{ét}}, \mathscr{O}_{\overline{X}_n}) \end{array} \tag{6.2.7.1}$$

is commutative up to canonical isomorphism. For every $\mathscr{O}_{\overline{X}'_n}$-module $\mathscr{F}'$ of $X'_{s,\text{ét}}$ and every integer $q \geq 0$, we have a canonical base change morphism ([1] (1.2.3.3))

$$\sigma_n^*(\mathrm{R}^q\overline{g}_{n*}(\mathscr{F}')) \to \mathrm{R}^q\mathfrak{g}_{n*}(\pi_n^*(\mathscr{F}')), \tag{6.2.7.2}$$

where σ_n^* and π_n^* denote the pullbacks in the sense of ringed topos.

Theorem 6.2.8 ([2] 6.5.31) *Assume that the morphism* $g\colon X' \to X$ *is proper. Then there exists an integer* $N \geq 0$ *such that for all integers* $n \geq 1$ *and* $q \geq 0$ *and every quasi-coherent* $\mathscr{O}_{\overline{X}'_n}$*-module* $\mathscr{F}'$ *of* $X'_{s,\text{zar}}$, *that we also consider as an* $\mathscr{O}_{\overline{X}'_n}$*-module of* $X'_{s,\text{ét}}$ (2.1.13), *the kernel and the cokernel of the base change morphism* (6.2.7.2)

$$\sigma_n^*(\mathrm{R}^q\overline{g}_{n*}(\mathscr{F}')) \to \mathrm{R}^q\mathfrak{g}_{n*}(\pi_n^*(\mathscr{F}')) \tag{6.2.8.1}$$

are annihilated by p^N.

6.2.9 The diagram of morphisms of ringed topos (6.1.22.1)

$$\begin{array}{ccc} (\widetilde{G}_s^{\mathbb{N}^\circ}, \overline{\breve{\mathscr{B}}}^!) & \xrightarrow{\breve{\pi}} & (X'^{\mathbb{N}^\circ}_{s,\text{ét}}, \mathscr{O}_{\breve{\overline{X}}'}) \\ \downarrow{\scriptstyle \breve{\mathfrak{g}}} & & \downarrow{\scriptstyle \breve{\overline{g}}} \\ (\widetilde{E}_s^{\mathbb{N}^\circ}, \overline{\breve{\mathscr{B}}}) & \xrightarrow{\breve{\sigma}} & (X^{\mathbb{N}^\circ}_{s,\text{ét}}, \mathscr{O}_{\breve{\overline{X}}}) \end{array} \tag{6.2.9.1}$$

is commutative up to canonical isomorphism. For every $\mathscr{O}_{\breve{\overline{X}}'}$-module $\mathscr{F}'$ of $X'^{\mathbb{N}^\circ}_{s,\text{ét}}$ and every integer $q \geq 0$, we have a canonical base change morphism ([1] (1.2.3.3))

$$\breve{\sigma}^*(\mathrm{R}^q\breve{\overline{g}}_*(\mathscr{F}')) \to \mathrm{R}^q\breve{\mathfrak{g}}_*(\breve{\pi}^*(\mathscr{F}')), \tag{6.2.9.2}$$

where $\breve{\sigma}^*$ and $\breve{\pi}^*$ denote the pullbacks in the sense of ringed topos.

Lemma 6.2.10 *For every $\mathscr{O}_{\breve{\overline{X}}'}$-module $\mathscr{F}' = (\mathscr{F}'_n)_{n\geq 0}$ of $X'^{\mathbb{N}^\circ}_{s,\text{ét}}$ and every $q \geq 0$, the base change morphism* (6.2.9.2)

$$\breve{\sigma}^*(\mathrm{R}^q\breve{\overline{g}}_*(\mathscr{F}')) \to \mathrm{R}^q\breve{\mathfrak{g}}_*(\breve{\pi}^*(\mathscr{F}')) \tag{6.2.10.1}$$

is induced by the base change morphisms (6.2.7.2), *for all integers $n \geq 0$,*

$$\sigma_n^*(\mathrm{R}^q\overline{g}_{n*}(\mathscr{F}'_n)) \to \mathrm{R}^q\mathfrak{g}_{n*}(\pi_n^*(\mathscr{F}'_n)). \tag{6.2.10.2}$$

Indeed, by ([3] III.7.1), for every integer $n \geq 0$, there exist two morphisms of topos $a_n : \widetilde{E}_s \to \widetilde{E}_s^{\mathbb{N}^\circ}$ and $b_n : \widetilde{G}_s \to \widetilde{G}_s^{\mathbb{N}^\circ}$ such that for all objects $M = (M_n)_{n\geq 0}$ of $\widetilde{E}_s^{\mathbb{N}^\circ}$ and $N = (N_n)_{n\geq 0}$ of $\widetilde{G}_s^{\mathbb{N}^\circ}$, we have $a_n^*(M) = M_n$ and $b_n^*(G) = G_n$. By ([3] (III.7.5.4)), the diagram of morphisms of topos

$$\begin{array}{ccc} \widetilde{G}_s & \xrightarrow{b_n} & \widetilde{G}_s^{\mathbb{N}^\circ} \\ {\scriptstyle g_s}\downarrow & & \downarrow{\scriptstyle \breve{\mathfrak{g}}} \\ \widetilde{E}_s & \xrightarrow{a_n} & \widetilde{E}_s^{\mathbb{N}^\circ} \end{array} \tag{6.2.10.3}$$

is commutative up to canonical isomorphism. Moreover, for every $\breve{\overline{\mathscr{B}}}^!$-module N of $\widetilde{G}_s^{\mathbb{N}^\circ}$, the base change morphism

$$a_n^*(\mathrm{R}^q\breve{\mathfrak{g}}_*(N)) \to \mathrm{R}^q\mathfrak{g}_{n*}(b_n^*(N)) \tag{6.2.10.4}$$

is an isomorphism ([3] (III.7.5.5)). The proposition then follows from ([1] 1.2.4(ii)).

Proposition 6.2.11 *Assume that the morphism $g : X' \to X$ is proper. Then there exists an integer $N \geq 0$ such that for every $\mathscr{O}_{\breve{\overline{X}}'}$-module $\mathscr{F}' = (\mathscr{F}'_n)_{n\geq 0}$ of $X'^{\mathbb{N}^\circ}_{s,\text{ét}}$, where the $\mathscr{O}_{\overline{X}'_n}$-modules $\mathscr{F}'_n$ are induced by quasi-coherent $\mathscr{O}_{\overline{X}'_n}$-modules of $X'_{s,\text{zar}}$* (2.1.13.3), *and every $q \geq 0$, the kernel and the cokernel of the base change morphism* (6.2.9.2)

$$\breve{\sigma}^*(\mathrm{R}^q\breve{\overline{g}}_*(\mathscr{F}')) \to \mathrm{R}^q\breve{\mathfrak{g}}_*(\breve{\pi}^*(\mathscr{F}')) \tag{6.2.11.1}$$

are annihilated by p^N.

This follows from 6.2.8, 6.2.10 and ([3] III.7.3(i)).

6.2.12 We check immediately that the two squares of the diagram of morphisms of ringed topos

$$\begin{array}{ccccc} (X'^{\mathbb{N}^\circ}_{s,\text{ét}}, \mathscr{O}_{\check{\overline{X}}'}) & \xrightarrow{\check{u}'} & (X'^{\mathbb{N}^\circ}_{s,\text{zar}}, \mathscr{O}_{\check{\overline{X}}'}) & \xrightarrow{\lambda'} & (X'_{s,\text{zar}}, \mathscr{O}_{\mathfrak{X}'}) \\ \downarrow{\scriptstyle \check{\overline{g}}} & & \downarrow{\scriptstyle \check{\overline{g}}} & & \downarrow{\scriptstyle \mathfrak{g}} \\ (X^{\mathbb{N}^\circ}_{s,\text{ét}}, \mathscr{O}_{\check{\overline{X}}}) & \xrightarrow{\check{u}} & (X^{\mathbb{N}^\circ}_{s,\text{zar}}, \mathscr{O}_{\check{\overline{X}}}) & \xrightarrow{\lambda} & (X_{s,\text{zar}}, \mathscr{O}_{\mathfrak{X}}), \end{array} \tag{6.2.12.1}$$

where λ and λ' (resp. $\check{u}$ and $\check{u}'$) are the canonical morphisms of ringed topos (2.1.9.1) (resp. (6.1.23.2)), are commutative up to canonical isomorphisms.

Lemma 6.2.13 ([2] 6.5.37) *Assume that the morphism $g\colon X' \to X$ is separated and quasi-compact. Let $\mathscr{F}' = (\mathscr{F}'_n)_{n\in\mathbb{N}}$ be an $\mathscr{O}_{\check{\overline{X}}'}$-module of $X'^{\mathbb{N}^\circ}_{s,\text{zar}}$ such that for every integer $n \geq 0$, the $\mathscr{O}_{\overline{X}'_n}$-module $\mathscr{F}'_n$ is quasi-coherent, q an integer ≥ 0. Then, the base change morphism with respect to the left square of the diagram* (6.2.12.1)

$$\check{u}^*(\mathrm{R}^q\check{\overline{g}}_{\text{zar}*}(\mathscr{F}')) \to \mathrm{R}^q\check{\overline{g}}_{\text{ét}*}(\check{u}'^*(\mathscr{F}')) \tag{6.2.13.1}$$

is an isomorphism.

Proposition 6.2.14 ([2] 6.5.38) *Assume the morphism $g\colon X' \to X$ is proper. Let $\mathscr{F}'$ be a coherent $\mathscr{O}_{\mathfrak{X}'}$-module of $X'_{s,\text{zar}}$, q an integer ≥ 0. Then, there exists an integer $N \geq 0$ such that the kernel and the cokernel of the base change morphism with respect to the right square of the diagram* (6.2.12.1)

$$\lambda^*(\mathrm{R}^q\mathfrak{g}_*(\mathscr{F}')) \to \mathrm{R}^q\check{\overline{g}}_{\text{zar}*}(\lambda'^*(\mathscr{F}')) \tag{6.2.14.1}$$

are annihilated by p^N.

6.2.15 The diagram of morphisms of ringed topos (6.1.23.6)

$$\begin{array}{ccc} (\widetilde{G}^{\mathbb{N}^\circ}_s, \check{\overline{\mathscr{B}}}^{!}) & \xrightarrow{\widehat{\pi}} & (X'_{s,\text{zar}}, \mathscr{O}_{\mathfrak{X}'}) \\ \downarrow{\scriptstyle \check{\mathfrak{g}}} & & \downarrow{\scriptstyle \mathfrak{g}} \\ (\widetilde{E}^{\mathbb{N}^\circ}_s, \check{\overline{\mathscr{B}}}) & \xrightarrow{\widehat{\sigma}} & (X_{s,\text{zar}}, \mathscr{O}_{\mathfrak{X}}) \end{array} \tag{6.2.15.1}$$

is commutative up to canonical isomorphism. For every $\mathscr{O}_{\mathfrak{X}'}$-module $\mathscr{F}'$ of $X'_{s,\text{zar}}$ and every integer $q \geq 0$, we have a canonical base change morphism ([1] (1.2.3.3))

$$\widehat{\sigma}^*(\mathrm{R}^q\mathfrak{g}_*(\mathscr{F}')) \to \mathrm{R}^q\check{\mathfrak{g}}_*(\widehat{\pi}^*(\mathscr{F}')), \tag{6.2.15.2}$$

where $\widehat{\sigma}^*$ and $\widehat{\pi}^*$ denote the pullbacks in the sense of ringed topos. In view of (2.9.6.4), this induces for every object F' of $\mathbf{Mod}_{\mathbb{Q}}(\mathscr{O}_{\mathfrak{X}'})$ and every integer $q \geq 0$, a canonical morphism, also called a base change morphism,

$$\widehat{\sigma}^*_{\mathbb{Q}}(\mathrm{R}^q\mathfrak{g}_{\mathbb{Q}*}(F')) \to \mathrm{R}^q\check{\mathfrak{g}}_{\mathbb{Q}*}(\widehat{\pi}^*_{\mathbb{Q}}(F')). \tag{6.2.15.3}$$

For every $\mathscr{O}_{\mathfrak{X}'}[\frac{1}{p}]$-coherent module $\mathscr{F}'$ of $X'_{s,\mathrm{zar}}$ and every integer $q \geq 0$ such that the $\mathscr{O}_{\mathfrak{X}}[\frac{1}{p}]$-module $\mathrm{R}^q\mathfrak{g}_*(\mathscr{F}')$ is coherent, the morphism (6.2.15.3) induces a canonical morphism, also called a base change morphism,

$$\widehat{\sigma}^*_{\mathbb{Q}}(\mathrm{R}^q\mathfrak{g}_*(\mathscr{F}')) \to \mathrm{R}^q\check{\mathfrak{g}}_{\mathbb{Q}*}(\widehat{\pi}^*_{\mathbb{Q}}(\mathscr{F}')), \tag{6.2.15.4}$$

where $\widehat{\sigma}^*_{\mathbb{Q}}$ and $\widehat{\pi}^*_{\mathbb{Q}}$ denote the functors (6.1.24.4) and (6.1.24.15).

Proposition 6.2.16 *Assume that the morphism $g\colon X' \to X$ is proper. Then there exists an integer $N \geq 0$ such that for every coherent $\mathscr{O}_{\mathfrak{X}'}$-module $\mathscr{F}'$ and every integer $q \geq 0$, the kernel and the cokernel of the base change morphism* (6.2.15.2)

$$\widehat{\sigma}^*(\mathrm{R}^q\mathfrak{g}_*(\mathscr{F}')) \to \mathrm{R}^q\check{\mathfrak{g}}_*(\widehat{\pi}^*(\mathscr{F}')) \tag{6.2.16.1}$$

are annihilated by p^N.

Indeed, we have $\widehat{\pi} = \check{\pi} \circ \check{u}' \circ \lambda'$ and $\widehat{\sigma} = \check{\sigma} \circ \check{u} \circ \lambda$. The proposition then follows from 6.2.11, 6.2.13 and 6.2.14, taking into account (2.1.13.6), ([2] 2.6.3) and ([1] 1.2.4(ii)). Note that the sheaf of rings $\mathscr{O}_{\mathfrak{X}'}$ of $X'_{s,\mathrm{zar}}$ is coherent ([1] 2.8.1).

Corollary 6.2.17 *Assume that the morphism $g\colon X' \to X$ is proper. Let $\mathscr{F}'$ be a coherent $\mathscr{O}_{\mathfrak{X}'}[\frac{1}{p}]$-module, q an integer ≥ 0. Then the $\mathscr{O}_{\mathfrak{X}}[\frac{1}{p}]$-module $\mathrm{R}^q\mathfrak{g}_*(\mathscr{F}')$ is coherent and the base change morphism* (6.2.15.4)

$$\widehat{\sigma}^*_{\mathbb{Q}}(\mathrm{R}^q\mathfrak{g}_*(\mathscr{F}')) \to \mathrm{R}^q\check{\mathfrak{g}}_{\mathbb{Q}*}(\widehat{\pi}^*_{\mathbb{Q}}(\mathscr{F}')) \tag{6.2.17.1}$$

is an isomorphism.

Indeed, the first assertion follows from ([1] 2.10.24 and 2.11.5). The second assertion then follows from 6.2.16. Note that the sheaf of rings $\mathscr{O}_{\mathfrak{X}'}[\frac{1}{p}]$ of $X'_{s,\mathrm{zar}}$ is coherent.

6.2.18 By ([59] 013K), for every complex of $\overline{\check{\mathscr{B}}}^!$-modules bounded from below $\mathscr{K}^\bullet$ of $\widetilde{G}_s^{\mathbb{N}^\circ}$, there exists a complex bounded from below of injective $\overline{\check{\mathscr{B}}}^!$-modules $\mathscr{L}^\bullet$ and a quasi-isomorphism $u\colon \mathscr{K}^\bullet \to \mathscr{L}^\bullet$. The latter induces a morphism of $\mathbf{D}^+(\mathbf{Mod}(\mathscr{O}_{\mathfrak{X}'}))$

$$\widehat{\pi}_*(\mathscr{K}^\bullet) \to \mathrm{R}\widehat{\pi}_*(\mathscr{K}^\bullet). \tag{6.2.18.1}$$

By ([59] 05TG), this morphism depends only on $\mathscr{K}^\bullet$, but not on u, on which it depends functorially.

Similarly, since $\mathbf{Mod}_{\mathbb{Q}}(\overline{\check{\mathscr{B}}}^!)$ has enough injectives in view of 2.9.3(iii), for every bounded from below complex $\mathscr{K}^\bullet$ of $\mathbf{Mod}_{\mathbb{Q}}(\overline{\check{\mathscr{B}}}^!)$, there exists a bounded from below complex of injective objects $\mathscr{L}^\bullet$ of $\mathbf{Mod}_{\mathbb{Q}}(\overline{\check{\mathscr{B}}}^!)$ and a quasi-isomorphism $u\colon \mathscr{K}^\bullet \to \mathscr{L}^\bullet$. The latter induces a morphism of $\mathbf{D}^+(\mathbf{Mod}_{\mathbb{Q}}(\mathscr{O}_{\mathfrak{X}'}))$,

$$\widehat{\pi}_{\mathbb{Q}*}(\mathscr{K}^\bullet) \to \mathrm{R}\widehat{\pi}_{\mathbb{Q}*}(\mathscr{K}^\bullet). \tag{6.2.18.2}$$

By ([59] 05TG), this morphism only depends on $\mathscr{K}^\bullet$, but not on u, and it functorially depends on it.

6.2.19 For every complex of $\mathscr{O}_{\mathfrak{X}'}$-modules $\mathscr{F}'^{\bullet}$ such that $\mathscr{F}'^{i} = 0$ for $i < 0$, and every integer $q \geq 0$, we have a canonical base change morphism with respect to the diagram (6.2.15.1),

$$\widehat{\sigma}^*(\mathrm{R}^q \mathfrak{g}_*(\mathscr{F}'^{\bullet})) \to \mathrm{R}^q \breve{\mathfrak{g}}_*(\widehat{\pi}^*(\mathscr{F}'^{\bullet})), \tag{6.2.19.1}$$

where $\widehat{\pi}^*(\mathscr{F}'^{\bullet})$ denotes the pullback of $\mathscr{F}'^{\bullet}$ defined term by term (not derived), and $\mathrm{R}^q \mathfrak{g}_*(-)$ and $\mathrm{R}^q \breve{\mathfrak{g}}_*(-)$ denote the hypercohomology modules. Indeed, this amounts to giving a morphism

$$\mathrm{R}^q \mathfrak{g}_*(\mathscr{F}'^{\bullet}) \to \widehat{\sigma}_*(\mathrm{R}^q \breve{\mathfrak{g}}_*(\widehat{\pi}^*(\mathscr{F}'^{\bullet}))), \tag{6.2.19.2}$$

and we take the composed morphism

$$\begin{aligned} &\mathrm{R}^q \mathfrak{g}_*(\mathscr{F}'^{\bullet}) \to \mathrm{R}^q \mathfrak{g}_*(\widehat{\pi}_*(\widehat{\pi}^*(\mathscr{F}'^{\bullet}))) \to \\ &\quad \mathrm{R}^q(\mathfrak{g} \circ \widehat{\pi})_*(\widehat{\pi}^*(\mathscr{F}'^{\bullet})) \xrightarrow{\sim} \mathrm{R}^q(\widehat{\sigma} \circ \breve{\mathfrak{g}})_*(\widehat{\pi}^*(\mathscr{F}'^{\bullet})) \to \widehat{\sigma}_*(\mathrm{R}^q \breve{\mathfrak{g}}_*(\widehat{\pi}^*(\mathscr{F}'^{\bullet}))), \end{aligned} \tag{6.2.19.3}$$

where the first arrow is induced by the adjunction morphism $\mathscr{F}'^{\bullet} \to \widehat{\pi}_*(\widehat{\pi}^*(\mathscr{F}'^{\bullet}))$, the second by (6.2.18.1), the third by the isomorphism underlying (6.2.15.1), and the fourth is the edge-homomorphism of the second hypercohomology spectral sequence of the functor $\widehat{\sigma}_*$ with respect to the complex $\mathrm{R}\breve{\mathfrak{g}}_*(\widehat{\pi}^*(\mathscr{F}'^{\bullet}))$ ([29] 0.11.4.3) and ([14] 1.4.5 and 1.4.6).

The base change morphism (6.2.19.1) is functorial in the complex $\mathscr{F}'^{\bullet}$. Note however that the functor $\widehat{\pi}^*$ does not a priori transform quasi-isomorphisms into quasi-isomorphisms. The morphism (6.2.19.1) cannot therefore be extended to the $\mathscr{F}'^{\bullet}$ object of $\mathbf{D}^+(\mathbf{Mod}(\mathscr{O}_{\mathfrak{X}'}))$.

Lemma 6.2.20 *For every $\mathscr{O}_{\mathfrak{X}'}$-module $\mathscr{F}'$ of $X'_{s,\mathrm{zar}}$ and every integer $q \geq 0$, the base change morphism*

$$\widehat{\sigma}^*(\mathrm{R}^q \mathfrak{g}_*(\mathscr{F}'))) \to \mathrm{R}^q \breve{\mathfrak{g}}_*(\widehat{\pi}^*(\mathscr{F}')) \tag{6.2.20.1}$$

defined in (6.2.19.1) *coincides with the base change morphism defined in* (6.2.15.2).

Indeed, the second hypercohomology spectral sequence of the functor $\widehat{\sigma}_*$ with respect to the complex $\mathrm{R}\breve{\mathfrak{g}}_*(\widehat{\pi}^*(\mathscr{F}'))$ coincides with the Cartan–Leray spectral sequence of the composed functor $\widehat{\sigma}_* \circ \breve{\mathfrak{g}}_*$ ([5] V 5.4). The fourth morphism of (6.2.19.3) thus coincides with that induced by this Cartan–Leray spectral sequence. Similarly, the second morphism (6.2.19.3) coincides with the edge-homomorphism of the Cartan–Leray spectral sequence of the composed functor $\mathfrak{g}_* \circ \widehat{\pi}_*$ by virtue of ([29] (0.11.3.4.2)). The proposition follows.

Lemma 6.2.21 *Let $h\colon T' \to T$ be a morphism of topos, $K^{\bullet}$ (resp. $L^{\bullet}$) a complex of abelian groups of T (resp. T') such that $K^i = 0$ (resp. $L^i = 0$) for $i < 0$, $u\colon h^{-1}(K^{\bullet}) \to L^{\bullet}$ a morphism of complexes of abelian groups, q an integer ≥ 0. We denote by $v\colon K^{\bullet} \to h_*(L^{\bullet})$ the adjoint morphism of u, defined term by term, by $u^q : h^{-1}(\mathrm{H}^q(K^{\bullet})) \to \mathrm{H}^q(L^{\bullet})$ the morphism induced by u and by $v^q : \mathrm{H}^q(K^{\bullet}) \to h_*(\mathrm{H}^q(L^{\bullet}))$ its adjoint. Then, v^q is the composition*

$$\mathrm{H}^q(K^{\bullet}) \xrightarrow{\mathrm{H}^q(v)} \mathrm{H}^q(h_*(L^{\bullet})) \longrightarrow \mathrm{R}^q h_*(L^{\bullet}) \longrightarrow h_*(\mathrm{H}^q(L^{\bullet})) , \tag{6.2.21.1}$$

where the second arrow is induced by the canonical morphism $h_*(L^\bullet) \to \mathrm{R}h_*(L^\bullet)$ *and the third arrow is the edge-homomorphism of the second hypercohomology spectral sequence of the functor* h_* *with respect to the complex* $L^\bullet$.

Indeed, the second and the third arrows of (6.2.21.1) being functorial in $L^\bullet$, we may reduce to the case where $L = h^{-1}(K^\bullet)$ and $u = \mathrm{id}$. Since v is the adjunction morphism $K^\bullet \to h_*(h^{-1}(K^\bullet))$ and u^q is the identity of $h^{-1}(\mathrm{H}^q(K^\bullet))$, it suffices to show that the composition

$$\mathrm{H}^q(K^\bullet) \xrightarrow{\mathrm{H}^q(v)} \mathrm{H}^q(h_*(h^{-1}(K^\bullet))) \longrightarrow \mathrm{R}^q h_*(h^{-1}(K^\bullet)) \longrightarrow h_*(h^{-1}(\mathrm{H}^q(K^\bullet))), \tag{6.2.21.2}$$

is the adjunction morphism of $\mathrm{H}^q(K^\bullet)$. By functoriality, considering the canonical morphism $\tau_{\leq q}(K^\bullet) \to K^\bullet$, we reduce to the case where $K^\bullet$ is concentrated in degrees $[0, q]$, then to the case where $K^\bullet = M[-q]$ for an abelian group M of T. The first morphism of (6.2.21.2) identifies then with the adjunction morphism $M \to h_*(h^{-1}(M))$, and the second and third morphisms identify with the identity morphism of $h_*(h^{-1}(M))$. The proposition follows.

6.2.22 Consider the commutative diagram of morphisms of ringed topos (6.1.23.6)

$$\begin{array}{ccc} (\widetilde{G}_s^{\mathbb{N}^\circ}, \widehat{\pi}^{-1}(\mathscr{O}_{\mathfrak{X}'})) & \xrightarrow{\widehat{\pi}} & (X'_{s,\mathrm{zar}}, \mathscr{O}_{\mathfrak{X}'}) \\ \Big\downarrow{\scriptstyle \breve{\mathfrak{g}}} & & \Big\downarrow{\scriptstyle \mathfrak{g}} \\ (\widetilde{E}_s^{\mathbb{N}^\circ}, \widehat{\sigma}^{-1}(\mathscr{O}_{\mathfrak{X}})) & \xrightarrow{\widehat{\sigma}} & (X_{s,\mathrm{zar}}, \mathscr{O}_{\mathfrak{X}}). \end{array} \tag{6.2.22.1}$$

We use the notation $\widehat{\pi}^{-1}$ and $\widehat{\sigma}^{-1}$ to denote pullbacks in the sense of abelian sheaves and we keep the notation $\widehat{\pi}^*$ and $\widehat{\sigma}^*$ for the pullbacks in the sense of the modules by the morphisms of ringed topos represented by the horizontal arrows of the diagram (6.2.15.1). Copying the construction of 6.2.19, for every complex of $\mathscr{O}_{\mathfrak{X}'}$-modules $\mathscr{F}'^\bullet$ such that $\mathscr{F}'^i = 0$ for $i < 0$, and all integers $q \geq 0$, we have a canonical base change morphism with respect to the diagram (6.2.22.1)

$$u^q \colon \widehat{\sigma}^{-1}(\mathrm{R}^q \mathfrak{g}_*(\mathscr{F}'^\bullet)) \to \mathrm{R}^q \breve{\mathfrak{g}}_*(\widehat{\pi}^{-1}(\mathscr{F}'^\bullet)). \tag{6.2.22.2}$$

We give an alternative construction. Let $K^{\bullet\bullet}$ (resp. $L^{\bullet\bullet}$) be an injective Cartan–Eilenberg resolution of $\mathscr{F}'^\bullet$ (resp. $\widehat{\pi}^{-1}(\mathscr{F}'^\bullet)$) such that $K^{ij} = 0$ (resp. $L^{ij} = 0$) for $i < 0$ ([29] 0.11.4.2). The base change morphism with respect to the diagram of (6.2.22.1) applied term by term induces a morphism of bicomplexes of $\widehat{\sigma}^{-1}(\mathscr{O}_{\mathfrak{X}})$-modules

$$\widehat{\sigma}^{-1}(\mathfrak{g}_*(K^{\bullet\bullet})) \to \breve{\mathfrak{g}}_*(\widehat{\pi}^{-1}(K^{\bullet\bullet})). \tag{6.2.22.3}$$

It is clear that $\widehat{\pi}^{-1}(K^{\bullet\bullet})$ is a Cartan–Eilenberg resolution of $\widehat{\pi}^{-1}(\mathscr{F}'^\bullet)$. By ([29] 0.11.4.2), there exists a morphism of bicomplexes of $\widehat{\pi}^{-1}(\mathscr{O}_{\mathfrak{X}'})$-modules $\widehat{\pi}^{-1}(K^{\bullet\bullet}) \to L^{\bullet\bullet}$. Taking the image by the functor $\breve{\mathfrak{g}}_*$ and composing with (6.2.22.3), we obtain a morphism of bicomplexes of $\widehat{\sigma}^{-1}(\mathscr{O}_{\mathfrak{X}})$-modules

$$\widehat{\sigma}^{-1}(\mathfrak{g}_*(K^{\bullet\bullet})) \to \breve{\mathfrak{g}}_*(L^{\bullet\bullet}). \tag{6.2.22.4}$$

This induces a morphism between the cohomology $\widehat{\sigma}^{-1}(\mathscr{O}_{\mathfrak{X}})$-modules of the associated simple complexes

$$v^q\colon \widehat{\sigma}^{-1}(\mathrm{R}^q\mathfrak{g}_*(\mathscr{F}'^\bullet)) \to \mathrm{R}^q\breve{\mathfrak{g}}_*(\widehat{\pi}^{-1}(\mathscr{F}'^\bullet)), \tag{6.2.22.5}$$

that is none other than the morphism u^q (6.2.22.2). Indeed, the adjoint of the morphism (6.2.22.3) is the composition

$$\mathfrak{g}_*(K^{\bullet\bullet}) \to \mathfrak{g}_*(\widehat{\pi}_*(\widehat{\pi}^{-1}(K^{\bullet\bullet}))) \to \mathfrak{g}_*(\widehat{\pi}_*(L^{\bullet\bullet})) \xrightarrow{\sim} \widehat{\sigma}_*(\breve{\mathfrak{g}}_*(L^{\bullet\bullet})), \tag{6.2.22.6}$$

where the first arrow is induced by the adjunction morphism $\mathrm{id} \to \widehat{\pi}_*\widehat{\pi}^{-1}$, the second arrow by the morphism $\widehat{\pi}^{-1}(K^{\bullet\bullet}) \to L^{\bullet\bullet}$, and the third arrow by the isomorphism underlying the diagram (6.2.22.1). By 6.2.21, the adjoint of the morphism v^q is the composition

$$\mathrm{R}^q\mathfrak{g}_*(\mathscr{F}'^\bullet) \to \mathrm{R}^q(\widehat{\sigma}\circ\breve{\mathfrak{g}})_*(\widehat{\pi}^{-1}(\mathscr{F}'^\bullet)) \to \widehat{\sigma}_*(\mathrm{R}^q\breve{\mathfrak{g}}_*(\widehat{\pi}^{-1}(\mathscr{F}'^\bullet))), \tag{6.2.22.7}$$

where the first arrow is induced by the composed morphism (6.2.22.6) and the second arrow is the edge-homomorphism of the second hypercohomology spectral sequence of the functor $\widehat{\sigma}_*$ with respect to the complex $\mathrm{R}\breve{\mathfrak{g}}_*(\widehat{\pi}^*(\mathscr{F}'^\bullet))$. Moreover, the diagram

$$\begin{array}{ccccc}
\mathrm{R}^q\mathfrak{g}_*(\mathscr{F}'^\bullet) & \xrightarrow{a} & \mathrm{R}^q\mathfrak{g}_*(\widehat{\pi}_*(\widehat{\pi}^{-1}(\mathscr{F}'^\bullet))) & \xrightarrow{b} & \mathrm{R}^q\mathfrak{g}_*(\mathrm{R}\widehat{\pi}_*(\widehat{\pi}^{-1}(\mathscr{F}'^\bullet))) \\
\downarrow & & \downarrow & & \downarrow \\
\mathrm{R}^q\mathfrak{g}_*(\mathrm{Tot}(K^{\bullet\bullet})) & \xrightarrow{a'} & \mathrm{R}^q\mathfrak{g}_*(\widehat{\pi}_*(\widehat{\pi}^{-1}(\mathrm{Tot}(K^{\bullet\bullet})))) & \xrightarrow{b'} & \mathrm{R}^q\mathfrak{g}_*(\widehat{\pi}_*(\mathrm{Tot}(L^{\bullet\bullet}))) \\
\downarrow & & & & \downarrow \\
\mathrm{H}^q(\mathfrak{g}_*(\mathrm{Tot}(K^{\bullet\bullet}))) & & \xrightarrow{a''} & & \mathrm{H}^q(\mathrm{Tot}(\mathfrak{g}_*(\widehat{\pi}_*(L^{\bullet\bullet})))),
\end{array}$$

where $\mathrm{Tot}(-)$ denotes the associated total complex, a and a' are induced by the adjunction morphism $\mathrm{id} \to \widehat{\pi}_*\widehat{\pi}^{-1}$, b is induced by (6.2.18.1), b' is induced by the morphism $\widehat{\pi}^{-1}(K^{\bullet\bullet}) \to L^{\bullet\bullet}$, a'' is induced by the composition of the first two arrows of (6.2.22.6) and the vertical arrows are the canonical morphisms, is commutative. The equality $u^q = v^q$ follows in view of the definition (6.2.19.3) of the adjoint of the morphism u^q (6.2.22.2).

6.2.23 For every complex $F'^\bullet$ of $\mathbf{Mod}_{\mathbb{Q}}(\mathscr{O}_{\mathfrak{X}'})$ such that $F'^i = 0$ for $i < 0$, and every integer q, we have a canonical base change morphism relative to the diagram (6.2.15.1),

$$\widehat{\sigma}^*_{\mathbb{Q}}(\mathrm{R}^q\mathfrak{g}_{\mathbb{Q}*}(F'^\bullet)) \to \mathrm{R}^q\breve{\mathfrak{g}}_{\mathbb{Q}*}(\widehat{\pi}^*_{\mathbb{Q}}(F'^\bullet)), \tag{6.2.23.1}$$

where $\widehat{\pi}^*_{\mathbb{Q}}(F'^\bullet)$ denotes the pullback of $F'^\bullet$ defined term by term (6.1.24.13), and $\mathrm{R}^q\mathfrak{g}_{\mathbb{Q}*}(-)$ and $\mathrm{R}^q\breve{\mathfrak{g}}_{\mathbb{Q}*}(-)$ denote the hypercohomology modules. Indeed, this amounts to giving a morphism

$$\mathrm{R}^q\mathfrak{g}_{\mathbb{Q}*}(F'^\bullet) \to \widehat{\sigma}_{\mathbb{Q}*}(\mathrm{R}^q\breve{\mathfrak{g}}_{\mathbb{Q}*}(\widehat{\pi}^*_{\mathbb{Q}}(F'^\bullet))), \tag{6.2.23.2}$$

and we take the composed morphism

$$\begin{aligned}&\mathrm{R}^q\mathfrak{g}_{\mathrm{Q}*}(F'^\bullet) \to \mathrm{R}^q\mathfrak{g}_{\mathrm{Q}*}(\widehat{\pi}_{\mathrm{Q}*}(\widehat{\pi}^*_{\mathrm{Q}}(F'^\bullet))) \to \\ &\quad \mathrm{R}^q(\mathfrak{g}\circ\widehat{\pi})_{\mathrm{Q}*}(\widehat{\pi}^*_{\mathrm{Q}}(F'^\bullet)) \xrightarrow{\sim} \mathrm{R}^q(\widehat{\sigma}\circ\breve{\mathfrak{g}})_{\mathrm{Q}*}(\widehat{\pi}^*_{\mathrm{Q}}(F'^\bullet)) \to \widehat{\sigma}_{\mathrm{Q}*}(\mathrm{R}^q\breve{\mathfrak{g}}_{\mathrm{Q}*}(\widehat{\pi}^*_{\mathrm{Q}}(F'^\bullet))),\end{aligned} \tag{6.2.23.3}$$

where the first arrow is induced by the adjunction morphism $F'^\bullet \to \widehat{\pi}_{\mathrm{Q}*}(\widehat{\pi}^*_{\mathrm{Q}}(F'^\bullet))$, the second by (6.2.18.2), the third by the isomorphism underlying (6.2.15.1), and the fourth is the edge-homomorphism of the second hypercohomology spectral sequence of the functor $\widehat{\sigma}_{\mathrm{Q}*}$ with respect to the complex $\mathrm{R}\breve{\mathfrak{g}}_{\mathrm{Q}*}(\widehat{\pi}^*_{\mathrm{Q}}(F'^\bullet))$ ([29] 0.11.4.3) and ([14] 1.4.5 and 1.4.6). Note that the category $\mathbf{Mod}_{\mathrm{Q}}(\breve{\overline{\mathscr{B}}})$ has enough injectives by 2.9.3(iii). In particular, every complex of $\mathbf{Mod}_{\mathrm{Q}}(\breve{\overline{\mathscr{B}}})$ admits an injective Cartan–Eilenberg resolution.

The base change morphism (6.2.23.1) is functorial in the complex $F'^\bullet$. Note however that the functor $\widehat{\pi}^*_{\mathrm{Q}}$ does not a priori transform quasi-isomorphisms into quasi-isomorphisms. The morphism (6.2.19.1) cannot therefore be extended to the $F'^\bullet$ object of $\mathbf{D}^+(\mathbf{Mod}_{\mathrm{Q}}(\mathscr{O}_{\mathfrak{X}'}))$.

Lemma 6.2.24

(i) *For every complex of $\mathscr{O}_{\mathfrak{X}'}$-modules $\mathscr{F}'^\bullet$ such that $\mathscr{F}'^i = 0$ for $i < 0$, and every integer q, the base change morphism* (6.2.23.1)

$$\widehat{\sigma}^*_{\mathrm{Q}}(\mathrm{R}^q\mathfrak{g}_{\mathrm{Q}*}(\mathscr{F}'^\bullet_{\mathrm{Q}})) \to \mathrm{R}^q\breve{\mathfrak{g}}_{\mathrm{Q}*}(\widehat{\pi}^*_{\mathrm{Q}}(\mathscr{F}'^\bullet_{\mathrm{Q}})) \tag{6.2.24.1}$$

is the canonical image of the base change morphism (6.2.19.1)

$$\widehat{\sigma}^*(\mathrm{R}^q\mathfrak{g}_*(\mathscr{F}'^\bullet)) \to \mathrm{R}^q\breve{\mathfrak{g}}_*(\widehat{\pi}^*(\mathscr{F}'^\bullet)). \tag{6.2.24.2}$$

(ii) *For every object F' of $\mathbf{Mod}_{\mathrm{Q}}(\mathscr{O}_{\mathfrak{X}'})$ and every integer $q \geq 0$, the base change morphism*

$$\widehat{\sigma}^*_{\mathrm{Q}}(\mathrm{R}^q\mathfrak{g}_{\mathrm{Q}*}(F'))) \to \mathrm{R}^q\breve{\mathfrak{g}}_{\mathrm{Q}*}(\widehat{\pi}^*_{\mathrm{Q}}(F')) \tag{6.2.24.3}$$

defined in (6.2.23.1) *coincides with the base change morphism defined in* (6.2.15.3).

(i) Indeed, each of the morphisms appearing in (6.2.23.3) is the canonical image of the corresponding morphism appearing in (6.2.19.3). This is obvious for the first and third arrows in view of (2.9.6.4), and follows from 2.9.3(iii) for the second arrow. The case of the fourth arrow follows from the fact that the canonical image of an injective Cartan–Eilenberg resolution of a complex of $\breve{\overline{\mathscr{B}}}$-modules $\mathscr{K}^\bullet$ is an injective Cartan–Eilenberg resolution of the complex $\mathscr{K}^\bullet_{\mathrm{Q}}$.

(ii) This follows from (i) and 6.2.20.

6.2.25 Copying the construction of 6.2.19, for every complex $F'^\bullet$ of $\mathbf{Mod}_{\mathrm{Q}}(\mathscr{O}_{\mathfrak{X}'})$ such that $F'^i = 0$ for $i < 0$, and every q, we have a canonical base change morphism with respect to the diagram (6.2.22.1)

$$\widehat{\sigma}^{-1}_{\mathrm{Q}}(\mathrm{R}^q\mathfrak{g}_{\mathrm{Q}*}(F'^\bullet)) \to \mathrm{R}^q\breve{\mathfrak{g}}_{\mathrm{Q}*}(\widehat{\pi}^{-1}_{\mathrm{Q}}(F'^\bullet)). \tag{6.2.25.1}$$

As in 6.2.22, we can give another construction using injective Cartan–Eilenberg resolutions of $F'^{\bullet}$ and $\widehat{\pi}_{\mathbb{Q}}^{-1}(F'^{\bullet})$, which exist since the categories $\mathbf{Mod}_{\mathbb{Q}}(\mathscr{O}_{\mathfrak{X}'})$ and $\mathbf{Mod}_{\mathbb{Q}}(\widehat{\pi}^{-1}(\mathscr{O}_{\mathfrak{X}'}))$ have enough injectives by 2.9.3(iii).

Theorem 6.2.26 *Assume that* $\mathfrak{g}\colon X' \to X$ *is proper. Let* $\mathscr{F}'^{\bullet}$ *be a complex of coherent* $\mathscr{O}_{\mathfrak{X}'}[\frac{1}{p}]$*-modules such that* $\mathscr{F}'^{i} = 0$ *for* $i < 0$, q *an integer* ≥ 0. *Then the* $\mathscr{O}_{\mathfrak{X}}[\frac{1}{p}]$*-module* $\mathrm{R}^q\mathfrak{g}_*(\mathscr{F}'^{\bullet})$ *is coherent and the base change morphism* (6.2.23.1)

$$\widehat{\sigma}_{\mathbb{Q}}^*(\mathrm{R}^q\mathfrak{g}_*(\mathscr{F}'^{\bullet})) \to \mathrm{R}^q\breve{\mathfrak{g}}_{\mathbb{Q}*}(\widehat{\pi}_{\mathbb{Q}}^*(\mathscr{F}'^{\bullet})), \tag{6.2.26.1}$$

where $\widehat{\sigma}_{\mathbb{Q}}^*$ *and* $\widehat{\pi}_{\mathbb{Q}}^*$ *denote the exact functors* (6.1.24.4) *and* (6.1.24.15) *(see* 6.2.5*), is an isomorphism.*

Indeed, the first assertion follows from ([1] 2.10.24 and 2.11.5) in view of the second hypercohomology spectral sequence of the functor $\mathfrak{g}_*$ with respect to the complex $\mathscr{F}'^{\bullet}$,

$$\mathrm{E}_2^{i,j} = \mathrm{R}^i\mathfrak{g}_*(\mathscr{H}^j(\mathscr{F}'^{\bullet})) \Rightarrow \mathrm{R}^{i+j}\mathfrak{g}_*(\mathscr{F}'^{\bullet}). \tag{6.2.26.2}$$

In view of 6.2.5, the latter induces a spectral sequence

$$\mathrm{E}_2^{i,j} = \widehat{\sigma}_{\mathbb{Q}}^*(\mathrm{R}^i\mathfrak{g}_*(\mathscr{H}^j(\mathscr{F}'^{\bullet}))) \Rightarrow \widehat{\sigma}_{\mathbb{Q}}^*(\mathrm{R}^{i+j}\mathfrak{g}_*(\mathscr{F}'^{\bullet})). \tag{6.2.26.3}$$

Moreover, again taking into account 6.2.5, the second hypercohomology spectral sequence of the functor $\breve{\mathfrak{g}}_{\mathbb{Q}*}$ with respect to the complex $\widehat{\pi}_{\mathbb{Q}}^*(\mathscr{F}'^{\bullet})$ is written as

$$\mathrm{E}_2^{i,j} = \mathrm{R}^i\breve{\mathfrak{g}}_{\mathbb{Q}*}(\widehat{\pi}_{\mathbb{Q}}^*(\mathscr{H}^j(\mathscr{F}'^{\bullet}))) \Rightarrow \mathrm{R}^{i+j}\breve{\mathfrak{g}}_{\mathbb{Q}*}(\widehat{\pi}_{\mathbb{Q}}^*(\mathscr{F}'^{\bullet})). \tag{6.2.26.4}$$

Recall that the category $\mathbf{Mod}_{\mathbb{Q}}(\overline{\breve{\mathscr{B}}}^{!})$ has enough injectives by 2.9.3(iii). It suffices to show that the base change morphisms (6.2.15.4)

$$\widehat{\sigma}_{\mathbb{Q}}^*(\mathrm{R}^i\mathfrak{g}_*(\mathscr{H}^j(\mathscr{F}'^{\bullet}))) \to \mathrm{R}^i\breve{\mathfrak{g}}_{\mathbb{Q}*}(\widehat{\pi}_{\mathbb{Q}}^*(\mathscr{H}^j(\mathscr{F}'^{\bullet}))) \tag{6.2.26.5}$$

and the base change morphisms (6.2.26.1)

$$\widehat{\sigma}_{\mathbb{Q}}^*(\mathrm{R}^{i+j}\mathfrak{g}_*(\mathscr{F}'^{\bullet})) \to \mathrm{R}^{i+j}\breve{\mathfrak{g}}_{\mathbb{Q}*}(\widehat{\pi}_{\mathbb{Q}}^*(\mathscr{F}'^{\bullet})) \tag{6.2.26.6}$$

define a morphism from the spectral sequence (6.2.26.3) to the spectral sequence (6.2.26.4). Indeed, since the morphisms (6.2.26.5) are isomorphisms by virtue of 6.2.17, we deduce from this that the morphism (6.2.26.1) is an isomorphism for every $q \geq 0$.

For every object F' of $\mathbf{Mod}_{\mathbb{Q}}(\mathscr{O}_{\mathfrak{X}'})$, we have a canonical base change morphism with respect to the diagram (6.2.22.1),

$$\widehat{\sigma}_{\mathbb{Q}}^{-1}(\mathrm{R}^q\mathfrak{g}_{\mathbb{Q}*}(F')) \to \mathrm{R}^q\breve{\mathfrak{g}}_{\mathbb{Q}*}(\widehat{\pi}_{\mathbb{Q}}^{-1}(F')). \tag{6.2.26.7}$$

Moreover, for every complex $F'^{\bullet}$ of $\mathbf{Mod}_{\mathbb{Q}}(\mathscr{O}_{\mathfrak{X}'})$ such that $F'^{i} = 0$ for $i < 0$, and every integer $q \geq 0$, we have a canonical base change morphism (6.2.25.1) with respect to the diagram (6.2.22.1),

$$\widehat{\sigma}_{\mathbb{Q}}^{-1}(\mathrm{R}^q \mathfrak{g}_{\mathbb{Q}*}(F'^{\bullet})) \to \mathrm{R}^q \breve{\mathfrak{g}}_{\mathbb{Q}*}(\widehat{\pi}_{\mathbb{Q}}^{-1}(F'^{\bullet})). \tag{6.2.26.8}$$

Moreover, we have two spectral sequences

$$\mathrm{E}_2^{i,j} = \widehat{\sigma}_{\mathbb{Q}}^{-1}(\mathrm{R}^i \mathfrak{g}_{\mathbb{Q}*}(\mathscr{H}^j(F'^{\bullet}))) \Rightarrow \widehat{\sigma}_{\mathbb{Q}}^{-1}(\mathrm{R}^{i+j} \mathfrak{g}_{\mathbb{Q}*}(F'^{\bullet})), \tag{6.2.26.9}$$

$$\mathrm{E}_2^{i,j} = \mathrm{R}^i \breve{\mathfrak{g}}_{\mathbb{Q}*}(\widehat{\pi}_{\mathbb{Q}}^{-1}(\mathscr{H}^j(F'^{\bullet}))) \Rightarrow \mathrm{R}^{i+j} \breve{\mathfrak{g}}_{\mathbb{Q}*}(\widehat{\pi}_{\mathbb{Q}}^{-1}(F'^{\bullet})). \tag{6.2.26.10}$$

Let us first show that the base change morphisms (6.2.26.7)

$$\widehat{\sigma}_{\mathbb{Q}}^{-1}(\mathrm{R}^i \mathfrak{g}_{\mathbb{Q}*}(\mathscr{H}^j(F'^{\bullet}))) \to \mathrm{R}^i \breve{\mathfrak{g}}_{\mathbb{Q}*}(\widehat{\pi}_{\mathbb{Q}}^{-1}(\mathscr{H}^j(F'^{\bullet}))) \tag{6.2.26.11}$$

and the morphisms (6.2.26.8) define a morphism from the spectral sequence (6.2.26.9) to the spectral sequence (6.2.26.10).

Let $K^{\bullet\bullet}$ (resp. $L^{\bullet\bullet}$) be an injective Cartan–Eilenberg resolution of $F'^{\bullet}$ (resp. $\widehat{\pi}_{\mathbb{Q}}^{-1}(F'^{\bullet})$) in the category $\mathbf{Mod}_{\mathbb{Q}}(\mathscr{O}_{\mathfrak{X}'})$ (resp. $\mathbf{Mod}_{\mathbb{Q}}(\widehat{\pi}_{\mathbb{Q}}^{-1}(\mathscr{O}_{\mathfrak{X}'}))$) such that $K^{ij} = 0$ (resp. $L^{ij} = 0$) for $i < 0$ ([29] 0.11.4.2). The spectral sequence (6.2.26.9) (resp. (6.2.26.10)) is by definition the second spectral sequence of the bicomplex $\widehat{\sigma}_{\mathbb{Q}}^{-1}(\mathfrak{g}_{\mathbb{Q}*}(K'^{\bullet\bullet}))$ (resp. $\breve{\mathfrak{g}}_{\mathbb{Q}*}(L^{\bullet\bullet})$) ([29] 0.11.3.2). The base change morphism (6.2.26.7) defines a morphism of bicomplexes

$$\widehat{\sigma}_{\mathbb{Q}}^{-1}(\mathfrak{g}_{\mathbb{Q}*}(K^{\bullet\bullet})) \to \breve{\mathfrak{g}}_{\mathbb{Q}*}(\widehat{\pi}_{\mathbb{Q}}^{-1}(K^{\bullet\bullet})). \tag{6.2.26.12}$$

Moreover, $\widehat{\pi}_{\mathbb{Q}}^{-1}(K^{\bullet\bullet})$ being a Cartan–Eilenberg resolution of $\widehat{\pi}_{\mathbb{Q}}^{-1}(F'^{\bullet})$, there exists a morphism of bicomplexes $\widehat{\pi}_{\mathbb{Q}}^{-1}(K^{\bullet\bullet}) \to L^{\bullet\bullet}$ ([29] 0.11.4.2), compatible with the morphisms from $\widehat{\pi}_{\mathbb{Q}}^{-1}(F'^{\bullet})$ to $\widehat{\pi}_{\mathbb{Q}}^{-1}(K^{\bullet 0})$ and $L^{\bullet 0}$. We deduce from this a morphism of bicomplexes

$$\widehat{\sigma}_{\mathbb{Q}}^{-1}(\mathfrak{g}_{\mathbb{Q}*}(K^{\bullet\bullet})) \to \breve{\mathfrak{g}}_{\mathbb{Q}*}(L^{\bullet\bullet}), \tag{6.2.26.13}$$

and consequently a morphism between the associated second spectral sequences (6.2.26.9) and (6.2.26.10). In view of 6.2.25, the latter is defined on the initial terms by (6.2.26.11) and on the abutments by (6.2.26.8).

We take for $F'^{\bullet}$ the complex $\mathscr{F}'^{\bullet}$ (4.3.16.2). By 6.2.5, the canonical morphism of complexes $\widehat{\pi}_{\mathbb{Q}}^{-1}(\mathscr{F}'^{\bullet}) \to \widehat{\pi}_{\mathbb{Q}}^{*}(\mathscr{F}'^{\bullet})$ induces a morphism from the spectral sequence (6.2.26.10) to the spectral sequence (6.2.26.4). Composing with the morphism defined above, we deduce a morphism from the spectral sequence (6.2.26.9) to the spectral sequence (6.2.26.4). By $\breve{\overline{\mathscr{B}}}_{\mathbb{Q}}$-linearization (6.2.5), we deduce the desired morphism from the spectral sequence (6.2.26.3) to the spectral sequence (6.2.26.4). The proposition follows.

6.3 Functoriality of Higgs–Tate algebras in Faltings Topos

In this section r denotes a rational number ≥ 0 and n an integer ≥ 0.

6.3.1 We denote by $\mathbf{P}$ the full subcategory of $\mathbf{\acute{E}t}_{/X}$ made up of *affine* schemes U such that one of the following conditions is satisfied:

(i) the scheme U_s is empty; or
(ii) the morphism $(U, \mathscr{M}_X|U) \to (S, \mathscr{M}_S)$ induced by f (6.1.3) admits an adequate chart (3.1.15) ([3] III.4.4).

We denote by **Q** the full subcategory of **P** made up of affine schemes U such that one of the following conditions is satisfied:

(iii) the scheme U_s is empty; or
(iv) there exists a fine and saturated chart $M \to \Gamma(U, \mathscr{M}_X)$ for $(U, \mathscr{M}_X|U)$ inducing an isomorphism

$$M \xrightarrow{\sim} \Gamma(U, \mathscr{M}_X)/\Gamma(U, \mathscr{O}_X^\times). \tag{6.3.1.1}$$

This chart is a priori independent of the adequate chart required in (ii).

We denote by

$$\pi_{\mathbf{P}} \colon E_{\mathbf{P}} \to \mathbf{P}, \tag{6.3.1.2}$$

$$\pi_{\mathbf{Q}} \colon E_{\mathbf{Q}} \to \mathbf{Q}, \tag{6.3.1.3}$$

the fibered sites deduced from π (6.1.8.1) by base change by the canonical injection functors from **P** and **Q** to $\mathbf{\acute{E}t}_{/X}$. Note that **P** and **Q** are topologically generating subcategories of $\mathbf{\acute{E}t}_{/X}$. Hence, $E_{\mathbf{P}}$ and $E_{\mathbf{Q}}$ are topologically generating subcategories of E. Moreover, **P** being stable by fiber products, $\pi_{\mathbf{P}}$ is a covanishing fibered site (2.10.1); but this is generally not the case for $E_{\mathbf{Q}}$ (see 4.4.1, 4.4.2 and 4.4.3).

We define in the same way the full subcategories $\mathbf{P}'$ and $\mathbf{Q}'$ of $\mathbf{\acute{E}t}_{/X'}$ with respect to the morphism f' (6.1.3).

6.3.2 Since X is Noetherian and therefore quasi-separated, every object of **P** is coherent over X. Let us consider the covanishing fibered site (6.1.8.2)

$$\pi'_{\mathrm{coh}} \colon E'_{\mathrm{coh}} \to \mathbf{\acute{E}t}_{\mathrm{coh}/X'}. \tag{6.3.2.1}$$

We denote by

$$\varphi \colon \mathbf{P} \to \mathbf{\acute{E}t}_{\mathrm{coh}/X'} \tag{6.3.2.2}$$

the pullback functor by $g \colon X' \to X$, and by

$$\Phi \colon E_{\mathbf{P}} \to E'_{\mathrm{coh}} \tag{6.3.2.3}$$

the functor induced by Θ^+ (6.1.10.1). The diagram of functors

$$\begin{array}{ccc} E_{\mathbf{P}} & \xrightarrow{\pi_{\mathbf{P}}} & \mathbf{P} \\ {\scriptstyle\Phi}\downarrow & & \downarrow{\scriptstyle\varphi} \\ E'_{\mathrm{coh}} & \xrightarrow{\pi'_{\mathrm{coh}}} & \mathbf{\acute{E}t}_{\mathrm{coh}/X'} \end{array} \tag{6.3.2.4}$$

is strictly commutative, i.e., we have $\varphi \circ \pi_{\mathbf{P}} = \pi'_{\mathrm{coh}} \circ \Phi$. We denote by

$$\pi'_{\mathbf{P}} \colon E'_{\mathbf{P}} \to \mathbf{P} \tag{6.3.2.5}$$

the fibered category deduced from π'_{coh} by base change by φ ([24] VI § 3), and by

$$\phi\colon E_{\mathbf{P}} \to E'_{\mathbf{P}} \qquad (6.3.2.6)$$

the **P**-functor induced by Φ (6.3.2.3). This functor ϕ is clearly Cartesian. The setting considered above is therefore a special case of the one considered in 2.10.7. Moreover, the assumptions of 2.10.11 are satisfied. The functor Φ is thus continuous for the covanishing topologies on $E_{\mathbf{P}}$ and E'_{coh} (2.10.7). It induces a morphism of topos which is canonically identified with $\Theta\colon \widetilde{E}' \to \widetilde{E}$ (6.1.10.2).

6.3.3 We denote by $\mathbb{I}$ the category of morphisms $U' \to U$ above the morphism $g\colon X' \to X$ such that the morphism $U' \to X'$ (resp. $U \to X$) is étale of finite presentation (resp. an object of **P**). This identifies with the category bearing the same name defined in 2.10.13 with respect to the functor φ (6.3.2.2). Consider the functors

$$\mathsf{s}\colon \mathbb{I} \to \mathbf{\acute{E}t}_{\mathrm{coh}/X'}, \quad (U' \to U) \mapsto U', \qquad (6.3.3.1)$$

$$\mathsf{b}\colon \mathbb{I} \to \mathbf{P}, \quad (U' \to U) \mapsto U. \qquad (6.3.3.2)$$

For any $U' \in \mathrm{Ob}(\mathbf{\acute{E}t}_{\mathrm{coh}/X'})$, we denote by $I_{\varphi}^{U'}$ the fiber category of s above U', i.e., the category of objects $(\mu\colon U' \to U)$ of $\mathbb{I}$; such an object of $I_{\varphi}^{U'}$ will also be denoted by (U, μ).

We denote by $\mathbb{J}$ the full subcategory of $\mathbb{I}$ made up of the morphisms $U' \to U$ such that U' is an object of $\mathbf{Q}'$. For any object $(U' \to U)$ of $\mathbb{I}$, we denote by $\mathbb{J}_{/(U' \to U)}$ the category of morphisms of $\mathbb{I}$ from an object of $\mathbb{J}$ to $(U' \to U)$.

We denote by J' the essential image of the category $\mathbb{J}$ by the functor s. For any object U' of $\mathbf{\acute{E}t}_{\mathrm{coh}/X'}$, we denote by $J'_{/U'}$ the category of X'-morphisms from an object of J' to U'.

Lemma 6.3.4

(i) *The functor* b *is essentially surjective, and the subcategory* J' *of* $\mathbf{\acute{E}t}_{\mathrm{coh}/X'}$ *is topologically generating.*

(ii) *Assume that the scheme* X *is separated. Then, for every object* $(U' \to U)$ *of* $\mathbb{I}$*, every object* V' *of* J' *and every* X'*-morphism* $u'\colon V' \to U'$*, there exist an object* $(V' \to W)$ *of* $\mathbb{J}$ *and a morphism* $(v', v)\colon (V' \to W) \to (U' \to U)$ *of* $\mathbb{I}$

$$\begin{array}{ccc} V' & \xrightarrow{v'} & U' \\ \downarrow & & \downarrow \\ W & \xrightarrow{v} & U. \end{array} \qquad (6.3.4.1)$$

(i) This follows from ([3] II.5.17).

(ii) Let $(U' \to U)$ be an object of $\mathbb{I}$, V' an object of J', $u'\colon V' \to U'$ an X'-morphism. Then there exists an object $V' \to V$ of $\mathbb{J}$. Consider the commutative diagram

$$\begin{array}{ccccc} V' & \xrightarrow{u'} & U' & \longrightarrow & X' \\ \downarrow & & \downarrow & & \downarrow \\ V \times_X U & \longrightarrow & U & & \\ \downarrow & & & \searrow & \downarrow \\ V & & \longrightarrow & & X. \end{array} \tag{6.3.4.2}$$

Since X is separated and U and V are affine, $V \times_X U$ is affine. Moreover, an adequate chart for $(V, \mathscr{M}_X|V)$ induces an adequate chart for $(V \times_X U, \mathscr{M}_X|V \times_X U)$. Hence, $V \times_X U$ is an object of $\mathbf{P}$, and the morphism $V' \to V \times_X U$ is an object of $\mathbb{J}$. The proposition follows.

Proposition 6.3.5 *Let $F = \{U \in \mathbf{P}^\circ \mapsto F_U\}$ be a v-presheaf on $E_{\mathbf{P}}$ (2.10.3). Then,*

(i) *For every $U' \in \mathrm{Ob}(\mathbf{\acute{E}t}_{\mathrm{coh}/X'})$, the sheaves $\overline{\mu}^*(F_U)$, for $(U, \mu\colon U' \to U) \in \mathrm{Ob}((I_\varphi^{U'})^\circ)$, where $\overline{\mu}\colon \overline{U'}^{\triangleright} \to \overline{U}^\circ$ abusively denotes the morphism induced by μ, naturally form a direct system of $\overline{U'}^{\triangleright}_{\mathrm{f\acute{e}t}}$. We define*

$$F'_{U'} = \varinjlim_{(U,\mu)\in(I_\varphi^{U'})^\circ} \overline{\mu}^*(F_U). \tag{6.3.5.1}$$

(ii) *The collection $F' = \{U' \in \mathbf{\acute{E}t}^\circ_{\mathrm{coh}/X'} \mapsto F'_{U'}\}$ naturally forms a v-presheaf on E'. For every morphism $\lambda\colon U'_1 \to U'_2$ of $\mathbf{\acute{E}t}_{\mathrm{coh}/X'}$ and every object $\mu_2\colon U'_2 \to U$ of $\mathbb{I}$ (6.3.3), setting $\mu_1 = \mu_2 \circ \lambda\colon U'_1 \to U$, the diagram*

$$\begin{array}{ccc} \overrightarrow{\lambda}^{\triangleright *}(\overline{\mu}_2^*(F_U)) & \longrightarrow & \overline{\mu}_1^*(F_U) \\ \downarrow & & \downarrow \\ \overrightarrow{\lambda}^{\triangleright *}(F'_{U'_2}) & \longrightarrow & F'_{U'_1}, \end{array} \tag{6.3.5.2}$$

where the vertical arrows are the canonical morphisms (6.3.5.1), the upper horizontal arrow is the canonical isomorphism and the lower horizontal arrow is the adjoint morphism of the morphism $F'_{U'_2} \to \overrightarrow{\lambda}^{\triangleright}_*(F'_{U'_1})$ defining the presheaf structure on F', is commutative.

(iii) *We have a canonical functorial isomorphism*

$$\Theta^*(F^a) \xrightarrow{\sim} F'^a, \tag{6.3.5.3}$$

where the exponent ^{a} denotes the associated sheaves.

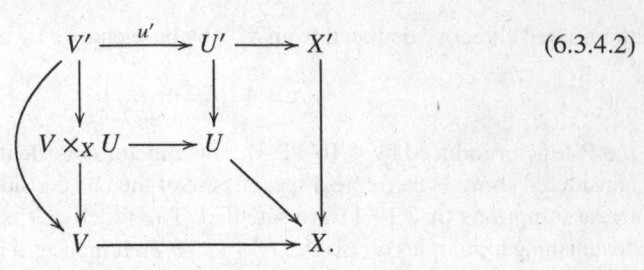

(iv) *With the notation of* 6.1.12, *for every object* $\mu\colon U' \to U$ *of* $\mathbb{I}$ (6.3.3), *the diagram*

$$\begin{array}{ccccc} \overline{\mu}^*(F_U) & \xrightarrow{\overline{\mu}^*(a)} & \overline{\mu}^*(\beta_{U*}(J_U^*(F^a))) & \xrightarrow{c} & \beta'_{U'*}(\Theta_\mu^*(J_U^*(F^a))) \\ \downarrow{b} & & & & \downarrow{d} \\ F'_{U'} & \xrightarrow{a'} & \beta'_{U'*}(J'^*_{U'}(F'^a)) & \xrightarrow{e} & \beta'_{U'*}(J'^*_{U'}(\Theta^*(F^a))), \end{array} \qquad (6.3.5.4)$$

where $a\colon F_U \to \beta_{U*}(J_U^*(F^a))$, a' *and* b *are the canonical morphisms,* c *is the base change morphism with respect to* (6.1.12.5), d *is the isomorphism underlying* (6.1.12.4) *and* e *is the isomorphism induced by* (6.3.5.3), *is commutative.*

Propositions (i), (ii) and (iii) are special cases of 2.10.12. We prove proposition (iv). By localization (6.1.12.4), we may reduce to the case where $\mu = g$. It then follows from 2.10.10(ii), (2.10.11.3) and (2.10.11.6) that the diagram

$$\begin{array}{ccccc} F_X & \xrightarrow{a} & \beta_*(F^a) & \xrightarrow{\beta_*(\mathrm{ad})} & \beta_*(\Theta_*(\Theta^*(F^a))) \\ \downarrow{b'} & & & & \downarrow{c'} \\ \gamma_*(F'_{X'}) & \xrightarrow{\gamma_*(a')} & \gamma_*(\beta'_*(F'^a)) & \xrightarrow{e'} & \gamma_*(\beta'_*(\Theta^*(F^a))), \end{array} \qquad (6.3.5.5)$$

where $\gamma\colon \overline{X}'^{\rhd} \to \overline{X}^{\circ}$ is the morphism induced by g, a and $a'\colon F'_{X'} \to \beta'_*(F'^a)$ are the canonical morphisms, b' is the adjoint of the canonical morphism $\gamma^*(F_X) \to F'_{X'}$ (6.3.5.1), $\mathrm{ad}\colon F^a \to \Theta_*(\Theta^*(F^a))$ is the adjunction morphism, e' is induced by the isomorphism (6.3.5.3) and c' is the isomorphism underlying (6.1.12.5), is commutative. The proposition follows by adjunction.

6.3.6 It immediately follows from 6.3.5 that the canonical homomorphism $\Theta^{-1}(\overline{\mathscr{B}}) \to \overline{\mathscr{B}}'$ (6.1.11.3) induces for every object $(\mu\colon U' \to U)$ of $\mathbb{I}$ (6.3.3) a homomorphism of $\overline{U}'^{\rhd}_{\text{fét}}$,

$$\overline{\mu}^*(\overline{\mathscr{B}}_U) \to \overline{\mathscr{B}}'_{U'}, \qquad (6.3.6.1)$$

where $\overline{\mu}\colon \overline{U}'^{\rhd} \to \overline{U}^{\circ}$ abusively denotes the morphism induced by μ. With the terminology of 2.10.15, these homomorphisms form an $\mathbb{I}$-system of Φ-compatible morphisms from $\overline{\mathscr{B}}$ to $\overline{\mathscr{B}}'$ (6.3.2.3).

6.3.7 Let Y be an object of $\mathbf{P}$ such that Y_s is non-empty, $((P,\gamma),(\mathbb{N},\iota),\vartheta)$ an adequate chart for the morphism $f|Y\colon (Y, \mathscr{M}_X|Y) \to (S, \mathscr{M}_S)$ induced by f, $\overline{y}$ a geometric point of $\overline{Y}^{\circ}$. The scheme $\overline{Y}$ being locally irreducible by ([2] 4.2.7 and [3] III.3.3), it is the sum of the schemes induced on its irreducible components. We denote by $\overline{Y}^{\star}$ the irreducible component of $\overline{Y}$ containing $\overline{y}$. Similarly, $\overline{Y}^{\circ}$ is the sum of the schemes induced on its irreducible components and $\overline{Y}^{\star\circ} = \overline{Y}^{\star} \times_X X^{\circ}$ is the irreducible component of $\overline{Y}^{\circ}$ containing $\overline{y}$. We use the notation introduced in 4.4.4, in particular, the discrete representation $\overline{R}_Y^{\overline{y}}$ of $\pi_1(\overline{Y}^{\star\circ}, \overline{y})$ defined in (4.3.7.2), the canonical exact sequence (4.4.4.6)

$$0 \to \widehat{\overline{R}}_Y^{\overline{y}} \to \mathscr{F}_Y^{\overline{y}} \to \widetilde{\xi}^{-1}\widetilde{\Omega}^1_{X/S}(Y) \otimes_{\mathscr{O}_X(Y)} \widehat{\overline{R}}_Y^{\overline{y}} \to 0, \tag{6.3.7.1}$$

and the $\widehat{\overline{R}}_Y^{\overline{y}}$-algebra (4.4.4.7)

$$\mathscr{C}_Y^{\overline{y}} = \varinjlim_{m\geq 0} \mathrm{S}^m_{\widehat{\overline{R}}_Y^{\overline{y}}}(\mathscr{F}_Y^{\overline{y}}). \tag{6.3.7.2}$$

Observe that these representations depend on the adequate chart $((P, \gamma), (\mathbb{N}, \iota), \vartheta)$. However, they do not depend on it if Y is an object of $\mathbf{Q}$ by 3.2.12.

We denote by $\mathscr{F}_Y^{\overline{y},(r)}$ the extension of $\widehat{\overline{R}}_Y^{\overline{y}}$-modules deduced from $\mathscr{F}_Y^{\overline{y}}$ (6.3.7.1) by pullback by the morphism of multiplication by p^r on $\widetilde{\xi}^{-1}\widetilde{\Omega}^1_{X/S}(Y) \otimes_{\mathscr{O}_X(Y)} \widehat{\overline{R}}_Y^{\overline{y}}$, so that we have an exact sequence of $\widehat{\overline{R}}_Y^{\overline{y}}$-modules

$$0 \to \widehat{\overline{R}}_Y^{\overline{y}} \to \mathscr{F}_Y^{\overline{y},(r)} \to \widetilde{\xi}^{-1}\widetilde{\Omega}^1_{X/S}(Y) \otimes_{\mathscr{O}_X(Y)} \widehat{\overline{R}}_Y^{\overline{y}} \to 0. \tag{6.3.7.3}$$

We denote by $\mathscr{C}_Y^{\overline{y},(r)}$ the $\widehat{\overline{R}}_Y^{\overline{y}}$-algebra (3.2.17.3)

$$\mathscr{C}_Y^{\overline{y},(r)} = \varinjlim_{m\geq 0} \mathrm{S}^m_{\widehat{\overline{R}}_Y^{\overline{y}}}(\mathscr{F}_Y^{\overline{y},(r)}). \tag{6.3.7.4}$$

We consider the analogous notation for f', that we endow with an exponent $'$.

6.3.8 For any object U of $\mathbf{\acute{E}t}_{/X}$, we set $\overline{\mathscr{B}}_U = \overline{\mathscr{B}} \circ \iota_U$ (4.3.6.3) and $\overline{\mathscr{B}}_{U,n} = \overline{\mathscr{B}}_U/p^n\overline{\mathscr{B}}_U$ (4.3.10.2). As in 4.4.7, with any object Y of $\mathbf{P}$ such that Y_s is not empty and any adequate chart $((P, \gamma), (\mathbb{N}, \iota), \vartheta)$ for the morphism $f|Y\colon (Y, \mathscr{M}_X|Y) \to (S, \mathscr{M}_S)$ induced by f, we associate a canonical exact sequence of $\overline{\mathscr{B}}_{Y,n}$-modules of $\overline{Y}^{\circ}_{\text{fét}}$ (4.4.7.1)

$$0 \to \overline{\mathscr{B}}_{Y,n} \to \mathscr{F}^{(r)}_{Y,n} \to \widetilde{\xi}^{-1}\widetilde{\Omega}^1_{X/S}(Y) \otimes_{\mathscr{O}_X(Y)} \overline{\mathscr{B}}_{Y,n} \to 0, \tag{6.3.8.1}$$

and a $\overline{\mathscr{B}}_{Y,n}$-algebra of $\overline{Y}^{\circ}_{\text{fét}}$ (4.4.7.2)

$$\mathscr{C}^{(r)}_{Y,n} = \varinjlim_{m\geq 0} \mathrm{S}^m_{\overline{\mathscr{B}}_{Y,n}}(\mathscr{F}^{(r)}_{Y,n}). \tag{6.3.8.2}$$

These objects are defined by reduction modulo p^n of those defined in 6.3.7 (see 4.4.7 for more details). Let $\mathscr{F}_{Y,n} = \mathscr{F}^{(0)}_{Y,n}$ and $\mathscr{C}_{Y,n} = \mathscr{C}^{(0)}_{Y,n}$ (see 4.4.5). Note that if Y is an object of $\mathbf{Q}$, $\mathscr{F}^{(r)}_{Y,n}$ and $\mathscr{C}^{(r)}_{Y,n}$ do not depend on the adequate chart.

For any object Y of $\mathbf{P}$ such that Y_s is empty, we set $\mathscr{C}^{(r)}_{Y,n} = \mathscr{F}^{(r)}_{Y,n} = 0$ (4.4.9).

As in 4.4.10, in view of 2.10.5, we consider the associated sheaves of $\widetilde{E}$

$$\mathscr{F}_n^{(r)} = \{Y \in \mathbf{Q}^\circ \mapsto \mathscr{F}_{Y,n}^{(r)}\}^a, \tag{6.3.8.3}$$

$$\mathscr{C}_n^{(r)} = \{Y \in \mathbf{Q}^\circ \mapsto \mathscr{C}_{Y,n}^{(r)}\}^a. \tag{6.3.8.4}$$

By 4.4.11, we have a canonical locally split exact sequence of $\overline{\mathscr{B}}_n$-modules

$$0 \to \overline{\mathscr{B}}_n \to \mathscr{F}_n^{(r)} \to \sigma_n^*(\widetilde{\xi}^{-1}\widetilde{\Omega}^1_{\overline{X}_n/\overline{S}_n}) \to 0, \tag{6.3.8.5}$$

and a canonical isomorphism of $\overline{\mathscr{B}}_n$-algebras

$$\mathscr{C}_n^{(r)} \xrightarrow{\sim} \varinjlim_{m\geq 0} \mathrm{S}^m_{\overline{\mathscr{B}}_n}(\mathscr{F}_n^{(r)}), \tag{6.3.8.6}$$

where the transition morphisms of the direct system are induced by (6.3.8.5). We set $\mathscr{F}_n = \mathscr{F}_n^{(0)}$ and $\mathscr{C}_n = \mathscr{C}_n^{(0)}$ (see 4.4.10).

For all rational numbers $r \geq r' \geq 0$, we have a canonical $\overline{\mathscr{B}}_n$-linear morphism (4.4.10.6)

$$\mathrm{a}_n^{r,r'} : \mathscr{F}_n^{(r)} \to \mathscr{F}_n^{(r')}, \tag{6.3.8.7}$$

and a canonical homomorphism of $\overline{\mathscr{B}}_n$-algebras (4.4.10.7)

$$\alpha_n^{r,r'} : \mathscr{C}_n^{(r)} \to \mathscr{C}_n^{(r')}. \tag{6.3.8.8}$$

For all rational numbers $r \geq r' \geq r'' \geq 0$, we have

$$\mathrm{a}_n^{r,r''} = \mathrm{a}_n^{r',r''} \circ \mathrm{a}_n^{r,r'} \quad \text{and} \quad \alpha_n^{r,r''} = \alpha_n^{r',r''} \circ \alpha_n^{r,r'}. \tag{6.3.8.9}$$

We refer to 4.4.11 for the properties of these morphisms.

We consider the analogous notation for f', that we endow with a $'$.

6.3.9 Let $(\mu\colon Y' \to Y)$ be an object of $\mathbb{J}$ (6.3.3) such that Y'_s is not empty, $((P,\gamma),(\mathbb{N},\iota),\vartheta)$ an adequate chart for the morphism $f|Y\colon (Y,\mathscr{M}_X|Y) \to (S,\mathscr{M}_S)$ induced by f, $\overline{y}'$ a geometric point of $\overline{Y}'^{\rhd}$ (6.1.3.3). We abusively denote by $\overline{\mu}\colon \overline{Y}'^{\rhd} \to \overline{Y}^\circ$ the morphism induced by μ and we set $\overline{y} = \overline{\mu}(\overline{y}')$. We use the notation of 6.3.7 for $(Y,\overline{y})$ and $(Y',\overline{y}')$. By 5.2.12, we have a $\pi_1(\overline{Y}'^{\star\rhd},\overline{y}')$-equivariant $\widehat{\overline{R}}{}^{\overline{y}}_Y$-linear morphism (5.2.12.3)

$$\mathscr{F}_Y^{\overline{y}} \to \mathscr{F}'^{\overline{y}'}_{Y'} \tag{6.3.9.1}$$

that fits into a commutative diagram

$$\begin{array}{ccccccccc}
0 & \longrightarrow & \widehat{\overline{R}}{}^{\overline{y}}_Y & \longrightarrow & \mathscr{F}_Y^{\overline{y}} & \longrightarrow & \widetilde{\xi}^{-1}\widetilde{\Omega}^1_{X/S}(Y) \otimes_{\mathscr{O}_X(Y)} \widehat{\overline{R}}{}^{\overline{y}}_Y & \longrightarrow & 0 \\
 & & \downarrow & & \downarrow & & \downarrow & & \\
0 & \longrightarrow & \widehat{\overline{R}}{}'^{\overline{y}'}_{Y'} & \longrightarrow & \mathscr{F}'^{\overline{y}'}_{Y'} & \longrightarrow & \widetilde{\xi}^{-1}\widetilde{\Omega}^1_{X'/S}(Y') \otimes_{\mathscr{O}_{X'}(Y')} \widehat{\overline{R}}{}'^{\overline{y}'}_{Y'} & \longrightarrow & 0
\end{array} \tag{6.3.9.2}$$

We deduce from this a $\pi_1(\overline{Y}'^{\star\triangleright}, \overline{y}')$-equivariant morphism of $\widehat{\overline{R}}_Y^{\overline{y}}$-algebras (5.2.12.4)

$$\mathscr{C}_Y^{\overline{y}} \to \mathscr{C}'^{\overline{y}'}_{Y'} \tag{6.3.9.3}$$

extending (6.3.9.1).

The morphism (6.3.9.1) induces a $\pi_1(\overline{Y}'^{\star\triangleright}, \overline{y}')$-equivariant $\widehat{\overline{R}}'^{\overline{y}'}_{Y'}$-linear morphism (6.3.7.3)

$$\mathscr{F}^{\overline{y},(r)} \otimes_{\widehat{\overline{R}}_Y^{\overline{y}}} \widehat{\overline{R}}'^{\overline{y}'}_{Y'} \to \mathscr{F}'^{\overline{y}',(r)} \tag{6.3.9.4}$$

and hence a $\pi_1(\overline{Y}'^{\star\triangleright}, \overline{y}')$-equivariant homomorphism of $\widehat{\overline{R}}_Y^{\overline{y}}$-algebras (6.3.7.4)

$$\mathscr{C}_Y^{\overline{y},(r)} \to \mathscr{C}'^{\overline{y}',(r)}_{Y'}. \tag{6.3.9.5}$$

We denote by $\Pi(\overline{Y}^{\star\circ})$ and $\Pi(\overline{Y}'^{\star\triangleright})$ the fundamental groupoids of $\overline{Y}^{\star\circ}$ and $\overline{Y}'^{\star\triangleright}$ and by

$$\overline{\mu}_*\colon \Pi(\overline{Y}'^{\star\triangleright}) \to \Pi(\overline{Y}^{\star\circ}) \tag{6.3.9.6}$$

the functor induced by the inverse image functor $\mathbf{\acute{E}t}_{\mathrm{f}/\overline{Y}^{\star\circ}} \to \mathbf{\acute{E}t}_{\mathrm{f}/\overline{Y}'^{\star\triangleright}}$ by $\overline{\mu}$. We denote by

$$F_{Y,n}\colon \Pi(\overline{Y}^{\star\circ}) \to \mathbf{Sets} \quad \text{and} \quad F'_{Y',n}\colon \Pi(\overline{Y}'^{\star\triangleright}) \to \mathbf{Sets} \tag{6.3.9.7}$$

the functors associated by ([3] VI.9.11) with the objects $\mathscr{F}_{Y,n}|\overline{Y}^{\star\circ}$ of $\overline{Y}^{\star\circ}_{\text{fét}}$ and $\mathscr{F}'_{Y',n}|\overline{Y}'^{\star\triangleright}$ of $\overline{Y}'^{\star\triangleright}_{\text{fét}}$, respectively (6.3.8). The morphism (6.3.9.1) clearly induces a morphism of functors

$$F_{Y,n} \circ \overline{\mu}_* \to F'_{Y',n}. \tag{6.3.9.8}$$

We deduce from this by ([3] VI.9.11) a $\overline{\mu}^*(\overline{\mathscr{B}}_{Y,n})$-linear morphism of $\overline{Y}'^{\triangleright}_{\text{fét}}$,

$$\nu_{\mu,n}\colon \overline{\mu}^*(\mathscr{F}_{Y,n}) \to \mathscr{F}'_{Y',n}. \tag{6.3.9.9}$$

It follows from (6.3.9.2) that the diagram

$$\begin{array}{ccccccccc}
0 & \longrightarrow & \overline{\mu}^*(\overline{\mathscr{B}}_{Y,n}) & \longrightarrow & \overline{\mu}^*(\mathscr{F}_{Y,n}) & \longrightarrow & \xi^{-1}\widetilde{\Omega}^1_{X/S}(Y) \otimes_{\mathscr{O}_X(Y)} \overline{\mu}^*(\overline{\mathscr{B}}_{Y,n}) & \longrightarrow & 0 \\
 & & \downarrow & & \downarrow{\scriptstyle \nu_{\mu,n}} & & \downarrow & & \\
0 & \longrightarrow & \overline{\mathscr{B}}'_{Y',n} & \longrightarrow & \mathscr{F}'_{Y',n} & \longrightarrow & \xi^{-1}\widetilde{\Omega}^1_{X'/S}(Y') \otimes_{\mathscr{O}_{X'}(Y')} \overline{\mathscr{B}}'_{Y',n} & \longrightarrow & 0
\end{array} \tag{6.3.9.10}$$

is commutative.

The morphism (6.3.9.9) induces a $\overline{\mu}^*(\overline{\mathscr{B}}_{Y,n})$-linear morphism of $\overline{Y}'^{\triangleright}_{\text{fét}}$,

$$\nu^{(r)}_{\mu,n}\colon \overline{\mu}^*(\mathscr{F}^{(r)}_{Y,n}) \to \mathscr{F}'^{(r)}_{Y',n} \tag{6.3.9.11}$$

that fits into a commutative diagram

$$\begin{array}{ccccccccc} 0 & \longrightarrow & \overline{\mu}^*(\overline{\mathscr{B}}_{Y,n}) & \longrightarrow & \overline{\mu}^*(\mathscr{F}^{(r)}_{Y,n}) & \longrightarrow & \widetilde{\xi}^{-1}\widetilde{\Omega}^1_{X/S}(Y)\otimes_{\mathscr{O}_X(Y)}\overline{\mu}^*(\overline{\mathscr{B}}_{Y,n}) & \longrightarrow & 0 \\ & & \downarrow & & \downarrow{\scriptstyle v^{(r)}_{\mu,n}} & & \downarrow & & \\ 0 & \longrightarrow & \overline{\mathscr{B}}'_{Y',n} & \longrightarrow & \mathscr{F}'^{(r)}_{Y',n} & \longrightarrow & \widetilde{\xi}^{-1}\widetilde{\Omega}^1_{X'/S}(Y')\otimes_{\mathscr{O}_{X'}(Y')}\overline{\mathscr{B}}'_{Y',n} & \longrightarrow & 0 \end{array} \tag{6.3.9.12}$$

We deduce from this a morphism of $\overline{\mu}^*(\overline{\mathscr{B}}_{Y,n})$-algebras of $\overline{Y}'^{\triangleright}_{\text{fét}}$

$$w^{(r)}_{\mu,n}\colon \overline{\mu}^*(\mathscr{C}^{(r)}_{Y,n}) \to \mathscr{C}'^{(r)}_{Y',n}. \tag{6.3.9.13}$$

Observe that the morphisms (6.3.9.11) and (6.3.9.13) depend on the choice of the adequate chart $((P,\gamma),(\mathbb{N},\iota),\vartheta)$. However, they do not depend on it if Y is an object of $\mathbf{Q}$ by 3.2.12.

Remark 6.3.10 Suppose that the morphism g (6.1.3.1) is étale and strict. Let $(\mu\colon Y' \to Y)$ be an object of $\mathbb{I}$ (6.3.3) such that Y' and Y are objects of $\mathbf{Q}$. It follows from 4.4.8 and 6.3.9 that the $\overline{\mu}^*(\overline{\mathscr{B}}_{Y,n})$-linear morphism of $\overline{Y}'^{\circ}_{\text{fét}}$ (6.3.9.11)

$$v^{(r)}_{\mu,n}\colon \overline{\mu}^*(\mathscr{F}^{(r)}_{Y,n}) \to \mathscr{F}'^{(r)}_{Y',n} \tag{6.3.10.1}$$

is none other than the morphism (4.4.8.1), i.e., the adjoint of the transition morphism of the presheaf $\{U \in \mathbf{Q}^\circ \mapsto \mathscr{F}^{(r)}_{U,n}\}$ on $E_{\mathbf{Q}}$ (4.4.10).

6.3.11 Suppose that the scheme X is separated and that the morphism $f\colon (X,\mathscr{M}_X) \to (S,\mathscr{M}_S)$ admits an adequate chart $((P,\gamma),(\mathbb{N},\iota),\vartheta)$ that we fix. For any object Y of $\mathbf{P}$, endowing the morphism $f|Y\colon (Y,\mathscr{M}_X|Y) \to (S,\mathscr{M}_S)$ induced by f with the adequate chart induced by $((P,\gamma),(\mathbb{N},\iota),\vartheta)$, we define a $\overline{\mathscr{B}}_{Y,n}$-module $\mathscr{F}^{(r)}_{Y,n}$ and a $\overline{\mathscr{B}}_{Y,n}$-algebra $\mathscr{C}^{(r)}_{Y,n}$ (6.3.8). By 6.3.9, the correspondences

$$\{Y \in \mathbf{P}^\circ \mapsto \mathscr{F}^{(r)}_{Y,n}\} \quad\text{and}\quad \{Y \in \mathbf{P}^\circ \mapsto \mathscr{C}^{(r)}_{Y,n}\} \tag{6.3.11.1}$$

define presheaves on $E_{\mathbf{P}}$ (6.3.1.2) of modules and algebras, respectively, relative to the ring $\{Y \in \mathbf{P}^\circ \mapsto \overline{\mathscr{B}}_{Y,n}\}$. By virtue of 2.10.6(i), we have canonical isomorphisms

$$\mathscr{F}^{(r)}_n \xrightarrow{\sim} \{Y \in \mathbf{P}^\circ \mapsto \mathscr{F}^{(r)}_{Y,n}\}^a, \tag{6.3.11.2}$$

$$\mathscr{C}^{(r)}_n \xrightarrow{\sim} \{Y \in \mathbf{P}^\circ \mapsto \mathscr{C}^{(r)}_{Y,n}\}^a. \tag{6.3.11.3}$$

For every object $(\mu\colon Y' \to Y)$ of $\mathbb{J}$ (6.3.3), denoting abusively by $\overline{\mu}\colon \overline{Y}'^{\triangleright} \to \overline{Y}^{\circ}$ the morphism induced by μ, we have a canonical $\overline{\mu}^*(\overline{\mathscr{B}}_{Y,n})$-linear morphism of $\overline{Y}'^{\triangleright}_{\text{fét}}$ (6.3.9.11)

$$v^{(r)}_{\mu,n}\colon \overline{\mu}^*(\mathscr{F}^{(r)}_{Y,n}) \to \mathscr{F}'^{(r)}_{Y',n}, \tag{6.3.11.4}$$

and a canonical morphism of $\overline{\mu}^*(\overline{\mathscr{B}}_{Y,n})$-algebras of $\overline{Y}'^{\triangleright}_{\text{fét}}$ (6.3.9.13)

$$w^{(r)}_{\mu,n}\colon \overline{\mu}^*(\mathscr{C}^{(r)}_{Y,n}) \to \mathscr{C}'^{(r)}_{Y',n}. \tag{6.3.11.5}$$

Let $(m', m) \colon (\mu_1 \colon Y_1' \to Y_1) \to (\mu_2 \colon Y_2' \to Y_2)$ be a morphism of $\mathbb{J}$.

$$\begin{array}{ccc} Y_1' & \xrightarrow{\mu_1} & Y_1 \\ {\scriptstyle m'}\downarrow & & \downarrow{\scriptstyle m} \\ Y_2' & \xrightarrow{\mu_2} & Y_2 \end{array} \tag{6.3.11.6}$$

With the notation of 5.2.3, we check immediately that the diagram

$$\begin{array}{ccccccc} (\widetilde{\mathbb{Y}}'_1, \mathscr{M}'_{\widetilde{\mathbb{Y}}'_1}) & & & \longrightarrow & & & (\widetilde{\mathbb{Y}}_1, \mathscr{M}_{\widetilde{\mathbb{Y}}_1}) \\ & \nwarrow^{i'_{Y'_1}} & & & & \nearrow^{i_{Y_1}} & \\ \downarrow & & (\widehat{\mathbb{Y}}'_1, \mathscr{M}_{\widehat{\mathbb{Y}}'_1}) & \longrightarrow & (\widehat{\mathbb{Y}}_1, \mathscr{M}_{\widehat{\mathbb{Y}}_1}) & & \downarrow \\ & & \downarrow & & \downarrow & & \\ & & (\widehat{\mathbb{Y}}'_2, \mathscr{M}_{\widehat{\mathbb{Y}}'_2}) & \longrightarrow & (\widehat{\mathbb{Y}}_2, \mathscr{M}_{\widehat{\mathbb{Y}}_2}) & & \\ & \swarrow^{i'_{Y'_2}} & & & & \searrow^{i_{Y_2}} & \\ (\widetilde{\mathbb{Y}}'_2, \mathscr{M}'_{\widetilde{\mathbb{Y}}'_2}) & & & \longrightarrow & & & (\widetilde{\mathbb{Y}}_2, \mathscr{M}_{\widetilde{\mathbb{Y}}_2}) \end{array} \tag{6.3.11.7}$$

is commutative (see (5.2.12.2)). We painfully deduce from this that the diagram

$$\begin{array}{ccccc} \overline{m}'^{\triangleright *}(\overline{\mu}_2^*(\mathscr{F}^{(r)}_{Y_2,n})) & & \xrightarrow{\overline{m}'^{\triangleright *}(v^{(r)}_{\mu_2,n})} & & \overline{m}'^{\triangleright *}(\mathscr{F}'^{(r)}_{Y'_2,n}) \\ {\scriptstyle c}\downarrow & & & & \downarrow{\scriptstyle a'} \\ \overline{\mu}_1^*(\overline{m}^{\circ *}(\mathscr{F}^{(r)}_{Y_2,n})) & \xrightarrow{\overline{\mu}_1^*(a)} & \overline{\mu}_1^*(\mathscr{F}^{(r)}_{Y_1,n}) & \xrightarrow{v^{(r)}_{\mu_1,n}} & \mathscr{F}'^{(r)}_{Y'_1,n}, \end{array} \tag{6.3.11.8}$$

where $a \colon \overline{m}^{\circ *}(\mathscr{F}^{(r)}_{Y_2,n}) \to \mathscr{F}^{(r)}_{Y_1,n}$ (resp. a') is the transition morphism of the v-presheaf $\{Y \in \mathbf{P}^\circ \mapsto \mathscr{F}^{(r)}_{Y,n}\}$ (resp. $\{Y' \in \mathbf{Q}'^\circ \mapsto \mathscr{F}'^{(r)}_{Y',n}\}$) (2.10.3) and c is the isomorphism induced by the commutative diagram (6.3.11.6), is commutative. We deduce from this that the diagram

$$\begin{array}{ccccc} \overline{m}'^{\triangleright *}(\overline{\mu}_2^*(\mathscr{C}^{(r)}_{Y_2,n})) & & \xrightarrow{\overline{m}'^{\triangleright *}(w^{(r)}_{\mu_2,n})} & & \overline{m}'^{\triangleright *}(\mathscr{C}'^{(r)}_{Y'_2,n}) \\ {\scriptstyle c}\downarrow & & & & \downarrow{\scriptstyle b'} \\ \overline{\mu}_1^*(\overline{m}^{\circ *}(\mathscr{C}^{(r)}_{Y_2,n})) & \xrightarrow{\overline{\mu}_1^*(b)} & \overline{\mu}_1^*(\mathscr{C}^{(r)}_{Y_1,n}) & \xrightarrow{w^{(r)}_{\mu_1,n}} & \mathscr{C}'^{(r)}_{Y'_1,n}, \end{array} \tag{6.3.11.9}$$

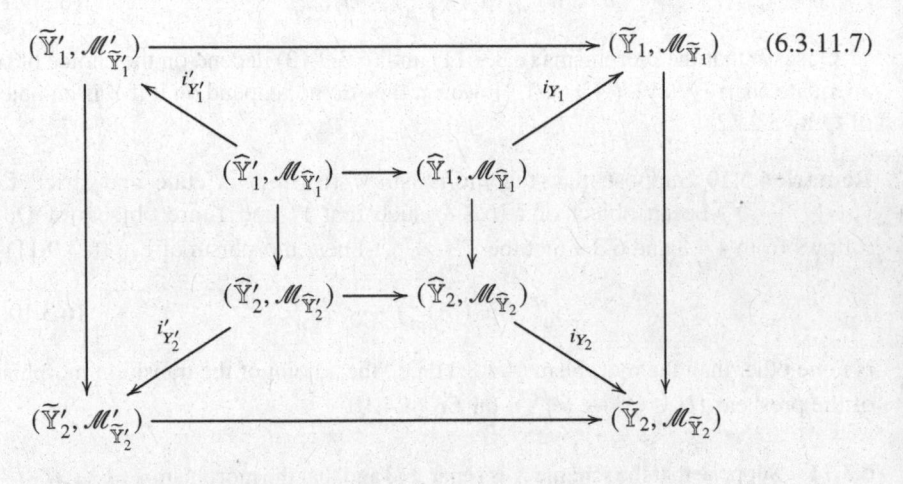

where $b\colon \overline{m}^{\circ *}(\mathscr{C}^{(r)}_{Y_2,n}) \to \mathscr{C}^{(r)}_{Y_1,n}$ (resp. b') is the transition morphism of the v-presheaf $\{Y \in \mathbf{P}^\circ \mapsto \mathscr{C}^{(r)}_{Y,n}\}$ (resp. $\{Y' \in \mathbf{Q}'^\circ \mapsto \mathscr{C}'^{(r)}_{Y',n}\}$) and c is the isomorphism induced by the commutative diagram (6.3.11.6), is commutative.

With the terminology of 2.10.15, the morphisms $v^{(r)}_{\mu,n}$ (6.3.11.4) (resp. $w^{(r)}_{\mu,n}$ (6.3.11.5)) therefore form a $\mathbb{J}$-system of compatible Φ-morphisms from $\{Y \in \mathbf{P}^\circ \mapsto \mathscr{F}^{(r)}_{Y,n}\}$ to $\{Y' \in \mathbf{Q}'^\circ \mapsto \mathscr{F}'^{(r)}_{Y',n}\}$ (resp. from $\{Y \in \mathbf{P}^\circ \mapsto \mathscr{C}^{(r)}_{Y,n}\}$ to $\{Y' \in \mathbf{Q}'^\circ \mapsto \mathscr{C}'^{(r)}_{Y',n}\}$) (6.3.2.3). Hence, by virtue of 2.10.19 and 6.3.4, they define morphisms of $\widetilde{E}'$

$$v^{(r)}_n : \Theta^{-1}(\mathscr{F}^{(r)}_n) \to \mathscr{F}'^{(r)}_n, \qquad (6.3.11.10)$$

$$w^{(r)}_n : \Theta^{-1}(\mathscr{C}^{(r)}_n) \to \mathscr{C}'^{(r)}_n, \qquad (6.3.11.11)$$

where Θ is the morphism of topos (6.1.10.2).

Remark 6.3.12 We keep the assumptions and notation of 6.3.11. It follows from 6.3.10 that if the morphism g (6.1.3.1) is étale and strict, the morphism $v^{(r)}_n$ (6.3.11.10) is none other than the isomorphism (4.4.16.10). Hence, $w^{(r)}_n$ (6.3.11.11) is also an isomorphism.

Lemma 6.3.13 *We take again the assumptions and notation of* 6.3.11. *Then,*

(i) *The morphism* $v^{(r)}_n$ (6.3.11.10) *is* $\Theta^{-1}(\overline{\mathscr{B}}_n)$*-linear. It therefore induces a* $\overline{\mathscr{B}}'_n$*-linear morphism of* $\widetilde{E}'_s$

$$v^{(r)}_n : \theta^*_n(\mathscr{F}^{(r)}_n) \to \mathscr{F}'^{(r)}_n, \qquad (6.3.13.1)$$

where θ_n *is the morphism of ringed topos* (6.1.11.4).

(ii) *We have a commutative diagram*

$$\begin{array}{ccccccccc} 0 & \longrightarrow & \overline{\mathscr{B}}'_n & \longrightarrow & \theta^*_n(\mathscr{F}^{(r)}_n) & \longrightarrow & \theta^*_n(\sigma^*_n(\widetilde{\xi}^{-1}\widetilde{\Omega}^1_{\overline{X}_n/\overline{S}_n})) & \longrightarrow & 0 \\ & & \| & & \downarrow{\scriptstyle v^{(r)}_n} & & \downarrow & & \\ 0 & \longrightarrow & \overline{\mathscr{B}}'_n & \longrightarrow & \mathscr{F}'^{(r)}_n & \longrightarrow & \sigma'^*_n(\widetilde{\xi}^{-1}\widetilde{\Omega}^1_{\overline{X}'_n/\overline{S}_n}) & \longrightarrow & 0 \end{array} \qquad (6.3.13.2)$$

where the horizontal lines are induced by the exact sequences (6.3.8.5) *and the right vertical arrow is induced by the canonical morphism*

$$\overline{g}^*_n(\widetilde{\Omega}^1_{\overline{X}_n/\overline{S}_n}) \to \widetilde{\Omega}^1_{\overline{X}'_n/\overline{S}_n} \qquad (6.3.13.3)$$

and the commutative diagram (6.1.11.5).

(iii) *The morphism* $w^{(r)}_n$ (6.3.11.11) *is a homomorphism of* $\Theta^{-1}(\overline{\mathscr{B}}_n)$*-algebras. It therefore induces a morphism of* $\overline{\mathscr{B}}'_n$*-algebras of* $\widetilde{E}'_s$

$$\omega^{(r)}_n : \theta^*_n(\mathscr{C}^{(r)}_n) \to \mathscr{C}'^{(r)}_n. \qquad (6.3.13.4)$$

(iv) *The diagram*

$$\begin{array}{ccc} \theta_n^*(\mathscr{F}_n^{(r)}) & \xrightarrow{\nu_n^{(r)}} & \mathscr{F}_n'^{(r)} \\ \downarrow & & \downarrow \\ \theta_n^*(\mathscr{C}_n^{(r)}) & \xrightarrow{\omega_n^{(r)}} & \mathscr{C}_n'^{(r)}, \end{array} \tag{6.3.13.5}$$

where the vertical arrows are the canonical morphisms, is commutative.

(i) We set $F = \{U \in \mathbf{P}^\circ \mapsto \mathscr{F}_{U,n}^{(r)}\}$ which is a v-presheaf on $E_{\mathbf{P}}$. By 2.10.17 and its proof, there exist a v-presheaf $F' = \{U' \in \mathbf{\acute{E}t}^\circ_{\mathrm{coh}/X'} \mapsto F'_{U'}\}$ on E'_{coh} and an $\mathbb{I}$-system of compatible Φ-morphisms from F to F',

$$\nu'_\mu : \overline{\mu}^*(\mathscr{F}_{U,n}^{(r)}) \to F'_{U'}, \quad \forall\, (\mu : U' \to U) \in \mathrm{Ob}(\mathbb{I}), \tag{6.3.13.6}$$

where $\overline{\mu}: \overline{U}'^{\triangleright} \to \overline{U}^\circ$ abusively denotes the morphism induced by μ, such that for all objects U' of $\mathbf{Q}'$ and $(\mu : U' \to U)$ of $\mathbb{J}$, we have $F'_{U'} = \mathscr{F}'^{(r)}_{U',n}$ and $\nu'_\mu = \nu^{(r)}_{\mu,n}$. Moreover, for all objects U' of $\mathbf{\acute{E}t}_{\mathrm{coh}/X'}$ and $(\mu : U' \to U)$ of $\mathbb{I}$, we may assume that that $F'_{U'}$ is a $\overline{\mathscr{B}}'_{U',n}$-module and that the morphism ν'_μ is $\overline{\mu}^*(\overline{\mathscr{B}}_{U,n})$-linear. The proposition follows in view of 6.3.5 (see 2.10.19).

(ii) We have a canonical isomorphism (4.3.3.4)

$$\sigma_n^*(\widetilde{\xi}^{-1}\widetilde{\Omega}^1_{\overline{X}_n/\overline{S}_n}) \xrightarrow{\sim} \sigma^{-1}(\widetilde{\xi}^{-1}\widetilde{\Omega}^1_{X/S}) \otimes_{\sigma^{-1}(\mathscr{O}_X)} \overline{\mathscr{B}}_n. \tag{6.3.13.7}$$

Then, by virtue of ([3] VI.5.34(ii), VI.8.9 and VI.5.17), $\sigma_n^*(\xi^{-1}\widetilde{\Omega}^1_{\overline{X}_n/\overline{S}_n})$ is the sheaf of $\widetilde{E}$ associated with the presheaf on E defined by the correspondence

$$\{U \in \mathbf{\acute{E}t}^\circ_{/X} \mapsto \xi^{-1}\widetilde{\Omega}^1_{X/S}(U) \otimes_{\mathscr{O}_X(U)} \overline{\mathscr{B}}_{U,n}\}. \tag{6.3.13.8}$$

By 6.3.6, for every object $(\mu : U' \to U)$ of $\mathbb{I}$ (6.3.3), denoting abusively by $\overline{\mu}: \overline{U}'^{\triangleright} \to \overline{U}^\circ$ the morphism induced by μ, we have a canonical morphism of $\overline{U}'^{\triangleright}_{\mathrm{f\acute{e}t}}$,

$$\widetilde{\xi}^{-1}\widetilde{\Omega}^1_{X/S}(U) \otimes_{\mathscr{O}_X(U)} \overline{\mu}^*(\overline{\mathscr{B}}_{U,n}) \to \widetilde{\xi}^{-1}\widetilde{\Omega}^1_{X'/S}(U') \otimes_{\mathscr{O}_{X'}(U')} \overline{\mathscr{B}}'_{U',n}. \tag{6.3.13.9}$$

These morphisms form an $\mathbb{I}$-system of Φ-compatible morphisms from

$$\{U \in \mathbf{P}^\circ \mapsto \xi^{-1}\widetilde{\Omega}^1_{X/S}(U) \otimes_{\mathscr{O}_X(U)} \overline{\mathscr{B}}_{U,n}\} \text{ to}$$
$$\{U' \in \mathbf{\acute{E}t}^\circ_{\mathrm{coh}/X'} \mapsto \xi^{-1}\widetilde{\Omega}^1_{X'/S}(U') \otimes_{\mathscr{O}_{X'}(U')} \overline{\mathscr{B}}_{U',n}\}.$$

Hence, by virtue of 2.10.6(i) and 2.10.19, they define a morphism of $\widetilde{E}'$

$$\Theta^{-1}(\sigma_n^*(\widetilde{\xi}^{-1}\widetilde{\Omega}^1_{\overline{X}_n/\overline{S}_n})) \to \sigma_n'^*(\widetilde{\xi}^{-1}\widetilde{\Omega}^1_{\overline{X}'_n/\overline{S}_n}), \tag{6.3.13.10}$$

which is none other than the morphism induced by the canonical morphism (6.3.13.3) and the commutative diagram (6.1.11.5). The proposition follows in view of (6.3.9.12).

(iii) We set $G = \{U \in \mathbf{P}^\circ \mapsto \mathscr{C}^{(r)}_{U,n}\}$, which is a v-presheaf on $E_{\mathbf{P}}$. By 2.10.17 and its proof, there exist a v-presheaf $G' = \{U' \in \mathbf{\acute{E}t}^\circ_{\mathrm{coh}/X'} \mapsto G'_{U'}\}$ on E'_{coh} and an $\mathbb{I}$-system of compatible Φ-morphisms from G to G',

$$\nu'_\mu \colon \overline{\mu}^*(\mathscr{C}^{(r)}_{U,n}) \to G'_{U'}, \quad \forall\, (\mu \colon U' \to U) \in \mathrm{Ob}(\mathbb{I}), \tag{6.3.13.11}$$

where $\overline{\mu} \colon \overline{U}'^{\triangleright} \to \overline{U}^\circ$ abusively denotes the morphism induced by μ, such that for all objects U' of $\mathbf{Q}'$ and $(\mu \colon U' \to U)$ of $\mathbb{J}$, we have $G'_{U'} = \mathscr{C}'^{(r)}_{U',n}$ and $\nu'_\mu = w^{(r)}_{\mu,n}$. Moreover, for all objects U' of $\mathbf{\acute{E}t}_{\mathrm{coh}/X'}$ and $(\mu \colon U' \to U)$ of $\mathbb{I}$, we may assume that $G'_{U'}$ is a $\overline{\mathscr{B}}'_{U',n}$-algebra and the morphism ν'_μ is a homomorphism of $\overline{\mu}^*(\overline{\mathscr{B}}_{U,n})$-algebras. The proposition follows in view of 6.3.5 (see 2.10.19).

(iv) This follows from 6.3.5 (see 2.10.19) in view of the definition of (6.3.9.13).

Lemma 6.3.14 *Suppose that the scheme X is separated and that the morphism $f \colon (X, \mathscr{M}_X) \to (S, \mathscr{M}_S)$ admits an adequate chart $((P,\gamma), (\mathbb{N}, \iota), \vartheta)$ that we fix. Let $(\mu \colon U' \to U)$ be an object of $\mathbb{I}$ (6.3.3). We abusively denote by $\overline{\mu} \colon \overline{U}'^{\triangleright} \to \overline{U}^\circ$ the morphism induced by μ and we take again the notation of 6.1.12. Then there exists a canonical morphism*

$$\overline{\mu}^*(\mathscr{F}^{(r)}_{U,n}) \to \beta'_{U'*}(J'^*_{U'}(\Theta^{-1}(\mathscr{F}^{(r)}_n))). \tag{6.3.14.1}$$

This is independent of the adequate chart $((P,\gamma), (\mathbb{N}, \iota), \vartheta)$ if U is an object of $\mathbf{Q}$.

If $(\mu \colon U' \to U)$ is an object of $\mathbb{J}$, this morphism fits into a commutative diagram

$$\begin{array}{ccc} \overline{\mu}^*(\mathscr{F}^{(r)}_{U,n}) & \xrightarrow{\nu^{(r)}_{\mu,n}} & \mathscr{F}'^{(r)}_{U',n} \\ \downarrow & & \downarrow \\ \beta'_{U'*}(J'^*_{U'}(\Theta^{-1}(\mathscr{F}^{(r)}_n))) & \xrightarrow{\beta'_{U'*}(J'^*_{U'}(\nu^{(r)}_n))} & \beta'_{U'*}(J'^*_{U'}(\mathscr{F}'^{(r)}_n)), \end{array} \tag{6.3.14.2}$$

where the right vertical arrow is the morphism (4.4.18.1) and $\nu^{(r)}_{\mu,n}$ (resp. $\nu^{(r)}_n$) is the morphism (6.3.11.4) (resp. (6.3.11.10)).

We set $F = \{Y \in \mathbf{P}^\circ \mapsto \mathscr{F}^{(r)}_{Y,n}\}$, which is a v-presheaf on $E_{\mathbf{P}}$, and for any $Y' \in \mathrm{Ob}(\mathbf{\acute{E}t}_{\mathrm{coh}/X'})$,

$$F'_{Y'} = \varinjlim_{(Y,\lambda) \in (I^{Y'}_\varphi)^\circ} \overline{\lambda}^*(\mathscr{F}^{(r)}_{Y,n}), \tag{6.3.14.3}$$

where for any object (Y, λ) of $I^{Y'}_\varphi$ (6.3.3), we abusively denoted by $\overline{\lambda} \colon \overline{Y}'^{\triangleright} \to \overline{Y}^\circ$ the morphism induced by $\lambda \colon Y' \to Y$. By 6.3.5, the collection $F' = \{Y' \in \mathbf{\acute{E}t}^\circ_{\mathrm{coh}/X'} \mapsto F'_{Y'}\}$ naturally forms a v- presheaf on E', and we have a canonical functorial isomorphism

$$\Theta^{-1}(\mathscr{F}^{(r)}_n) \xrightarrow{\sim} F'^a. \tag{6.3.14.4}$$

We then take for morphism (6.3.14.1) the composed morphism

$$\overline{\mu}^*(\mathscr{F}_{U,n}^{(r)}) \to F'_{U'} \to \beta'_{U'*}(j'^*_{U'}(F'^a)) \xrightarrow{\sim} \beta'_{U'*}(j'^*_{U'}(\Theta^{-1}(\mathscr{F}_n^{(r)}))), \qquad (6.3.14.5)$$

where the first and the second arrows are the canonical morphisms and the last arrow is induced by the isomorphism (6.3.14.4). It follows from 6.3.5(iv) and 4.4.18 that this morphism is independent of the adequate chart $((P,\gamma),(\mathbb{N},\iota),\vartheta)$ if U is an object of $\mathbf{Q}$.

The second assertion follows from the definition of the morphism $\nu_n^{(r)}$ (6.3.11.10) (see 2.10.19).

6.3.15 Suppose that the scheme X is separated and that the morphism $f: (X,\mathscr{M}_X) \to (S,\mathscr{M}_S)$ admits an adequate chart $((P,\gamma),(\mathbb{N},\iota),\vartheta)$ that we fix. Let $\overline{x}'$ be a geometric point of X', $\underline{X}'$ the strict localization of X' at $\overline{x}'$. We denote by $\underline{\widetilde{E}}'$ the Faltings topos associated with the canonical morphism $\underline{\overline{X}}'^{\rhd} \to \underline{X}'$, by

$$\Phi': \underline{\widetilde{E}}' \to \widetilde{E}' \qquad (6.3.15.1)$$

the functoriality morphism induced by the canonical morphism $\underline{X}' \to X'$, by

$$\underline{\beta}': \underline{\widetilde{E}}' \to \underline{\overline{X}}'^{\rhd}_{\text{fét}} \qquad (6.3.15.2)$$

the canonical morphism (4.3.2.13) and by

$$\underline{\theta}': \underline{\overline{X}}'^{\rhd}_{\text{fét}} \to \underline{\widetilde{E}}' \qquad (6.3.15.3)$$

the canonical section of $\underline{\beta}'$ ([3] VI.10.23). We denote by

$$\varphi'_{\overline{x}'}: \widetilde{E}' \to \underline{\overline{X}}'^{\rhd}_{\text{fét}} \qquad (6.3.15.4)$$

the composed functor $\theta'^* \circ \Phi'^*$.

We denote by $\mathfrak{V}'_{\overline{x}'}$ (resp. $\mathbf{P}'_{\overline{x}'}$, resp. $\mathbf{Q}'_{\overline{x}'}$) the category of neighborhoods of $\overline{x}'$ in the site $\mathbf{\acute{E}t}_{/X'}$ (resp. $\mathbf{P}'$, resp. $\mathbf{Q}'$) (4.4.19), and by $\mathbb{I}_{\overline{x}'}$ the category of pairs of morphisms $(\iota': \overline{x}' \to U', \mu: U' \to U)$, where ι' is an X'-morphism and μ is an object of $\mathbb{I}$ (6.3.3).

Let $(\iota': \overline{x}' \to U', \mu: U' \to U)$ be an object of $\mathbb{I}_{\overline{x}'}$. We also denote by $\iota': \underline{X}' \to U'$ the X'-morphism deduced from ι' ([5] VIII 7.3) and we abusively denote by $\overline{\iota}': \underline{\overline{X}}'^{\rhd} \to \overline{U}'^{\rhd}$ the induced morphism. We take again the notation of 6.1.12. The morphism $\iota': \underline{X}' \to U'$ induces by functoriality a morphism

$$\Phi'_{\iota'}: \underline{\widetilde{E}}' \to \widetilde{E}'_{/\sigma'^*(U')} \qquad (6.3.15.5)$$

that fits into a commutative diagram up to canonical isomorphism

$$\begin{array}{ccc} \underline{\widetilde{E}}' & \xrightarrow{\Phi'_{\iota'}} & \widetilde{E}'_{/\sigma'^*(U')} \\ {\scriptstyle \underline{\beta}'}\downarrow & & \downarrow{\scriptstyle \beta'_{U'}} \\ \underline{\overline{X}}'^{\rhd}_{\text{fét}} & \xrightarrow{\overline{\iota}'} & \overline{U}'^{\rhd}_{\text{fét}}. \end{array} \qquad (6.3.15.6)$$

Applying the composed functor $\theta'^* \circ \Phi'^*_{\iota'}$ to the morphism $\beta'^*_{U'}(\overline{\mu}^*(\mathscr{F}_A^{(r)})) \to J'^*_{U'}(\Theta^{-1}(\mathscr{F}_n^{(r)}))$ adjoint of (6.3.14.1) and taking into account the commutative diagram (6.3.15.6), we get a morphism

$$a' \colon \overline{\iota}'^*(\overline{\mu}^*(\mathscr{F}_{U,n}^{(r)})) \to \varphi'_{\overline{x}'}(\Theta^{-1}(\mathscr{F}_n^{(r)})). \tag{6.3.15.7}$$

This is independent of the adequate chart $((P,\gamma),(\mathbb{N},\iota),\vartheta)$ if U is an object of $\mathbf{Q}$ (6.3.14).

We set $\overline{x} = g(\overline{x}')$ and denote by $\underline{X}$ the strict localization of X at $\overline{x}$. The X'-morphism ι' induces an X-morphism $\iota = \iota' \circ \mu \colon \overline{x} \to U$. We take again the notation of 4.4.19 and 6.1.13. Since (U,ι) is an object of $\mathbf{P}_{\overline{x}}$, we can consider the morphism (4.4.19.7)

$$a \colon \overline{\iota}^*(\mathscr{F}_{U,n}^{(r)}) \to \varphi_{\overline{x}}(\mathscr{F}_n^{(r)}). \tag{6.3.15.8}$$

The diagram (6.3.5.4) induces a commutative diagram

$$\begin{array}{ccc} \underline{\gamma}^*(\overline{\iota}^*(\mathscr{F}_{U,n}^{(r)})) & \xrightarrow{\underline{\gamma}^*(a)} & \underline{\gamma}^*(\varphi_{\overline{x}}(\mathscr{F}_n^{(r)})) \\ \downarrow & & \downarrow \\ \overline{\iota}'^*(\overline{\mu}^*(\mathscr{F}_{U,n}^{(r)})) & \xrightarrow{a'} & \varphi'_{\overline{x}'}(\Theta^{-1}(\mathscr{F}_n^{(r)})), \end{array} \tag{6.3.15.9}$$

where the right (resp. left) vertical arrow is the isomorphism (6.1.13.11) (resp. induced by the relation $\overline{\iota} \circ \underline{\gamma} = \overline{\mu} \circ \overline{\iota}'$).

If $(\mu \colon U' \to U)$ is an object of $\mathbb{J}$ (6.3.3), the diagram (6.3.14.2) induces a commutative diagram

$$\begin{array}{ccc} \overline{\iota}'^*(\overline{\mu}^*(\mathscr{F}_{U,n}^{(r)})) & \xrightarrow{\overline{\iota}'^*(\nu_{\mu,n}^{(r)})} & \overline{\iota}'^*(\mathscr{F}'^{(r)}_{U',n}) \\ {\scriptstyle a'}\downarrow & & \downarrow{\scriptstyle b'} \\ \varphi'_{\overline{x}'}(\Theta^{-1}(\mathscr{F}_n^{(r)})) & \xrightarrow{\varphi'_{\overline{x}'}(\nu_n^{(r)})} & \varphi'_{\overline{x}'}(\mathscr{F}'^{(r)}_n), \end{array} \tag{6.3.15.10}$$

where b' is the morphism (4.4.19.7) for $\mathscr{F}'^{(r)}_n$, and $\nu_n^{(r)}$ (resp. $\nu_{\mu,n}^{(r)}$) is the morphism (6.3.11.10) (resp. (6.3.11.4)).

Proposition 6.3.16 *Suppose that the scheme X is separated and that the morphism $f \colon (X,\mathscr{M}_X) \to (S,\mathscr{M}_S)$ admits an adequate chart $((P,\gamma),(\mathbb{N},\iota),\vartheta)$. Then, the morphism*

$$\nu_n^{(r)} \colon \Theta^{-1}(\mathscr{F}_n^{(r)}) \to \mathscr{F}'^{(r)}_n \tag{6.3.16.1}$$

defined in (6.3.11.10) *does not depend on the choice of the adequate chart* $((P,\gamma),(\mathbb{N},\iota),\vartheta)$.

Let $\overline{x}'$ be a geometric point of X', $\overline{x} = g(\overline{x}')$. We take again the notation of 4.4.19 and 6.3.15. We denote by $\mathbb{K}_{\overline{x}'}$ the subcategory of $\mathbb{I}_{\overline{x}'}$ made up of the objects $(\iota' \colon \overline{x}' \to U', \mu \colon U' \to U)$ such that U (resp. U') is an object of $\mathbf{Q}$ (resp. $\mathbf{Q}'$). Note that finite inverse limits in $\mathbb{I}_{\overline{x}'}$ are representable ([5] I 2.3). Consequently, the category $\mathbb{I}_{\overline{x}'}$ is cofiltered ([5] I 2.7.1). The canonical injection functor $\mathbb{K}_{\overline{x}'} \to \mathbb{I}_{\overline{x}'}$ is initial and the category $\mathbb{K}_{\overline{x}'}$ is cofiltered by virtue of ([5] I 8.1.3(c)) and ([3] II.5.17).

The functors s (6.3.3.1) and b (6.3.3.2) induce functors

$$\mathsf{s}_{\overline{x}'} : \mathbb{K}_{\overline{x}'} \to \mathbf{Q}'_{\overline{x}'}, (\iota' : \overline{x}' \to U', \mu : U' \to U) \mapsto (U', \iota'), \tag{6.3.16.2}$$

$$\mathsf{b}_{\overline{x}'} : \mathbb{K}_{\overline{x}'} \to \mathbf{Q}_{\overline{x}}, (\iota' : \overline{x}' \to U', \mu : U' \to U) \mapsto (U, \iota = \iota' \circ \mu : \overline{x} \to U). \tag{6.3.16.3}$$

It follows from ([5] I 8.1.3(b)), ([3] II.5.17) and from the fact that the categories $\mathbf{Q}'_{\overline{x}'}$ and $\mathbf{Q}_{\overline{x}}$ are cofiltered that the functors $\mathsf{s}^{\circ}_{\overline{x}'}$ and $\mathsf{b}^{\circ}_{\overline{x}'}$ are cofinal. Hence, in view of 4.4.20(i), (6.3.15.9) and (6.3.15.10), the morphism $\varphi'_{\overline{x}'}(v_n^{(r)})$ identifies with the morphism obtained by taking the direct limit on the category $\mathbb{K}^{\circ}_{\overline{x}'}$ of the composed morphism

$$\underline{\gamma}^*(\overline{\iota}^*(\mathscr{F}^{(r)}_{U,n})) \longrightarrow \overline{\iota}'^*(\overline{\mu}^*(\mathscr{F}^{(r)}_{U,n})) \xrightarrow{\overline{\iota}'^*(v^{(r)}_{\mu,n})} \overline{\iota}'^*(\mathscr{F}'^{(r)}_{U',n}), \tag{6.3.16.4}$$

where the first arrow is the isomorphism induced by the relation $\overline{\iota} \circ \underline{\gamma} = \overline{\mu} \circ \overline{\iota}'$. Since U (resp. U') is an object of $\mathbf{Q}$ (resp. $\mathbf{Q}'$), the morphism $v^{(r)}_{\mu,n}$ does not depend on the adequate chart $((P, \gamma), (\mathbb{N}, \iota), \vartheta)$. The proposition follows since the family of functors $\varphi'_{\overline{x}'}$ when $\overline{x}'$ goes through the set of geometric points of X' is conservative ([3] VI.10.32).

6.3.17 Consider a commutative diagram of morphisms of schemes

$$\begin{array}{ccc} U' & \xrightarrow{\mu} & U \\ {\scriptstyle u'}\downarrow & & \downarrow{\scriptstyle u} \\ X' & \xrightarrow{g} & X \end{array} \tag{6.3.17.1}$$

such that u and u' are étale of finite presentation. We endow U (resp. U') with the logarithmic structure $\mathscr{M}_U$ (resp. $\mathscr{M}_{U'}$) pullback of $\mathscr{M}_X$ (resp. $\mathscr{M}_{X'}$). We associate with it the commutative diagram of morphisms of schemes

$$\begin{array}{ccccccc} (\check{\overline{U}}', \mathscr{M}_{\check{\overline{U}}'}) & & & \xrightarrow{\check{\overline{u}}'} & & & (\check{\overline{X}}', \mathscr{M}_{\check{\overline{X}}'}) \\ & \searrow & & & & \swarrow & \\ & & (\widetilde{U}', \mathscr{M}_{\widetilde{U}'}) & \xrightarrow{\widetilde{u}'} & (\widetilde{X}', \mathscr{M}_{\widetilde{X}'}) & & \\ {\scriptstyle \check{\overline{\mu}}}\downarrow & & {\scriptstyle \widetilde{\mu}}\downarrow & & \downarrow{\scriptstyle \widetilde{g}} & & \downarrow{\scriptstyle \check{\overline{g}}} \\ & & (\widetilde{U}, \mathscr{M}_{\widetilde{U}}) & \xrightarrow{\widetilde{u}} & (\widetilde{X}, \mathscr{M}_{\widetilde{X}}) & & \\ & \nearrow & & & & \nwarrow & \\ (\check{\overline{U}}, \mathscr{M}_{\check{\overline{U}}}) & & & \xrightarrow{\check{\overline{u}}} & & & (\check{\overline{X}}, \mathscr{M}_{\check{\overline{X}}}) \end{array} \tag{6.3.17.2}$$

where $\mathscr{M}_{\check{\overline{U}}}$ (resp. $\mathscr{M}_{\check{\overline{U}}'}$) denotes the pullback of the logarithmic structure $\mathscr{M}_X$ (resp. $\mathscr{M}_{X'}$) on $\check{\overline{U}}$ (resp. $\check{\overline{U}}'$) (6.1.1.2), $\widetilde{u}$ (resp. $\widetilde{u}'$) is the unique strict étale lifting of $\check{\overline{u}}$ (resp. $\check{\overline{u}}'$) and $\widetilde{\mu}$ is the unique lifting of $\check{\overline{\mu}}$ that makes the inner square commutative. We take again the notation of 6.1.12. By 4.4.16, more precisely (4.4.16.10), $j_U^*(\mathscr{F}_n^{(r)})$ identifies canonically with the analogous sheaf for the adequate morphism $f \circ u\colon (U, \mathscr{M}_U) \to (S, \mathscr{M}_S)$ endowed with the $(\widetilde{S}, \mathscr{M}_{\widetilde{S}})$-deformation $(\widetilde{U}, \mathscr{M}_{\widetilde{U}})$ of $(\check{\overline{U}}, \mathscr{M}_{\check{\overline{U}}})$. Similarly, $j_{U'}^{\prime *}(\mathscr{F}_n^{\prime(r)})$ identifies canonically with the analogous sheaf for the adequate morphism $f' \circ u'\colon (U', \mathscr{M}_{U'}) \to (S, \mathscr{M}_S)$ endowed with the $(\widetilde{S}, \mathscr{M}_{\widetilde{S}})$-deformation $(\widetilde{U}', \mathscr{M}_{\widetilde{U}'})$ of $(\check{\overline{U}}', \mathscr{M}_{\check{\overline{U}}'})$.

Suppose that the scheme X is separated and that the morphism f admits an adequate chart. We then have the morphism (6.3.11.10)

$$\nu_n^{(r)}\colon \Theta^{-1}(\mathscr{F}_n^{(r)}) \to \mathscr{F}_n^{\prime(r)}, \tag{6.3.17.3}$$

that we proved in 6.3.16 does not depend on the adequate chart for f. Suppose moreover that the scheme U is separated and that the morphism $f \circ u$ admits an adequate chart. The composed morphism

$$\Theta_\mu^{-1}(j_U^*(\mathscr{F}_n^{(r)})) \xrightarrow{\sim} j_{U'}^{\prime *}(\Theta^{-1}(\mathscr{F}_n^{(r)})) \xrightarrow{j_{U'}^{\prime *}(\nu_n^{(r)})} j_{U'}^{\prime *}(\mathscr{F}_n^{\prime(r)}), \tag{6.3.17.4}$$

where the first arrow is the isomorphism underlying the diagram (6.1.12.4), then identifies with the morphism analogous to $\nu_n^{(r)}$ defined with respect to the morphisms μ and $\widetilde{\mu}$. Indeed, choosing an adequate chart for f and taking for $f \circ u$ the induced map, the assertion easily follows from the definitions.

Proposition 6.3.18 *There exists a canonical $\Theta^{-1}(\overline{\mathscr{B}}_n)$-linear morphism of $\widetilde{E}'_s$,*

$$\nu_n^{(r)}\colon \Theta^{-1}(\mathscr{F}_n^{(r)}) \to \mathscr{F}_n^{\prime(r)}, \tag{6.3.18.1}$$

satisfying the following property: for every commutative diagram of morphisms of schemes

$$\begin{array}{ccc} U' & \xrightarrow{\mu} & U \\ {\scriptstyle u'}\downarrow & & \downarrow{\scriptstyle u} \\ X' & \xrightarrow{g} & X \end{array} \tag{6.3.18.2}$$

such that the morphisms u and u' are étale of finite presentation, the scheme U is separated and the morphism $f|U\colon (U, \mathscr{M}_X|U) \to (S, \mathscr{M}_S)$ induced by f admits an adequate chart, considering the diagram (6.3.17.2) associated with (6.3.18.2) and using the notation of 6.1.12, the composed morphism

$$\Theta_\mu^{-1}(j_U^*(\mathscr{F}_n^{(r)})) \xrightarrow{\sim} j_{U'}^{\prime *}(\Theta^{-1}(\mathscr{F}_n^{(r)})) \xrightarrow{j_{U'}^{\prime *}(\nu_n^{(r)})} j_{U'}^{\prime *}(\mathscr{F}_n^{\prime(r)}), \tag{6.3.18.3}$$

where the first arrow is the isomorphism underlying the diagram (6.1.12.4), *identifies, in view of* (4.4.16.10), *with the morphism defined in* (6.3.11.10) *with respect to the morphisms* μ *and* $\widetilde{\mu}$ *(and to any adequate chart for* $f|U$*).*

Indeed, the morphism $\nu_n^{(r)}$ (6.3.18.1) is defined by descent from the morphism defined in (6.3.11.10), in view of 6.3.16 and 6.3.17.

6.3.19 The morphism $\nu_n^{(r)}$ (6.3.18.1) induces a $\overline{\mathscr{B}}'_n$-linear morphism of $\widetilde{E}'_s$

$$\nu_n^{(r)} : \theta_n^*(\mathscr{F}_n^{(r)}) \to \mathscr{F}_n'^{(r)}, \tag{6.3.19.1}$$

where θ_n is the morphism of ringed topos (6.1.11.4). It immediately follows from 6.3.13(ii) that we have a commutative diagram

$$\begin{array}{ccccccccc} 0 & \longrightarrow & \overline{\mathscr{B}}'_n & \longrightarrow & \theta_n^*(\mathscr{F}_n^{(r)}) & \longrightarrow & \theta_n^*(\sigma_n^*(\widetilde{\xi}^{-1}\widetilde{\Omega}^1_{\overline{X}_n/\overline{S}_n})) & \longrightarrow & 0 \\ & & \| & & \downarrow{\scriptstyle \nu_n^{(r)}} & & \downarrow & & \\ 0 & \longrightarrow & \overline{\mathscr{B}}'_n & \longrightarrow & \mathscr{F}_n'^{(r)} & \longrightarrow & \sigma_n'^*(\widetilde{\xi}^{-1}\widetilde{\Omega}^1_{\overline{X}'_n/\overline{S}_n}) & \longrightarrow & 0 \end{array} \tag{6.3.19.2}$$

where the horizontal lines are induced by the exact sequences (6.3.8.5) and the right vertical arrow is induced by the canonical morphism

$$\overline{g}_n^*(\widetilde{\Omega}^1_{\overline{X}_n/\overline{S}_n}) \to \widetilde{\Omega}^1_{\overline{X}'_n/\overline{S}_n} \tag{6.3.19.3}$$

and the commutative diagram (6.1.11.5). In view of (6.3.8.6), the morphism $\nu_n^{(r)}$ induces a morphism of $\overline{\mathscr{B}}'_n$-algebras of $\widetilde{E}'_s$

$$\omega_n^{(r)} : \theta_n^*(\mathscr{C}_n^{(r)}) \to \mathscr{C}_n'^{(r)}. \tag{6.3.19.4}$$

In the following, we consider $\mathscr{C}_n'^{(r)}$ as a $\theta_n^*(\mathscr{C}_n^{(r)})$-algebra via $\omega_n^{(r)}$.

For any rational number $t \geq r$, we consider the $\theta_n^*(\mathscr{C}_n^{(r)})$-algebra

$$\mathscr{C}_n'^{(t,r)} = \mathscr{C}_n'^{(t)} \otimes_{\theta_n^*(\mathscr{C}_n^{(t)})} \theta_n^*(\mathscr{C}_n^{(r)}) \tag{6.3.19.5}$$

deduced from $\mathscr{C}_n'^{(t)}$ by base change by the homomorphism $\theta_n^*(\alpha_n^{t,r})$, where $\alpha_n^{t,r} : \mathscr{C}_n^{(t)} \to \mathscr{C}_n^{(r)}$ is the canonical homomorphism (4.4.10.7). Taking again the notation of (6.1.21.10), the canonical homomorphism $\theta_n^*(\mathscr{C}_n^{(r)}) \to \mathscr{C}_n'^{(t,r)}$ induces a morphism of $\overline{\mathscr{B}}^!_n$-algebras of $\widetilde{G}_s$

$$\mathrm{g}_n^*(\mathscr{C}_n^{(r)}) \to \tau_{n*}(\mathscr{C}_n'^{(t,r)}). \tag{6.3.19.6}$$

For all rational numbers t, t', r, r' such that $t' \geq t \geq r$ and $t' \geq r' \geq r$, the diagrams

$$\begin{array}{ccccccc} \mathscr{C}_n^{(r')} & \xleftarrow{\alpha_n^{t',r'}} & \mathscr{C}_n^{(t')} & & \theta_n^*(\mathscr{C}_n^{(t')}) & \xrightarrow{\omega_n^{(t')}} & \mathscr{C}_n'^{(t')} \\ \downarrow{\scriptstyle \alpha_n^{r',r}} & & \downarrow{\scriptstyle \alpha_n^{t',t}} & & \downarrow{\scriptstyle \theta_n^*(\alpha_n^{t',t})} & & \downarrow{\scriptstyle \alpha_n'^{t',t}} \\ \mathscr{C}_n^{(r)} & \xleftarrow[\alpha_n^{t,r}]{} & \mathscr{C}_n^{(t)} & & \theta_n^*(\mathscr{C}_n^{(t)}) & \xrightarrow[\omega_n^{(t)}]{} & \mathscr{C}_n'^{(t)} \end{array} \qquad (6.3.19.7)$$

are commutative. We deduce from this a canonical homomorphism of $\theta_n^*(\mathscr{C}_n^{(r')})$-algebras

$$\alpha_n'^{t',t,r',r} : \mathscr{C}_n'^{(t',r')} \to \mathscr{C}_n'^{(t,r)}. \qquad (6.3.19.8)$$

It follows from (4.4.10.8) that for all rational numbers t, t', t'', r, r', r'' such that $t'' \geq t' \geq t \geq r \geq 0$, $t'' \geq t' \geq r' \geq r$ and $t'' \geq r'' \geq r' \geq r$, we have

$$\alpha_n'^{t',t,r',r} \circ \alpha_n'^{t'',t',r'',r'} = \alpha_n'^{t'',t,r'',r}. \qquad (6.3.19.9)$$

6.3.20 We take again the notation of 6.1.22. We denote by $\breve{\mathscr{F}}^{(r)} = (\mathscr{F}_{m+1}^{(r)})_{m\in\mathbb{N}}$ the Higgs–Tate $\breve{\overline{\mathscr{B}}}$-extension of thickness r and by $\breve{\mathscr{C}}^{(r)} = (\mathscr{C}_{m+1}^{(r)})_{m\in\mathbb{N}}$ the Higgs–Tate $\breve{\overline{\mathscr{B}}}$-algebra of thickness r, associated with $(f, \widetilde{X}, \mathscr{M}_{\widetilde{X}})$ (see 4.4.22). We also consider analogous objects and notation for f', that we endow with an exponent $'$.

By ([3] III.7.5 and III.7.12), the homomorphisms $(\omega_{m+1}^{(r)})_{m\in\mathbb{N}}$ (6.3.19.4) define a homomorphism

$$\breve{\omega}^{(r)} : \breve{\theta}^*(\breve{\mathscr{C}}^{(r)}) \to \breve{\mathscr{C}}'^{(r)}. \qquad (6.3.20.1)$$

In the following, we consider $\breve{\mathscr{C}}'^{(r)}$ as a $\breve{\theta}^*(\breve{\mathscr{C}}^{(r)})$-algebra via $\breve{\omega}^{(r)}$.

For any rational number $t \geq r$, we consider the $\breve{\theta}^*(\breve{\mathscr{C}}^{(r)})$-algebra

$$\breve{\mathscr{C}}'^{(t,r)} = \breve{\mathscr{C}}'^{(t)} \otimes_{\breve{\theta}^*(\breve{\mathscr{C}}^{(t)})} \breve{\theta}^*(\breve{\mathscr{C}}^{(r)}) \qquad (6.3.20.2)$$

deduced from $\breve{\mathscr{C}}'^{(t)}$ by base change by the homomorphism $\breve{\theta}^*(\breve{\alpha}^{t,r})$, where $\breve{\alpha}^{t,r} : \breve{\mathscr{C}}^{(t)} \to \breve{\mathscr{C}}^{(r)}$ is the canonical homomorphism (4.4.22.6). We have a canonical isomorphism (6.3.19.5)

$$\breve{\mathscr{C}}'^{(t,r)} \xrightarrow{\sim} (\mathscr{C}_{m+1}'^{(t,r)})_{m\in\mathbb{N}}. \qquad (6.3.20.3)$$

The canonical homomorphism $\breve{\theta}^*(\breve{\mathscr{C}}^{(r)}) \to \breve{\mathscr{C}}'^{(t,r)}$ induces a homomorphism

$$\breve{\mathfrak{g}}^*(\breve{\mathscr{C}}^{(r)}) \to \breve{\tau}_*(\breve{\mathscr{C}}'^{(t,r)}), \qquad (6.3.20.4)$$

that identifies with that induced by the homomorphisms (6.3.19.6).

For all rational numbers t, t', r, r' such that $t' \geq t \geq r \geq 0$ and $t' \geq r' \geq r$, we have a canonical homomorphism of $\breve{\theta}^*(\breve{\mathscr{C}}^{(r')})$-algebras

$$\breve{\alpha}'^{t',t,r',r} : \breve{\mathscr{C}}'^{(t',r')} \to \breve{\mathscr{C}}'^{(t,r)}, \qquad (6.3.20.5)$$

that identifies with the homomorphism $(\alpha_{m+1}'^{t',t,r',r})_{m\in\mathbb{N}}$ (6.3.19.8).

It follows from (6.3.19.9) that for all rational numbers t, t', t'', r, r', r'' such that $t'' \geq t' \geq t \geq r \geq 0$, $t'' \geq t' \geq r' \geq r$ and $t'' \geq r'' \geq r' \geq r$, we have

$$\breve{\alpha}'^{t',t,r',r} \circ \breve{\alpha}'^{t'',t',r'',r'} = \breve{\alpha}'^{t'',t,r'',r}. \tag{6.3.20.6}$$

We denote by $\mathbf{Mod}(\breve{\mathfrak{g}}^*(\breve{\mathscr{C}}^{(r)}))$ the category of $\breve{\mathfrak{g}}^*(\breve{\mathscr{C}}^{(r)})$-modules of $\widetilde{G}_s^{\mathbb{N}^\circ}$, by $\mathbf{Mod}_{\mathbb{Q}}(\breve{\mathfrak{g}}^*(\breve{\mathscr{C}}^{(r)}))$ the category of $\breve{\mathfrak{g}}^*(\breve{\mathscr{C}}^{(r)})$-modules up to isogeny (2.9.2) and by $\mathbf{Ind\text{-}Mod}(\breve{\mathfrak{g}}^*(\breve{\mathscr{C}}^{(r)}))$ the category of ind-$\breve{\mathfrak{g}}^*(\breve{\mathscr{C}}^{(r)})$-modules (2.7.1). We consider the analogous notation for the category of $\breve{\theta}^*(\breve{\mathscr{C}}^{(r)})$-modules of $\widetilde{E}_s'^{\mathbb{N}^\circ}$. The morphism of ringed topos $\breve{\tau}$ (6.1.23.6) induces a morphism of ringed topos that we still abusively denote by

$$\breve{\tau}\colon (\widetilde{E}_s'^{\mathbb{N}^\circ}, \breve{\theta}^*(\breve{\mathscr{C}}^{(r)})) \to (\widetilde{G}_s^{\mathbb{N}^\circ}, \breve{\mathfrak{g}}^*(\breve{\mathscr{C}}^{(r)})). \tag{6.3.20.7}$$

This notation does not induce any risk of ambiguity since the pullback functor for the $\breve{\mathfrak{g}}^*(\breve{\mathscr{C}}^{(r)})$-modules coincides with the restriction of the pullback functor for $\overline{\breve{\mathscr{B}}}^!$-modules (6.1.22.1). We use for $\breve{\tau}$ (6.3.20.7) the notation introduced in 2.7.10 and 2.9.8. In particular, we have two adjoint additive functors

$$\mathrm{I}\breve{\tau}^*\colon \mathbf{Ind\text{-}Mod}(\breve{\mathfrak{g}}^*(\breve{\mathscr{C}}^{(r)})) \to \mathbf{Ind\text{-}Mod}(\breve{\theta}^*(\breve{\mathscr{C}}^{(r)})), \tag{6.3.20.8}$$

$$\mathrm{I}\breve{\tau}_*\colon \mathbf{Ind\text{-}Mod}(\breve{\theta}^*(\breve{\mathscr{C}}^{(r)})) \to \mathbf{Ind\text{-}Mod}(\breve{\mathfrak{g}}^*(\breve{\mathscr{C}}^{(r)})). \tag{6.3.20.9}$$

The functor $\mathrm{I}\breve{\tau}^*$ (resp. $\mathrm{I}\breve{\tau}_*$) is right-exact (resp. left). The functor $\mathrm{I}\tau_*$ admits a right derived functor

$$\mathrm{RI}\breve{\tau}_*\colon \mathbf{D}^+(\mathbf{Ind\text{-}Mod}(\breve{\theta}^*(\breve{\mathscr{C}}^{(r)}))) \to \mathbf{D}^+(\mathbf{Ind\text{-}Mod}(\breve{\mathfrak{g}}^*(\breve{\mathscr{C}}^{(r)}))). \tag{6.3.20.10}$$

6.4 Cohomological Computations

6.4.1 For local cohomological computations, it is convenient to introduce the following notation. Let $(\overline{y}' \rightsquigarrow \overline{x}')$ be a point of $X'_{\text{ét}} \overleftarrow{\times}_{X'_{\text{ét}}} \overline{X}'^{\rhd}_{\text{ét}}$ ([2] 3.4.23) such that $\overline{x}'$ is above s, $\underline{X}'$ the strict localization of X' at $\overline{x}'$. By ([3] III.3.7), $\underline{\overline{X}}'$ is normal and strictly local (and in particular integral); it can therefore be identified with the strict localization of $\overline{X}'$ at $\overline{a}'(\overline{x}')$ (6.1.6). The X'-morphism $\overline{y}' \to \underline{X}'$ defining $(\overline{y}' \rightsquigarrow \overline{x}')$ lifts to an $\underline{\overline{X}}'^{\rhd}$-morphism $v'\colon \overline{y}' \to \underline{\overline{X}}'^{\rhd}$ and therefore induces a geometric point of $\underline{\overline{X}}'^{\rhd}$ that we (abusively) denote also by $\overline{y}'$. We set $\underline{\Delta}' = \pi_1(\underline{\overline{X}}'^{\rhd}, \overline{y}')$.

Let $\overline{x} = g(\overline{x}')$ and $\overline{y} = \gamma(\overline{y}')$ (6.1.10.3), which are therefore geometric points of X and $\overline{X}^\circ$ respectively, and let $(\overline{y} \rightsquigarrow \overline{x}')$ (resp. $(\overline{y} \rightsquigarrow \overline{x})$) be the image of $(\overline{y}' \rightsquigarrow \overline{x}')$ by the first (resp. the composition) of the canonical morphisms

$$X'_{\text{ét}} \overleftarrow{\times}_{X'_{\text{ét}}} \overline{X}'^{\rhd}_{\text{ét}} \to X'_{\text{ét}} \overleftarrow{\times}_{X_{\text{ét}}} \overline{X}^\circ_{\text{ét}} \to X_{\text{ét}} \overleftarrow{\times}_{X_{\text{ét}}} \overline{X}^\circ_{\text{ét}}. \tag{6.4.1.1}$$

We denote by $\underline{X}$ the strict localization of X at $\overline{x}$. By ([3] III.3.7), $\underline{\overline{X}}$ is normal and strictly local (and in particular integral); it can therefore be identified with the strict localization of $\overline{X}$ at $\overline{a}(\overline{x})$ (6.1.6). The X-morphism $\overline{y} \to \underline{X}$ defining $(\overline{y} \rightsquigarrow \overline{x})$ lifts to an $\overline{X}^{\circ}$-morphism $v\colon \overline{y} \to \underline{\overline{X}}^{\circ}$ and therefore induces a geometric point of $\underline{\overline{X}}^{\circ}$ which we (abusively) denote also by $\overline{y}$. We set $\underline{\Delta} = \pi_1(\underline{\overline{X}}^{\circ}, \overline{y})$. We denote by

$$\underline{\gamma}\colon \underline{\overline{X}}'^{\rhd} \to \underline{\overline{X}}^{\circ} \tag{6.4.1.2}$$

the morphism induced by g. Since $\underline{\gamma}(\overline{y}') = \overline{y}$, this induces a homomorphism

$$\underline{\Delta}' \to \underline{\Delta}. \tag{6.4.1.3}$$

We denote by $\underline{\Pi}$ its kernel. We denote by $\mathbf{B}_{\underline{\Delta}'}$ (resp. $\mathbf{B}_{\underline{\Delta}}$) the classifying topos of the profinite group $\underline{\Delta}'$ (resp. $\underline{\Delta}$) and by

$$\psi_{\underline{X}',\overline{y}'}\colon \underline{\overline{X}}'^{\rhd}_{\text{fét}} \xrightarrow{\sim} \mathbf{B}_{\underline{\Delta}'}, \tag{6.4.1.4}$$

$$\psi_{\underline{X},\overline{y}}\colon \underline{\overline{X}}^{\circ}_{\text{fét}} \xrightarrow{\sim} \mathbf{B}_{\underline{\Delta}}, \tag{6.4.1.5}$$

the fiber functors ([3] (VI.9.8.4)).

Recall that giving a neighborhood of the point of $X_{\text{ét}}$ associated with $\overline{x}$ in the site $\mathbf{\acute{E}t}_{/X}$ (resp. $\mathbf{P}$, resp. $\mathbf{Q}$ (6.3.1)) is equivalent to giving an $\overline{x}$-pointed étale X-scheme (resp. of $\mathbf{P}$, resp. of $\mathbf{Q}$) ([5] IV 6.8.2). These objects naturally form a cofiltered category, that we denote by $\mathfrak{V}_{\overline{x}}$ (resp. $\mathbf{P}_{\overline{x}}$, resp. $\mathbf{Q}_{\overline{x}}$). The categories $\mathbf{P}_{\overline{x}}$ and $\mathbf{Q}_{\overline{x}}$ are $\mathbb{U}$-small, and the canonical injection functors $\mathbf{Q} \to \mathbf{P} \to \mathbf{\acute{E}t}_{/X}$ induce fully faithful and cofinal functors $\mathbf{Q}_{\overline{x}} \to \mathbf{P}_{\overline{x}} \to \mathfrak{V}_{\overline{x}}$. The categories $\mathfrak{V}'_{\overline{x}'}$, $\mathbf{P}'_{\overline{x}'}$ and $\mathbf{Q}'_{\overline{x}'}$ are defined in the same way.

Consider $\mathfrak{C}_{\overline{x}'}$, the category associated with $\overline{x}'$ in 6.1.16. Let $(\overline{x}' \to U' \to U)$ be an object of $\mathfrak{C}_{\overline{x}'}$; we will omit $\overline{x}'$ to lighten the notation. We have a canonical X'-morphism $\underline{X}' \to U'$ and a canonical X-morphism $\underline{X} \to U$ that fit into a commutative diagram

$$\begin{array}{ccccc} & & \underline{X}' & \xrightarrow{\underline{g}} & \underline{X} \\ & \nearrow & \downarrow & & \downarrow \\ \overline{x}' & \longrightarrow & U' & \longrightarrow & U \end{array} \tag{6.4.1.6}$$

where $\underline{g}$ is the morphism induced by g. We deduce from this morphisms $\underline{\overline{X}}' \to \overline{U}'$ and $\underline{\overline{X}} \to \overline{U}$. The morphism $v'\colon \overline{y}' \to \underline{\overline{X}}'^{\rhd}$ induces a geometric point of $\overline{U}'^{\rhd}$ which we also denote by $\overline{y}'$. Similarly, the morphism $v\colon \overline{y} \to \underline{\overline{X}}^{\circ}$ induces a geometric point of $\overline{U}^{\circ}$ which we also denote by $\overline{y}$. Observe that the diagram

$$\begin{array}{ccc} \overline{y}' & \longrightarrow & \overline{y} \\ \downarrow & & \downarrow \\ \overline{U}'^{\rhd} & \longrightarrow & \overline{U}^{\circ} \end{array} \tag{6.4.1.7}$$

is commutative.

The schemes $\overline{U}$ and $\overline{U}'$ being locally irreducible by ([3] III.3.3 and III.4.2(iii)), they are the sums of the schemes induced on their irreducible components. We denote by $\overline{U}^{\star}$ (resp. $\overline{U}'^{\star}$) the irreducible component of $\overline{U}$ (resp. $\overline{U}'$) containing $\overline{y}$ (resp. $\overline{y}'$). Similarly, $\overline{U}^{\circ}$ (resp. $\overline{U}'^{\triangleright}$) is the sum of the schemes induced on its irreducible components and $\overline{U}^{\star\circ} = \overline{U}^{\star} \times_X X^{\circ}$ (resp. $\overline{U}'^{\star\triangleright} = \overline{U}'^{\star} \times_{X'} X'^{\triangleright}$) is the irreducible component of $\overline{U}^{\circ}$ (resp. $\overline{U}'^{\triangleright}$) containing $\overline{y}$ (resp. $\overline{y}'$). We set $\Delta_U = \pi_1(\overline{U}^{\star\circ}, \overline{y})$ and $\Delta'_{U'} = \pi_1(\overline{U}'^{\star\triangleright}, \overline{y}')$. By (6.4.1.7), we have a canonical homomorphism $\Delta'_{U'} \to \Delta_U$. We denote by $\Pi_{U' \to U}$ its kernel. We denote by $\mathbf{B}_{\Delta'_{U'}}$ (resp. $\mathbf{B}_{\Delta_U}$) the classifying topos of the profinite group $\Delta'_{U'}$ (resp. Δ_U) and by

$$\psi_{U',\overline{y}'} : \overline{U}'^{\triangleright}_{\text{fét}} \xrightarrow{\sim} \mathbf{B}_{\Delta'_{U'}}, \tag{6.4.1.8}$$

$$\psi_{U,\overline{y}} : \overline{U}^{\circ}_{\text{fét}} \xrightarrow{\sim} \mathbf{B}_{\Delta_U}, \tag{6.4.1.9}$$

the fiber functors.

The functors

$$\mathfrak{C}_{\overline{x}'} \to \mathfrak{V}'_{\overline{x}'}, \quad (\overline{x}' \to U' \to U) \mapsto (\overline{x}' \to U'), \tag{6.4.1.10}$$

$$\mathfrak{C}_{\overline{x}'} \to \mathfrak{V}_{\overline{x}}, \quad (\overline{x}' \to U' \to U) \mapsto (\overline{x} \to U), \tag{6.4.1.11}$$

where the second is defined by composition and the fact that $\overline{x} = g(\overline{x}')$, are initial by virtue of ([5] I 8.1.3(b)). Consequently, the canonical morphisms

$$\underline{\overline{X}}' \to \lim_{\substack{\longleftarrow \\ (U' \to U) \in \mathfrak{C}_{\overline{x}'}}} \overline{U}'^{\star}, \tag{6.4.1.12}$$

$$\underline{\overline{X}} \to \lim_{\substack{\longleftarrow \\ (U' \to U) \in \mathfrak{C}_{\overline{x}'}}} \overline{U}^{\star}, \tag{6.4.1.13}$$

are isomorphisms. By ([3] VI.11.8), the canonical morphisms

$$\underline{\Delta}' \to \lim_{\substack{\longleftarrow \\ (U' \to U) \in \mathfrak{C}_{\overline{x}'}}} \Delta'_{U'}, \tag{6.4.1.14}$$

$$\underline{\Delta} \to \lim_{\substack{\longleftarrow \\ (U' \to U) \in \mathfrak{C}_{\overline{x}'}}} \Delta_U \tag{6.4.1.15}$$

are therefore isomorphisms. We deduce from this that the canonical morphism

$$\underline{\Pi} \to \lim_{\substack{\longleftarrow \\ (U' \to U) \in \mathfrak{C}_{\overline{x}'}}} \Pi_{U' \to U} \tag{6.4.1.16}$$

is an isomorphism.

We denote by

$$\varphi'_{\overline{x}'} : \widetilde{E}' \to \underline{\overline{X}}'^{\triangleright}_{\text{fét}}, \tag{6.4.1.17}$$

$$\phi_{\overline{x}'} : \widetilde{G} \to \underline{\overline{X}}^{\circ}_{\text{fét}}, \tag{6.4.1.18}$$

$$\varphi_{\overline{x}} : \widetilde{E} \to \underline{\overline{X}}^{\circ}_{\text{fét}}, \tag{6.4.1.19}$$

the canonical functors defined in (6.1.13.10) and (6.1.18.3), It follows from 6.1.19(i), ([2] 6.6.7 and (6.5.11.1)), that for every abelian group F of $\widetilde{E}'$ and every integer $i \geq 0$, we have a canonical isomorphism

$$(\mathrm{R}^i \tau_*(F))_{\varrho(\overline{y} \rightsquigarrow \overline{x}')} \xrightarrow{\sim} \mathrm{H}^i(\underline{\Pi}, \psi'_{\underline{X}', \overline{y}'}(\varphi'_{\overline{x}'}(F))), \tag{6.4.1.20}$$

where ϱ is the morphism (6.1.14.8).

6.4.2 We keep the assumptions and notation of 6.4.1. For any object $(\overline{x}' \to U' \to U)$ of $\mathfrak{C}_{\overline{x}'}$, we set

$$\overline{R}^{\prime\overline{y}'}_{U'} = \psi'_{U',\overline{y}'}(\overline{\mathscr{B}}'_{U'} | \overline{U}^{\prime\star\triangleright}), \tag{6.4.2.1}$$

$$\overline{R}^{\overline{y}}_{U} = \psi_{U,\overline{y}}(\overline{\mathscr{B}}_{U} | \overline{U}^{\star\circ}), \tag{6.4.2.2}$$

$$\overline{R}^{!\overline{y}}_{U'\to U} = \psi_{U,\overline{y}}(\overline{\mathscr{B}}^{!}_{U'\to U} | \overline{U}^{\star\circ}), \tag{6.4.2.3}$$

where $\overline{\mathscr{B}}_U$ and $\overline{\mathscr{B}}'_{U'}$ (resp. $\overline{\mathscr{B}}^{!}_{U'\to U}$) are the sheaves defined in (4.3.6.3) (resp. (6.1.15.2)).

Explicitly, let $(V_i)_{i\in I}$ be the normalized universal cover of $\overline{U}^{\star\circ}$ at $\overline{y}$ ([3] VI.9.8). For every $i \in I$, $(V_i \to U)$ (resp. $(U' \to U \leftarrow V_i)$) is naturally an object of E (resp. G). We then have

$$\overline{R}^{\overline{y}}_{U} = \lim_{\substack{\longrightarrow \\ i\in I}} \overline{\mathscr{B}}(V_i \to U), \tag{6.4.2.4}$$

$$\overline{R}^{!\overline{y}}_{U'\to U} = \lim_{\substack{\longrightarrow \\ i\in I}} \overline{\mathscr{B}}^{!}(U' \to U \leftarrow V_i). \tag{6.4.2.5}$$

The canonical homomorphism $\mathrm{g}^*(\overline{\mathscr{B}}) \to \overline{\mathscr{B}}^{!}$ (6.1.21.3) induces for every $i \in I$ a morphism (functorial in i)

$$\overline{\mathscr{B}}(V_i \to U) \to \overline{\mathscr{B}}^{!}(U' \to U \leftarrow V_i). \tag{6.4.2.6}$$

We deduce from this, by taking the direct limit, a homomorphism

$$\overline{R}^{\overline{y}}_{U} \to \overline{R}^{!\overline{y}}_{U'\to U}. \tag{6.4.2.7}$$

Let $(W_j)_{j\in J}$ be the normalized universal cover of $\overline{U}^{\prime\star\triangleright}$ at $\overline{y}'$. For every $j \in J$, $(W_j \to U')$ is naturally an object of E'. We then have

$$\overline{R}^{\prime\overline{y}'}_{U'} = \lim_{\substack{\longrightarrow \\ j\in J}} \overline{\mathscr{B}}'(W_j \to U'). \tag{6.4.2.8}$$

For every $i \in I$, we have a canonical $\overline{U}^{\circ}$-morphism $\overline{y} \to V_i$. We deduce from this a $\overline{U}^{\prime\triangleright}$-morphism $\overline{y}' \to V_i \times_{\overline{U}^{\circ}} \overline{U}^{\prime\triangleright}$. The scheme $V_i \times_{\overline{U}^{\circ}} \overline{U}^{\prime\triangleright}$ being locally irreducible, it is the sum of the schemes induced on its irreducible components. We denote by V'_i

the irreducible component of $V_i \times_{\overline{U}^\circ} \overline{U}'^{\rhd}$ containing the image of $\overline{y}'$. The schemes $(V_i')_{i \in I}$ naturally form an inverse system of connected finite étale $\overline{y}'$-pointed covers of $\overline{U}'^{\star\rhd}$. The canonical morphism

$$\overline{U}'^{\rhd} \times_{\overline{U}^\circ} V_i \to U' \times_{(U \times_X X')} (V_i \times_{X^\circ} X'^{\rhd}) \tag{6.4.2.9}$$

is an isomorphism. We therefore have a canonical isomorphism of $\widetilde{E}'$

$$\tau^*((U' \to U \leftarrow V_i)^a) \xrightarrow{\sim} (V_i \times_{\overline{U}^\circ} \overline{U}'^{\rhd} \to U')^a. \tag{6.4.2.10}$$

Since the canonical homomorphism $\overline{\mathscr{B}}^! \to \tau_*(\overline{\mathscr{B}}')$ (6.1.21.2) is an isomorphism, we deduce from this a canonical homomorphism (functorial in i)

$$\overline{\mathscr{B}}^!(U' \to U \leftarrow V_i) = \overline{\mathscr{B}}'(V_i \times_{\overline{U}^\circ} \overline{U}'^{\rhd} \to U') \to \overline{\mathscr{B}}'(V_i' \to U'). \tag{6.4.2.11}$$

We set

$$\overline{R}^{\circ\overline{y}'}_{U' \to U} = \varinjlim_{i \in I} \overline{\mathscr{B}}'(V_i' \to U'), \tag{6.4.2.12}$$

we therefore have a canonical homomorphism

$$\overline{R}^{!\overline{y}}_{U' \to U} \to \overline{R}^{\circ\overline{y}'}_{U' \to U}. \tag{6.4.2.13}$$

For all $i \in I$ and $j \in J$, there exists at most one morphism of pointed $\overline{U}'^{\star\rhd}$-schemes $W_j \to V_i'$. Moreover, for every $i \in I$, there exist $j \in J$ and a morphism of pointed $\overline{U}'^{\star\rhd}$-schemes $W_j \to V_i'$. Then, we have a homomorphism

$$\overline{R}^{\circ\overline{y}'}_{U' \to U} = \varinjlim_{i \in I} \overline{\mathscr{B}}'(V_i' \to U') \to \varinjlim_{j \in J} \overline{\mathscr{B}}'(W_j \to U') = \overline{R}^{\prime\overline{y}'}_{U'}. \tag{6.4.2.14}$$

We therefore have three canonical $\Delta'_{U'}$-equivariant homomorphisms

$$\overline{R}^{\overline{y}}_U \to \overline{R}^{!\overline{y}}_{U' \to U} \to \overline{R}^{\circ\overline{y}'}_{U' \to U} \to \overline{R}^{\prime\overline{y}'}_{U'}. \tag{6.4.2.15}$$

These rings and these homomorphisms being functorial in $(\overline{x}' \to U' \to U) \in \mathrm{Ob}(\mathfrak{C}_{\overline{x}'})$, let

$$\overline{R}^{\prime\overline{y}'}_{\underline{X}'} = \varinjlim_{(\overline{x}' \to U' \to U) \in \mathrm{Ob}(\mathfrak{C}^\circ_{\overline{x}'})} \overline{R}^{\prime\overline{y}'}_{U'}, \tag{6.4.2.16}$$

$$\overline{R}^{\circ\overline{y}}_{\underline{X}' \to \underline{X}} = \varinjlim_{(\overline{x}' \to U' \to U) \in \mathrm{Ob}(\mathfrak{C}^\circ_{\overline{x}'})} \overline{R}^{\circ\overline{y}'}_{U' \to U}, \tag{6.4.2.17}$$

$$\overline{R}^{!\overline{y}}_{\underline{X}' \to \underline{X}} = \varinjlim_{(\overline{x}' \to U' \to U) \in \mathrm{Ob}(\mathfrak{C}^\circ_{\overline{x}'})} \overline{R}^{!\overline{y}}_{U' \to U}, \tag{6.4.2.18}$$

$$\overline{R}^{\overline{y}}_{\underline{X}} = \varinjlim_{(\overline{x}' \to U' \to U) \in \mathrm{Ob}(\mathfrak{C}^{\circ}_{\overline{x}'})} \overline{R}^{\overline{y}}_{U}, \tag{6.4.2.19}$$

of which the first two are rings of $\mathbf{B}_{\underline{\Delta}'}$ and the other two are rings of $\mathbf{B}_{\underline{\Delta}}$.

By (4.3.9.4), as the functors (6.4.1.10) and (6.4.1.11) are initial, we have canonical isomorphisms

$$\psi_{\underline{X}', \overline{y}'}(\varphi'_{\overline{x}'}(\overline{\mathscr{B}}')) \xrightarrow{\sim} \overline{R}'^{\overline{y}'}_{\underline{X}'}, \tag{6.4.2.20}$$

$$\psi_{\underline{X}, \overline{y}}(\varphi_{\overline{x}}(\overline{\mathscr{B}})) \xrightarrow{\sim} \overline{R}^{\overline{y}}_{\underline{X}}. \tag{6.4.2.21}$$

These induce canonical isomorphisms (4.3.8.2)

$$\overline{\mathscr{B}}'_{\rho'(\overline{y}' \rightsquigarrow \overline{x}')} \xrightarrow{\sim} \overline{R}'^{\overline{y}'}_{\underline{X}'}, \tag{6.4.2.22}$$

$$\overline{\mathscr{B}}_{\rho(\overline{y} \rightsquigarrow \overline{x})} \xrightarrow{\sim} \overline{R}^{\overline{y}}_{\underline{X}}. \tag{6.4.2.23}$$

Moreover, by (6.1.17.3), we have a canonical isomorphism

$$\overline{\mathscr{B}}^{!}_{\varrho(\overline{y} \rightsquigarrow \overline{x}')} \xrightarrow{\sim} \overline{R}^{!\overline{y}}_{\underline{X}' \to \underline{X}}. \tag{6.4.2.24}$$

The homomorphisms (6.4.2.15) induce by taking the direct limit canonical $\underline{\Delta}'$-equivariant homomorphisms

$$\overline{R}^{\overline{y}}_{\underline{X}} \to \overline{R}^{!\overline{y}}_{\underline{X}' \to \underline{X}} \to \overline{R}^{\circ\overline{y}'}_{\underline{X}' \to \underline{X}} \to \overline{R}'^{\overline{y}'}_{\underline{X}'}. \tag{6.4.2.25}$$

By virtue of ([2] 6.5.26), the homomorphism in the middle is an isomorphism. By the above, the first homomorphism and the composition of the other two are identified with the homomorphisms

$$\overline{\mathscr{B}}_{\rho(\overline{y} \rightsquigarrow \overline{x})} \to \overline{\mathscr{B}}^{!}_{\varrho(\overline{y} \rightsquigarrow \overline{x}')} \to \overline{\mathscr{B}}'_{\rho'(\overline{y}' \rightsquigarrow \overline{x}')} \tag{6.4.2.26}$$

induced by the adjoints of the homomorphisms (6.1.21.3) and (6.1.21.2), respectively.

Remark 6.4.3 Under the assumptions of 6.4.2, for every object $(\overline{x}' \to U' \to U)$ of $\mathfrak{C}_{\overline{x}'}$ such that the restriction $(U', \mathscr{M}_{X'}|U') \to (U, \mathscr{M}_X|U)$ of g satisfies the assumptions of 5.1.3, the rings $\overline{R}^{\overline{y}}_{U} \to \overline{R}^{\circ\overline{y}'}_{U' \to U} \to \overline{R}'^{\overline{y}'}_{U'}$ coincide with the rings $\overline{R} \to \overline{R}^{\circ} \to \overline{R}'$ defined in 5.1.13.

6.4.4 We keep the assumptions and notation of 6.4.1 and 6.4.2. We denote by W the ring of Witt vectors with coefficients in k with respect to p, K_0 the field of the fractions of W and $\mathfrak{d}$ the different of the extension K/K_0. We fix a uniformizer ϖ of $\mathscr{O}_K$. We set $\delta = 0$ in the absolute case (3.1.14) and $\delta = v(\varpi\mathfrak{d})$ in the relative case. Notice the change of notation compared to (3.4.1.1) to avoid confusion with the morphism ρ (6.1.8.6). By (3.1.10.1) and (3.1.5.3), we have a canonical $\mathscr{O}_C$-linear isomorphism

$$\mathscr{O}_C(1) \xrightarrow{\sim} p^{\delta + \frac{1}{p-1}} \xi \mathscr{O}_C. \tag{6.4.4.1}$$

We denote by $\mathfrak{C}'_{\overline{x}'}$ the full subcategory of the category $\mathfrak{C}_{\overline{x}'}$ (6.1.16) made up of the objects $(\overline{x}' \to U' \to U)$ such that the following conditions are satisfied:

(i) The schemes U and U' are affine and connected, and the restriction $(U', \mathscr{M}_{X'}|U') \to (U, \mathscr{M}_X|U)$ of the morphism g (6.1.4.1) admits a relative adequate chart ([2] 5.1.11). These conditions correspond to those (for g) set in 5.1.3.
(ii) There exists a fine and saturated chart $M \to \Gamma(U', \mathscr{M}_{X'})$ for $(U', \mathscr{M}_{X'}|U')$ inducing an isomorphism

$$M \xrightarrow{\sim} \Gamma(U', \mathscr{M}_{X'})/\Gamma(U', \mathscr{O}_{X'}^{\times}). \tag{6.4.4.2}$$

The category $\mathfrak{C}'_{\overline{x}'}$ is cofiltered, and the canonical injection functor $\mathfrak{C}'_{\overline{x}'} \to \mathfrak{C}_{\overline{x}'}$ is initial by ([3] II.5.17) and ([5] I 8.1.3(c)). Similarly, the functor $\mathfrak{C}'_{\overline{x}'} \to \mathbf{Q}'_{\overline{x}'}$ defined by $(\overline{x}' \to U' \to U) \mapsto (\overline{x}' \to U')$ is initial (6.4.1).

For every object $(\overline{x}' \to U' \to U)$ of $\mathfrak{C}'_{\overline{x}'}$ and every integer $n \geq 0$, we have a canonical exact sequence of $\overline{R}^{\overline{y}'}_{U'}$-representations of $\Delta'_{U'}$

$$0 \to \overline{R}^{\prime\overline{y}'}_{U'}/p^n\overline{R}^{\prime\overline{y}'}_{U'} \to \mathscr{F}^{\prime\overline{y}'}_{U'}/p^n\mathscr{F}^{\prime\overline{y}'}_{U'} \to \widetilde{\xi}^{-1}\widetilde{\Omega}^1_{\overline{X}'_n/\overline{S}_n}(\overline{U}'^{\star}) \otimes_{\mathscr{O}_{\overline{X}'_n}(\overline{U}'^{\star})} \overline{R}^{\prime\overline{y}'}_{U'} \to 0, \tag{6.4.4.3}$$

deduced from the exact sequence (4.4.4.6) (for f' and U'). We denote by

$$\partial_{U'\to U}\colon \widetilde{\xi}^{-1}\widetilde{\Omega}^1_{\overline{X}'_n/\overline{S}_n}(\overline{U}'^{\star}) \otimes_{\mathscr{O}_{\overline{X}'}(\overline{U}'^{\star})} \overline{R}^{\circ\overline{y}'}_{U'\to U} \to \mathrm{H}^1(\Pi_{U'\to U}, \overline{R}^{\prime\overline{y}'}_{U'}/p^n\overline{R}^{\prime\overline{y}'}_{U'}) \tag{6.4.4.4}$$

the boundary map of the long exact sequence of cohomology deduced from the short exact sequence (6.4.4.3), where $\Pi_{U'\to U}$ is the kernel of the canonical homomorphism $\Delta'_{U'} \to \Delta_U$. We then have a commutative diagram

$$\begin{array}{ccc}
\widetilde{\xi}^{-1}\widetilde{\Omega}^1_{\overline{X}'_n/\overline{S}_n}(\overline{U}'^{\star}) \otimes_{\mathscr{O}_{\overline{X}'}(\overline{U}'^{\star})} \overline{R}^{\circ\overline{y}'}_{U'\to U} & \xrightarrow{\partial_{U'\to U}} & \mathrm{H}^1(\Pi_{U'\to U}, \overline{R}^{\prime\overline{y}'}_{U'}/p^n\overline{R}^{\prime\overline{y}'}_{U'}) \\
 & & \uparrow{\scriptstyle c} \\
\downarrow{\scriptstyle u} & & \mathrm{H}^1(\Pi_{U'\to U}, p^{\delta+\frac{1}{p-1}}\overline{R}^{\prime\overline{y}'}_{U'}/p^{\delta+\frac{1}{p-1}+n}\overline{R}^{\prime\overline{y}'}_{U'}) \\
 & & \uparrow{\scriptstyle -b} \\
\widetilde{\xi}^{-1}\widetilde{\Omega}^1_{\overline{X}'_n/\overline{X}_n}(\overline{U}'^{\star}) \otimes_{\mathscr{O}_{\overline{X}'}(\overline{U}'^{\star})} \overline{R}^{\circ\overline{y}'}_{U'\to U} & \xrightarrow{a} & \mathrm{H}^1(\Pi_{U'\to U}, \widetilde{\xi}^{-1}(\overline{R}^{\prime\overline{y}'}_{U'}/p^n\overline{R}^{\prime\overline{y}'}_{U'})(1))
\end{array} \tag{6.4.4.5}$$

where u is the canonical morphism, a is the component in degree one of the morphism (5.1.21.1), b is induced by the isomorphism (6.4.4.1) and c is induced by the canonical injection $p^{\delta+\frac{1}{p-1}}\mathscr{O}_C \to \mathscr{O}_C$. Indeed, we may assume $U = X$ and $U' = X'$. The morphism $\partial_{U'\to U}$ does not depend on the chosen deformation $(\widetilde{X}', \mathscr{M}_{\widetilde{X}'})$ (3.2.20). We can therefore reduce to the case where $(\widetilde{X}', \mathscr{M}_{\widetilde{X}'})$ is the deformation defined by the adequate chart of f' fixed in (i) (see 3.4.6). Setting $\overline{R}' = \overline{R}^{\prime\overline{y}'}_{U'}$, the same chart determines a section of the Higgs–Tate torsor over $\mathrm{Spec}(\widehat{\overline{R}'})$ (5.2.6), and consequently a splitting of the extension (5.2.6.1) (see [3] II.10.11). The assertion then follows from 3.4.8; see ([2] 6.6.12), which treats the absolute case (3.1.14).

Lemma 6.4.5 *We keep the assumptions and notation of 6.4.4. Moreover, let $(\overline{x}' \to U' \to U)$ be an object of $\mathfrak{C}'_{\overline{x}'}$ (6.4.4), n an integer ≥ 0. Then,*

(i) *The morphism $\partial_{U'\to U}$ factors through u (6.4.4.5) and it induces a morphism*

$$\partial'_{U'\to U}\colon \widetilde{\xi}^{-1}\widetilde{\Omega}^1_{\overline{X}'_n/\overline{X}_n}(\overline{U}'^{\star}) \otimes_{\mathscr{O}_{\overline{X}'}(\overline{U}'^{\star})} \overline{R}^{\circ\overline{y}'}_{U'\to U} \to \mathrm{H}^1(\Pi_{U'\to U}, \overline{R}^{\prime\overline{y}'}_{U'}/p^n\overline{R}^{\prime\overline{y}'}_{U'}). \tag{6.4.5.1}$$

(ii) *There exists one and only one homomorphism of graded $\overline{R}^{\circ\overline{y}'}_{U'\to U}$-algebras*

$$\wedge\left(\widetilde{\xi}^{-1}\widetilde{\Omega}^1_{\overline{X}'_n/\overline{X}_n}(\overline{U}'^{\star}) \otimes_{\mathscr{O}_{\overline{X}'}(\overline{U}'^{\star})} \overline{R}^{\circ\overline{y}'}_{U'\to U}\right) \to \oplus_{i\geq 0}\mathrm{H}^i(\Pi_{U'\to U}, \overline{R}^{\prime\overline{y}'}_{U'}/p^n\overline{R}^{\prime\overline{y}'}_{U'}) \tag{6.4.5.2}$$

whose component in degree one is the morphism $\partial'_{U'\to U}$ (6.4.5.1). Furthermore, its kernel is annihilated by $p^{2\ell(\delta+\frac{1}{p-1})}\mathfrak{m}_{\overline{K}}$ and its cokernel is annihilated by $p^{2\ell\delta+\frac{2\ell+1}{p-1}}\mathfrak{m}_{\overline{K}}$, where $\ell = \dim(X'/X)$.

(iii) *For every $i \geq \ell+1$, $\mathrm{H}^i(\Pi_{U'\to U}, \overline{R}^{\prime\overline{y}'}_{U'}/p^n\overline{R}^{\prime\overline{y}'}_{U'})$ is α-zero.*

(i) This immediately follows from the diagram (6.4.4.5).

(ii) By virtue of 5.1.21(i), there exists one and only one homomorphism of graded $\overline{R}^{\circ\overline{y}'}_{U'\to U}$-algebras

$$\wedge(\xi^{-1}\widetilde{\Omega}^1_{\overline{X}'_n/\overline{X}_n}(\overline{U}'^{\star}) \otimes_{\mathscr{O}_{\overline{X}'}(\overline{U}'^{\star})} \overline{R}^{\circ\overline{y}'}_{U'\to U}) \to \oplus_{i\geq 0}\mathrm{H}^i(\Pi_{U'\to U}, \xi^{-i}(\overline{R}^{\prime\overline{y}'}_{U'}/p^n\overline{R}^{\prime\overline{y}'}_{U'})(i)) \tag{6.4.5.3}$$

whose component in degree one is the morphism a of the diagram (6.4.4.5). Its kernel is α-zero and its cokernel is annihilated by $p^{\frac{1}{p-1}}\mathfrak{m}_{\overline{K}}$. We deduce from this that there exists one and only one homomorphism of graded $\overline{R}^{\circ\overline{y}'}_{U'\to U}$-algebras

$$\wedge\left(\widetilde{\xi}^{-1}\widetilde{\Omega}^1_{\overline{X}'_n/\overline{X}_n}(\overline{U}'^{\star}) \otimes_{\mathscr{O}_{\overline{X}'}(\overline{U}'^{\star})} \overline{R}^{\circ\overline{y}'}_{U'\to U}\right) \to \oplus_{i\geq 0}\mathrm{H}^i(\Pi_{U'\to U}, \overline{R}^{\prime\overline{y}'}_{U'}/p^n\overline{R}^{\prime\overline{y}'}_{U'}) \tag{6.4.5.4}$$

whose component in degree one is the morphism $\partial'_{U'\to U}$. By chasing diagram (6.4.4.5), we see that the kernel of (6.4.5.4) is annihilated by $p^{2\ell(\delta+\frac{1}{p-1})}\mathfrak{m}_{\overline{K}}$. In view of (iii) below, the cokernel of (6.4.5.4) is annihilated by $p^{2\ell\delta+\frac{2\ell+1}{p-1}}\mathfrak{m}_{\overline{K}}$.

(iii) This is the statement 5.1.21(ii), mentioned as a reminder.

6.4.6 Let n be an integer ≥ 0. We take again the notation of 6.1.7 and (6.1.21.10) and consider the exact sequence of $\overline{\mathscr{B}}'_n$-modules

$$0 \to \overline{\mathscr{B}}'_n \to \mathscr{F}'_n \to \sigma'^{*}_n(\widetilde{\xi}^{-1}\widetilde{\Omega}^1_{\overline{X}'_n/\overline{S}_n}) \to 0, \tag{6.4.6.1}$$

the analogue of the sequence (6.3.8.5) for f' with $r = 0$. Consider the $\overline{\mathscr{B}}^{!}_n$-linear morphism of $\widetilde{G}_s$

$$\pi_n^*(\widetilde{\xi}^{-1}\widetilde{\Omega}^1_{\overline{X}'_n/\overline{S}_n}) \to \mathrm{R}^1\tau_{n*}(\overline{\mathscr{B}}'_n), \tag{6.4.6.2}$$

composed of the adjunction morphism

$$\pi_n^*(\widetilde{\xi}^{-1}\widetilde{\Omega}^1_{\overline{X}'_n/\overline{S}_n}) \to \tau_{n*}(\tau_n^*(\pi_n^*(\widetilde{\xi}^{-1}\widetilde{\Omega}^1_{\overline{X}'_n/\overline{S}_n}))) \tag{6.4.6.3}$$

and of the boundary map of the long exact sequence of cohomology deduced from the short exact sequence (6.4.6.1), taking into account the isomorphism $\sigma'_n \xrightarrow{\sim} \pi_n\tau_n$ (6.1.21.10).

Proposition 6.4.7 *We keep the assumptions and notation of* 6.4.6. *Then,*

(i) *The morphism* (6.4.6.2) *factors through the canonical surjective morphism*

$$\pi_n^*(\widetilde{\xi}^{-1}\widetilde{\Omega}^1_{\overline{X}'_n/\overline{S}_n}) \to \pi_n^*(\widetilde{\xi}^{-1}\widetilde{\Omega}^1_{\overline{X}'_n/\overline{X}_n}), \tag{6.4.7.1}$$

and it induces a $\overline{\mathscr{B}}^!_n$-linear morphism of $\widetilde{G}_s$

$$\pi_n^*(\widetilde{\xi}^{-1}\widetilde{\Omega}^1_{\overline{X}'_n/\overline{X}_n}) \to \mathrm{R}^1\tau_{n*}(\overline{\mathscr{B}}'_n). \tag{6.4.7.2}$$

(ii) *There exists one and only one homomorphism of graded $\overline{\mathscr{B}}^!_n$-algebras of $\widetilde{G}_s$*

$$\wedge(\pi_n^*(\widetilde{\xi}^{-1}\widetilde{\Omega}^1_{\overline{X}'_n/\overline{X}_n})) \to \oplus_{i\geq 0}\mathrm{R}^i\tau_{n*}(\overline{\mathscr{B}}'_n) \tag{6.4.7.3}$$

whose component in degree one is the morphism (6.4.7.2). *Furthermore, its kernel is annihilated by $p^{2\ell(\delta+\frac{1}{p-1})}\mathfrak{m}_{\overline{K}}$ and its cokernel is annihilated by $p^{2\ell\delta+\frac{2\ell+1}{p-1}}\mathfrak{m}_{\overline{K}}$, where δ is defined in* 6.4.4 *and $\ell = \dim(X'/X)$.*

(iii) *For every integer $i \geq \ell+1$, $\mathrm{R}^i\tau_{n*}(\overline{\mathscr{B}}'_n)$ is α-zero.*

Let $(\overline{y} \rightsquigarrow \overline{x}')$ be a point of $X'_{\text{ét}} \overleftarrow{\times}_{X_{\text{ét}}} \overline{X}^\circ_{\text{ét}}$ such that $\overline{x}'$ is above s ([2] 3.4.23), $\underline{X}'$ the strict localization of X' at $\overline{x}'$, $\overline{x} = g(\overline{x}')$, $\underline{X}$ the strict localization of X at $\overline{x}$. The morphism $\underline{\gamma}\colon \overline{\underline{X}}'^{\triangleright} \to \overline{\underline{X}}^\circ$ induced by g is faithfully flat by virtue of ([2] 2.4.1). Then, there exists a point $(\overline{y}' \rightsquigarrow \overline{x}')$ of $X'_{\text{ét}} \overleftarrow{\times}_{X'_{\text{ét}}} \overline{X}'^{\triangleright}_{\text{ét}}$ lifting the point $(\overline{y} \rightsquigarrow \overline{x}')$ of $X'_{\text{ét}} \overleftarrow{\times}_{X_{\text{ét}}} \overline{X}^\circ_{\text{ét}}$ (6.1.14.9). We take again the notation introduced in 6.4.1, 6.4.2 and 6.4.4.

By (4.3.5.7) and 4.4.20, the image of the exact sequence (6.4.6.1) by the composed functor $\psi_{\underline{X}',\overline{y}'} \circ \varphi'_{\overline{x}'}$ (6.4.1) identifies with the direct limit of the sequence (6.4.4.3), when $(\overline{x}' \to U' \to U)$ goes through the category $\mathfrak{C}'^{\circ}_{\overline{x}'}$ (6.4.4).

By virtue of (6.1.20.6) and (6.4.1.20), for every integer $i \geq 0$, we have a canonical isomorphism

$$\mathrm{R}^i\tau_{n*}(\overline{\mathscr{B}}'_n)_{\varrho(\overline{y}\rightsquigarrow\overline{x}')} \xrightarrow{\sim} \mathrm{H}^i(\underline{\Pi}, \psi_{\underline{X}',\overline{y}'}(\varphi'_{\overline{x}'}(\overline{\mathscr{B}}'_n))). \tag{6.4.7.4}$$

Moreover, by (4.3.9.4), we have a canonical isomorphism of $\mathbf{B}_{\underline{\Delta}'}$

$$\psi_{\underline{X}',\overline{y}'}(\varphi'_{\overline{x}'}(\overline{\mathscr{B}}')) \xrightarrow{\sim} \overline{R}^{\overline{y}'}_{\underline{X}'}. \tag{6.4.7.5}$$

By (6.4.1.16) and ([49] I prop. 8), the canonical morphism

$$\varinjlim_{(U'\to U)\in \mathfrak{C}'^{\circ}_{\overline{x}'}} \mathrm{H}^i(\Pi_{U'\to U}, \overline{R}'^{\overline{y}'}_{U'}/p^n\overline{R}'^{\overline{y}'}_{U'}) \to \mathrm{H}^i(\underline{\Pi}, \overline{R}'^{\overline{y}'}_{\underline{X}'}/p^n\overline{R}'^{\overline{y}'}_{\underline{X}'}) \tag{6.4.7.6}$$

is an isomorphism.

By virtue of (6.4.2.24) and ([2] (3.4.23.1) and 6.5.26), we have a canonical isomorphism

$$\pi_n^*(\xi^{-1}\widetilde{\Omega}^1_{\overline{X}'_n/\overline{S}_n})_{\varrho(\overline{y}\rightsquigarrow\overline{x}')} \xrightarrow{\sim} \xi^{-1}\widetilde{\Omega}^1_{\overline{X}'_n/\overline{S}_n,\overline{x}'} \otimes_{\mathscr{O}_{\overline{X}',\overline{x}'}} \overline{R}^{\circ\overline{y}'}_{\underline{X}'\to\underline{X}}, \tag{6.4.7.7}$$

and similarly if we replace $\widetilde{\Omega}^1_{\overline{X}'_n/\overline{S}_n}$ by $\widetilde{\Omega}^1_{\overline{X}'_n/\overline{X}_n}$.

By 6.1.19(ii), the stalk of the morphism (6.4.6.2) at $\varrho(\overline{y}\rightsquigarrow\overline{x}')$ therefore identifies with the direct limit, when $(\overline{x}'\to U'\to U)$ goes through the category $\mathfrak{C}'^{\circ}_{\overline{x}'}$, of the morphism (6.4.4.4)

$$\partial_{U'\to U}\colon \xi^{-1}\widetilde{\Omega}^1_{\overline{X}'_n/\overline{S}_n}(\overline{U}'^{\star})\otimes_{\mathscr{O}_{\overline{X}'}(\overline{U}'^{\star})} \overline{R}^{\circ\overline{y}'}_{U'\to U} \to \mathrm{H}^1(\Pi_{U'\to U}, \overline{R}'^{\overline{y}'}_{U'}/p^n\overline{R}'^{\overline{y}'}_{U'}). \tag{6.4.7.8}$$

The proposition then follows from 6.4.5 in view of ([2] 6.5.3).

6.4.8 We keep the assumptions and notation of 6.4.1 and 6.4.2. Suppose, moreover, that the scheme X is separated and that the morphism $f\colon (X,\mathscr{M}_X)\to(S,\mathscr{M}_S)$ admits an adequate chart $((P,\gamma),(\mathbb{N},\iota),\vartheta)$, and we take again the notation of 6.3.11.

We denote by $\mathbb{J}_{\overline{x}'}$ the full subcategory of the category $\mathfrak{C}_{\overline{x}'}$ defined in 6.1.16 made up of the objects $(\iota'\colon\overline{x}'\to U', \mu\colon U'\to U)$ such that μ is an object of the category $\mathbb{J}$ defined in 6.3.3, i.e., such that U' is an object of $\mathbf{Q}'$ and U is an object of $\mathbf{P}$ (6.3.1). The canonical injection functor $\mathbb{J}_{\overline{x}'}\to\mathfrak{C}_{\overline{x}'}$ is initial and the category $\mathbb{J}_{\overline{x}'}$ is cofiltered by virtue of ([5] I 8.1.3(c)) and ([3] II.5.17). Moreover, the functors

$$\mathfrak{C}_{\overline{x}'}\to\mathbf{Q}'_{\overline{x}'},\quad (\overline{x}'\to U'\to U)\mapsto(\overline{x}'\to U'), \tag{6.4.8.1}$$
$$\mathfrak{C}_{\overline{x}'}\to\mathbf{P}_{\overline{x}},\quad (\overline{x}'\to U'\to U)\mapsto(\overline{x}\to U), \tag{6.4.8.2}$$

where the cofiltered categories $\mathbf{P}_{\overline{x}}$ and $\mathbf{Q}'_{\overline{x}'}$ are defined in 6.4.1 and the second functor is defined by composition and the fact that $\overline{x}=g(\overline{x}')$, are initial by virtue of ([5] I 8.1.3(b)) and ([3] II.5.17).

For any rational number $r\geq 0$ and any object $(\iota'\colon\overline{x}'\to U'\to U)$ of $\mathbb{J}_{\overline{x}'}$, we endow the morphism $f|U\colon (U,\mathscr{M}_X|U)\to(S,\mathscr{M}_S)$ induced by f with the adequate chart induced by $((P,\gamma),(\mathbb{N},\iota),\vartheta)$, and we use the notation introduced in 6.3.9. We then have a canonical $\pi_1(\overline{U}'^{\star\triangleright},\overline{y}')$-equivariant morphism of $\widehat{\overline{R}}^{\overline{y}}_U$-algebras (6.3.9.5)

$$\mathscr{C}^{\overline{y},(t)}_U \to \mathscr{C}'^{\overline{y}',(t)}_{U'}. \tag{6.4.8.3}$$

We consider the $\mathscr{C}^{\overline{y},(r)}_U$-algebra (5.2.14.5)

$$\mathscr{C}'^{\overline{y}',(t,r)}_{U'\to U} = \mathscr{C}'^{\overline{y}',(t)}_{U'} \otimes_{\mathscr{C}^{\overline{y},(t)}_U} \mathscr{C}^{\overline{y},(r)}_U \tag{6.4.8.4}$$

deduced from $\mathscr{C}_{U'}^{\prime\overline{y}',(t)}$ by base change by the canonical homomorphism $\mathscr{C}_U^{\overline{y},(t)} \to \mathscr{C}_U^{\overline{y},(r)}$ (5.2.13).

By virtue of 4.4.20, (6.1.13.11) and (6.4.2.20), the image of the homomorphism $\omega_n^{(t)}\colon \theta_n^*(\mathscr{C}_n^{(t)}) \to \mathscr{C}_n^{\prime(t)}$ (6.3.19.4) by the composed functor $\psi_{\underline{X}',\overline{y}'} \circ \varphi'_{\overline{x}'}$ (6.4.1) identifies with the direct limit of the homomorphisms

$$(\mathscr{C}_U^{\overline{y},(r)}/p^n\mathscr{C}_U^{\overline{y},(t)}) \otimes_{\overline{R}_U^{\overline{y}}} \overline{R}_{U'}^{\prime\overline{y}'} \to \mathscr{C}_{U'}^{\prime\overline{y}',(r)}/p^n\mathscr{C}_{U'}^{\prime\overline{y}',(t)} \tag{6.4.8.5}$$

induced by (6.4.8.3), when $(\overline{x}' \to U' \to U)$ goes through the category $\mathbb{J}^{\circ}_{\overline{x}'}$ (see the proof of 6.3.16). We deduce from this a canonical isomorphism

$$\psi_{\underline{X}',\overline{y}'}(\varphi'_{\overline{x}'}(\mathscr{C}_n^{\prime(t,r)})) \xrightarrow{\sim} \lim_{\substack{\longrightarrow \\ (\overline{x}'\to U'\to U)\in \mathbb{J}^{\circ}_{\overline{x}'}}} \mathscr{C}_{U'\to U}^{\prime\overline{y}',(t,r)}/p^n\mathscr{C}_{U'\to U}^{\prime\overline{y}',(t,r)}. \tag{6.4.8.6}$$

Proposition 6.4.9 *Let r, t, t' be three rational numbers such that $t' > t > r \geq 0$. Then,*

(i) *For every integer $n \geq 1$, the canonical homomorphism* (6.3.19.6)

$$\mathrm{g}_n^*(\mathscr{C}_n^{(r)}) \to \tau_{n*}(\mathscr{C}_n^{\prime(t,r)}) \tag{6.4.9.1}$$

is α-injective. Let $\mathscr{H}_n^{(t,r)}$ be its cokernel.

(ii) *There exists a rational number $a > 0$ such that for every integer $n \geq 1$, the morphism*

$$\mathscr{H}_n^{(t',r)} \to \mathscr{H}_n^{(t,r)} \tag{6.4.9.2}$$

induced by canonical homomorphism $\mathscr{C}_n^{\prime(t',r)} \to \mathscr{C}_n^{\prime(t,r)}$ (6.3.19.8) is annihilated by p^a.

(iii) *There exists a rational number $b > 0$ such that for all integers $n, q \geq 1$, the canonical morphism*

$$\mathrm{R}^q\tau_{n*}(\mathscr{C}_n^{\prime(t',r)}) \to \mathrm{R}^q\tau_{n*}(\mathscr{C}_n^{\prime(t,r)}) \tag{6.4.9.3}$$

is annihilated by p^b.

The question being local on X ([3] VI.10.14 and [2] 3.4.12), we may assume that the scheme X is separated and that the morphism $f\colon (X, \mathscr{M}_X) \to (S, \mathscr{M}_S)$ admits an adequate chart $((P,\gamma), (\mathbb{N}, \iota), \vartheta)$.

Let $(\overline{y} \rightsquigarrow \overline{x}')$ be a point of $X'_{\text{ét}} \overleftarrow{\times}_{X_{\text{ét}}} \overline{X}^{\circ}_{\text{ét}}$ such that $\overline{x}'$ is above s, $\underline{X}'$ the strict localization of X' at $\overline{x}'$, $\overline{x} = g(\overline{x}')$, $\underline{X}$ the strict localization of X at $\overline{x}$. The morphism $\underline{\gamma}\colon \underline{\overline{X}}'^{\rhd} \to \underline{\overline{X}}^{\circ}$ induced by g being faithfully flat by virtue of ([2] 2.4.1), there exists a point $(\overline{y}' \rightsquigarrow \overline{x}')$ of $X'_{\text{ét}} \overleftarrow{\times}_{X'_{\text{ét}}} \overline{X}'^{\rhd}_{\text{ét}}$ lifting the point $(\overline{y} \rightsquigarrow \overline{x}')$ of $X'_{\text{ét}} \overleftarrow{\times}_{X_{\text{ét}}} \overline{X}^{\circ}_{\text{ét}}$ (6.4.1.1). We take again the notation introduced in 6.4.1, 6.4.2 and 6.4.8.

By virtue of (6.1.20.6) and (6.4.1.20), for every integer $i \geq 0$, we have a canonical isomorphism

$$\mathrm{R}^i\tau_{n*}(\mathscr{C}_n^{\prime(t,r)})_{\varrho(\overline{y}\rightsquigarrow\overline{x}')} \xrightarrow{\sim} \mathrm{H}^i(\underline{\Pi}, \psi_{\underline{X}',\overline{y}'}(\varphi'_{\overline{x}'}(\mathscr{C}_n^{\prime(t,r)}))). \tag{6.4.9.4}$$

In view of (6.4.1.16), (6.4.8.6) and ([49] I prop. 8), the canonical morphism

$$\lim_{\substack{\longrightarrow \\ (\overline{x}' \to U' \to U) \in \mathbb{J}^{\circ}_{\overline{x}'}}} \mathrm{H}^i(\Pi_{U' \to U}, \mathscr{C}'^{\overline{y}',(t,r)}_{U' \to U} / p^n \mathscr{C}'^{\overline{y}',(t,r)}_{U' \to U}) \to \mathrm{H}^i(\underline{\Pi}, \psi_{\underline{X}',\overline{y}'}(\varphi'_{\overline{x}'}(\mathscr{C}'^{(t,r)}_n))) \tag{6.4.9.5}$$

is an isomorphism. In view of 4.4.20, (6.4.2.24) and (6.4.8.5), we deduce from this that the stalk of the morphism (6.4.9.1) at $\varrho(\overline{y} \rightsquigarrow \overline{x}')$ identifies with the direct limit of the morphism (5.2.15.1)

$$(\mathscr{C}^{\overline{y},(r)}_U / p^n \mathscr{C}^{\overline{y},(r)}_U) \otimes_{\widehat{\overline{R}}^{\overline{y}}_U} \overline{R}^{\diamond \overline{y}'}_{U' \to U} \to (\mathscr{C}'^{\overline{y}',(t,r)}_{U' \to U} / p^n \mathscr{C}'^{\overline{y}',(t,r)}_{U' \to U})^{\Pi_{U' \to U}} \tag{6.4.9.6}$$

where $(\overline{x}' \to U' \to U)$ goes through the category $\mathbb{J}^{\circ}_{\overline{x}'}$. The proposition then follows from 5.2.15 in view of ([2] 6.5.3).

Corollary 6.4.10 *Let r, t, t' be three rational numbers such that $t' > t > r \geq 0$. Then,*

(i) *The canonical homomorphism of $\widetilde{G}^{\mathbb{N}^{\circ}}_s$ (6.3.20.4)*

$$\breve{\mathfrak{g}}^*(\breve{\mathscr{C}}^{(r)}) \to \breve{\tau}_*(\breve{\mathscr{C}}'^{(t,r)}) \tag{6.4.10.1}$$

is α-injective. Let $\breve{\mathscr{H}}^{(t,r)}$ be its cokernel.

(ii) *There exists a rational number $a > 0$ such that the morphism*

$$\breve{\mathscr{H}}^{(t',r')} \to \breve{\mathscr{H}}^{(t,r)} \tag{6.4.10.2}$$

induced by the canonical homomorphism $\breve{\alpha}'^{t',t,r,r}$ (6.3.20.5) is annihilated by p^a.

(iii) *There exists a rational number $b > 0$ such that for every integer $q \geq 1$, the canonical morphism of $\widetilde{G}^{\mathbb{N}^{\circ}}_s$*

$$\mathrm{R}^q \breve{\tau}_*(\breve{\mathscr{C}}'^{(t',r)}) \to \mathrm{R}^q \breve{\tau}_*(\breve{\mathscr{C}}^{(t,r)}) \tag{6.4.10.3}$$

is annihilated by p^b.

This follows from 6.4.9 and ([3] III.7.3(i) and (III.7.5.5)).

Corollary 6.4.11 *Let r, t, t' be three rational numbers such that $t' > t > r \geq 0$. Then,*

(i) *The canonical morphism of $\breve{\mathfrak{g}}^*(\breve{\mathscr{C}}^{(r)})_{\mathbb{Q}}$-modules (6.3.20.4)*

$$\mathrm{u}^{t,r} : \breve{\mathfrak{g}}^*(\breve{\mathscr{C}}^{(r)})_{\mathbb{Q}} \to \breve{\tau}_*(\breve{\mathscr{C}}'^{(t,r)})_{\mathbb{Q}} \tag{6.4.11.1}$$

admits a canonical left inverse

$$\mathrm{v}^{t,r} : \breve{\tau}_*(\breve{\mathscr{C}}'^{(t,r)})_{\mathbb{Q}} \to \breve{\mathfrak{g}}^*(\breve{\mathscr{C}}^{(r)})_{\mathbb{Q}}. \tag{6.4.11.2}$$

(ii) *The composition*

$$\breve{\tau}_*(\breve{\mathscr{C}}'^{(t',r)})_{\mathbb{Q}} \xrightarrow{\mathrm{v}^{t',r}} \breve{\mathfrak{g}}^*(\breve{\mathscr{C}}^{(r)})_{\mathbb{Q}} \xrightarrow{\mathrm{u}^{t,r}} \breve{\tau}_*(\breve{\mathscr{C}}'^{(t,r)})_{\mathbb{Q}} \tag{6.4.11.3}$$

is the canonical homomorphism.

(iii) *For every integer $q \geq 1$, the canonical morphism*

$$\mathrm{R}^q \check{\tau}_*(\breve{\mathscr{C}}'^{(t',r)})_{\mathbb{Q}} \to \mathrm{R}^q \check{\tau}_*(\breve{\mathscr{C}}'^{(t,r)})_{\mathbb{Q}} \tag{6.4.11.4}$$

vanishes.

Indeed, by 6.4.10(i)-(ii), $\mathrm{u}^{t,r}$ is injective and there exists one and only one $\breve{\mathfrak{g}}^*(\breve{\mathscr{C}}^{(r)})_{\mathbb{Q}}$-linear morphism

$$\mathrm{v}^{t',t,r} : \check{\tau}_*(\breve{\mathscr{C}}'^{(t',r)})_{\mathbb{Q}} \to \breve{\mathfrak{g}}^*(\breve{\mathscr{C}}^{(r)})_{\mathbb{Q}} \tag{6.4.11.5}$$

such that $\mathrm{u}^{t,r} \circ \mathrm{v}^{t',t,r}$ is the canonical homomorphism $\check{\tau}_*(\breve{\mathscr{C}}'^{(t',r)})_{\mathbb{Q}} \to \check{\tau}_*(\breve{\mathscr{C}}'^{(t,r)})_{\mathbb{Q}}$. Since we have $\mathrm{u}^{t,r} \circ \mathrm{v}^{t',t,r} \circ \mathrm{u}^{t',r} = \mathrm{u}^{t,r}$, we deduce from this that $\mathrm{v}^{t',t,r}$ is a left inverse of $\mathrm{u}^{t',r}$. We immediately check that it does not depend on t; the propositions (i) and (ii) follow. Proposition (iii) immediately follows from 6.4.10(iii).

Corollary 6.4.12 *For every rational number $r \geq 0$, the canonical homomorphism*

$$\breve{\mathfrak{g}}^*(\breve{\mathscr{C}}^{(r)})_{\mathbb{Q}} \to \varinjlim_{t \in \mathbb{Q}_{>r}} \check{\tau}_*(\breve{\mathscr{C}}'^{(t,r)})_{\mathbb{Q}} \tag{6.4.12.1}$$

is an isomorphism, and for every integer $q \geq 1$,

$$\varinjlim_{t \in \mathbb{Q}_{>r}} \mathrm{R}^q \check{\tau}_*(\breve{\mathscr{C}}'^{(t,r)})_{\mathbb{Q}} = 0, \tag{6.4.12.2}$$

where the limits are taken in $\mathbf{Mod}_{\mathbb{Q}}(\breve{\mathfrak{g}}^*(\breve{\mathscr{C}}^{(r)}))$. *These are, in particular, representable.*

The proposition is still true if we consider the different $\breve{\mathfrak{g}}^*(\breve{\mathscr{C}}^{(r)})_{\mathbb{Q}}$-modules as ind-$\breve{\mathfrak{g}}^*(\breve{\mathscr{C}}^{(r)})$-modules via the functor $\alpha_{\breve{\mathfrak{g}}^*(\breve{\mathscr{C}}^{(r)})}$ (2.9.7.3) and if we replace $\varinjlim_{t \in \mathbb{Q}_{>r}}$ by "$\varinjlim_{t \in \mathbb{Q}_{>r}}$".

6.4.13 Let t, r be two rational numbers such that $t \geq r \geq 0$, n an integer ≥ 1. We denote by

$$d_{\mathscr{C}_n'^{(t)}} : \mathscr{C}_n'^{(t)} \to \sigma_n'^*(\widetilde{\xi}^{-1}\widetilde{\Omega}^1_{\overline{X}'_n/\overline{S}_n}) \otimes_{\overline{\mathscr{B}}'_n} \mathscr{C}_n'^{(t)} \tag{6.4.13.1}$$

the universal $\overline{\mathscr{B}}'_n$-derivation of $\mathscr{C}_n'^{(t)}$ (see 4.4.21). It is a Higgs $\overline{\mathscr{B}}'_n$-field with coefficients in $\sigma_n'^*(\widetilde{\xi}^{-1}\widetilde{\Omega}^1_{\overline{X}'_n/\overline{S}_n})$ by 4.4.23(ii). We denote by

$$\underline{d}_{\mathscr{C}_n'^{(t)}} : \mathscr{C}_n'^{(t)} \to \sigma_n'^*(\widetilde{\xi}^{-1}\widetilde{\Omega}^1_{\overline{X}'_n/\overline{X}_n}) \otimes_{\overline{\mathscr{B}}'_n} \mathscr{C}_n'^{(t)} \tag{6.4.13.2}$$

the $\overline{\mathscr{B}}'_n$-derivation induced by $d_{\mathscr{C}_n'^{(t)}}$. This is actually the universal $\theta_n^*(\mathscr{C}_n^{(t)})$-derivation of $\mathscr{C}_n'^{(t)}$. This follows from the exact sequence

$$\begin{aligned}\sigma_n'^*(\overline{g}_n^*(\widetilde{\xi}^{-1}\widetilde{\Omega}^1_{\overline{X}_n/\overline{S}_n})) \otimes_{\overline{\mathscr{B}}'_n} \mathscr{C}_n'^{(t)} &\to \sigma_n'^*(\widetilde{\xi}^{-1}\widetilde{\Omega}^1_{\overline{X}'_n/\overline{S}_n}) \otimes_{\overline{\mathscr{B}}'_n} \mathscr{C}_n'^{(t)} \\ &\to \Omega^1_{\mathscr{C}_n'^{(t)}/\theta_n^*(\mathscr{C}_n^{(t)})} \to 0,\end{aligned} \tag{6.4.13.3}$$

in view of 4.4.21, (6.1.21.10) and (see [37] II 1.1.2). It is therefore a Higgs $\theta_n^*(\mathscr{C}_n^{(t)})$-field with coefficients in $\sigma_n'^*(\widetilde{\xi}^{-1}\widetilde{\Omega}^1_{\overline{X}'_n/\overline{X}_n}) \otimes_{\overline{\mathscr{B}}'_n} \theta_n^*(\mathscr{C}_n^{(t)})$ by 2.5.26(i).

The morphism

$$\underline{d}_{\mathscr{C}_n'^{(t,r)}} : \mathscr{C}_n'^{(t,r)} \to \sigma_n'^*(\widetilde{\xi}^{-1}\widetilde{\Omega}^1_{\overline{X}'_n/\overline{X}_n}) \otimes_{\overline{\mathscr{B}}'_n} \mathscr{C}_n'^{(t,r)}, \tag{6.4.13.4}$$

deduced from $\underline{d}_{\mathscr{C}_n'^{(t)}}$ by extension of scalars, is the universal $\theta_n^*(\mathscr{C}_n^{(r)})$-derivation of $\mathscr{C}_n'^{(t,r)}$ (6.3.19.5). It is a Higgs $\theta_n^*(\mathscr{C}_n^{(r)})$-field with coefficients in $\sigma_n'^*(\widetilde{\xi}^{-1}\widetilde{\Omega}^1_{\overline{X}'_n/\overline{X}_n}) \otimes_{\overline{\mathscr{B}}'_n} \theta_n^*(\mathscr{C}_n^{(r)})$. We denote by $\underline{\mathbb{K}}^\bullet(\mathscr{C}_n'^{(t,r)})$ the Dolbeault complex of the Higgs $\theta_n^*(\mathscr{C}_n^{(r)})$-module $(\mathscr{C}_n'^{(t,r)}, p^t\underline{d}_{\mathscr{C}_n'^{(t,r)}})$ (2.5.1.2) and by $\underline{\widetilde{\mathbb{K}}}^\bullet(\mathscr{C}_n'^{(t,r)})$ the augmented Dolbeault complex

$$\theta_n^*(\mathscr{C}_n^{(r)}) \to \underline{\mathbb{K}}^0(\mathscr{C}_n'^{(t,r)}) \to \underline{\mathbb{K}}^1(\mathscr{C}_n'^{(t,r)}) \to \underline{\mathbb{K}}^2(\mathscr{C}_n'^{(t,r)}) \to \dots, \tag{6.4.13.5}$$

where $\theta_n^*(\mathscr{C}_n^{(r)})$ is placed in degree -1 and the first differential is the canonical homomorphism.

For all rational numbers t', r' such that $t' \geq t$ and $t' \geq r' \geq r$, we have (4.4.21.3)

$$p^{t'}(\mathrm{id} \otimes \alpha_n'^{t',t,r',r}) \circ \underline{d}_{\mathscr{C}_n'^{(t',r')}} = p^t\underline{d}_{\mathscr{C}_n'^{(t,r)}} \circ \alpha_n'^{t',t,r',r}, \tag{6.4.13.6}$$

where $\alpha_n'^{t',t,r',r}$ is the homomorphism (6.3.19.8). Consequently, $\alpha_n'^{t',t,r,r}$ induces a morphism

$$\iota_n^{t',t,r} : \underline{\widetilde{\mathbb{K}}}^\bullet(\mathscr{C}_n'^{(t',r)}) \to \underline{\widetilde{\mathbb{K}}}^\bullet(\mathscr{C}_n'^{(t,r)}). \tag{6.4.13.7}$$

Proposition 6.4.14 *For all rational numbers r, t, t' such that $t' > t > r \geq 0$, there exists a rational number $a \geq 0$ such that for all integers n and q with $n \geq 0$, the morphism*

$$\mathrm{H}^q(\iota_n^{t',t,r}) : \mathrm{H}^q(\underline{\widetilde{\mathbb{K}}}^\bullet(\mathscr{C}_n'^{(t',r)})) \to \mathrm{H}^q(\underline{\widetilde{\mathbb{K}}}^\bullet(\mathscr{C}_n'^{(t,r)})) \tag{6.4.14.1}$$

is annihilated by p^a.

The question being local on X ([3] VI.10.14 and [2] 3.4.10), we may assume that the scheme X is separated and that the morphism $f : (X, \mathscr{M}_X) \to (S, \mathscr{M}_S)$ admits an adequate chart $((P, \gamma), (\mathbb{N}, \iota), \vartheta)$.

Let $(\overline{y}' \rightsquigarrow \overline{x}')$ be a point of $X'_{\text{ét}} \overleftarrow{\times}_{X'_{\text{ét}}} \overline{X}'^{\triangleright}_{\text{ét}}$ such that $\overline{x}'$ is above s, n an integer ≥ 0. We take again the notation introduced in 6.4.1, 6.4.2 and 6.4.8. It follows from ([3] III.10.30) that the stalk of the derivation $\underline{d}_{\mathscr{C}_n'^{(t,r)}}$ (6.4.13.4) at $\rho'(\overline{y}' \rightsquigarrow \overline{x}')$ identifies with the direct limit of the universal $(\mathscr{C}_U^{\overline{y},(r)}/p^n\mathscr{C}_U^{\overline{y},(r)})$-derivations

$$\mathscr{C}_{U'\to U}'^{\overline{y}',(t,r)}/p^n\mathscr{C}_{U'\to U}'^{\overline{y}',(t,r)} \to \widetilde{\xi}^{-1}\widetilde{\Omega}^1_{X'/X}(U') \otimes_{\mathscr{O}_{X'}(U')} (\mathscr{C}_{U'\to U}'^{\overline{y}',(t,r)}/p^n\mathscr{C}_{U'\to U}'^{\overline{y}',(t,r)})$$

of $\mathscr{C}_{U'\to U}'^{\overline{y}',(t,r)}/p^n\mathscr{C}_{U'\to U}'^{\overline{y}',(t,r)}$, when $(\overline{x}' \to U' \to U)$ goes through the category $\mathbb{J}^\circ_{\overline{x}'}$ (see 5.3.1). The proposition then follows from 5.3.2(i) in view of ([3] III.9.5).

6.4.15 Let r, t be two rational numbers such that $t \geq r \geq 0$. We denote by

$$d_{\breve{\mathscr{C}}'^{(t)}} : \breve{\mathscr{C}}'^{(t)} \to \widehat{\sigma}'^*(\widetilde{\xi}^{-1}\widetilde{\Omega}^1_{\mathfrak{X}'/\mathscr{S}}) \otimes_{\overline{\breve{\mathscr{B}}}'} \breve{\mathscr{C}}'^{(t)} \tag{6.4.15.1}$$

the universal $\overline{\breve{\mathscr{B}}}'$-derivation of $\breve{\mathscr{C}}'^{(t)}$ (4.4.33.1), where $\widehat{\sigma}'$ is the morphism of ringed topos defined in (6.1.23.6). It is a Higgs $\overline{\breve{\mathscr{B}}}'$-field with coefficients in $\breve{\sigma}^*(\widetilde{\xi}^{-1}\widetilde{\Omega}^1_{\mathfrak{X}'/\mathscr{S}})$ by 4.4.23(iii). We denote by

$$\underline{d}_{\breve{\mathscr{C}}'^{(t)}} : \breve{\mathscr{C}}'^{(t)} \to \widehat{\sigma}'^*(\widetilde{\xi}^{-1}\widetilde{\Omega}^1_{\mathfrak{X}'/\mathfrak{X}}) \otimes_{\overline{\breve{\mathscr{B}}}'} \breve{\mathscr{C}}'^{(t)} \tag{6.4.15.2}$$

the $\overline{\breve{\mathscr{B}}}'$-derivation induced by $d_{\breve{\mathscr{C}}'^{(t)}}$. This is actually the universal $\breve{\theta}^*(\breve{\mathscr{C}}^{(t)})$-derivation of $\breve{\mathscr{C}}'^{(t)}$. This follows from the exact sequence

$$\widehat{\sigma}'^*(\mathfrak{g}^*(\widetilde{\xi}^{-1}\widetilde{\Omega}^1_{\mathfrak{X}/\mathscr{S}})) \otimes_{\overline{\breve{\mathscr{B}}}'} \breve{\mathscr{C}}'^{(t)} \to \widehat{\sigma}'^*(\widetilde{\xi}^{-1}\widetilde{\Omega}^1_{\mathfrak{X}'/\mathscr{S}}) \otimes_{\overline{\breve{\mathscr{B}}}'} \breve{\mathscr{C}}'^{(t)} \to \Omega^1_{\breve{\mathscr{C}}'^{(t)}/\breve{\theta}^*(\breve{\mathscr{C}}^{(t)})} \to 0, \tag{6.4.15.3}$$

taking into account (4.4.33.1) and (6.1.23.6) (see [37] II 1.1.2). Observe that $\underline{d}_{\breve{\mathscr{C}}'^{(t)}}$ identifies with the morphism $(\underline{d}_{\mathscr{C}'^{(t)}_{n+1}})_{n\in\mathbb{N}}$ (6.4.13.2), and that it is a Higgs $\breve{\theta}^*(\breve{\mathscr{C}}^{(t)})$-field with coefficients in $\widehat{\sigma}'^*(\widetilde{\xi}^{-1}\widetilde{\Omega}^1_{\mathfrak{X}'/\mathfrak{X}}) \otimes_{\overline{\breve{\mathscr{B}}}'} \breve{\theta}^*(\breve{\mathscr{C}}^{(t)})$.

The morphism

$$\underline{d}_{\breve{\mathscr{C}}'^{(t,r)}} : \breve{\mathscr{C}}'^{(t,r)} \to \widehat{\sigma}'^*(\widetilde{\xi}^{-1}\widetilde{\Omega}^1_{\mathfrak{X}'/\mathfrak{X}}) \otimes_{\overline{\breve{\mathscr{B}}}'} \breve{\mathscr{C}}'^{(t,r)}, \tag{6.4.15.4}$$

deduced by extension of scalars of $\underline{d}_{\breve{\mathscr{C}}'^{(t)}}$, is the universal $\breve{\theta}^*(\breve{\mathscr{C}}^{(r)})$-derivation of $\breve{\mathscr{C}}'^{(t,r)}$. It is a Higgs $\breve{\theta}^*(\breve{\mathscr{C}}^{(r)})$-field with coefficients in $\widehat{\sigma}'^*(\widetilde{\xi}^{-1}\widetilde{\Omega}^1_{\mathfrak{X}'/\mathfrak{X}}) \otimes_{\overline{\breve{\mathscr{B}}}'} \breve{\theta}^*(\breve{\mathscr{C}}^{(r)})$. We denote by $\underline{\mathbb{K}}^\bullet(\breve{\mathscr{C}}'^{(t,r)})$ the Dolbeault complex of the Higgs $\breve{\theta}^*(\breve{\mathscr{C}}^{(r)})$-module $(\breve{\mathscr{C}}'^{(t,r)}, p^t \underline{d}_{\breve{\mathscr{C}}'^{(t,r)}})$, and by $\underline{\widetilde{\mathbb{K}}}^\bullet(\breve{\mathscr{C}}'^{(t,r)})$ the augmented Dolbeault complex

$$\breve{\theta}^*(\breve{\mathscr{C}}^{(r)}) \to \underline{\mathbb{K}}^0(\breve{\mathscr{C}}'^{(t,r)}) \to \underline{\mathbb{K}}^1(\breve{\mathscr{C}}'^{(t,r)}) \to \cdots \to \underline{\mathbb{K}}^n(\breve{\mathscr{C}}'^{(t,r)}) \to \ldots, \tag{6.4.15.5}$$

where $\breve{\theta}^*(\breve{\mathscr{C}}^{(r)})$ is placed in degree -1 and the first differential is the canonical homomorphism.

For all rational numbers t', r' such that $t' \geq t$ and $t' \geq r' \geq r$, we have (6.4.13.6)

$$p^{t'}(\mathrm{id} \otimes \breve{\alpha}'^{t',t,r',r}) \circ \underline{d}_{\breve{\mathscr{C}}'^{(t',r')}} = p^t \underline{d}_{\breve{\mathscr{C}}'^{(t,r)}} \circ \breve{\alpha}'^{t',t,r',r}, \tag{6.4.15.6}$$

where $\breve{\alpha}'^{t',t,r',r}$ is the homomorphism (6.3.20.5). Consequently, $\breve{\alpha}'^{t',t,r,r}$ induces a morphism of complexes

$$\mathfrak{i}^{t',t,r} : \underline{\widetilde{\mathbb{K}}}^\bullet(\breve{\mathscr{C}}'^{(t',r)}) \to \underline{\widetilde{\mathbb{K}}}^\bullet(\breve{\mathscr{C}}'^{(t,r)}). \tag{6.4.15.7}$$

We denote by $\underline{\mathbb{K}}^\bullet_{\mathbb{Q}}(\breve{\mathscr{C}}'^{(t,r)})$ and $\underline{\widetilde{\mathbb{K}}}^\bullet_{\mathbb{Q}}(\breve{\mathscr{C}}'^{(t,r)})$ the images of the complexes $\underline{\mathbb{K}}^\bullet(\breve{\mathscr{C}}'^{(t,r)})$ and $\underline{\widetilde{\mathbb{K}}}^\bullet(\breve{\mathscr{C}}'^{(t,r)})$ in $\mathbf{Mod}_{\mathbb{Q}}(\breve{\theta}^*(\breve{\mathscr{C}}^{(r)}))$ (6.3.20). These complexes will also be considered as complexes of the category $\mathbf{Ind\text{-}Mod}(\breve{\theta}^*(\breve{\mathscr{C}}^{(r)}))$ via the functor $\alpha_{\breve{\theta}^*(\breve{\mathscr{C}}^{(r)})}$ (2.9.7.3).

Proposition 6.4.16 *For all rational numbers r, t, t' such that $t' > t > r \geq 0$ and every integer q, the canonical morphism of $\breve{\theta}^*(\breve{\mathscr{C}}^{(r)})_{\mathbb{Q}}$-modules* (6.4.15.7)

$$\mathrm{H}^q(\breve{\iota}^{t',t,r}_{\mathbb{Q}})\colon \mathrm{H}^q(\widetilde{\underline{\mathbb{K}}}^\bullet_{\mathbb{Q}}(\breve{\mathscr{C}}'^{(t',r)})) \to \mathrm{H}^q(\widetilde{\underline{\mathbb{K}}}^\bullet_{\mathbb{Q}}(\breve{\mathscr{C}}'^{(t,r)})) \qquad (6.4.16.1)$$

vanishes.

This follows from 6.4.14 and ([3] III.7.3(i)).

Corollary 6.4.17 *Let r, t, t' be three rational numbers such that $t' > t > r \geq 0$. Then,*

(i) *The canonical morphism of $\breve{\theta}^*(\breve{\mathscr{C}}^{(r)})_{\mathbb{Q}}$-modules*

$$u^{t,r}\colon \breve{\theta}^*(\breve{\mathscr{C}}^{(r)})_{\mathbb{Q}} \to \mathrm{H}^0(\underline{\mathbb{K}}^\bullet_{\mathbb{Q}}(\breve{\mathscr{C}}'^{(t,r)})) \qquad (6.4.17.1)$$

admits a canonical left inverse

$$v^{t,r}\colon \mathrm{H}^0(\underline{\mathbb{K}}^\bullet_{\mathbb{Q}}(\breve{\mathscr{C}}'^{(t,r)})) \to \breve{\theta}^*(\breve{\mathscr{C}}^{(r)})_{\mathbb{Q}}. \qquad (6.4.17.2)$$

(ii) *The composition*

$$\mathrm{H}^0(\underline{\mathbb{K}}^\bullet_{\mathbb{Q}}(\breve{\mathscr{C}}'^{(t',r)})) \xrightarrow{v^{t',r}} \breve{\theta}^*(\breve{\mathscr{C}}^{(r)})_{\mathbb{Q}} \xrightarrow{u^{t,r}} \mathrm{H}^0(\underline{\mathbb{K}}^\bullet_{\mathbb{Q}}(\breve{\mathscr{C}}'^{(t,r)})) \qquad (6.4.17.3)$$

is the canonical morphism.

(iii) *For every integer $q \geq 1$, the canonical morphism*

$$\mathrm{H}^q(\underline{\mathbb{K}}^\bullet_{\mathbb{Q}}(\breve{\mathscr{C}}'^{(t',r)})) \to \mathrm{H}^q(\underline{\mathbb{K}}^\bullet_{\mathbb{Q}}(\breve{\mathscr{C}}'^{(t,r)})) \qquad (6.4.17.4)$$

vanishes.

Indeed, consider the canonical commutative diagram (without the dotted arrow)

$$\begin{array}{ccccc}
\breve{\theta}^*(\breve{\mathscr{C}}^{(r)})_{\mathbb{Q}} & \xrightarrow{u^{t',r}} & \mathrm{H}^0(\underline{\mathbb{K}}^\bullet_{\mathbb{Q}}(\breve{\mathscr{C}}'^{(t',r)})) & \twoheadrightarrow & \mathrm{H}^0(\widetilde{\underline{\mathbb{K}}}^\bullet_{\mathbb{Q}}(\breve{\mathscr{C}}'^{(t',r)})) \\
\| & \swarrow{\scriptstyle v^{t',t,r}} & \downarrow{\scriptstyle \varpi^{t',t,r}} & & \downarrow{\scriptstyle \mathrm{H}^0(\breve{\iota}^{t',t,r}_{\mathbb{Q}})} \\
\breve{\theta}^*(\breve{\mathscr{C}}^{(r)})_{\mathbb{Q}} & \xrightarrow[u^{t,r}]{} & \mathrm{H}^0(\underline{\mathbb{K}}^\bullet_{\mathbb{Q}}(\breve{\mathscr{C}}'^{(t,r)})) & \twoheadrightarrow & \mathrm{H}^0(\widetilde{\underline{\mathbb{K}}}^\bullet_{\mathbb{Q}}(\breve{\mathscr{C}}'^{(t,r)})).
\end{array} \qquad (6.4.17.5)$$

It follows from 6.4.16 that $u^{t',r}$ and consequently $u^{t,r}$ are injective, and that there exists one and only one morphism $v^{t',t,r}$ as above such that $\varpi^{t',t,r} = u^{t,r} \circ v^{t',t,r}$. Since we have $u^{t,r} \circ v^{t',t,r} \circ u^{t',r} = u^{t,r}$, we deduce from this that $v^{t',t,r}$ is a left inverse of $u^{t',r}$. We immediately check that it does not depend on t; the propositions (i) and (ii) follow. Proposition (iii) immediately follows from 6.4.16.

Corollary 6.4.18 *For every rational number $r \geq 0$, the canonical morphism of complexes of ind-$\breve{\theta}^*(\breve{\mathscr{C}}^{(r)})$-modules*

$$\breve{\theta}^*(\breve{\mathscr{C}}^{(r)})_{\mathbb{Q}}[0] \to \text{“}\varinjlim_{r\in\mathbb{Q}_{>0}}\text{”}\ \underline{\mathbb{K}}^{\bullet}_{\mathbb{Q}}(\breve{\mathscr{C}}'^{(t,r)}) \tag{6.4.18.1}$$

is a quasi-isomorphism.

Recall first that **Ind-Mod**$(\breve{\theta}^*(\breve{\mathscr{C}}^{(r)}))$ admits small direct limits and that small filtered direct limits are exact (2.6.7.5). The proposition is therefore equivalent to the fact that the canonical morphism

$$\breve{\theta}^*(\breve{\mathscr{C}}^{(r)})_{\mathbb{Q}} \to \text{“}\varinjlim_{t\in\mathbb{Q}_{>r}}\text{”}\ \mathrm{H}^0(\underline{\mathbb{K}}^{\bullet}_{\mathbb{Q}}(\breve{\mathscr{C}}'^{(t,r)})) \tag{6.4.18.2}$$

is an isomorphism, and that for every integer $q \geq 1$,

$$\text{“}\varinjlim_{t\in\mathbb{Q}_{>r}}\text{”}\ \mathrm{H}^q(\underline{\mathbb{K}}^{\bullet}_{\mathbb{Q}}(\breve{\mathscr{C}}'^{(t,r)})) = 0. \tag{6.4.18.3}$$

These statements follow immediately from 6.4.17.

Remark 6.4.19 Let r be a rational number ≥ 0. Although filtered direct limits are not a priori representable in the category $\mathbf{Mod}_{\mathbb{Q}}(\breve{\theta}^*(\breve{\mathscr{C}}^{(r)}))$, it follows from 6.4.17 that in this category, the canonical morphism

$$\breve{\theta}^*(\breve{\mathscr{C}}^{(r)})_{\mathbb{Q}} \to \varinjlim_{t\in\mathbb{Q}_{>r}} \mathrm{H}^0(\underline{\mathbb{K}}^{\bullet}_{\mathbb{Q}}(\breve{\mathscr{C}}'^{(t,r)})) \tag{6.4.19.1}$$

is an isomorphism, and that for every integer $q \geq 1$,

$$\varinjlim_{t\in\mathbb{Q}_{>r}} \mathrm{H}^q(\underline{\mathbb{K}}^{\bullet}_{\mathbb{Q}}(\breve{\mathscr{C}}'^{(t,r)})) = 0. \tag{6.4.19.2}$$

6.4.20 Let t, r be two rational numbers such that $t \geq r \geq 0$. We denote by

$$v\colon \breve{\theta}^*(\widehat{\sigma}^*(\widetilde{\xi}^{-1}\widetilde{\Omega}^1_{\mathfrak{X}/\mathscr{S}})) \to \widehat{\sigma}'^*(\widetilde{\xi}^{-1}\widetilde{\Omega}^1_{\mathfrak{X}'/\mathscr{S}}) \tag{6.4.20.1}$$

the morphism induced by the canonical morphism $\mathfrak{g}^*(\widetilde{\Omega}^1_{\mathfrak{X}/\mathscr{S}}) \to \widetilde{\Omega}^1_{\mathfrak{X}'/\mathscr{S}}$ and the commutative diagram (6.1.23.6). By 6.3.13(ii), the diagram

$$\begin{array}{ccc}
\breve{\theta}^*(\breve{\mathscr{C}}^{(t)}) & \xrightarrow{\breve{\theta}^*(d_{\breve{\mathscr{C}}^{(t)}})} & \breve{\theta}^*(\widehat{\sigma}^*(\widetilde{\xi}^{-1}\widetilde{\Omega}^1_{\mathfrak{X}/\mathscr{S}})) \otimes_{\overline{\mathscr{B}}'_n} \breve{\theta}^*(\breve{\mathscr{C}}^{(t)}) \\
\Big\downarrow{\scriptstyle \breve{\omega}^{(t)}} & & \Big\downarrow{\scriptstyle v\otimes\breve{\omega}^{(t)}} \\
\breve{\mathscr{C}}'^{(t)} & \xrightarrow{d_{\breve{\mathscr{C}}'^{(t)}}} & \widehat{\sigma}'^*(\widetilde{\xi}^{-1}\widetilde{\Omega}^1_{\mathfrak{X}'/\mathscr{S}}) \otimes_{\breve{\overline{\mathscr{B}}}'} \breve{\mathscr{C}}'^{(t)},
\end{array} \tag{6.4.20.2}$$

where $\breve{\omega}^{(t)}$ is the homomorphism (6.3.20.1), is commutative. In view of (4.4.33.3), there thus exists a $\overline{\breve{\mathscr{B}}}'$-derivation of the algebra $\breve{\mathscr{C}}'^{(t,r)} = \breve{\mathscr{C}}'^{(t)} \otimes_{\breve{\theta}^*(\breve{\mathscr{C}}^{(t)})} \breve{\theta}^*(\breve{\mathscr{C}}^{(r)})$,

$$\delta_{\breve{\mathscr{C}}'^{(t,r)}} : \breve{\mathscr{C}}'^{(t,r)} \to \widehat{\sigma}'^*(\widetilde{\xi}^{-1}\widetilde{\Omega}^1_{\mathfrak{X}'/\mathcal{S}}) \otimes_{\overline{\breve{\mathscr{B}}}'} \breve{\mathscr{C}}'^{(t,r)}, \tag{6.4.20.3}$$

defined by

$$\delta_{\breve{\mathscr{C}}'^{(t,r)}} = p^t d_{\breve{\mathscr{C}}'^{(t)}} \otimes \mathrm{id}_{\breve{\theta}^*(\breve{\mathscr{C}}^{(r)})} + (\nu \otimes_{\overline{\breve{\mathscr{B}}}'} \mathrm{id}_{\breve{\mathscr{C}}'^{(t,r)}}) \circ (\mathrm{id}_{\breve{\mathscr{C}}'^{(t)}} \otimes \breve{\theta}^*(p^r d_{\breve{\mathscr{C}}^{(r)}})). \tag{6.4.20.4}$$

Since $d_{\breve{\mathscr{C}}'^{(t)}}$ and $d_{\breve{\mathscr{C}}^{(r)}}$ are Higgs fields by 4.4.23(ii), we check immediately that $\delta_{\breve{\mathscr{C}}'^{(t,r)}}$ is a Higgs $\overline{\breve{\mathscr{B}}}'$-field with coefficients in $\widehat{\sigma}'^*(\widetilde{\xi}^{-1}\widetilde{\Omega}^1_{\mathfrak{X}'/\mathcal{S}})$. The morphism

$$\underline{\delta}_{\breve{\mathscr{C}}'^{(t,r)}} : \breve{\mathscr{C}}'^{(t,r)} \to \widehat{\sigma}'^*(\widetilde{\xi}^{-1}\widetilde{\Omega}^1_{\mathfrak{X}'/\mathfrak{X}}) \otimes_{\overline{\breve{\mathscr{B}}}'} \breve{\mathscr{C}}'^{(t,r)} \tag{6.4.20.5}$$

induced by $\delta_{\breve{\mathscr{C}}'^{(t,r)}}$ is none other than $p^t \underline{d}_{\breve{\mathscr{C}}'^{(t,r)}}$ (6.4.15.4). It is thus a $\breve{\theta}^*(\breve{\mathscr{C}}^{(r)})$-derivation and a Higgs $\breve{\theta}^*(\breve{\mathscr{C}}^{(r)})$-field with coefficients in $\widehat{\sigma}'^*(\widetilde{\xi}^{-1}\widetilde{\Omega}^1_{\mathfrak{X}'/\mathfrak{X}}) \otimes_{\overline{\breve{\mathscr{B}}}'} \breve{\theta}^*(\breve{\mathscr{C}}^{(r)})$.

For all rational numbers t, t', r, r' such that $t' \geq t \geq r \geq 0$ and $t' \geq r' \geq r$, we have (4.4.21.3)

$$(\mathrm{id} \otimes \breve{\alpha}'^{t',t,r',r}) \circ \delta_{\breve{\mathscr{C}}'^{(t',r')}} = \delta_{\breve{\mathscr{C}}'^{(t,r)}} \circ \breve{\alpha}'^{t',t,r',r}, \tag{6.4.20.6}$$

where $\breve{\alpha}'^{t',t,r',r}$ is the homomorphism (6.4.15.6).

6.4.21 Let t, r be two rational numbers such that $t \geq r \geq 0$. We denote by $\mathbb{K}^\bullet(\breve{\mathscr{C}}'^{(t,r)})$ the Dolbeault complex of the Higgs $\overline{\breve{\mathscr{B}}}'$-module $(\breve{\mathscr{C}}'^{(t,r)}, \delta_{\breve{\mathscr{C}}'^{(t,r)}})$ (6.4.20.3) and by $\underline{\mathbb{K}}^\bullet(\breve{\mathscr{C}}'^{(t,r)})$ the Dolbeault complex of the Higgs $\breve{\theta}^*(\breve{\mathscr{C}}^{(r)})$-module $(\breve{\mathscr{C}}'^{(t,r)}, p^t \underline{d}_{\breve{\mathscr{C}}'^{(t,r)}})$ (6.4.15.4). We endow $\mathbb{K}^\bullet(\breve{\mathscr{C}}'^{(t,r)})$ with the Koszul filtration (2.5.11.6) associated with the image by the functor $\widehat{\sigma}'^*$ of the exact sequence (6.1.25.4), and we denote by

$$\partial^{(t,r)} : \underline{\mathbb{K}}^\bullet(\breve{\mathscr{C}}'^{(t,r)}) \to \widehat{\sigma}'^*(\mathfrak{g}^*(\widetilde{\xi}^{-1}\widetilde{\Omega}^1_{\mathfrak{X}/\mathcal{S}})) \otimes_{\overline{\breve{\mathscr{B}}}'} \underline{\mathbb{K}}^\bullet(\breve{\mathscr{C}}'^{(t,r)}) \tag{6.4.21.1}$$

the associated boundary map in $\mathbf{D}^+(\mathbf{Mod}(\overline{\breve{\mathscr{B}}}'))$ (2.5.11.9). The diagram of morphisms in $\mathbf{D}^+(\mathbf{Mod}(\overline{\breve{\mathscr{B}}}'))$

$$\begin{array}{ccc} \breve{\theta}^*(\breve{\mathscr{C}}^{(r)})[0] & \xrightarrow{p^r \breve{\theta}^*(d_{\breve{\mathscr{C}}^{(r)}})} & \breve{\theta}^*(\widehat{\sigma}^*(\widetilde{\xi}^{-1}\widetilde{\Omega}^1_{\mathfrak{X}/\mathcal{S}}) \otimes_{\overline{\breve{\mathscr{B}}}} \breve{\mathscr{C}}^{(r)})[0] \\ \downarrow & & \downarrow \\ \underline{\mathbb{K}}^\bullet(\breve{\mathscr{C}}'^{(t,r)}) & \xrightarrow{\partial^{(t,r)}} & \widehat{\sigma}'^*(\mathfrak{g}^*(\widetilde{\xi}^{-1}\widetilde{\Omega}^1_{\mathfrak{X}/\mathcal{S}})) \otimes_{\overline{\breve{\mathscr{B}}}'} \underline{\mathbb{K}}^\bullet(\breve{\mathscr{C}}'^{(t,r)}), \end{array} \tag{6.4.21.2}$$

where the vertical arrows are the canonical morphisms, see (6.4.15.5) and (6.1.23.6), is commutative. This follows from 2.5.14 taking for exact sequence (2.5.11.1) the image by the functor $\widehat{\sigma}'^*$ of the exact sequence (6.1.25.4), for (M, θ) the trivial Higgs

$\overline{\breve{\mathscr{B}}}'$-module $(\overline{\breve{\mathscr{B}}}', 0)$ and for (N, κ) the Higgs $\overline{\breve{\mathscr{B}}}'$-module $(\breve{\theta}^*(\breve{\mathscr{C}}^{(r)}), \breve{\theta}^*(d_{\breve{\mathscr{C}}^{(r)}}))$. With the notation of 2.5.12, we have $\underline{\theta}' = 0$, so that the differentials of $\underline{\mathbb{K}}'^{\bullet}$ are zero. The desired assertion then follows from 2.5.14 by functoriality of the boundary map (2.5.11.9).

6.5 Relative Cohomology of Dolbeault ind-modules

6.5.1 We denote by $\mathbf{Ind\text{-}Mod}^{\mathrm{Dolb}}(\overline{\breve{\mathscr{B}}}')$ the category of Dolbeault ind-$\overline{\breve{\mathscr{B}}}'$-modules (4.5.5) and by $\mathbf{HM}^{\mathrm{sol}}(\mathscr{O}_{\mathfrak{X}'}[\frac{1}{p}], \widetilde{\xi}^{-1}\widetilde{\Omega}^1_{\mathfrak{X}'/\mathscr{S}})$ the category of solvable Higgs $\mathscr{O}_{\mathfrak{X}'}[\frac{1}{p}]$-bundles with coefficients in $\widetilde{\xi}^{-1}\widetilde{\Omega}^1_{\mathfrak{X}'/\mathscr{S}}$, with respect to the deformation $(\widetilde{X}', \mathscr{M}_{\widetilde{X}'})$ fixed in 6.1.4. We denote by

$$\mathscr{H}' : \mathbf{Ind\text{-}Mod}(\overline{\breve{\mathscr{B}}}') \to \mathbf{HM}(\mathscr{O}_{\mathfrak{X}'}, \widetilde{\xi}^{-1}\widetilde{\Omega}^1_{\mathfrak{X}'/\mathscr{S}}) \tag{6.5.1.1}$$

the functor defined in (4.5.7.3), with respect to the deformation $(\widetilde{X}', \mathscr{M}_{\widetilde{X}'})$. By 4.5.20, this induces an equivalence of categories that we denote again by

$$\mathscr{H}' : \mathbf{Ind\text{-}Mod}^{\mathrm{Dolb}}(\overline{\breve{\mathscr{B}}}') \xrightarrow{\sim} \mathbf{HM}^{\mathrm{sol}}(\mathscr{O}_{\mathfrak{X}'}[\frac{1}{p}], \widetilde{\xi}^{-1}\widetilde{\Omega}^1_{\mathfrak{X}'/\mathscr{S}}). \tag{6.5.1.2}$$

We denote by

$$\underline{\mathscr{H}}' : \mathbf{Ind\text{-}Mod}(\overline{\breve{\mathscr{B}}}') \to \mathbf{HM}(\mathscr{O}_{\mathfrak{X}'}, \widetilde{\xi}^{-1}\widetilde{\Omega}^1_{\mathfrak{X}'/\mathfrak{X}}) \tag{6.5.1.3}$$

the composition of the functor $\mathscr{H}'$ with the canonical functor (6.1.25.4)

$$\mathbf{HM}(\mathscr{O}_{\mathfrak{X}'}, \widetilde{\xi}^{-1}\widetilde{\Omega}^1_{\mathfrak{X}'/\mathscr{S}}) \to \mathbf{HM}(\mathscr{O}_{\mathfrak{X}'}, \widetilde{\xi}^{-1}\widetilde{\Omega}^1_{\mathfrak{X}'/\mathfrak{X}}). \tag{6.5.1.4}$$

6.5.2 Let t, r be two rational numbers such that $t \geq r \geq 0$. We associate with $(f', \widetilde{X}', \mathscr{M}_{\widetilde{X}'})$ objects analogous to those associated with $(f, \widetilde{X}, \mathscr{M}_{\widetilde{X}})$ in 4.5.1-4.5.3 denoted by the same symbols endowed with an exponent $'$. In particular, we have the category $\mathbf{Ind\text{-}MC}(\breve{\mathscr{C}}'^{(t)}/\overline{\breve{\mathscr{B}}}')$ and the functors $\mathrm{I}\mathfrak{S}'^{(t)}$ (4.5.1.2), $\mathrm{I}\widehat{\sigma}'^{(t)*}$ (4.5.1.6) and $\mathrm{I}\varepsilon'^{t,t'}$ (4.5.3.1) for every rational number t' such that $t \geq t' \geq 0$.

We introduce relative variants. We denote by $\mathbf{Ind\text{-}MC}(\breve{\mathscr{C}}'^{(t,r)}/\breve{\theta}^*(\breve{\mathscr{C}}^{(r)}))$ the category of ind-$\breve{\mathscr{C}}'^{(t,r)}$-modules with integrable p^t-connection with respect to the extension $\breve{\mathscr{C}}'^{(t,r)}/\breve{\theta}^*(\breve{\mathscr{C}}^{(r)})$ (see 2.8.4 and 2.8.8). Each object in this category is a Higgs $\breve{\theta}^*(\breve{\mathscr{C}}^{(r)})$-ind-module with coefficients in $\widehat{\sigma}'^*(\widetilde{\xi}^{-1}\widetilde{\Omega}^1_{\mathfrak{X}'/\mathfrak{X}}) \otimes_{\overline{\breve{\mathscr{B}}}'} \breve{\theta}^*(\breve{\mathscr{C}}^{(r)})$ by 2.8.11(i). We can therefore associate with it a Dolbeault complex in $\mathbf{Ind\text{-}Mod}(\breve{\theta}^*(\breve{\mathscr{C}}^{(r)}))$.

We have the functor

$$\begin{aligned} \mathrm{I}\underline{\mathfrak{S}}'^{(t,r)} : \mathbf{Ind\text{-}Mod}(\overline{\breve{\mathscr{B}}}') &\to \mathbf{Ind\text{-}MC}(\breve{\mathscr{C}}'^{(t,r)}/\breve{\theta}^*(\breve{\mathscr{C}}^{(r)})) \\ \mathscr{M} &\mapsto (\breve{\mathscr{C}}'^{(t,r)} \otimes_{\overline{\breve{\mathscr{B}}}'} \mathscr{M}, p^t \underline{d}_{\breve{\mathscr{C}}'^{(t,r)}} \otimes_{\overline{\breve{\mathscr{B}}}'} \mathrm{id}_{\mathscr{M}}), \end{aligned} \tag{6.5.2.1}$$

where $\underline{d}_{\breve{\mathscr{C}}'^{(t,r)}}$ is the universal $\breve{\theta}^*(\breve{\mathscr{C}}^{(r)})$-derivation of $\breve{\mathscr{C}}'^{(t,r)}$ (6.4.15.4). In view of 2.8.11(ii) and with the notation of 6.1.26, the functor $\mathrm{I}\widehat{\sigma}'^*$ (6.1.23.6) induces a functor

$$\begin{aligned} \mathrm{I}\underline{\widehat{\sigma}}'^{(t,r)*}\colon \mathbf{Ind\text{-}HM}(\mathscr{O}_{\mathfrak{X}'}, \widetilde{\xi}^{-1}\widetilde{\Omega}^1_{\mathfrak{X}'/\mathfrak{X}}) &\to \mathbf{Ind\text{-}MC}(\breve{\mathscr{C}}'^{(t,r)}/\breve{\theta}^*(\breve{\mathscr{C}}^{(r)})), \quad (6.5.2.2)\\ (\mathscr{N}, \theta) &\mapsto (\breve{\mathscr{C}}'^{(t,r)} \otimes_{\overline{\breve{\mathscr{B}}}'} \mathrm{I}\widehat{\sigma}'^*(\mathscr{N}), p^t \underline{d}_{\breve{\mathscr{C}}'^{(t,r)}} \otimes_{\overline{\breve{\mathscr{B}}}'} \mathrm{id} + \mathrm{id} \otimes_{\overline{\breve{\mathscr{B}}}'} \mathrm{I}\widehat{\sigma}'^*(\theta)). \end{aligned}$$

Let $(\mathscr{F}, \nabla)$ be an ind-$\breve{\mathscr{C}}'^{(t)}$-module with integrable p^t-connection with respect to the extension $\breve{\mathscr{C}}'^{(t)}/\overline{\breve{\mathscr{B}}}'$. We denote by

$$\underline{\nabla}\colon \mathscr{F} \to \widehat{\sigma}'^*(\widetilde{\xi}^{-1}\widetilde{\Omega}^1_{\mathfrak{X}'/\mathfrak{X}}) \otimes_{\overline{\breve{\mathscr{B}}}'} \mathscr{F} \tag{6.5.2.3}$$

the morphism of ind-$\overline{\breve{\mathscr{B}}}'$-modules deduced from ∇. Since the universal $\breve{\theta}^*(\breve{\mathscr{C}}^{(t)})$-derivation $\underline{d}_{\breve{\mathscr{C}}'^{(t)}}$ of $\breve{\mathscr{C}}'^{(t)}$ (6.4.15.2) is induced by $d_{\breve{\mathscr{C}}'^{(t)}}$ (6.4.15.1), we can canonically consider $\underline{\nabla}$ as a morphism of ind-$\breve{\theta}^*(\breve{\mathscr{C}}^{(t)})$-modules by 2.7.6 and 2.7.7, and it is then an integrable p^t-connection on $\mathscr{F}$ with respect to the extension $\breve{\mathscr{C}}'^{(t)}/\breve{\theta}^*(\breve{\mathscr{C}}^{(t)})$. The morphism

$$\underline{\nabla}^{(t,r)}\colon \mathscr{F} \otimes_{\breve{\theta}^*(\breve{\mathscr{C}}^{(t)})} \breve{\mathscr{C}}^*(\breve{\mathscr{C}}^{(r)}) \to \widehat{\sigma}'^*(\widetilde{\xi}^{-1}\widetilde{\Omega}^1_{\mathfrak{X}'/\mathfrak{X}}) \otimes_{\overline{\breve{\mathscr{B}}}'} \mathscr{F} \otimes_{\breve{\theta}^*(\breve{\mathscr{C}}^{(t)})} \breve{\theta}^*(\breve{\mathscr{C}}^{(r)}), \tag{6.5.2.4}$$

deduced from $\underline{\nabla}$ by extension of scalars, is an integrable p^t-connection on the ind-$\breve{\mathscr{C}}'^{(t,r)}$-module $\mathscr{F} \otimes_{\breve{\mathscr{C}}'^{(t)}} \breve{\mathscr{C}}'^{(t,r)}$ with respect to the extension $\breve{\mathscr{C}}'^{(t,r)}/\breve{\theta}^*(\breve{\mathscr{C}}^{(r)})$ (2.8.10). We thus obtain a functor

$$\begin{aligned} \mathrm{I}\underline{\lambda}^{t,r}\colon \mathbf{Ind\text{-}MC}(\breve{\mathscr{C}}'^{(t)}/\overline{\breve{\mathscr{B}}}') &\to \mathbf{Ind\text{-}MC}(\breve{\mathscr{C}}'^{(t,r)}/\breve{\theta}^*(\breve{\mathscr{C}}^{(r)})) \quad (6.5.2.5)\\ (\mathscr{F}, \nabla) &\mapsto (\mathscr{F} \otimes_{\breve{\mathscr{C}}'^{(t)}} \breve{\mathscr{C}}'^{(t,r)}, \underline{\nabla}^{(t,r)}). \end{aligned}$$

We immediately check that we have a canonical isomorphism of functors

$$\mathrm{I}\underline{\mathfrak{S}}'^{(t,r)} \xrightarrow{\sim} \mathrm{I}\underline{\lambda}^{t,r} \circ \mathrm{I}\mathfrak{S}'^{(t)}. \tag{6.5.2.6}$$

We also check that the diagram

$$\begin{array}{ccc} \mathbf{Ind\text{-}HM}(\mathscr{O}_{\mathfrak{X}'}, \widetilde{\xi}^{-1}\widetilde{\Omega}^1_{\mathfrak{X}'/S}) & \xrightarrow{\mathrm{I}\widehat{\sigma}'^{(t)*}} & \mathbf{Ind\text{-}MC}(\breve{\mathscr{C}}'^{(t)}/\overline{\breve{\mathscr{B}}}') \\ \downarrow & & \downarrow{\scriptstyle \mathrm{I}\underline{\lambda}^{t,r}} \\ \mathbf{Ind\text{-}HM}(\mathscr{O}_{\mathfrak{X}'}, \widetilde{\xi}^{-1}\widetilde{\Omega}^1_{\mathfrak{X}'/\mathfrak{X}}) & \xrightarrow{\mathrm{I}\underline{\widehat{\sigma}}'^{(t,r)*}} & \mathbf{Ind\text{-}MC}(\breve{\mathscr{C}}'^{(t,r)}/\breve{\theta}^*(\breve{\mathscr{C}}^{(r)})), \end{array} \tag{6.5.2.7}$$

where the unlabeled arrow is the canonical functor, is commutative up to canonical isomorphism.

Let t', r' be two rational numbers such that $t \geq t' \geq r' \geq 0$ and $r \geq r'$. By 2.8.9 and (6.4.15.6), for every ind-$\breve{\mathscr{C}}'^{(t,r)}$-module with integrable p^t-connection $(\mathscr{F}, \nabla)$ with respect to the extension $\breve{\mathscr{C}}'^{(t,r)}/\breve{\theta}^*(\breve{\mathscr{C}}^{(r)})$, the ind-$\breve{\mathscr{C}}'^{(t',r')}$-module $\mathscr{F} \otimes_{\breve{\mathscr{C}}'^{(t,r)}} \breve{\mathscr{C}}'^{(t',r')}$ defined by extension of scalars (6.3.20.5) is canonically endowed with an integrable $p^{t'}$-connection ∇' with respect to the extension $\breve{\mathscr{C}}'^{(t',r')}/\breve{\theta}^*(\breve{\mathscr{C}}^{(r')})$. We thus define a functor

$$\begin{array}{rccc} \mathrm{I}\underline{\varepsilon}'^{t,t',r,r'}: & \mathbf{Ind\text{-}MC}(\breve{\mathscr{C}}'^{(t,r)}/\check{\theta}^*(\breve{\mathscr{C}}^{(r)})) & \to & \mathbf{Ind\text{-}MC}(\breve{\mathscr{C}}'^{(t',r')}/\check{\theta}^*(\breve{\mathscr{C}}^{(r')})) \\ & (\mathscr{F},\nabla) & \mapsto & (\mathscr{F}\otimes_{\breve{\mathscr{C}}'^{(t,r)}}\breve{\mathscr{C}}'^{(t',r')},\nabla'). \end{array} \tag{6.5.2.8}$$

We have canonical isomorphisms of functors

$$\mathrm{I}\underline{\varepsilon}'^{t,t',r,r'}\circ \mathrm{I}\underline{\mathfrak{S}}'^{(t,r)} \xrightarrow{\sim} \mathrm{I}\underline{\mathfrak{S}}'^{(t',r')}, \tag{6.5.2.9}$$

$$\mathrm{I}\underline{\varepsilon}'^{t,t',r,r'}\circ \mathrm{I}\underline{\widehat{\sigma}}'^{(t,r)*} \xrightarrow{\sim} \mathrm{I}\underline{\widehat{\sigma}}'^{(t',r')*}. \tag{6.5.2.10}$$

For all rational numbers t',r',t'',r'' such $t\geq t'\geq t''\geq r''\geq 0$, $t'\geq r'\geq r''$ and $r\geq r'$, we have a canonical isomorphism of functors

$$\mathrm{I}\underline{\varepsilon}'^{t',t'',r',r''}\circ \mathrm{I}\underline{\varepsilon}'^{t,t',r,r'} \xrightarrow{\sim} \mathrm{I}\underline{\varepsilon}'^{t,t'',r,r''}. \tag{6.5.2.11}$$

6.5.3 Let t,r be two rational numbers such that $t\geq r\geq 0$. We associate with $(f',\widetilde{X}',\mathscr{M}_{\widetilde{X}'})$ objects analogous to those associated with $(f,\widetilde{X},\mathscr{M}_{\widetilde{X}})$ in 4.6.1–4.6.3 denoted by the same symbols endowed with an exponent $'$. In particular, we have the category $\mathbf{IC}_{\mathbb{Q}}(\breve{\mathscr{C}}'^{(t)}/\overline{\breve{\mathscr{B}}}')$ and the functor $\widehat{\sigma}'^{(t)*}$ (4.6.1.7).

We introduce relative variants. We will simply denote by $\mathbf{IC}(\breve{\mathscr{C}}'^{(t,r)}/\check{\theta}^*(\breve{\mathscr{C}}^{(r)}))$ the category of integrable p^t-isoconnections with respect to the extension $\breve{\mathscr{C}}'^{(t,r)}/\check{\theta}^*(\breve{\mathscr{C}}^{(r)})$ (see 2.9.11); we therefore omit the exponent p^t from the notation introduced in 2.9.11 considering that it is redundant with the exponent of $\breve{\mathscr{C}}'^{(t,r)}$. It is an additive category. We denote by $\mathbf{IC}_{\mathbb{Q}}(\breve{\mathscr{C}}'^{(t,r)}/\check{\theta}^*(\breve{\mathscr{C}}^{(r)}))$ the category of objects of $\mathbf{IC}(\breve{\mathscr{C}}'^{(t,r)}/\check{\theta}^*(\breve{\mathscr{C}}^{(r)}))$ up to isogeny. By 2.9.14(i), every object of $\mathbf{IC}(\breve{\mathscr{C}}'^{(t,r)}/\check{\theta}^*(\breve{\mathscr{C}}^{(r)}))$ is a Higgs $\check{\theta}^*(\breve{\mathscr{C}}^{(r)})$-isogeny with coefficients in $\widehat{\sigma}'^*(\widetilde{\xi}^{-1}\widetilde{\Omega}^1_{\mathfrak{X}'/\mathfrak{X}})$ (6.4.15.4). In particular, one can functorially associate with every object of $\mathbf{IC}_{\mathbb{Q}}(\breve{\mathscr{C}}'^{(t,r)}/\check{\theta}^*(\breve{\mathscr{C}}^{(r)}))$ a Dolbeault complex in $\mathbf{Mod}_{\mathbb{Q}}(\check{\theta}^*(\breve{\mathscr{C}}^{(r)}))$ (see 2.9.9).

If $(\mathscr{N},\mathscr{N}',v,\theta)$ is a Higgs $\mathscr{O}_{\mathfrak{X}'}$-isogeny with coefficients in $\widetilde{\xi}^{-1}\widetilde{\Omega}^1_{\mathfrak{X}'/\mathfrak{X}}$,

$$(\breve{\mathscr{C}}'^{(t,r)}\otimes_{\overline{\breve{\mathscr{B}}}'}\widehat{\sigma}'^*(\mathscr{N}),\breve{\mathscr{C}}'^{(t,r)}\otimes_{\overline{\breve{\mathscr{B}}}'}\widehat{\sigma}'^*(\mathscr{N}'),\mathrm{id}\otimes_{\overline{\breve{\mathscr{B}}}'}\widehat{\sigma}'^*(v),p^t\underline{d}_{\breve{\mathscr{C}}'^{(t,r)}}\otimes\widehat{\sigma}'^*(v)+\mathrm{id}\otimes\widehat{\sigma}'^*(\theta))$$

is an object of $\mathbf{IC}(\breve{\mathscr{C}}'^{(t,r)}/\check{\theta}^*(\breve{\mathscr{C}}^{(r)}))$. With the notation of 6.1.26, we thus obtain a functor

$$\underline{\widehat{\sigma}}'^{(t,r)*}\colon \mathbf{HI}(\mathscr{O}_{\mathfrak{X}'},\widetilde{\xi}^{-1}\widetilde{\Omega}^1_{\mathfrak{X}'/\mathfrak{X}}) \to \mathbf{IC}(\breve{\mathscr{C}}'^{(t,r)}/\check{\theta}^*(\breve{\mathscr{C}}^{(r)})). \tag{6.5.3.1}$$

In view of (6.1.26.1), this induces a functor that we denote again by

$$\underline{\widehat{\sigma}}'^{(t,r)*}_{\mathbb{Q}}\colon \mathbf{HM}^{\mathrm{coh}}(\mathscr{O}_{\mathfrak{X}'}[\frac{1}{p}],\widetilde{\xi}^{-1}\widetilde{\Omega}^1_{\mathfrak{X}'/\mathfrak{X}}) \to \mathbf{IC}_{\mathbb{Q}}(\breve{\mathscr{C}}'^{(t,r)}/\check{\theta}^*(\breve{\mathscr{C}}^{(r)})). \tag{6.5.3.2}$$

6.5.4 Let t,r be two rational numbers such that $t\geq r\geq 0$, $(\mathscr{F},\nabla)$ an ind-$\breve{\mathscr{C}}'^{(t)}$-module with integrable p^t-connection with respect to the extension $\breve{\mathscr{C}}'^{(t)}/\overline{\breve{\mathscr{B}}}'$. Recall that the $\overline{\breve{\mathscr{B}}}'$-derivation $\delta_{\breve{\mathscr{C}}'^{(t,r)}}$ (6.4.20.3) of the algebra $\breve{\mathscr{C}}'^{(t,r)}$ is a Higgs $\overline{\breve{\mathscr{B}}}'$-field with coefficients in $\widehat{\sigma}'^*(\widetilde{\xi}^{-1}\widetilde{\Omega}^1_{\mathfrak{X}'/\mathscr{S}})$. By 2.8.12, ∇ and $\delta_{\breve{\mathscr{C}}'^{(t,r)}}$ define on the $\breve{\mathscr{C}}'^{(t,r)}$-module

$\mathscr{F} \otimes_{\breve{\mathscr{C}}'^{(t)}} \breve{\mathscr{C}}'^{(t,r)}$ a Higgs $\overline{\breve{\mathscr{B}}}'$-field with coefficients in $\widehat{\sigma}'^*(\widetilde{\xi}^{-1}\widetilde{\Omega}^1_{\mathfrak{X}'/\mathscr{S}})$ denoted by $\nabla^{(t,r)}$.

For $\mathscr{Z} = \mathfrak{X}$ or $\mathscr{S}$, we denote by $\mathbf{Ind\text{-}HM}(\overline{\breve{\mathscr{B}}}', \widehat{\sigma}'^*(\widetilde{\xi}^{-1}\widetilde{\Omega}^1_{\mathfrak{X}'/\mathscr{Z}}))$ the category of Higgs ind-$\overline{\breve{\mathscr{B}}}'$-modules with coefficients in $\widehat{\sigma}'^*(\widetilde{\xi}^{-1}\widetilde{\Omega}^1_{\mathfrak{X}'/\mathscr{Z}})$. We then get a functor

$$\begin{aligned} \mathrm{I}\lambda^{t,r}\colon \mathbf{Ind\text{-}MC}(\breve{\mathscr{C}}'^{(t)}/\overline{\breve{\mathscr{B}}}') &\to \mathbf{Ind\text{-}HM}(\overline{\breve{\mathscr{B}}}', \widehat{\sigma}'^*(\widetilde{\xi}^{-1}\widetilde{\Omega}^1_{\mathfrak{X}'/\mathscr{S}})) \\ (\mathscr{F}, \nabla) &\mapsto (\mathscr{F} \otimes_{\breve{\mathscr{C}}'^{(t)}} \breve{\mathscr{C}}'^{(t,r)}, \nabla^{(t,r)}). \end{aligned} \tag{6.5.4.1}$$

We check immediately that the diagram

$$\begin{array}{ccc} \mathbf{Ind\text{-}MC}(\breve{\mathscr{C}}'^{(t)}/\overline{\breve{\mathscr{B}}}') & \xrightarrow{\mathrm{I}\lambda^{t,r}} & \mathbf{Ind\text{-}HM}(\overline{\breve{\mathscr{B}}}', \widehat{\sigma}'^*(\widetilde{\xi}^{-1}\widetilde{\Omega}^1_{\mathfrak{X}'/\mathscr{S}})) \\ \Big\downarrow{\scriptstyle \mathrm{I}\underline{\lambda}^{t,r}} & & \Big\downarrow \\ \mathbf{Ind\text{-}MC}(\breve{\mathscr{C}}'^{(t,r)}/\breve{\theta}^*(\breve{\mathscr{C}}^{(r)})) & \longrightarrow & \mathbf{Ind\text{-}HM}(\overline{\breve{\mathscr{B}}}', \widehat{\sigma}^*(\widetilde{\xi}^{-1}\widetilde{\Omega}^1_{\mathfrak{X}'/\mathfrak{X}})), \end{array} \tag{6.5.4.2}$$

where $\mathrm{I}\underline{\lambda}^{t,r}$ is the functor (6.5.2.5), and the unlabeled functors are the canonical functors, is commutative.

We denote by $\mathrm{I}\widehat{\sigma}'^{(t,r)*}$ the composed functor $\mathrm{I}\lambda^{t,r} \circ \mathrm{I}\widehat{\sigma}'^{(t)*}$ (4.5.1.6). In view of 2.8.12 and 6.4.20, this one is defined by

$$\begin{aligned} \mathrm{I}\widehat{\sigma}'^{(t,r)*}\colon \mathbf{Ind\text{-}HM}(\mathscr{O}_{\mathfrak{X}'}, \widetilde{\xi}^{-1}\widetilde{\Omega}^1_{\mathfrak{X}'/\mathscr{S}}) &\to \mathbf{Ind\text{-}HM}(\overline{\breve{\mathscr{B}}}', \widehat{\sigma}'^*(\widetilde{\xi}^{-1}\widetilde{\Omega}^1_{\mathfrak{X}'/\mathscr{S}})), \\ (\mathscr{N}, \theta) \mapsto (\breve{\mathscr{C}}'^{(t,r)} \otimes_{\underline{\breve{\mathscr{B}}}'} \mathrm{I}\widehat{\sigma}'^*(\mathscr{N}), \delta_{\breve{\mathscr{C}}'^{(t,r)}} &\otimes_{\underline{\breve{\mathscr{B}}}'} \mathrm{id} + \mathrm{id} \otimes_{\underline{\breve{\mathscr{B}}}'} \mathrm{I}\widehat{\sigma}'^*(\theta)). \end{aligned} \tag{6.5.4.3}$$

It follows from (6.5.2.7) and (6.5.4.2) that the diagram

$$\begin{array}{ccc} \mathbf{Ind\text{-}HM}(\mathscr{O}_{\mathfrak{X}'}, \widetilde{\xi}^{-1}\widetilde{\Omega}^1_{\mathfrak{X}'/\mathscr{S}}) & \xrightarrow{\mathrm{I}\widehat{\sigma}'^{(t,r)*}} & \mathbf{Ind\text{-}HM}(\overline{\breve{\mathscr{B}}}', \widehat{\sigma}^*(\widetilde{\xi}^{-1}\widetilde{\Omega}^1_{\mathfrak{X}'/\mathscr{S}})) \\ \Big\downarrow & & \Big\downarrow \\ \mathbf{Ind\text{-}HM}(\mathscr{O}_{\mathfrak{X}'}, \widetilde{\xi}^{-1}\widetilde{\Omega}^1_{\mathfrak{X}'/\mathfrak{X}}) & & \\ \Big\downarrow{\scriptstyle \mathrm{I}\underline{\widehat{\sigma}}'^{(t,r)*}} & & \\ \mathbf{Ind\text{-}MC}(\breve{\mathscr{C}}'^{(t,r)}/\breve{\theta}^*(\breve{\mathscr{C}}^{(r)})) & \longrightarrow & \mathbf{Ind\text{-}HM}(\overline{\breve{\mathscr{B}}}', \widehat{\sigma}^*(\widetilde{\xi}^{-1}\widetilde{\Omega}^1_{\mathfrak{X}'/\mathfrak{X}})), \end{array} \tag{6.5.4.4}$$

where $\mathrm{I}\underline{\widehat{\sigma}}'^{(t,r)*}$ is the functor (6.5.2.2) and the unlabeled functors are the canonical functors, is commutative.

We denote by $\mathrm{I}\mathfrak{S}'^{(t,r)}$ the composed functor $\mathrm{I}\lambda^{t,r} \circ \mathrm{I}\mathfrak{S}'^{(t)}$ (4.5.1.2). In view of 2.8.12 and 6.4.20, this one is defined by

$$\begin{aligned} \mathrm{I}\mathfrak{S}'^{(t,r)}\colon \mathbf{Ind\text{-}Mod}(\overline{\breve{\mathscr{B}}}') &\to \mathbf{Ind\text{-}HM}(\overline{\breve{\mathscr{B}}}', \widehat{\sigma}^*(\widetilde{\xi}^{-1}\widetilde{\Omega}^1_{\mathfrak{X}'/\mathscr{S}})), \\ \mathscr{M} &\mapsto (\breve{\mathscr{C}}'^{(t,r)} \otimes_{\underline{\breve{\mathscr{B}}}'} \mathscr{M}, \delta_{\breve{\mathscr{C}}'^{(t,r)}} \otimes_{\underline{\breve{\mathscr{B}}}'} \mathrm{id}). \end{aligned} \tag{6.5.4.5}$$

Lemma 6.5.5 *Let $\mathscr{M}$ be a rational* (4.3.15) *and flat* (2.7.9) *ind-$\overline{\breve{\mathscr{B}}}'$-module, r a rational number ≥ 0, q an integer ≥ 0. We then have a canonical functorial isomorphism of ind-$\breve{\mathfrak{g}}^*(\breve{\mathscr{C}}^{(r)})$-modules* (6.1.22.1)

$$\mathrm{R}^q\mathrm{I}\breve{\tau}_*(\mathscr{M}\otimes_{\overline{\breve{\mathscr{B}}}'}\breve{\theta}^*(\breve{\mathscr{C}}^{(r)}))\xrightarrow{\sim}\text{“}\varinjlim_{t\in\mathbb{Q}_{>r}}\text{”}\,\mathrm{R}^q\mathrm{I}\breve{\tau}_*(\mathscr{M}\otimes_{\overline{\breve{\mathscr{B}}}'}\underline{\mathbb{K}}^\bullet(\breve{\mathscr{C}}'^{(t,r)})),\qquad(6.5.5.1)$$

where $\mathrm{I}\breve{\tau}_*$ is the functor (6.3.20.9) *and $\underline{\mathbb{K}}^\bullet(\breve{\mathscr{C}}'^{(t,r)})$ is the Dolbeault complex of the Higgs $\breve{\theta}^*(\breve{\mathscr{C}}^{(r)})$-module $(\breve{\mathscr{C}}'^{(t,r)}, p^t\underline{d}_{\breve{\mathscr{C}}'^{(t,r)}})$* (6.4.15).

Indeed, by virtue of 6.4.18, the canonical morphism of complexes of ind-$\breve{\theta}^*(\breve{\mathscr{C}}^{(r)})$-modules

$$\breve{\theta}^*(\breve{\mathscr{C}}^{(r)})_{\mathbb{Q}}[0]\to\text{“}\varinjlim_{t\in\mathbb{Q}_{>r}}\text{”}\,\underline{\mathbb{K}}^\bullet_{\mathbb{Q}}(\breve{\mathscr{C}}'^{(t,r)})\qquad(6.5.5.2)$$

is a quasi-isomorphism. Moreover, $\mathscr{M}$ being rational, for every $\overline{\breve{\mathscr{B}}}'$-module $\mathscr{F}$, the canonical morphism $\mathscr{M}\otimes_{\overline{\breve{\mathscr{B}}}'}\mathscr{F}\to\mathscr{M}\otimes_{\overline{\breve{\mathscr{B}}}'}\mathscr{F}_{\mathbb{Q}}$ (2.6.6.7) is an isomorphism. Since $\mathscr{M}$ is flat, we deduce from this that the canonical morphism of complexes of ind-$\breve{\theta}^*(\breve{\mathscr{C}}^{(r)})$-modules

$$\mathscr{M}\otimes_{\overline{\breve{\mathscr{B}}}'}\breve{\theta}^*(\breve{\mathscr{C}}^{(r)})[0]\to\text{“}\varinjlim_{t\in\mathbb{Q}_{>r}}\text{”}\,\mathscr{M}\otimes_{\overline{\breve{\mathscr{B}}}'}\underline{\mathbb{K}}^\bullet(\breve{\mathscr{C}}'^{(t,r)})\qquad(6.5.5.3)$$

is a quasi-isomorphism. Since $\mathrm{R}^q\mathrm{I}\breve{\tau}_*$ commutes with small filtered direct limits (2.6.9.4), we deduce from this the isomorphism (6.5.5.1).

6.5.6 Let $\mathbf{HI}(\overline{\breve{\mathscr{B}}}^!,\widehat{\pi}^*(\widetilde{\xi}^{-1}\widetilde{\Omega}^1_{\mathfrak{X}'/\mathfrak{X}}))$ be the category of Higgs $\overline{\breve{\mathscr{B}}}^!$-isogenies with coefficients in $\widehat{\pi}^*(\widetilde{\xi}^{-1}\widetilde{\Omega}^1_{\mathfrak{X}'/\mathfrak{X}})$ (2.9.9) and $\mathbf{HI}_{\mathbb{Q}}(\overline{\breve{\mathscr{B}}}^!,\widehat{\pi}^*(\widetilde{\xi}^{-1}\widetilde{\Omega}^1_{\mathfrak{X}'/\mathfrak{X}}))$ the category of objects of $\mathbf{HI}(\overline{\breve{\mathscr{B}}}^!,\widehat{\pi}^*(\widetilde{\xi}^{-1}\widetilde{\Omega}^1_{\mathfrak{X}'/\mathfrak{X}}))$ up to isogeny ([3] III.6.1.1). We associate with every object $\mathscr{F}$ of $\mathbf{HI}_{\mathbb{Q}}(\overline{\breve{\mathscr{B}}}^!,\widehat{\pi}^*(\widetilde{\xi}^{-1}\widetilde{\Omega}^1_{\mathfrak{X}'/\mathfrak{X}}))$ a Dolbeault complex $\underline{\mathbb{K}}(\mathscr{F})$ in $\mathbf{Mod}_{\mathbb{Q}}(\overline{\breve{\mathscr{B}}}^!)$ (2.9.9.4).

With the notation of 6.1.26, the morphism $\widehat{\pi}$ (6.1.23.5) induces a functor

$$\widehat{\pi}^*\colon\mathbf{HI}(\mathscr{O}_{\mathfrak{X}'},\widetilde{\xi}^{-1}\widetilde{\Omega}^1_{\mathfrak{X}'/\mathfrak{X}})\to\mathbf{HI}(\overline{\breve{\mathscr{B}}}^!,\widehat{\pi}^*(\widetilde{\xi}^{-1}\widetilde{\Omega}^1_{\mathfrak{X}'/\mathfrak{X}})),\qquad(6.5.6.1)$$

and hence a functor

$$\widehat{\pi}^*_{\mathbb{Q}}\colon\mathbf{HI}_{\mathbb{Q}}(\mathscr{O}_{\mathfrak{X}'},\widetilde{\xi}^{-1}\widetilde{\Omega}^1_{\mathfrak{X}'/\mathfrak{X}})\to\mathbf{HI}_{\mathbb{Q}}(\overline{\breve{\mathscr{B}}}^!,\widehat{\pi}^*(\widetilde{\xi}^{-1}\widetilde{\Omega}^1_{\mathfrak{X}'/\mathfrak{X}})).\qquad(6.5.6.2)$$

Taking into account (6.1.26.1), we deduce from this a functor that we denote again by

$$\widehat{\pi}^*_{\mathbb{Q}}\colon\mathbf{HM}^{\mathrm{coh}}(\mathscr{O}_{\mathfrak{X}'}[\tfrac{1}{p}],\widetilde{\xi}^{-1}\widetilde{\Omega}^1_{\mathfrak{X}'/\mathfrak{X}})\to\mathbf{HI}_{\mathbb{Q}}(\overline{\breve{\mathscr{B}}}^!,\widehat{\pi}^*(\widetilde{\xi}^{-1}\widetilde{\Omega}^1_{\mathfrak{X}'/\mathfrak{X}})).\qquad(6.5.6.3)$$

For every object $\mathscr{N}$ of $\mathbf{HM}^{\mathrm{coh}}(\mathscr{O}_{\mathfrak{X}'}[\frac{1}{p}], \widetilde{\xi}^{-1}\widetilde{\Omega}^1_{\mathfrak{X}'/\mathfrak{X}})$, we immediately see that the Dolbeault complex $\underline{\mathbb{K}}^\bullet(\widehat{\pi}^*_{\mathbb{Q}}(\mathscr{N}))$ of $\widehat{\pi}^*_{\mathbb{Q}}(\mathscr{N})$ is deduced from that of $\mathscr{N}$ by applying the functor $\widehat{\pi}^*_{\mathbb{Q}}$ (6.1.24.15) term by term.

Lemma 6.5.7 *Let $\mathscr{N}$ be a Higgs $\mathscr{O}_{\mathfrak{X}'}[\frac{1}{p}]$-bundle with coefficients in $\widetilde{\xi}^{-1}\widetilde{\Omega}^1_{\mathfrak{X}'/\mathfrak{X}}$ (6.1.27), r a rational number ≥ 0. We denote by $\underline{\mathbb{K}}^\bullet(\mathscr{N})$ the Dolbeault complex of $\mathscr{N}$ (2.5.1.2), that we consider as a complex of ind-$\mathscr{O}_{\mathfrak{X}'}$-modules (6.1.24.1), by $\mathrm{I}\widehat{\pi}^*(\underline{\mathbb{K}}^\bullet(\mathscr{N}))$ its image by the functor $\mathrm{I}\widehat{\pi}^*$ (6.2.5.3), and for any rational number $t \geq r$, by $\mathrm{I}\underline{\widehat{\sigma}}'^{(t,r)*}(\mathscr{N})$ the object of $\mathbf{Ind\text{-}MC}(\breve{\mathscr{C}}'^{(t,r)}/\breve{\theta}^*(\breve{\mathscr{C}}^{(r)}))$ associated with $\mathscr{N}$ (6.5.2.2) and by $\underline{\mathbb{K}}^\bullet(\mathrm{I}\underline{\widehat{\sigma}}'^{(t,r)*}(\mathscr{N}))$ its Dolbeault complex (6.5.2). Then, for every rational number $t > r$, there exists a canonical morphism of $\mathbf{D}^+(\mathbf{Ind\text{-}Mod}(\breve{\mathfrak{g}}^*(\breve{\mathscr{C}}^{(r)})))$*

$$\mathrm{I}\widehat{\pi}^*(\underline{\mathbb{K}}^\bullet(\mathscr{N})) \otimes_{\underline{\breve{\mathscr{B}}}} \breve{\mathfrak{g}}^*(\breve{\mathscr{C}}^{(r)}) \to \mathrm{R}\mathrm{I}\breve{\tau}_*(\underline{\mathbb{K}}^\bullet(\mathrm{I}\underline{\widehat{\sigma}}'^{(t,r)*}(\mathscr{N}))), \qquad (6.5.7.1)$$

where the left tensor product is defined term by term (not derived) and $\mathrm{I}\breve{\tau}_*$ is the functor (6.3.20.9). It is in fact a morphism of direct systems indexed by $t \in \mathbb{Q}_{>r}$, where the transition morphisms of the right-hand side are induced by the homomorphisms $\breve{\alpha}''^{t',t,r,r}$ (6.3.20.5) for $t' \geq t > r$. Moreover, the induced morphism of $\mathbf{D}^+(\mathbf{Ind\text{-}Mod}(\breve{\mathfrak{g}}^*(\breve{\mathscr{C}}^{(r)})))$

$$\mathrm{I}\widehat{\pi}^*(\underline{\mathbb{K}}^\bullet(\mathscr{N})) \otimes_{\underline{\breve{\mathscr{B}}}} \breve{\mathfrak{g}}^*(\breve{\mathscr{C}}^{(r)}) \to \mathrm{R}\mathrm{I}\breve{\tau}_*(\text{“}\varinjlim_{t\in\mathbb{Q}_{>r}}\text{”}\, \underline{\mathbb{K}}^\bullet(\mathrm{I}\underline{\widehat{\sigma}}'^{(t,r)*}(\mathscr{N}))) \qquad (6.5.7.2)$$

is an isomorphism. In particular, for every integer $q \geq 0$, the induced morphism

$$\mathrm{H}^q(\mathrm{I}\widehat{\pi}^*(\underline{\mathbb{K}}^\bullet(\mathscr{N})) \otimes_{\underline{\breve{\mathscr{B}}}} \breve{\mathfrak{g}}^*(\breve{\mathscr{C}}^{(r)})) \to \text{“}\varinjlim_{t\in\mathbb{Q}_{>r}}\text{”}\, \mathrm{R}^q\mathrm{I}\breve{\tau}_*(\underline{\mathbb{K}}^\bullet(\mathrm{I}\underline{\widehat{\sigma}}'^{(t,r)*}(\mathscr{N}))) \qquad (6.5.7.3)$$

is an isomorphism.

Let t be a rational number $> r$. We denote by $\underline{\widehat{\sigma}}'^{(t,r)*}_{\mathbb{Q}}(\mathscr{N})$ the object of $\mathbf{IC}_{\mathbb{Q}}(\breve{\mathscr{C}}'^{(t,r)}/\breve{\theta}^*(\breve{\mathscr{C}}^{(r)}))$ associated with $\mathscr{N}$ (6.5.3.2), and by $\underline{\mathbb{K}}^\bullet(\underline{\widehat{\sigma}}'^{(t,r)*}_{\mathbb{Q}}(\mathscr{N}))$ its Dolbeault complex (6.5.3). Since $\underline{d}_{\breve{\mathscr{C}}'^{(t,r)}}$ is a $\breve{\theta}^*(\breve{\mathscr{C}}^{(r)})$-derivation (6.4.15.4), we have a canonical morphism of complexes of $\breve{\mathfrak{g}}^*(\breve{\mathscr{C}}^{(r)})_{\mathbb{Q}}$-modules

$$\widehat{\pi}^*_{\mathbb{Q}}(\underline{\mathbb{K}}^\bullet(\mathscr{N})) \otimes_{\underline{\breve{\mathscr{B}}}} \breve{\mathfrak{g}}^*(\breve{\mathscr{C}}^{(r)}) \to \breve{\tau}_{\mathbb{Q}*}(\underline{\mathbb{K}}^\bullet(\underline{\widehat{\sigma}}'^{(t,r)*}_{\mathbb{Q}}(\mathscr{N}))). \qquad (6.5.7.4)$$

By ([59] 013K), since the abelian category $\mathbf{Mod}_{\mathbb{Q}}(\breve{\theta}^*(\breve{\mathscr{C}}^{(r)}))$ has enough injectives (2.9.6), there exist a bounded from below complex of $\breve{\theta}^*(\breve{\mathscr{C}}^{(r)})_{\mathbb{Q}}$-injective modules $\mathscr{L}^\bullet$ and a quasi-isomorphism $u\colon \underline{\mathbb{K}}^\bullet(\underline{\widehat{\sigma}}'^{(t,r)*}_{\mathbb{Q}}(\mathscr{N})) \to \mathscr{L}^\bullet$. The latter induces a morphism of $\mathbf{D}^+(\mathbf{Mod}_{\mathbb{Q}}(\breve{\mathfrak{g}}^*(\breve{\mathscr{C}}^{(r)})))$

$$\breve{\tau}_{\mathbb{Q}*}(\underline{\mathbb{K}}^\bullet(\underline{\widehat{\sigma}}'^{(t,r)*}_{\mathbb{Q}}(\mathscr{N}))) \to \mathrm{R}\breve{\tau}_{\mathbb{Q}*}(\underline{\mathbb{K}}^\bullet(\underline{\widehat{\sigma}}'^{(t,r)*}_{\mathbb{Q}}(\mathscr{N}))), \qquad (6.5.7.5)$$

that depends only on $\underline{\mathbb{K}}^\bullet(\underline{\widehat{\sigma}}'^{(t,r)*}_{\mathbb{Q}}(\mathscr{N}))$, but not on u ([59] 05TG).

In view of (2.9.5.1), (2.9.6.7) and (2.9.7.4), we take for morphism (6.5.7.1) the image by the functor $\alpha_{\check{\mathfrak{g}}^*(\check{\mathscr{C}}^{(r)})}$ (2.9.7.3) of the morphism composed of (6.5.7.4) and (6.5.7.5). It is clearly a morphism of direct systems indexed by $t \in \mathbb{Q}_{>r}$, where the transition morphisms of the right-hand side are induced by the homomorphisms $\check{\alpha}'^{t',t,r,r}$ (6.3.20.5) for $t' \geq t > r$.

By 6.2.2(i), for every integer $j \geq 0$, $\underline{d}_{\check{\mathscr{C}}'^{(t,r)}}$ (6.4.15.4) induces a $\check{\mathfrak{g}}^*(\check{\mathscr{C}}^{(r)})$-linear morphism

$$\theta^{j,(t,r)} \colon \mathrm{R}^j\check{\tau}_*(\check{\mathscr{C}}'^{(t,r)}) \to \widehat{\pi}^*(\widetilde{\xi}^{-1}\widetilde{\Omega}^1_{\mathfrak{X}'/\mathfrak{X}}) \otimes_{\check{\overline{\mathscr{B}}}^!} \mathrm{R}^j\check{\tau}_*(\check{\mathscr{C}}'^{(t,r)}), \tag{6.5.7.6}$$

which is clearly a Higgs $\check{\mathfrak{g}}^*(\check{\mathscr{C}}^{(r)})$-field on $\mathrm{R}^j\check{\tau}_*(\check{\mathscr{C}}'^{(t,r)})$ with coefficients in $\widehat{\pi}^*(\widetilde{\xi}'^{-1}\widetilde{\Omega}^1_{\mathfrak{X}'/\mathfrak{X}}) \otimes_{\check{\overline{\mathscr{B}}}^!} \check{\mathfrak{g}}^*(\check{\mathscr{C}}^{(r)})$. Let $\widehat{\pi}^*_{\mathbb{Q}}(\mathscr{N})$ be the object of $\mathbf{HI}_{\mathbb{Q}}(\check{\overline{\mathscr{B}}}^!, \widehat{\pi}^*(\widetilde{\xi}^{-1}\widetilde{\Omega}^1_{\mathfrak{X}'/\mathfrak{X}}))$ associated with $\mathscr{N}$ (6.5.6.3). We can then consider $\widehat{\pi}^*_{\mathbb{Q}}(\mathscr{N}) \otimes_{\check{\overline{\mathscr{B}}}^!_{\mathbb{Q}}} \mathrm{R}^j\check{\tau}_*(\check{\mathscr{C}}'^{(t,r)})_{\mathbb{Q}}$, the tensor product (2.9.9.7) of $\widehat{\pi}^*_{\mathbb{Q}}(\mathscr{N})$ and the Higgs isogeny

$$(\mathrm{R}^j\check{\tau}_*(\check{\mathscr{C}}'^{(t,r)}), \mathrm{R}^j\check{\tau}_*(\check{\mathscr{C}}'^{(t,r)}), \mathrm{id}, p^t\theta^{j,(t,r)}), \tag{6.5.7.7}$$

which is naturally a Higgs $\check{\mathfrak{g}}^*(\check{\mathscr{C}}^{(r)})$-isogeny with coefficients in the module $\widehat{\pi}^*(\widetilde{\xi}^{-1}\widetilde{\Omega}^1_{\mathfrak{X}'/\mathfrak{X}}) \otimes_{\check{\overline{\mathscr{B}}}^!} \check{\mathfrak{g}}^*(\check{\mathscr{C}}^{(r)})$, defined up to isogeny (see 2.9.9). We denote by $\underline{\mathbb{K}}^\bullet(\widehat{\pi}^*_{\mathbb{Q}}(\mathscr{N}) \otimes_{\check{\overline{\mathscr{B}}}^!_{\mathbb{Q}}} \mathrm{R}^j\check{\tau}_*(\check{\mathscr{C}}'^{(t,r)})_{\mathbb{Q}})$ its Dolbeault complex, which is a complex of $\mathbf{Mod}_{\mathbb{Q}}(\check{\mathfrak{g}}^*(\check{\mathscr{C}}^{(r)}))$ (2.9.9.4).

By 6.2.2(ii), for every integer $i \geq 0$, we have a canonical isomorphism of $\check{\mathfrak{g}}^*(\check{\mathscr{C}}^{(r)})_{\mathbb{Q}}$-modules

$$\mathrm{R}^j\check{\tau}_{\mathbb{Q}*}(\underline{\mathbb{K}}^i(\underline{\widehat{\sigma}}'^{(t,r)*}_{\mathbb{Q}}(\mathscr{N}))) \xrightarrow{\sim} \underline{\mathbb{K}}^i(\widehat{\pi}^*_{\mathbb{Q}}(\mathscr{N}) \otimes_{\check{\overline{\mathscr{B}}}^!_{\mathbb{Q}}} \mathrm{R}^j\check{\tau}_{\mathbb{Q}*}(\check{\mathscr{C}}'^{(t,r)}_{\mathbb{Q}})), \tag{6.5.7.8}$$

compatible with the morphisms induced by the differentials of the two Dolbeault complexes.

The abelian category $\mathbf{Mod}_{\mathbb{Q}}(\check{\theta}^*(\check{\mathscr{C}}^{(r)}))$ having enough injectives, we have a canonical spectral sequence, functorial in t,

$${}^t\mathrm{E}_1^{i,j} = \mathrm{R}^j\check{\tau}_{\mathbb{Q}*}(\underline{\mathbb{K}}^i(\underline{\widehat{\sigma}}'^{(t,r)*}_{\mathbb{Q}}(\mathscr{N}))) \Rightarrow \mathrm{R}^{i+j}\check{\tau}_{\mathbb{Q}*}(\underline{\mathbb{K}}^\bullet(\underline{\widehat{\sigma}}'^{(t,r)*}_{\mathbb{Q}}(\mathscr{N}))). \tag{6.5.7.9}$$

This induces, for every integer $q \geq 0$, a canonical morphism

$$\mathrm{H}^q({}^t\mathrm{E}_1^{\bullet,0}) \to \mathrm{R}^q\check{\tau}_{\mathbb{Q}*}(\underline{\mathbb{K}}^\bullet(\underline{\widehat{\sigma}}'^{(t,r)*}_{\mathbb{Q}}(\mathscr{N}))), \tag{6.5.7.10}$$

which is none other than the morphism induced by (6.5.7.5) by ([29] 0.11.3.4).

In view of (2.9.5.1), (2.9.6.7) and (2.9.7.4), applying the exact functor $\alpha_{\check{\mathfrak{g}}^*(\check{\mathscr{C}}^{(r)})}$ to the spectral sequence (6.5.7.9), we obtain a spectral sequence of ind-$\check{\mathfrak{g}}^*(\check{\mathscr{C}}^{(r)})$-modules

$${}^t\mathscr{E}_1^{i,j} = \mathrm{R}^j\mathrm{I}\check{\tau}_*(\underline{\mathbb{K}}^i(\mathrm{I}\underline{\widehat{\sigma}}'^{(t,r)*}(\mathscr{N}))) \Rightarrow \mathrm{R}^{i+j}\mathrm{I}\check{\tau}_*(\underline{\mathbb{K}}^\bullet(\mathrm{I}\underline{\widehat{\sigma}}'^{(t,r)*}(\mathscr{N}))). \tag{6.5.7.11}$$

By 6.4.11 and (6.5.7.8), for every $i \geq 0$, we have a canonical isomorphism

$$\text{“}\varinjlim_{t \in \mathbb{Q}_{>r}}\text{”}\, {}^t\mathscr{C}_1^{i,0} \xrightarrow{\sim} \mathrm{I}\widehat{\pi}^*(\underline{\mathbb{K}}^i(\mathscr{N})) \otimes_{\underline{\breve{\mathscr{B}}}'} \breve{\mathfrak{g}}^*(\breve{\mathscr{C}}^{(r)}), \qquad (6.5.7.12)$$

and for every $j \geq 1$, we have

$$\text{“}\varinjlim_{t \in \mathbb{Q}_{>r}}\text{”}\, {}^t\mathscr{C}_1^{i,j} = 0. \qquad (6.5.7.13)$$

Moreover, the isomorphisms (6.5.7.12) (for $i \in \mathbb{N}$) form an isomorphism of complexes. Since small filtered direct limits exist and are exact in **Ind-Mod**$(\breve{\mathfrak{g}}^*(\breve{\mathscr{C}}^{(r)}))$ (2.6.7.5), we deduce from this that the morphisms (6.5.7.3) are isomorphisms. Consequently, the morphism (6.5.7.2) is an isomorphism since the functors $\mathrm{R}^q\mathrm{I}\breve{\tau}_*$ commute with small filtered direct limits (2.6.9.4).

Proposition 6.5.8 *Let t, r be two rational numbers such that $t > r \geq 0$, $\mathscr{M}$ a rational* (4.3.15) *and flat* (2.7.9) $\underline{\breve{\mathscr{B}}}'$*-module, $\mathscr{N}$ a Higgs $\mathscr{O}_{\mathfrak{X}'}[\frac{1}{p}]$-bundle with coefficients in $\widetilde{\xi}^{-1}\widetilde{\Omega}^1_{\mathfrak{X}'/\mathscr{X}'}$*

$$\underline{\alpha} \colon \mathrm{I}\underline{\mathfrak{S}}'^{(t,r)}(\mathscr{M}) \xrightarrow{\sim} \mathrm{I}\underline{\widehat{\sigma}}'^{(t,r)*}(\mathscr{N}) \qquad (6.5.8.1)$$

an isomorphism of **Ind-MC**$(\breve{\mathscr{C}}'^{(t,r)}/\breve{\theta}^*(\breve{\mathscr{C}}^{(r)}))$*, where $\mathrm{I}\underline{\mathfrak{S}}'^{(t,r)}$ is the functor* (6.5.2.1) *and $\mathrm{I}\underline{\widehat{\sigma}}'^{(t,r)*}$ is the functor* (6.5.2.2)*. We denote by $\underline{\mathbb{K}}^\bullet(\mathscr{N})$ the Dolbeault complex of $\mathscr{N}$* (2.5.1.2) *and by $\mathrm{I}\widehat{\pi}^*(\underline{\mathbb{K}}^\bullet(\mathscr{N}))$ its image by the functor $\mathrm{I}\widehat{\pi}^*$* (6.2.5.3)*. We then have a canonical isomorphism of* $\mathbf{D}^+$(**Ind-Mod**$(\breve{\mathfrak{g}}^*(\breve{\mathscr{C}}^{(r)})))$

$$\mathrm{RI}\breve{\tau}_*(\mathscr{M} \otimes_{\underline{\breve{\mathscr{B}}}'} \breve{\theta}^*(\breve{\mathscr{C}}^{(r)})) \to \mathrm{I}\widehat{\pi}^*(\underline{\mathbb{K}}^\bullet(\mathscr{N})) \otimes_{\underline{\breve{\mathscr{B}}}'} \breve{\mathfrak{g}}^*(\breve{\mathscr{C}}^{(r)}), \qquad (6.5.8.2)$$

where $\mathrm{I}\breve{\tau}_*$ is the functor (6.3.20.9) *and the tensor product on the right-hand side is defined term by term (not derived).*

Indeed, for any rational number t' such that $t \geq t' \geq r$, we denote by

$$\underline{\alpha}^{(t',r)} \colon \mathrm{I}\underline{\mathfrak{S}}'^{(t',r)}(\mathscr{M}) \xrightarrow{\sim} \mathrm{I}\underline{\widehat{\sigma}}'^{(t',r)*}(\mathscr{N}) \qquad (6.5.8.3)$$

the isomorphism of **Ind-MC**$(\breve{\mathscr{C}}'^{(t',r)}/\breve{\theta}^*(\breve{\mathscr{C}}^{(r)}))$ induced by $\mathrm{I}\underline{\varepsilon}'^{t,t',r}(\underline{\alpha})$ (6.5.2.8) and the isomorphisms (6.5.2.9) and (6.5.2.10). This induces an isomorphism of complexes of ind-$\breve{\theta}^*(\breve{\mathscr{C}}^{(r)})$-modules

$$\mathscr{M} \otimes_{\underline{\breve{\mathscr{B}}}'} \underline{\mathbb{K}}^\bullet(\breve{\mathscr{C}}'^{(t',r)}) \xrightarrow{\sim} \underline{\mathbb{K}}^\bullet(\mathrm{I}\underline{\widehat{\sigma}}'^{(t',r)*}(\mathscr{N})), \qquad (6.5.8.4)$$

where $\underline{\mathbb{K}}^\bullet(\breve{\mathscr{C}}'^{(t',r)})$ is the Dolbeault complex of the Higgs $\breve{\theta}^*(\breve{\mathscr{C}}^{(r)})$-module $(\breve{\mathscr{C}}'^{(t',r)}, p^{t'}\underline{d}_{\breve{\mathscr{C}}'^{(t',r)}})$ (6.4.15). These isomorphisms form an isomorphism of direct systems (for $r < t' < t$) and they therefore induce an isomorphism of complexes of ind-$\breve{\theta}^*(\breve{\mathscr{C}}^{(r)})$-modules

$$\text{“}\varinjlim_{t' \in \mathbb{Q}_{>r}}\text{”}\, \mathscr{M} \otimes_{\underline{\breve{\mathscr{B}}}'} \underline{\mathbb{K}}^\bullet(\breve{\mathscr{C}}'^{(t',r)}) \xrightarrow{\sim} \text{“}\varinjlim_{t' \in \mathbb{Q}_{>r}}\text{”}\, \underline{\mathbb{K}}^\bullet(\mathrm{I}\underline{\widehat{\sigma}}'^{(t',r)*}(\mathscr{N})). \qquad (6.5.8.5)$$

The proposition follows in view of (6.5.5.3) and 6.5.7.

Proposition 6.5.9 *Let* $\mathscr{M}$ *be a Dolbeault ind-*$\overline{\breve{\mathscr{B}}}'$*-module* (6.5.1), $\underline{\mathscr{H}}'(\mathscr{M})$ *the associated Higgs* $\mathscr{O}_{\mathfrak{X}'}[\frac{1}{p}]$*-bundle with coefficients in* $\widetilde{\xi}^{-1}\widetilde{\Omega}^1_{\mathfrak{X}'/\mathfrak{X}}$ (6.5.1.3), $\underline{\mathbb{K}}^\bullet(\underline{\mathscr{H}}'(\mathscr{M}))$ *its Dolbeault complex,* $\mathrm{I}\widehat{\pi}^*(\underline{\mathbb{K}}^\bullet(\underline{\mathscr{H}}'(\mathscr{M})))$ *the image of the latter by the functor* $\mathrm{I}\widehat{\pi}^*$ (6.2.6.3)*. Then, there exist a rational number* $r > 0$ *and an isomorphism of* $\mathbf{D}^+(\mathbf{Ind\text{-}Mod}(\breve{\mathfrak{g}}^*(\breve{\mathscr{C}}^{(r)})))$

$$\mathrm{RI}\breve{\tau}_*(\mathscr{M} \otimes_{\overline{\breve{\mathscr{B}}}'} \breve{\theta}^*(\breve{\mathscr{C}}^{(r)})) \to \mathrm{I}\widehat{\pi}^*(\underline{\mathbb{K}}^\bullet(\underline{\mathscr{H}}'(\mathscr{M}))) \otimes_{\overline{\breve{\mathscr{B}}}'} \breve{\mathfrak{g}}^*(\breve{\mathscr{C}}^{(r)}), \tag{6.5.9.1}$$

where $\mathrm{I}\breve{\tau}_*$ *is the functor* (6.3.20.9) *and the tensor product on the right-hand side is defined term by term (not derived).*

Indeed, there exist a rational number $t > 0$ and an isomorphism of $\mathbf{Ind\text{-}MC}(\breve{\mathscr{C}}'^{(t)}/\overline{\breve{\mathscr{B}}}')$,

$$\alpha\colon \mathrm{I}\mathfrak{S}'^{(t)}(\mathscr{M}) \xrightarrow{\sim} \mathrm{I}\widehat{\sigma}'^{(t)*}(\mathscr{H}'(\mathscr{M})), \tag{6.5.9.2}$$

where $\mathscr{H}'(\mathscr{M})$ is the Higgs $\mathscr{O}_{\mathfrak{X}'}[\frac{1}{p}]$-bundle with coefficients in $\widetilde{\xi}^{-1}\widetilde{\Omega}^1_{\mathfrak{X}'/\mathscr{S}}$ associated with $\mathscr{M}$ (6.5.1.1). For every rational number $0 < r < t$, the isomorphism $\mathrm{I}\underline{\lambda}^{t,r}(\alpha)$ (6.5.2.5) induces, in view of the isomorphism (6.5.2.6) and the commutative diagram (6.5.2.7), an isomorphism of $\mathbf{Ind\text{-}MC}(\breve{\mathscr{C}}'^{(t,r)}/\breve{\theta}^*(\breve{\mathscr{C}}^{(r)}))$,

$$\underline{\alpha}\colon \mathrm{I}\underline{\mathfrak{S}}'^{(t,r)}(\mathscr{M}) \xrightarrow{\sim} \mathrm{I}\underline{\widehat{\sigma}}'^{(t,r)*}(\underline{\mathscr{H}}'(\mathscr{M})). \tag{6.5.9.3}$$

Since the ind-$\overline{\breve{\mathscr{B}}}'$-module $\mathscr{M}$ is rational and flat by 4.5.6, the proposition then follows from 6.5.8.

Remark 6.5.10 Under the assumptions of 6.5.9, there exists a rational number $r > 0$ such that the isomorphism (6.5.9.1) is canonical and it depends functorially on $\mathscr{M}$. Indeed, by virtue of 4.5.13, there exist a rational number $t > 0$ and an isomorphism of $\mathbf{Ind\text{-}MC}(\breve{\mathscr{C}}'^{(t)}/\overline{\breve{\mathscr{B}}}')$,

$$\alpha\colon \mathrm{I}\mathfrak{S}'^{(t)}(\mathscr{M}) \xrightarrow{\sim} \mathrm{I}\widehat{\sigma}'^{(t)*}(\mathscr{H}'(\mathscr{M})), \tag{6.5.10.1}$$

satisfying the properties 4.5.13(i)–(ii), where $\mathscr{H}'(\mathscr{M})$ is the Higgs $\mathscr{O}_{\mathfrak{X}'}[\frac{1}{p}]$-bundle with coefficients in $\widetilde{\xi}^{-1}\widetilde{\Omega}^1_{\mathfrak{X}'/\mathscr{S}}$ associated with $\mathscr{M}$ (6.5.1.1). For any rational number t' such that $0 \leq t' < t$, we denote by

$$\alpha^{(t')}\colon \mathrm{I}\mathfrak{S}'^{(t')}(\mathscr{M}) \xrightarrow{\sim} \mathrm{I}\widehat{\sigma}'^{(t')*}(\mathscr{H}'(\mathscr{M})) \tag{6.5.10.2}$$

the isomorphism of $\mathbf{Ind\text{-}MC}(\breve{\mathscr{C}}'^{(t')}/\overline{\breve{\mathscr{B}}}')$ induced by $\mathrm{I}\varepsilon'^{t,t'}(\alpha)$ and the analogue of the isomorphisms (4.5.3.2) and (4.5.3.3). By the proof of 4.5.20, $\alpha^{(t')}$ depends only on $\mathscr{M}$ (but not on α), on which it depends functorially. For all rational numbers t', r such that $0 \leq r < t' < t$, we denote by

$$\underline{\alpha}^{(t',r)}\colon \mathrm{I}\underline{\mathfrak{S}}'^{(t',r)}(\mathscr{M}) \xrightarrow{\sim} \mathrm{I}\underline{\widehat{\sigma}}'^{(t',r)*}(\underline{\mathscr{H}}'(\mathscr{M})) \tag{6.5.10.3}$$

the isomorphism of **Ind-MC**($\breve{\mathscr{C}}'^{(t',r)}/\breve{\theta}^*(\breve{\mathscr{C}}^{(r)})$) induced by $\mathrm{I}\underline{\lambda}^{t',r}(\alpha^{(t')})$, the isomorphism (6.5.2.6) and the commutative diagram (6.5.2.7). Fixing r such that $0 < r < t$, the isomorphisms $\underline{\alpha}^{(t',r)}$ form an isomorphism of direct systems, for $r < t' < t$. The assertion follows in view of 6.5.5 and 6.5.7 (see the proof of 6.5.8).

Lemma 6.5.11 *Let r be a rational number ≥ 0, n, q two integers ≥ 0, $\mathscr{M}$ a $\overline{\mathscr{B}}'_n$-module of $\widetilde{E}'_s$, $\mathscr{N}$ a $\overline{\mathscr{B}}^!_n$-module of $\widetilde{G}_s$. Then, the canonical morphisms*

$$\mathrm{R}^q\tau_{n*}(\mathscr{M}) \otimes_{\overline{\mathscr{B}}^!_n} \mathfrak{g}^*_n(\mathscr{C}^{(r)}_n) \to \mathrm{R}^q\tau_{n*}(\mathscr{M} \otimes_{\overline{\mathscr{B}}'_n} \theta^*_n(\mathscr{C}^{(r)}_n)), \qquad (6.5.11.1)$$

$$\mathrm{R}^q\mathfrak{g}_{n*}(\mathscr{N}) \otimes_{\overline{\mathscr{B}}_n} \mathscr{C}^{(r)}_n \to \mathrm{R}^q\mathfrak{g}_{n*}(\mathscr{N} \otimes_{\overline{\mathscr{B}}^!_n} \mathfrak{g}^*_n(\mathscr{C}^{(r)}_n)), \qquad (6.5.11.2)$$

$$\mathrm{R}^q\theta_{n*}(\mathscr{M}) \otimes_{\overline{\mathscr{B}}_n} \mathscr{C}^{(r)}_n \to \mathrm{R}^q\theta_{n*}(\mathscr{M} \otimes_{\overline{\mathscr{B}}'_n} \theta^*_n(\mathscr{C}^{(r)}_n)), \qquad (6.5.11.3)$$

where the morphisms of ringed topos τ_n, $\mathfrak{g}_n$ and θ_n are defined in (6.1.21.10) *and* (6.1.21.9), *are isomorphisms.*

Indeed, since the $\overline{\mathscr{B}}_n$-module $\mathscr{F}^{(r)}_n$ is locally free of finite type (6.3.8.5), for every integer $m \geq 0$, the canonical morphisms

$$\mathrm{R}^q\tau_{n*}(\mathscr{M}) \otimes_{\overline{\mathscr{B}}^!_n} \mathfrak{g}^*_n(\mathrm{S}^m_{\overline{\mathscr{B}}_n}(\mathscr{F}^{(r)}_n)) \to \mathrm{R}^q\tau_{n*}(\mathscr{M} \otimes_{\overline{\mathscr{B}}'_n} \theta^*_n(\mathrm{S}^m_{\overline{\mathscr{B}}_n}(\mathscr{F}^{(r)}_n))), \quad (6.5.11.4)$$

$$\mathrm{R}^q\mathfrak{g}_{n*}(\mathscr{N}) \otimes_{\overline{\mathscr{B}}_n} \mathrm{S}^m_{\overline{\mathscr{B}}_n}(\mathscr{F}^{(r)}_n) \to \mathrm{R}^q\mathfrak{g}_{n*}(\mathscr{N} \otimes_{\overline{\mathscr{B}}^!_n} \mathfrak{g}^*_n(\mathrm{S}^m_{\overline{\mathscr{B}}_n}(\mathscr{F}^{(r)}_n))), \quad (6.5.11.5)$$

$$\mathrm{R}^q\theta_{n*}(\mathscr{M}) \otimes_{\overline{\mathscr{B}}_n} \mathrm{S}^m_{\overline{\mathscr{B}}_n}(\mathscr{F}^{(r)}_n) \to \mathrm{R}^q\theta_{n*}(\mathscr{M} \otimes_{\overline{\mathscr{B}}'_n} \theta^*_n(\mathrm{S}^m_{\overline{\mathscr{B}}_n}(\mathscr{F}^{(r)}_n))), \quad (6.5.11.6)$$

are isomorphisms (2.1.10). Moreover, the topos $\widetilde{E}$ and $\widetilde{G}$ are coherent by ([3] VI.10.5(ii)) and ([2] 3.4.22(ii)). The morphism of schemes $g: X' \to X$ being coherent, the morphisms of topos τ and $\mathfrak{g}$ (6.1.14.6) are coherent by virtue of ([3] VI.10.4 and VI.10.5(i)), ([2] 3.4.21(i) and 3.4.22(ii)) and ([5] VI 3.3). Consequently, the functors $\mathrm{R}^q\tau_*$, $\mathrm{R}^q\mathfrak{g}_*$ and $\mathrm{R}^q\Theta_*$ commute with filtered direct limits of abelian sheaves by virtue of ([5] VI 5.1). The proposition follows in view of (6.3.8.6), (6.1.20.6) and (6.1.20.10).

Lemma 6.5.12 *Let r be a rational number ≥ 0, q an integer ≥ 0, $\mathscr{M}$ an ind-$\breve{\overline{\mathscr{B}}}'$-module of $\widetilde{E}'^{\mathbb{N}^\circ}$, $\mathscr{N}$ an ind-$\breve{\overline{\mathscr{B}}}^!$-module of $\widetilde{G}^{\mathbb{N}^\circ}$. Then, we have canonical isomorphisms*

$$\mathrm{R}^q\mathrm{I}\breve{\tau}_*(\mathscr{M}) \otimes_{\breve{\overline{\mathscr{B}}}^!} \breve{\mathfrak{g}}^*(\breve{\mathscr{C}}^{(r)}) \xrightarrow{\sim} \mathrm{R}^q\mathrm{I}\breve{\tau}_*(\mathscr{M} \otimes_{\breve{\overline{\mathscr{B}}}'} \breve{\theta}^*(\breve{\mathscr{C}}^{(r)})), \qquad (6.5.12.1)$$

$$\mathrm{R}^q\mathrm{I}\breve{\mathfrak{g}}_*(\mathscr{N}) \otimes_{\breve{\overline{\mathscr{B}}}} \breve{\mathscr{C}}^{(r)} \xrightarrow{\sim} \mathrm{R}^q\mathrm{I}\breve{\mathfrak{g}}_*(\mathscr{N} \otimes_{\breve{\overline{\mathscr{B}}}^!} \breve{\mathfrak{g}}^*(\breve{\mathscr{C}}^{(r)})), \qquad (6.5.12.2)$$

$$\mathrm{R}^q\mathrm{I}\breve{\theta}_*(\mathscr{M}) \otimes_{\breve{\overline{\mathscr{B}}}} \breve{\mathscr{C}}^{(r)} \xrightarrow{\sim} \mathrm{R}^q\mathrm{I}\breve{\theta}_*(\mathscr{M} \otimes_{\breve{\overline{\mathscr{B}}}'} \breve{\theta}^*(\breve{\mathscr{C}}^{(r)})), \qquad (6.5.12.3)$$

where the morphism of ringed topos $\breve{\tau}$, $\breve{\mathfrak{g}}$ and $\breve{\theta}$ are defined in (6.1.22.1).

Indeed, the case where $\mathscr{M}$ is a $\breve{\overline{\mathscr{B}}}'$-module of $\widetilde{E}'^{\mathbb{N}^\circ}$ and $\mathscr{N}$ is a $\breve{\overline{\mathscr{B}}}^!$-module of $\widetilde{G}^{\mathbb{N}^\circ}$, follows from (2.7.10.6), 6.5.11 and ([3] III.7.5). It implies the general case in view of 2.7.3 and (2.7.10.5).

Proposition 6.5.13 *Suppose that* $g\colon X' \to X$ *is proper. Let* $\mathscr{M}$ *be a Dolbeault ind-*$\overline{\breve{\mathscr{B}}}'$*-module* (6.5.1), q *an integer* $q \geq 0$. *We denote by* $\underline{\mathscr{H}}'(\mathscr{M})$ *the Higgs* $\mathscr{O}_{\mathfrak{X}'}[\frac{1}{p}]$*-bundle with coefficients in* $\widetilde{\xi}^{-1}\widetilde{\Omega}^1_{\mathfrak{X}'/\mathfrak{X}}$ *associated with* $\mathscr{M}$ (6.5.1.3), *and by* $\underline{\mathbb{K}}^\bullet(\underline{\mathscr{H}}'(\mathscr{M}))$ *its Dolbeault complex. Then, the* $\mathscr{O}_{\mathfrak{X}}[\frac{1}{p}]$*-module* $\mathrm{R}^q\mathfrak{g}_*(\underline{\mathbb{K}}^\bullet(\underline{\mathscr{H}}'(\mathscr{M})))$ *is coherent, and there exist a rational number* $r > 0$, *independent of* q, *and an isomorphism of ind-*$\breve{\mathscr{C}}^{(r)}$*-modules*

$$\mathrm{R}^q\mathrm{I}\breve{\theta}_*(\mathscr{M}) \otimes_{\overline{\breve{\mathscr{B}}}} \breve{\mathscr{C}}^{(r)} \xrightarrow{\sim} \mathrm{I}\widehat{\sigma}^*(\mathrm{R}^q\mathfrak{g}_*(\underline{\mathbb{K}}^\bullet(\underline{\mathscr{H}}'(\mathscr{M})))) \otimes_{\overline{\breve{\mathscr{B}}}} \breve{\mathscr{C}}^{(r)}, \tag{6.5.13.1}$$

where $\mathrm{I}\widehat{\sigma}^*$ *is the functor* (6.1.24.2).

This follows from 6.5.9, 6.5.12, 6.2.26, (2.9.6.8) and (2.9.8.5).

6.5.14 Let $\mathscr{K}^\bullet$ be a bounded from below complex of $\mathbf{Mod}_{\mathbb{Q}}(\overline{\breve{\mathscr{B}}}^!)$. We denote by $\check{\tau}^*_{\mathbb{Q}}(\mathscr{K}^\bullet)$ the pullback complex defined term by term (6.1.23.6). We then have a canonical morphism of complexes of $\overline{\breve{\mathscr{B}}}^!_{\mathbb{Q}}$-modules

$$\mathscr{K}^\bullet \to \check{\tau}_{\mathbb{Q}*}(\check{\tau}^*_{\mathbb{Q}}(\mathscr{K}^\bullet)). \tag{6.5.14.1}$$

By ([59] 013K), since the abelian category $\mathbf{Mod}_{\mathbb{Q}}(\overline{\breve{\mathscr{B}}}')$ has enough injectives (2.9.6), there exist a bounded from below complex of injective $\overline{\breve{\mathscr{B}}}'_{\mathbb{Q}}$-modules $\mathscr{L}^\bullet$ and a quasi-isomorphism $u\colon \check{\tau}^*_{\mathbb{Q}}(\mathscr{K}^\bullet) \to \mathscr{L}^\bullet$. The latter induces a morphism of $\mathbf{D}^+(\mathbf{Mod}_{\mathbb{Q}}(\overline{\breve{\mathscr{B}}}^!))$

$$\check{\tau}_{\mathbb{Q}*}(\check{\tau}^*_{\mathbb{Q}}(\mathscr{K}^\bullet)) \to \mathrm{R}\check{\tau}_{\mathbb{Q}*}(\check{\tau}^*_{\mathbb{Q}}(\mathscr{K}^\bullet)). \tag{6.5.14.2}$$

By ([59] 05TG), this morphism depends only on $\mathscr{K}^\bullet$, but not on u, on which it depends functorially. We obtain by composition a canonical morphism of $\mathbf{D}^+(\mathbf{Mod}_{\mathbb{Q}}(\overline{\breve{\mathscr{B}}}^!))$, functorial in $\mathscr{K}^\bullet$,

$$\mathscr{K}^\bullet \to \mathrm{R}\check{\tau}_{\mathbb{Q}*}(\check{\tau}^*_{\mathbb{Q}}(\mathscr{K}^\bullet)). \tag{6.5.14.3}$$

By (2.9.5.1) and (2.9.6.7), this induces a canonical morphism of $\mathbf{D}^+(\mathbf{Ind}\text{-}\mathbf{Mod}(\overline{\breve{\mathscr{B}}}^!))$, functorial in $\mathscr{K}^\bullet$,

$$\mathscr{K}^\bullet \to \mathrm{RI}\check{\tau}_*(\mathrm{I}\check{\tau}^*(\mathscr{K}^\bullet)). \tag{6.5.14.4}$$

Note that the functors $\check{\tau}^*_{\mathbb{Q}}$ and $\mathrm{I}\check{\tau}^*$ do not transform a priori quasi-isomorphisms into quasi-isomorphisms. The morphisms (6.5.14.3) and (6.5.14.4) cannot therefore be extended to the $\mathscr{K}^\bullet$ object of $\mathbf{D}^+(\mathbf{Mod}_{\mathbb{Q}}(\overline{\breve{\mathscr{B}}}^!))$. However, we have the statement 6.5.17.

We will apply the above constructions in particular to the case where $\mathscr{K}^\bullet = \widehat{\pi}^*_{\mathbb{Q}}(\mathbb{K}^\bullet) \otimes_{\overline{\breve{\mathscr{B}}}^!} \mathscr{E}$, $\mathbb{K}^\bullet$ being a bounded from below complex of $\mathbf{Mod}^{\mathrm{coh}}(\mathscr{O}_{\mathfrak{X}'}[\frac{1}{p}])$ and $\mathscr{E}$ a flat $\overline{\breve{\mathscr{B}}}^!$-module (see 6.1.24).

6.5.15 Let $(\mathscr{N},\theta)$ be a Higgs $\mathscr{O}_{\mathfrak{X}'}[\frac{1}{p}]$-bundle with coefficients in $\widetilde{\xi}^{-1}\widetilde{\Omega}^1_{\mathfrak{X}'/\mathscr{S}}$. To lighten the notation, we will omit θ when there is no risk of ambiguity. We denote by

$$\underline{\theta}\colon \mathscr{N} \to \widetilde{\xi}^{-1}\widetilde{\Omega}^1_{\mathfrak{X}'/\mathfrak{X}} \otimes_{\mathscr{O}_{\mathfrak{X}'}} \mathscr{N} \tag{6.5.15.1}$$

the Higgs $\mathscr{O}_{\mathfrak{X}'}[\frac{1}{p}]$-field induced by θ and by $\mathbb{K}^\bullet(\mathscr{N})$ (resp. $\underline{\mathbb{K}}^\bullet(\mathscr{N})$) the Dolbeault complex of $(\mathscr{N},\theta)$ (resp. $(\mathscr{N},\underline{\theta})$) (2.5.1.4). We endow the complex $\mathbb{K}^\bullet(\mathscr{N})$ with the Koszul filtration associated with the exact sequence (6.1.25.4), see (2.5.11.6), and we denote by

$$\partial\colon \underline{\mathbb{K}}^\bullet(\mathscr{N}) \to \mathfrak{g}^*(\widetilde{\xi}^{-1}\widetilde{\Omega}^1_{\mathfrak{X}/\mathscr{S}}) \otimes_{\mathscr{O}_{\mathfrak{X}'}} \underline{\mathbb{K}}^\bullet(\mathscr{N}) \tag{6.5.15.2}$$

the associated boundary map in $\mathbf{D}^+(\mathbf{Mod}(\mathscr{O}_{\mathfrak{X}'}[\frac{1}{p}]))$ (2.5.11.9).

In view of 6.2.5, the Koszul filtration of $\mathbb{K}^\bullet(\mathscr{N})$ induces a filtration of the complex $\widehat{\sigma}'^*_{\mathrm{Q}}(\mathbb{K}^\bullet(\mathscr{N}))$ (6.2.5.2) and an associated boundary map

$$\widehat{\sigma}'^\star_{\mathrm{Q}}(\partial)\colon \widehat{\sigma}'^*_{\mathrm{Q}}(\underline{\mathbb{K}}^\bullet(\mathscr{N})) \to \widehat{\sigma}'^*_{\mathrm{Q}}(\mathfrak{g}^*(\widetilde{\xi}^{-1}\widetilde{\Omega}^1_{\mathfrak{X}/\mathscr{S}}) \otimes_{\mathscr{O}_{\mathfrak{X}'}} \underline{\mathbb{K}}^\bullet(\mathscr{N})) \tag{6.5.15.3}$$

in $\mathbf{D}^+(\mathbf{Mod}_{\mathrm{Q}}(\breve{\overline{\mathscr{B}}}'))$, defined similarly to ∂. We use the notation $\widehat{\sigma}'^\star_{\mathrm{Q}}(\partial)$ to distinguish it from the functor $\widehat{\sigma}'^*_{\mathrm{Q}}$, which we do not extend to the considered derived categories.

We denote by

$$\mathrm{I}\widehat{\sigma}'^\star(\partial)\colon \mathrm{I}\widehat{\sigma}'^*(\underline{\mathbb{K}}^\bullet(\mathscr{N})) \to \mathrm{I}\widehat{\sigma}'^*(\mathfrak{g}^*(\widetilde{\xi}^{-1}\widetilde{\Omega}^1_{\mathfrak{X}/\mathscr{S}}) \otimes_{\mathscr{O}_{\mathfrak{X}'}} \underline{\mathbb{K}}^\bullet(\mathscr{N})) \tag{6.5.15.4}$$

the canonical image of $\widehat{\sigma}'^\star_{\mathrm{Q}}(\partial)$ in $\mathbf{D}^+(\mathbf{Ind\text{-}Mod}(\breve{\overline{\mathscr{B}}}'))$ (2.9.7.3). This morphism is none other than the boundary map associated with the Koszul filtration of the Dolbeault complex $\mathrm{I}\widehat{\sigma}'^*(\mathbb{K}^\bullet(\mathscr{N}))$ of the Higgs ind-$\breve{\overline{\mathscr{B}}}'$-module $(\mathrm{I}\widehat{\sigma}'^*(\mathscr{N}), \mathrm{I}\widehat{\sigma}'^*(\theta))$ (6.2.6.1), associated with the image by the functor $\widehat{\sigma}'^*$ of the exact sequence (6.1.25.4), see (2.8.2.5). We use the notation $\mathrm{I}\widehat{\sigma}'^\star(\partial)$ to distinguish it from the functor $\mathrm{I}\widehat{\sigma}'^*$, which we do not extend to the considered derived categories.

Similarly, the Koszul filtration of $\mathbb{K}^\bullet(\mathscr{N})$ induces a filtration of the complex $\widehat{\pi}^*_{\mathrm{Q}}(\mathbb{K}^\bullet(\mathscr{N}))$ (6.2.5.3) and an associated boundary map

$$\widehat{\pi}^\star_{\mathrm{Q}}(\partial)\colon \widehat{\pi}^*_{\mathrm{Q}}(\underline{\mathbb{K}}^\bullet(\mathscr{N})) \to \widehat{\pi}^*_{\mathrm{Q}}(\mathfrak{g}^*(\widetilde{\xi}^{-1}\widetilde{\Omega}^1_{\mathfrak{X}/\mathscr{S}}) \otimes_{\mathscr{O}_{\mathfrak{X}'}} \underline{\mathbb{K}}^\bullet(\mathscr{N})) \tag{6.5.15.5}$$

in $\mathbf{D}^+(\mathbf{Mod}_{\mathrm{Q}}(\breve{\overline{\mathscr{B}}}^!))$, defined analogously to ∂. We denote by

$$\mathrm{I}\widehat{\pi}^\star(\partial)\colon \mathrm{I}\widehat{\pi}^*(\underline{\mathbb{K}}^\bullet(\mathscr{N})) \to \mathrm{I}\widehat{\pi}^*(\mathfrak{g}^*(\widetilde{\xi}^{-1}\widetilde{\Omega}^1_{\mathfrak{X}/\mathscr{S}}) \otimes_{\mathscr{O}_{\mathfrak{X}'}} \underline{\mathbb{K}}^\bullet(\mathscr{N})) \tag{6.5.15.6}$$

its canonical image in $\mathbf{D}^+(\mathbf{Ind\text{-}Mod}(\breve{\overline{\mathscr{B}}}^!))$.

For any integer $q \geq 0$, we denote by

$$\kappa^q\colon \mathrm{R}^q\mathfrak{g}_*(\underline{\mathbb{K}}^\bullet(\mathscr{N})) \to \widetilde{\xi}^{-1}\widetilde{\Omega}^1_{\mathfrak{X}/\mathscr{S}} \otimes_{\mathscr{O}_{\mathfrak{X}}} \mathrm{R}^q\mathfrak{g}_*(\underline{\mathbb{K}}^\bullet(\mathscr{N})) \tag{6.5.15.7}$$

the Katz–Oda $\mathscr{O}_{\mathfrak{X}}$-field on the Dolbeault cohomology of $\mathscr{N}$ (2.5.16), which is none other than $\mathrm{R}^q\mathfrak{g}_*(\partial)$.

Proposition 6.5.16 *We keep the assumptions and notation of* 6.5.15. *We further suppose that the morphism* $g\colon X' \to X$ *is proper and that the Higgs* $\mathscr{O}_{\mathfrak{X}'}[\frac{1}{p}]$*-module* $(\mathscr{N}, \theta)$ *is nilpotent* (4.2.5). *Then, for every integer* $q \geq 0$, *the Higgs* $\mathscr{O}_{\mathfrak{X}}[\frac{1}{p}]$*-module* $(\mathrm{R}^q\mathfrak{g}_*(\underline{\mathbb{K}}^\bullet(\mathscr{N})), \kappa^q)$, *where* κ^q *is the Katz–Oda field* (6.5.15.7), *is nilpotent.*

The proof is identical to that of 2.5.22 using 4.2.8 instead of 2.5.4. Note that the functor $\mathrm{R}^q\mathfrak{g}_*$ transforms coherent $\mathscr{O}_{\mathfrak{X}'}[\frac{1}{p}]$-modules into coherent $\mathscr{O}_{\mathfrak{X}}[\frac{1}{p}]$-modules by ([1] 2.10.24 and 2.11.5).

Lemma 6.5.17 *We keep the assumptions and notation of* 6.5.15. *Let, moreover,* $\mathscr{E}$ *be a flat* $\overline{\breve{\mathscr{B}}}^!$*-module. Then, the diagram of morphisms in* $\mathbf{D}^+(\mathbf{Ind\text{-}Mod}(\overline{\breve{\mathscr{B}}}^!))$

$$\begin{array}{ccc}
\mathrm{I}\widehat{\pi}^*(\underline{\mathbb{K}}^\bullet(\mathscr{N})) \otimes_{\overline{\breve{\mathscr{B}}}^!} \mathscr{E} & \xrightarrow{\mathrm{I}\widehat{\pi}^*(\partial)\otimes\mathrm{id}} & \mathrm{I}\widehat{\pi}^*(\mathfrak{g}^*(\widetilde{\xi}^{-1}\widetilde{\Omega}^1_{\mathfrak{X}/\mathscr{S}}) \otimes_{\mathscr{O}_{\mathfrak{X}'}} \underline{\mathbb{K}}^\bullet(\mathscr{N})) \otimes_{\overline{\breve{\mathscr{B}}}^!} \mathscr{E} \\
\downarrow & & \downarrow \\
\mathrm{RI}\breve{\tau}_*(\mathrm{I}\widehat{\sigma}'^*(\underline{\mathbb{K}}^\bullet(\mathscr{N})) \otimes_{\overline{\breve{\mathscr{B}}}'} \breve{\tau}^*(\mathscr{E})) & \xrightarrow{\mathrm{RI}\breve{\tau}_*(\mathrm{I}\widehat{\sigma}'^*(\partial)\otimes\mathrm{id})} & \mathrm{RI}\breve{\tau}_*(\mathrm{I}\widehat{\sigma}'^*(\mathfrak{g}^*(\widetilde{\xi}^{-1}\widetilde{\Omega}^1_{\mathfrak{X}/\mathscr{S}}) \otimes_{\mathscr{O}_{\mathfrak{X}'}} \underline{\mathbb{K}}^\bullet(\mathscr{N})) \otimes_{\overline{\breve{\mathscr{B}}}'} \breve{\tau}^*(\mathscr{E})),
\end{array}$$

where the vertical arrows are the morphisms (6.5.14.4), *is commutative.*

To lighten the notation, we omit $\mathscr{N}$ from the notation of the complexes $\mathbb{K}^\bullet(\mathscr{N})$ and $\underline{\mathbb{K}}^\bullet(\mathscr{N})$, and denote their differentials by $\theta^\bullet$ and $\underline{\theta}^\bullet$. Let $\Omega = \mathfrak{g}^*(\widetilde{\xi}^{-1}\widetilde{\Omega}^1_{\mathfrak{X}/\mathscr{S}})$ and

$$\mathbb{G}^\bullet = \mathbb{K}^\bullet/\mathrm{W}^2\mathbb{K}^\bullet, \tag{6.5.17.1}$$

where $\mathrm{W}^\bullet\mathbb{K}^\bullet$ denotes the Koszul filtration of $\mathbb{K}^\bullet$ associated with the exact sequence (6.1.25.4), see (2.5.11.6). We also denote by $\theta^\bullet$ the differentials of $\mathbb{G}^\bullet$. We have a canonical exact sequence of complexes of $\mathscr{O}_{\mathfrak{X}'}[\frac{1}{p}]$-modules (2.5.11.8)

$$0 \longrightarrow \Omega \otimes_{\mathscr{O}_{\mathfrak{X}'}} \underline{\mathbb{K}}^\bullet[-1] \xrightarrow{u^\bullet} \mathbb{G}^\bullet \xrightarrow{v^\bullet} \underline{\mathbb{K}}^\bullet \longrightarrow 0. \tag{6.5.17.2}$$

We denote by $\mathrm{C}^\bullet$ the mapping cone of $u^\bullet$. For every integer i, we therefore have $\mathrm{C}^i = (\Omega \otimes_{\mathscr{O}_{\mathfrak{X}'}} \underline{\mathbb{K}}^i) \oplus \mathbb{G}^i$ and the differential $c^i\colon \mathrm{C}^i \to \mathrm{C}^{i+1}$ is defined by the matrix

$$\begin{pmatrix} \mathrm{id} \otimes \underline{\theta}^i & 0 \\ u^{i+1} & \theta^i \end{pmatrix}. \tag{6.5.17.3}$$

Let $\pi_1^\bullet$ and $\pi_2^\bullet$ denote the canonical projections of $\mathrm{C}^\bullet$ on $\Omega \otimes_{\mathscr{O}_{\mathfrak{X}'}} \underline{\mathbb{K}}^\bullet$ and $\mathbb{G}^\bullet$, respectively. The composition $v^\bullet \circ \pi_2^\bullet\colon \mathrm{C}^\bullet \to \underline{\mathbb{K}}^\bullet$ is then a quasi-isomorphism, and $-\partial$ (6.5.15.2) is the composition in $\mathbf{D}^+(\mathbf{Mod}(\mathscr{O}_{\mathfrak{X}'}[\frac{1}{p}]))$ of the inverse of $v^\bullet \circ \pi_2^\bullet$ and $\pi_1^\bullet$ ([59] 09KF).

By 6.2.6, the sequence of complexes of ind-$\overline{\breve{\mathscr{B}}}'$-modules

$$0 \to \mathrm{I}\widehat{\sigma}'^*(\Omega \otimes_{\mathscr{O}_{\mathfrak{X}'}} \underline{\mathbb{K}}^\bullet)[-1] \to \mathrm{I}\widehat{\sigma}'^*(\mathbb{G}^\bullet) \to \mathrm{I}\widehat{\sigma}'^*(\underline{\mathbb{K}}^\bullet) \to 0 \tag{6.5.17.4}$$

is exact. The mapping cone of $\mathrm{I}\widehat{\sigma}'^*(u^\bullet)$ can be identified with $\mathrm{I}\widehat{\sigma}'^*(\mathrm{C}^\bullet)$, and the morphism $\mathrm{I}\widehat{\sigma}'^\star(\partial)$ of $\mathbf{D}^+(\mathbf{Ind\text{-}Mod}(\overline{\breve{\mathscr{B}}}'))$ is then defined analogously to ∂. We describe similarly the morphism $\mathrm{I}\widehat{\pi}^\star(\partial)$ of $\mathbf{D}^+(\mathbf{Ind\text{-}Mod}(\overline{\breve{\mathscr{B}}}^!))$.

Since the morphism (6.5.14.4) is functorial, the diagrams of $\mathbf{D}^+(\mathbf{Ind\text{-}Mod}(\breve{\overline{\mathscr{B}}}^!))$

$$\begin{array}{ccccc}
\mathrm{I}\widehat{\pi}^*(\underline{\mathbb{K}}^\bullet)\otimes_{\breve{\overline{\mathscr{B}}}^!}\mathscr{E} & \xleftarrow{a_2} & \mathrm{I}\widehat{\pi}^*(\mathrm{C}^\bullet)\otimes_{\breve{\overline{\mathscr{B}}}^!}\mathscr{E} & \xrightarrow{a_1} & \mathrm{I}\widehat{\pi}^*(\Omega\otimes_{\mathscr{O}_{\mathfrak{X}'}}\underline{\mathbb{K}}^\bullet)\otimes_{\breve{\overline{\mathscr{B}}}^!}\mathscr{E}\\
\downarrow & & \downarrow & & \downarrow\\
\mathrm{RI}\check{\tau}_*(\mathrm{I}\widehat{\sigma}'^*(\underline{\mathbb{K}}^\bullet)\otimes_{\breve{\overline{\mathscr{B}}}}\check{\tau}^*(\mathscr{E})) & \xleftarrow{b_2} & \mathrm{RI}\check{\tau}_*(\mathrm{I}\widehat{\sigma}'^*(\mathrm{C}^\bullet)\otimes_{\breve{\overline{\mathscr{B}}}}\check{\tau}^*(\mathscr{E})) & \xrightarrow{b_1} & \mathrm{RI}\check{\tau}_*(\mathrm{I}\widehat{\sigma}'^*(\Omega\otimes_{\mathscr{O}_{\mathfrak{X}'}}\underline{\mathbb{K}}^\bullet)\otimes_{\breve{\overline{\mathscr{B}}}}\check{\tau}^*(\mathscr{E})),
\end{array}$$

where the vertical arrows are the morphisms (6.5.14.4), $a_1=\mathrm{I}\widehat{\pi}^*(\pi_1^\bullet)\otimes\mathrm{id}$, $a_2=\mathrm{I}\widehat{\pi}^*(\nu^\bullet\circ\pi_2^\bullet)\otimes\mathrm{id}$ $b_1=\mathrm{RI}\check{\tau}_*(\mathrm{I}\widehat{\sigma}'^*(\pi_1^\bullet)\otimes\mathrm{id})$ and $b_2=\mathrm{RI}\check{\tau}_*(\mathrm{I}\widehat{\sigma}'^*(\nu^\bullet\circ\pi_2^\bullet)\otimes\mathrm{id})$, are commutative. Since $\mathscr{E}$ is $\breve{\overline{\mathscr{B}}}^!$-flat, a_2 is a quasi-isomorphism (6.2.6), and $-\mathrm{I}\widehat{\pi}^\star(\partial)\otimes\mathrm{id}$ is the composition of the inverse of a_2 and a_1. Similarly, $\mathrm{I}\widehat{\sigma}'^*(\nu^\bullet\circ\pi_2^\bullet)\otimes\mathrm{id}$ is a quasi-isomorphism, and so is b_2, and $-\mathrm{RI}\check{\tau}_*(\mathrm{I}\widehat{\sigma}'^\star(\partial)\otimes\mathrm{id})$ is the composition of the inverse of b_2 and b_1. The proposition follows.

Lemma 6.5.18 *Under the assumptions of* 6.5.15 *and with the same notation, for every integer* $q\geq 0$, *the diagram*

$$\begin{array}{ccc}
\widehat{\sigma}_{\mathrm{Q}}^*(\mathrm{R}^q\mathfrak{g}_*(\underline{\mathbb{K}}(\mathscr{N}))) & \longrightarrow & \mathrm{R}^q\check{\mathfrak{g}}_{\mathrm{Q}*}(\widehat{\pi}_{\mathrm{Q}}^*(\underline{\mathbb{K}}(\mathscr{N})))\\
\downarrow{\scriptstyle \widehat{\sigma}_{\mathrm{Q}}^*(\kappa^q)} & & \\
\widehat{\sigma}_{\mathrm{Q}}^*(\widetilde{\xi}^{-1}\widetilde{\Omega}^1_{\mathfrak{X}/\mathscr{S}}\otimes_{\mathscr{O}_{\mathfrak{X}}}\mathrm{R}^q\mathfrak{g}_*(\underline{\mathbb{K}}(\mathscr{N}))) & & \downarrow{\scriptstyle \mathrm{R}^q\check{\mathfrak{g}}_{\mathrm{Q}*}(\widehat{\pi}_{\mathrm{Q}}^\star(\partial))}\\
\downarrow & & \\
\widehat{\sigma}_{\mathrm{Q}}^*(\mathrm{R}^q\mathfrak{g}_*(\mathfrak{g}^*(\widetilde{\xi}^{-1}\widetilde{\Omega}^1_{\mathfrak{X}/\mathscr{S}})\otimes_{\mathscr{O}_{\mathfrak{X}'}}\underline{\mathbb{K}}(\mathscr{N}))) & \longrightarrow & \mathrm{R}^q\check{\mathfrak{g}}_{\mathrm{Q}*}(\widehat{\pi}_{\mathrm{Q}}^*(\mathfrak{g}^*(\widetilde{\xi}^{-1}\widetilde{\Omega}^1_{\mathfrak{X}/\mathscr{S}})\otimes_{\mathscr{O}_{\mathfrak{X}'}}\underline{\mathbb{K}}(\mathscr{N}))),
\end{array}$$

where the horizontal arrows are induced by the base change morphism (6.2.23.1) *and the unlabeled vertical arrow is the canonical isomorphism, is commutative.*

This amounts to saying that the diagram

$$\begin{array}{ccc}
\widehat{\sigma}_{\mathrm{Q}}^*(\mathrm{R}^q\mathfrak{g}_*(\underline{\mathbb{K}}(\mathscr{N}))) & \longrightarrow & \mathrm{R}^q\check{\mathfrak{g}}_{\mathrm{Q}*}(\widehat{\pi}_{\mathrm{Q}}^*(\underline{\mathbb{K}}(\mathscr{N})))\\
\downarrow{\scriptstyle \widehat{\sigma}_{\mathrm{Q}}^*(\mathrm{R}^q\mathfrak{g}_*(\partial))} & & \downarrow{\scriptstyle \mathrm{R}^q\check{\mathfrak{g}}_{\mathrm{Q}*}(\widehat{\pi}_{\mathrm{Q}}^\star(\partial))}\\
\widehat{\sigma}_{\mathrm{Q}}^*(\mathrm{R}^q\mathfrak{g}_*(\mathfrak{g}^*(\widetilde{\xi}^{-1}\widetilde{\Omega}^1_{\mathfrak{X}/\mathscr{S}})\otimes_{\mathscr{O}_{\mathfrak{X}'}}\underline{\mathbb{K}}(\mathscr{N}))) & \longrightarrow & \mathrm{R}^q\check{\mathfrak{g}}_{\mathrm{Q}*}(\widehat{\pi}_{\mathrm{Q}}^*(\mathfrak{g}^*(\widetilde{\xi}^{-1}\widetilde{\Omega}^1_{\mathfrak{X}/\mathscr{S}})\otimes_{\mathscr{O}_{\mathfrak{X}'}}\underline{\mathbb{K}}(\mathscr{N})))
\end{array}$$

is commutative. We take again the notation of the proof of 6.5.17. By 6.2.5, the sequence of complexes of ind-$\breve{\overline{\mathscr{B}}}^!$-modules

$$0\to\widehat{\pi}_{\mathrm{Q}}^*(\Omega\otimes_{\mathscr{O}_{\mathfrak{X}'}}\underline{\mathbb{K}}^\bullet)[-1]\to\widehat{\pi}_{\mathrm{Q}}^*(\mathbb{G}^\bullet)\to\widehat{\pi}_{\mathrm{Q}}^*(\underline{\mathbb{K}}^\bullet)\to 0 \qquad (6.5.18.1)$$

is exact. The mapping cone of $\widehat{\pi}_{\mathrm{Q}}^*(u^\bullet)$ can be identified with $\widehat{\pi}_{\mathrm{Q}}^*(\mathrm{C}^\bullet)$, and the morphism $\widehat{\pi}_{\mathrm{Q}}^\star(\partial)$ of $\mathbf{D}^+(\mathbf{Mod}_{\mathrm{Q}}(\breve{\overline{\mathscr{B}}}^!))$ is then defined analogously to ∂.

Since the base change morphism (6.2.23.1) is functorial, the diagrams of $\mathbf{D}^+(\mathbf{Ind\text{-}Mod}(\overline{\breve{\mathscr{B}}}))$

$$\begin{array}{ccccc}
\widehat{\sigma}^*_{\mathbb{Q}}(\mathrm{R}^q \mathfrak{g}_*(\underline{\mathbb{K}}^\bullet)) & \xleftarrow{a_2} & \widehat{\sigma}^*_{\mathbb{Q}}(\mathrm{R}^q \mathfrak{g}_*(\mathrm{C}^\bullet)) & \xrightarrow{a_1} & \widehat{\sigma}^*_{\mathbb{Q}}(\mathrm{R}^q \mathfrak{g}_*(\Omega \otimes_{\mathscr{O}_{\mathfrak{X}'}} \underline{\mathbb{K}}^\bullet)) \\
\downarrow & & \downarrow & & \downarrow \\
\mathrm{R}^q \breve{\mathfrak{g}}_{\mathbb{Q}*}(\widehat{\pi}^*_{\mathbb{Q}}(\underline{\mathbb{K}}^\bullet)) & \xleftarrow{b_2} & \mathrm{R}^q \breve{\mathfrak{g}}_{\mathbb{Q}*}(\widehat{\pi}^*_{\mathbb{Q}}(\mathrm{C}^\bullet)) & \xrightarrow{b_1} & \mathrm{R}^q \breve{\mathfrak{g}}_{\mathbb{Q}*}(\widehat{\pi}^*_{\mathbb{Q}}(\Omega \otimes_{\mathscr{O}_{\mathfrak{X}'}} \underline{\mathbb{K}}^\bullet)),
\end{array}$$

where the vertical arrows are the morphisms (6.2.23.1), $a_1 = \widehat{\sigma}^*_{\mathbb{Q}}(\mathrm{R}^q \mathfrak{g}_*(\pi_1^\bullet))$, $a_2 = \widehat{\sigma}^*_{\mathbb{Q}}(\mathrm{R}^q \mathfrak{g}_*(\nu^\bullet \circ \pi_2^\bullet))$, $b_1 = \mathrm{R}^q \breve{\mathfrak{g}}_{\mathbb{Q}*}(\widehat{\pi}^*_{\mathbb{Q}}(\pi_1^\bullet))$ and $b_2 = \mathrm{R}^q \breve{\mathfrak{g}}_{\mathbb{Q}*}(\widehat{\pi}^*_{\mathbb{Q}}(\nu^\bullet \circ \pi_2^\bullet))$, are commutative. Note that a_2 is an isomorphism, and that $-\widehat{\sigma}^*_{\mathbb{Q}}(\mathrm{R}^q \mathfrak{g}_*(\partial))$ is the composition of the inverse of a_2 and a_1. Moreover, $\widehat{\pi}^*_{\mathbb{Q}}(\nu^\bullet \circ \pi_2^\bullet)$ is a quasi-isomorphism (6.2.5). Hence, b_2 is an isomorphism, and $-\mathrm{R}^q \breve{\mathfrak{g}}_{\mathbb{Q}*}(\widehat{\pi}^\star_{\mathbb{Q}}(\partial))$ is the composition of the inverse of b_2 and b_1. The proposition follows.

6.5.19 Let $(\mathscr{N}, \theta)$ be a Higgs $\mathscr{O}_{\mathfrak{X}'}[\frac{1}{p}]$-bundle with coefficients in $\widetilde{\xi}^{-1}\widetilde{\Omega}^1_{\mathfrak{X}'/\mathscr{S}}$, t, r two rational numbers such that $t \geq r \geq 0$. We denote by $\mathrm{I}\widehat{\sigma}'^{(t,r)*}(\mathscr{N})$ the image of $(\mathscr{N}, \theta)$ by the functor (6.5.4.3), by $\underline{\theta}$ the Higgs $\mathscr{O}_{\mathfrak{X}'}[\frac{1}{p}]$-field on $\mathscr{N}$ with coefficients in $\widetilde{\xi}^{-1}\widetilde{\Omega}^1_{\mathfrak{X}'/\mathfrak{X}}$ induced by θ, by $\mathrm{I}\underline{\widehat{\sigma}}'^{(t,r)*}(\mathscr{N})$ the image of $(\mathscr{N}, \underline{\theta})$ by the functor (6.5.2.2), and by $\mathscr{F}^{(r)}$ the ind-$\breve{\theta}^*(\breve{\mathscr{C}}^{(r)})$-module

$$\mathscr{F}^{(r)} = \text{“}\varinjlim_{t \in \mathbb{Q}_{>r}}\text{''}\ \mathrm{I}\widehat{\sigma}'^*(\mathscr{N}) \otimes_{\overline{\breve{\mathscr{B}}}'} \breve{\mathscr{C}}'^{(t,r)}. \tag{6.5.19.1}$$

The ind-Higgs $\overline{\breve{\mathscr{B}}}'$-modules $\mathrm{I}\widehat{\sigma}'^{(t,r)*}(\mathscr{N})$ with coefficients in $\widehat{\sigma}'^*(\xi^{-1}\widetilde{\Omega}^1_{\mathfrak{X}'/\mathscr{S}})$, for $t \in \mathbb{Q}_{>r}$, form a direct system (6.4.20.6). We denote by

$$\vartheta^{(r)} \colon \mathscr{F}^{(r)} \to \widehat{\sigma}'^*(\xi^{-1}\widetilde{\Omega}^1_{\mathfrak{X}'/\mathscr{S}}) \otimes_{\overline{\breve{\mathscr{B}}}'} \mathscr{F}^{(r)} \tag{6.5.19.2}$$

the Higgs $\overline{\breve{\mathscr{B}}}'$-field induced by those of the $\mathrm{I}\widehat{\sigma}'^{(t,r)*}(\mathscr{N})$'s for $t \in \mathbb{Q}_{>r}$, and by $\mathbb{K}^\bullet(\mathscr{F}^{(r)})$ the Dolbeault complex of $(\mathscr{F}^{(r)}, \vartheta^{(r)})$. Similarly, the Higgs ind-$\breve{\theta}^*(\breve{\mathscr{C}}^{(r)})$-modules $\mathrm{I}\underline{\widehat{\sigma}}'^{(t,r)*}(\mathscr{N})$ with coefficients in $\widehat{\sigma}'^*(\xi^{-1}\widetilde{\Omega}^1_{\mathfrak{X}'/\mathfrak{X}}) \otimes_{\overline{\breve{\mathscr{B}}}'} \breve{\theta}^*(\breve{\mathscr{C}}^{(r)})$, for $t \in \mathbb{Q}_{>r}$, form a direct system. We denote by

$$\underline{\vartheta}^{(r)} \colon \mathscr{F}^{(r)} \to \widehat{\sigma}'^*(\xi^{-1}\widetilde{\Omega}^1_{\mathfrak{X}'/\mathfrak{X}}) \otimes_{\overline{\breve{\mathscr{B}}}'} \mathscr{F}^{(r)} \tag{6.5.19.3}$$

the Higgs $\breve{\theta}^*(\breve{\mathscr{C}}^{(r)})$-field induced by those of the $\mathrm{I}\underline{\widehat{\sigma}}'^{(t,r)*}(\mathscr{N})$'s for $t \in \mathbb{Q}_{>r}$, and by $\underline{\mathbb{K}}^\bullet(\mathscr{F}^{(r)})$ the Dolbeault complex of $(\mathscr{F}^{(r)}, \underline{\vartheta}^{(r)})$. Note that morphism of the ind-$\overline{\breve{\mathscr{B}}}'$-modules underlying $\underline{\vartheta}^{(r)}$ is induced by $\vartheta^{(r)}$. We denote by

$$\partial^{(r)} \colon \underline{\mathbb{K}}^\bullet(\mathscr{F}^{(r)}) \to \widehat{\sigma}'^*(\mathfrak{g}^*(\widetilde{\xi}^{-1}\widetilde{\Omega}^1_{\mathfrak{X}/\mathscr{S}})) \otimes_{\overline{\breve{\mathscr{B}}}'} \underline{\mathbb{K}}^\bullet(\mathscr{F}^{(r)}) \tag{6.5.19.4}$$

the morphism of $\mathbf{D}^+(\mathbf{Ind\text{-}Mod}(\overline{\check{\mathscr{B}}}'))$, the boundary map associated with the Koszul filtration of $\mathbb{K}^\bullet(\mathscr{F}^{(r)})$ which is associated with the image by the functor $\widehat{\sigma}'^*$ of the exact sequence (6.1.25.4), see (2.8.2.5).

By 6.5.7, there exists a canonical isomorphism of $\mathbf{D}^+(\mathbf{Ind\text{-}Mod}(\check{\mathfrak{g}}^*(\check{\mathscr{C}}^{(r)})))$

$$\mathrm{I}\widehat{\pi}^*(\underline{\mathbb{K}}^\bullet(\mathscr{N}))\otimes_{\overline{\check{\mathscr{B}}}^!}\check{\mathfrak{g}}^*(\check{\mathscr{C}}^{(r)})\to \mathrm{RI}\check{\tau}_*(\underline{\mathbb{K}}^\bullet(\mathscr{F}^{(r)})). \qquad (6.5.19.5)$$

In view of (6.1.23.6), we identify $\check{\mathfrak{g}}^*(d_{\check{\mathscr{C}}^{(r)}})$ (4.4.33.1) with a $\overline{\check{\mathscr{B}}}^!$-derivation

$$\check{\mathfrak{g}}^*(d_{\check{\mathscr{C}}^{(r)}})\colon \check{\mathfrak{g}}^*(\check{\mathscr{C}}^{(r)})\to \widehat{\pi}^*(\mathfrak{g}^*(\widetilde{\xi}^{-1}\widetilde{\Omega}^1_{\mathfrak{X}/\mathscr{S}}))\otimes_{\overline{\check{\mathscr{B}}}^!}\check{\mathfrak{g}}^*(\check{\mathscr{C}}^{(r)}). \qquad (6.5.19.6)$$

Proposition 6.5.20 *Under the assumptions of* 6.5.19 *and with the same notation, for every rational number $r\geq 0$, the diagram of* $\mathbf{D}^+(\mathbf{Ind\text{-}Mod}(\overline{\check{\mathscr{B}}}'))$

$$\begin{array}{ccc}
\mathrm{I}\widehat{\sigma}'^*(\underline{\mathbb{K}}^\bullet(\mathscr{N}))\otimes_{\overline{\check{\mathscr{B}}}^!}\check{\theta}^*(\check{\mathscr{C}}^{(r)}) & \xrightarrow{\ a\ } & \mathrm{I}\widehat{\sigma}'^*(\mathfrak{g}^*(\widetilde{\xi}^{-1}\widetilde{\Omega}^1_{\mathfrak{X}/\mathscr{S}})\otimes_{\mathscr{O}_{\mathfrak{X}'}}\underline{\mathbb{K}}^\bullet(\mathscr{N}))\otimes_{\overline{\check{\mathscr{B}}}'}\check{\theta}^*(\check{\mathscr{C}}^{(r)}) \\
\downarrow & & \downarrow \\
\underline{\mathbb{K}}^\bullet(\mathscr{F}^{(r)}) & \xrightarrow{\ \partial^{(r)}\ } & \widehat{\sigma}'^*(\mathfrak{g}^*(\widetilde{\xi}^{-1}\widetilde{\Omega}^1_{\mathfrak{X}/\mathscr{S}}))\otimes_{\overline{\check{\mathscr{B}}}'}\underline{\mathbb{K}}^\bullet(\mathscr{F}^{(r)}),
\end{array}$$

where the vertical arrows are the canonical morphisms and

$$a=\mathrm{I}\widehat{\sigma}'^\star(\partial)\otimes\mathrm{id}+p^r\mathrm{id}\otimes\check{\theta}^*(d_{\check{\mathscr{C}}^{(r)}})$$

is commutative (6.5.15.4).

This follows from 2.8.3 by functoriality of the boundary map (2.8.2.5).

Corollary 6.5.21 *Under the assumptions of* 6.5.19 *and with the same notation, for every rational number $r\geq 0$, the diagram of* $\mathbf{D}^+(\mathbf{Ind\text{-}Mod}(\overline{\check{\mathscr{B}}}^!))$

$$\begin{array}{ccc}
\mathrm{I}\widehat{\pi}^*(\underline{\mathbb{K}}^\bullet(\mathscr{N}))\otimes_{\overline{\check{\mathscr{B}}}^!}\check{\mathfrak{g}}^*(\check{\mathscr{C}}^{(r)}) & \xrightarrow{\ a\ } & \mathrm{I}\widehat{\pi}^*(\mathfrak{g}^*(\widetilde{\xi}^{-1}\widetilde{\Omega}^1_{\mathfrak{X}/\mathscr{S}})\otimes_{\mathscr{O}_{\mathfrak{X}'}}\underline{\mathbb{K}}^\bullet(\mathscr{N}))\otimes_{\overline{\check{\mathscr{B}}}^!}\check{\mathfrak{g}}^*(\check{\mathscr{C}}^{(r)}) \\
\downarrow & & \downarrow \\
\mathrm{RI}\check{\tau}_*(\underline{\mathbb{K}}^\bullet(\mathscr{F}^{(r)})) & \xrightarrow{\ b\ } & \widehat{\pi}^*(\mathfrak{g}^*(\widetilde{\xi}^{-1}\widetilde{\Omega}^1_{\mathfrak{X}/\mathscr{S}}))\otimes_{\overline{\check{\mathscr{B}}}^!}\mathrm{RI}\check{\tau}_*(\underline{\mathbb{K}}^\bullet(\mathscr{F}^{(r)})),
\end{array}$$

where the vertical arrows are induced by the isomorphism (6.5.19.5), $a=\mathrm{I}\widehat{\pi}^\star(\partial)\otimes\mathrm{id}+p^r\mathrm{id}\otimes\check{\mathfrak{g}}^*(d_{\check{\mathscr{C}}^{(r)}})$ (6.5.15.4) *and b is induced by* $\mathrm{RI}\check{\tau}_*(\partial^{(r)})$ *and* 2.7.14, *is commutative.*

This follows from 6.5.20, 6.5.17 and from the functoriality of the morphism (6.5.14.4).

6.5.22 Let $\mathscr{M}$ be a Dolbeault ind-$\overline{\check{\mathscr{B}}}'$-module (6.5.1). We denote by $\mathscr{H}'(\mathscr{M})$ (resp. $\underline{\mathscr{H}}'(\mathscr{M})$) the Higgs $\mathscr{O}_{\mathfrak{X}'}[\frac{1}{p}]$-module with coefficients in $\widetilde{\xi}^{-1}\widetilde{\Omega}^1_{\mathfrak{X}'/\mathscr{S}}$ (resp. $\widetilde{\xi}^{-1}\widetilde{\Omega}^1_{\mathfrak{X}'/\mathfrak{X}}$) associated with $\mathscr{M}$ (6.5.1.1) (resp. (6.5.1.3)), by $\mathbb{K}^\bullet(\mathscr{H}'(\mathscr{M}))$ (resp. $\underline{\mathbb{K}}^\bullet(\underline{\mathscr{H}}'(\mathscr{M}))$) its

Dolbeault complex, and by $\mathrm{I}\widehat{\pi}^*(\underline{\mathbb{K}}^\bullet(\underline{\mathscr{H}}'(\mathscr{M})))$ the image of the latter by the functor $\mathrm{I}\widehat{\pi}^*$ (6.2.6.3). By 6.5.15, the Koszul filtration of $\mathbb{K}^\bullet(\mathscr{H}'(\mathscr{M}))$ associated with the exact sequence (6.1.25.4) determines a boundary map in $\mathbf{D}^+(\mathbf{Mod}(\mathscr{O}_{\mathfrak{X}'}[\frac{1}{p}]))$ (6.5.15.2)

$$\partial_{\mathscr{M}} \colon \underline{\mathbb{K}}^\bullet(\underline{\mathscr{H}}'(\mathscr{M})) \to \mathfrak{g}^*(\widetilde{\xi}^{-1}\widetilde{\Omega}^1_{\mathfrak{X}/\mathscr{S}}) \otimes_{\mathscr{O}_{\mathfrak{X}'}} \underline{\mathbb{K}}^\bullet(\underline{\mathscr{H}}'(\mathscr{M})), \tag{6.5.22.1}$$

and a boundary map in $\mathbf{D}^+(\mathbf{Ind\text{-}Mod}(\overline{\breve{\mathscr{B}}}^!))$ (6.5.15.6)

$$\mathrm{I}\widehat{\pi}^\star(\partial_{\mathscr{M}}) \colon \mathrm{I}\widehat{\pi}^*(\underline{\mathbb{K}}^\bullet(\underline{\mathscr{H}}'(\mathscr{M}))) \to \mathrm{I}\widehat{\pi}^*(\mathfrak{g}^*(\widetilde{\xi}^{-1}\widetilde{\Omega}^1_{\mathfrak{X}/\mathscr{S}}) \otimes_{\mathscr{O}_{\mathfrak{X}'}} \underline{\mathbb{K}}^\bullet(\underline{\mathscr{H}}'(\mathscr{M}))). \tag{6.5.22.2}$$

For any integer $q \geq 0$, we denote by

$$\kappa^q_{\mathscr{M}} \colon \mathrm{R}^q\mathfrak{g}_*(\underline{\mathbb{K}}^\bullet(\underline{\mathscr{H}}'(\mathscr{M}))) \to \widetilde{\xi}^{-1}\widetilde{\Omega}^1_{\mathfrak{X}/\mathscr{S}} \otimes_{\mathscr{O}_{\mathfrak{X}}} \mathrm{R}^q\mathfrak{g}_*(\underline{\mathbb{K}}^\bullet(\underline{\mathscr{H}}'(\mathscr{M}))) \tag{6.5.22.3}$$

the Katz–Oda $\mathscr{O}_{\mathfrak{X}}$-field (2.5.16), which is none other than the morphism $\mathrm{R}^q\mathfrak{g}_*(\partial_{\mathscr{M}})$.

Proposition 6.5.23 *Under the assumptions of* 6.5.22, *there exist a rational number* $r > 0$ *and an isomorphism of* $\mathbf{D}^+(\mathbf{Ind\text{-}Mod}(\breve{\mathfrak{g}}^*(\breve{\mathscr{C}}^{(r)})))$

$$\mathrm{RI}\breve{\tau}_*(\mathscr{M} \otimes_{\overline{\breve{\mathscr{B}}}'} \breve{\theta}^*(\breve{\mathscr{C}}^{(r)})) \xrightarrow{\sim} \mathrm{I}\widehat{\pi}^*(\underline{\mathbb{K}}^\bullet(\underline{\mathscr{H}}'(\mathscr{M}))) \otimes_{\overline{\breve{\mathscr{B}}}^!} \breve{\mathfrak{g}}^*(\breve{\mathscr{C}}^{(r)}), \tag{6.5.23.1}$$

where $\mathrm{I}\breve{\tau}_*$ *is the functor* (6.3.20.9) *and the tensor product on the right-hand side is defined term by term (not derived), such that the diagram of* $\mathbf{D}^+(\mathbf{Ind\text{-}Mod}(\overline{\breve{\mathscr{B}}}^!))$

$$\begin{array}{ccc}
\mathrm{RI}\breve{\tau}_*(\mathscr{M} \otimes_{\overline{\breve{\mathscr{B}}}'} \breve{\theta}^*(\breve{\mathscr{C}}^{(r)})) & \longrightarrow & \mathrm{I}\widehat{\pi}^*(\underline{\mathbb{K}}^\bullet(\underline{\mathscr{H}}'(\mathscr{M}))) \otimes_{\overline{\breve{\mathscr{B}}}^!} \breve{\mathfrak{g}}^*(\breve{\mathscr{C}}^{(r)}) \\
\Big\downarrow a & & \Big\downarrow b \\
\mathrm{RI}\breve{\tau}_*(\mathscr{M} \otimes_{\overline{\breve{\mathscr{B}}}'} \breve{\theta}^*(\breve{\mathscr{C}}^{(r)} \otimes_{\overline{\breve{\mathscr{B}}}} \widehat{\sigma}^*(\widetilde{\xi}^{-1}\widetilde{\Omega}^1_{\mathfrak{X}/\mathscr{S}}))) & \longrightarrow & \mathrm{I}\widehat{\pi}^*(\mathfrak{g}^*(\widetilde{\xi}^{-1}\widetilde{\Omega}^1_{\mathfrak{X}/\mathscr{S}}) \otimes_{\mathscr{O}_{\mathfrak{X}'}} \underline{\mathbb{K}}^\bullet(\underline{\mathscr{H}}'(\mathscr{M}))) \otimes_{\overline{\breve{\mathscr{B}}}^!} \breve{\mathfrak{g}}^*(\breve{\mathscr{C}}^{(r)}),
\end{array} \tag{6.5.23.2}$$

where the horizontal arrows are induced by the isomorphism (6.5.23.1) *and* 2.7.14, $a = p^r\mathrm{RI}\breve{\tau}_*(\mathrm{id} \otimes \breve{\theta}^*(d_{\breve{\mathscr{C}}^{(r)}}))$ (4.4.33.1) *and* $b = \mathrm{I}\widehat{\pi}^\star(\partial_{\mathscr{M}}) \otimes \mathrm{id} + p^r\mathrm{id} \otimes \breve{\mathfrak{g}}^*(d_{\breve{\mathscr{C}}^{(r)}})$ (6.5.22.2), *is commutative.*

To lighten the notation, we set $(\mathscr{N}, \theta) = \mathscr{H}'(\mathscr{M})$ and $(\underline{\mathscr{N}}, \underline{\theta}) = \underline{\mathscr{H}}'(\mathscr{M})$; the Higgs fields will be omitted when there is no risk of ambiguity. By 4.5.12, there exist a rational number $t > 0$ and an isomorphism of $\mathbf{Ind\text{-}MC}(\breve{\mathscr{C}}'^{(t)}/\overline{\breve{\mathscr{B}}}')$

$$\alpha \colon \mathrm{I}\mathfrak{S}'^{(t)}(\mathscr{M}) \xrightarrow{\sim} \mathrm{I}\widehat{\sigma}'^{(t)*}(\mathscr{N}). \tag{6.5.23.3}$$

For any rational number t' such that $0 \leq t' \leq t$, we denote by

$$\alpha^{(t')} \colon \mathrm{I}\mathfrak{S}'^{(t')}(\mathscr{M}) \xrightarrow{\sim} \mathrm{I}\widehat{\sigma}'^{(t')*}(\mathscr{N}) \tag{6.5.23.4}$$

the isomorphism of $\mathbf{Ind\text{-}MC}(\breve{\mathscr{C}}'^{(t')}/\overline{\breve{\mathscr{B}}}')$ induced by $\mathrm{I}\varepsilon'^{t,t'}(\alpha)$ (6.5.2) and the analogue of the isomorphisms (4.5.3.2) and (4.5.3.3). For any rational numbers t', r such that $0 \leq r \leq t' \leq t$, we denote by

$$\alpha^{(t',r)} \colon \mathrm{I}\mathfrak{S}'^{(t',r)}(\mathscr{M}) \xrightarrow{\sim} \mathrm{I}\widehat{\sigma}'^{(t',r)*}(\mathscr{N}) \qquad (6.5.23.5)$$

the isomorphism of $\mathbf{Ind\text{-}HM}(\breve{\overline{\mathscr{B}}}', \widehat{\sigma}'^{*}(\widetilde{\xi}^{-1}\widetilde{\Omega}^{1}_{\mathfrak{X}'/\mathscr{S}}))$ induced by $\mathrm{I}\lambda^{t',r}(\alpha^{(t')})$ (6.5.4.1), where $\mathrm{I}\mathfrak{S}'^{(t',r)}$ is the functor (6.5.4.5) and $\mathrm{I}\widehat{\sigma}'^{(t',r)*}$ is the functor (6.5.4.3). Let r be a rational number such that $0 < r < t$. Since the $\alpha^{(t')}$'s (6.5.23.4), for $r < t' \leq t$, form an isomorphism of direct systems of Higgs ind-$\breve{\overline{\mathscr{B}}}'$-modules with coefficients in $\widehat{\sigma}'^{*}(\widetilde{\xi}^{-1}\widetilde{\Omega}^{1}_{\mathfrak{X}'/\mathscr{S}})$, the same is true for the $\alpha^{(t',r)}$'s (6.5.23.5) in view of (6.4.20.6). Using the notation of 6.5.19, we deduce from this, by taking the direct limit in $\mathbf{Ind\text{-}HM}(\breve{\overline{\mathscr{B}}}', \widehat{\sigma}'^{*}(\widetilde{\xi}^{-1}\widetilde{\Omega}^{1}_{\mathfrak{X}'/\mathscr{S}}))$, a morphism

$$\mathscr{M} \otimes_{\breve{\overline{\mathscr{B}}}} \breve{\mathscr{C}}'^{(t',r)} \to \mathscr{F}^{(r)}, \qquad (6.5.23.6)$$

where $\mathscr{M} \otimes_{\breve{\overline{\mathscr{B}}}} \breve{\mathscr{C}}'^{(t',r)}$ is endowed with the Higgs $\breve{\overline{\mathscr{B}}}'$-field $\mathrm{id} \otimes \delta_{\breve{\mathscr{C}}'^{(t',r)}}$ (6.4.20.3), and hence a morphism of complexes of ind-$\breve{\overline{\mathscr{B}}}'$-modules

$$\mathscr{M} \otimes_{\breve{\overline{\mathscr{B}}}} \mathbb{K}^{\bullet}(\breve{\mathscr{C}}'^{(t',r)}) \to \mathbb{K}^{\bullet}(\mathscr{F}^{(r)}), \qquad (6.5.23.7)$$

$$\mathscr{M} \otimes_{\breve{\overline{\mathscr{B}}}} \underline{\mathbb{K}}^{\bullet}(\breve{\mathscr{C}}'^{(t',r)}) \to \underline{\mathbb{K}}^{\bullet}(\mathscr{F}^{(r)}), \qquad (6.5.23.8)$$

where the Dolbeault complexes $\mathbb{K}^{\bullet}(\breve{\mathscr{C}}'^{(t',r)})$ and $\underline{\mathbb{K}}^{\bullet}(\breve{\mathscr{C}}'^{(t',r)})$ are defined in 6.4.21. By 4.5.6, the ind-$\breve{\overline{\mathscr{B}}}$-module $\mathscr{M}$ is rational and flat. Taking the boundary maps associated with the Koszul filtrations (2.8.2.5) of the Dolbeault complexes appearing in (6.5.23.7) with respect to the image by the functor $\widehat{\sigma}'^{*}$ of the exact sequence (6.1.25.4), we obtain by functoriality a commutative diagram of $\mathbf{D}^{+}(\mathbf{Ind\text{-}Mod}(\breve{\overline{\mathscr{B}}}'))$

$$\begin{array}{ccc}
\mathscr{M} \otimes_{\breve{\overline{\mathscr{B}}}} \underline{\mathbb{K}}^{\bullet}(\breve{\mathscr{C}}'^{(t',r)}) & \longrightarrow & \underline{\mathbb{K}}^{\bullet}(\mathscr{F}^{(r)}) \\
\Big\downarrow {\scriptstyle \mathrm{id} \otimes \partial^{(t',r)}} & & \Big\downarrow {\scriptstyle \partial^{(r)}} \\
\widehat{\sigma}'^{*}(\mathfrak{g}^{*}(\widetilde{\xi}^{-1}\widetilde{\Omega}^{1}_{\mathfrak{X}/\mathscr{S}})) \otimes_{\breve{\overline{\mathscr{B}}}} \mathscr{M} \otimes_{\breve{\overline{\mathscr{B}}}} \underline{\mathbb{K}}^{\bullet}(\breve{\mathscr{C}}'^{(t',r)}) & \longrightarrow & \widehat{\sigma}'^{*}(\mathfrak{g}^{*}(\widetilde{\xi}^{-1}\widetilde{\Omega}^{1}_{\mathfrak{X}/\mathscr{S}})) \otimes_{\breve{\overline{\mathscr{B}}}} \underline{\mathbb{K}}^{\bullet}(\mathscr{F}^{(r)})
\end{array} \qquad (6.5.23.9)$$

where the horizontal arrows are induced by (6.5.23.8), and $\partial^{(t',r)}$ (resp. $\partial^{(r)}$) is the boundary map (6.4.21.1) (resp. (6.5.19.4)).

By (6.5.5.3) and (6.5.19.1), the morphisms (6.5.23.8) induce a quasi-isomorphism

$$\mathscr{M} \otimes_{\breve{\overline{\mathscr{B}}}} \check{\theta}^{*}(\breve{\mathscr{C}}^{(r)})[0] \to \underline{\mathbb{K}}^{\bullet}(\mathscr{F}^{(r)}). \qquad (6.5.23.10)$$

In view of (6.4.21.2) and (6.5.23.9), the diagram

$$\begin{array}{ccc}
\mathscr{M} \otimes_{\breve{\overline{\mathscr{B}}}} \check{\theta}^{*}(\breve{\mathscr{C}}^{(r)})[0] & \longrightarrow & \underline{\mathbb{K}}^{\bullet}(\mathscr{F}^{(r)}) \\
\Big\downarrow {\scriptstyle p^{r}\mathrm{id} \otimes \check{\theta}^{*}(d_{\breve{\mathscr{C}}^{(r)}})} & & \Big\downarrow {\scriptstyle \partial^{(r)}} \\
\check{\theta}^{*}(\widehat{\sigma}^{*}(\widetilde{\xi}^{-1}\widetilde{\Omega}^{1}_{\mathfrak{X}/\mathscr{S}})) \otimes_{\breve{\overline{\mathscr{B}}}} \mathscr{M} \otimes_{\breve{\overline{\mathscr{B}}}} \check{\theta}^{*}(\breve{\mathscr{C}}^{(r)})[0] & \longrightarrow & \widehat{\sigma}'^{*}(\mathfrak{g}^{*}(\widetilde{\xi}^{-1}\widetilde{\Omega}^{1}_{\mathfrak{X}/\mathscr{S}})) \otimes_{\breve{\overline{\mathscr{B}}}} \underline{\mathbb{K}}^{\bullet}(\mathscr{F}^{(r)}),
\end{array} \qquad (6.5.23.11)$$

where the horizontal arrows are induced by (6.5.23.10) and (6.1.23.6), is commutative. The proposition follows by virtue of 6.5.7 and 6.5.21.

Theorem 6.5.24 *Assume that $g\colon X' \to X$ is proper. Let $\mathscr{M}$ be a Dolbeault ind-$\overline{\breve{\mathscr{B}}}'$-module (6.5.1), q an integer ≥ 0. We denote by $\underline{\mathscr{H}}'(\mathscr{M})$ the Higgs $\mathscr{O}_{\mathfrak{X}'}[\frac{1}{p}]$-module with coefficients in $\widetilde{\xi}^{-1}\widetilde{\Omega}^1_{\mathfrak{X}'/\mathfrak{X}}$ associated with $\mathscr{M}$ (6.5.1.3) and by $\underline{\mathbb{K}}^\bullet(\underline{\mathscr{H}}'(\mathscr{M}))$ its Dolbeault complex. We endow the $\mathscr{O}_{\mathfrak{X}}[\frac{1}{p}]$-module $\mathrm{R}^q\mathfrak{g}_*(\underline{\mathbb{K}}^\bullet(\underline{\mathscr{H}}'(\mathscr{M})))$ with the Katz–Oda field $\kappa^q_{\mathscr{M}}$ (6.5.22.3). Then, the $\mathscr{O}_{\mathfrak{X}}[\frac{1}{p}]$-module $\mathrm{R}^q\mathfrak{g}_*(\underline{\mathbb{K}}^\bullet(\underline{\mathscr{H}}'(\mathscr{M})))$ is coherent, and there exist a rational number $r > 0$, independent of q, and an isomorphism of* **Ind-MC**$(\breve{\mathscr{C}}^{(r)}/\breve{\overline{\mathscr{B}}})$

$$\mathrm{I}\mathfrak{S}^{(r)}(\mathrm{R}^q\mathrm{I}\breve{\theta}_*(\mathscr{M})) \xrightarrow{\sim} \mathrm{I}\widehat{\sigma}^{(r)*}(\mathrm{R}^q\mathfrak{g}_*(\underline{\mathbb{K}}^\bullet(\underline{\mathscr{H}}'(\mathscr{M}))), \kappa^q_{\mathscr{M}}), \tag{6.5.24.1}$$

where the functors $\mathrm{I}\mathfrak{S}^{(r)}$ and $\mathrm{I}\widehat{\sigma}^{(r)*}$ are defined in (4.5.1.2) *and* (4.5.1.6) *in view of* (4.5.2.3).

Indeed, by virtue of 6.5.13, the $\mathscr{O}_{\mathfrak{X}}[\frac{1}{p}]$-module $\mathrm{R}^q\mathfrak{g}_*(\underline{\mathbb{K}}^\bullet(\underline{\mathscr{H}}'(\mathscr{M})))$ is coherent, and there exist a rational number $r > 0$ and an isomorphism of ind-$\breve{\mathscr{C}}^{(r)}$-modules

$$\mathrm{R}^q\mathrm{I}\breve{\theta}_*(\mathscr{M}) \otimes_{\breve{\overline{\mathscr{B}}}} \breve{\mathscr{C}}^{(r)} \xrightarrow{\sim} \mathrm{I}\widehat{\sigma}^*(\mathrm{R}^q\mathfrak{g}_*(\underline{\mathbb{K}}^\bullet(\underline{\mathscr{H}}'(\mathscr{M})))) \otimes_{\breve{\overline{\mathscr{B}}}} \breve{\mathscr{C}}^{(r)}. \tag{6.5.24.2}$$

Moreover, it follows from 6.5.23, 6.5.12, 6.2.26, 6.5.18, (2.9.6.8) and (2.9.8.5) that we can find such an isomorphism compatible with the Higgs fields. The left-hand and the right-hand sides being endowed with the total Higgs fields, the proposition follows.

Corollary 6.5.25 *Let $\mathscr{M}$ be a Dolbeault ind-$\overline{\breve{\mathscr{B}}}'$-module (6.5.1), q an integer ≥ 0. We denote by $\underline{\mathscr{H}}'(\mathscr{M})$ the Higgs $\mathscr{O}_{\mathfrak{X}'}[\frac{1}{p}]$-module with coefficients in $\widetilde{\xi}^{-1}\widetilde{\Omega}^1_{\mathfrak{X}'/\mathfrak{X}}$ associated with $\mathscr{M}$ (6.5.1.3) and by $\underline{\mathbb{K}}^\bullet(\underline{\mathscr{H}}'(\mathscr{M}))$ its Dolbeault complex. We endow the $\mathscr{O}_{\mathfrak{X}}[\frac{1}{p}]$-module $\mathrm{R}^q\mathfrak{g}_*(\underline{\mathbb{K}}^\bullet(\underline{\mathscr{H}}'(\mathscr{M})))$ with the Katz–Oda field $\kappa^q_{\mathscr{M}}$ (6.5.22.3). Suppose that the morphism $g\colon X' \to X$ is proper and the $\mathscr{O}_{\mathfrak{X}}[\frac{1}{p}]$-module $\mathrm{R}^q\mathfrak{g}_*(\underline{\mathbb{K}}^\bullet(\underline{\mathscr{H}}'(\mathscr{M})))$ is locally projective of finite type. Then, the ind-$\breve{\overline{\mathscr{B}}}$-module $\mathrm{R}^q\mathrm{I}\breve{\theta}_*(\mathscr{M})$ is Dolbeault, the Higgs $\mathscr{O}_{\mathfrak{X}}[\frac{1}{p}]$-module $(\mathrm{R}^q\mathfrak{g}_*(\underline{\mathbb{K}}^\bullet(\underline{\mathscr{H}}'(\mathscr{M}))), \kappa^q_{\mathscr{M}})$ is solvable (4.5.5), and we have an isomorphism of Higgs $\mathscr{O}_{\mathfrak{X}}[\frac{1}{p}]$-bundles*

$$\mathscr{H}(\mathrm{R}^q\mathrm{I}\breve{\theta}_*(\mathscr{M})) \xrightarrow{\sim} (\mathrm{R}^q\mathfrak{g}_*(\underline{\mathbb{K}}^\bullet(\underline{\mathscr{H}}'(\mathscr{M}))), \kappa^q_{\mathscr{M}}), \tag{6.5.25.1}$$

where $\mathscr{H}$ is the functor (4.5.7.3).

This follows from 6.5.24 and 4.5.11.

6.5.26 We take the notation of 4.12.2 for f and we consider the analogous notation for f', that we equip with an exponent $'$ (6.1.3). We then have the commutative diagram of morphisms of topos

$$\begin{array}{ccc} (\overline{X}'^{\rhd}_{\text{ét}})^{\mathbb{N}^\circ} & \xrightarrow{\check{\psi}'} & \widetilde{E}'^{\mathbb{N}^\circ} \\ \Big\downarrow{\scriptstyle \check{\gamma}} & & \Big\downarrow{\scriptstyle \check{\Theta}} \\ (\overline{X}^\circ_{\text{ét}})^{\mathbb{N}^\circ} & \xrightarrow{\check{\psi}} & \widetilde{E}^{\mathbb{N}^\circ} \end{array} \tag{6.5.26.1}$$

induced by ψ, ψ' (6.1.8.5), γ and Θ (6.1.10.3). We will naturally consider them as morphisms of topos ringed by the rings $\check{\mathbb{Z}}_p$ (4.12.1).

We denote by

$$\mathscr{H}'_{\mathbb{Q}}\colon \mathbf{Mod}_{\mathbb{Q}}(\overline{\breve{\mathscr{B}}}') \to \mathbf{HM}(\mathscr{O}_{\mathfrak{X}'}[\frac{1}{p}], \widetilde{\xi}^{-1}\widetilde{\Omega}^1_{\mathfrak{X}'/\mathscr{S}}) \tag{6.5.26.2}$$

the functor defined in (4.6.9.2), and by

$$\underline{\mathscr{H}'}_{\mathbb{Q}}\colon \mathbf{Mod}_{\mathbb{Q}}(\overline{\breve{\mathscr{B}}}') \to \mathbf{HM}(\mathscr{O}_{\mathfrak{X}'}[\frac{1}{p}], \widetilde{\xi}^{-1}\widetilde{\Omega}^1_{\mathfrak{X}'/\mathfrak{X}}) \tag{6.5.26.3}$$

the composition of the functor $\mathscr{H}'_{\mathbb{Q}}$ and the canonical functor (6.5.1.4). These functors are compatible with the functors (6.5.1.1) and (6.5.1.3); see (4.6.9.3).

Proposition 6.5.27 *Assume that $g\colon X' \to X$ is proper. Let $\mathscr{M}$ be a Dolbeault $\overline{\breve{\mathscr{B}}}'_{\mathbb{Q}}$-module (4.6.6), q an integer ≥ 0. We denote by $\underline{\mathscr{H}'}_{\mathbb{Q}}(\mathscr{M})$ the Higgs $\mathscr{O}_{\mathfrak{X}'}[\frac{1}{p}]$-bundle with coefficients in $\widetilde{\xi}^{-1}\widetilde{\Omega}^1_{\mathfrak{X}'/\mathfrak{X}}$ associated with $\mathscr{M}$ (6.5.26.3) and by $\underline{\mathbb{K}}^\bullet$ its Dolbeault complex. Then, the $\mathscr{O}_{\mathfrak{X}}[\frac{1}{p}]$-module $\mathrm{R}^q\mathfrak{g}_*(\underline{\mathbb{K}}^\bullet)$ is coherent, and there exist a rational number $r > 0$, independent of q, and an isomorphism of $\check{\mathscr{C}}^{(r)}_{\mathbb{Q}}$-modules*

$$\mathrm{R}^q\check{\theta}_{\mathbb{Q}*}(\mathscr{M}) \otimes_{\breve{\overline{\mathscr{B}}}_{\mathbb{Q}}} \check{\mathscr{C}}^{(r)}_{\mathbb{Q}} \xrightarrow{\sim} \widehat{\sigma}^*_{\mathbb{Q}}(\mathrm{R}^q\mathfrak{g}_*(\underline{\mathbb{K}}^\bullet)) \otimes_{\breve{\overline{\mathscr{B}}}_{\mathbb{Q}}} \check{\mathscr{C}}^{(r)}_{\mathbb{Q}}, \tag{6.5.27.1}$$

where $\widehat{\sigma}^*_{\mathbb{Q}}$ is the functor (6.1.24.4).

This follows from 6.5.13, in view of (2.9.5.1), (2.9.6.8) and the fact that the functor $\alpha_{\check{\mathscr{C}}^{(r)}}$ is fully faithful (2.9.2.3).

Corollary 6.5.28 *Let $M = (M_n)_{n\geq 0}$ be a $\check{\mathbb{Z}}_p$-local system of $(\overline{X}'^{\rhd}_{\text{ét}})^{\mathbb{N}^\circ}$ (4.12.3). We set $\mathscr{M} = \check{\psi}'_*(M) \otimes_{\check{\mathbb{Z}}_p} \overline{\breve{\mathscr{B}}}'$, and denote by $\underline{\mathscr{H}'}_{\mathbb{Q}}(\mathscr{M}_{\mathbb{Q}})$ the Higgs $\mathscr{O}_{\mathfrak{X}'}[\frac{1}{p}]$-bundle with coefficients in $\widetilde{\xi}^{-1}\widetilde{\Omega}^1_{\mathfrak{X}'/\mathfrak{X}}$ associated with $\mathscr{M}_{\mathbb{Q}}$ (6.5.26.3) and by $\underline{\mathbb{K}}^\bullet$ its Dolbeault complex. Assume that the following conditions are satisfied:*

(i) *the morphism $g\colon X' \to X$ is proper;*

(ii) *the $\check{\mathbb{Z}}_{p,\mathbb{Q}}$-local system $M_{\mathbb{Q}}$ is Dolbeault (4.12.4), i.e., the $\overline{\breve{\mathscr{B}}}'_{\mathbb{Q}}$-module $\mathscr{M}_{\mathbb{Q}}$ is Dolbeault.*

Then the $\mathscr{O}_{\mathfrak{X}}[\frac{1}{p}]$-modules $\mathrm{R}^i\mathfrak{g}_*(\mathrm{H}^j(\underline{\mathbb{K}}^\bullet))$ are coherent for all $i, j \geq 0$, and there exist *a rational number $r > 0$ and a spectral sequence of $\breve{\mathscr{C}}^{(r)}_{\mathbb{Q}}$-modules*

$$\mathrm{E}_2^{i,j} = \widehat{\sigma}^*_{\mathbb{Q}}(\mathrm{R}^i \mathfrak{g}_*(\mathrm{H}^j(\underline{\mathbb{K}}^\bullet))) \otimes_{\overline{\breve{\mathscr{B}}}_{\mathbb{Q}}} \breve{\mathscr{C}}^{(r)}_{\mathbb{Q}} \Rightarrow (\breve{\psi}_*(\mathrm{R}^{i+j}\breve{\gamma}_*(M)) \otimes_{\breve{\mathbb{Z}}_p} \breve{\mathscr{C}}^{(r)})_{\mathbb{Q}}, \quad (6.5.28.1)$$

where $\widehat{\sigma}^*_{\mathbb{Q}}$ is the functor (6.1.24.4).

Indeed, the first assertion follows from ([1] 2.10.24 and 2.11.5). By 6.2.5, for every rational number $r \geq 0$, since $\breve{\mathscr{C}}^{(r)}$ is $\overline{\breve{\mathscr{B}}}$-flat, the second spectral sequence of hypercohomology

$$\mathrm{E}_2^{i,j} = \mathrm{R}^i \mathfrak{g}_*(\mathrm{H}^j(\underline{\mathbb{K}}^\bullet)) \Rightarrow \mathrm{R}^{i+j}\mathfrak{g}_*(\underline{\mathbb{K}}^\bullet) \quad (6.5.28.2)$$

induces a spectral sequence of $\breve{\mathscr{C}}^{(r)}_{\mathbb{Q}}$-modules

$$\mathrm{E}_2^{i,j} = \widehat{\sigma}^*_{\mathbb{Q}}(\mathrm{R}^i \mathfrak{g}_*(\mathrm{H}^j(\underline{\mathbb{K}}^\bullet))) \otimes_{\overline{\breve{\mathscr{B}}}_{\mathbb{Q}}} \breve{\mathscr{C}}^{(r)}_{\mathbb{Q}} \Rightarrow \widehat{\sigma}^*_{\mathbb{Q}}(\mathrm{R}^{i+j}\mathfrak{g}_*(\underline{\mathbb{K}}^\bullet)) \otimes_{\overline{\breve{\mathscr{B}}}_{\mathbb{Q}}} \breve{\mathscr{C}}^{(r)}_{\mathbb{Q}}. \quad (6.5.28.3)$$

The proposition follows by virtue of 6.5.27, ([2] 5.7.3) and ([3] III.7.5).

Corollary 6.5.29 *Assume that the morphism $g: X' \to X$ is proper. Then, there exist a rational number $r > 0$ and for every integer $n \geq 0$, a canonical isomorphism of $\breve{\mathscr{C}}^{(r)}_{\mathbb{Q}}$-modules*

$$\breve{\psi}_*(\mathrm{R}^n\breve{\gamma}_*(\breve{\mathbb{Z}}_p)) \otimes_{\breve{\mathbb{Z}}_p} \breve{\mathscr{C}}^{(r)}_{\mathbb{Q}} \xrightarrow{\sim} \oplus_{0 \leq i \leq n} \widehat{\sigma}^*(\mathrm{R}^i \mathfrak{g}_*(\widetilde{\Omega}^{n-i}_{\mathfrak{X}'/\mathfrak{X}})) \otimes_{\overline{\breve{\mathscr{B}}}} \breve{\mathscr{C}}^{(r)}_{\mathbb{Q}}(i-n), \quad (6.5.29.1)$$

where $\widehat{\sigma}$ is the morphism of ringed topos (4.3.11.4).

This follows from 6.5.27 applied to $\mathscr{M} = \overline{\breve{\mathscr{B}}}'_{\mathbb{Q}}$, ([2] 5.7.4) and ([3] III.7.5), since the $\overline{\breve{\mathscr{B}}}'_{\mathbb{Q}}$-module $\overline{\breve{\mathscr{B}}}'_{\mathbb{Q}}$ is Dolbeault and $\underline{\mathscr{H}}'_{\mathbb{Q}}(\overline{\breve{\mathscr{B}}}'_{\mathbb{Q}})$ is the trivial bundle $\mathscr{O}_{\mathfrak{X}'}[\frac{1}{p}]$ endowed with the zero Higgs field (4.6.10).

Remark 6.5.30 The corollary 6.5.29 applies in particular to the relative case (3.1.14), where $\widetilde{g}$ is the trivial deformation (6.1.5).

Theorem 6.5.31 ([2] 6.7.5) *If the morphism $g: X' \to X$ is proper, there exists a canonical spectral sequence of $\overline{\breve{\mathscr{B}}}_{\mathbb{Q}}$-modules*

$$\mathrm{E}_2^{i,j} = \widehat{\sigma}^*(\mathrm{R}^i \mathfrak{g}_*(\widetilde{\Omega}^j_{\mathfrak{X}'/\mathfrak{X}})) \otimes_{\overline{\breve{\mathscr{B}}}} \overline{\breve{\mathscr{B}}}_{\mathbb{Q}}(-j) \Rightarrow \breve{\psi}_*(\mathrm{R}^{i+j}\breve{\gamma}_*(\breve{\mathbb{Z}}_p)) \otimes_{\mathbb{Z}_p} \overline{\breve{\mathscr{B}}}_{\mathbb{Q}}, \quad (6.5.31.1)$$

where $\widehat{\sigma}$ is the morphism of ringed topos (4.3.11.4).

In the statement ([2] 6.7.5), we actually require g to be projective. However, as pointed in ([2] 6.7.6), the result holds under the more general assumption that g is proper. Indeed, the projectivity assumption on g is used in the proof of ([2] 5.7.4), which also extends to proper morphisms (see [2] 5.7.6).

This spectral sequence does not require the consideration of any deformation (6.1.4). It is G_K-equivariant ([2] 6.7.10) and it degenerates at E_2 ([2] 6.7.13). Nevertheless the abutment filtration is not split in general (see [2] 1.3.2 and 1.3.3). However, we can check that it splits after base change from $\overline{\breve{\mathscr{B}}}$ to $\breve{\mathscr{C}}^{(r)}$ for a rational number $r > 0$, and that it corresponds to the decomposition (6.5.29.1).

Proposition 6.5.32 *Assume that $g\colon X' \to X$ is proper. Let $\mathscr{M}$ be a Dolbeault $\breve{\overline{\mathscr{B}}}'_{\mathbb{Q}}$-module (4.6.6), q an integer ≥ 0. We denote by $\underline{\mathscr{H}'_{\mathbb{Q}}}(\mathscr{M})$ the Higgs $\mathscr{O}_{\mathfrak{X}'}[\frac{1}{p}]$-bundle with coefficients in $\widetilde{\xi}^{-1}\widetilde{\Omega}^1_{\mathfrak{X}'/\mathfrak{X}}$ associated with $\mathscr{M}$ (6.5.26.3), by $\underline{\mathbb{K}}^\bullet$ its Dolbeault complex and by*

$$\kappa^q_{\mathscr{M}}\colon \mathrm{R}^q\mathfrak{g}_*(\underline{\mathbb{K}}^\bullet) \to \widetilde{\xi}^{-1}\widetilde{\Omega}^1_{\mathfrak{X}/\mathscr{S}} \otimes_{\mathscr{O}_{\mathfrak{X}}} \mathrm{R}^q\mathfrak{g}_*(\underline{\mathbb{K}}^\bullet) \qquad (6.5.32.1)$$

the Katz–Oda field (2.5.16). Then, the $\mathscr{O}_{\mathfrak{X}}[\frac{1}{p}]$-module $\mathrm{R}^q\mathfrak{g}_*(\underline{\mathbb{K}}^\bullet)$ is coherent, and there exist a rational number $r > 0$, independent of q, and an isomorphism of $\mathbf{IC}_{\mathbb{Q}}(\breve{\mathscr{C}}^{(r)}/\breve{\overline{\mathscr{B}}})$

$$\mathfrak{S}^{(r)}(\mathrm{R}^q\breve{\theta}_{\mathbb{Q}*}(\mathscr{M})) \xrightarrow{\sim} \widehat{\sigma}^{(r)*}(\mathrm{R}^q\mathfrak{g}_*(\underline{\mathbb{K}}^\bullet), \kappa^q_{\mathscr{M}}), \qquad (6.5.32.2)$$

where the functors $\mathfrak{S}^{(r)}$ and $\widehat{\sigma}^{(r)*}$ are defined in (4.6.1.2) and (4.6.1.7).

This follows from 6.5.24 by means of the fully faithful functor $\alpha_{\breve{\mathscr{C}}^{(r)}/\breve{\overline{\mathscr{B}}}}$ (4.6.3.3), in view of (4.6.3.4), (4.6.3.5) and (4.6.9.3).

Corollary 6.5.33 *Let $\mathscr{M}$ be a Dolbeault $\breve{\overline{\mathscr{B}}}'_{\mathbb{Q}}$-module (4.6.6), q an integer ≥ 0. We denote by $\underline{\mathscr{H}'_{\mathbb{Q}}}(\mathscr{M})$ the Higgs $\mathscr{O}_{\mathfrak{X}'}[\frac{1}{p}]$-bundle with coefficients in $\widetilde{\xi}^{-1}\widetilde{\Omega}^1_{\mathfrak{X}'/\mathfrak{X}}$ associated with $\mathscr{M}$ (6.5.26.3), by $\underline{\mathbb{K}}^\bullet$ its Dolbeault complex and by*

$$\kappa^q_{\mathscr{M}}\colon \mathrm{R}^q\mathfrak{g}_*(\underline{\mathbb{K}}^\bullet) \to \widetilde{\xi}^{-1}\widetilde{\Omega}^1_{\mathfrak{X}/\mathscr{S}} \otimes_{\mathscr{O}_{\mathfrak{X}}} \mathrm{R}^q\mathfrak{g}_*(\underline{\mathbb{K}}^\bullet) \qquad (6.5.33.1)$$

the Katz–Oda field (2.5.16). Assume that the morphism $g\colon X' \to X$ is proper and the $\mathscr{O}_{\mathfrak{X}}[\frac{1}{p}]$-module $\mathrm{R}^q\mathfrak{g}_*(\underline{\mathbb{K}}^\bullet)$ is locally projective of finite type. Then, the $\breve{\overline{\mathscr{B}}}_{\mathbb{Q}}$-module $\mathrm{R}^q\breve{\theta}_{\mathbb{Q}*}(\mathscr{M})$ is Dolbeault, the Higgs $\mathscr{O}_{\mathfrak{X}}[\frac{1}{p}]$-module $(\mathrm{R}^q\mathfrak{g}_*(\underline{\mathbb{K}}^\bullet), \kappa^q_{\mathscr{M}})$ is rationally solvable (4.6.6), and we have an isomorphism of Higgs $\mathscr{O}_{\mathfrak{X}}[\frac{1}{p}]$-bundles

$$\mathscr{H}_{\mathbb{Q}}(\mathrm{R}^q\breve{\theta}_{\mathbb{Q}*}(\mathscr{M})) \xrightarrow{\sim} (\mathrm{R}^q\mathfrak{g}_*(\underline{\mathbb{K}}^\bullet), \kappa^q_{\mathscr{M}}), \qquad (6.5.33.2)$$

where $\mathscr{H}_{\mathbb{Q}}$ is the functor (4.6.9.2).

This follows from 6.5.32 and 4.6.12.

Corollary 6.5.34 *Assume that the morphism $g\colon X' \to X$ is proper. For any integer $q \geq 0$, set $\mathscr{M}^q = \breve{\psi}_*(\mathrm{R}^q\breve{\gamma}_*(\breve{\mathbb{Z}}_p)) \otimes_{\breve{\mathbb{Z}}_p} \breve{\overline{\mathscr{B}}}$. Then, the $\breve{\overline{\mathscr{B}}}_{\mathbb{Q}}$-module $\mathscr{M}^q_{\mathbb{Q}}$ is Hodge–Tate (4.12.4), and we have an isomorphism of Higgs $\mathscr{O}_{\mathfrak{X}}[\frac{1}{p}]$-bundles*

$$\mathscr{H}_{\mathbb{Q}}(\mathscr{M}^q_{\mathbb{Q}}) \xrightarrow{\sim} \oplus_{0\leq i\leq q}\mathrm{R}^i\mathfrak{g}_*(\widetilde{\xi}^{i-q}\widetilde{\Omega}^{q-i}_{\mathfrak{X}'/\mathfrak{X}}) \otimes_{\mathscr{O}_{\mathfrak{X}}} \mathscr{O}_{\mathfrak{X}}[\frac{1}{p}], \qquad (6.5.34.1)$$

where $\mathscr{H}_{\mathbb{Q}}$ is the functor (4.6.9.2), the Higgs field on the right-hand side being induced by the Kodaira–Spencer maps of $\mathfrak{g}$ (6.1.25.6).

Indeed, it follows from ([2] 5.7.5 and 5.7.6) and ([3] III.7.5) that we have a canonical isomorphism

$$\mathscr{M}^q_{\mathbb{Q}} \xrightarrow{\sim} \mathrm{R}^q \breve{\theta}_{\mathbb{Q}*}(\overline{\breve{\mathscr{B}}}'_{\mathbb{Q}}). \tag{6.5.34.2}$$

By 4.6.10, the $\overline{\breve{\mathscr{B}}}'_{\mathbb{Q}}$-module $\overline{\breve{\mathscr{B}}}'_{\mathbb{Q}}$ is Dolbeault and the Higgs module $\underline{\mathscr{H}}'_{\mathbb{Q}}(\overline{\breve{\mathscr{B}}}'_{\mathbb{Q}})$ (6.5.26.3) is the trivial bundle $\mathscr{O}_{\mathfrak{X}'}[\frac{1}{p}]$ endowed with the zero Higgs field. Its Dolbeault complex $\underline{\mathbb{K}}^\bullet$ is therefore given by

$$\underline{\mathbb{K}}^\bullet = \oplus_{i\geq 0}\widetilde{\xi}^{-i}\widetilde{\Omega}^i_{\mathfrak{X}'/\mathfrak{X}} \otimes_{\mathscr{O}_{\mathfrak{X}'}} \mathscr{O}_{\mathfrak{X}'}[\frac{1}{p}][-i]. \tag{6.5.34.3}$$

Consequently, we have

$$\mathrm{R}^q \mathfrak{g}_*(\underline{\mathbb{K}}^\bullet) = \oplus_{0\leq i\leq q}\mathrm{R}^i \mathfrak{g}_*(\widetilde{\xi}^{i-q}\widetilde{\Omega}^{q-i}_{\mathfrak{X}'/\mathfrak{X}}) \otimes_{\mathscr{O}_{\mathfrak{X}}} \mathscr{O}_{\mathfrak{X}}[\frac{1}{p}]. \tag{6.5.34.4}$$

The Katz–Oda field (2.5.16)

$$\kappa^q : \mathrm{R}^q \mathfrak{g}_*(\underline{\mathbb{K}}^\bullet) \to \widetilde{\xi}^{-1}\widetilde{\Omega}^1_{\mathfrak{X}/\mathscr{S}} \otimes_{\mathscr{O}_{\mathfrak{X}}} \mathrm{R}^q \mathfrak{g}_*(\underline{\mathbb{K}}^\bullet) \tag{6.5.34.5}$$

is induced by the Kodaira–Spencer maps of $\mathfrak{g}$ (6.1.25.6). It is therefore nilpotent (4.2.5).

By virtue of 6.1.28 and 6.5.33, the $\breve{\overline{\mathscr{B}}}_{\mathbb{Q}}$-module $\mathrm{R}^q \breve{\theta}_{\mathbb{Q}*}(\overline{\breve{\mathscr{B}}}'_{\mathbb{Q}})$ is Hodge–Tate, and we have an isomorphism of Higgs $\mathscr{O}_{\mathfrak{X}}[\frac{1}{p}]$-bundles

$$\mathscr{H}_{\mathbb{Q}}(\mathrm{R}^q \breve{\theta}_{\mathbb{Q}*}(\overline{\breve{\mathscr{B}}}'_{\mathbb{Q}})) \xrightarrow{\sim} (\mathrm{R}^q \mathfrak{g}_*(\underline{\mathbb{K}}^\bullet), \kappa^q). \tag{6.5.34.6}$$

The proposition follows.

Remark 6.5.35 The corollary 6.5.34 applies in particular to the relative case (3.1.14), where $\widetilde{g}$ is the trivial deformation (6.1.5).

Errata and Addenda to "The p-adic Simpson Correspondence"

by A. Abbes, M. Gros and T. Tsuji, Ann. of Math. Stud. **193**, Princeton Univ. Press (2016)

A) Misprints

(I.1.2). Line 1, replace *"valuation ring"* by *"valuation field"*.

(I.2.1). Line 1, replace *"valuation ring"* by *"valuation field"*.

(I.4.6.1). One line under (I.4.6.1), replace $\Gamma = \pi_1(X, \overline{y})$ by $\Gamma = \pi_1(X_\eta, \overline{y})$.

(I.4.7.4). Two lines above (I.4.7.4), replace $\alpha^{r,r'} : \mathscr{C}^{(r')} \to \mathscr{C}^{(r)}$ by $\alpha^{r,r'} : \mathscr{C}^{(r)} \to \mathscr{C}^{(r')}$.

(I.4.7.4). One line above (I.4.7.4), replace $h_\alpha^{r,r'} : \widehat{\mathscr{C}}^{(r')} \to \widehat{\mathscr{C}}^{(r)}$ by $h_\alpha^{r,r'} : \widehat{\mathscr{C}}^{(r)} \to \widehat{\mathscr{C}}^{(r')}$.

(I.4.7.7). Replace $p^{r'}(\mathrm{id} \times \alpha^{r,r'}) \circ d_{\mathscr{C}^{(r')}} = p^r d_{\mathscr{C}^{(r)}} \circ \alpha^{r,r'}$ by $p^r(\mathrm{id} \times \alpha^{r,r'}) \circ d_{\mathscr{C}^{(r)}} = p^{r'} d_{\mathscr{C}^{(r')}} \circ \alpha^{r,r'}$.

(I.4.13.1). Replace the target $\oplus_{i \in \mathbb{Z}} \mathrm{D}^i(V) \otimes_{\widehat{R}} \widehat{R_1}(-1)$ by $\oplus_{i \in \mathbb{Z}} \mathrm{D}^i(V) \otimes_{\widehat{R}} \widehat{R_1}(-i)$.

(I.5.12.7). One line under (I.5.12.7), replace *"multiplication by $p^{r'-r}$"* by *"multiplication by $p^{r-r'}$"*.

(II.2.1). Line 1-2, replace *"valuation ring"* by *"valuation field"*.

(II.3.9). Line 5, replace *"has a left inverse"* by *"has a right inverse"*.

(II.3.12). Line 1-2, replace *"M a topological A-G-module, and N a topological A-H-module"* by *"M a linearly topologized A-G-module, and N a linearly topologized A-H-module"*.

(II.3.34). Line 4-5, replace *"with values in* $\mathrm{id}_r + a^m \mathrm{Mat}_r(A/a^q A)$ *associated with N and N', respectively"* by *"with values in* $\mathrm{id}_r + a^m \mathrm{Mat}_r(A/a^q A)$ *associated with ρ and ρ', respectively"*.

A. Abbes, M. Gros, *The p-adic Simpson Correspondence and Hodge-Tate Local Systems*, Lecture Notes in Mathematics 2345, https://doi.org/10.1007/978-3-031-55914-3

(II.5.11). Line -2, replace "*... under f, or, equivalently ...*" by "*... under f. If f is strict, then the canonical homomorphism $f^{-1}(\mathscr{M}_Y^\sharp) \to \mathscr{M}_X^\sharp$ is an isomorphism and the converse holds if, moreover, $\mathscr{M}_X$ is u-integral in the sense of [58], I. Def. 1.3.1, 3. (cf. [58], III. Cor. 1.2.11)*".

(II.6.20.1). Replace $E_1^{i,j}$ by $E_2^{i,j}$.

(II.8.1.10). Replace the H_1's by the H_0's without changing the conclusion.

(II.8.1.11). Replace the H_0's by the H_1's without changing the conclusion.

(II.8.21). Replace in the last line "*whose kernel is annihilated by $p^{\frac{1}{p-1}}$.*" by "*whose kernel is annihilated by $\mathfrak{m}_{\overline{K}} p^{\frac{1}{p-1}}$.*"

(II.9.3). Line 5-6, replace the sentence "*For every integer $n \geq 1$, the canonical projection $\mathscr{R}_A \to A/pA$ onto the $(n+1)$th component of the inverse system $(A/pA)_{\mathbb{N}}$ (that is, the component of index n)*" by the sentence "*For every integer $n \geq 1$, the canonical projection $\mathscr{R}_A \to A/pA$ onto the nth component of the inverse system $(A/pA)_{\mathbb{N}}$ (that is, the component of index $n-1$)*".

(II.11.1). Line 4-5, we could have warn the reader that the notation $\alpha^{r,r'}$ and $h_\alpha^{r,r'}$ is not compatible with that used in (I.4.7). Same remark for (II.12.1.6).

(II.11.13). Line 3, replace "*for every $1 \leq i \leq n$*" by "*for every $1 \leq i \leq d$*".

(II.11.14). Line -3 of the proof, replace "*by virtue of (II.11.14.6) and II.11.12(ii)*" by "*by virtue of (II.11.14.6) and II.11.12(iii)*".

(II.12.1.6). Line -2, we could have warn the reader that the notation $\alpha^{r,r'}$ and $h_\alpha^{r,r'}$ are not compatible with those used in (I.4.7).

(II.13.15.7). Replace $k\varphi(g) = \exp(\sum_{i=1}^d \xi^{-1}\theta_i \otimes \chi_i(g))$ by

$$\varphi(g) = \exp(\sum_{i=1}^{d} \xi^{-1}\theta_i \otimes \chi_i(g)).$$

(II.14.2). Line 1, replace "*Let α be rational number $> \frac{1}{p-1}$*" by "*Let α be a rational number $> \frac{1}{p-1}$*".

(II.14.4.4). Four lines before (II.14.4.4), replace "*By II.14.3 ... and a Δ_{p^∞}-equivariant $\overline{R}$-linear isomorphism*" by "*By II.14.3 .. and a Δ-equivariant $\overline{R}$-linear isomorphism*".

(II.14.4.4). One line under (II.14.4.4), replace the sentences:
"*By virtue of II.14.1, for all integers $n \geq m > \alpha$, there exists a unique Δ_{p^∞}-equivariant R_1-linear isomorphism*

$$N_n/p^{m-\alpha}N_n \xrightarrow{\sim} N_m/p^{m-\alpha}N_m \tag{6.5.35.1}$$

that is compatible with the isomorphisms (II.14.4.4). Consequently, the R_1-modules $(N_n)_{n>\alpha}$

$$(N_n/p^{n-\alpha-\beta-\frac{1}{p-1}}N_n)_{n>\alpha+\beta+\frac{1}{p-1}}$$

form an inverse system..."

by the sentences:

"By virtue of II.14.1, for all integers $n \geq m > \alpha+\beta+\frac{1}{p-1}$, there exists a unique Δ_{p^∞}-equivariant R_1-linear isomorphism

$$N_n/p^{m-\alpha-\beta-\frac{1}{p-1}}N_n \xrightarrow{\sim} N_m/p^{m-\alpha-\beta-\frac{1}{p-1}}N_m \qquad (6.5.35.2)$$

that is compatible with the isomorphisms (II.14.4.4). Consequently, the R_1-modules

$$(N_n/p^{n-\alpha-\beta-\frac{1}{p-1}}N_n)_{n>\alpha+\beta+\frac{1}{p-1}}$$

form an inverse system..."

(III.1). Line 11, replace *"valuation ring"* by *"valuation field"*.

(III.2.1). Line 1-2, replace *"valuation ring"* by *"valuation field"*.

(III.10.16.11). Replace the $\times$ by $\otimes$ on each of the 2 horizontal lines.

(III.10.19.2). Replace the $\times$ by $\otimes$ on each of the 2 horizontal lines.

(III.10.30 (iii)). Line 2, replace *"inverse limit"* by *"direct limit"*.

(III.12.7). Line 8, replace *"every object of Ξ^r is a Higgs $\overline{\check{\mathscr{B}}}$-isogeny"* by *"every object of $\Xi^r_{\mathbb{Q}}$ is a Higgs $\overline{\check{\mathscr{B}}}$-isogeny"*.

(III.14.7.5). In the descending left arrow, replace Φ^* by $\check{\Phi}^*$.

(III.14.7.11). Replace the middle term $\top'^r_+ \circ \check{\Phi}^* \circ \mathfrak{S}$ by $\top'^r_+ \circ \check{\Phi}^* \circ \mathfrak{S}^r$.

(IV.5.1). Line 21 of page 384, replace $[\ ] : \overline{\mathscr{A}} \to W(R_{\overline{\mathscr{A}}})$ by $[\ ] : R_{\overline{\mathscr{A}}} \to W(R_{\overline{\mathscr{A}}})$.

(IV.6.1). The last line of page 411, replace $\Gamma(\overline{s}, O_{\overline{s}})$ by $\kappa(\overline{s}) = \Gamma(\overline{s}, O_{\overline{s}})$.

(IV.6.2). Line 20 of page 414, replace $\mathrm{Gal}(\kappa(\overline{s})/\kappa(s^g))$ by $\mathrm{Gal}(\kappa(s)^{\mathrm{ur}}/\kappa(s^g))$.

(IV.6.2). Line 20 of page 414, replace $\mathrm{Aut}_{C_{\mathrm{gpt}}}((U,\overline{s}))^\circ$ by $\mathrm{Aut}_{(\mathcal{U}_{\overline{K},\mathrm{triv}})_{\mathrm{gpt}}}(\overline{s})^\circ$.

(V.11.7). Replace $E_1^{a,b}$ by $E_2^{a,b}$.

(V.12.1). Line 3 of page 482, replace *"the multiplication by a"* by *"the multiplication by $\mathrm{tr}_G(b)$"*.

(VI.1.13). Line 19, replace *"valuation ring"* by *"valuation field"*.

(VI.3.4). Line 2, Line 3 and 2 times Line -3, replace e_Z by e_S.

(VI.3.5). Line 5, replace 2 times e_Z by e_S.

(VI.5.9). Line 4 : replace $(V_{n,m} \to V_n)_{n\in M_n}$ by $(V_{n,m} \to V_n)_{m\in M_n}$

(VI.6.5.1). Two lines under (VI.6.5.1), replace $,\!\to$ by $\to$.

B) Errata

(II.9.5). Replace the proof of the proposition by the following:

Indeed, we clearly have $\theta(\xi)=0$. By ([73] A.2.3), since $\mathrm{W}(A^\flat)$ and $\widehat{A}$ are $\mathbb{Z}_p$-flat and complete and separated for the p-adic topologies, it is enough to prove that the sequence

$$0 \longrightarrow A^\flat \xrightarrow{\cdot \underline{p}} A^\flat \xrightarrow{\nu_1} A/pA \longrightarrow 0, \qquad \text{(II.9.5.3)}$$

where ν_1 is the homomorphism induced by the projection on the first component of the projective system $(A/pA)_{\mathbb{N}}$ (II.9.3.1), is exact. By (iii), ν_1 is surjective.

Let $y=(y_n)_{n\in\mathbb{N}}\in A^\flat$ such that $\underline{p}y=0$. For every $n\geq 0$, let $\widetilde{y}_n$ be a lift of y_n in A. We have $p_n\widetilde{y}_n\in pA$. Consequently, $\widetilde{y}_n\in p_n^{p^n-1}A$ because p is not a zero divisor in A. It follows that

$$y_n=y_{n+1}^p=(\widetilde{y}_{n+1}^p \mod pA)=0 \qquad \text{(II.9.5.4)}$$

because $p^{n+2}-p\geq p^{n+1}$. Then, $\underline{p}$ is not a zero divisor in $A^\flat$.

It is clear that $\nu_1(\underline{p}y)=0$ for every $y\in A^\flat$. Conversely, let $x=(x_n)_{n\in\mathbb{N}}\in A^\flat$ such that $x_0=0$. For every $n\geq 0$, let $\widetilde{x}_n$ be a lift of x_n in A. By (i) and (ii), there exists $\widetilde{y}_n\in A$ such that $\widetilde{x}_n=p_n\widetilde{y}_n$. From the relation $\widetilde{x}_{n+1}^p\equiv\widetilde{x}_n \mod pA$, we deduce that $\widetilde{y}_{n+1}^p\equiv\widetilde{y}_n \mod p_n^{p^n-1}A$. For every $n\geq 1$, we have

$$\widetilde{y}_{n+1}^{p^2}\equiv\widetilde{y}_n^p \mod pA, \qquad \text{(II.9.5.5)}$$

because $p^{n+1}-p\geq p^n$. Hence, $y=(\widetilde{y}_{n+1}^p \mod pA)_{n\geq 0}\in A^\flat$. Since $x_n=x_{n+1}^p=p_n\widetilde{y}_{n+1}^p \mod p$, we have $x=\underline{p}y$. Therefore, the sequence (II.9.5.3) is exact in the center.

(II.11.12). Proof of (ii): The sufficiency of the required exactness of (II.11.12.5) is missing there. One can rather argue as follows:

(ii) Since for every integer $n\geq 0$, we clearly have

$$p^n\cdot{}_i\mathscr{S}_{p^\infty}^{(r)}={}_i\mathscr{S}_{p^\infty}^{(r)}\cap(p^n\cdot{}_{(i-1)}\mathscr{S}_{p^\infty}^{(r)}), \qquad \text{(II.11.12.4)}$$

the canonical morphism $u\colon {}_i\widehat{\mathscr{S}}_{p^\infty}^{(r)}\to{}_{(i-1)}\widehat{\mathscr{S}}_{p^\infty}^{(r)}$ is injective.

For every $\nu\in{}_{(i-1)}\Xi_{p^\infty}$, we denote by $(R_{p^\infty}^{(\nu)})^\wedge$ and $({}_{(i-1)}\mathscr{S}_{p^\infty}^{(r)}(\nu))^\wedge$ the p-adic Hausdorff completions of $R_{p^\infty}^{(\nu)}$ and ${}_{(i-1)}\mathscr{S}_{p^\infty}^{(r)}(\nu)$, respectively. Note that $({}_{(i-1)}\mathscr{S}_{p^\infty}^{(r)}(\nu))^\wedge$ is identified with a sub-R_1-module of $({}_{(i-1)}\mathscr{S}_{p^\infty}^{(r)})^\wedge$ stable by γ_i. By (II.11.11.3), every element x of $({}_{(i-1)}\mathscr{S}_{p^\infty}^{(r)})^\wedge$ can be written as the sum of a series

$$\sum_{\nu\in{}_{(i-1)}\Xi_{p^\infty}} x_\nu, \qquad \text{(II.11.12.5)}$$

where $x_\nu\in({}_{(i-1)}\mathscr{S}_{p^\infty}^{(r)}(\nu))^\wedge$ and, for every integer $n\geq 0$, except for finitely many $\nu\in{}_{(i-1)}\Xi_{p^\infty}$, $x_\nu\in p^n({}_{(i-1)}\mathscr{S}_{p^\infty}^{(r)}(\nu))^\wedge$. Such an element x is zero if and only if x_ν is zero for every $\nu\in{}_{(i-1)}\Xi_{p^\infty}$.

Let $\nu \in {}_{(i-1)}\Xi_{p^\infty}$. For every

$$z = \sum_{\underline{n} \in J_{i-1}} p^{r|\underline{n}|} a_{\underline{n}} y^{\underline{n}} \in {}_{(i-1)}\mathscr{S}^{(r)}_{p^\infty}(\nu), \tag{II.11.12.6}$$

we have (II.11.6.1)

$$(\gamma_i - \mathrm{id})(z) = \sum_{\underline{n} \in J_{i-1}} p^{r|\underline{n}|} b_{\underline{n}} y^{\underline{n}} \in {}_{(i-1)}\mathscr{S}^{(r)}_{p^\infty}(\nu), \tag{II.11.12.7}$$

where, for every $\underline{n} = (n_1, \ldots, n_{d'}) \in J_{i-1}$,

$$b_{\underline{n}} = (\nu(\gamma_i) - 1)a_{\underline{n}} + \sum_{\underline{m}=(m_1,\ldots,m_{d'}) \in J_{i-1}(\underline{n})} p^{r(m_i - n_i)} \binom{m_i}{n_i} \nu(\gamma_i) a_{\underline{m}} w^{m_i - n_i},$$

$J_{i-1}(\underline{n})$ denotes the subset J_{i-1} made up of elements $\underline{m} = (m_1, \ldots, m_{d'})$ such that $m_j = n_j$ for $j \neq i$ and $m_i > n_i$, and $w = \xi^{-1} \log([\zeta])$ is an element of valuation $\frac{1}{p-1}$ of $\mathscr{O}_C$ (II.9.18).

Let z be an element of $({}_{(i-1)}\mathscr{S}^{(r)}_{p^\infty}(\nu))^\wedge$. Then, z can be written as the sum of a series

$$z = \sum_{\underline{n} \in J_{i-1}} p^{r|\underline{n}|} a_{\underline{n}} y^{\underline{n}},$$

where $a_{\underline{n}} \in (R^{(\nu)}_{p^\infty})^\wedge$ and $a_{\underline{n}}$ tends to 0 when $|\underline{n}|$ tends to infinity. As $R^{(\nu)}_{p^\infty}$ is $\mathscr{O}_C$-flat, the same therefore holds for $(R^{(\nu)}_{p^\infty})^\wedge$ (cf. the proof of II.6.14). Consequently, z is zero if and only if $a_{\underline{n}}$ is zero for every $\underline{n} \in J_{i-1}$. We immediately see that $(\gamma_i - \mathrm{id})(z)$ is also given by the formula (II.11.12.7).

Suppose that $\gamma_i(z) = z$ and $\nu(\gamma_i) \neq 1$. As $(R^{(\nu)}_{p^\infty})^\wedge$ is $\mathscr{O}_C$-flat and $v(\nu(\gamma_i) - 1) \leq \frac{1}{p-1}$, for every $\underline{n} = (n_1, \ldots, n_{d'}) \in J_{i-1}$, we have

$$a_{\underline{n}} = -(\nu(\gamma_i) - 1)^{-1} \sum_{\underline{m}=(m_1,\ldots,m_{d'}) \in J_{i-1}(\underline{n})} p^{r(m_i - n_i)} \binom{m_i}{n_i} \nu(\gamma_i) a_{\underline{m}} w^{m_i - n_i}.$$

We deduce from this that for every $\alpha \in \mathbb{N}$ and every $\underline{n} \in J_{i-1}$, we have $a_{\underline{n}} \in p^{r\alpha}(R'^{(\nu)}_{p^\infty})^\wedge$ (this is proved by induction on α); therefore $z = 0$ as $(R'^{(\nu)}_{p^\infty})^\wedge$ is separated for the p-adic topology. Consequently, $\gamma_i - \mathrm{id}$ is injective on $({}_{(i-1)}\mathscr{S}^{(r)}_{p^\infty}(\nu))^\wedge$.

Suppose that $\gamma_i(z) = z$ and $\nu(\gamma_i) = 1$, so that $\nu \in {}_{(i)}\Xi_{p^\infty}$. Then, for every $\underline{n} = (n_1, \ldots, n_{d'}) \in J_{i-1}$, if we set $\underline{n}' = (n'_1, \ldots, n'_{d'}) \in J_{i-1}(\underline{n})$ with $n'_i = n_i + 1$, we have

$$(n_i + 1)! a_{\underline{n}'} = - \sum_{\underline{m}=(m_1,\ldots,m_{d'}) \in J_{i-1}(\underline{n}')} p^{r(m_i - n_i - 1)} m_i! a_{\underline{m}} \frac{w^{m_i - n_i - 1}}{(m_i - n_i)!}.$$

We have $w^{m-1}/m! \in \mathscr{O}_C$ for every integer $m \geq 1$. We deduce from this that for every $\alpha \in \mathbb{N}$ and every $\underline{n} = (n_1, \ldots, n_{d'}) \in J_{i-1}$ such that $n_i \geq 1$, we have $n_i! a_{\underline{n}} \in p^{r\alpha} R'^{(\nu)}_{p^\infty}$ (this is proved by induction on α); therefore $a_{\underline{n}} = 0$. Consequently $z \in ({}_{i}\mathscr{S}^{(r)}_{p^\infty}(\nu))^\wedge$.

We deduce from this that the sequence

$$0 \longrightarrow ({}_i\mathscr{S}^{(r)}_{p^\infty}(\nu))^\wedge \xrightarrow{u_\nu} ({}_{(i-1)}\mathscr{S}^{(r)}_{p^\infty}(\nu)^\wedge \xrightarrow{\gamma_i - \mathrm{id}} ({}_{(i-1)}\mathscr{S}^{(r)}_{p^\infty}(\nu))^\wedge , \quad \text{(II.11.12.8)}$$

where u_ν is the canonical morphism, is exact. The proposition follows.

(II.13.9). Modify the beginning of (II.13.9) as follows:

Let M be an $\widehat{R_1}$-module of finite type which is $\mathscr{O}_C$-flat. By ([1] 1.10.2), M is complete and separated for the p-adic topology. We deduce from this that the canonical morphism

$$\mathrm{End}_{\widehat{R_1}}(M) \to \lim_{\substack{\longleftarrow \\ n\geq 0}} \mathrm{Hom}_{\widehat{R_1}}(M, M/p^n M) \quad \text{(II.13.9.a)}$$

is an isomorphism. On the other hand, for every integer $n \geq 0$, the exact sequence

$$0 \to M \xrightarrow{p^n} M \to M/p^n M \to 0 \quad \text{(II.13.9.b)}$$

shows that the canonical map

$$\mathrm{End}_{\widehat{R_1}}(M)/p^n \mathrm{End}_{\widehat{R_1}}(M) \to \mathrm{Hom}_{\widehat{R_1}}(M, M/p^n M) \quad \text{(II.13.9.c)}$$

is injective. We deduce from (II.13.9.a) and (II.13.9.c) that $\mathrm{End}_{\widehat{R_1}}(M)$ is complete and separated for the p-adic topology. Moreover, it is $\mathscr{O}_C$-flat and for every rational number $\alpha \geq 0$, the canonical homomorphism $p^\alpha \mathrm{End}_{\widehat{R_1}}(M) \to \mathrm{Hom}_{\widehat{R_1}}(M, p^\alpha M)$ is an isomorphism. Let u...

(III.10.3). The category **P** should moreover be required to contain the affine schemes U such that the special fiber U_s is empty in order to have covering families.

(III.10.5). The category **Q** should moreover be required to contain the affine schemes U such that the special fiber U_s is empty in order to have covering families.

(V.12.1). Lemma V.12.1(2) is wrong. There is an obvious counterexample: $A \neq 0$, $B = A[X]$, $G = \{1\}$, $M = B$, $a = 1$, $r = 1$, $b_1 = c_1 = 1$. The claim $\varphi \circ \psi = a \cdot 1_{B\otimes_A M^G}$ in the proof of Lemma V.12.1(2) is not true in general unless the homomorphism $A \to B^G$ is surjective.

Replace Lemma V.12.1(2) and its proof by the following:

Assume that $a \in B$ and $a' \in B^G$ satisfy the following conditions:

(i) *There exist $b_i, c_i \in B$ $(1 \leq i \leq r)$ such that $\sum_{i=1}^r b_i c_i = a$ and $\sum_{i=1}^r b_i g(c_i) = 0$ for all $g \in G \backslash \{1\}$.*

(ii) *$a' B^G$ is contained in the image of the homomorphism $A \to B^G$.*

Then, the kernel and the cokernel of $\psi\colon B \otimes_A M^G \to M; b \otimes x \mapsto bx$ are killed by aa'.

Proof. We define a map $\varphi\colon M \to B \otimes_A M^G$ by $\varphi(m) = \sum_{i=1}^r b_i \otimes a' \mathrm{tr}_G(c_i m)$, which is A-linear. We assert $\varphi \circ \psi = aa' \cdot 1_{B\otimes_A M^G}$ and $\psi \circ \varphi = aa' \cdot 1_M$. The image of $b \otimes m \in B \otimes_A M^G$ $(b \in B, m \in M^G)$ under $\varphi \circ \psi$ is $\sum_{i=1}^r b_i \otimes a' \mathrm{tr}_G(c_i bm) = \sum_{i=1}^r b_i \otimes (a' \mathrm{tr}_G(c_i b))m$ by the definition of φ and ψ, and $m \in M^G$. By the condition (ii), we obtain

$$\begin{aligned}\varphi\circ\psi(b\otimes m) &= \sum_{i=1}^{r}(b_i a'\mathrm{tr}_G(c_i b)\otimes m)\\ &= \left(a'\sum_{g\in G}\left(\sum_{i=1}^{r} b_i g(c_i)\right)g(b)\right)\otimes m = a'ab\otimes m.\end{aligned}$$

The second equality is shown as

$$\psi\circ\varphi(m) = \sum_{i=1}^{r} b_i a'\mathrm{tr}_G(c_i m) = a'\sum_{g\in G}\left(\sum_{i=1}^{r} b_i g(c_i)\right)g(m) = a'am. \qquad \square$$

(V.12.5) and (V.12.6). Replace the proof of Corollary V.12.6 by the direct proof below. This replacement allows us to apply the corrected Lemma V.12.1(2) above in the proof of Proposition V.12.5.

PROOF. By Lemma V.12.4, $R \to S$ is almost faithfully flat and the homomorphism $S\otimes_R S \to \prod_{g\in G} S; x\otimes y \mapsto (xg(y))_{g\in G}$ is an almost isomorphism. By Proposition V.9.1, we have an almost exact sequence $R \to S \xrightarrow{d} S\otimes_R S$, where d is defined by $d(s) = 1\otimes s - s\otimes 1$. The composition of d with the almost isomorphism $S\otimes_R S \xrightarrow{\approx} \prod_{g\in G} S$ sends s to $(g(s)-s)_{g\in G}$. Hence $R\to S^G$ is an almost isomorphism.
$\square$

(VI.10.40). The introduction of $X_{(\overline{x})}$ in the proposition is useless. A better way to formulate the conclusion is to say that the map

$$\mathscr{H}^i(F) \to \mathrm{R}^i\sigma_*(F^a)$$

is an isomorphism. It follows immediately from the corresponding isomorphism on the stalks given in the proposition and its proof.

References

1. A. Abbes, *Éléments de géométrie rigide. Volume I. Construction et étude géométrique des espaces rigides*, Progress in Mathematics Vol. **286**, Birkhäuser (2010).
2. A. Abbes, M. Gros, Les suites spectrales de Hodge–Tate, *Astérisque* **448** (2024), arXiv:2003.04714.
3. A. Abbes, M. Gros, T. Tsuji, *The p-adic Simpson correspondence*, Ann. of Math. Stud., **193**, Princeton Univ. Press (2016).
4. P. Achinger, $K(\pi, 1)$-neighborhoods and comparison theorems, *Compositio Math.* **151** (2015), 1945-1964.
5. M. Artin, A. Grothendieck, J. L. Verdier, *Théorie des topos et cohomologie étale des schémas, SGA 4*, Lecture Notes in Math. Tome 1, **269** (1972); Tome 2, **270** (1972); Tome 3, **305** (1973), Springer-Verlag.
6. P. Berthelot, A. Grothendieck, L. Illusie, *Théorie des intersections et théorème de Riemann–Roch, SGA 6*, Lecture Notes in Math. **225** (1971), Springer-Verlag.
7. O. Biquard, Fibrés de Higgs et connexions intégrables : le cas logarithmique (diviseur lisse), *Ann. scient. de l'E.N.S.*, **30** (1997), 41–96.
8. N. Bourbaki, *Algèbre*, Chapitres 1-3, Masson (1970).
9. N. Bourbaki, *Algèbre*, Chapitre 10, Masson (1980).
10. N. Bourbaki, *Algèbre commutative*, Chapitres 1-9, Hermann (1985).
11. N. Bourbaki, *Topologie générale*, Chapitres 1-4, Hermann (1971).
12. P. Deligne, Théorème de Lefschetz et critères de dégénérescence de suites spectrales, *Pub. Math. IHÉS* **35** (1968), 107–126.
13. R. Donagi, T. Pantev, C. Simpson, Direct Images in Non Abelian Hodge Theory, preprint (2016), arXiv:1612.06388.
14. P. Deligne, Théorie de Hodge II, *Pub. Math. IHÉS* **40** (1971), 5–57.
15. P. Deligne, *Cohomologie Étale, SGA* $4\frac{1}{2}$, Lecture Notes in Math. **569** (1977), Springer-Verlag.
16. G. Faltings, p-adic Hodge theory, *J. Amer. Math. Soc.* **1** (1988), 255–299.
17. G. Faltings, Almost étale extensions, dans Cohomologies p-adiques et applications arithmétiques. II, *Astérisque* **279** (2002), 185–270.
18. G. Faltings, A p-adic Simpson correspondence, *Adv. Math.* **198** (2005), 847-862.
19. J.-M. Fontaine, Sur certains types de représentations p-adiques du groupe de Galois d'un corps local; construction d'un anneau de Barsotti-Tate, *Annals of Math.* **115** (1982), 529–577.
20. J.-M. Fontaine, Le corps des périodes p-adiques, dans Périodes p-adiques, Séminaire de Bures, 1988, *Astérisque* **223** (1994), 59–111.
21. O. Gabber, L. Ramero, *Almost Ring Theory*, Springer-Verlag, Lecture Notes in Math. **1800** (2003), Springer-Verlag.
22. P. Gabriel, Des catégories abéliennes, *Bull. Soc. Math. France* **90** (1962) 323–448.
23. J. Giraud, *Cohomologie non abélienne*, Springer-Verlag (1971).
24. A. Grothendieck, *Revêtements étales et groupe fondamental, SGA 1*, Lecture Notes in Math. **224** (1971), Springer-Verlag; Édition recomposée et annotée par la SMF, Documents mathématiques **3** (2003).

A. Abbes, M. Gros, *The p-adic Simpson Correspondence and Hodge-Tate Local Systems*,
Lecture Notes in Mathematics 2345, https://doi.org/10.1007/978-3-031-55914-3

25. A. Grothendieck, *Cohomologie ℓ-adique et fonctions L, SGA 5*, Lecture Notes in Math. **589** (1977), Springer-Verlag.
26. A. Grothendieck, *Groupes de Barsotti-Tate et cristaux de Dieudonné*, SMS (1974) Montréal.
27. A. Grothendieck, J.A. Dieudonné, *Éléments de Géométrie Algébrique I*, Seconde édition, Grundlehren der mathematischen Wissenschaften **166**, Springer-Verlag (1971).
28. A. Grothendieck, J.A. Dieudonné, Éléments de Géométrie Algébrique, II Étude globale élémentaire de quelques classes de morphismes, *Pub. Math. IHÉS* **8** (1961).
29. A. Grothendieck, J.A. Dieudonné, Éléments de Géométrie Algébrique, III Étude cohomologique des faisceaux cohérents, *Pub. Math. IHÉS* **11** (1961), **17** (1963).
30. A. Grothendieck, J.A. Dieudonné, Éléments de Géométrie Algébrique, IV Étude locale des schémas et des morphismes de schémas, *Pub. Math. IHÉS* **20** (1964), **24** (1965), **28** (1966), **32** (1967).
31. T. He, Sen operators and Lie algebras arising from Galois representations over p-adic varieties, preprint (2022), arXiv:2208.07519.
32. T. He, Almost coherence of higher direct images, preprint (2022), arXiv:2212.01797.
33. B. Heuer, A p-adic Simpson correspondence for smooth proper rigid varieties, preprint (2023), arXiv:2212.01797.
34. N. Hitchin, Stable bundles and integrable systems, *Duke Math.* **54** (1987), 91–114.
35. N. Hitchin, The self-duality equations on a Riemann surface, *Proceedings of the London Mathematical Society* **3** (1987), 59–126.
36. O. Hyodo, On variation of Hodge–Tate structures, *Math. Ann.* **284** (1989), 7–22.
37. L. Illusie, *Complexe cotangent et déformations. I*, Lecture Notes in Math. **239** (1971), Springer-Verlag.
38. L. Illusie, K. Kato, C. Nakayama, Quasi-unipotent Logarithmic Riemann–Hilbert Correspondences, *J. Math. Sci. Univ. Tokyo* **12** (2005), 1-66.
39. U. Jannsen, Continuous étale Cohomology, *Math. Ann.* **280** (1988), 207–245.
40. M. Kashiwara, P. Schapira, *Categories and sheaves*, Grundlehren der mathematischen Wissenschaften **332**, Springer-Verlag (2006).
41. K. Kato, Logarithmic structures of Fontaine–Illusie, in *Algebraic analysis, geometry, and number theory*, Johns Hopkins UP, Baltimore (1989), 191–224.
42. N. Katz, Algebraic solutions of differential equations (p-curvature and the Hodge filtration), *Invent. math.* **18** (1972), 1–118.
43. N. Katz, T. Oda, On the differentiation of De Rham cohomology classes with respect to parameters, *J. Math. Kyoto Univ.* **8** (1968), 199–213.
44. R. Liu, X. Zhu, Rigidity and a Riemann–Hilbert correspondence for p-adic local systems, *Invent. math.* **207** (2017), 291–343.
45. A. Ogus, *Lectures on Logarithmic Algebraic Geometry*, Cambridge University Press (2018).
46. A. Ogus, V. Vologodsky, Nonabelian Hodge theory in characteristic p, *Pub. Math. IHES* **106** (2007), 1–138.
47. L. Ribes, P. Zalesskii, *Profinite Groups*, Ergebnisse der Mathematik, Vol. **40** (2010), Springer-Verlag.
48. J.-E. Roos, Caractérisation des catégories qui sont quotients des catégories de modules par des sous-catégories bilocalisantes, *C.R. Acad. Sc. Paris* **261** (1965), 4954–4957.
49. J.-P. Serre, *Cohomologie galoisienne*, Cinquième édition révisée et complétée, Lecture Notes in Math. **5**, Springer-Verlag (1997).
50. C. Simpson, Constructing variations of Hodge structure using Yang-Mills theory and applications to uniformization, *J. Amer. Math. Soc.* **1** (1988), no. 4, 867–918.
51. C. Simpson, Higgs bundles and local systems, *Pub. Math. IHÉS* **75** (1992), 5–95.
52. C. Simpson, Moduli of representations of the fundamental group of a smooth variety I, *Pub. Math. IHÉS* **79** (1994), 47–129.
53. C. Simpson, Moduli of representations of the fundamental group of a smooth projective variety II, *Publ. Math. IHÉS* **80** (1994), 5–79.
54. J. Tate, p-divisible groups, in *Proceedings of a Conference on Local Fields (Driebergen, 1966)*, Springer (1967), Berlin, 158–183.
55. T. Tsuji, p-adic étale cohomology and crystalline cohomology in the semi-stable reduction case, *Invent. math.* **137** (1999), 233–411.

56. T. Tsuji, Saturated morphisms of logarithmic schemes, *Tunisian Journal of Mathematics* **1** (2019), 185–220.
57. T. Tsuji, Notes on the local p-adic Simpson correspondence, *Math. Annalen* (2018) **371** 795–881.
58. D. Xu, Parallel transport for Higgs bundles over p-adic curves, preprint (2023), arXiv:2201.06697.
59. Authors of the Stacks Project, Stacks Project, 2018.

Index

A. Abbes, M. Gros, *The p-adic Simpson Correspondence and Hodge-Tate Local Systems*, Lecture Notes in Mathematics 2345, https://doi.org/10.1007/978-3-031-55914-3

Chapter VI, 331

Printed in the United States
by Baker & Taylor Publisher Services